Atom–Photon Interactions

Atom–Photon Interactions

Basic Processes and Applications

Claude Cohen-Tannoudji
Jacques Dupont-Roc
Gilbert Grynberg

A WILEY-INTERSCIENCE PUBLICATION

JOHN WILEY & SONS, INC.

New York   Chichester   Brisbane   Toronto   Singapore

Copyright © 1992 by John Wiley & Sons, Inc.
This is a translation of *Processus d'interaction entre photons et atomes.*
Copyright © 1988, InterEditions and Editions du CNRS.
Translation by Patricia Thickstun.

Library of Congress Cataloging in Publication Data:

Cohen-Tannoudji, Claude, 1933–
    [Processus d'interaction entre photons et atomes. English]
    Atom-photon interactions: basic processes and applications/
  Claude Cohen-Tannoudji, Jacques Dupont-Roc, Gilbert Grynberg.
      p.   cm.
    Translation of: Processus d'interaction entre photons et atomes.
    "A Wiley-Interscience publication."
    Includes bibliographical references and index.
    ISBN 0-471-62556-6 (cloth: acid-free paper)
    1. Photonuclear reactions. 2. Quantum theory. 3. Statistical
  physics. I. Dupont-Roc, Jacques. II. Grynberg, Gilbert.
  III. Title.
QC794.8.P4C6413 1992   91-40587
539.7′56–dc20   CIP

Printed in the United States of America

10 9 8 7 6 5 4 3 2 1

Contents

COMPLEMENT B₁—DESCRIPTION OF THE EFFECT OF A
PERTURBATION BY AN EFFECTIVE HAMILTONIAN

COMPLEMENT C₁—DISCRETE LEVEL COUPLED TO A BROAD
CONTINUUM: A SIMPLE MODEL

II
A SURVEY OF SOME INTERACTION PROCESSES
BETWEEN PHOTONS AND ATOMS

COMPLEMENT A_{II}—PHOTODETECTION SIGNALS AND
CORRELATION FUNCTIONS

COMPLEMENT A_{III}—ANALYTIC PROPERTIES OF THE
RESOLVENT

COMPLEMENT B_{III}—NONPERTURBATIVE EXPRESSIONS FOR
THE SCATTERING AMPLITUDES OF A PHOTON BY AN ATOM

COMPLEMENT C_{III}—DISCRETE STATE COUPLED TO A
FINITE-WIDTH CONTINUUM: FROM THE WEISSKOPF–WIGNER
EXPONENTIAL DECAY TO THE RABI OSCILLATION

IV
RADIATION CONSIDERED AS A RESERVOIR: MASTER
EQUATION FOR THE PARTICLES

COMPLEMENT A_{IV}—FLUCTUATIONS AND LINEAR RESPONSE
APPLICATION TO RADIATIVE PROCESSES

1. Statistical Functions and Physical Interpretation of the Master
   Equation—*a. Symmetric Correlation Function. b. Linear Sus-*

COMPLEMENT B_{IV}—MASTER EQUATION FOR A DAMPED HARMONIC OSCILLATOR

COMPLEMENT C_{IV}—QUANTUM LANGEVIN EQUATIONS FOR A SIMPLE PHYSICAL SYSTEM

V
OPTICAL BLOCH EQUATIONS

COMPLEMENT A_V—BLOCH–LANGEVIN EQUATIONS AND
QUANTUM REGRESSION THEOREM

VI
THE DRESSED ATOM APPROACH

COMPLEMENT A_{VI}—THE DRESSED ATOM IN THE
RADIO-FREQUENCY DOMAIN

COMPLEMENT B_{VI}—COLLISIONAL PROCESSES IN THE
PRESENCE OF LASER IRRADIATION

EXERCISES

APPENDIX
QUANTUM ELECTRODYNAMICS IN THE COULOMB
GAUGE—SUMMARY OF THE ESSENTIAL RESULTS

Preface

The spectacular development of new sources of electromagnetic radiation, covering a range of frequencies from radio waves to far ultraviolet (lasers, masers, synchrotron radiation, microwave sources), has resulted in considerably renewed interest in photon-atom interactions. New methods have appeared for obtaining more precise information about the structure and dynamics of atoms and molecules, for controlling their internal and external degrees of freedom, and for generating new types of radiation. These developments have caused a growing number of physicists, chemists, researchers, and engineers to become interested in interactions occurring between matter and low-energy radiation. With these two books on photons and atoms, our aim is to provide the theoretical bases necessary for undertaking the study of these processes beginning at a level of quantum mechanics and classical electromagnetism corresponding to that of first-year graduate course.

Such a program is naturally composed of two parts. First, one must introduce a theoretical framework that can be used to describe the quantum dynamics of the global system "electromagnetic field + nonrelativistic charged particles" and discuss the physical content of the theory, as well as its different possible formulations. These problems have been studied in a previous volume entitled *Photons and Atoms—Introduction to Quantum Electrodynamics*. Second, one must show how such a theoretical framework can be used to analyze the interactions between photons and atoms as they appear in atomic and molecular physics, quantum optics, and laser physics. This is the goal of the present volume entitled *Atom–Photon Interactions: Basic Processes and Applications*. The goals of these two volumes are thus clearly distinct and, depending on the concerns and needs of the reader, one or the other or both volumes of this work may be used.

It is, of course, impossible to present in a single volume an exhaustive study of the interaction between matter and radiation and of all the related physical phenomena. We have thus emphasized the aspects that we consider to be essential. First we will analyze in detail the elementary processes in which photons are emitted, absorbed, scattered, emitted and reabsorbed, or exchanged between atoms. Extensive use of diagrammatic representations will allow us to visualize the processes being described. A knowledge of these elementary processes is, nevertheless, not always sufficient for analyzing in simple terms the extremely large variety of

xxi

phenomena which may result from the interplay of these processes. Thus we thought it important to bring together in this book different theoretical approaches, which are usually dispersed in more specialized works, and which are more particularly adapted to one aspect or another of the phenomena being discussed (perturbative methods, resolvent method, master equation, Langevin equation, optical Bloch equations, dressed atom method, etc.). Finally, we have decided to illustrate each of these methods in simple systems, so as to be able to show as clearly as possible their significance and their limitations. Our hope is to have integrated in this volume the basic elements allowing the physics of the matter-radiation interaction be mastered in all its different aspects.

ACKNOWLEDGMENTS

This work is based on teaching and research that has been carried out over several years at Collège de France and the Physics Department of the Ecole Normale Supérieure. We would like to express our thanks to our colleagues and friends, in particular Serge Reynaud and Jean Dalibard, who directly participated in this research and from whose comments we benefited.

We are also especially grateful to Martine Guillaume, Patricia Bouniol, and Catherine Emo who were in charge of preparing the manuscript, and to Christophe Salomon, who checked the translation.

Atom–Photon Interactions

Introduction

Electromagnetic interaction governs the motion of electrons and nuclei, which are, on the electron-volt scale, the elementary constituents of matter. Understanding the mechanisms of interaction between the electromagnetic field and these particles is thus fundamental to the interpretation of many phenomena in our environment. Responsible for the cohesion of atoms and molecules and their mutual interactions, the electromagnetic interaction is also at the origin of the emission and absorption of radiation by these systems. The study of the light emitted or absorbed by these atoms and molecules provides an essential source of information on the structure and dynamics of these systems. Finally, it is also possible to use photons to act on the atoms, to control their internal as well as external degrees of freedom, and to place them in situations far from thermodynamic equilibrium.

The Coulomb gauge is particularly useful to describe the processes which involve the electromagnetic field and atoms or molecules. Recall that, in this representation, the Coulomb interaction between charged particles, which is related to the longitudinal component of the electromagnetic field and which predominates at low energy, appears directly in the particle Hamiltonian. The coupling of the charged particles with the quantized transverse field can then be analyzed by using a perturbative method. Such a formulation presents several advantages. First, because of the Coulomb interaction, the particle Hamiltonian may have bound states. Atoms, molecules, and matter in general may thus be described to a first approximation in a simple way. Next, we have to consider the coupling with the quantum field which is characterized by its elementary excita-

1

tions, photons. Exchanges of energy, momentum, and angular momentum between the systems of particles and the transverse field are thus described by the absorptions and emissions of photons. These processes constitute the elementary interactions between matter and the transverse field.

Let us first of all point out that identifying the elementary processes does not exhaust the subject. In fact, the interplay of several of these processes may result in qualitatively new effects. For example, the large number of degrees of freedom of the field is at the origin of the appearance of irreversible evolutions for the atoms. The field may appear in this case as a reservoir, and statistical mechanics methods prove to be especially convenient for describing the effect of the field on the particles. Another aspect of the electromagnetic field requires a special approach: the electromagnetic field is, in fact, one of the rare fields that we know how to produce in states having a nonzero average amplitude. The evolution of atomic systems in such coherent fields gives rise to a wealth of phenomena which were first explored in magnetic resonance experiments before being studied in the field of optics.

This variety of phenomena and the appropriate theoretical methods required for understanding them are discussed in the six chapters of this book. The first chapter reviews the concept of transition amplitude, which is essential for the quantum description of photon-atom interactions. The important states of the particles and of the field are identified, and we introduce diagrammatic representations to visualize the processes under study. Chapter II presents an overview of photon-atom interactions. The different elementary processes are reviewed and analyzed in terms of transition amplitudes: absorption, emission, and scattering of photons; radiative corrections, and photon exchanges. These processes are illustrated in several examples selected from various contexts. Chapter III takes up the problem of calculating transition amplitudes with the resolvent, a tool more powerful than perturbation theory. We demonstrate how nonperturbative expressions for the transition amplitudes can be obtained, particularly for the decay of an unstable state or for resonant scattering. Starting in Chapter IV, we no longer attempt to determine the evolution of the "field + particle" system, but only that of the atom. We show in Chapter IV that, for several field states, the evolution of the atom can be described by a relaxation equation for the density matrix of the atom (or by a Langevin equation for its observables). Several results are also established with regard to the damping and fluctuations of internal and external atomic observables. In Chapter V, we consider how the effect of coherent, monochromatic radiation, which may be intense, can perturb the evolution of the atom. Using some approximations, such an evolution may be described by optical Bloch equations, in which the effect of the coherent field is considered as a time-dependent perturbation. The optical Bloch equations constitute, along with the Maxwell equations, basic equations that can be used for analyzing many physical problems. Finally, Chapter VI describes the same problem from the point of view of the "dressed atom", which uses a quantum description of the field mode having a macroscopic amplitude. This point of view clearly emphasizes the role played by the fundamental processes of absorption, stimulated emission, and spontaneous emission discussed in Chapter II and proves to be quite convenient for interpreting many phenomena that occur in the presence of intense radiation.

As in the other volume, *Photons and Atoms—Introduction to Quantum Electrodynamics*, each chapter is supplemented by complements which have various functions; they may clarify ideas introduced in the chapter or extend the chapter by presenting examples of applications; they may also propose other points of view or treat additional problems. A succinct, nonexhaustive bibliography is presented, either in the form of general references at the end of the chapter and the complements, or in the form of more specific references mentioned in notes at the bottom of the page. The books are cited by author's name, except for *Photons and Atoms— Introduction to Quantum Electrodynamics*, which is referred to by its title. To avoid excessive references to this book, we have assembled in an Appendix the essential elements of the description of Coulomb gauge electrodynamics. However, this Appendix is extremely succinct, and the reader is invited to refer to the previously mentioned work for questions concerning foundations of the theory. Finally, we have assembled at the end of this volume 20 corrected exercises that illustrate several aspects of the interaction between matter and radiation. We made them essentially independent so that they may, for the most part, be undertaken independently of a detailed reading of any given chapter.

For a simple first approach, the reader can cover just Chapters I, II and V, and accept the existence of relaxation rates after a look at Section A of Chapter IV. This reader will thus have an overall view of elementary processes and will have become familiar with simple methods for calculating transition amplitudes and with optical Bloch equations as well.

NOTE

According to current usage, the term "radiation" is used in general to designate the transverse field, although they are identical only in the absence of charged particles.

CHAPTER I

Transition Amplitudes in Electrodynamics

The goal of this first chapter is to introduce the concept of transition amplitude, which is essential for the quantum description of interactions between atoms and photons.

We begin in Section A by recalling that the transition amplitude associated with a physical process is the evolution-operator matrix element between the initial and final states of the process under study. The calculation of these amplitudes frequently uses perturbation theory and is based on the splitting of the total Hamiltonian H into an unperturbed part H_0 and a coupling V.

We then discuss in Section B the basic ideas of quantum mechanics concerning the time dependence of transition amplitudes. Distinctions are made among several cases according to whether the initial and final states of the process under study belong to the discrete or to the continuous spectrum of H_0. The three complements explore this problem in greater detail. Complement A_I assembles several important results concerning the perturbative calculation of transition amplitudes and physical quantities that can be deduced from these amplitudes (transition rates, cross-sections, etc.). Complement B_I introduces the concept of effective Hamiltonian, which is useful for describing situations where several energy levels of H_0 forming a well-isolated manifold are indirectly coupled through other levels of H_0. Complement C_I presents a very simple model of a discrete state coupled to a continuum, which allows one to exactly calculate the transition amplitudes and to understand the way the discrete state of H_0 can be traced in the eigenstates of H.

5

In Section C, we apply these ideas to a system of charged particles interacting with the electromagnetic field. Starting with the Hamiltonian H of quantum electrodynamics in the Coulomb gauge, we consider several possible splittings of this Hamiltonian into an unperturbed part H_0 and a coupling V. We emphasize the advantages of the Coulomb gauge, which allows the Coulomb interaction to be included in the particle Hamiltonian, and the bound states of charged particles, such as atoms, molecules, or ions to be considered as "unperturbed". We also introduce diagrammatic representations of interaction processes that allow the evolution of the global system to be simply visualized. These are the interaction processes (absorption, emission, scattering, etc.) that we will review in Chapter II.

A—PROBABILITY AMPLITUDE ASSOCIATED
WITH A PHYSICAL PROCESS

The idea of probability amplitude plays a central role in the quantum description of the time evolution of a physical process. The system under study is prepared at an instant t_i in a given state $|\psi_i\rangle$. The probability amplitude of finding it, at another instant t_f, in the state $|\psi_f\rangle$ is given, in the Schrödinger representation, by

$$\langle \psi_f | U(t_f, t_i) | \psi_i \rangle, \tag{A.1}$$

where $U(t_f, t_i)$ is the evolution operator between t_i and t_f. The main advantage of these amplitudes (A.1) is that they can be multiplied: the amplitude for going from $|\psi_1\rangle$ to $|\psi_2\rangle$ between t_1 and t_2, and then from $|\psi_2\rangle$ to $|\psi_3\rangle$ between t_2 and t_3 is given by the product

$$\langle \psi_3 | U(t_3, t_2) | \psi_2 \rangle \langle \psi_2 | U(t_2, t_1) | \psi_1 \rangle. \tag{A.2}$$

Another interesting property of the amplitudes (A.1) is that they interfere. If the system is not observed at an intermediate instant t_2, the amplitudes associated with all the possible intermediate states must be summed over. It is in fact well known that

$$\langle \psi_3 | U(t_3, t_1) | \psi_1 \rangle = \sum_n \langle \psi_3 | U(t_3, t_2) | \varphi_n \rangle \langle \varphi_n | U(t_2, t_1) | \psi_1 \rangle \tag{A.3}$$

where the $\{|\varphi_n\rangle\}$ form an orthonormal basis of states.

The calculation of the amplitude (A.1) assumes, of course, that we already know how to determine $|\psi_i\rangle$ and $|\psi_f\rangle$. Generally the initial state and the final state are characterized by well-defined values of some physical variables. Thus we must be able to calculate the eigenvalues and eigenstates of the observables that represent these physical variables. We must also know the evolution operator $U(t_f, t_i)$ which is determined by diagonalizing the Hamiltonian H of the system. However, in most cases, and particularly in electrodynamics, we do not know how to exactly calculate the eigenstates and eigenvalues of H. It is thus necessary to resort to approximation methods.

The perturbative calculation of amplitudes (A.1) depends in general on the splitting of the Hamiltonian H into an "unperturbed" part H_0, for which the eigenstates $|\varphi_n\rangle$ and eigenvalues E_n are known, and a perturba

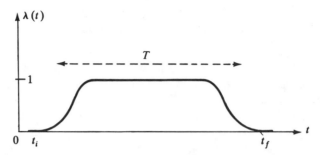

Figure 1. Temporal variation of the parameter $\lambda(t)$ allowing the perturbation V to be adiabatically switched on and switched off. This simulates a collision of duration T.

tion $V = H - H_0$:

$$H = H_0 + V. \tag{A.4}$$

Since any state $|\psi_i\rangle$ or $|\psi_f\rangle$ may always be expanded on the basis $|\varphi_n\rangle$ of eigenstates of H_0, the amplitudes (A.1) can be expressed as a function of the quantities

$$\langle\varphi_p|U(t_f, t_i)|\varphi_n\rangle, \tag{A.5}$$

which represent the transition amplitudes induced by the perturbation V between unperturbed states. In this chapter, we will concern ourselves with these transition amplitudes and with their calculation in the form of a perturbative expansion in powers of V.

When the problem under study can be stated in terms of collisions, it is quite convenient to adiabatically "switch on" and "switch off" the perturbation V by formally multiplying V by a parameter $\lambda(t)$, whose time variations are represented in Figure 1. In this way we can "simulate" the collision of two wave packets, which initially ($t \le t_i$) do not interact because they are too distant from each other, and then, after the collision ($t_f \le t$), are again separated and no longer interact. The limit of the expression (A.5), when the duration T of the collision (see Figure 1) tends to infinity, is simply an element of the scattering matrix S. (The evolution operator U must then be taken in interaction representation with respect to H_0 to eliminate the free evolution exponentials in t_i and in t_f due to H_0; see Complement A_I, §1).

B—TIME DEPENDENCE OF TRANSITION AMPLITUDES

To give some idea of the type of physical information that may be extracted from the preceding transition amplitudes, we are going to distinguish among different cases, depending on whether the initial and final states $|\varphi_i\rangle$ and $|\varphi_f\rangle$ of the process belong to the discrete or to the continuous spectrum of the unperturbed Hamiltonian H_0. For each of these cases, we review the time dependence that quantum mechanics predicts for the matrix elements of the evolution operator, and we show how it is possible to connect the transition amplitudes to measurable physical quantities such as level shifts, lifetimes, cross sections, etc. Such general ideas will be useful for analyzing the physical processes to be reviewed in Chapter II.

1. Coupling between Discrete Isolated States

We begin by considering the case in which the unperturbed Hamiltonian H_0 has one or several discrete eigenstates that are well isolated from all other eigenstates of H_0.

One particularly simple case is that of a single discrete state $|\varphi_1\rangle$, well isolated, having unperturbed energy E_1. Let us now consider the amplitude

$$\tilde{U}_{11}(T) = \langle \varphi_1 | \tilde{U}(T) | \varphi_1 \rangle \tag{B.1.a}$$

where

$$\tilde{U}(T) = e^{iH_0T/2\hbar} U(T) e^{iH_0T/2\hbar} \tag{B.1.b}$$

is the evolution operator between $t_i = -T/2$ and $t_f = +T/2$ in the interaction representation with respect to H_0 (see Complement A_I, §1). $\tilde{U}_{11}(T)$ represents the probability amplitude that the system, prepared in state $|\varphi_1\rangle$ at t_i, is still there a time T later. By inserting on the right or on the left of $U(T) = \exp(-iHT/2\hbar)$ the closure relation on the eigenstates of H, a superposition of exponentials of T appears, one of which has a clearly preponderant weight (zero order in V):

$$\tilde{U}_{11}(T) \simeq |\langle \varphi_1 | \psi_1 \rangle|^2 e^{-i\delta E_1 T/\hbar}. \tag{B.2}$$

In (B.2), $|\psi_1\rangle$ is the eigenstate of H that approaches $|\varphi_1\rangle$ when V tends to zero, and δE_1 is the shift of state $|\varphi_1\rangle$ due to the coupling V, given by the well-known perturbative expansion:

$$\delta E_1 = \langle \varphi_1 | V | \varphi_1 \rangle + \sum_{n \neq 1} \frac{\langle \varphi_1 | V | \varphi_n \rangle \langle \varphi_n | V | \varphi_1 \rangle}{E_1 - E_n} + \cdots . \tag{B.3}$$

The study of the amplitude (B.1.a) allows the level shifts to be calculated. Note that, in the expansion of the amplitude (B.2), it is sufficient to know the term linear in T to obtain δE_1. We will encounter situations of this type in the study of the radiative shift of atomic states caused by the virtual emission and reabsorption of a photon.

Another interesting case is the one in which H_0 has two discrete states, $|\varphi_1\rangle$ and $|\varphi_2\rangle$, well isolated from the others, having the same unperturbed energy $E_1 = E_2$. Consider the amplitude

$$\tilde{U}_{21}(T) = \langle \varphi_2 | \tilde{U}(T) | \varphi_1 \rangle \tag{B.4}$$

which allows us to calculate the probability

$$P_{21}(T) = |\tilde{U}_{21}(T)|^2 \tag{B.5}$$

that the system, initially in the state $|\varphi_1\rangle$, passes after a time T to the state $|\varphi_2\rangle$. It is well known in quantum mechanics (the two-level problem) that $P_{21}(T)$ has an oscillating character. The frequency of this reversible oscillation of the system between $|\varphi_2\rangle$ and $|\varphi_1\rangle$, called the "Rabi nutation frequency", is proportional to the coupling introduced by V between $|\varphi_1\rangle$ and $|\varphi_2\rangle$, either directly (if $\langle \varphi_2 | V | \varphi_1 \rangle$ is nonzero), or indirectly (if $\langle \varphi_2 | V | \varphi_1 \rangle$ is zero) via other levels far from $|\varphi_1\rangle$ and $|\varphi_2\rangle$ (see Complement B_I). Such situations occur, for example, when an atomic system interacts with an intense monochromatic wave (see Complement A_{VI}).

If several discrete eigenstates of H_0 that are close to each other form a group sufficiently well isolated from all the other levels of H_0, then the transition amplitudes between two levels of the group are superpositions of Rabi oscillations with different amplitudes and frequencies. At the limit where the number of coupled states becomes extremely large, the interferences between these different Rabi oscillations eventually give an irreversible character to the evolution of the system. This is what we will study now by considering couplings involving a continuum of eigenstates of H_0.

2. Resonant Coupling between a Discrete Level and a Continuum

In this paragraph we assume that a discrete state $|\varphi_i\rangle$, having energy E_i, is coupled by V to a continuum $|\varphi_f\rangle$ of eigenstates of H_0 (the energy E_f of $|\varphi_f\rangle$ varies continuously).

The calculation of the transition amplitude $\langle \varphi_f | \tilde{U}(T) | \varphi_i \rangle$ is well-known in quantum mechanics. At the lowest order in V, that is, first order, we find (see Complement A_I, §2-b)

$$\tilde{U}_{fi}(T) = \langle \varphi_f | \tilde{U}(T) | \varphi_i \rangle \simeq \delta_{fi} - 2\pi i \delta^{(T)}(E_f - E_i) V_{fi} + \cdots \quad (B.6)$$

where $V_{fi} = \langle \varphi_f | V | \varphi_i \rangle$ is the matrix element of V between $\langle \varphi_f |$ and $| \varphi_i \rangle$ and where $\delta^{(T)}(E_f - E_i)$ is a delta function of width $\hbar/T$ that expresses the conservation of the unperturbed energy to within $\hbar/T$ (uncertainty related to the interaction duration T). In fact, $\delta(T)(E_f - E_i)$ is the Fourier transform of the product of an exponential with frequency $(E_f - E_i)/\hbar$ and a square function of width T. To obtain the transition probability from $| \varphi_i \rangle$ to $| \varphi_f \rangle$, we must take the square of the modulus of (B.6). Because states $| \varphi_i \rangle$ and $| \varphi_f \rangle$ are assumed to be different, the first term of (B.6), δ_{fi}, is zero. In addition, the square of $\delta^{(T)}$ is proportional to $T\delta^{(T)}$ [see Complement A_I, Equation (49)]. The probability of transition from $| \varphi_i \rangle$ to $| \varphi_f \rangle$ is thus proportional to the duration of the interaction, which allows us to define a transition rate equal to

$$w_{fi} = \frac{1}{T} |\tilde{U}_{fi}(T)|^2 \simeq \frac{2\pi}{\hbar} |V_{fi}|^2 \delta^{(T)}(E_f - E_i). \quad (B.7)$$

In fact, the final state $| \varphi_f \rangle$, which belongs to a continuous spectrum, is not normalizable. The quantity that does have a physical meaning is the transition rate toward a group of final states. For example, the sum of (B.7) over all the states $| \varphi_f \rangle$ gives the transition rate Γ of the discrete state $| \varphi_i \rangle$ to any state of the continuum:

$$\Gamma = \sum_f w_{fi} = \frac{2\pi}{\hbar} \sum_f |V_{fi}|^2 \delta^{(T)}(E_f - E_i)$$

$$= \frac{2\pi}{\hbar} |V_{fi}|^2 \rho(E_f = E_i). \quad (B.8)$$

In the second line of (B.8), it is assumed that $|V_{fi}|^2$ depends only on E_f and ρ is the density of final states evaluated at $E_f = E_i$ (Fermi's golden rule).

Another interesting quantity is the probability $|\tilde{U}_{ii}(T)|^2$ that the system will remain in the discrete state $| \varphi_i \rangle$ after a time interval T. The preceding perturbative calculation gives (conservation of the norm):

$$|\tilde{U}_{ii}(T)|^2 = 1 - \sum_f |\tilde{U}_{fi}(T)|^2 \simeq 1 - \Gamma T. \quad (B.9)$$

The probability of finding the system in state $|\varphi_i\rangle$ thus decreases proportionally to Γ. In fact, a nonperturbative calculation of the amplitude $\tilde{U}_{ii}(T)$, to which we will return in Chapter III, is possible (*). It gives

$$\tilde{U}_{ii}(T) = e^{-\Gamma T/2} e^{-i\delta E_i T/\hbar} \tag{B.10}$$

which shows that the discrete state decays exponentially over time with a "lifetime":

$$\tau = \frac{1}{\Gamma}. \tag{B.11}$$

The probability $|\tilde{U}_{ii}(T)|^2$ decays indeed as $\exp(-\Gamma T) = \exp(-T/\tau)$. A shift δE_i of the discrete state also appears as a result of its coupling with the continuum. The expression for δE_i is

$$\delta E_i = \mathscr{P} \sum_f \frac{|V_{fi}|^2}{E_i - E_f} \tag{B.12}$$

where $\mathscr{P}$ denotes the principal part. Thus it is clear that the study of transition amplitudes involving a discrete state and a continuum gives access to important physical quantities, such as lifetimes or level shifts. In Chapter II we will discuss an important example of this type of situation, the spontaneous emission of radiation by a discrete excited atomic state.

Remark

The exponential decay of $|\tilde{U}_{ii}(T)|^2$ is a simple example of irreversible evolution resulting from the superposition of an extremely large number of Rabi oscillations having different frequencies (see end of preceding subsection). It is important, however, to note that such a result holds only if the continuum to which the discrete state $|\varphi_i\rangle$ is coupled is extremely flat, more precisely, if the quantity $|V_{fi}|^2\rho(E_f)$ varies extremely slowly with E_f. If the continuum has structures responsible for rapid variations in $|V_{fi}|^2\rho(E_f)$, damped oscillations may persist in $|\tilde{U}_{ii}(T)|^2$ (see Complement C_{III}).

3. Couplings inside a Continuum or between Continua

It remains for us to consider the case in which both the initial state $|\varphi_i\rangle$ and the final state $|\varphi_f\rangle$ of the physical process being studied belong to the same continuum or to two different continua of eigenstates of H_0. Situa-

---

(*) A simple model of a discrete state coupled to a continuum is also analyzed in Complement C_I. It allows the exponential decay described by expression (B.10) to be simply obtained.

tions of this type are frequently encountered in the study of the scattering of photons by atoms. In this case, the state $|\varphi_i\rangle$ ($|\varphi_f\rangle$) represents the atom in a given energy level in the presence of an incident (scattered) photon. Because the energy of incident and scattered photons may vary in a continuous fashion, $|\varphi_i\rangle$ and $|\varphi_f\rangle$ do indeed belong to a continuum.

For sufficiently large T, the transition amplitude $\tilde{U}_{fi}(T)$ is just an element of the S matrix. The quantum calculation of $\tilde{U}_{fi}(T)$ gives (see Complements A_I and B_{III}):

$$\tilde{U}_{fi}(T) = \delta_{fi} - 2\pi i \delta^{(T)}(E_f - E_i)\mathcal{T}_{fi} \qquad (B.13)$$

where $\delta^{(T)}$ is a delta function of width $\hbar/T$, and where $\mathcal{T}_{fi}$ is the transition matrix which can be expanded in powers of V (Born expansion) as

$$\mathcal{T}_{fi} = \langle\varphi_f|V|\varphi_i\rangle + \langle\varphi_f|V\frac{1}{E_i - H_0 + i\eta}V|\varphi_i\rangle +$$

$$+ \langle\varphi_f|V\frac{1}{E_i - H_0 + i\eta}V\frac{1}{E_i - H_0 + i\eta}V|\varphi_i\rangle + \cdots \qquad (B.14)$$

η being an infinitely small and positive quantity.

Remark

If the Born expansion does not converge, then expression (B.14) becomes meaningless. This occurs, for instance, when H_0 has a discrete eigenstate $|\varphi_k\rangle$ whose energy E_k coincides with that of the initial $|\varphi_i\rangle$ and final $|\varphi_f\rangle$ states ("resonant scattering" case). However, a compact expression of the transition matrix $\mathcal{T}_{fi}$ can be determined (see complement B_{III}, §1.b):

$$\mathcal{T}_{fi} = \langle\varphi_f|V|\varphi_i\rangle + \langle\varphi_f|V\frac{1}{E_i - H + i\eta}V|\varphi_i\rangle \qquad (B.15)$$

where it is H instead of H_0 that appears in the denominator of the second term. We can verify that the formal expansion of the expression (B.15) actually results in the expansion (B.14). Indeed, the identity

$$\frac{1}{A} = \frac{1}{B} + \frac{1}{B}(B - A)\frac{1}{A} \qquad (B.16)$$

applied to $A = E_i - H + i\eta$, $B = E_i - H_0 + i\eta$ gives

$$\frac{1}{E_i - H + i\eta} = \frac{1}{E_i - H_0 + i\eta} + \frac{1}{E_i - H_0 + i\eta} V \frac{1}{E_i - H + i\eta}. \quad (B.17)$$

The iteration of (B.17), when substituted into the second term of (B.15), gives (B.14).

As in subsection B.2, if $|\varphi_f\rangle \neq |\varphi_i\rangle$, the calculation of the transition probability $|\varphi_i\rangle \rightarrow |\varphi_f\rangle$ causes the square of the $\delta^{(T)}$ function, which is proportional to $T\delta^{(T)}$, to appear. It is thus possible to define a transition rate

$$w_{fi} = \frac{1}{T} |\tilde{U}_{fi}(T)|^2 = \frac{2\pi}{\hbar} |\mathcal{T}_{fi}|^2 \delta^{(T)}(E_f - E_i). \quad (B.18)$$

Since the final state $|\varphi_f\rangle$ is not normalizable, only the sum of w_{fi} over a group of final states has a physical meaning. This summation causes the density of final states $\rho(E_f = E_i)$ to appear. Finally, the initial state $|\varphi_i\rangle$ is also nonnormalizable because it belongs to a continuum. However, an incident flux can be associated with such a state $|\varphi_i\rangle$, and it is well known that the ratio of the transition rate from state $|\varphi_i\rangle$ toward a group of final states and the incident flux associated with $|\varphi_i\rangle$ is simply a *scattering cross-section*. We can thus see how it is possible to derive a measurable physical quantity, such as a cross-section, from the transition amplitudes $\tilde{U}_{fi}(T)$ between two states belonging to two continua. In the following, we will give several examples of scattering processes (Rayleigh, Raman, and Compton scattering, photoionization, bremsstrahlung, etc.).

Remark

Expression (B.18) appears as a generalized Fermi golden rule, where the coupling $\mathcal{T}_{fi}$ between $|\varphi_i\rangle$ and $|\varphi_f\rangle$ contains all the orders in V. However, it should be emphasized that expression (B.13), and expression (B.18), which results from it, are valid at all orders in V only if $|\varphi_i\rangle$ and $|\varphi_f\rangle$ both belong to a continuum. In addition, the exact expression (B.15) for $\mathcal{T}_{fi}$ must be used if intermediate resonant states exist. In the case where either one of the two states $|\varphi_i\rangle$ or $|\varphi_f\rangle$ is discrete, the difference between $\tilde{U}_{fi}(T)$ and δ_{fi} is proportional to $\delta^{(T)}(E_f - E_i)$ only at the lowest order in V.

C—APPLICATION TO ELECTRODYNAMICS

1. Coulomb Gauge Hamiltonian

We will now consider the electrodynamics case. The system under study consists of an ensemble of charged particles α, having charge q_α and mass m_α, interacting with an electromagnetic field. Let $\mathbf{r}_\alpha$ and $\mathbf{p}_\alpha = (\hbar/i)\nabla_\alpha$ be the position and momentum of the particle α, and $\mathbf{A}(\mathbf{r})$ be the potential vector of the field which, in the Coulomb gauge, is transverse ($\mathbf{A} = \mathbf{A}_\perp$). The Hamiltonian that describes the dynamics of this system is written, in the Coulomb gauge (see Appendix, §3),

$$H = \sum_\alpha \frac{1}{2m_\alpha} [\mathbf{p}_\alpha - q_\alpha \mathbf{A}(\mathbf{r}_\alpha)]^2 - \sum_\alpha \frac{g_\alpha q_\alpha}{2m_\alpha} \mathbf{S}_\alpha \cdot \mathbf{B}(\mathbf{r}_\alpha) + V_{\text{Coul}} + H_R.$$

$$(\text{C.1})$$

The first term of (C.1) represents the kinetic energies of the particles, since the velocity $\dot{\mathbf{r}}_\alpha$ of the particle α is

$$\dot{\mathbf{r}}_\alpha = \frac{1}{i\hbar}[\mathbf{r}_\alpha, H] = \frac{\partial H}{\partial \mathbf{p}_\alpha} = \frac{1}{m_\alpha}[\mathbf{p}_\alpha - q_\alpha \mathbf{A}(\mathbf{r}_\alpha)]. \qquad (\text{C.2})$$

The second term of (C.1) represents the interaction of the spin magnetic moments of the particles ($\mathbf{S}_\alpha$ is the spin of the particle α; g_α is its Landé factor) with the magnetic field $\mathbf{B}$ of the radiation field evaluated at the points where the particles are located. The third term, V_{Coul}, is the Coulomb energy of the system of particles. It is the sum of the Coulomb interaction energies between pairs of particles (α, β) and of the Coulomb self-energies $\varepsilon_{\text{Coul}}^\alpha$ of each particle

$$V_{\text{Coul}} = \sum_{\alpha \neq \beta} \frac{q_\alpha q_\beta}{8\pi\varepsilon_0} \frac{1}{|\mathbf{r}_\alpha - \mathbf{r}_\beta|} + \sum_\alpha \varepsilon_{\text{Coul}}^\alpha. \qquad (\text{C.3})$$

$\varepsilon_{\text{Coul}}^\alpha$ is a constant given by formula (43) in the Appendix. Finally, the last term H_R is the energy of the transverse field (electric $\mathbf{E}_\perp$ and magnetic $\mathbf{B}$):

$$H_R = \frac{\varepsilon_0}{2} \int d^3r \left[\mathbf{E}_\perp^2(\mathbf{r}) + c^2 \mathbf{B}^2(\mathbf{r}) \right] \qquad (\text{C.4.a})$$

which can also be expressed simply as function of the annihilation and creation operators a_j and a_j^+ of a photon in the normal vibrational "mode" j of the field (identified by the wave vector $\mathbf{k}_j$, the polarization $\boldsymbol{\varepsilon}_j$, and the frequency $\omega_j = ck_j$)

$$H_R = \sum_j \hbar\omega_j \left(a_j^+ a_j + \tfrac{1}{2}\right). \qquad \text{(C.4.b)}$$

Each mode j is thus associated with a one-dimensional harmonic oscillator. The eigenstate $|n_j\rangle$ of such an oscillator (with $n_j = 0, 1, 2 \dots$) represents a state in which the mode j contains n_j photons having energy $\hbar\omega_j$, momentum $\hbar\mathbf{k}_j$ and polarization $\boldsymbol{\varepsilon}_j$. Finally, recall that the different transverse fields $\mathbf{E}_\perp$ and $\mathbf{B}$, as well as $\mathbf{A}$, can be expressed as a linear combination of operators a_j and a_j^+ (see expressions (29), (30), and (28) in the Appendix).

For what follows, it is useful to rewrite the Hamiltonian H in the form

$$H = \sum_\alpha \frac{\mathbf{p}_\alpha^2}{2m_\alpha} + H_R + V_{\text{Coul}} + H_{I1} + H_{I2} + H_{I1}^S \qquad \text{(C.5.a)}$$

with

$$H_{I1} = -\sum_\alpha \frac{q_\alpha}{m_\alpha} \mathbf{p}_\alpha \cdot \mathbf{A}(\mathbf{r}_\alpha) \qquad \text{(C.5.b)}$$

$$H_{I1}^S = -\sum_\alpha g_\alpha \frac{q_\alpha}{2m_\alpha} \mathbf{S}_\alpha \cdot \mathbf{B}(\mathbf{r}_\alpha) \qquad \text{(C.5.c)}$$

$$H_{I2} = \sum_\alpha \frac{q_\alpha^2}{2m_\alpha} [\mathbf{A}(\mathbf{r}_\alpha)]^2. \qquad \text{(C.5.d)}$$

The splitting of H into $H_0 + V$ may be achieved in several ways. We will now discuss two that lead to different types of perturbative expansions for the transition amplitudes.

2. Expansion in Powers of the Charges q_α

A first possibility consists of gathering in H_0 all the terms independent of q_α:

$$H_0 = \sum_\alpha \frac{\mathbf{p}_\alpha^2}{2m_\alpha} + H_R. \qquad \text{(C.6.a)}$$

The charges q_α thus appear as coupling parameters characterizing the strength of the perturbation:

$$V = V_{\text{Coul}} + H_{I1} + H_{I1}^S + H_{I2}. \tag{C.6.b}$$

The eigenstates of H_0 thus simply represent the free particles (eigenstates of $\mathbf{p}_\alpha^2/2m_\alpha$) in the presence of transverse photons (eigenstates of H_R). No bound state of the particles can appear in H_0, because no coupling, either direct or indirect, can exist between the particles.

All terms contained in V—the Coulomb interaction as well as the interaction between the particles and the transverse field—will then be treated in a perturbative fashion. The physics problems that can be studied simply with such an approach are thus essentially particle-particle or particle-photon scattering problems (Coulomb scattering, Compton scattering, bremsstrahlung emission, etc.).

3. Expansion in Powers of the Interaction with the Transverse Field

The second splitting of H that we will consider here consists of collecting in H_0 all the terms that depend either on the dynamical variables of the particles, or on the dynamical variables of the transverse field, but not on both at the same time:

$$H_0 = H_P + H_R \tag{C.7.a}$$

where

$$H_P = \sum_\alpha \frac{\mathbf{p}_\alpha^2}{2m_\alpha} + V_{\text{Coul}} \tag{C.7.b}$$

is a particle Hamiltonian. The perturbation V thus contains all terms containing both the particle operators and transverse field operators:

$$V = H_{I1} + H_{I1}^S + H_{I2}. \tag{C.7.c}$$

The perturbative expansion in powers of V is thus an expansion in powers of the interaction between the particles and the transverse field.

Since the Coulomb interaction V_{Coul} is included in the particle Hamiltonian, such a Hamiltonian can now describe bound states. In fact, the eigenstates of H_P are characterized by two types of quantum numbers: on the one hand, the *external* quantum numbers, describing the motion of the center of mass of the group of particles, and on the other hand, the *internal* quantum numbers, describing the excitation state of the system of particles in the center-of-mass reference frame. For example, if the particles under study form an atom or a molecule, the internal quantum numbers indicate the energy levels of this atom or this molecule. The eigenstates of H_0 thus describe situations in which systems of particles, such as atoms, molecules, ions, etc. in well-defined internal states and with a well-defined global momentum, are in the presence of a certain number of transverse photons characterized by well-defined energy, momentum, and polarization.

The perturbative treatment of V then allows us to describe the processes of absorption or emission of photons by such systems of particles. Further on we will give a diagrammatic representation of the transition amplitudes induced by V that allow these different processes to be visualized.

Remark

When the system of particles consists of several distinct systems $A, B, \ldots$ (atoms, molecules, ions, electrons, etc.) well separated from each other, it may be advantageous to retain in H_0 only the Coulomb energies inside each system, V_{Coul}^{AA}, V_{Coul}^{BB}, and to include in the interaction Hamiltonian V the Coulomb interaction energy between different systems, $V_{\text{Coul}}^{AB} \ldots$. The eigenstates of H_0 thus represent situations in which the different systems $A, B, \ldots$ are in well-defined external and internal states, without mutual interactions, and in the presence of a certain number of photons.

4. Advantages of Including the Coulomb Interaction in the Particle Hamiltonian

The first advantage of the splitting (C.7) as compared to (C.6) is that it allows us to consider states where charged particles are bound by the Coulomb interaction (atoms, molecules, ions, etc.) as unperturbed states. Such states can indeed exist before the switching on and survive after the switching off of V. (See Figure 1.) With the other splitting (C.6), no bound state could exist in the absence of V (*). Atoms, molecules, ions, etc. would dissociate into nuclei and electrons before the switching on and after the switching off of V, and it would be much more difficult to describe the emission, absorption, or scattering of photons by such systems.

(*) Of course, we ignore here any interactions other than electromagnetic ones, such as, for example, strong interactions responsible for the cohesion of nuclei.

The second advantage of the splitting (C.7) is that V_{Coul} is, in the nonrelativistic domain, much more important than the interaction with the transverse field. It is thus natural to start with the eigenstates of H_0 and deal with the effect of interactions with the transverse field as a perturbation.

Remarks

(i) The elimination of the scalar potential and the longitudinal vector potential causes the Coulomb interaction to appear as a purely particle term in the Coulomb gauge representation. In another gauge, for example, in the Lorentz gauge, the Coulomb interaction appears only in a second-order treatment of the interaction between particles and scalar and longitudinal potentials (*). The advantages of the splitting (C.7) that we just discussed are thus closely connected to the choice of the Coulomb gauge.

(ii) Such advantages also exist in other formulations of quantum electrodynamics equivalent to Coulomb gauge electrodynamics, such as, for example, the one leading to the electric dipole Hamiltonian (see Appendix, §5). The new Hamiltonian H', which is the unitary transform of $H = H_0 + V$, may also be written in the form $H' = H_0 + V'$, where H_0 has the same expression as in (C.7) and contains in particular the Coulomb interaction.

5. Diagrammatic Representation of Transition Amplitudes

Let us use, for example, the splitting (C.7). The perturbative expansion of the evolution operator causes the transition amplitudes to appear as products of matrix elements of V between eigenstates of H_0 and of free evolution exponentials between two interactions, these products being summed over all times and over all intermediate states (see Complement A_I, §2.a):

$$\langle \varphi_f | U(t_f, t_i) | \varphi_i \rangle = \delta_{fi}\, e^{-iE_i(t_f - t_i)/\hbar} +$$

$$+ \sum_{n=1}^{\infty} \left(\frac{1}{i\hbar} \right)^n \int_{t_f \geq \tau_n \cdots \tau_2 \geq \tau_1 \geq t_i} d\tau_n \cdots d\tau_2\, d\tau_1 \times$$

$$\times \sum_{\varphi_{n-1} \cdots \varphi_1} e^{iE_f(t_f - \tau_n)/\hbar} \langle \varphi_f | V | \varphi_{n-1} \rangle e^{-iE_{n-1}(\tau_n - \tau_{n-1})/\hbar} \cdots$$

$$\cdots \langle \varphi_2 | V | \varphi_1 \rangle e^{-iE_1(\tau_2 - \tau_1)/\hbar} \langle \varphi_1 | V | \varphi_i \rangle e^{-iE_i(\tau_1 - t_i)/\hbar}. \qquad \text{(C.8)}$$

In what follows, it will be convenient to use diagrammatic representations

(*) See, for example, Complement B_V of *Photons and Atoms—Introduction to Quantum Electrodynamics*.

of these products, because such representations allow one to visualize the physical processes involved in the transition amplitude.

The free propagation of particles will be represented by straight lines, while that of photons will be represented by wavy lines. Next to each line will be quantum numbers indicating the corresponding free state ($a, b, \ldots$ for the state of particles, wave vector $\mathbf{k}$, and polarization $\boldsymbol{\varepsilon}$ for photons). Each of these lines has associated with it exponential factors of the type $\exp[-iE(\tau_k - \tau_l)/\hbar]$, describing the free evolution of the state of energy E between times τ_k and τ_l associated with the ends of the line (these times are not explicitly shown on the diagram to keep it as simple as possible). Each diagram is read going from bottom to top (following the time course of the process). The lines for particles and for photons coming from the bottom correspond to the free state in the ket $|\varphi_i\rangle$ of the matrix element; the lines for particles and for photons going toward the top correspond to the free state in the bra $\langle\varphi_f|$. The matrix elements of V are represented by the points at which these lines intersect (vertex), with the matrix element being taken between the state described by the lines which arrive at the vertex and that described by those which leave from it. A photon line arriving and disappearing at this point (Figure 2a) corresponds to the annihilation of a photon (a term of H_{I1} and H_{I1}^S). A photon line leaving from this point (Figure 2b) corresponds to the creation of a photon (a^+ term of H_{I1} and H_{I1}^S). The two-photon terms of H_{I2} corre-

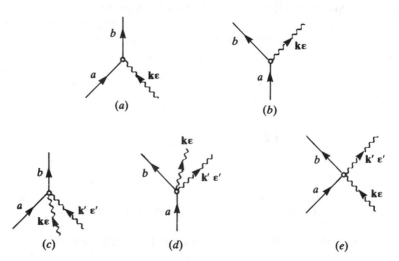

Figure 2. Different types of vertex corresponding to different matrix elements of V.

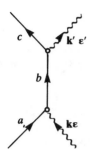

Figure 3. Diagrammatic representation of transition amplitude (C.9).

spond to the three Figures $2c$ (a^2 term annihilating two photons), $2d$ (a^{+2} term creating two photons), and $2e$ (aa^+ or a^+a terms creating one photon while annihilating another).

In many processes, there is more than one interaction between atoms and fields. The corresponding diagram thus contains several vertices and several lines associated with the free propagation between interactions. It symbolizes the product of all the quantities (matrix elements of V, free evolution exponentials) represented in this way. For example, the diagram in Figure 3 symbolizes the quantity

$$\exp\left[-i(E_c + \hbar\omega')(t_f - \tau_2)/\hbar\right]\langle c; \mathbf{k}'\boldsymbol{\varepsilon}'|V|b; 0\rangle \times$$
$$\times \exp\left[-iE_b(\tau_2 - \tau_1)/\hbar\right]\langle b; 0|V|a; \mathbf{k}\boldsymbol{\varepsilon}\rangle \times \qquad \text{(C.9)}$$
$$\times \exp\left[-i(E_a + \hbar\omega)(\tau_1 - t_i)/\hbar\right]$$

and represents the amplitude that the system of particles, initially in state a, absorbs at instant τ_1 an incident photon $\mathbf{k}\boldsymbol{\varepsilon}$ and passes into state b, then finishes at instant τ_2 in state c by emitting a photon $\mathbf{k}'\boldsymbol{\varepsilon}'$. Other types of diagrams will be introduced later on in this book.

Remark

In other formulations of quantum electrodynamics equivalent to Coulomb gauge electrodynamics, the interaction Hamiltonian V' may have matrix elements that are more simple than V. For example, in the electric dipole point of view (see Appendix, §5), the interaction Hamiltonian (of the form $-q\mathbf{E}_\perp \cdot \mathbf{r}$) contains only one-photon terms that are linear in a and a^+. There are no more two-photon terms such as those coming from $\mathbf{A}^2$ and resulting in matrix elements of the same type as those represented in Figures $2c$, $2d$, and $2e$. This results in a great simplification in the higher order terms of the perturbative expansion of the transition amplitudes.

GENERAL REFERENCES

For the importance of transition amplitudes, see Feynman, Volume III, Chapters 3 and 7, Cohen-Tannoudji, Diu, and Laloë, Chapter III, Section E; Levy-Leblond and Balibar, Chapters 4 and 5.

The perturbative calculation of transition amplitudes may be found in many quantum mechanics books. See, for example, Messiah, Chapter XVII; Cohen-Tannoudji, Diu, and Laloë, Chapter XIII; Merzbacher, Chapter XVIII; Schiff, Chapters 8 and 9; Feynman and Hibbs, Chapter VI. References concerning transition amplitudes between two continua (collision problems) will be given in Chapter III.

For Coulomb gauge electrodynamics, see the Appendix and the references therein.

COMPLEMENT A$_I$

PERTURBATIVE CALCULATION OF TRANSITION AMPLITUDES—SOME USEFUL RELATIONS

In Chapter I, without presenting any formal proofs, we used several results concerning transition amplitudes. The goal of this complement is to sketch a brief derivation of these results and to gather several useful relations to which we will continue to refer throughout this book.

We will begin (§1) by introducing the interaction representation and by emphasizing its advantages. We will then (§2) proceed to the perturbative calculation of transition amplitudes and to the determination of the lowest-order terms (order $0, 1, 2$) in the expansion of these amplitudes in powers of the coupling V. Finally, we will study (§3) the transition probability from an initial state toward a final state by distinguishing several cases according to the discrete or continuous nature of the energy spectrum.

1. Interaction Representation

As in Chapter I, the Hamiltonian H of the system under study is split into

$$H = H_0 + V. \tag{1}$$

The unperturbed Hamiltonian H_0 is assumed to be time independent. The coupling V may or may not depend on time [for example, if V is multiplied by the parameter $\lambda(t)$, whose temporal variations are shown in Figure 1 of the chapter].

Going from the usual Schrödinger representation to the interaction representation with respect to H_0 is achieved by applying the unitary transformation

$$T(t) = e^{iH_0(t-t_0)/\hbar} \tag{2}$$

to the vectors $|\psi(t)\rangle$ and operators A of the Schrödinger representation, t_0 being a reference instant that we take as the origin of time ($t_0 = 0$). If

$|\tilde{\psi}(t)\rangle$ and $\tilde{A}(t)$ represent the vectors and operators in the new representation, then we have

$$\left|\tilde{\psi}(t)\right\rangle = e^{iH_0t/\hbar}\left|\psi(t)\right\rangle \tag{3.a}$$

$$\tilde{A}(t) = e^{iH_0t/\hbar}\, A\, e^{-iH_0t/\hbar}. \tag{3.b}$$

If V were zero, the interaction representation would be identical to the Heisenberg representation, and $|\tilde{\psi}(t)\rangle$ would remain fixed over time. It follows that, in the general case where V is nonzero, $|\tilde{\psi}(t)\rangle$ evolves only as a result of the presence of the coupling V. To see this more precisely, let us determine the evolution equation of $|\tilde{\psi}(t)\rangle$ by applying $i\hbar d/dt$ to (3.a) and by using the Schrödinger equation for $i\hbar d|\psi(t)\rangle/dt$. We get

$$i\hbar\frac{d}{dt}\left|\tilde{\psi}(t)\right\rangle = -H_0\left|\tilde{\psi}(t)\right\rangle + e^{iH_0t/\hbar}(H_0 + V)\left|\psi(t)\right\rangle$$

$$= \tilde{V}(t)\left|\tilde{\psi}(t)\right\rangle \tag{4}$$

where

$$\tilde{V}(t) = e^{iH_0t/\hbar}\, V\, e^{-iH_0t/\hbar}. \tag{5}$$

Equation (4) clearly shows that the rate of variation of $|\tilde{\psi}(t)\rangle$ is at least first order in V. In particular, in the study of a collision process, where the coupling V can be switched off in the remote past and the far future (see Figure 1 in Chapter I), the state vector does not evolve in the interaction representation before the collision begins and after it ends. This allows us to understand how the matrix elements of the evolution operator $\tilde{U}(t_f, t_i)$ in interaction representation have a well-defined limit when t_f and t_i tend, respectively, to $+\infty$ and $-\infty$ (scattering matrix). It is useful also for what follows to determine the relation that exists between $\tilde{U}(t_f, t_i)$ and the evolution operator $U(t_f, t_i)$ in the Schrödinger representation. The equation

$$\left|\psi(t_f)\right\rangle = U(t_f, t_i)\left|\psi(t_i)\right\rangle \tag{6}$$

gives, taking in account (3.a)

$$\left|\tilde{\psi}(t_f)\right\rangle = \tilde{U}(t_f, t_i)\left|\tilde{\psi}(t_i)\right\rangle \tag{7}$$

with

$$\tilde{U}(t_f, t_i) = e^{iH_0t_f/\hbar} U(t_f, t_i) e^{-iH_0t_i/\hbar}. \tag{8}$$

2. Perturbative Expansion of Transition Amplitudes

a) PERTURBATIVE EXPANSION OF THE EVOLUTION OPERATOR

The evolution operator $U(t_f, t_i)$ of the Schrödinger representation is defined by (6), and satisfies the initial condition

$$U(t_i, t_i) = \mathbb{1}. \tag{9}$$

Taking into account the Schrödinger equation satisfied by $|\psi(t_f)\rangle$, Equation (6) is equivalent to the integral equation

$$U(t_f, t_i) = U_0(t_f, t_i) + \frac{1}{i\hbar} \int_{t_i}^{t_f} dt \, U_0(t_f, t) V U(t, t_i) \tag{10}$$

where

$$U_0(t_f, t_i) = e^{-iH_0(t_f - t_i)/\hbar} \tag{11}$$

is the unperturbed evolution operator associated with H_0. To prove this equivalence, it is sufficient to verify that the operator U defined by (10) actually satisfies (9) and the evolution equation

$$i\hbar \frac{d}{dt_f} U(t_f, t_i) = (H_0 + V) U(t_f, t_i). \tag{12}$$

By successive iterations, Equation (10) thus leads to the well-known perturbative expansion of the evolution operator

$$U(t_f, t_i) = U_0(t_f, t_i) + \sum_{n=1}^{\infty} U^{(n)}(t_f, t_i) \tag{13.a}$$

with

$$U^{(n)}(t_f, t_i) = \left(\frac{1}{i\hbar}\right)^n \int_{t_f \geq \tau_n \cdots \tau_2 \geq \tau_1 \geq t_i} d\tau_n \cdots d\tau_2 \, d\tau_1 \times$$
$$\times e^{-iH_0(t_f - \tau_n)/\hbar} V \cdots V e^{-iH_0(\tau_2 - \tau_1)/\hbar} V e^{-iH_0(\tau_1 - t_i)/\hbar}. \tag{13.b}$$

The structure of the term (13.b) is that of a product of $(n + 1)$ unperturbed evolution operators separated by n interaction operators V. By taking the matrix elements of (13.b) between the eigenstates $\langle \varphi_f |$ and $| \varphi_i \rangle$ of H_0 and by inserting $(n - 1)$ times the closure relation on the eigenstates of H_0, between two successive V operators, formula (C.8) of the chapter is obtained.

Equation (8) finally allows us to obtain the perturbative expansion of the evolution operator in the interaction representation

$$\tilde{U}(t_f, t_i) = \mathbb{1} + \sum_{n=1}^{\infty} \tilde{U}^{(n)}(t_f, t_i) \tag{14.a}$$

$$\tilde{U}^{(n)}(t_f, t_i) = \left(\frac{1}{i\hbar}\right)^n \int_{t_f \geq \tau_n \cdots \tau_2 \geq \tau_1 \geq t_i} d\tau_n \cdots d\tau_2 \, d\tau_1 \tilde{V}(\tau_n) \cdots \tilde{V}(\tau_2)\tilde{V}(\tau_1). \tag{14.b}$$

Comparison of (14.b) and (13.b) demonstrates that using the interaction representation eliminates the free evolution exponentials $\exp(-iH_0 t_f/\hbar)$ and $\exp(iH_0 t_i/\hbar)$ relative to the initial and final times. Let $\mathscr{S}_{fi}$ be the matrix element of $\tilde{U}(t_f, t_i)$ between the eigenstates $\langle \varphi_f |$ and $| \varphi_i \rangle$ of H_0

$$\mathscr{S}_{fi} = \langle \varphi_f | \tilde{U}(t_f, t_i) | \varphi_i \rangle. \tag{15}$$

The perturbative expansion (14) thus yields

$$\mathscr{S}_{fi} = \delta_{fi} + \sum_{n=1}^{\infty} \mathscr{S}_{fi}^{(n)} \tag{16.a}$$

$$\mathscr{S}_{fi}^{(n)} = \langle \varphi_f | \tilde{U}^{(n)}(t_f, t_i) | \varphi_i \rangle. \tag{16.b}$$

We will now calculate the first- and second-order terms in V of this perturbative expansion of the transition amplitude $\mathscr{S}_{fi}$.

b) FIRST-ORDER TRANSITION AMPLITUDE

Using (16.b), (14.b), and (5), we get

$$\mathscr{S}_{fi}^{(1)} = \frac{1}{i\hbar} \int_{t_i}^{t_f} d\tau_1 \, V_{fi} \, e^{i(E_f - E_i)\tau_1/\hbar} \tag{17}$$

where we have set $V_{fi} = \langle \varphi_f | V | \varphi_i \rangle$. We assume here that V is time independent and select the origin of time such that

$$t_i = -T/2 \qquad t_f = +T/2 \tag{18}$$

where T is the duration of the interaction. The integral over τ_1 of (17) can thus be calculated and it yields

$$\mathcal{S}_{fi}^{(1)} = -2\pi i V_{fi} \delta^{(T)}(E_f - E_i) \tag{19}$$

where

$$\delta^{(T)}(E_f - E_i) = \frac{1}{2\pi} \int_{-T/2}^{+T/2} \frac{d\tau_1}{\hbar} \, e^{i(E_f - E_i)\tau_1 / \hbar}$$

$$= \frac{1}{\pi} \frac{\sin(E_f - E_i)T/2\hbar}{(E_f - E_i)}. \tag{20}$$

According to the first equality in (20), $\delta^{(T)}(E_f - E_i)$ tends to $\delta(E_f - E_i)$ when $T \to \infty$. In fact, according to the second equality in (20), $\delta^{(T)}(E_f - E_i)$ is a diffraction function, represented in Figure 1. Its maximal amplitude $T/2\pi\hbar$ is obtained for $E_f - E_i = 0$, and its width is on the order of $4\pi\hbar/T$ (distance between the first two zeros on either side of the

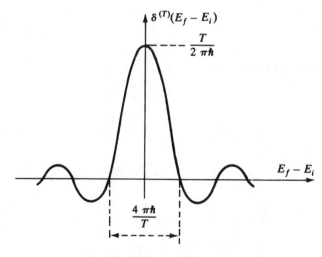

Figure 1. Variations of the function $\delta^{(T)}(E_f - E_i)$ versus $(E_f - E_i)$.

maximum). Its integral equals 1. Hence this function is an approximate delta function expressing the conservation of energy with an uncertainty $\hbar/T$ due to the finite duration of the interaction.

Remark

If V is multiplied by the function $\lambda(t)$ shown in Figure 1 of Chapter I, relation (19) remains valid but $\delta^{(T)}$ is then the Fourier transform of a square function with a width on the order of T, with smooth edges. The function $\delta^{(T)}$ is in this case an apodized diffraction function for which the integral is always equal to 1 and which can still be considered as a delta function of width $\hbar/T$.

c) SECOND-ORDER TRANSITION AMPLITUDE

For second order, Equations (16.b), (14.b), and (5) yield

$$\mathscr{S}_{fi}^{(2)} = \left(\frac{1}{i\hbar}\right)^2 \int_{+T/2 \geq \tau_2 \geq \tau_1 \geq -T/2} d\tau_1\, d\tau_2 \quad \times$$
$$\times \sum_k V_{fk} V_{ki}\, e^{i(E_f - E_k)\tau_2/\hbar}\, e^{i(E_k - E_i)\tau_1/\hbar}. \tag{21}$$

The sum over the intermediate states $|\varphi_k\rangle$ represents a sum over the eigenstates of H_0. To eliminate the restriction $\tau_2 \geq \tau_1$, the integrand in (21) is multiplied by the Heaviside function $\theta(\tau_2 - \tau_1)$, which is equal to 1 for $\tau_2 > \tau_1$ and equal to 0 for $\tau_2 < \tau_1$. We then use the identity

$$e^{-iE_k(\tau_2 - \tau_1)/\hbar}\, \theta(\tau_2 - \tau_1) = \lim_{\eta \to 0_+} -\frac{1}{2\pi i} \int_{-\infty}^{+\infty} \frac{e^{-iE(\tau_2 - \tau_1)/\hbar}}{E + i\eta - E_k}\, dE \tag{22}$$

which can be easily checked by calculating the integral of (22) by the method of residues. Equation (21) may thus be rewritten

$$\mathscr{S}_{fi}^{(2)} = \left(\frac{1}{i\hbar}\right)^2 \left(\frac{-1}{2\pi i}\right) \int_{-T/2}^{+T/2} d\tau_2 \int_{-T/2}^{+T/2} d\tau_1 \times$$
$$\times \int_{-\infty}^{+\infty} dE\, e^{i(E_f - E)\tau_2/\hbar}\, e^{-i(E_i - E)\tau_1/\hbar}\, W_{fi}(E) \tag{23}$$

where

$$W_{fi}(E) = \lim_{\eta \to 0_+} \sum_k \frac{V_{fk} V_{ki}}{E + i\eta - E_k}. \tag{24}$$

From subsection 2-b above, the integrals over τ_2 and τ_1 give, respectively, $2\pi\hbar\delta^{(T)}(E_f - E)$ and $2\pi\hbar\delta^{(T)}(E_i - E)$. These two functions are practically zero as soon as E differs from E_f for the former and from E_i for the latter by more than $\hbar/T$. Their product is thus a function of E which is nonzero if E_f and E_i are equal within $\hbar/T$, and if E also coincides with this common value of E_i and E_f within $\hbar/T$.

To carry out the latter integral over E which remains to be done in (23), we must now compare the width in E of $\delta^{(T)}(E_f - E)\delta^{(T)}(E_i - E)$ which is on the order of $\hbar/T$ about $E_i \simeq E_f$, to the width of the other function $W_{fi}(E)$ which appears in (23). In many cases, which we will describe further on, $W_{fi}(E)$ is a function of E that varies only slightly with E over an interval of width $\hbar/T$. It is thus possible to replace $W_{fi}(E)$ by $W_{fi}(E_i)$ in (23) and to take $W_{fi}(E_i)$ out of the integral, which gives

$$\mathscr{S}_{fi}^{(2)} = -\frac{1}{2\pi i}\frac{4\pi^2\hbar^2}{(i\hbar)^2}\left[\lim_{\eta\to 0_+}\sum_k \frac{V_{fk}V_{ki}}{E_i - E_k + i\eta}\right] \times$$

$$\times \int_{-\infty}^{+\infty} dE\, \delta^{(T)}(E - E_i)\delta^{(T)}(E - E_f). \tag{25}$$

The integral over E of (25) can be easily calculated from the integral form (20) of $\delta(T)$

$$\int_{-\infty}^{+\infty} dE\, \delta^{(T)}(E - E_i)\,\delta^{(T)}(E - E_f) =$$

$$= \frac{1}{4\pi^2\hbar^2}\int_{-\infty}^{+\infty} dE \int_{-T/2}^{T/2} d\tau \int_{-T/2}^{T/2} d\tau'\, e^{i(E-E_f)\tau/\hbar}\, e^{i(E-E_i)\tau'/\hbar}. \tag{26}$$

(Note that there is no restriction on the temporal order of τ and τ'.) The integral over E of (26) gives $2\pi\hbar\delta(\tau + \tau')$, so that (26) reduces to

$$\int_{-\infty}^{+\infty} dE\, \delta^{(T)}(E - E_i)\,\delta^{(T)}(E - E_f) =$$

$$= \frac{1}{2\pi\hbar}\int_{-T/2}^{+T/2} d\tau\, e^{i(E_i-E_f)\tau/\hbar} = \delta^{(T)}(E_i - E_f). \tag{27}$$

Finally, by substituting (27) into (25), we obtain

$$\mathscr{S}_{fi}^{(2)} = -2\pi i\left[\lim_{\eta\to 0_+}\sum_k \frac{V_{fk}V_{ki}}{E_i - E_k + i\eta}\right]\delta^{(T)}(E_i - E_f). \tag{28}$$

Let us now specify the cases for which $W_{fi}(E)$ may be indeed considered to be a function of E varying slowly on the scale of $\hbar/T$. Essentially, there must not be any discrete eigenstates $|\varphi_k\rangle$ of H_0 for which the energy E_k is extremely close to E_i and E_f. In such a case, the function $(E - E_k + i\eta)^{-1}$ would vary quite rapidly with E near $E = E_i$, because it diverges for $E = E_k$. By contrast, note that H_0 may have a continuous spectrum E_k near $E_i \simeq E_f$. In this case, the sum over k in (24) becomes actually an integral over E_k. The fraction $1/(E + i\eta - E_k)$ is then expressed as a function of $\delta(E - E_k)$ and $\mathcal{P}[1/(E - E_k)]$ and yields, after integration over E_k, functions of E which have no reason to diverge near $E_i \simeq E_f$. For sufficiently large T, these functions may thus be considered as varying slowly on the scale of $\hbar/T$.

Finally, summarizing the results obtained in this paragraph, we may write

$$\mathcal{S}_{fi} = \delta_{fi} - 2\pi i \delta^{(T)}(E_f - E_i)\left[V_{fi} + \lim_{\eta\to 0_+}\sum_k \frac{V_{fk}V_{ki}}{E_i - E_k + i\eta}\right] + 0(V^3).$$

(29)

Remarks

(i) If either the initial state $|\varphi_i\rangle$ or the final state $|\varphi_f\rangle$ or both states are discrete, the intermediate state $|\varphi_k\rangle$ may be the same as this discrete state in the sum (24), which leads to a divergence of $W_{fi}(E)$ near E_i or E_f. This may, however, occur only if the diagonal element of V in the discrete state, V_{ii} (or V_{ff}), is nonzero. We can always put these diagonal elements back into H_0 (which amounts to replacing E_i by $E_i + V_{ii}$). It is then possible to use expression (28) for $\mathcal{S}_{fi}^{(2)}$. Nevertheless, for orders higher than 2, it is, in general, impossible to prevent $|\varphi_i\rangle$ (or $|\varphi_f\rangle$) from appearing as an intermediate state.

(ii) In Chapter III, we will introduce methods for calculating transition amplitudes that are more powerful than those discussed in this complement. These methods allow us to sum up all the terms of the perturbation expansion in which one or more discrete states, which appear as intermediate states, might lead to divergences of the transition amplitudes because their energy is extremely close to that of the initial and final states. We will thus obtain nonperturbative expressions for these transition amplitudes that are valid under conditions for which expression (29) is not. The same approach allows us to generalize expression (29) of $\mathcal{S}_{fi}$ for higher orders in the case of a scattering problem where $|\varphi_i\rangle$ and $|\varphi_f\rangle$ both belong to the continuous spectrum of H_0. We obtain in this case

$$\mathcal{S}_{fi} = \delta_{fi} - 2\pi i \delta(E_f - E_i)\mathcal{T}_{fi}$$

(30)

where the transition matrix $\mathscr{T}_{fi}$ is given by expression (B.15) in Chapter I. When it is valid, the expansion in powers of V of $\mathscr{T}_{fi}$ is the Born expansion, the first two terms of which appear inside the brackets in (29).

3. Transition Probability

a) CALCULATION OF THE TRANSITION PROBABILITY TO A FINAL STATE DIFFERENT FROM THE INITIAL STATE

Assume that $|\varphi_f\rangle$ is different from $|\varphi_i\rangle$. The first term, δ_{fi}, of (29) is thus zero. The transition probability $\mathscr{P}_{fi}(T)$ from $|\varphi_i\rangle$ to $|\varphi_f\rangle$ at the end of time T is

$$\mathscr{P}_{fi}(T) = |\mathscr{S}_{fi}|^2 = 4\pi^2 \left[\delta^{(T)}(E_i - E_f)\right]^2 \times$$

$$\times \left| V_{fi} + \lim_{\eta \to 0_+} \sum_k \frac{V_{fk}V_{ki}}{E_i - E_k + i\eta} \right|^2 + \cdots . \quad (31)$$

From (20), the function

$$\left[\delta^{(T)}(E_f - E_i)\right]^2 = \frac{1}{\pi^2} \frac{\sin^2(E_f - E_i)T/2\hbar}{(E_f - E_i)^2} \quad (32)$$

is the square of a diffraction function. For $E_f = E_i$, $[\delta^{(T)}(E_f - E_i)]^2$ reaches its maximum value, $T^2/4\pi^2\hbar^2$. The distance between the first two zeros of this function on either side of its maximum is, as for $\delta^{(T)}$, equal to $4\pi\hbar/T$ and gives some idea of its width. Later on, we will need the integral over E_f of this function, which is on the order of $(T^2/4\pi^2\hbar^2) \times (4\pi\hbar/T)$, that is, on the order of $T/\pi\hbar$. In fact, from (27), we obtain

$$\int_{-\infty}^{+\infty} dE_f \left[\delta^{(T)}(E_f - E_i)\right]^2 = \frac{T}{2\pi\hbar}. \quad (33)$$

b) TRANSITION PROBABILITY BETWEEN TWO DISCRETE STATES. LOWEST-ORDER CALCULATION

Let us begin by assuming that $|\varphi_i\rangle$ and $|\varphi_f\rangle$ are two discrete eigenstates of H_0, having energies E_i and E_f that are close to each other and far from all other eigenstates of H_0. To lowest order in V, the transition probability from $|\varphi_i\rangle$ to $|\varphi_f\rangle$ is written, using (31) and (32)

$$\mathscr{P}_{fi}(T) = \frac{4|V_{fi}|^2}{(E_f - E_i)^2} \sin^2(E_f - E_i)T/2\hbar. \quad (34)$$

This relation is, of course, valid only if T is sufficiently small so that the perturbative treatment of V is justified. It is interesting to compare it with the exact formula obtained in the simple case where $|\varphi_i\rangle$ and $|\varphi_f\rangle$ are the only eigenstates of H_0, and where as a consequence the exact diagonalization of $H_0 + V$ is possible (*):

$$\mathscr{P}_{fi}(T) = \frac{4|V_{fi}|^2}{(E_f - E_i)^2 + 4|V_{fi}|^2} \sin^2\left[\frac{T}{2\hbar} \sqrt{(E_f - E_i)^2 + 4|V_{fi}|^2}\right]. \quad (35)$$

Expressions (34) and (35) are identical to lowest order in V_{fi}. In particular, for $E_f = E_i$, Equation (34) gives

$$\mathscr{P}_{fi}(T) = |V_{fi}|^2 T^2/\hbar^2 \qquad (36)$$

which is the first term in the expansion in powers of V_{fi} of the function $\sin^2[|V_{fi}|T/\hbar]$ describing the resonant Rabi nutation between $|\varphi_i\rangle$ and $|\varphi_f\rangle$.

c) CASE WHERE THE FINAL STATE BELONGS TO AN ENERGY CONTINUUM. DENSITY OF STATES

When the final state belongs to an energy continuum, $\mathscr{P}_{fi}(T)$ is no longer a transition probability, but rather a *transition probability density*. The quantity which has a physical meaning is thus the probability that the system will reach a group of final states characterized by eigenvalues falling within a certain domain D_f. To be specific, let us first consider two concrete examples which will allow us to introduce the concept of density of states.

α) *Density of States for a Free Massive Particle*

We consider a free particle with mass M. The final states $|\varphi_f\rangle$ are momentum states $|\mathbf{p}\rangle$ satisfying the classical orthonormalization relation

$$\langle \mathbf{p}|\mathbf{p}'\rangle = \delta(\mathbf{p} - \mathbf{p}'). \qquad (37)$$

The final state $|\mathbf{p}_f\rangle$ has no physical meaning (its norm is infinite). By contrast, we can consider the probability that in the final state, the momentum of the particle points into the solid angle $\delta\Omega_f$ about the direction of $\mathbf{p}_f$ and its energy is within the interval δE_f about $E_f = p_f^2/2M$.

(*) See, for example, Cohen-Tannoudji, Diu, and Laloë, Chapter IV, §C-3. We also assume here that $V_{ii} = V_{ff} = 0$.

These conditions define a domain D_f in momentum space and the corresponding probability equals

$$\delta\mathscr{P}(\mathbf{p}_f, T) = \int_{\mathbf{p}\in D_f} d^3p \left| \langle \mathbf{p} | \tilde{U}(T) | \varphi_i \rangle \right|^2. \tag{38}$$

Instead of using p_x, p_y, p_z to characterize the final state $|\mathbf{p}\rangle$, it is also possible to use other variables such as the energy E and the polar angles θ and φ characterizing the direction of $\mathbf{p}$. If $d\Omega$ is the solid angle corresponding to $d\theta$ and $d\varphi$, one has

$$d^3p = p^2\, dp\, d\Omega = \rho(E)\, dE\, d\Omega \tag{39}$$

where $d\Omega = \sin\theta\, d\theta\, d\varphi$ and where

$$\rho(E) = p^2 \frac{dp}{dE} = p^2 \frac{M}{p} = M\sqrt{2ME} \tag{40}$$

is by definition the density of final states. Expression (38) is thus written

$$\delta\mathscr{P}(\mathbf{p}_f, T) = \int_{\substack{\Omega\in\delta\Omega_f \\ E\in\delta E_f}} d\Omega\, dE\, \rho(E) \left| \langle \mathbf{p} | \tilde{U}(T) | \varphi_i \rangle \right|^2. \tag{41}$$

Remark

It is possible to discretize the continuum and to obtain final states of norm 1 by enclosing the particles in a cubic box with sides of length L and by imposing periodic boundary conditions to obtain final states having the same spatial dependence as the states $|\mathbf{p}\rangle$. The sum over the final states included in a domain D_f is thus a true discrete sum. When L tends to infinity, it is convenient to replace this discrete sum by an integral. The density of states $\rho(E)$ is thus defined such that $\rho(E)\, dE\, d\Omega$ is the number of discretized states contained in the domain associated with dE and $d\Omega$. As a result of the periodic boundary conditions, the possible values of the wave vector $\mathbf{k}$ (which is equal to $\mathbf{p}/\hbar$) form a regular lattice of points in k space with one point per elementary volume $(2\pi/L)^3$. The number of states in d^3k is thus $(L/2\pi)^3 k^2\, dk\, d\Omega$, which for $\rho(E)$ gives the value

$$\rho(E) = \frac{L^3}{(2\pi)^3} \frac{1}{\hbar^3} M\sqrt{2ME}. \tag{42}$$

The final result for the transition amplitude (41) certainly does not depend on L. To see this, it is sufficient to note that the presence of $\langle p|$ in the matrix element of the evolution operator implies that the square of the modulus of this matrix element contains a factor $1/L^3$ which compensates for the factor L^3 appearing in the density of states (42).

β) *Density of States for a Photon*

Another important example of energy continuum is that of one-photon states $|\mathbf{k}\varepsilon\rangle$ satisfying the orthonormalization relation

$$\langle \mathbf{k}'\varepsilon'|\mathbf{k}\varepsilon\rangle = \delta_{\varepsilon\varepsilon'}\delta(\mathbf{k} - \mathbf{k}'). \tag{43}$$

Recall that the states $|\mathbf{k}\varepsilon\rangle$ result from the action of creation operators $a_\varepsilon^+(\mathbf{k})$ on the vacuum. It is thus possible to write an equation analogous to (38) where it is also necessary to sum over the polarizations and to replace d^3p by

$$d^3k = k^2\, dk\, d\Omega = \rho(E)\, dE\, d\Omega \tag{44}$$

with

$$\rho(E) = k^2\frac{dk}{dE} = \frac{E^2}{\hbar^2 c^2}\frac{1}{\hbar c} = \frac{E^2}{\hbar^3 c^3}. \tag{45}$$

To determine (45), we have used the relation $E = \hbar\omega = \hbar c k$ between the photon energy E and the modulus k of its wave vector (instead of $E = p^2/2M$).

In the case where the radiation is confined in a box of volume L^3, (45) must be replaced by

$$\rho(E) = \frac{L^3}{(2\pi)^3}\frac{E^2}{\hbar^3 c^3}. \tag{46}$$

γ) *General Case*

In the general case, we assume that the final state $|\varphi_f\rangle$, belonging to an energy continuum, is characterized by its energy E and a group of other physical variables designated by β, and we get

$$\delta\mathscr{P}(E_f, \beta_f, T) = \int_{\substack{E\in\delta E_f \\ \beta\in\delta\beta_f}} dE\, d\beta\, \rho(E, \beta)\big|\langle E, \beta|\tilde{U}(T)|\varphi_i\rangle\big|^2 \tag{47}$$

for the probability that, starting from the normalized state $|\varphi_i\rangle$, the system will arrive after time T in one of the final states of the domain D_f characterized by δE_f and $\delta\beta_f$. In (47), $\rho(E, \beta)$ is the density of final states which, in the general case, depends on both E and β.

d) TRANSITION RATE TOWARD A CONTINUUM OF FINAL STATES

In expression (47), $|\langle E, \beta|\tilde{U}(T)|\varphi_i\rangle|^2$, is simply the square of the modulus of the transition amplitude $\mathcal{S}_{fi}$ given in (29) with $\delta_{fi} = 0$. At the lowest order in V we obtain, by writing $v(E, \beta; \varphi_i)$, the matrix element V_{fi}:

$$\delta\mathcal{P}(E_f, \beta_f, T) =$$

$$\int_{\substack{E \in \delta E_f \\ \beta \in \delta\beta_f}} dE\, d\beta\, \rho(E, \beta) 4\pi^2 |v(E, \beta; \varphi_i)|^2 [\delta^{(T)}(E - E_i)]^2. \quad (48)$$

In general, $\rho(E, \beta)|v(E, \beta; \varphi_i)|^2$ is a function of E varying with E much more slowly than $[\delta^{(T)}(E - E_i)]^2$ which, for sufficiently large T, has an extremely small width, on the order of $\hbar/T$. If this is the case, we can then replace $[\delta^{(T)}(E - E_i)]^2$ by a "delta function" centered on E_i. Because the integral over E of $[\delta^{(T)}(E - E_i)]^2$ is, according to (33), equal to $T/2\pi\hbar$, we are justified to write (with regard to slowly varying functions of E)

$$[\delta^{(T)}(E - E_i)]^2 \simeq \frac{T}{2\pi\hbar}\delta^{(T)}(E - E_i). \quad (49)$$

The substitution of (49) into (48) then shows that $\delta\mathcal{P}$ is proportional to the duration T of the interaction, which allows us to define a transition probability *per unit time* $\delta w(E_f, \beta_f)$

$$\delta w(E_f, \beta_f) = \frac{1}{T}\delta\mathcal{P}(E_f, \beta_f, T) =$$

$$= \frac{2\pi}{\hbar}\int_{\substack{E \in \delta E_f \\ \beta \in \delta\beta_f}} dE\, d\beta\, \rho(E, \beta)|v(E, \beta; \varphi_i)|^2 \delta^{(T)}(E - E_i).$$

$$(50)$$

Assume that the interval δE_f contains E_i and that δE_f is greater than the width $\hbar/T$ of $\delta^{(T)}(E - E_i)$. The integral over E is thus straightforward. If, in addition, $\delta\beta_f$ is sufficiently small so that the integral over β becomes unnecessary, we finally get

$$\frac{\delta w(E_f, \beta_f)}{\delta \beta_f} = \frac{2\pi}{\hbar} \left| v(E_f = E_i, \beta_f; \varphi_i) \right|^2 \rho(E_f = E_i, \beta_f) \qquad (51)$$

which is simply the Fermi golden rule for the transition probability per unit time and per unit interval $\delta\beta$.

Remark: Sum over Polarizations

Let us return to the photon example. Frequently the squared matrix element of (50) has the form $|\boldsymbol{\varepsilon} \cdot \mathbf{X}|^2$, where $\mathbf{X}$ is a vectorial quantity and where $\boldsymbol{\varepsilon}$ is the polarization vector of the photon. If we do not observe the polarization of the emitted photon, then we must, for a given emission direction $\mathbf{k}$, sum over the two states of polarization $\boldsymbol{\varepsilon}$ and $\boldsymbol{\varepsilon}'$ orthogonal to $\mathbf{k}$ and orthogonal to each other; that is, calculate

$$\sum_{\boldsymbol{\varepsilon} \perp \mathbf{k}} |\boldsymbol{\varepsilon} \cdot \mathbf{X}|^2 = \sum_{i,j=x,y,z} \left(\sum_{\boldsymbol{\varepsilon} \perp \mathbf{k}} \varepsilon_i \varepsilon_j \right) X_i X_j^*. \qquad (52)$$

To evaluate the sum between parentheses in (52), it is sufficient to note that $\boldsymbol{\varepsilon}$, $\boldsymbol{\varepsilon}'$, and $\boldsymbol{\kappa} = \mathbf{k}/\kappa$ form an orthonormal basis for which the closure relation is written

$$\varepsilon_i \varepsilon_j + \varepsilon_i' \varepsilon_j' + \kappa_i \kappa_j = \delta_{ij} \qquad (53)$$

from which we deduce

$$\sum_{\boldsymbol{\varepsilon} \perp \mathbf{k}} \varepsilon_i \varepsilon_j = \varepsilon_i \varepsilon_j + \varepsilon_i' \varepsilon_j' = \delta_{ij} - \kappa_i \kappa_j = \delta_{ij} - \frac{k_i k_j}{k^2} \qquad (54)$$

and consequently

$$\sum_{\boldsymbol{\varepsilon} \perp \mathbf{k}} |\boldsymbol{\varepsilon} \cdot \mathbf{X}|^2 = \mathbf{X} \cdot \mathbf{X}^* - \frac{(\mathbf{k} \cdot \mathbf{X})(\mathbf{k} \cdot \mathbf{X}^*)}{k^2}. \qquad (55)$$

e) CASE WHERE BOTH THE INITIAL AND FINAL STATES BELONG
     TO A CONTINUUM

In this case, the initial state is itself also not physical (because it has an infinite norm). It is nevertheless possible to derive from it a quantity having a physical meaning, such as a particle flux, if $|\varphi_i\rangle$ represents an incident free particle having a well-defined momentum.

Let us calculate such an incident flux for a particle having mass M and for a photon. It is thus particularly convenient to use discretized states in a cube having sides of length L. In the discretized state $|\mathbf{p}_i\rangle$ or $|\mathbf{k}_i\rangle$, having norm 1, the particle density is $1/L^3$ (one particle in a volume L^3) and the velocity equals $\mathbf{p}_i/M = \hbar\mathbf{k}_i/M$ for the particle of mass M, or $c\boldsymbol{\kappa}_i$ for the photon (where $\boldsymbol{\kappa}_i = \mathbf{k}_i/k_i$). From that we deduce that the incident flux $\boldsymbol{\Phi}_i$ equals

$$\boldsymbol{\Phi}_i = \frac{\mathbf{p}_i}{M}\frac{1}{L^3} = \frac{\hbar\mathbf{k}_i}{M}\frac{1}{L^3} \tag{56}$$

for the free particle with mass M and

$$\boldsymbol{\Phi}_i = \frac{c}{L^3}\boldsymbol{\kappa}_i \tag{57}$$

for the photon. Dividing by $|\boldsymbol{\Phi}_i|$ the transition probability per unit time and per unit solid angle yields the differential scattering cross-section from $\mathbf{k}_i$ toward $\mathbf{k}_f$.

Remark

If both $|\varphi_i\rangle$ and $|\varphi_f\rangle$ belong to a continuum, two factors $1/L^3$ appear in the squared matrix element of (51). One factor $1/L^3$ is compensated for by the factor L^3 that appears in the final-state density (see Remark in paragraph 3-c above). The other factor $1/L^3$ compensates for the one appearing in expression (56) or (57) of the flux, when $\delta w/\delta\Omega$ is divided by this flux. It is therefore clear that the scattering cross-section does not depend on L.

GENERAL REFERENCES

Same bibliography as for Chapter I.

COMPLEMENT B_I

DESCRIPTION OF THE EFFECT OF A PERTURBATION BY AN EFFECTIVE HAMILTONIAN

1. Introduction—Motivation

For many systems, the eigenstates of the Hamiltonian H cannot be exactly determined. In contrast, those of an approximate Hamiltonian H_0 are sometimes known. Thus, in the case of electrodynamics, we saw in Section C of the chapter that the Hamiltonian is diagonalizable in the absence of coupling between the particles and the field. In this case, perturbation theory can be used to determine the eigenstates of H, by taking as a perturbation the difference between H and H_0.

In this complement, we will consider the case of a Hamiltonian H_0 having energy levels $E_{i\alpha}$ which are grouped into manifolds $\mathscr{E}_\alpha^0, \mathscr{E}_\beta^0, \ldots$ that are well separated from each other. The subscript i denotes the different levels $|i, \alpha\rangle$ of the same manifold, and P_α is the projector over the manifold $\mathscr{E}_\alpha^0$:

$$H_0|i, \alpha\rangle = E_{i\alpha}|i, \alpha\rangle \tag{1}$$

$$P_\alpha = \sum_i |i, \alpha\rangle\langle i, \alpha|. \tag{2}$$

To say that the manifolds are well separated signifies that the spectrum of H_0 has the shape indicated in Figure 1. More precisely, we assume

$$|E_{i\alpha} - E_{j\alpha}| \ll |E_{i\alpha} - E_{j\beta}| \qquad \text{with } \alpha \neq \beta. \tag{3}$$

Figure 1. Manifolds $\mathscr{E}_\alpha^0, \mathscr{E}_\beta^0, \ldots$ of the Hamiltonian H_0.

In physical terms, the quantum number i characterizes the degrees of freedom for which the Bohr frequencies $(E_{i\alpha} - E_{j\alpha})/\hbar$ (intervals between levels of a same manifold) are small. In contrast, the index α is a quantum number relative to quantities for which the Bohr frequencies $(E_{i\alpha} - E_{i\beta})/\hbar$ (intervals between two different manifolds) are much larger. Thus, the existence of well-separated manifolds reveals the presence within the system of two types of degrees of freedom: fast degrees of freedom characterized by Greek indices such as α, and slow degrees of freedom characterized by Roman subscripts such as i.

Situations of this type are encountered in many physics problems, especially in the study of interactions between matter and radiation. Let us consider, for example, the system made up, on the one hand, of an electron in a external static potential and, on the other hand, of a mode $\mathbf{k}\varepsilon$ of the radiation field with frequency ω. In the absence of interaction between the electron and the radiation, the energy levels of the overall system are designated by the quantum numbers i of the electron in the external potential and the number $\alpha = N$ of photons in the mode. If the frequency ω of the mode under consideration is extremely large as compared to the frequencies $(E_i - E_j)/\hbar$ characterizing the movement of the electron in the external potential, the situation is analogous to that in Figure 1. The manifold $\mathscr{E}_\alpha^0 = \mathscr{E}_N^0$ thus consists of the energy levels of the electron in the external potential in the presence of N photons and the other manifolds corresponding to a number $N' \neq N$ of photons are at a distance $(N' - N)\hbar\omega$.

To obtain the total Hamiltonian H, let us now add to H_0 the perturbation or the coupling which we write in the form λV, where λ is a dimensionless parameter:

$$H = H_0 + \lambda V. \tag{4}$$

The operator V has matrix elements inside a manifold as well as between two different manifolds. For example, for the system mentioned above, the interaction between the electron and the mode $\mathbf{k}\varepsilon$ couples the manifold $\mathscr{E}_N^0$ to manifolds $\mathscr{E}_{N+1}^0$ (and $\mathscr{E}_{N-1}^0$), the corresponding physical processes being the emission (and absorption) of a photon $\mathbf{k}\varepsilon$ by the electron. If λ is sufficiently small, more precisely, if

$$|\langle i, \alpha | \lambda V | j, \beta \rangle| \ll |E_{i\alpha} - E_{j\beta}| \qquad (\beta \neq \alpha) \tag{5}$$

the energy levels of the Hamiltonian H are clustered, as are those of H_0, in manifolds $\mathscr{E}_\alpha, \mathscr{E}_\beta, \ldots$ well separated from each other, the levels of $\mathscr{E}_\alpha$ tending to those of $\mathscr{E}_\alpha^0$ when $\lambda \to 0$.

The physical effects of the coupling λV are of two types. On the one hand, the wave functions are modified. In particular, the wave functions of the manifold $\mathscr{E}_\alpha^0$ are "contaminated" by the wave functions of other manifolds $\mathscr{E}_\beta^0$ with $\beta \neq \alpha$. On the other hand, the energies are modified. In particular, the slow Bohr frequencies (within one manifold) are changed. Let us now study in more detail these two types of effect in the case of the physical example introduced above. As a result of the contamination of the states of $\mathscr{E}_N^0$ by those of $\mathscr{E}_{N\pm1}^0$, N is no longer a good quantum number, and the electron observables, which commute with the photon number operator, may thus have nonzero matrix elements between perturbed states of $\mathscr{E}_N$ and perturbed states of $\mathscr{E}_{N\pm1}$. Physically, rapid components at frequency ω appear in the electron motion, which, in fact, correspond to the vibration of the electron in the electric field of mode $\mathbf{k}\boldsymbol{\varepsilon}$. In addition, the nonresonant coupling between $\mathscr{E}_N^0$ and $\mathscr{E}_{N\pm1}^0$ shifts the states of $\mathscr{E}_N^0$ to second order in λV. Physically, the virtual emission and reabsorption (or virtual absorption and reemission) of one photon by the electron changes the slow electron motion in the external potential.

In this complement, we are essentially interested in the modification made to the slow motion by the coupling λV and not to the contamination of wave functions. Our goal is to attempt to construct a Hamiltonian acting only within each manifold $\mathscr{E}_\alpha^0$ such that its eigenvalues in $\mathscr{E}_\alpha^0$ are identical to those of H in $\mathscr{E}_\alpha$. Such a Hamiltonian, called the effective Hamiltonian, acts only on the slow degrees of freedom because its matrix elements between $\langle i, \alpha |$ and $| j, \beta \rangle$ are zero if $\alpha \neq \beta$. Because it correctly describes the slow motion, it incorporates the effect on the slow degrees of freedom of the coupling of the latter with the fast degrees of freedom. In the case of the physical system described above, the effective Hamiltonian is a *purely electronic* Hamiltonian which describes the perturbed slow motion of the electron by means of corrective electron terms: correction to the kinetic energy of the electron (due to the fact that its inertia is modified by the virtual photon cloud which surrounds it) and correction to the potential energy (due to the fact that the electron vibrating in the field of the mode ω averages the static external potential over the extent of its vibrational mode) (*). Other important physical examples may be given, e.g., the *effective magnetic interaction* between two electrons associated with the virtual emission of a transverse photon by an electron and the

---

(*) Such an effective Hamiltonian is derived in P. Avan, C. Cohen-Tannoudji, J. Dupont-Roc, and C. Fabre, *J. Physique*, **37**, 993 (1976).

reabsorption of this photon by the other electron (see Chapter II, Section F).

We will show in this complement how an effective Hamiltonian can be constructed by means of a unitary transformation applied to the total Hamiltonian H (*). The principle of the method is described in Section 2. We will then determine (§3) the unitary transformation and the expression of the effective Hamiltonian. Finally, we will examine (§4) the case in which the system under study is an ensemble of two subsystems whose interaction is described by the coupling λV. We will demonstrate in particular how it is possible to obtain an operator expression for the effective Hamiltonian which involves only observables of the system that evolve with slow frequencies.

2. Principle of the Method

Thus we are seeking an effective Hamiltonian H' having the following properties:

a) H' is Hermitian

b) H' has the same eigenvalues as H, with the same degeneracy

c) H' has no matrix elements between *unperturbed* manifolds $\mathscr{E}_\alpha^0, \mathscr{E}_\beta^0, \ldots$

Properties a) and b) result in the fact that there is a unitary transformation

$$T = e^{iS} \tag{6}$$

$$S = S^+ \tag{7}$$

which allows us to go from H to H'

$$H' = THT^+. \tag{8}$$

Property c) is expressed by the equation

$$P_\alpha H' P_\beta = 0 \qquad \text{for } \alpha \neq \beta. \tag{9}$$

(*) A review of the different ways of formally constructing effective Hamiltonians may be found in D. J. Klein, *J. Chem. Phys.*, **61**, 786 (1974). A more recent reference explicitly gives the terms of the effective Hamiltonian up to the fifth order: I. Shavitt and L. T. Redmon, *J. Chem. Phys.*, **73**, 5711 (1980).

The matrix representing H' in the unperturbed initial basis $|i, \alpha\rangle$ is block diagonal. Each block is relative to a manifold $\mathscr{E}_\alpha^0$ and represents an effective Hamiltonian H_{eff}^α which describes the perturbed levels of this manifold

$$H' = \sum_\alpha P_\alpha H_{\text{eff}}^\alpha. \tag{10}$$

Equations (8) and (9) are not sufficient to entirely determine the transformation T. In fact, if T is a solution, we can construct an infinite number of other solutions of the form UT, where U is an arbitrary unitary transformation acting only inside manifolds $\mathscr{E}_\alpha^0$. One way to remove this uncertainty is to impose on S the condition that it does not have matrix elements inside each manifold:

$$P_\alpha S P_\alpha = 0 \qquad \text{for any } \alpha. \tag{11}$$

To explicitly calculate S, it is convenient to write it in the form of an expansion in powers of λ

$$S = \lambda S_1 + \lambda^2 S_2 + \cdots + \lambda^n S_n + \cdots. \tag{12}$$

It is obvious that the zero-order term is zero, because H_0 is itself diagonal in the basis $\{|i, \alpha\rangle\}$. Equation (8) can then be expanded in the form

$$H' = H + [iS, H] + \frac{1}{2!}[iS, [iS, H]] + \frac{1}{3!}[iS, [iS, [iS, H]]] + \cdots. \tag{13}$$

By substituting S from expression (12), we obtain an expansion of the effective Hamiltonian.

$$H' = H_0 + \lambda H_1' + \lambda^2 H_2' + \cdots + \lambda^p H_p' + \cdots. \tag{14}$$

Each term H_p' can be expressed as a function of S_n, of H_0 and of V. Conditions (9) and (11), applied step by step, determine S_n, and consequently H'.

It is helpful to write the effective Hamiltonian H' thus determined in the form

$$H' = H_0 + W \tag{15}$$

where

$$W = \lambda H_1' + \lambda^2 H_2' + \cdots + \lambda^p H_p' + \cdots. \tag{16}$$

W is called the "level-shift" operator. According to Property b) of H', W produces exactly the same effect over the energy levels of H_0 as the perturbation λV. In contrast, according to Property c), it has the advantage of acting only inside manifolds. If we seek only new energies in the manifold $\mathscr{E}_\alpha$, it is much simpler to use W than to use λV, which couples $\mathscr{E}_\alpha^0$ to all other manifolds.

3. Determination of the Effective Hamiltonian

a) ITERATIVE CALCULATION OF S

Let us order the expression (13) for H' in increasing powers of λ, after having replaced S by its expansion (12) and H by its expression (4). Thus

$$H' = H_0 + [i\lambda S_1, H_0] + \lambda V +$$

$$+ [i\lambda^2 S_2, H_0] + [i\lambda S_1, \lambda V] + \frac{1}{2}[i\lambda S_1, [i\lambda S_1, H_0]] +$$

$$\vdots$$

$$+ [i\lambda^n S_n, H_0] + [i\lambda^{n-1} S_{n-1}, \lambda V] +$$

$$+ \frac{1}{2}[i\lambda^{n-1} S_{n-1}, [i\lambda S_1, H_0]] +$$

$$+ \frac{1}{2}[i\lambda S_1, [i\lambda^{n-1} S_{n-1}, H_0]] + \cdots +$$

$$+ \frac{1}{n!}[i\lambda S_1, [i\lambda S_1, \cdots [i\lambda S_1, H_0] \cdots] +$$

$$\vdots \tag{17}$$

The last term of (17) contains n stacked commutators.

Let us first consider first-order terms in λ:

$$\lambda H_1' = [i\lambda S_1, H_0] + \lambda V \tag{18}$$

and write that the matrix element of H_1' between two different manifolds is zero. We obtain the equation

$$\langle i, \alpha | i\lambda S_1 | j, \beta \rangle (E_{j\beta} - E_{i\alpha}) + \langle i, \alpha | \lambda V | j, \beta \rangle = 0 \qquad (19)$$

which determines the matrix elements of S_1 between two different manifolds, since the other elements are zero according to (11).

$$\langle i, \alpha | i\lambda S_1 | j, \beta \rangle = \frac{\langle i, \alpha | \lambda V | j, \beta \rangle}{E_{i\alpha} - E_{j\beta}}, \qquad \text{for } \alpha \neq \beta \qquad (20.\text{a})$$

$$\langle i, \alpha | i\lambda S_1 | j, \alpha \rangle = 0. \qquad (20.\text{b})$$

Let us now consider the nth-order term, $\lambda^n H'_n$. It involves all the operators S_p, p ranging from 1 to n. The fact that all matrix elements of $\lambda^n H'_n$ between two different manifolds α and β are zero allows us to express the matrix elements of $\lambda^n S_n$ between these same manifolds as a function of those of the operators S_p of lower order than n. Indeed, $\lambda^n S_n$ appears only in a single term, that of the commutator with H_0. We thus obtain an equation of the type

$$\langle i, \alpha | i\lambda^n S_n | j, \beta \rangle (E_{j\beta} - E_{i\alpha}) =$$

$$\mathscr{F}(\lambda V; \lambda S_1, \dots, \lambda^{n-1} S_{n-1}) \qquad \text{for } \alpha \neq \beta \qquad (21.\text{a})$$

additionally with

$$\langle i, \alpha | i\lambda^n S_n | j, \alpha \rangle = 0 \qquad (21.\text{b})$$

Step by step, S is thus entirely determined.

b) Expression of the Second-Order Effective Hamiltonian

Note first of all that the expression for the nth-order effective Hamiltonian does not involve $\lambda^n S_n$. In fact, we have previously seen that, in the expression for the nth-order term of (17), S_n is involved only in the simple commutator with H_0. Because H_0 is diagonal with regard to α and since S_n is nondiagonal, such a commutator is nondiagonal and thus does not contribute to the expression of H' within each manifold. Hence, to determine the effective Hamiltonian up to order 2, it is thus sufficient to know λS_1, which is given by (20.a). The matrix elements of H' within the manifold α are written

$$\langle i|H_{\text{eff}}^{\alpha}|j\rangle = \langle i, \alpha|H'|j, \alpha\rangle =$$
$$= \langle i, \alpha|H_0 + \lambda V + [i\lambda S_1, \lambda V] + \tfrac{1}{2}[i\lambda S_1, [i\lambda S_1, H_0]] + \cdots |j, \alpha\rangle.$$
$$(22)$$

Let us show that the last term of (22) is identical, except for one factor, to the next to last term. In fact, $[i\lambda S_1, H_0]$ is purely nondiagonal, and according to (19), its matrix elements are opposite to those of λV between different manifolds. Let us call λV^{nd} the nondiagonal part (which connects the different manifolds) of λV. Hence we have

$$[i\lambda S_1, H_0] + \lambda V^{\text{nd}} = 0 \qquad (23)$$

and the last term of (22) is reduced to $-[i\lambda S_1, \lambda V^{\text{nd}}]/2$. The next-to-last term also involves only the nondiagonal part of λV, so that the product by λS_1 yields a diagonal term. Hence, up to second order, H_{eff}^{α} is reduced to

$$H_{\text{eff}}^{\alpha} = H_0 P_{\alpha} + P_{\alpha} \lambda V P_{\alpha} + \tfrac{1}{2} P_{\alpha}[i\lambda S_1, \lambda V] P_{\alpha} + \cdots. \qquad (24)$$

Similar simplifications occur in all orders.

Let us now explicitly calculate the matrix elements of the last term of (24) by using the matrix elements (20.a) of λS_1:

$$\langle i, \alpha|[i\lambda S_1, \lambda V]|j, \alpha\rangle = \sum_{k, \gamma \neq \alpha} \langle i, \alpha|i\lambda S_1|k, \gamma\rangle\langle k\gamma|\lambda V|j, \alpha\rangle -$$
$$- \langle i, \alpha|\lambda V|k, \gamma\rangle\langle k, \gamma|i\lambda S_1|j, \alpha\rangle$$
$$= \sum_{k, \gamma \neq \alpha} \langle i, \alpha|\lambda V|k, \gamma\rangle\langle k, \gamma|\lambda V|j, \alpha\rangle \times$$
$$\times \left[\frac{1}{E_{i\alpha} - E_{k\gamma}} + \frac{1}{E_{j\alpha} - E_{k\gamma}} \right]. \qquad (25)$$

We finally get, for $\langle i|H_{\text{eff}}^{\alpha}|j\rangle$ up to second order in λ:

$$\langle i|H_{\text{eff}}^{\alpha}|j\rangle = E_{i\alpha}\delta_{ij} + \langle i, \alpha|\lambda V|j, \alpha\rangle +$$
$$+ \frac{1}{2} \sum_{k, \gamma \neq \alpha} \langle i, \alpha|\lambda V|k, \gamma\rangle\langle k, \gamma|\lambda V|j, \alpha\rangle \times$$
$$\times \left[\frac{1}{E_{i\alpha} - E_{k\gamma}} + \frac{1}{E_{j\alpha} - E_{k\gamma}} \right] + \cdots. \qquad (26)$$

The first term of (26) represents the unperturbed energy of levels of $\mathscr{E}_\alpha^0$, the second term represents the *direct* coupling between the levels i and j of $\mathscr{E}_\alpha^0$, and the third term represents the *indirect* coupling between these two levels through all the levels $k\gamma$ of other manifolds $\mathscr{E}_\gamma^0$. This last term has the structure of a second-order perturbation term, that is, a product of two matrix elements of λV divided by an unperturbed energy denominator (the two energies $E_{i\alpha}$ and $E_{j\alpha}$ of the two levels of $\mathscr{E}_\alpha^0$ appearing symmetrically if they are different).

c) HIGHER-ORDER TERMS

We have seen how, step by step, the higher-order terms may be explicitly calculated. In fact, at least from a formal point of view, the expression for the effective Hamiltonian may also be given in compact algebraic forms. These two types of expressions may be found in the references quoted at the beginning of this complement.

4. Case of Two Interacting Systems

We will now consider the case of two systems $\mathscr{A}$ and $\mathscr{R}$, with respective Hamiltonians H_A and H_R, that interact by means of a Hamiltonian λV. The spacings between energy levels of $\mathscr{R}$ are assumed to be large as compared to those of $\mathscr{A}$. This is shown by the following equations:

$$H_0 = H_A + H_R \tag{27}$$

$$H_A|i\rangle = E_i|i\rangle \tag{28}$$

$$H_R|\alpha\rangle = E_\alpha|\alpha\rangle \tag{29}$$

$$|E_i - E_j| \ll |E_\alpha - E_\beta|, \qquad \alpha \neq \beta. \tag{30}$$

We assume that the coupling between the two systems is written in the form

$$\lambda V = \lambda \sum_\mu A_\mu R_\mu \tag{31}$$

where the operators A_μ (resp. R_μ) are relative to the subsystem $\mathscr{A}$ (resp. $\mathscr{R}$). Finally, we assume A_μ and R_μ to be purely *nondiagonal* in the bases $\{|i\rangle\}$ and $\{|\alpha\rangle\}$.

To determine H_{eff}^α, let us replace $E_{i\alpha}$ by $E_i + E_\alpha$, in (26), and λV by its expression (31). Because R_μ is nondiagonal in α, the second term of (26) is zero. Hence we get

$$\langle i|H_{\text{eff}}^{\alpha}|j\rangle = E_{\alpha} + E_{i}\delta_{ij} +$$

$$+ \frac{1}{2}\sum_{\mu\mu'}\sum_{k,\gamma\neq\alpha}(\langle\alpha|\lambda R_{\mu}|\gamma\rangle\langle\gamma|\lambda R_{\mu'}|\alpha\rangle)(\langle i|A_{\mu}|k\rangle\langle k|A_{\mu'}|j\rangle) \times$$

$$\times\left[\frac{1}{E_{\alpha} - E_{\gamma} + E_{i} - E_{k}} + \frac{1}{E_{\alpha} - E_{\gamma} + E_{j} - E_{k}}\right]. \tag{32}$$

By using the inequality (30), we can expand the fractions appearing in (32)

$$\frac{1}{E_{\alpha} - E_{\gamma} + E_{i} - E_{k}} = \frac{1}{E_{\alpha} - E_{\gamma}} - \frac{E_{i} - E_{k}}{(E_{\alpha} - E_{\gamma})^{2}} + \cdots \tag{33}$$

and the equivalent by replacing i by j. The second order term becomes

$$\sum_{\mu\mu'}\left(\sum_{\gamma\neq\alpha}\langle\alpha|\lambda R_{\mu}|\gamma\rangle\frac{1}{E_{\alpha} - E_{\gamma}}\langle\gamma|\lambda R_{\mu'}|\alpha\rangle\right)\times$$

$$\times\left(\sum_{k}\langle i|A_{\mu}|k\rangle\langle k|A_{\mu'}|j\rangle\right) +$$

$$+ \sum_{\mu\mu'}\left(\sum_{\gamma\neq\alpha}\frac{\langle\alpha|\lambda R_{\mu}|\gamma\rangle\langle\gamma|\lambda R_{\mu'}|\alpha\rangle}{(E_{\alpha} - E_{\gamma})^{2}}\right)\times$$

$$\times\left(\sum_{k}\langle i|A_{\mu}|k\rangle\langle k|A_{\mu'}|j\rangle\left(\frac{E_{k} - E_{i} + E_{k} - E_{j}}{2}\right)\right). \tag{34}$$

Note that the sum over γ may be extended to the case $\gamma = \alpha$, because the corresponding terms $\langle\alpha|R_{\mu}|\alpha\rangle$ are zero. The closure relations over states $|\gamma\rangle$ on one hand, and over $|k\rangle$ on the other hand thus allow us to rewrite these two terms in operator forms. Actually

$$\sum_{k}\langle i|A_{\mu}|k\rangle\langle k|A_{\mu'}|j\rangle(E_{k} - E_{i} + E_{k} - E_{j})$$

$$= \langle i|[A_{\mu}, H_{A}]A_{\mu'} - A_{\mu}[A_{\mu'}, H_{A}]|j\rangle. \tag{35}$$

Finally, the effective Hamiltonian may be written in operator form:

$$H_{\text{eff}}^{\alpha} = E_{\alpha}\mathbb{1}_A + H_A + \sum_{\mu\mu'} \langle \alpha | \lambda R_{\mu} \frac{1}{E_{\alpha} - H_R} \lambda R_{\mu'} | \alpha \rangle A_{\mu} A_{\mu'} +$$

$$+ \frac{1}{2} \sum_{\mu\mu'} \langle \alpha | \lambda R_{\mu} \frac{1}{(E_{\alpha} - H_R)^2} \lambda R_{\mu'} | \alpha \rangle ([A_{\mu}, H_A] A_{\mu'} - A_{\mu}[A_{\mu'}, H_A])$$

$$+ \cdots. \tag{36}$$

Hence it does indeed appear as a sum of operators relative to $\mathscr{A}$, with coefficients that are the average value in the state $|\alpha\rangle$ of operators relative to $\mathscr{R}$. The first two terms represent the free evolution in the manifold $\mathscr{E}_{\alpha}^0$. The third term describes the effect of virtual transitions to other manifolds. The last term of (36) is a correction to the third term which takes into account the fact that a virtual transition to the manifold β is not infinitely short but rather lasts a time which is on the order of $\hbar/|E_{\alpha} - E_{\beta}|$. The operators A_{μ} and $A_{\mu'}$ have the time to evolve when acted upon by H_A during this period. This is expressed by the last term of (36).

COMPLEMENT C_I

DISCRETE LEVEL COUPLED TO A BROAD CONTINUUM: A SIMPLE MODEL

Many interaction processes between atoms and photons may be analyzed in terms of a discrete level coupled to a continuum. Such an analysis can provide good insights into these processes, but it often requires a detailed knowledge of the essential characteristics of the new eigenstates resulting from the coupling between the discrete state and the continuum. Hence we consider it important to devote a complement to the analysis of this problem. The results obtained will allow us to clarify several of the discussions in Chapter II.

Rather than considering the most general situation, we prefer to limit ourselves in the majority of this complement to a sufficiently simple model so that formalism will not be a major obstacle. This model cannot, of course, include all the details of the phenomena, but it will allow us to point out and to explain their essential features.

We begin (§1) by describing the model and the simplifying assumptions, which consist of taking a continuum extending from $-\infty$ to $+\infty$ on the energy axis and a coupling with the discrete state independent of the energy. "Seen" from the discrete state, the continuum appears in this case to be completely "flat" and structureless. It is then possible, after "discretization" of the continuum, to perform an exact calculation of the eigenstates and eigenvalues of the total Hamiltonian. The essential result is that the contamination of the new eigenstates by the discrete state is significant only in an interval centered on the energy of the discrete state, the width of this interval being on the order of $\hbar\Gamma$, where Γ is the transition rate of the discrete state to the continuum calculated by using the Fermi golden rule (§2). We then demonstrate (§3) how this "dissolution" of the discrete state, over an interval of width $\hbar\Gamma$ in the new continuum, allows us to quantitatively understand several important physical phenomena, such as the exponential decay of the discrete state, the excitation of this discrete state starting from another state of the system, the resonant scattering through this discrete state, and Fano profiles. The last subsection (§4) provides some indications concerning the way eigenstates of the new continuum may be calculated in more general situations (nonflat continuum) and without discretization of the continuum.

1. Description of the Model (*)

a) THE DISCRETE STATE AND THE CONTINUUM

Let us consider a Hamiltonian H_0 which has as eigenstates a discrete state $|\varphi\rangle$ and a continuum of states $|E, \beta\rangle$. For the state $|E, \beta\rangle$, E denotes the eigenvalue associated with H_0 (unperturbed energy) and β denotes other quantum numbers allowing one to distinguish $|E, \beta\rangle$ among orthogonal states having the same energy E.

Assume that the discrete state $|\varphi\rangle$ is coupled to the continuum $|E, \beta\rangle$ by a coupling Hamiltonian V. It is always possible to change the basis within each subspace of energy E of the continuum in order to single out the linear combination of states $|E, \beta\rangle$ which is coupled to $|\varphi\rangle$, and which is denoted by $|E\rangle$, among all other orthogonal linear combinations which are not coupled to $|\varphi\rangle$. We use v_E to denote the matrix element $\langle E|V|\varphi\rangle$. Knowing eigenvalues and eigenvectors of the total Hamiltonian allows us to determine the dynamics of the system. It is thus appropriate to diagonalize the total Hamiltonian in the base $\{|\varphi\rangle, |E\rangle \cdots\}$ (**). We begin by introducing some simplifications which will allow us to reduce the calculations as much as possible while retaining the essential physical results.

b) DISCRETIZATION OF THE CONTINUUM

First we discretize the continuum. Such a step was presented for the electromagnetic field in the chapter, with the introduction of a fictitious box with periodic boundary conditions. If the dimension L of the box is large compared to all other characteristic lengths of the problem, the physical results will not depend on the volume L^3 and will be obtained at the limit $1/L^3 \to 0$.

We thus proceed similarly and replace the states of the continuum $|E\rangle$ by discrete states $|k\rangle$ spaced by δ in energy. The density of states is thus $1/\delta$. The matrix element $\langle k|V|\varphi\rangle$ is designated v_k. The physical results are obtained at the limit $\delta \to 0$.

(*) Such a model was introduced by U. Fano, *Nuovo Cimento*, **12**, 156 (1935).

(**) We assume that V has no matrix elements within the continuum. Then the linear combinations of states $|E, \beta\rangle$, of the same energy E, which are not coupled to the discrete state $|\varphi\rangle$ remain eigenstates of the total Hamiltonian, with the eigenvalue E. If $\langle E, \beta|V|E', \beta'\rangle$ was different from zero, a prediagonalization of the Hamiltonian inside the continuum would result in the previous situation.

The application of Fermi's golden rule for a system initially in state $|\varphi\rangle$ leads to a transition rate Γ to the continuum equal to

$$\Gamma = \frac{2\pi}{\hbar} v^2 \frac{1}{\delta} \tag{1}$$

where v is the matrix element (assumed to be real) of V between $|\varphi\rangle$ and the state $|k\rangle$ of the same energy as $|\varphi\rangle$. Relation (1) clearly implies that, at the limit $\delta \to 0$, v^2/δ must remain constant and equal to $\hbar\Gamma/2\pi$.

c) SIMPLIFYING ASSUMPTIONS

First we assume that the discretized continuum extends from $-\infty$ to $+\infty$ with equidistant levels separated by the quantity δ. The unperturbed energy of the level $|k\rangle$ is thus

$$\langle k|H_0|k\rangle = E_k = k\delta \tag{2.a}$$

where k is any integer (positive, negative, or zero). In what follows, the energy E_φ of the discrete level is taken as the origin of energies ($E_\varphi = 0$) and thus coincides with the energy of the level $k = 0$ of the quasi-continuum.

$$\langle \varphi|H_0|\varphi\rangle = E_\varphi = 0. \tag{2.b}$$

For the coupling, we assume that all the matrix elements of V between the level $|\varphi\rangle$ and the states $|k\rangle$ are equal and real

$$v_k = \langle k|V|\varphi\rangle = \langle \varphi|V|k\rangle = v. \tag{2.c}$$

Finally, all the other matrix elements of V are assumed to be zero.

$$\langle \varphi|V|\varphi\rangle = \langle k|V|k'\rangle = 0. \tag{2.d}$$

2. Stationary States of the System. Traces of the Discrete State in the New Continuum

a) THE EIGENVALUE EQUATION

Let E_μ and $|\psi_\mu\rangle$ be the eigenvalues and eigenvectors of the total Hamiltonian $H = H_0 + V$.

$$H|\psi_\mu\rangle = E_\mu|\psi_\mu\rangle. \tag{3}$$

Let us project Equation (3) respectively onto $\langle k|$ and $\langle \varphi|$. Using the assumptions made concerning H_0 and V [see relations (2)], we get

$$E_k\langle k|\psi_\mu\rangle + v\langle\varphi|\psi_\mu\rangle = E_\mu\langle k|\psi_\mu\rangle \tag{4.a}$$

$$\sum_k v\langle k|\psi_\mu\rangle = E_\mu\langle\varphi|\psi_\mu\rangle. \tag{4.b}$$

Equation (4.a) gives (*):

$$\langle k|\psi_\mu\rangle = v\frac{\langle\varphi|\psi_\mu\rangle}{E_\mu - E_k}. \tag{5}$$

This expression, when substituted into (4.b), yields the eigenvalue equation

$$\sum_k \frac{v^2}{E_\mu - E_k} = E_\mu. \tag{6}$$

In addition, by using (5) and the normalization condition

$$\sum_k |\langle k|\psi_\mu\rangle|^2 + |\langle\varphi|\psi_\mu\rangle|^2 = 1 \tag{7}$$

we find for the components of $|\psi_\mu\rangle$ on $|\varphi\rangle$ and $|k\rangle$ (with an appropriate choice of phase):

$$\langle\varphi|\psi_\mu\rangle = \frac{1}{\left[1 + \sum_{k'}\left(\dfrac{v}{E_\mu - E_{k'}}\right)^2\right]^{1/2}} \tag{8.a}$$

$$\langle k|\psi_\mu\rangle = \frac{v/(E_\mu - E_k)}{\left[1 + \sum_{k'}\left(\dfrac{v}{E_\mu - E_{k'}}\right)^2\right]^{1/2}}. \tag{8.b}$$

Let us return to the eigenvalue equation (6). It involves a series of the form $\sum_k(z - k)^{-1}$ with $z = E_\mu/\delta$. Similarly, the sum $\sum_k(z - k)^{-2}$ appears in the components (8.a) and (8.b) of the eigenvectors. It is possible

(*) We will see further on that $E_\mu - E_k$ is always nonzero.

to show that (*)

$$\sum_k (z - k)^{-2} = \frac{\pi^2}{\sin^2 \pi z}. \tag{9}$$

By integration, we obtain

$$\sum_k (z - k)^{-1} = \frac{\pi}{\tan \pi z}. \tag{10}$$

It follows that the eigenvalue equation (6) takes the simpler form

$$\frac{\pi v^2}{\delta \tan(\pi E_\mu/\delta)} = E_\mu \tag{11}$$

that we will transform, by using (1), into

$$\frac{1}{\tan(\pi E_\mu/\delta)} = \frac{2 E_\mu}{\hbar \Gamma}. \tag{12}$$

Finally, we introduce the angle defined by

$$\varphi_\mu = \tan^{-1} \frac{\hbar \Gamma}{2 E_\mu} \tag{13.a}$$

which allows us to write the solution of (12) in the form

$$\frac{E_\mu}{\delta} = m + \frac{\varphi_\mu}{\pi} \tag{13.b}$$

where m is an integer (≥ 0 or ≤ 0). Since φ_μ is (by definition of the $\tan^{-1}$ function) between $-\pi/2$ and $\pi/2$, φ_μ/π is the difference between E_μ/δ and the integer closest to E_μ/δ. We will use this angle φ_μ later on.

b) GRAPHIC DETERMINATION OF THE NEW EIGENVALUES

To determine the eigenvalues of the total Hamiltonian, we must solve Equation (12); that is, find the intersections of the line $y = ax$ with the curve $y = 1/\tan bx$ (a and b being respectively equal to $2/\hbar\Gamma$ and π/δ). This may be done graphically (see Figure 1).

(*) See, for example, Cartan (Chapter V).

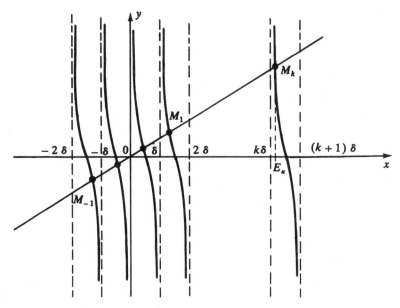

Figure 1. Graphic determination of the eigenvalues of H. The abscissa of each point of intersection between the line $y = 2x/\hbar\Gamma$ and the curve $y = 1/\tan(\pi x/\delta)$ is an eigenvalue of the total Hamiltonian. Let M_k be the point of intersection whose abscissa is between $k\delta$ and $(k + 1)\delta$ (the abscissa of M_{-k} being between $-(k + 1)\delta$ and $-k\delta$). The associated eigenvalue is denoted E_κ (Greek index corresponding to the Roman index k).

An eigenvalue $E_{\pm\kappa}$ is associated with each intersection M_k between the line $y = ax$ and the curve $y = 1/\tan bx$ (the unperturbed levels are indicated by a Roman index and the perturbed levels by corresponding Greek index). It is clear from Figure 1 that the new eigenvalues $E_{\pm\kappa}$ are interspersed between the old ones ($E_k < E_\kappa < E_{k+1}$ and $E_{-(k+1)} < E_{-\kappa} < E_{-k}$). The eigenstates of H thus form a quasi-continuum for which the density of states is extremely close to $1/\delta$ (there is one eigenstate per interval of energy δ).

We also note in Figure 1 that, for sufficiently large values of k, the abscissa of point M_k differs very little from $k\delta$, so that, to a first approximation, we have $E_\kappa \simeq E_k$. This relation is satisfied when, at the point of intersection, the curve is sufficiently close to its asymptote. According to (12), this occurs when

$$E_\kappa \gg \hbar\Gamma. \tag{14}$$

The presence of the discrete level thus significantly modifies the eigenvalues and eigenvectors of the quasi-continuum only over an energy interval on the order of $\hbar\Gamma$ around the energy of the discrete level.

c) PROBABILITY DENSITY OF THE DISCRETE STATE IN THE NEW CONTINUUM

In the presence of the coupling V, the discrete state $|\varphi\rangle$ is found diluted in the different states $\{|\psi_\mu\rangle\}$ of the quasi-continuum of H, the component of $|\varphi\rangle$ in the state $|\psi_\mu\rangle$ being given by the square of the expression (8.a). To transform the denominator of this expression, we use (9). We get

$$1 + v^2 \sum_k \left(\frac{1}{E_\mu - E_k} \right)^2 = 1 + \frac{v^2}{\delta^2} \sum_k \left(\frac{E_\mu}{\delta} - k \right)^{-2}$$

$$= 1 + \frac{\pi^2 v^2}{\delta^2} \left(1 + \left(\tan \pi \frac{E_\mu}{\delta} \right)^{-2} \right) \quad (15)$$

that is, using (11) and (1)

$$1 + v^2 \sum_k \left(\frac{1}{E_\mu - E_k} \right)^2 = 1 + \frac{\pi^2 v^2}{\delta^2} + \frac{E_\mu^2}{v^2} = \frac{1}{v^2} \left[v^2 + \left(\frac{\hbar\Gamma}{2} \right)^2 + E_\mu^2 \right] \quad (16)$$

which finally yields

$$\langle \varphi | \psi_\mu \rangle = \frac{v}{\left[v^2 + \left(\frac{\hbar\Gamma}{2} \right)^2 + E_\mu^2 \right]^{1/2}} . \quad (17)$$

Let us consider an interval $[E, E + dE]$ with dE large compared to δ, but small compared to $\hbar\Gamma$. The probability dN_φ of finding the discrete state $|\varphi\rangle$ in this interval equals

$$dN_\varphi = \sum_{E < E_\mu < E + dE} |\langle \varphi | \psi_\mu \rangle|^2 \simeq \frac{dE}{\delta} |\langle \varphi | \psi_\mu \rangle|^2 \quad (18)$$

which yields, taking into account (17) and (1),

$$\frac{dN_\varphi}{dE} = \frac{v^2/\delta}{v^2 + \left(\dfrac{\hbar\Gamma}{2}\right)^2 + E^2} = \frac{\hbar\Gamma/2\pi}{v^2 + \left(\dfrac{\hbar\Gamma}{2}\right)^2 + E^2}. \tag{19}$$

At the limit $\delta \to 0$, v^2 tends to 0 and expression (19) becomes

$$\frac{dN_\varphi}{dE} = \frac{\hbar\Gamma/2\pi}{\left(\dfrac{\hbar\Gamma}{2}\right)^2 + E^2} \tag{20}$$

which is a Lorentzian curve of width $\hbar\Gamma$, centered on $E = E_\varphi = 0$ and having an integral over E equal to 1.

Such a result demonstrates that, after coupling, the discrete level $|\varphi\rangle$ is spread over an interval of width $\hbar\Gamma$ in the new continuum. In other words, only the levels of the new continuum located in an interval on the order of $\hbar\Gamma$ about E_φ retain the memory of the level $|\varphi\rangle$ in their wave function.

3. A Few Applications of This Simple Model

The results obtained above concerning the new continuum of states $\{|\psi_\mu\rangle\}$ and the density dN_φ/dE characterizing the traces of the discrete state $|\varphi\rangle$ in this new continuum are helpful for a quantitative treatment of several problems. We review now some of them.

a) Decay of the Discrete Level (*)

Let us first attempt to calculate the probability that the system, initially prepared (at $t = 0$) in the discrete state $|\varphi\rangle$, still remains in the same state an instant t later. Let us use the relation (17) and expand $|\varphi\rangle$ on the basis of eigenstates $|\psi_\mu\rangle$ of the Hamiltonian. We get

$$|\psi(0)\rangle = |\varphi\rangle = \sum_\mu \frac{v}{\left[v^2 + \left(\dfrac{\hbar\Gamma}{2}\right)^2 + E_\mu^2\right]^{1/2}} |\psi_\mu\rangle. \tag{21}$$

(*) This problem may also be treated by using the Weisskopf-Wigner method. See, for example Cohen-Tannoudji, Diu, and Laloë, Complement D_{XIII}.

At time t, this state $|\psi(t)\rangle$ becomes

$$|\psi(t)\rangle = \sum_\mu \frac{v \, e^{-iE_\mu t/\hbar}}{\left[v^2 + \left(\frac{\hbar\Gamma}{2} \right)^2 + E_\mu^2 \right]^{1/2}} |\psi_\mu\rangle \tag{22}$$

and the probability amplitude of finding the system in the state $|\varphi\rangle$ is, according to (21) and (22), equal to

$$\langle\varphi|\psi(t)\rangle = \sum_\mu \frac{v^2 \, e^{-iE_\mu t/\hbar}}{v^2 + \left(\frac{\hbar\Gamma}{2} \right)^2 + E_\mu^2} \tag{23}$$

which we may rewrite by replacing v^2 by $(\hbar\Gamma/2\pi)\delta$:

$$\langle\varphi|\psi(t)\rangle = \delta \sum_\mu \frac{(\hbar\Gamma/2\pi)e^{-iE_\mu t/\hbar}}{v^2 + \left(\frac{\hbar\Gamma}{2} \right)^2 + E_\mu^2}. \tag{24}$$

At the limit $\delta \to 0$, the sum $\delta\sum_\mu f(E_\mu)$ tends to an integral $\int dE f(E)$, so that

$$\langle\varphi|\psi(t)\rangle = \int_{-\infty}^{+\infty} \frac{\hbar\Gamma}{2\pi} \frac{e^{-iEt/\hbar}}{\left(\frac{\hbar\Gamma}{2} \right)^2 + E^2} \, dE. \tag{25}$$

This integral is calculated by the method of residues and reduces to

$$\langle\varphi|\psi(t)\rangle = e^{-\Gamma|t|/2}. \tag{26}$$

The probability of finding the system at the instant t in the level $|\varphi\rangle$ is thus

$$|\langle\varphi|\psi(t)\rangle|^2 = e^{-\Gamma|t|}. \tag{27}$$

It decays exponentially with a time constant Γ^{-1}.

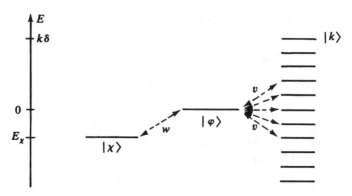

Figure 2. Energy-level scheme considered in this paragraph. The arrows represent the coupling between the levels.

b) EXCITATION OF THE SYSTEM IN THE DISCRETE LEVEL FROM ANOTHER STATE

Let us now assume that the system under study has another discrete level $|\chi\rangle$, with unperturbed energy E_χ, orthogonal to the discrete state $|\varphi\rangle$ and to the quasi-continuum of states $\{|k\rangle\}$. This level $|\chi\rangle$ is coupled directly only to the state $|\varphi\rangle$ by a coupling Hamiltonian W.

$$\langle\varphi|W|\chi\rangle = w \tag{28.a}$$

$$\langle k|W|\chi\rangle = 0. \tag{28.b}$$

All the other matrix elements of W are assumed to be zero.

We show in this section that, if w is sufficiently small, the system, initially in state $|\chi\rangle$, leaves this state with a well-defined rate. In other words, as a consequence of its coupling with a discrete level $|\varphi\rangle$, which is itself coupled to a continuum, the level $|\chi\rangle$ acquires a finite lifetime.

We have shown previously (§2) that the coupling of $|\varphi\rangle$ with the states $|k\rangle$ gives rise to a new quasi-continuum of states $|\psi_\mu\rangle$. Let us now calculate $\langle\psi_\mu|W|\chi\rangle$. Because W couples $|\chi\rangle$ only to $|\varphi\rangle$, we have

$$\langle\psi_\mu|W|\chi\rangle = \langle\psi_\mu|\varphi\rangle\langle\varphi|W|\chi\rangle = \langle\psi_\mu|\varphi\rangle w. \tag{29}$$

It thus appears that W couples the level $|\chi\rangle$ to the quasi-continuum of states $\{|\psi_\mu\rangle\}$. If the coupling w remains small, the probability per unit time that the system leaves the state $|\chi\rangle$ is given by Fermi's golden rule:

$$\Gamma_\chi = \frac{2\pi}{\hbar} |\langle\psi_\mu|W|\chi\rangle|^2 \frac{1}{\delta}. \tag{30}$$

(The density of states $\{|\psi_\mu\rangle\}$, for $E_\mu = E_\chi$, is equal to $1/\delta$.) By using (29) and (17), we find at the limit $\delta \to 0$:

$$\Gamma_\chi = w^2 \frac{\Gamma}{\left(\dfrac{\hbar\Gamma}{2}\right)^2 + E_\chi^2}. \tag{31}$$

In the particular case where $|\chi\rangle$ and $|\varphi\rangle$ have the same energy ($E_\chi = E_\varphi = 0$), expression (31) becomes

$$\Gamma_\chi = \frac{4w^2}{\hbar^2\Gamma}. \tag{32}$$

Hence, a discrete state $|\chi\rangle$ coupled to another discrete state $|\varphi\rangle$ which is unstable will itself decay irreversibly, with a rate Γ_χ given by (32).

Remarks

(i) The treatment presented above is valid only for sufficiently small w. More precisely the coupling w must be smaller than the width $\hbar\Gamma$ of the interval over which $|\langle\psi_\mu|W|\chi\rangle|^2$ is significant, or equivalently, the width of the interval of the new continuum $\{|\psi_\mu\rangle\}$ in which the probability that $|\varphi\rangle$ is present is important. Such a result may also be understood by comparing the period of the Rabi oscillation between $|\chi\rangle$ and $|\varphi\rangle$, which is on the order of $\hbar/w$, and the lifetime of the system in the level $|\varphi\rangle$, which is on the order of Γ^{-1}. If this lifetime is shorter than the period of the Rabi oscillation, a system initially in the state $|\chi\rangle$ will evolve irreversibly to the continuum because, once it has passed through $|\varphi\rangle$, it will immediately decay in the continuum $\{|k\rangle\}$ and will have an extremely low probability of returning to $|\chi\rangle$. Note that the condition $w \ll \hbar\Gamma$ implies that $\Gamma_\chi \ll \Gamma$. The lifetime associated with the level $|\chi\rangle$ is much longer than that associated with the level $|\varphi\rangle$.

(ii) The level $|\chi\rangle$ introduced above may itself be part of the continuum $\{|k\rangle\}$ (this is in particular the case for the absorption of a photon between the ground state a and a discrete excited state b of an atom, a problem that we will consider in subsection B-4 of Chapter II). To apply the preceding treatment, we must remove the state $|\chi\rangle$ from the continuum. Clearly, if δ is sufficiently small, the new continuum resulting from the coupling between the discrete state $|\varphi\rangle$ and the continuum of states $\{|k\rangle\}$ from which we have subtracted $|\chi\rangle$

will be sufficiently close to the continuum $\{|\psi_\mu\rangle\}$ studied above, so that all the preceding results remain valid.

c) RESONANT SCATTERING THROUGH A DISCRETE LEVEL

Let us now consider two levels $|\chi_i\rangle$ and $|\chi_j\rangle$ having the same energy E_χ. For example, in the problem of the resonant scattering of a photon by an atom, $|\chi_i\rangle$ and $|\chi_j\rangle$ are states which represent the atom in the ground state in the presence of the incident or scattered photon, respectively (see §C-3 of Chapter II). These two levels $|\chi_i\rangle$ and $|\chi_j\rangle$ are coupled to $|\varphi\rangle$ by a coupling term W ($\langle\varphi|W|\chi_i\rangle = w_i$ and $\langle\varphi|W|\chi_j\rangle = w_j$). On the other hand, we assume that $|\chi_i\rangle$ and $|\chi_j\rangle$ are neither coupled to each other ($\langle\chi_i|W|\chi_j\rangle = 0$) nor to the levels of the quasi-continuum $\{|k\rangle\}$ ($\langle k|W|\chi_i\rangle = 0$). We will study the scattering from $|\chi_i\rangle$ to $|\chi_j\rangle$ and show that, even if $E_\chi = E_\varphi$, the scattering amplitude does not diverge. This result is, of course, related to the fact that the discrete level $|\varphi\rangle$ is spread by the coupling V in the quasi-continuum of states $\{|\psi_\mu\rangle\}$.

First, we recall how the scattering amplitude diverges when the coupling between the discrete state $|\varphi\rangle$ and the continuum $\{|k\rangle\}$ is not taken into consideration. To lowest order in V, the transition matrix element is written

$$\mathcal{T}_{ji} \simeq \lim_{\eta\to 0_+} \langle\chi_j|W\frac{1}{E_\chi - H_0 + i\eta}W|\chi_i\rangle \tag{33}$$

that is, again

$$\mathcal{T}_{ji} \simeq \lim_{\eta\to 0_+} \frac{\langle\chi_j|W|\varphi\rangle\langle\varphi|W|\chi_i\rangle}{E_\chi - E_\varphi + i\eta} \tag{34}$$

an expression that diverges if $E_\chi = E_\varphi$.

The coupling between $|\varphi\rangle$ and $|k\rangle$ appears only to higher orders in V. In fact it is possible to write the transition matrix element to all orders in V. It is sufficient to replace H_0 by $H = H_0 + V$ in (33) [see expression (B.15) in this chapter]. Because W couples $|\chi_i\rangle$ and $|\chi_j\rangle$ only to $|\varphi\rangle$, we get

$$\mathcal{T}_{ji} = \lim_{\eta\to 0_+} \langle\chi_j|W|\varphi\rangle\langle\varphi|\frac{1}{E_\chi - H + i\eta}|\varphi\rangle\langle\varphi|W|\chi_i\rangle. \tag{35}$$

Let us now introduce the closure relation over the eigenstates $|\psi_\mu\rangle$ of H in the central matrix element of (35). The density of the state $|\varphi\rangle$ in the new continuum then appears explicitly, and expression (35) becomes, taking (17) and (1) into consideration,

$$\mathscr{T}_{ji} = \lim_{\eta \to 0_+} w_j^* w_i \delta \sum_\mu \frac{\hbar\Gamma/2\pi}{(E_\chi - E_\mu + i\eta)\left(E_\mu^2 + \left(\dfrac{\hbar\Gamma}{2}\right)^2 + v^2\right)}. \qquad (36)$$

In the limit $\delta \to 0$, this sum tends to the following integral:

$$\mathscr{T}_{ji} = w_j^* w_i \lim_{\eta \to 0_+} \int_{-\infty}^{+\infty} dE\, \frac{\hbar\Gamma/2\pi}{(E_\chi - E + i\eta)\left(E^2 + \left(\dfrac{\hbar\Gamma}{2}\right)^2\right)} \qquad (37)$$

which may be calculated by the residue method and gives

$$\mathscr{T}_{ji} = w_j^* w_i \frac{1}{E_\chi + i\hbar(\Gamma/2)}. \qquad (38)$$

The scattering amplitude thus does not diverge any more when $E_\chi = E_\varphi = 0$. Note that everything happens as if, in the lowest-order expression, the energy of the discrete level $|\varphi\rangle$ had been replaced by a complex energy $E_\varphi - i\hbar\Gamma/2 = -i\hbar\Gamma/2$. While remaining finite, the transition matrix element thus varies in a resonant fashion when E_χ is swept over an interval of width $\hbar\Gamma$ about $E_\varphi = 0$.

d) FANO PROFILES

We come back to the case where the system has another state $|\chi\rangle$ coupled to $|\varphi\rangle$ by a coupling term $W(\langle\varphi|W|\chi\rangle = w)$, but we now assume that $|\chi\rangle$ is also directly coupled by W to the states $\{|k\rangle\}$ of the quasi-continuum. We also make a simplifying assumption concerning the matrix elements $\langle k|W|\chi\rangle$ which are assumed to be independent of $|k\rangle$ ($\langle k|W|\chi\rangle = w'$). What is the probability that a level $|\psi_\mu\rangle$ will be excited from $|\chi\rangle$? This probability is proportional to the square of the matrix element $\langle\psi_\mu|W|\chi\rangle$ which may be calculated from (17) and (5):

$$\langle\psi_\mu|W|\chi\rangle = \frac{\langle\varphi|W|\chi\rangle v + \sum_k \langle k|W|\chi\rangle v^2/(E_\mu - E_k)}{\left[v^2 + \left(\dfrac{\hbar\Gamma}{2}\right)^2 + E_\mu^2\right]^{1/2}} \qquad (39)$$

an expression which is written, taking (6) into consideration,

$$\langle \psi_\mu | W | \chi \rangle = \frac{wv + w'E_\mu}{\left[v^2 + \left(\frac{\hbar \Gamma}{2} \right)^2 + E_\mu^2 \right]^{1/2}}. \tag{40}$$

When the energy of the state E_μ varies, the two terms of the numerator of the right-hand side of (40) add up or subtract depending on the respective signs of E_μ and of vw/w'. Hence the lineshape (probability of excitation as a function of E_μ) is generally asymmetric.

Relation (40) is frequently rewritten as a function of the reduced variables

$$\varepsilon_\mu = \frac{E_\mu}{\hbar \Gamma / 2} \tag{41}$$

$$q = \frac{\delta}{\pi v} \frac{w}{w'} \tag{42}$$

ε_μ is the energy in units of $\hbar \Gamma / 2$. The parameter q defined in (42) characterizes the ratio between the coupling to the discrete state $|\varphi\rangle$ and the coupling to the quasi-continuum $\{|k\rangle\}$. Let us also introduce the parameter ξ

$$\xi = \frac{4v^2}{\hbar^2 \Gamma^2} = \frac{2}{\pi} \frac{\delta}{\hbar \Gamma}. \tag{43}$$

ξ^{-1} represents the number of levels of the discretized continuum within the natural width ($\xi \to 0$ when $\delta \to 0$). By using these definitions, we obtain from (40)

$$\frac{|\langle \psi_\mu | W | \chi \rangle|^2}{w'^2} = \frac{|q + \varepsilon_\mu|^2}{1 + \varepsilon_\mu^2 + \xi}. \tag{44}$$

In Figure 3 we have represented several possible excitation profiles obtained for different values of q. They are called Fano profiles (*). Note

(*) Such profiles are encountered in several physical situations, for example in photoionization near an autoionizing state.

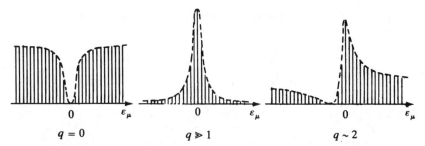

Figure 3. Fano profiles obtained for different values of q. The intensity of the transitions to the states $|\psi_\mu\rangle$ of the quasi-continuum is represented by solid lines. The dashed line corresponding to the envelope of these intensities allows us to visualize the profile of excitation at the continuous limit ($\delta \to 0$). Note that except for the limiting situations $q = 0$ and $q \gg 1$, the profiles are asymmetric.

that the situation $q \gg 1$ (case where the coupling between $|\chi\rangle$ and $|\varphi\rangle$ is much larger than the coupling between $|\chi\rangle$ and $|k\rangle$) corresponds to the situation considered in subsection 3-b of this complement.

Remark

In all the situations considered previously, the limit $\delta \to 0$ did not pose any problem. This is not always the case, and certain precautions must be taken when the problem under study (*) involves quantities such as $\langle k|\psi_\mu\rangle$ [relation (8.b)] in which the factor $1/(E_\mu - E_k)$ appears. Actually, in the discrete case $(E_\mu - E_k)$ never vanishes (see Figure 1), and thus division by $(E_\mu - E_k)$ presents no difficulties. This is what we did, for example, to deduce (5) from (4.a). By contrast, more precautions must be taken in determining the limit of $1/(E_\mu - E_k)$. In order to determine such a limit considered as a distribution, we reexpress $(E_\mu - E_k)^{-1}$ as a function of the angle φ_μ defined by Equation (13.a) and of the integer m closest to E_μ/δ (see Figure 1). By using Equation (13.b) and the fact that $E_k = k\delta$, we obtain

$$\frac{1}{E_\mu - E_k} = \frac{1}{\left(m - k + \dfrac{\varphi_\mu}{\pi}\right)\delta}. \tag{45}$$

(*) This is the case, for example, when one studies the energy distribution of the final states resulting from the disintegration of the discrete state.

Let us rewrite (45) as the sum of an odd function of $(m - k)$ and an even function.

$$\frac{1}{E_\mu - E_k} = \frac{(m - k)\delta}{(m - k)^2\delta^2 - \left(\varphi_\mu\delta/\pi\right)^2} - \frac{\varphi_\mu\delta/\pi}{(m - k)^2\delta^2 - \left(\varphi_\mu\delta/\pi\right)^2}. \quad (46)$$

Since φ_μ/π is, in modulus, less than $\frac{1}{2}$, $\varphi_\mu\delta/\pi$ tends to zero when δ tends to zero. The first term of the right-hand side of (46) is odd in $(m - k)\delta$, and hence tends, when $\delta \to 0$, to $\mathscr{P}(1/(E' - E))$, where E' and E are associated, respectively, with E_μ and E_k and $\mathscr{P}$ denotes the principal part. Let us now analyze the second term. It has significant values only for m close to k. In particular, for $E_m = E_k$, its value is equal to $\pi/(\varphi_\mu\delta)$ and thus tends to infinity when $\delta \to 0$. When $E_m \neq E_k$, this term has the opposite sign but its value remains on the order of $1/\delta$ for small $(m - k)$. Its width in E_k is of the order of δ. The second term of (46) thus tends to a distribution localized in $E_m = E_k$ when $\delta \to 0$. Let us now calculate the sum.

$$-\delta \sum_k \frac{\varphi_\mu\delta/\pi}{(m - k)^2\delta^2 - \left(\varphi_\mu\delta/\pi\right)^2} = \sum_p \frac{\varphi_\mu/\pi}{\left(\varphi_\mu/\pi\right)^2 - p^2}. \quad (47)$$

The sum of this series is known (*) to be $\pi/\tan \varphi_\mu$; that is, again according to (13.a),

$$\frac{2\pi}{\hbar\Gamma}E_\mu. \quad (48)$$

It follows that the limit $\delta \to 0$ of the second term of (46) is

$$(2\pi/\hbar\Gamma)E\delta(E - E').$$

Finally,

$$\lim_{\delta \to 0} \frac{1}{E_\mu - E_k} = \mathscr{P}\left(\frac{1}{E' - E}\right) + \frac{2\pi}{\hbar\Gamma}E\delta(E - E'). \quad (49)$$

4. Generalization to More Realistic Continua. Diagonalization of the Hamiltonian without Discretization

We return in this last subsection to the problem of a discrete state $|\varphi\rangle$ coupled to a continuum of states $\{|E\rangle\}$, to directly study (without discretizing the continuum) the eigenstates $|\psi(E)\rangle$ of the total Hamiltonian. In contrast to the situation in subsections 2 and 3, we will not make any

(*) See for example, Cartan, corrected exercise 25, p. 226.

restrictive assumptions on the continuum and on its coupling with the state $|\varphi\rangle$. In particular, the matrix element $\langle E|V|\varphi\rangle = v(E)$ is now an arbitrary function of E. The orthonormalization relation between states of the continuum is

$$\langle E'|E\rangle = \delta(E - E') \tag{50}$$

and the matrix elements of the Hamiltonian H_0 and of the coupling V are

$$\langle\varphi|H_0|\varphi\rangle = E_\varphi \tag{51.a}$$

$$\langle E|H_0|E'\rangle = E\delta(E - E') \tag{51.b}$$

$$\langle E|V|\varphi\rangle = v(E) \tag{51.c}$$

$$\langle E'|V|E\rangle = \langle\varphi|V|\varphi\rangle = 0 \tag{51.d}$$

Let $|\psi(E')\rangle$ be an eigenstate of H_0 with eigenvalue E'

$$H|\psi(E')\rangle = E'|\psi(E')\rangle. \tag{52}$$

To find the expansion of $|\psi(E')\rangle$ over the unperturbed states $|\varphi\rangle$ and $|E\rangle$, we proceed as in subsection 2-a and first project (52) onto $|\varphi\rangle$, then onto $|E\rangle$. We thus obtain the equations

$$E_\varphi\langle\varphi|\psi(E')\rangle + \int dE\, v(E)^*\langle E|\psi(E')\rangle = E'\langle\varphi|\psi(E')\rangle \tag{53.a}$$

$$E\langle E|\psi(E')\rangle + v(E)\langle\varphi|\psi(E')\rangle = E'\langle E|\psi(E')\rangle. \tag{53.b}$$

Equation (53.b) is transformed to

$$(E' - E)\langle E|\psi(E')\rangle = v(E)\langle\varphi|\psi(E')\rangle. \tag{54}$$

Distribution, theory allows the general solution of (54) to be written in the form

$$\langle E|\psi(E')\rangle = \left[\mathscr{P}\frac{1}{E' - E} + z(E')\delta(E - E')\right]v(E)\langle\varphi|\psi(E')\rangle \tag{55}$$

where $z(E')$ is an arbitrary function of E'. To determine this function $z(E')$, we substitute (55) into (53.a), and we obtain

$$E_\varphi + \mathscr{P}\int dE\,\frac{|v(E)|^2}{E' - E} + z(E')|v(E')|^2 = E' \tag{56}$$

that is

$$E_\varphi + \hbar\Delta(E') + \frac{1}{\pi}\hbar\frac{\Gamma(E')}{2}z(E') = E' \tag{57}$$

with

$$\Gamma(E') = \frac{2\pi}{\hbar}|v(E')|^2 \tag{58.a}$$

$$\Delta(E') = \frac{1}{\hbar}\mathscr{P}\int dE\,\frac{|v(E)|^2}{E'-E} = \frac{1}{2\pi}\mathscr{P}\int dE\,\frac{\Gamma(E)}{E'-E}. \tag{58.b}$$

From this we deduce

$$z(E') = 2\pi\frac{E'-E_\varphi-\hbar\Delta(E')}{\hbar\Gamma(E')}. \tag{59}$$

Equations (55) and (59) allow us to find the new eigenvectors to within a normalization coefficient, which is determined by the equation

$$\langle\psi(E')|\psi(E)\rangle = \delta(E'-E). \tag{60}$$

We do not give here the calculation of this normalization coefficient, but we will simply quote the result (*):

$$
\begin{aligned}
|\psi(E')\rangle = {} & \frac{1}{\left\{[E'-E_\varphi-\hbar\Delta(E')]^2 + [\hbar\Gamma(E')/2]^2\right\}^{1/2}} \\
& \times \left\{v(E')\left[|\varphi\rangle + \mathscr{P}\int dE\,\frac{V(E)}{E'-E}|E\rangle\right]\right. \\
& \left. + [E'-E_\varphi-\hbar\Delta(E')]|E'\rangle\right\}.
\end{aligned} \tag{61}
$$

No restrictive assumption having been made on the variation of the coupling $v(E)$ with E, the vectors thus obtained may be applied to the study of many problems.

(*) See U. Fano, *Phys. Rev.*, **124**, 1866 (1961).

A Survey of Some Interaction Processes between Photons and Atoms

Chapter II presents an overall view of the interactions between atoms and photons. Our main goal is to introduce in the simplest possible way some fundamental processes, to specify the vocabulary used to describe them, and to analyze the associated physical phenomena. The emphasis will be placed more on the physical content of the processes under study than on the calculation methods. During this discussion, we will use only simple ideas concerning the transition amplitudes introduced in Chapter I, reserving for subsequent chapters the presentation of more powerful and more precise calculation methods.

To classify the different processes to be studied, we find it convenient to consider the number of photons involved in a given process (to lowest order where this process appears). Hence, the simplest processes are those during which a new photon appears (*emission processes* studied in Section A), or during which a photon, initially present, disappears (*absorption processes* studied in Section B). Distinctions are made among several cases, depending on the discrete or continuous nature of the initial and final internal atomic states. This allows us to review several important phenomena such as photoionization, Bremsstrahlung, photodissociation, etc. We also show in Section B how the dynamics of the absorption process (or stimulated emission) may be sensitive to the state of the incident field. This discussion is extended in Complement A_{II} where the probability of absorption is described in terms of correlation functions for the two interacting systems, the atom and the field.

We then discuss, in Section C, a more complex process involving several photons, the *scattering* process during which one photon disappears and a new photon appears. Distinctions are made among several types of scattering processes, depending on whether the incident photon has an energy that is small or large compared to the excitation energies of the atom, a special treatment being devoted to the case of resonant scattering.

The following section, Section D, deals with *multiphoton processes* during which several photons appear or disappear. The variety of these processes is extremely large, and we review only some significant examples.

Finally, in the last two sections, Sections E and F, we study processes for which the state of the field is the same in the initial and final states, but during which photons may be emitted or reabsorbed, or absorbed and reemitted. We first consider in Section E spontaneous (or stimulated) *radiative corrections* resulting from the emission and the reabsorption (or the absorption and reemission) of a photon by a charged particle or by an atom. The case in which photons are emitted by a particle (or an atom) and reabsorbed by another particle (or another atom) is then analyzed in Section F. Such *photon exchanges between particles* or atoms give rise to effective interactions for which two examples are given (magnetic interactions and Van der Waals interactions). Finally, the problem of radiative corrections is described in Complement B_{II} using another approach (Pauli–Fierz approach), which consists of carrying out on the quantum electrodynamic Hamiltonian a unitary transformation that subtracts the transverse field "bound" to the particles from the total field.

A—EMISSION PROCESS: A NEW PHOTON APPEARS

The emission of a photon with wave vector $\mathbf{k}$ and polarization $\boldsymbol{\varepsilon}$ is a process in which the number of photons in the mode $\mathbf{k}\boldsymbol{\varepsilon}$ increases by one unit, passing from n in the initial state to $n + 1$ in the final state, while all the other modes remain in the same state. We will assume in this subsection that, in the initial state, all the modes are empty. The emission is then called *spontaneous emission*. We will subsequently return to the case in which photons are present in the initial state. Distinctions are made depending on whether the initial or final (internal) atomic states are discrete or continuous.

1. Spontaneous Emission between Two Discrete Atomic Levels. Radiative Decay of an Excited Atomic State

a) DIAGRAMMATIC REPRESENTATION

Let us first consider the spontaneous emission of a photon by an atom passing from one discrete internal state $|b\rangle$ to another discrete internal state $|a\rangle$. More precisely, we take, for the initial state of the global system, the state

$$|\varphi_i\rangle = |b, \mathbf{K}; 0\rangle \tag{A.1}$$

which represents an atom in the internal state b, with global momentum $\hbar\mathbf{K}$, the field being in the vacuum state, and for the final state

$$|\varphi_f\rangle = |a, \mathbf{K}'; \mathbf{k}\boldsymbol{\varepsilon}\rangle \tag{A.2}$$

which represents the atom in the internal state a, with global momentum $\hbar\mathbf{K}'$, in the presence of a photon $\mathbf{k}\boldsymbol{\varepsilon}$. The diagrammatic representation of such a process (see Chapter I, §C-5) is given in Figure 1.

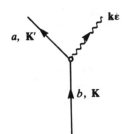

Figure 1. Diagrammatic representation of spontaneous emission between two discrete states b and a.

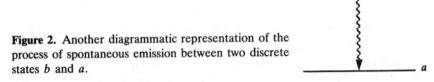

Figure 2. Another diagrammatic representation of the process of spontaneous emission between two discrete states b and a.

The conservation of energy, expressed by the function $\delta^{(T)}(E_f - E_i)$ appearing in the transition amplitude [relation (29) of Complement A_1], leads to

$$E_b + \frac{\hbar^2 K^2}{2M} = E_a + \frac{\hbar^2 K'^2}{2M} + \hbar\omega. \tag{A.3}$$

Similarly, the conservation of the total momentum, due to the translation invariance of the interaction Hamiltonian (*), results in

$$\hbar\mathbf{K} = \hbar\mathbf{K}' + \hbar\mathbf{k}. \tag{A.4}$$

The internal state b necessarily has an energy greater than that of the state a. This is obvious when (A.3) is written in the center-of-mass reference frame ($\mathbf{K} = \mathbf{0}$), but it does not clearly appear in the diagram of Figure 1, which does not show the internal energies. For this reason, another type of diagrammatic representation (Figure 2) is sometimes used, where the internal states are represented by horizontal lines with ordinates proportional to the internal energy, and the emission of a photon is represented by a wavy arrow going from b to a (the atom "falls" from b to a by emitting a photon).

b) SPONTANEOUS EMISSION RATE

Because total momentum is conserved, the problem of spontaneous emission can be studied separately in each subspace corresponding to a given value of this total momentum, for example, in the center-of-mass system where $\mathbf{K} = \mathbf{0}$ and where, as a result of (A.4), $\mathbf{K}' = -\mathbf{k}$. In this subspace, $|\varphi_i\rangle = |b, \mathbf{0}; 0\rangle$ is a discrete state, because we have fixed the value $\mathbf{K} = \mathbf{0}$ and because both the internal state b and the vacuum state $|0\rangle$ of radiation are discrete. By contrast, there exists an infinite number of

(*) See, for example, *Photons and Atoms—Introduction to Quantum Electrodynamics*, Chapter III, §D-3.

final states $|\varphi_f\rangle = |a, -\mathbf{k}; \mathbf{k}\varepsilon\rangle$, corresponding to all the possible values of $\mathbf{k}$, for which the energies $E_a + \hbar\omega + (\hbar^2 k^2/2M)$ vary continuously. These states are coupled to $|\varphi_i\rangle$ by H_{I1} [see (C.5.b) of Chapter I)]: the problem of spontaneous emission between two discrete atomic states is thus finally, with regard to the *global* system atom + radiation, a problem involving a discrete state coupled to a continuum. The evolution of the system thus will present an irreversible character. There will be no reversible oscillations between $|\varphi_i\rangle$ and $|\varphi_f\rangle$, but rather an irreversible departure from the discrete state $|\varphi_i\rangle$ to the continuum $|\varphi_f\rangle$ with a transition rate Γ given, according to expression (B.8) of Chapter I, by (*)

$$\Gamma = \frac{2\pi}{\hbar} \sum_{\substack{\mathbf{k}, \varepsilon \\ E_f = E_i}} |\langle a, -\mathbf{k}; \mathbf{k}\varepsilon|H_{I1}|b, 0; 0\rangle|^2 \rho(E_f = E_i). \qquad (\text{A.5})$$

On the other hand, the energy conservation relation (A.3), combined with (A.4), determines the energy of the final states, and consequently the emitted frequencies. In the center-of-mass system $(\mathbf{K} = 0, \mathbf{K}' = -\mathbf{k})$ we obtain

$$\hbar\omega = \hbar\omega_0 - E_{\text{rec}} \qquad (\text{A.6})$$

where

$$\hbar\omega_0 = E_b - E_a \qquad (\text{A.7})$$

is the energy of the atomic transition $b \rightarrow a$ and where

$$E_{\text{rec}} = \frac{\hbar^2 k^2}{2M} = \frac{\hbar^2 \omega^2}{2Mc^2} \qquad (\text{A.8})$$

is the recoil kinetic energy of the atom associated with the emission of the photon. Because $\hbar\omega \ll Mc^2$, this recoil energy is extremely small compared to $\hbar\omega$ in the optical (and a fortiori microwave) range, and is frequently neglected. This is equivalent to considering $M = \infty$, and thus an infinitely heavy atom. The frequency ω of the spontaneously emitted photons in the center-of-mass system is thus equal, to within $E_{\text{rec}}/\hbar$, to the atomic transition frequency ω_0.

(*) Exercise 1 presents a calculation of Γ for a two-level atom.

Remarks

(i) Consider a reference frame, other than the center-of-mass system ($\mathbf{K} \neq 0$), in which the atom moves with a velocity $\mathbf{v} = \hbar\mathbf{K}/M$. The simultaneous resolution of (A.3) and (A.4) yields

$$\hbar\omega = \hbar\omega_0 - E_{rec} + \hbar\mathbf{k} \cdot \mathbf{v}. \tag{A.9}$$

For each direction $\mathbf{k}$ of emission, there is thus a well-defined emission frequency, depending on the velocity $\mathbf{v}$ of the atom and on the angle between $\mathbf{v}$ and $\mathbf{k}$. This is the well-known Doppler effect.

(ii) In all the foregoing, we have implicitly assumed that the atom's center of mass moves freely. When the atom evolves in an external potential, the total momentum is no longer a constant of the motion [Equation (A.4) is no longer valid]. Expression (A.3) then becomes

$$E_b + E_{c.m.}^{(i)} = E_a + E_{c.m.}^{(f)} + \hbar\omega \tag{A.3'}$$

where $E_{c.m.}^{(i)}$ and $E_{c.m.}^{(f)}$ are, respectively, the initial energy (before emission) and final energy (after emission) associated with the motion of the center of mass in the external potential. Under certain circumstances, the probability that this energy does not change during the emission process ($E_{c.m.}^{(i)} = E_{c.m.}^{(f)}$) may be significant. In this case, the emitted photon has a frequency that is shifted neither by the recoil effect nor by the Doppler effect. Such a situation corresponds to the Dicke and Mössbauer effects, and is studied in greater detail in Exercise 2. The analysis made here on the emission process may be easily generalized to the absorption process considered in Section B. When the atom is placed in an external potential, it may happen that the frequency of an absorbed photon exactly corresponds to the atomic frequency.

c) Nonperturbative Results

All the previous considerations result from a calculation of the lowest-order transition amplitude. As we have already indicated in Chapter I (§B-2), it is possible to go further and to obtain expressions containing terms to all orders in H_{I1} for the amplitudes $\langle \varphi_i | \tilde{U}(T) | \varphi_i \rangle$ and $\langle \varphi_f | \tilde{U}(T) | \varphi_i \rangle$ (see also Complement C_I or Chapter III, §C-1). The results of such a nonperturbative treatment are thus as follows:

(i) The probability $P_b(T)$ that the atom is still present in the highest state b at time T decreases exponentially with a time constant:

$$\tau = \frac{1}{\Gamma} \tag{A.10}$$

where Γ is given by (A.5) (*). τ is called the *radiative lifetime* of the state b. The spontaneous emission is thus responsible for a radiative decay of the excited atomic levels.

(ii) As a result of its finite lifetime τ, the excited state b has a certain width in energy, characterized by $\hbar/\tau = \hbar\Gamma$. Γ is called the *natural width* of the level b (in angular frequency units). The spontaneous emission lines are thus not infinitely narrow, as predicted by the first-order theory, but rather have a certain width (equal to the sum of the natural widths of the two levels involved in the transition).

(iii) Finally, it is found that the state b is shifted in energy. This is a radiative correction to which we will return later in paragraph E-1-b.

Remark

Spontaneous emission from a discrete excited atomic level may be profoundly modified if, instead of being in free space, the atom is enclosed in a cavity (**). Actually, the modes of the electromagnetic field in the cavity form a discrete ensemble (if the damping of the cavity is negligible). First of all, suppose that one of the eigenfrequencies of the cavity coincides with the atomic frequency ω_0, all the other eigenfrequencies being sufficiently far from ω_0 so that we may ignore the nonresonant coupling of the atom with these other modes (for this it is necessary that the cavity be small enough). We will also ignore the external quantum numbers (negligible recoil energy). The two states $|\varphi_i\rangle = |b;0\rangle$, atom in the state b, resonant mode in the state $|0\rangle$, and $|\varphi_f\rangle = |a;1\rangle$, atom in the state a with one photon in the resonant mode, thus form two discrete states of the global system, degenerate in energy and far from all the other states. The interaction between the atom and the radiation therefore gives rise to a Rabi oscillation: the atom emits a photon into the cavity, then reabsorbs it, then reemits it, and so on. In the case where the eigenfrequencies of the cavity do not coincide with ω_0, the evolution is completely different: the atom cannot emit a photon into the cavity, because the total energy would not be conserved.

2. Spontaneous Emission between a Continuum State and a Discrete State

We now assume that one of the two atomic states, a or b, belongs to a continuum, the other being discrete.

(*) We assume that the only energy level lower than E_b to which the atom can decay from b by spontaneous emission is the level a.

(**) See for example D. Kleppner, *Phys. Rev. Lett.*, **47**, 233 (1981); S. Haroche and J. M. Raimond, in *Advances in Atomic and Molecular Physics*, edited by D. R. Bates and B. Bederson, Academic Press, New York, 1985, Vol. 20, p. 347; P. Filipowicz, P. Meystre, G. Rempe and H. Walther, *Opt. Acta*, **32**, 1105 (1985).

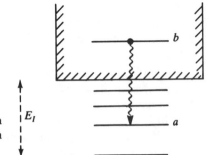

Figure 3. Spontaneous emission from a state b of the ionization continuum to a discrete atomic state a.

a) First Example: Radiative Capture

In the first example we study, the atomic state b belongs to the ionization continuum of the atom (*). This continuum is represented by hatched lines in Figure 3 (the distance E_I between the beginning of the continuum and the ground state is the ionization energy). Spontaneous emission causes the atom to pass from a state b of this continuum to a discrete state a located below the ionization limit.

Physically, a state such as b describes a collision state between an electron and an ion. The electron is not bound to the ion in the state b. By linearly superposing states such as b, wave packets can be constructed which describe an electron arriving from infinity, interacting with the ion, and then going far away. If a spontaneous emission process of the type shown in Figure 3 occurs, the electron loses energy radiatively and is then bound to the ion in a state a, thus forming a neutral atom. This is known as radiative capture of the electron by the ion. Radiative capture of electrons by protons occurs in interstellar space and is at the origin of the formation of hydrogen atoms in Rydberg states n, which are detected in radio astronomy by the microwave radiation that these atoms then emit in transitions $n \to n - 1, n - 1 \to n - 2 \ldots$.

Because the states $|\varphi_i\rangle = |b; 0\rangle$ and $|\varphi_f\rangle = |a; \mathbf{k}\varepsilon\rangle$ both belong to continua (the first because of b, the second because of $\mathbf{k}\varepsilon$), it is possible to define a transition rate and an incident flux, and, consequently, a radiative capture cross-section of the electron in the state a. Such cross-sections play an important role in astrophysics and plasma physics.

(*) The atom is assumed to be infinitely heavy, so that the external quantum numbers are ignored.

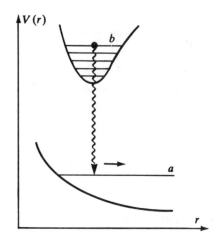

Figure 4. Interaction potentials and energy levels of a diatomic molecule. The figure corresponds to an excimer or exciplex, the ground state being dissociative. An excited electronic level where the two atoms are bound is considered.

b) SECOND EXAMPLE: RADIATIVE DISSOCIATION OF A MOLECULE

We now consider a second example, borrowed from molecular physics, where it is the higher state b that is discrete, and the lower state a that belongs to a continuum.

The state b is assumed to belong to an excited electronic state of a diatomic molecule, in which the effective interaction potential $V(r)$ between the two atoms has a minimum as a function of the internuclear distance r (Figure 4). The state b is a state of vibration-rotation in this potential well.

The state a, on the other hand, belongs to a dissociative electronic state, that is, to a state for which the potential $V(r)$ between the two atoms does not have a minimum (or for which the minimum is extremely shallow). In such an electronic state, the two atoms cannot form a stable molecule, and the state a actually describes a collision state between the two atoms. Molecules having configurations of the same type as those shown in Figure 4 are called excimers, for homonuclear molecules (such as Xe_2), or exciplexes, for heteronuclear molecules (such as XeF).

In a state such as b, the two atoms vibrate about their equilibrium position. The spontaneous emission of a photon $\hbar\omega$ causes the molecule to pass into the state a located at $\hbar\omega$ below b, then the two atoms rapidly move away from each other. The spontaneous emission has thus dissociated the molecule. This is why the name "radiative dissociation" is given to such a process. Because state a belongs to a continuum, the frequency of the emitted photon can vary within a certain interval and the spectrum

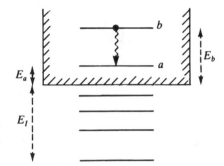

Figure 5. Spontaneous emission between two states of the ionization continuum of an atom.

of the light emitted is a continuous spectrum and not a spectrum of discrete lines.

Remark

Such systems are well suited to the realization of laser sources (*). Because the lower state a dissociates extremely rapidly, with characteristic times (on the order of the molecular dimensions divided by the relative velocity of the two separating atoms) that are much shorter than the radiative lifetime of the higher state b, the state a may be less populated than the state b (for example, in a discharge where the state b is permanently populated by collisions). It is therefore possible to realize significant population inversions between b and a leading to a high gain.

3. Spontaneous Emission between Two States of the Ionization Continuum—Bremsstrahlung

We finally consider, in this last subsection, the case in which the two states a and b both belong to a continuum. For example, a and b may both be located in the ionization continuum of an atom (Figure 5).

Physically, an electron moves from infinity toward an ion with an initial kinetic energy E_b. Arriving in the vicinity of the ion, it spontaneously emits a photon, which causes it to lose the energy $\hbar\omega$. However, this energy loss is insufficient to allow it to be captured by the ion, and the electron again moves away toward infinity with a final kinetic energy $E_a = E_b - \hbar\omega$ lower than E_b. The spontaneous emission of a photon has thus slowed down the electron, hence the name Bremsstrahlung (deceleration-induced

(*) See, for example, *IEEE J. of Quantum. Electron.*, **QE-15**, 265 (1979); *Excimer Lasers*, Vol. 30 of *Topics in Applied Physics*, edited by C. K. Rhodes, Springer Verlag, Berlin, 1979; M. H. R. Hutchinson, *Appl. Phys.*, **21**, 95 (1980).

radiation) given to the process. Because the energy E_a of the final state a may vary continuously, the Bremsstrahlung spectrum is continuous.

We must emphasize the importance of the presence of the ion. Actually, the real emission of a photon by a free electron is forbidden (*). To see this, it is sufficient to consider the reference frame in which the electron is at rest. The emission of a photon and the recoil of the electron to conserve global momentum necessarily result in a state for which the energy is greater than that of the initial state. Hence there is no final state of the global system, coupled by H_{I_1} to the initial state, that has the same energy. A third partner is required, in this case the ion, which may provide the momentum necessary to achieve a final state of the global system having the same energy and the same momentum as the initial state (**).

(*) On the other hand, a free electron may virtually emit a photon and reabsorb it (see §E-1 below).

(**) Exercise 8 presents a calculation of the cross-section for the emission of a photon by Bremsstrahlung. This calculation is a perturbative one with regard to the potential created by the ion and it uses the Pauli–Fierz representation introduced in Complement B_{II}.

B—ABSORPTION PROCESS: A PHOTON DISAPPEARS

During an absorption process, a photon disappears. The corresponding energy is gained by the atom in the form of recoil kinetic energy and an increase in internal energy. As for spontaneous emission, distinctions can be made among several cases, depending on the discrete or continuous character of the internal atomic energy in the initial and the final states. Some examples of these different situations are briefly summarized in subsections 1–3.

A new element appears in the study of absorption processes. Whereas in the study of spontaneous emission, the initial state of the field is always the vacuum, several types of initial states may now be envisioned for the field: monochromatic or broadband incident radiation, weak or strong radiation, with constant or pulsed intensity, etc. We will show in subsection 4, by using qualitative arguments, how the dynamics of the absorption process between two discrete states may be sensitive to the properties of the incident radiation.

Finally, note that there is an emission process known as *stimulated emission* that has many similarities to the process of absorption. As with absorption, stimulated emission is caused by photons present in the initial state. On the other hand, when the atom undergoes a transition to a lower state, the energy lost by the atom is gained by the field that induced this transition. Instead of being depleted, the incident radiation is thus amplified. Most of the results we establish in this subsection concerning absorption may thus be adapted to stimulated emission.

1. Absorption between Two Discrete States

This involves a process which is exactly the reverse of the emission process discussed previously in subsection A-1. By absorbing a photon $\hbar\omega$ the atom passes from a discrete internal state a to another discrete internal state b having a higher energy. Two possible diagrammatic representations of such a process are shown in Figure 6.

The conservation of global energy and momentum implies that the frequency ω of the photon absorbed by an atom initially at rest ($\mathbf{K} = \mathbf{0}$) in the state a, is given by

$$\hbar\omega = \hbar\omega_0 + E_{\text{rec}}. \tag{B.1}$$

Thus, for an *atom initially at rest*, the frequencies of emission and absorption differ, respectively, from the atomic frequency ω_0 by the quantities $-E_{\text{rec}}/\hbar$ and $+E_{\text{rec}}/\hbar$ [compare (A.6) to (B.1)].

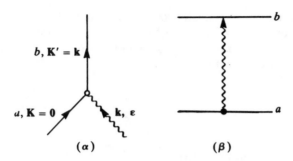

Figure 6. Two diagrammatic representations of the absorption process (compare Figure 1 to 6α, and 2 to 6β).

We will return later (§4) to the conditions under which it is possible to define a probability of absorption (or of stimulated emission) per unit time.

2. Absorption between a Discrete State and a Continuum State

a) FIRST EXAMPLE: PHOTOIONIZATION

This process is the reverse of radiative capture. An atom, initially in a discrete internal state *a*, for example, the ground state, absorbs a photon having an energy $\hbar\omega$ higher than the ionization energy E_I of the atom, and ends up in the state *b* of the ionization continuum (see Figure 7).

This disappearance of a photon, accompanied by the appearance of a photoelectron, is simply the well-known photoelectric effect. The conservation of global energy at the end of the absorption process results in the

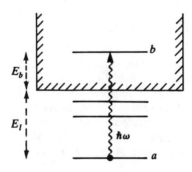

Figure 7. Photoionization of an atom by the absorption of a photon ($\hbar\omega > E_I$).

fact that the energy of state b above the ionization threshold, that is also the final kinetic energy $mv^2/2$ of the photoelectron, once it is separated from the ion, is related to the energy $\hbar\omega$ of the incident photon and to the ionization energy E_I by the equation:

$$\hbar\omega = E_I + \frac{1}{2}mv^2 \tag{B.2}$$

which is just the Einstein relation.

Because the final state belongs necessarily to a continuum (as a result of the continuous character of E_b), the Fermi golden rule allows us to calculate a photoionization rate and, consequently, after division by the flux associated with the incident photon, a photoionization cross-section.

Remarks

(i) The preceding discussion seems to indicate that the probability of photoionization is zero for $\hbar\omega < E_I$ because the absorption of a photon $\hbar\omega$ cannot bring the atom into the ionization continuum. Actually, at high intensities, the atom can be photoionized even if $\hbar\omega < E_I$. This occurs through a multiphoton ionization process, during which the atom absorbs n photons (with $n > 1$), bringing to it an energy $n\hbar\omega$ sufficient to pass into the ionization continuum (see Section D below).

(ii) For $\hbar\omega$ sufficiently large compared to E_I, the probability of photoionization decreases extremely rapidly as ω increases. The physical interpretation of this rapid decrease is simple. Actually, the greater $\hbar\omega$ is compared to E_I, the more the atomic electron appears free to the incident photon. But we know that the absorption of a photon by a free electron is not allowed as a real process. To see this, consider Figure 8a, where the relation between the energy E and the momentum p for a free electron is shown (parabola $E = p^2/2m$). A free electron initially at rest (point O) that absorbs a photon having energy $E = \hbar\omega$ and momentum $p = E/c$ goes from point O to point O' (the line OO' has a slope c that is nearly vertical compared to the slope of the tangents to the parabola, which is equal to $p/m = v \ll c$). The electron "thus leaves the parabola", so that the corresponding transition cannot be real. This remains true even if the electron is not initially at rest.

On the other hand, in a bound state, the momentum of the electron is no longer perfectly defined. For example, in the ground state $|\varphi_a\rangle$, having energy $-E_I$, the wave function has a certain width in momentum space represented by segment MN in Figure 8β. The length of MN is a characteristic distance over which the wave function $\varphi_a(\mathbf{p})$ decreases. Hence, the amplitude of absorption of a photon $\mathbf{k}\varepsilon$ is proportional to the matrix element $\langle \varphi_f | \boldsymbol{\varepsilon} \cdot \mathbf{p} \exp(i\mathbf{k} \cdot \mathbf{r}) | \varphi_a \rangle$ where $|\varphi_a\rangle$ is the ground state and $|\varphi_f\rangle$ is the state of the continuum with

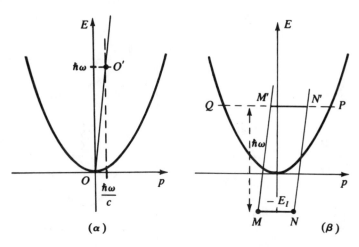

Figure 8. Diagrams describing the role of energy and momentum conservation in the absorption of a photon by a free (α) or bound (β) electron.

energy $\hbar\omega - E_I$. The operator $\exp(i\mathbf{k} \cdot \mathbf{r})$ is a translation operator in momentum space translating $\mathbf{p}$ by $\hbar\mathbf{k}$. The amplitude of absorption of the photon is thus proportional to the overlap integral, in momentum space, of the wave function $\boldsymbol{\varepsilon} \cdot \mathbf{p}\varphi_a(\mathbf{p} - \hbar\mathbf{k})$ with the wave functions $\varphi_f(\mathbf{p})$ of the states of the continuum having energy $\hbar\omega - E_I$. In Figure 8β, the width of $\boldsymbol{\varepsilon} \cdot \mathbf{p}\varphi_a(\mathbf{p} - \hbar\mathbf{k})$ is represented by the segment $M'N'$ obtained starting from MN by a translation $\hbar\omega/c$ on the p axis and $\hbar\omega$ on the E axis. The functions $\varphi_f(\mathbf{p})$ are schematically represented by the points P and Q. The overlap of these functions can thus be nonzero and the transition is therefore allowed. When $\hbar\omega$ increases, the segment $M'N'$ moves away from the points P and Q and the overlap integral for the corresponding wave functions decreases. It is thus possible to qualitatively understand (*) why the photoionization cross-section decreases with $\hbar\omega$ for $\hbar\omega \gg E_I$.

b) Second Example: Photodissociation

In the second example, borrowed from molecular physics, the initial state a is a vibration-rotation sublevel of the electronic ground state of a stable molecule (the effective potential has a minimum in this state—see Figure 9). The absorption of a photon $\hbar\omega$ brings the molecule to a dissociative excited electronic state b.

(*) For more details, see Bethe and Salpeter, Section 69.

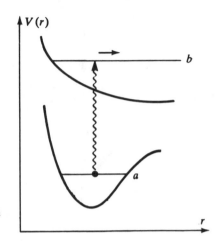

Figure 9. Photodissociation of a molecule by absorption of a photon.

Hence, the absorption of a photon $\hbar\omega$ breaks apart the molecule. Such a process is called photodissociation (*).

3. Absorption between Two States of the Ionization Continuum: Inverse Bremsstrahlung

Figure 10 shows a photon absorption between two states a and b of the ionization continuum. Such a transition, known as a "*free-free*" transition, is accompanied by an acceleration of the electron. The corresponding process, which is the reverse of that in Figure 5, is called "inverse

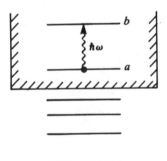

Figure 10. Photon absorption between two states of the ionization continuum.

(*) See for example Herzberg, Chapter VII.

Bremsstrahlung". It plays an important role in the heating of electrons in a laser-induced plasma.

4. Influence of the Initial State of the Field on the Dynamics of the Absorption Process

We return to the question of absorption between two discrete internal atomic states, the ground state a and an excited state b (for the sake of simplicity, we assume the atom to be infinitely heavy and disregard the external quantum numbers). We will attempt to identify the discrete or continuous nature of the final state of the global atom + radiation system, and show that several time evolutions are possible, depending on the nature of the state of the incident radiation.

Let us first assume that the incident radiation is monochromatic and resonant. For example, a single mode $\mathbf{k}_0\mathbf{\varepsilon}_0$ is populated and contains N photons of frequency $\omega_0 = ck_0$ equal to the frequency of the atomic transition $(E_b - E_a)/\hbar$. The initial state of the total system is thus $|\varphi_i\rangle = |a; N\mathbf{k}_0\mathbf{\varepsilon}_0\rangle$ (atom in the state a in the presence of N photons $\mathbf{k}_0\mathbf{\varepsilon}_0$, all the other modes being empty). The interaction Hamiltonian H_{I1} couples this state $|\varphi_i\rangle$ to another state $|\varphi_f\rangle = |b; (N-1)\mathbf{k}_0\mathbf{\varepsilon}_0\rangle$ representing the atom in the excited state b in the presence of $N-1$ photons $\mathbf{k}_0\mathbf{\varepsilon}_0$, all the other modes being empty. As a result of the resonant character of the excitation, the loss of energy due to photon annihilation is compensated for by the gain in the internal energy of the atom, and the two states $|\varphi_i\rangle$ and $|\varphi_f\rangle$ have the same energy (left-hand side of Figure 11). We now study the states to which the state $|\varphi_f\rangle$ is coupled. In addition to the coupling with the state $|\varphi_i\rangle$, we must also consider the possibility that the atom will return to the state a by emitting a photon in a mode $\mathbf{k}\mathbf{\varepsilon}$ that was initially empty. The $N-1$ photons $\mathbf{k}_0\mathbf{\varepsilon}_0$ remain spectators in such a transition, which is simply the transition studied in subsection A-1 to describe the spontaneous emission of a photon $\mathbf{k}\mathbf{\varepsilon}$ from the state b. The state $|\varphi_f\rangle$ is thus also coupled to all the states $|\varphi_g\rangle =$

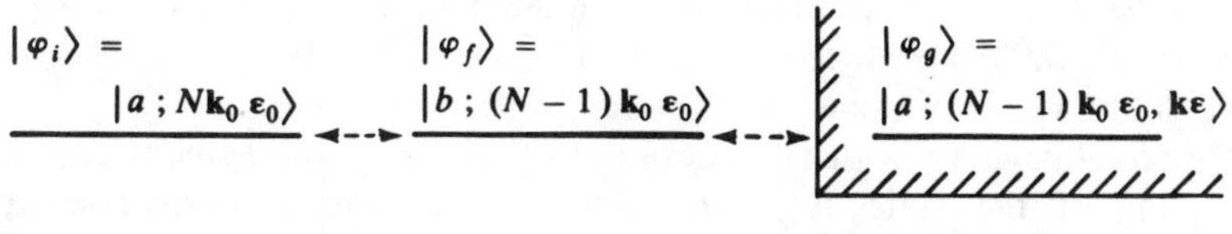

Figure 11. Some important states of the global "atom + radiation" system appearing in the study of the absorption process.

$|a; (N - 1)\mathbf{k}_0\boldsymbol{\varepsilon}_0, \mathbf{k}\boldsymbol{\varepsilon}\rangle$ (atom in the ground state a in the presence of $N - 1$ photons $\mathbf{k}_0\boldsymbol{\varepsilon}_0$ and one photon $\mathbf{k}\boldsymbol{\varepsilon}$) which form a continuum (right-hand side of Figure 11). The preceding analysis thus demonstrates that the state $|\varphi_f\rangle$ is not a "true final state" for the atom leaving $|\varphi_i\rangle$ because it is also coupled to the continuum of states $|\varphi_g\rangle$. Moreover, the transition from $|\varphi_i\rangle$ to $|\varphi_g\rangle$ through $|\varphi_f\rangle$, that is, the appearance of a new photon $\mathbf{k}\boldsymbol{\varepsilon}$ accompanied by the disappearance of an incident photon $\mathbf{k}_0\boldsymbol{\varepsilon}_0$, is simply a scattering process. We will return to this important process in Section C that follows.

The temporal evolution of the global system starting from the initial state $|\varphi_i\rangle$, which may be studied using the amplitude $\langle\varphi_i|\tilde{U}(T)|\varphi_i\rangle$, is simple in several cases which we will now discuss. First consider the evolution of the state $|\varphi_f\rangle$ in the two limiting cases where either the coupling on the right with the states $|\varphi_g\rangle$ or the coupling on the left with the state $|\varphi_i\rangle$ can be neglected. If the state $|\varphi_f\rangle$ were coupled only to the state $|\varphi_i\rangle$, the problem would then be reduced to that of two discrete states of the same energy coupled to each other (see Chapter I, §B-1). The evolution of the system would thus be a Rabi nutation between $|\varphi_i\rangle$ and $|\varphi_f\rangle$ occurring at the Rabi frequency Ω_1 proportional to the product of the dipole moment d_{ab} of the transition $a \leftrightarrow b$ by $\sqrt{\langle E^2\rangle}$, where $\langle E^2\rangle$ is the mean value of the square of the electric field at the atom location, evaluated in the state $|N\mathbf{k}_0\boldsymbol{\varepsilon}_0\rangle$. This Rabi frequency, proportional to $\sqrt{N}$, increases with N. On the other hand, if the state $|\varphi_f\rangle$ were coupled only to the continuum $|\varphi_g\rangle$, it would disintegrate exponentially toward the continuum with a lifetime $\tau = 1/\Gamma$ (see §A-1). The two limiting cases thus correspond to the two situations $\Omega_1 \gg \Gamma$ or $\Omega_1 \ll \Gamma$. In the first case (intense monochromatic wave), we can, to a first approximation, neglect the spontaneous decay of b. The atom oscillates between a and b at the Rabi frequency Ω_1, and this rapid nutation is slowly damped with a time constant on the order of Γ^{-1}. In this case, we obviously cannot describe the process in terms of transition rate or absorption cross-section. In the second case (weak monochromatic excitation), we can disregard in a first approximation the coupling of $|\varphi_f\rangle$ with $|\varphi_i\rangle$. The diagonalization of the coupling on the right in Figure 11 in the subspace formed by the discrete state $|\varphi_f\rangle$ and the continuum $\{|\varphi_g\rangle\}$ yields a single continuum of new states $|\psi_\mu\rangle$ in which the density of the state $|\varphi_f\rangle$ (near the unperturbed energy of $|\varphi_f\rangle$) is on the order of $1/\hbar\Gamma$ (one state over an interval $\hbar\Gamma$): The state $|\varphi_f\rangle$ is in some way "diluted" within the continuum $\{|\psi_\mu\rangle\}$ over an interval having a width on the order of $\hbar\Gamma$ (see Complement C_1). The coupling of the state $|\varphi_i\rangle$ with this new continuum gives rise to a disintegration rate of $|\varphi_i\rangle$, which, according to Fermi's golden rule, is on the order of $(1/\hbar) \times$ (square of the coupling between $|\varphi_i\rangle$ and $|\varphi_f\rangle$) $\times$

(density of the state $|\varphi_f\rangle$ in the new continuum), that is on the order of $(1/\hbar) \times \hbar^2\Omega_1^2 \times (1/\hbar\Gamma) = \Omega_1^2/\Gamma$. It is thus possible in this case to define an absorption rate, proportional to Ω_1^2, thus to N, i.e., to the incident flux, and consequently to derive an absorption cross section.

All the preceding considerations concern monochromatic incident radiation. Let us now assume that, in the initial state, the atom is in the presence of N_1 photons $\mathbf{k}_1\boldsymbol{\varepsilon}_1$, N_2 photons $\mathbf{k}_2\boldsymbol{\varepsilon}_2, \ldots, N_i$ photons $\mathbf{k}_i\boldsymbol{\varepsilon}_i \ldots$ the frequencies $\omega_1 = ck_1$, $\omega_2 = ck_2 \cdots \omega_i = ck_i \cdots$ forming an extremely dense ensemble, that we can consider as a continuous spectrum. The initial state $|\varphi_i\rangle = |a; N_1\mathbf{k}_1\boldsymbol{\varepsilon}_1 \cdots N_i\mathbf{k}_i\boldsymbol{\varepsilon}_i \cdots \rangle$ is thus coupled to a large number of states $|\varphi_f\rangle = |b; N_1\mathbf{k}_1\boldsymbol{\varepsilon}_1 \cdots (N_i - 1)\mathbf{k}_i\boldsymbol{\varepsilon}_i \cdots \rangle$ corresponding to the atom in the state b, a photon $\mathbf{k}_i\boldsymbol{\varepsilon}_i$ having been absorbed. Because $\omega_i = ck_i$ may take a large number of values, the states $|\varphi_f\rangle$ now form a continuum, instead of being reduced to a single discrete state as in the monochromatic case displayed in Figure 11. Even if the coupling of these states $|\varphi_f\rangle$ with the states $|\varphi_g\rangle$ associated with spontaneous emission of a photon $\mathbf{k}\boldsymbol{\varepsilon}$ from b is disregarded, the temporal evolution of $|\varphi_i\rangle$ is thus that of a discrete state coupled to a continuum, and it is possible to define an absorption rate which can easily be seen to be proportional to the value of the incident intensity for $\omega = \omega_0$. It is thus possible to recover in this way the Einstein B coefficients that describe absorption and stimulated emission (see also Chapter IV, Section E).

We have limited ourselves, in the preceding examples, to stationary states for the incident field (eigenstates of H_R). It is easy to envision the large variety of situations which may be realized with nonstationary incident radiation (pulsed excitation).

C—SCATTERING PROCESS: A PHOTON DISAPPEARS AND ANOTHER PHOTON APPEARS

1. Scattering Amplitude—Diagrammatic Representation

In a scattering process, the global system passes from an initial state $|\varphi_i\rangle$, in which the atom is in a state a in the presence of a photon $\mathbf{k}\boldsymbol{\varepsilon}$, to a final state $|\varphi_f\rangle$ in which the atom is in the state a' in the presence of another photon $\mathbf{k}'\boldsymbol{\varepsilon}'$. Three possible pathways allow $|\varphi_i\rangle$ to be connected to $|\varphi_f\rangle$: the atom absorbs the photon $\mathbf{k}\boldsymbol{\varepsilon}$ and passes into the state b, then emits the photon $\mathbf{k}'\boldsymbol{\varepsilon}'$ and passes into the state a' (Figure 12α); it may also first emit the photon $\mathbf{k}'\boldsymbol{\varepsilon}'$ before absorbing the photon $\mathbf{k}\boldsymbol{\varepsilon}$ (Figure 12β); finally, the absorption of $\mathbf{k}\boldsymbol{\varepsilon}$ and the emission of $\mathbf{k}'\boldsymbol{\varepsilon}'$ may occur simultaneously (Figure 12γ). The two processes 12α and 12β correspond to the action of H_{I1} to the second order with an intermediate state, whereas the process 12γ is associated with the action of H_{I2} to first order (see Chapter I, §C-5 and Figure 2).

According to the relation (B.14) of Chapter I, the three amplitudes $\mathscr{T}_{fi}$ associated with these three pathways equal, respectively, in the lowest order,

$$\mathscr{T}_{fi}^{\alpha} = \sum_b \lim_{\eta \to 0_+} \frac{\langle a'; \mathbf{k}'\boldsymbol{\varepsilon}'|H_{I1}|b; 0\rangle\langle b; 0|H_{I1}|a; \mathbf{k}\boldsymbol{\varepsilon}\rangle}{E_a + \hbar\omega - E_b + i\eta} \qquad \text{(C.1.a)}$$

$$\mathscr{T}_{fi}^{\beta} = \sum_b \lim_{\eta \to 0_+} \frac{\langle a'; \mathbf{k}'\boldsymbol{\varepsilon}'|H_{I1}|b; \mathbf{k}\boldsymbol{\varepsilon}, \mathbf{k}'\boldsymbol{\varepsilon}'\rangle\langle b; \mathbf{k}\boldsymbol{\varepsilon}, \mathbf{k}'\boldsymbol{\varepsilon}'|H_{I1}|a; \mathbf{k}\boldsymbol{\varepsilon}\rangle}{E_a - \hbar\omega' - E_b + i\eta} \qquad \text{(C.1.b)}$$

$$\mathscr{T}_{fi}^{\gamma} = \langle a'; \mathbf{k}'\boldsymbol{\varepsilon}'|H_{I2}|a; \mathbf{k}\boldsymbol{\varepsilon}\rangle. \qquad \text{(C.1.c)}$$

H_{I1} and H_{I2} are the one-photon and two-photon interaction Hamiltonians given by the expressions (C.5.b) and (C.5.d) of Chapter I. Here we neglect H_{I1}^S and the effects due to the spins.

Remark

In the electric dipole approach (see Appendix, §5), the interaction Hamiltonian between the atom and the radiation contains only linear terms in a and a^+ so that, in this approach, the scattering always appears as a second-order process for the interaction Hamiltonian H_I'. Hence there are only two possible pathways associated with diagrams of the type 12α and 12β.

The intermediate state of process 12β is never discrete, due to the presence of the photons $\hbar\omega$ and $\hbar\omega'$. By contrast, the intermediate state

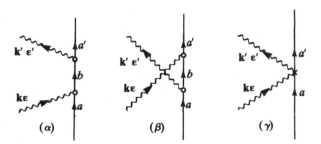

Figure 12. Possible pathways connecting the initial state and the final state of a scattering process.

of process 12α may be discrete if the atomic state b is discrete. In such a case, the energy denominator of (C.1.a) may lead to a divergence (*) of $\mathcal{T}_{fi}^{\alpha}$ if $\hbar\omega$ is equal to $E_b - E_a$, that is, if the frequency ω of the incident photon is resonant with the frequency $(E_b - E_a)/\hbar$ of the transition $b \to a$. This is a resonant scattering process that we exclude for the moment and to which we will return later (§C-3).

The conservation of total energy between $|\varphi_i\rangle$ and $|\varphi_f\rangle$ results in (for an infinitely heavy atom for which recoil is neglected)

$$E_a + \hbar\omega = E_{a'} + \hbar\omega'. \tag{C.2}$$

If the atom is found in the same state at the end of the scattering process ($a' = a$), we have $E_{a'} = E_a$, and consequently, according to (C.2), $\omega' = \omega$. The scattering is known as elastic. If a' differs from a, the scattering is inelastic and the change in energy of the photon $\hbar(\omega' - \omega)$ reflects the change in atomic energy $E_a - E_{a'}$.

Finally, we note the existence of another diagrammatic representation of process α that is frequently used (Figure 13). The absorption of the photon $\hbar\omega$ is represented by an upward arrow joining the state a to a dashed line located at a distance $\hbar\omega$ above a. The emission of $\hbar\omega'$ is represented by a downward arrow connecting this dashed line to the state a'. The advantage of this representation is that the distance between the state b and the dashed line in Figure 13 represents clearly the energy

(*) If state b belongs to a continuous atomic spectrum, the presence of $i\eta$ in the denominator of (C.1.a) results in the appearance of a principal part and a delta function that yield finite results when the integration over the energy E_b of b is performed, even if $E_b - E_a$ may be equal to $\hbar\omega$.

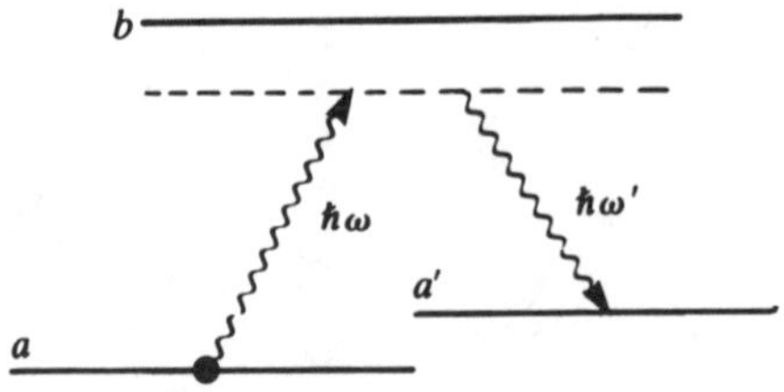

Figure 13. Another diagrammatic representation of process α shown in Figure 12.

defect of the intermediate state of the scattering process. The scattering is resonant when the dashed line coincides with the state b. One might well wonder whether such a dashed line represents an energy level. Actually, what is involved is not an atomic level, but rather an energy level of the global system: atom in the state a + one photon $\mathbf{k}\varepsilon$, or also atom in the state a' + one photon $\mathbf{k}'\varepsilon'$ [these two states have the same energy according to (C.2)]. The state b of Figure 13 may also be considered as a state of the global system, atom in the state b without any photon. Hence we can clearly understand how the distance between the level b and the dashed line in Figure 13 represents the energy defect of the global system in the intermediate state of process α.

In the preceding, we have considered the scattering of a photon by an atom. A photon can also be scattered by a free electron (recall that a photon cannot be really absorbed or emitted by a free electron). Such a process is just *Compton scattering*. Diagrams α, β, γ in Figure 12 remain valid provided that a, b, and a' are considered as the quantum numbers specifying the free-electron state (for example, momentum and spin). The resolution of the equations expressing the conservation of total energy and total momentum between $|\varphi_i\rangle$ and $|\varphi_f\rangle$ yields the well-known expression for the Compton frequency shift.

2. Different Types of Photon Scattering by an Atomic or Molecular System

Depending on the energy of the incident photon and the elastic or inelastic character of the scattering, different names are used to designate the scattering of a photon by an atom or a molecule. We will now review some examples of such processes.

a) Low-Energy Elastic Scattering: Rayleigh Scattering

The energy $\hbar\omega$ of the incident photon is assumed to be very small compared to the ionization energy of the atom E_I, and thus compared to the energy differences $E_b - E_a$ separating the initial atomic level a, which

b ————————————————

$\hbar\omega$ $\hbar\omega' = \hbar\omega$

a ————————•————————   **Figure 14.** Rayleigh scattering.

is generally the ground state, from excited electronic states b appearing in the intermediate state:

$$\hbar\omega \ll E_I, |E_b - E_a|. \tag{C.3}$$

Since the scattering is elastic, we have $a' = a$, and $\hbar\omega' = \hbar\omega$. The Rayleigh scattering is thus the scattering, without change in frequency, by an atom or a molecule irradiated by a radiation having a frequency much lower than its resonance frequencies (Figure 14).

It is possible to show that the Rayleigh scattering cross-section is proportional to ω^4 (it increases extremely rapidly with the frequency of the incident photon), and to the static polarizability of the atomic system in state a (see Exercise 3).

b) LOW-ENERGY INELASTIC SCATTERING: RAMAN SCATTERING

The energy $\hbar\omega$ of the photon is always assumed to be small compared with the ionization energy E_I, but the final state a' of the atomic system is now different from a. The scattering, which is then called Raman scattering, is thus accompanied by a change in frequency, $\omega - \omega'$, which according to (C.2) is equal to $(E_{a'} - E_a)/\hbar$. In general, the levels a and a' are different rovibrational levels of a molecule, and the measure of the frequency shift between the incident light and the scattered light allows us to determine the frequencies of vibration and rotation of the molecule (*).

Depending on whether the initial molecular level a is lower or higher in energy than the final level a', the scattering process is called *Stokes* (Figure 15α) or *anti-Stokes* (Figure 15β) Raman scattering. The scattered light has a frequency ω' lower than ω in the first case, and higher in the second case.

(*) See, for example, Herzberg, Chapter II, §4.

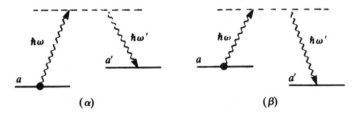

Figure 15. Stokes (α) and anti-Stokes (β) Raman scattering.

Figure 16. Inverse Raman process of that in Figure 15α.

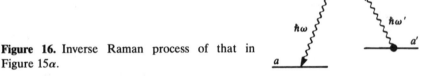

Remark

Assume that the modes ω and ω' (Figure 15α) are both initially populated. It is then possible to consider the *inverse* process of process 15α, where the molecule, initially in the state a', ends in the state a by absorbing a photon $\hbar\omega'$ and by emitting a photon $\hbar\omega$ (Figure 16). The processes in Figure 15α amplify the radiation at the frequency ω' (by absorption of a photon $\hbar\omega$ and induced emission of a photon $\hbar\omega'$), whereas the processes in Figure 16 deplete it (by absorption of a photon $\hbar\omega'$ and induced emission of a photon $\hbar\omega$). However, the initial molecular state is not the same in the two cases, which leads to amplification and depletion processes respectively proportional to the numbers N_a and $N_{a'}$ of molecules in states a and a'. If $N_a > N_{a'}$, the molecular medium presents a gain at frequency ω'. Thus, a laser oscillation at frequency ω' can be obtained if the gain of the medium enclosed in a cavity tuned around ω' is larger than the losses. This is the principle of Raman lasers (*).

c) HIGH-ENERGY ELASTIC SCATTERING: THOMSON SCATTERING

The energy $\hbar\omega$ of the incident photon is assumed to be large compared with the ionization energy of the atom E_I. The elastic scattering, with $a' = a$ and $\omega' = \omega$ (Figure 17) is, in this case, called *Thomson* scattering.

(*) See, for example, Bloembergen, §4 and 5; Shen, Chapter 10; W. Kaiser and M. Maier in *Laser Handbook*, edited by F. T. Arrecchi and E. O. Schultz-Dubois, Vol. 2, p. 1077, North-Holland, Amsterdam, 1972; Hanna, Yuratich and Cotter, Chapters 5 and 6.

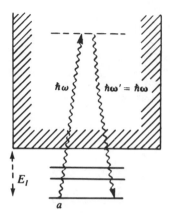

Figure 17. Thomson scattering.

Such a process plays a preponderant role in the scattering of soft x rays by atoms.

The quantum calculation of the total Thomson-scattering cross-section σ_T (see Exercise 4) yields, in the limit where the wavelength λ of incident radiation is large compared with the atomic dimensions a_0 (*), the well-known result $\sigma_T = 8\pi r_0^2/3$ (where r_0 is the classical electron radius), that is, the scattering cross section of a monochromatic wave by a *classical free electron*. Such a result shows that Thomson scattering by an atomic electron may be interpreted in semi-classical terms. The incident wave causes the electron to vibrate at a frequency ω much higher than those of its motion around the nucleus (on the order of $E_I/\hbar$). To first approximation, everything happens as if the atomic electron were free. Moreover, when $\hbar\omega \gg E_I$, it is the amplitude (C.1.c) that is the most significant, as for the free electron and not (C.1.a). The contribution of the latter is actually decreased by the factor $\hbar\omega$ in the denominator.

d) HIGH-ENERGY INELASTIC SCATTERING WITH THE FINAL ATOMIC STATE IN THE IONIZATION CONTINUUM: COMPTON SCATTERING

Finally, we assume that $\hbar\omega$ is very large compared with E_I, and even sufficiently large so that the wavelength λ is small compared with the atomic dimensions a_0. The binding of the electron to the nucleus thus

---

(*) For a hydrogen atom, the conditions $\hbar\omega \gg E_I$ and $\lambda \gg a_0$, considered in this paragraph, correspond to the energy interval $[\alpha^2 mc^2 - \alpha mc^2]$, where α is the fine structure constant. See *Photons and Atoms—Introduction to Quantum Electrodynamics*, Chapter III, Figure 2.

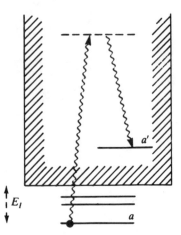

Figure 18. Compton scattering by an atomic electron.

plays only a small role in the scattering that greatly resembles Compton scattering by a free electron. One then expects the important scattering processes to be inelastic processes where the atom is ionized in the final state (Figure 18).

The momenta $\hbar\mathbf{k}$ and $\hbar\mathbf{k}'$ of the incident and scattered photons and the momentum $\mathbf{p}_0$ of the electron in the final state a' are thus very large compared with that of the electron in the initial state a, which is distributed over an interval having a width on the order of $\hbar/a_0$ about a zero mean value. To first approximation, we can thus consider the process in Figure 18 as a Compton-scattering process by a free electron initially at rest. The corrections to this approximation are due to the distribution of electron momentum in the initial state and in the final state and thus yield information about the atomic wave functions. As for Thomson scattering, it is the amplitude (C.1.c) that is preponderant.

Remark

It is possible to analyze Compton scattering by a free or bound electron by means of energy-momentum diagrams analogous to those in Figure 8. Thus, Figure 19α represents the Compton scattering by a free electron initially at rest. In order to be able to draw two-dimensional figures, with the energy on the ordinate and the momentum on the abscissa, we will limit ourselves to backward scattering, the momenta $\hbar\mathbf{k}$ and $\hbar\mathbf{k}'$ of the incident and scattered photons being in opposite directions. It can be seen from Figure 19α that it is possible, starting from point O of the parabola, to arrive at another point O'' on the parabola by successively carrying out the translation $\mathbf{00}'$ corresponding to the

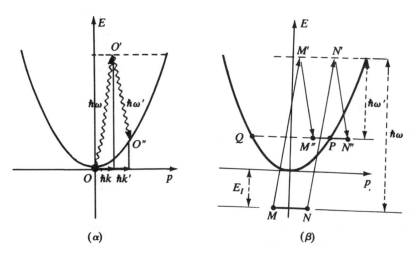

Figure 19. Energy-momentum diagrams associated with Compton scattering by a free electron initially at rest (α) and by a bound electron (β).

absorption of the incident photon $\hbar\omega$, $\hbar\mathbf{k}$ and the translation $0'0''$ corresponding to the emission of the scattered photon $\hbar\omega'$, $\hbar\mathbf{k}'$. The scattering process may therefore be real. For a bound electron, it is necessary to start from the ground state, for which the wave function $\varphi_a(\mathbf{p})$ is, as in Figure 8, represented by a horizontal segment MN, located at the ordinate $-E_I$. At high energy $(\hbar\omega \gg E_I)$, the preponderant term of the amplitude for Compton scattering is given by the expression (C.1.c), which is, according to the expression for H_{12} [see relation (C.5.d) in Chapter I], proportional to $\langle \varphi_{a'} | \exp i(\mathbf{k} - \mathbf{k}') \cdot \mathbf{r} | \varphi_a \rangle$. Thus we obtain the overlap integral of the ground-state wave function $\varphi_a(\mathbf{p})$, translated by $\hbar(\mathbf{k} - \mathbf{k}')$, with the wave function $\varphi_{a'}^*(\mathbf{p})$ of the state of the continuum with energy $-E_I + \hbar(\omega - \omega')$. In Figure 19$\beta$, these two wave functions are represented, respectively, by the segment $M''N''$ (obtained starting from MN as a result of two translations $\hbar\omega$, $\hbar\mathbf{k}$ and $-\hbar\omega'$, $-\hbar\mathbf{k}'$) and by the points P and Q (corresponding to the intersection of the parabola by the horizontal line of energy $-E_I + \hbar\omega - \hbar\omega'$). Because of the finite width of $M''N''$, the preceding overlap integral may be important for a range of values of ω', ω being fixed. Thus, the Compton scattering by a bound electron does not give rise to an infinitely narrow line (in a given scattering direction), unlike what occurs for a free electron.

3. Resonant Scattering

We now return to the case in which the frequency ω of the incident photon is very close to the frequency $\omega_0 = (E_b - E_a)/\hbar$ of an atomic

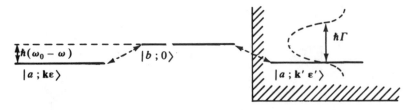

Figure 20. Some important states of the total atom + radiation system involved in the study of resonant scattering.

transition between two *discrete* states a and b. The process α in Figure 12 is thus preponderant compared with the others, because the amplitude (C.1.a) that is associated with it diverges at resonance. Consequently, we concentrate in this subsection on the study of this process. We assume that $a' = a$ (elastic scattering), and that there is only a single incident photon $\mathbf{k}\varepsilon$.

In Figure 20, we have represented the states of the total "atom + radiation" system that play an important role in the scattering process. The final state of the scattering process, $|a; \mathbf{k}'\varepsilon'\rangle$, has the same energy as the initial state $|a; \mathbf{k}\varepsilon\rangle$. Moreover, these two states both belong to the same continuum of states (atom in the state a in the presence of one photon). At the lowest order in H_{I1}, the initial and final states of the scattering process are indirectly coupled through the discrete state $|b; 0\rangle$. We can vary at will the difference $\hbar(\omega_0 - \omega)$ between the energy of the state $|b; 0\rangle$ and that of the initial state, by scanning ω about ω_0.

To qualitatively understand how we can resolve the difficulty connected with the divergence of the amplitude (C.1.a), let us leave the perturbative point of view that is used to determine (C.1.a) and attempt to understand the dynamics of the system, keeping only the single states of Figure 20 and their mutual couplings (represented by dashes and arrows). Hence we replace the transition matrix element (C.1.a), valid for the lowest order in H_{I1}, by

$$\mathcal{T}_{a\mathbf{k}'\varepsilon', a\mathbf{k}\varepsilon} = \langle a; \mathbf{k}'\varepsilon'|H_{I1}\frac{1}{E_a + \hbar\omega - H + i\eta}H_{I1}|a; \mathbf{k}\varepsilon\rangle \quad (C.4)$$

valid to all others in H_{I1} (it is H and not H_0 that appears in the energy denominator associated with the intermediate state [see equation (B.15) in Chapter I]).

Thus it appears that, in a theory to all orders in H_{I1}, it is not the discrete state $|b; 0\rangle$ that must be considered as the intermediate state of the scattering process, as is suggested by Figure 20, but rather the ensemble of all the eigenstates of the total Hamiltonian H. When we limit ourselves to the states of Figure 20, the eigenstates of H form a continuum (see Complement C_I: the discrete state $|b; 0\rangle$ "dissolves" in the continuum of eigenstates of H). The sum over the intermediate states appearing in the expression (C.4) is thus a sum over a continuum that does not lead to any divergence (due to the $+i\eta$ of the denominator, which introduces a principal part and a delta function). In addition, the interaction Hamiltonian H_{I1} couples the initial state and the final state to the eigenstates of H insofar as they have a nonzero component on the discrete state $|b; 0\rangle$. Hence, in the expression (C.4), the sum over the intermediate states involves the density of the discrete state $|b; 0\rangle$ in the new continuum of eigenstates of H, a density that varies resonantly over an interval of width $\hbar\Gamma$ about the energy E_b (see Complement C_I). Thus one expects that, while remaining finite, the scattering amplitude exhibits a resonance when the frequency ω of the incident photon is swept about $\omega_0 = (E_b - E_a)/\hbar$.

The preceding qualitative discussion is confirmed by the simple calculation presented in Complement C_I. Recall here the result: it is sufficient to replace the nonperturbed energy E_b of $|b; 0\rangle$ in the denominator of (C.1.a) with $E_b - i\hbar(\Gamma/2)$. Note that such a modification applied to the evolution exponential $\exp(-iE_b t/\hbar)$ of the state $|b; 0\rangle$ transforms this exponential into $\exp(-iE_b t/\hbar)\exp(-\Gamma t/2)$, and accounts for the well-known exponential decrease of the probability amplitude of being in the state b. The nonperturbative expression for the transition matrix near resonance is thus finally:

$$\mathcal{T}_{fi}^{\text{res}} = \frac{\langle a; \mathbf{k'\varepsilon'}|H_{I1}|b; 0\rangle\langle b; 0|H_{I1}|a; \mathbf{k\varepsilon}\rangle}{\hbar\omega + E_a - E_b + i\hbar(\Gamma/2)} \tag{C.5}$$

that is an expression for which the modulus and the phase vary extremely rapidly over an interval of width Γ about $\omega = \omega_0 = (E_b - E_a)/\hbar$. The nonperturbative character of (C.5) clearly appears if one uses the following power-series expansion:

$$\frac{1}{\hbar[\omega - \omega_0 + i(\Gamma/2)]} = \frac{1}{\hbar}\sum_{n=0}^{\infty}(-1)^n\frac{(i\Gamma/2)^n}{(\omega - \omega_0)^{n+1}}. \tag{C.6}$$

The presence of Γ, which is second order in H_{I1} (see A.5), in the denominator of (C.5) is thus equivalent, according to (C.6), to the sum of an infinite number of terms of increasing powers of H_{I1}.

Expression (C.6) may also be interpreted in semiclassical terms. Let us consider a classical oscillator of eigenfrequency ω_0, damped with a time constant Γ^{-1}. If this oscillator is subjected to a monochromatic excitation of frequency ω, its "response" will have a modulus and a phase varying in a resonant fashion over an interval of width Γ about $\omega = \omega_0$, and described by the same factor $[\omega - \omega_0 + i(\Gamma/2)]^{-1}$ as that appearing in (C.6).

The preceding analogy allows us to very simply understand the essential characteristics of the scattered light close to resonance, which is also called "*resonance fluorescence*". If the incident light is monochromatic, the scattered light is also monochromatic (*) with the same frequency, and its intensity varies with the detuning from resonance $\omega - \omega_0$ as a Lorentzian of width Γ. If, on the other hand, the incident radiation has a very flat spectrum near $\omega = \omega_0$, the different components of this spectrum will be scattered with an efficiency proportional to $[(\omega - \omega_0)^2 + (\Gamma^2/4)]^{-1}$ and the scattered light will have a Lorentzian spectrum of width Γ centered on $\omega = \omega_0$.

Similar considerations may also be used to understand the scattering of a wave packet. If the wave packet is very long in time ($\Delta t \gg \Gamma^{-1}$), and as a result very narrow in frequency ($\Delta \omega \ll \Gamma$), the different waves making up the packet will be scattered in practically the same way because $\Delta \omega \ll \Gamma$, and the scattered wave packet will have the same form as that of the incident wave packet. On the other hand, if the wave packet is very short in time ($\Delta t \ll \Gamma^{-1}$—see Figure 21α), and consequently very wide in frequency ($\Delta \omega \gg \Gamma$), the scattered wave packet will have a different shape (Figure 21β), because only the components of the wave packet having a frequency within an interval of width Γ about $\omega = \omega_0$ will be efficiently scattered. More precisely, the scattered wave packet has a time dependence given by the convolution product of the incident wave packet by the Fourier transform of the scattering amplitude $[\omega - \omega_0 + i(\Gamma/2)]^{-1}$. After a rising edge analogous to that of the incident wave packet, the scattered wave packet decays exponentially over a time on the order of Γ^{-1}. It can also be said that, in this second case ($\Delta t \ll \Gamma^{-1}$), the resonant scattering may be decomposed into two steps: an excitation of the atom during the very brief passage time Δt of the wave packet, followed by the exponential decay by spontaneous emission of the excited state prepared in such a quasi-percussional manner (see also Exercise 13).

(*) We disregard here the nonlinear scattering processes that appear at high intensity [see Remark (iii) at the end of this paragraph].

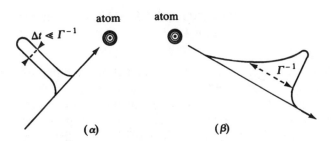

Figure 21. α—Representation of an incident short wave packet $(\Delta t \ll \Gamma^{-1})$. β—Shape of the scattered wave packet.

Remarks

(i) Every discrete atomic state b, except for the ground state, "disappears" when the coupling between the discrete state $|b; 0\rangle$ and the continuum of spontaneous emission $|a; \mathbf{k}'\varepsilon'\rangle$ is taken into consideration. Its "trace" manifests itself only in the form of a resonance in the scattering amplitude. Strictly speaking, the problem of the spontaneous emission from an excited state prepared at $t = 0$, (studied in subsection A-1 above), should be analyzed in terms of the scattering of a short wave packet, which "prepares percussionally the excited state".

(ii) In this entire subsection we have ignored any possible degeneracy in the higher state b and lower state a. Some interesting effects connected with the existence of such manifolds may be observed in resonant scattering (level-crossing resonance, quantum beats, optical pumping, etc.). See Exercise 6.

(iii) We have also limited ourselves in this subsection to the case of low-intensity incident radiation (a single incident photon). At higher intensity, new effects appear, such as nonlinear scattering processes involving many incident photons (see, for example, Figure 29β). More generally, the Rabi nutation induced by an intense resonant laser beam can considerably modify the spectrum of scattered light. We will return to this question later (see Chapters V and VI).

D—MULTIPHOTON PROCESSES: SEVERAL PHOTONS APPEAR OR DISAPPEAR

The elementary processes of emission and absorption that we studied in Sections A and B concerned only a single photon, which appeared or disappeared during the transition. More complex processes may occur in which the number of photons may increase or decrease by several units, the total energy of the final state of the particles + radiation system being of course equal to that of the initial state (likewise for the total momentum and total angular momentum). Such processes are called multiphoton processes. Their variety is so great that it is impossible to present an exhaustive review here. Instead, we will confine ourselves to qualitatively describing a small number of simple examples. Before beginning, note that scattering, studied in Section C above, may also be considered as a multiphoton process, in the sense that two photons are "involved" during the process, one that disappears ($\hbar\omega$) and another that appears ($\hbar\omega'$). Although the total number of photons is the same in the final state and in the initial state, nothing prevents a scattering process from being considered as a two-photon process, sharing many properties with the absorption (or emission) of two photons.

1. Spontaneous Emission of Two Photons

Let us consider the transition $2s_{1/2} \to 1s_{1/2}$ in the hydrogen atom. As a result of the Wigner-Eckart theorem and the space reflection symmetry, the only allowed electromagnetic transition between these two levels is a $M1$ magnetic dipole transition. Actually, this transition is extremely weak for the following reasons. First, it can result only from the spin magnetic moment $\mathbf{M}_S$, the orbital quantum number being zero in the two states. Furthermore, $\mathbf{M}_S$, which acts only on the spin degrees of freedom, has a zero matrix element between the nonrelativistic wave functions of the two states (the orbital and spin components separate, and the scalar product of the orbital components is zero). Hence, the spontaneous emission of a magnetic dipole photon on the transition $2s_{1/2} \to 1s_{1/2}$ can result only from relativistic corrections to the wave functions, and is quite negligible for the hydrogen atom (*).

In the next order of perturbation theory (second order), it is possible to get a radiative disintegration of state $2s$ to state $1s$ by the spontaneous

(*) This would no longer be true for highly charged hydrogenlike ions for which the relativistic effects are much greater (the transition probability $M1$ varies as a function of Z^{10}).

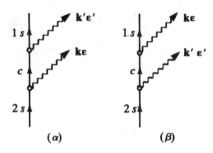

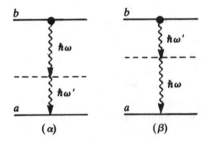

Figure 22. Spontaneous emission of two photons $\mathbf{k}\varepsilon$ and $\mathbf{k}'\varepsilon'$ associated with the transition $2s \to 1s$.

Figure 23. Another possible diagrammatic representation of the spontaneous emission of two photons on the transition $2s \to 1s$.

emission of two electric dipole photons $E1$. The diagrammatic representation of the corresponding transition amplitudes is given in Figure 22. The atom emits a photon $\mathbf{k}\varepsilon$ (Figure 22α), and goes from state $2s$ to an intermediate state c (which, according to selection rules for electric dipole transitions, can only be a state p ($l = 1$) of the discrete or continuous spectrum), then emits a second photon $\mathbf{k}'\varepsilon'$ to end in state $1s$. Diagram 22β corresponds to the other possible order of emission of the two photons (*). The conservation of total energy at the end of the process implies that

$$E_{2s} - E_{1s} = \hbar\omega + \hbar\omega' \qquad (D.1)$$

and appears more clearly in the diagrammatic representation in Figure 23.

The calculation of the spontaneous emission rate for two photons $\mathbf{k}\varepsilon$ and $\mathbf{k}'\varepsilon'$, summed over all pairs of photons satisfying (D.1), results in a radiative lifetime of the $2s$ state that is much shorter than that associated

(*) We have not represented the amplitude associated with H_{I2}, analogous to that in Figure 2d in Chapter I (with $a = 2s$ and $b = 1s$), because its value is zero within the long-wavelength approximation.

with the spontaneous emission of a $M1$ photon on the transition $2s_{1/2} \rightarrow 1s_{1/2}$, or an $E1$ photon on the transition $2s_{1/2} \rightarrow 2p_{1/2}$ (*). It can thus be said that the radiative disintegration of the state $2s$ is dominated by a spontaneous emission process involving two photons. The lifetime is on the order of 0.1 sec.

2. Multiphoton Absorption (and Stimulated Emission) between Two Discrete Atomic States

An atom submitted to sufficiently intense radiation can absorb several incident photons and go from a discrete level a to another discrete level b, located at a higher energy, at a distance equal to the sum of the energies of the absorbed photons. Such a process is called multiphoton absorption, the reverse process being multiphoton stimulated emission. Figure 24 represents an example of the two amplitudes associated with the absorption of two photons between two discrete states a and b.

Multiphoton absorption processes can appear in a large number of atomic or molecular systems, in all frequency ranges (hertzian, microwave, infrared, visible, etc.). The photons that are absorbed can have the same or different frequencies (these frequencies may even be very different, such as, for example, in multiphoton transitions simultaneously involving optical and microwave photons).

As previously mentioned above, the total momentum and total angular momentum must be conserved at the end of the process. Such laws of conservation (in combination with the conservation of parity) explain several important properties of multiphoton transitions. Two simple examples are analyzed in the following remarks.

Remarks

(i) As a first example of the application of the laws of conservation, we will consider a two-photon absorption process, in which both photons have the same frequency ω and opposite wave vectors $+\mathbf{k}$ and $-\mathbf{k}$, thus opposite momenta $+\hbar\mathbf{k}$ and $-\hbar\mathbf{k}$. When the atom goes from state a to state b by absorbing these two photons having zero total momentum, its momentum does not change, and hence its kinetic energy does not change. Consequently, to the first order in v/c, where v is the atomic velocity, the resonance condition (conservation of

---

(*) The interval between the $2s_{1/2}$ and $2p_{1/2}$ energy levels (Lamb shift) is quite small (on the order of 1000 MHz), and spontaneous emission is negligible in this frequency range (it varies as ω^3). See the discussion of the various deexcitation processes of the $2s$ state in Bethe and Salpeter, Section 7.

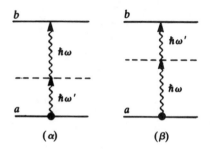

Figure 24. Absorption of two photons between two discrete states.

total energy) is written

$$E_b - E_a = 2\hbar\omega. \tag{D.2}$$

In a two-photon absorption process in which the photons have the same frequency but opposite directions of propagation, there is thus neither a first-order Doppler effect nor a recoil effect. We can also understand the origin of this result by going from the laboratory reference frame to the atom rest frame. In this rest frame, the first-order Doppler shifts of the two photons, $+\mathbf{k} \cdot \mathbf{v}$ and $-\mathbf{k} \cdot \mathbf{v}$, are opposite and compensate for each other exactly in the energy balance. All these considerations are easily generalized to transitions involving more than two photons. If the sum of the wave vectors of photons absorbed during the transition is zero ($\sum_i \mathbf{k}_i = \mathbf{0}$), any first-order Doppler effect disappears in the resonance condition (*).

(ii) The second example we will consider concerns an atomic level having a $J = \frac{1}{2}$ angular momentum. The atom is subjected to a static magnetic field $\mathbf{B}_0$, parallel to $0z$, which splits the two Zeeman sublevels $m = -\frac{1}{2}$ and $m = +\frac{1}{2}$ by a quantity $\hbar\omega_0$, proportional to B_0, and to a radio-frequency field $\mathbf{B}_1 \cos \omega t$, having a fixed frequency ω and linear polarization perpendicular to $\mathbf{B}_0$. When the field B_0 is swept, resonant transitions occur between the two Zeeman sublevels when the splitting $\hbar\omega_0$ between these sublevels is equal to an odd number times $\hbar\omega$:

$$\omega_0 = (2n + 1)\omega. \tag{D.3}$$

The *odd* character of the spectrum is a direct consequence of the conservation of total angular momentum. The linear radio-frequency field may actually be considered as the superposition of two rotating fields, right-hand and left-hand circularly polarized, with which are associated σ^+ and σ^- photons, having

(*) See, for example, G. Grynberg and B. Cagnac, *Rep. Prog. Phys.*, **40**, 791 (1977); V. S. Letokhov and V. P. Chebotayev, in *Non-linear Laser Spectroscopy*, Vol. 4 of *Springer Series in Optical Sciences*, Springer-Verlag, Berlin, 1977, Chapter 4.

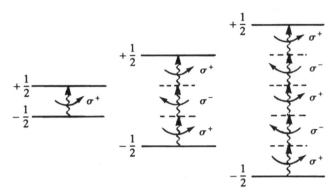

Figure 25. One, three, or five photon resonances obtained with a linearly polarized radio-frequency field orthogonal to the static magnetic field.

respectively angular momenta $+\hbar$ and $-\hbar$ with respect to $0z$. Thus, to go from sublevel $m = -\frac{1}{2}$ to $m = +\frac{1}{2}$, the atom needs to increase its angular momentum along $0z$ by $+\hbar$. It can do this only by absorbing either one photon σ^+, or two photons σ^+ and one photon σ^- (Figure 25) and more generally $n + 1$ photons σ^+ and n photons σ^-. The conservation of energy thus immediately yields the resonance condition (D.3) (*).

3. Multiphoton Ionization

Multiphoton ionization processes may also cause an atom to go from a discrete level to a level of the continuum. For example, an atom, initially in the ground state, can be ionized by laser radiation of frequency ω, even if the energy $\hbar\omega$ of the incident photons is lower than the ionization energy E_I. Actually, by absorbing a sufficiently large number of photons, the atom may always acquire the necessary energy to reach the ionization continuum (Figure 26) (**). The probability of multiphoton ionization increases significantly when an intermediate excitation step involves a quasi-resonant transition between discrete levels. For example, in the three-photon ionization process represented in Figure 26, the discrete

(*) For the study of multiphoton processes in the radiofrequency range, see J. M. Winter, *Ann. Phys.* (*Paris*), **4**, 745 (1959); Brossel; Cohen-Tannoudji, Diu, and Laloë, Complement B_{XIII}.

(**) For a review of the work on these processes, see, for example, G. Mainfray and C. Manus, in *Multiphoton Ionization of Atoms*, edited by S. Chin and P. Lambropoulos, Academic Press, New York, 1984, p. 7; Shen, Chapter 22.

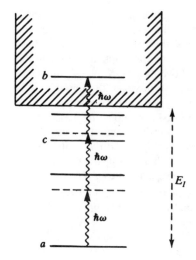

Figure 26. Ionization of an atom by the absorption of three photons (the energy of state a increased by $2\hbar\omega$ is quasi-resonant with the energy of the discrete state c).

level c has an energy close to that of level a increased by $2\hbar\omega$, and the two-photon transition $a \to c$ is quasi-resonant. The state of the global system "atom in the state a + 2 photons" (represented by a dotted line) has an energy close to that of the discrete state c.

Remarks

(i) The variation of the photoionization probability versus $(2\omega - \omega_{ca})$ for the process shown in Figure 26 is generally not symmetric around the resonance frequency $(2\omega = \omega_{ca})$. We can actually distinguish two contributions to the transition amplitude to the continuum: that in which the intermediate level attained after absorption of two photons differs from c and that in which the intermediate level is c. The first contribution generally varies little when $2\omega - \omega_{ca}$ varies about 0, whereas the second has a resonant behavior and changes sign for $2\omega = \omega_{ca}$. Depending on the sign of the detuning $(2\omega - \omega_{ca})$, the two contributions add together or cancel each other out, which explains the asymmetry in the variation of the probability of photoionization. Actually, we find again in this problem the Fano profiles considered in Complement C_I (§3-d). To be specific, the level $|a; N\mathbf{k}\boldsymbol{\varepsilon}\rangle$ is, on the one hand, coupled directly to the continuum $|b; (N - 3)\mathbf{k}\boldsymbol{\varepsilon}\rangle$ through nonresonant terms, and on the other hand to the discrete level $|c; (N - 2)\mathbf{k}\boldsymbol{\varepsilon}\rangle$, which is itself coupled to the continuum.

(ii) The probability of multiphoton ionization also depends on the radiation state. Consider, for instance, a two-photon ionization process obtained either with a monochromatic field in the state $|N\mathbf{k}\boldsymbol{\varepsilon}\rangle$, or with a state $|N_1\mathbf{k}_1\boldsymbol{\varepsilon}_1; N_2\mathbf{k}_2\boldsymbol{\varepsilon}_2\rangle$

in which two field modes are filled. The wave vectors $\mathbf{k}$, $\mathbf{k}_1$, and $\mathbf{k}_2$ are assumed to be sufficiently close so that the atomic matrix elements involved in the transition amplitude are the same regardless of the type of photons absorbed. In the case in which the initial state is $|a; N\mathbf{k}\varepsilon\rangle$, the transition amplitude to the continuum is proportional to $\sqrt{N(N-1)} \simeq N$ (we assume $N, N_1, N_2 \gg 1$) and the transition probability varies as N^2. In the case in which the initial state is $|a; N_1\mathbf{k}_1\varepsilon_1, N_2\mathbf{k}_2\varepsilon_2\rangle$, the probability of absorbing two photons of mode $\mathbf{k}_1\varepsilon_1$ is proportional to N_1^2 and that of absorbing two photons of mode $\mathbf{k}_2\varepsilon_2$ is proportional to N_2^2. Let us now consider the transition amplitude associated with the absorption of a photon in each mode. We must add the contributions of the two possible orders of absorption "$\mathbf{k}_1\varepsilon_1$ then $\mathbf{k}_2\varepsilon_2$" and "$\mathbf{k}_2\varepsilon_2$ then $\mathbf{k}_1\varepsilon_1$" which result in a transition to the same final state $|b; (N_1 - 1)\mathbf{k}_1\varepsilon_1, (N_2 - 1)\mathbf{k}_2\varepsilon_2\rangle$. The transition amplitude is thus proportional to $2\sqrt{N_1 N_2}$ and the transition probability varies as $4N_1 N_2$. Finally, the probability of ionization from the state $|a; N_1\mathbf{k}_1\varepsilon_1, N_2\mathbf{k}_2\varepsilon_2\rangle$ is proportional to $N_1^2 + N_2^2 + 4N_1 N_2$, which is always greater than $(N_1 + N_2)^2$. For the same incident radiation energy, the multiphoton ionization is thus more efficient when the photons are taken from different modes.

4. Harmonic Generation

An atom or a molecule, starting from a discrete state a, for example, the ground state, can absorb p photons $\hbar\omega$ (with $p > 1$), then return to state a by emitting a single photon of energy $p\hbar\omega$. Figure 27 shows such a process for $p = 3$. It correspond to "third harmonic generation". Here again, intermediate quasi-resonances can increase the efficiency of such a process.

Remark

The generation of a harmonic field may be considered to be a scattering process with the absorption of p photons $\mathbf{k}\varepsilon$ and the emission of a photon $\mathbf{k}'\varepsilon'$, the conservation of energy resulting in $\omega' = p\omega$.

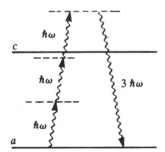

Figure 27. Third harmonic generation (with quasi-resonance at the discrete level c).

Let us now analyze the contributions of two atoms A and B to such a process. We assume that these atoms are initially in the ground state a and that their centers of mass have momenta $\hbar \mathbf{K}_A$ and $\hbar \mathbf{K}_B$. The initial radiation state corresponds to N photons in mode $\mathbf{k}\boldsymbol{\varepsilon}$ and 0 photons in mode $\mathbf{k}'\boldsymbol{\varepsilon}'$, such that for the global photon + atom system, the initial state is

$$|\varphi_i\rangle = |N\mathbf{k}\boldsymbol{\varepsilon}, 0\mathbf{k}'\boldsymbol{\varepsilon}'\rangle \otimes |a, \mathbf{K}_A\rangle \otimes |b, \mathbf{K}_B\rangle.$$

If the creation of a harmonic photon results from the interaction with atom A, the final state of the process is

$$|\varphi_f\rangle = {}_{1}(N-p)\mathbf{k}\boldsymbol{\varepsilon}, \mathbf{k}'\boldsymbol{\varepsilon}'\rangle \otimes |a, \mathbf{K}_A + \delta\mathbf{k}\rangle \otimes |b, \mathbf{K}_B\rangle$$

where $\hbar\delta\mathbf{k} = \hbar(p\mathbf{k} - \mathbf{k}')$ represents the modification of the momentum of the atom A resulting from the conservation of total momentum for the atom + field system. If, on the other hand, the emission of the harmonic photon results from the interaction with atom B, it is the latter that undergoes the recoil effect and the final state of the process is

$$|\varphi'_f\rangle = {}_{1}(N-p)\mathbf{k}\boldsymbol{\varepsilon}, \mathbf{k}'\boldsymbol{\varepsilon}'\rangle \otimes |a, \mathbf{K}_A\rangle \otimes |b, \mathbf{K}_B + \delta\mathbf{k}\rangle.$$

When $\delta\mathbf{k} \neq \mathbf{0}$, the two final states $|\varphi_f\rangle$ and $|\varphi'_f\rangle$ are different, and the probability of emission of a harmonic photon is the sum of the probabilities obtained for each atom. On the other hand, when $\delta\mathbf{k} = \mathbf{0}$, the final state of the process is the same whether the scattering atom is A or B: it is not possible to specify which atom emitted the harmonic photon. To find the probability of emission, the scattering amplitudes for each atom must be added before the sum is squared. If the scattering amplitudes are in phase, the resulting probability of emission is double in the direction for which $\delta\mathbf{k} = \mathbf{0}$, which is just $\mathbf{k}' = p\mathbf{k}$, i.e., the forward direction.

The foregoing reasoning may be extended to any number N_a of atoms. The emission of the harmonic wave in the direction $\mathbf{k}' = p\mathbf{k}$ results from the coherent contribution of different atoms and increases as N_a^2, whereas the emission in other directions increases linearly with N_a. The situation obtained in this way is characteristic of forward scattering.

Let us finish by emphasizing that the foregoing analysis is applicable only for diluted media, the dimensions of which are sufficiently small so that the effects of the refraction index may be ignored. If this is not the case, the incident photons as well as the scattered photons may interact several times with the atoms of the medium, which modifies their propagation. The phase velocities of the incident wave and of the harmonic wave are no longer necessarily equal. During propagation, dephasing may thus appear between the two waves, resulting in destructive interference between the contributions of successive planes of atoms. To obtain a coherent emission of the harmonic wave that is as intense as possible, a phase-matching condition must be realized between the incident wave and the harmonic wave (*).

5. Multiphoton Processes and Quasi-Resonant Scattering

The foregoing example may be considered to be multiphoton scattering in the sense that some photons disappear and others appear (the arrows point in different directions, either up or down). Another example of such a situation is the resonant excitation of a level b located at a distance $\hbar\omega_0$ above level a, an excitation that can appear at high intensity, even if the frequency ω of incident radiation is not exactly equal to ω_0. The atom, leaving a, absorbs an incident photon $\hbar\omega$, spontaneously emits (**) a photon $\hbar\omega'$, then absorbs a second incident photon $\hbar\omega$ to end up in level b (Figure 28).

The conservation of energy yields

$$\hbar\omega_0 = E_b - E_a = \hbar\omega - \hbar\omega' + \hbar\omega \qquad (D.4)$$

or equivalently

$$\omega' = 2\omega - \omega_0. \qquad (D.5)$$

Because the two intermediate states are quasi-resonant (if $\omega_0 - \omega$ is not too large), the probability of such a process is sufficiently large to be observable.

Actually, we saw in subsection B-4 above that an excited state is never the actual final state of a physical process, because this excited state can always disintegrate by spontaneous emission. In addition to the quasi-reso-nant scattering process to lowest order in the incident field (Figure 29α), we are thus naturally led to now consider the scattering process to second order in the incident field (Figure 29β), in which the atom absorbs a photon $\hbar\omega$, emits a photon $h\omega'$, absorbs a second photon $\hbar\omega$, and finally emits a photon $\hbar\omega''$ to end up in a.

Whereas the conservation of energy implies for process α that the scattered photon has the same frequency as the incident photon (as for

(*) For more details on harmonic wave generation and frequency mixing, see Bloember-gen, §5 and 6; Shen, Chapters 6, 7, and 8; Hanna, Yuratich, and Cotter, Chapter 4 and the included references.

(**) If radiation is already present at frequency ω', it can be amplified through the process in Figure 28 by stimulated emission [see, for example, F. Y. Wu, S. Ezekiel, M. Ducloy, and B. R. Mollow, *Phys. Rev. Lett.*, **38**, 1077 (1977)].

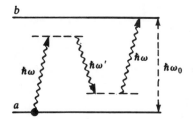

Figure 28. Three-photon process allowing level b to be excited, even if the incident photons do not have the exact resonance energy ($\omega \neq \omega_0$).

Rayleigh scattering), for process β, we must have only

$$2\omega = \omega' + \omega'' \tag{D.6}$$

which shows that ω' and ω'' may vary continuously. Thus, at high intensity, the light scattered by an excited atom near resonance is no longer necessarily monochromatic. Actually, the amplitude associated with process β exhibits a resonance when the energy attained after absorption of the second photon $\hbar\omega$ (third dotted line in Figure 29) varies about the energy of state b, within an interval of width $\hbar\Gamma$. Thus, it follows that ω'' is distributed over an interval of width Γ about ω_0, and consequently, ω' over an interval symmetric with respect to ω, thus centered on $2\omega - \omega_0$ (Figure 30).

Figure 30 was drawn by assuming that $|\omega - \omega_0| > \Gamma$. If this is not the case, and if, in addition, the incident intensity is sufficiently large to obtain a Rabi frequency larger than Γ, then the preceding perturbative approach is no longer valid because it does not converge. We will see later in

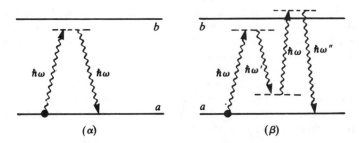

Figure 29. Quasi-resonant scattering, first order in incident field (α) and second order (β).

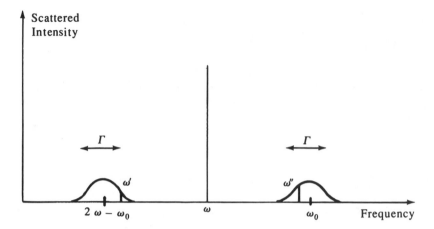

Figure 30. Spectrum of the light scattered by an atom excited near resonance. The central monochromatic line and the two sidebands of width Γ correspond respectively to processes α and β in Figure 29.

Chapters V and VI how other approaches (for example, that of the "dressed" atom) allow the essential properties of the scattered light to be understood.

Remark

In the process shown in Figure 29β, photons of frequency ω' and $\omega'' = 2\omega - \omega'$ are strongly correlated because the emission of a photon ω' is necessarily associated with the emission of a photon of frequency $2\omega - \omega'$ so that energy is conserved at the end of the process. Moreover, for a group of N atoms ($N \gg 1$), the emission of such a pair of photons occurs preferentially in directions $\mathbf{k}'$ and $\mathbf{k}''$ satisfying the phase matching relation $2\mathbf{k} = \mathbf{k}' + \mathbf{k}''$. Actually, when photons are emitted in these directions, the momentum of the atom does not change at the end of the process, and the transition amplitudes relative to different atoms must be added together (rather than the probabilities —see remark in subsection D-4). The emission of two photons in directions satisfying the phase-matching condition is thus N times more probable than the emission in two arbitrary directions.

Experimentally, a standing wave of frequency ω is often used. In this case, the absorption of two photons propagating in opposite directions leads to the phase-matching condition $\mathbf{k}' + \mathbf{k}'' = \mathbf{0}$. The two correlated photons are thus emitted in opposite directions (*).

(*) Cf. P. Grangier, G. Roger, A. Aspect, A. Heidmann, and S. Reynaud, *Phys. Rev. Lett.*, **57**, 687 (1986).

E—RADIATIVE CORRECTIONS: PHOTONS ARE EMITTED
AND REABSORBED (OR ABSORBED AND REEMITTED)

In all the examples considered above, the state of the radiation field is not the same in the initial state and the final state of the process being studied. Even in the scattering process, where the number of photons remains constant, the direction of the scattered photon is usually different from that of the incident photon (except for forward scattering). We will now consider processes for which the state of the radiation field is the same in the final state and the initial state, while changing in the intermediate state.

1. Spontaneous Radiative Corrections

a) CASE OF A FREE ELECTRON: MASS CORRECTION

Let us begin with the most simple case, that of a free electron with momentum $\mathbf{p}$, initially in the vacuum state $|0\rangle$ of the radiation field. The initial state is thus

$$|\varphi_i\rangle = |\mathbf{p}; 0\rangle. \tag{E.1}$$

This electron may always emit a photon $\mathbf{k}\varepsilon$, its momentum changing from $\mathbf{p}$ to $\mathbf{p} - \hbar\mathbf{k}$ (Figure 31α). However, as we mentioned above (§A-3), the new state thus obtained, $|\mathbf{p} - \hbar\mathbf{k}; \mathbf{k}\varepsilon\rangle$, can never have the same energy as the initial state $|\mathbf{p}; 0\rangle$, so that it cannot be the final state of a real process conserving total momentum and energy. On the other hand, the electron may always reabsorb the emitted photon and thus reach a final state which exactly coincides with the initial state (Figure 31α). Thus, although a free electron may not emit a photon in a real fashion, it nevertheless may always emit and reabsorb it virtually, as is sometimes said to emphasize the fact that such an emission of a photon can be only transient. The process in Figure 31α is order 2 in H_{I1}, hence order 2 with regard to the charge q. At the same order q^2, we must also take into account the effect of H_{I2} to order 1. The $a_{\mathbf{k}\varepsilon} a_{\mathbf{k}\varepsilon}^+$ term of H_{I2} may in fact simultaneously create and annihilate a photon $\mathbf{k}\varepsilon$, with the electron remaining in the same state $\mathbf{p}$ (Figure 31β).

In the subspace of total momentum $\mathbf{p}$, the initial state $|\varphi_i\rangle = |\mathbf{p}; 0\rangle$ of processes 31α and 31β is a discrete state. Because it is not coupled to any other state with the same energy by the interaction Hamiltonian, it is relatively well isolated from the other unperturbed states of H_0. The amplitude $\langle\varphi_i|\tilde{U}(T)|\varphi_i\rangle$, for which Figure 31 shows the lowest-order terms

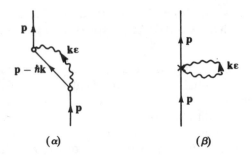

Figure 31. Second-order processes in q occurring for a free electron with momentum $\mathbf{p}$, in the vacuum state of the radiation field.

in q, should thus allow us to determine the shift of the state $|\mathbf{p}; 0\rangle$ due to the coupling between the electron and the transverse radiation field (see §B-1 in Chapter I). In fact, the shifts associated with processes 31α and 31β are, respectively,

$$\delta E_\alpha = \sum_{\mathbf{k}\varepsilon} \frac{|\langle \mathbf{p} - \hbar\mathbf{k}; \mathbf{k}\varepsilon|H_{I1}|\mathbf{p}; 0\rangle|^2}{\dfrac{\mathbf{p}^2}{2m} - \dfrac{(\mathbf{p} - \hbar\mathbf{k})^2}{2m} - \hbar\omega} \tag{E.2}$$

$$\delta E_\beta = \langle \mathbf{p}; 0|H_{I2}|\mathbf{p}; 0\rangle \tag{E.3}$$

and are just the shifts of the states $|\mathbf{p}; 0\rangle$, obtained to second order in H_{I1} and first order in H_{I2}, from perturbation theory.

Because the denominator of (E.2) is always negative, the shift δE_α is negative. To lowest order in $1/c^2$ (we perform a nonrelativistic calculation), $-\hbar\omega$ is the predominant term in the denominator. Because $H_{I1} = -q\mathbf{A} \cdot \mathbf{p}/m$ is proportional to the operator $\mathbf{p}$, and because $|\mathbf{p}; 0\rangle$ is an eigenstate of this operator, the numerator of (E.2) is proportional to $\mathbf{p}^2$. Finally, δE_α may be expressed in the form

$$\delta E_\alpha = -\frac{\mathbf{p}^2}{2m} \frac{\delta m}{m} \tag{E.4}$$

where δm is a constant given by (32) in Complement B_{II}. Process 31α thus describes a decrease in the kinetic energy of the electron caused by an increase in its mass from m to $m + \delta m$ (according to (E.4), δE_α is actually a term of order 1 in the expansion of $\mathbf{p}^2/2(m + \delta m)$ in powers of $\delta m/m$).

The physical origin of this effect is discussed in Complement B_{II}. An electron, having a velocity $\mathbf{p}/m$, creates a transverse field proportional to its velocity, which is carried along by the electron (one can also say that the electron moves surrounded by a cloud of "virtual photons"). The energy δE_α is the sum of the energy of the transverse field bound to the electron (which is positive), and the coupling energy of the electron with this transverse field (which is negative and greater in absolute value).

Expression (E.3) is easy to calculate from the expression of H_{I2} and from the mode expansion of $\mathbf{A}$. We find

$$\delta E_\beta = \sum_{\mathbf{k}\varepsilon} \frac{q^2 \mathscr{A}_\omega^2}{2m} = \sum_{\mathbf{k}\varepsilon} \frac{q^2 \mathscr{E}_\omega^2}{2m\omega^2}, \tag{E.5}$$

an expression that represents the vibrational kinetic energy of the electron in vacuum fluctuations since $\mathscr{E}_\omega^2$ is simply the contribution of the mode $\mathbf{k}\varepsilon$ to $\langle 0|\mathbf{E}_\perp^2|0\rangle$. With this energy δE_β is associated another mass correction $\delta m' = \delta E_\beta/c^2$, which has to be added to the correction appearing in (E.4). By extending the calculation to higher orders in $1/c^2$, one would also find a term analogous to (E.4) with $\delta m'$ in the numerator. Finally, processes 31α and 31β are responsible for a correction to the electron mass.

Remarks

(i) The transverse field associated with virtual photons in Figure 31 may be considered, to a first approximation, to be bound to the electron. Another approach that can be used to study radiative corrections consists of carrying out a unitary transformation (Pauli–Fierz transformation) on the quantum electrodynamics Hamiltonian for the purpose of subtracting this bound transverse field from the total field. The lowest-order radiative corrections would thus appear directly in the particle Hamiltonian. This approach is discussed in Complement B_{II}.

(ii) Expressions (E.2) and (E.5), which result from a nonrelativistic treatment of the electron, are certainly incorrect with regard to the contributions of the relativistic modes $\mathbf{k}\varepsilon$ ($\hbar\omega \geq mc^2$). In fact, the intermediate state, $\mathbf{p} - \hbar\mathbf{k}$, of the electron in process 31α may be relativistic for sufficiently large $\hbar\omega$, although the initial (and final) states are not (we assume $p/mc \ll 1$). In addition, we did not consider here the possibility that electron-positron pairs could appear in the intermediate state. In fact, a completely relativistic calculation leads to high-frequency terms different from those of the sums (E.2) and (E.5), and yields, for the total mass correction δm, an integral over ω that diverges logarithmically (whereas the integrals (E.2) and (E.5) diverge linearly and quadratically). We will not consider here the problems involved in the appearance of divergent

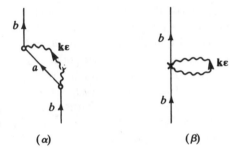

Figure 32. Emission and reabsorption of a photon by an atom initially in internal state b.

expressions in the radiative correction calculations and the way these problems are solved (renormalization of the electron mass and charge).

b) CASE OF AN ATOMIC ELECTRON: NATURAL WIDTH AND RADIATIVE SHIFT

Let us now consider the case of a bound electron, for example, an atomic electron bound to a nucleus by the Coulomb interaction. Processes completely analogous to those in Figure 31 may be considered. For example, an atom, initially in the discrete state b (*), may emit a photon $\mathbf{k}\varepsilon$ and move to another state a, then reabsorb the photon $\mathbf{k}\varepsilon$ and return to state b (Figure 32α). The summation over the intermediate states should now concern not only all the photons $\mathbf{k}\varepsilon$, but also all the internal atomic states a. Likewise, process 32β directly generalizes 31β.

The essential difference between processes 32α and 32β is that intermediate states $|a; \mathbf{k}\varepsilon\rangle$ having the same energy as the initial state $|b; 0\rangle$ may appear when b is an excited atomic state, and may thus be the final states of a real process. Such a situation occurs when b can disintegrate radiatively by spontaneous emission toward a state a with energy $E_a < E_b$ (see §A-1). In other words, the transition to the intermediate state in Figure 32α is not always virtual as it is for 31α. It may also be real.

Mathematically, the possibility of real transitions starting from the initial state $|\varphi_i\rangle = |b; 0\rangle$ is manifested by the fact that the amplitude $\langle \varphi_i | \tilde{U}(T) | \varphi_i \rangle$ for the system to remain in the same state after a time T is now not only oscillating, but it is also damped:

$$\langle \varphi_i | \tilde{U}(T) | \varphi_i \rangle = \mathrm{e}^{-i[\Delta - i(\Gamma/2)]T}. \tag{E.6}$$

(*) For the sake of simplicity, we do not write the external quantum numbers, which, in the center-of-mass system are $\mathbf{K} = 0$ in the initial and final states, $\mathbf{K}' = -\mathbf{k}$ in the intermediate state.

The diagrams in Figure 32 allow us to calculate Γ and δE to lowest order in q. Γ is the probability per unit time of leaving state b for a lower energy level by real spontaneous emission of a photon. The expression for Γ was given previously [see (A.5)]. $\hbar\Delta$ is a shift of state b, generalizing (E.2) and (E.3):

$$\hbar\Delta = \hbar\Delta_1 + \hbar\Delta_2 \tag{E.7}$$

$$\hbar\Delta_1 = \mathscr{P} \sum_a \sum_{\mathbf{k}\varepsilon} \frac{|\langle b; 0|H_{I1}|a; \mathbf{k}\varepsilon\rangle|^2}{E_b - E_a - \hbar\omega} \tag{E.8}$$

$$\hbar\Delta_2 = \langle b; 0|H_{I2}|b; 0\rangle. \tag{E.9}$$

In (E.8), $\mathscr{P}$ denotes principal part. Thus, for an excited atomic state, the spontaneous radiative corrections are manifested by a radiative shift and a radiative broadening, appearing as real and imaginary parts of a complex correction to the energy. For the ground state, only the radiative shift remains, because no real spontaneous emission process can occur starting from the ground state.

Let us return to expression (E.7) for the shift $\hbar\Delta$. It is easy to see that (E.9) coincides with (E.5) in the long-wavelength approximation and thus represents the vibrational kinetic energy of the electron in vacuum fluctuations. As for $\hbar\Delta_1$, it is possible, for the high-frequency part of the sum over ω [i.e., for $\hbar\omega$ large compared to $|E_b - E_a|$, but small compared to mc^2 in order for the nonrelativistic expression (E.8) to be valid], to expand $(E_b - E_a - \hbar\omega)^{-1}$ in powers of $(E_b - E_a)/\hbar\omega$. The zero-order approximation amounts to neglecting $E_b - E_a$ in the denominator of (E.8): one can show (see Exercise 7) that it gives the contribution, to the shift $\hbar\Delta_1$ of state b, of the variation in kinetic energy of the electron due to the mass correction δm previously introduced for a free electron. The following term, of order 1 in $(E_b - E_a)/\hbar\omega$, is related to a correction of the potential energy $V(r)$, proportional to the Laplacian $\Delta V(r)$ of $V(r)$. For a Coulomb potential [$V(r) \sim 1/r$], this correction, which is proportional to $\delta(\mathbf{r})$, thus particularly affects the s states and removes the degeneracy between the $2s_{1/2}$ and $2p_{1/2}$ states of the hydrogen atom predicted by the Dirac equation (Lamb shift). Physically, this correction to V, proportional to ΔV, may be interpreted as resulting from the averaging of the potential "seen" by the electron in its vibrational motion under the influence of vacuum fluctuations (see Complement B_{II} and Exercise 7).

Remark

It is also possible to consider the energy levels of an electron in a uniform static magnetic field, and to calculate the radiative shifts of these levels, as we did

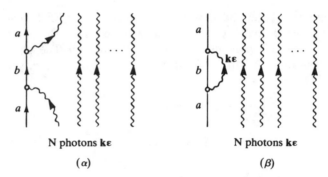

N photons kε N photons kε

(α)   (β)

Figure 33. α: Absorption and reemission of an incident photon. β: Induced emission of a photon kε identical to an incident photon, and reabsorption of this photon.

above for the levels of an atomic electron. Such a calculation shows that the frequencies ω_c of the cyclotron motion of the charge and ω_L of the Larmor precession motion of the spin are both decreased, ω_c being decreased more than ω_L. The g factor of the electron, which may be defined as the ratio $2\omega_L/\omega_c$, is thus increased by radiative corrections above the unperturbed value of 2 predicted by the Dirac equation ($g - 2$ anomaly—see Exercises 10 and 12).

2. Stimulated Radiative Corrections

In the previous subsection, the radiation field state is the vacuum $|0\rangle$ in the initial state and in the final state. We will now suppose that several incident photons are initially present, and we will study the process by which an atom *absorbs* one of these photons, then reemits it (Figure 33α), or emits in a stimulated fashion a photon identical to one of the incident photons and then reabsorbs it (Figure 33β), returning, in these two cases, to a final atomic state a identical to the initial state. The intermediate atomic state b is arbitrary. To the same order q^2, we must also take into account the effect of H_{I2} to the first order, which can be represented by the diagrams in Figure 34.

In many cases, the probability amplitude $\langle \varphi_i' | \tilde{U}(T) | \varphi_i' \rangle$ for the system to remain in the same state $|\varphi_i'\rangle = |a; N\mathbf{k}\varepsilon$, other incident photons$\rangle$ after a time T is, to a very good approximation, an oscillating and damped exponential, as in (E.6):

$$\langle \varphi_i' | \tilde{U}(T) | \varphi_i' \rangle = e^{-i[\Delta' - i(\Gamma'/2)]T}. \tag{E.10}$$

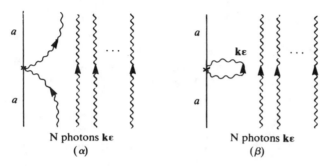

Figure 34. Diagrams representing the effect of H_{I2} to first order. α: Terms in a^+a; β: Terms in aa^+.

For example, such a situation occurs when the incident radiation is monochromatic, but of low intensity, or sufficiently far from resonance, or if it has a large spectral width (see discussion of subsection B-4 above). In this case, what is the physical significance of the parameters Γ' and Δ' appearing in the amplitude (E.10)?

The damping rate Γ' of the state $|\varphi_i'\rangle$ represents the probability per unit time for the global system to leave state $|\varphi_i'\rangle$. We can also say that Γ' is the probability per unit time that the atom in state a absorbs an incident photon, or emits in a stimulated way a photon identical to the incident photons. Because the intermediate state in Figure 33 may not be resonant and serves only as an intermediate state in the scattering process, it is actually more correct to consider that Γ' is a rate for any incident photon to be scattered. This probability is directly related to the total scattering cross-section (*).

The physical effect associated with the damping rate Γ' may be considered either from the point of view of the atom or from the point of view of the field. From the atom's point of view, Γ' appears as a *radiative broadening* of the level a caused by incident photons. In other words, Γ' is analogous, for the absorption and stimulated emission processes, to the natural width Γ associated with spontaneous emission processes. In contrast with Γ, which is an "intrinsic" quantity, Γ' depends on characteristics of the incident beam (intensity, direction, polarization, spectral

(*) The fact that the amplitudes associated with the processes in Figures 33 and 34 allow us to calculate a rate Γ' simply related to the total scattering cross-section is not surprising. The processes in Figures 33 and 34 are indeed *forward-scattering processes*, and it is well-known that there exists a relation between the total scattering cross-section and the imaginary part of the forward-scattering amplitude (Bohr–Peierls–Placzek relation; see, for instance, Messiah, Chapter XIX, Section 31).

distribution, etc.). In addition, in contrast with Γ, Γ' may be nonzero for the ground state (there is no real spontaneous emission process starting from the ground state a, whereas there may be real absorption processes, and more generally, scattering processes starting from a). With regard to the field, Γ' is associated with the attenuation of the incident beam caused by absorption or with its amplification caused by stimulated emission. In fact, the processes in Figures 33 and 34 are forward-scattering processes. Recall that, classically, the attenuation (or amplification) of the incident field is related to the interference between the incident field and the forward-scattered field.

Let us now consider the other parameter Δ' appearing in (E.10). From the atom's point of view, $\hbar\Delta'$ may be considered to be a shift of state a caused by incident photons (light shift). Such an effect is sometimes called the "dynamic Stark effect" by analogy with the ordinary Stark effect, which is associated with the polarization energy of an atom in a static electric field. In the case studied here, the energy $\hbar\Delta'$ may be considered to be the polarization energy of the atom in an electric field with a nonzero frequency, a field to which the atom "responds" with its "dynamic" rather than static polarizability. The shift $\hbar\Delta'$ is the equivalent, for absorption and stimulated emission, of the shift $\hbar\Delta$ associated with spontaneous emission and discussed above in subsection E-1-b. The symmetry of the exciting light is involved in the properties of $\hbar\Delta'$. Because this symmetry is no longer the spherical symmetry of spontaneous emission (where the atom interacts with *all* the modes), it is clear that the light shifts of different Zeeman sublevels of an atom may be different (*). From the point of view of the radiation field, the forward-scattered field is, in classical terms, phase-shifted with respect to the incident field, so that the total field undergoes not only in a change in amplitude (this is the effect of Γ'), but also a change in phase (effect of Δ'). Such a phase change, summed over all the atoms of a medium, is at the origin of the refraction index of this medium.

Remark

To keep this discussion brief, we will not give the expression for Δ' here. Let us say only that it is considerably simplified in two extreme cases. If the excitation is quasi-resonant, that is, if the energy of the incident photons is close to that

---

(*) It was in this way that light shifts were observed for the first time as a modification, caused by light, of the Zeeman splittings in the atomic ground state. The magnetic resonance lines in the ground state are actually very narrow, and differential shifts of the Zeeman sublevels, although very small, may be easily detected. See, for instance, C. Cohen-Tannoudji and A. Kastler, *Progress in Optics*, (edited by E. Wolf), North-Holland, Amsterdam, 1966, Vol. V, p. 1.

allowing a transition from level a to another level b, then process 33α predominates, with, in addition, a single significant intermediate atomic state, that is quasi-resonant. The other simple extreme case corresponds to a high-frequency excitation, for which $\hbar\omega$ is large compared to all the energies $(E_b - E_a)$. An expansion in powers of $(E_b - E_a)/\hbar\omega$, similar to the one we introduced above for (E.8), allows $\hbar\Delta'$ to be expressed as a sum of terms each having a simple physical interpretation: vibrational kinetic energy of the electron in the incident wave, variation in potential energy resulting from the averaging of external fields by this vibrational motion, etc. The comparison of radiative corrections induced by an incident field with spontaneous radiative corrections allows one to identify in the latter the part which may be attributed to vacuum fluctuations (they have the same form as radiative corrections that would be induced by a fluctuating field having a spectral density of $\hbar\omega/2$ per mode), and the part that may be attributed to the interaction of the electron with its own field (radiation reaction) (*).

(*) See Complement A_{IV} and J. Dupont-Roc, C. Fabre, and C. Cohen-Tannoudji, *J. Phys.* B **11**, 563 (1978).

F—INTERACTION BY PHOTON EXCHANGE

In Section E, we considered a situation in which the same particle (or the same group of particles) virtually emits and reabsorbs a photon. In the presence of several particles (or groups of particles), the photon emitted by one may be reabsorbed by another. Conservation of energy between the initial state and final state holds only for the global system, each subsystem being possibly modified. Such a modification may be considered as resulting from an interaction between the two subsystems. In this section we illustrate this type of process with two examples.

1. Exchange of Transverse Photons between Two Charged Particles: First Correction to the Coulomb Interaction

Consider two charged particles α and β, with respective momenta $\mathbf{p}_\alpha$ and $\mathbf{p}_\beta$, in the vacuum state of the radiation field. Let us take as the unperturbed Hamiltonian H_0 that of the free particles and the free field (point of view described in subsection C-2 of Chapter I). The interaction Hamiltonian thus includes the Coulomb interaction between the two particles and the interaction with the transverse field. Using the notation in subsection E-1-a, the initial state is written

$$|\varphi_i\rangle = |\mathbf{p}_\alpha, \mathbf{p}_\beta; 0\rangle. \tag{F.1}$$

The processes affecting these particles to the second order in electric charges may be divided into two classes. We first rediscover those described in the diagrams in Figure 31, in which one of the particles emits and reabsorbs a photon, while the other remains a spectator. These are the radiative corrections specific to each of the particles. In addition, we have the Coulomb interaction between the two particles (Figure 35a) and the exchange of transverse photons (Figures 35b and c).

Figures 35b and 35c describe a scattering process in which the pair of particles moves from state $\mathbf{p}_\alpha, \mathbf{p}_\beta$ to state $\mathbf{p}'_\alpha, \mathbf{p}'_\beta$ by exchanging a transverse photon. In the space of the states of the global system particles + photons, Figures 35b and 35c demonstrate the existence of an indirect coupling between two states belonging to the 0 photon manifold (states $|\mathbf{p}_\alpha, \mathbf{p}_\beta; 0\rangle$ and $|\mathbf{p}'_\alpha, \mathbf{p}'_\beta; 0\rangle$) via an intermediate state belonging to the one-photon manifold (which is separated from the zero-photon manifold by a large energy gap if $\hbar\omega$ is sufficiently large). Such an indirect coupling may also be described by a particle "effective Hamiltonian" δV (acting only inside the zero-photon manifold) which is to be added to V_{Coul}.

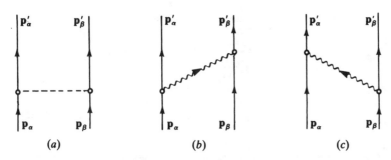

Figure 35. Electromagnetic interactions between two charged particles α and β. (*a*) Coulomb interaction. (*b*) Emission of a transverse photon by α and reabsorption of this photon by β. (*c*) Emission of a transverse photon by β and reabsorption of this photon by α.

According to the results in Complement B_I, the matrix elements of δV are given by the expression

$$\langle \mathbf{p}'_\alpha, \mathbf{p}'_\beta | \delta V | \mathbf{p}_\alpha, \mathbf{p}_\beta \rangle =$$

$$= \sum_{\mathbf{k}\varepsilon} \sum_{\mathbf{p}''_\alpha \mathbf{p}''_\beta} \frac{1}{2} \left[\frac{1}{E_p - E_{p''} - \hbar\omega} + \frac{1}{E_{p'} - E_{p''} - \hbar\omega} \right] \times$$

$$\times \langle \mathbf{p}'_\alpha, \mathbf{p}'_\beta; 0 | H_{I1} | \mathbf{p}''_\alpha, \mathbf{p}''_\beta; \mathbf{k}\varepsilon \rangle \langle \mathbf{p}''_\alpha, \mathbf{p}''_\beta; \mathbf{k}\varepsilon | H_{I1} | \mathbf{p}_\alpha, \mathbf{p}_\beta; 0 \rangle. \quad (F.2)$$

On the right-hand side of (F.2), E_p, $E_{p'}$, $E_{p''}$ are the unperturbed energies for particles in states $|\mathbf{p}_\alpha, \mathbf{p}_\beta\rangle$, $|\mathbf{p}'_\alpha, \mathbf{p}'_\beta\rangle$, $|\mathbf{p}''_\alpha, \mathbf{p}''_\beta\rangle$. For sufficiently large $\hbar\omega$, $E_p - E_{p''}$ and $E_{p'} - E_{p''}$ can be neglected in the denominators, while only $\hbar\omega$ is retained. The summation over $\mathbf{p}''_\alpha$ and $\mathbf{p}''_\beta$ introduces a closure relation, which yields an explicit expression for δV:

$$\delta V = -\sum_{\mathbf{k}\varepsilon} \frac{1}{2\varepsilon_0 L^3 \omega^2} \frac{q_\alpha q_\beta}{m_\alpha m_\beta} (\boldsymbol{\varepsilon} \cdot \mathbf{p}_\alpha)(\boldsymbol{\varepsilon} \cdot \mathbf{p}_\beta) e^{i\mathbf{k} \cdot (\mathbf{r}_\alpha - \mathbf{r}_\beta)} + (\alpha \leftrightarrow \beta)$$

$$= -\frac{q_\alpha q_\beta}{m_\alpha m_\beta \varepsilon_0 c^2} \sum_{i,j=x,y,z} (\mathbf{p}_\alpha)_i (\mathbf{p}_\beta)_j S_{ij}(\mathbf{r}_\alpha - \mathbf{r}_\beta) \quad (F.3.a)$$

where

$$S_{ij}(\boldsymbol{\rho}) = \frac{1}{L^3} \sum_{\mathbf{k}\varepsilon} \varepsilon_i \varepsilon_j \frac{e^{i\mathbf{k}\cdot\boldsymbol{\rho}}}{k^2}$$

$$= \int \frac{d^3k}{(2\pi)^3} \left(\delta_{ij} - \frac{k_i k_j}{k^2} \right) \frac{e^{i\mathbf{k}\cdot\boldsymbol{\rho}}}{k^2}. \qquad \text{(F.3.b)}$$

To derive (F.3.b), we used Equation (54) of Complement A_I and replaced the discrete sum by an integral. The Fourier transform appearing in (F.3.b) can be calculated (*). One finds

$$S_{ij}(\rho) = \frac{1}{8\pi} \left(\frac{\delta_{ij}}{\rho} + \frac{\rho_i \rho_j}{\rho^3} \right). \qquad \text{(F.3.c)}$$

Inserting (F.3.c) into (F.3.a) finally gives

$$\delta V = - \frac{q_\alpha q_\beta}{8\pi\varepsilon_0 m_\alpha m_\beta c^2} \left\{ \mathbf{p}_\alpha \cdot \frac{1}{|\mathbf{r}_\alpha - \mathbf{r}_\beta|} \mathbf{p}_\beta + \right.$$

$$\left. + [\mathbf{p}_\alpha \cdot (\mathbf{r}_\alpha - \mathbf{r}_\beta)] \frac{1}{|\mathbf{r}_\alpha - \mathbf{r}_\beta|^3} [(\mathbf{r}_\alpha - \mathbf{r}_\beta) \cdot \mathbf{p}_\beta] \right\}. \qquad \text{(F.4)}$$

We thus see in this example how the effect of photon exchange between two systems may be described, to a certain approximation, by an effective Hamiltonian acting only on these systems. One can then use this effective Hamiltonian as an "interaction Hamiltonian" while "forgetting" the field producing this interaction.

Expression (F.4) describes the first correction to the Coulomb interaction (**). When added to the particle Hamiltonian, it gives rise to two types of forces in the equations of motion: on the one hand, the magnetic force between the currents associated with the particle velocities; on the other hand, a delay correction of order v^2/c^2 to the Coulomb force related to the propagation time of the real interaction between the two particles.

(*) For such calculation, it is convenient to regularize the functions to be integrated by replacing $1/k^2$ by $1/(k^2 + \eta^2)$, where η is an infinitely small positive quantity.
(**) The absence of $\hbar$ indicates that it will be identical in classical theory. Cf. Landau and Lifchitz, *The Classical Theory of Fields*, §8–4.

Remarks

(i) Let us emphasize the fact that δV is an effective Hamiltonian derived from the real interaction H_{I1}. It must be used only for calculating energies. We can also make it appear directly in the particle Hamiltonian by carrying out a unitary transformation on the Hamiltonian (C.1) of Chapter I.

(ii) Instead of being eliminated to cause the Coulomb interaction to appear, the longitudinal part of the vector potential and the scalar potential may be quantized, which gives rise to longitudinal and scalar photons. In this case, the Coulomb interaction itself is described as an exchange of longitudinal and scalar photons between particles (*).

(iii) The relativistic dynamics of spin-$\frac{1}{2}$ particles are described by the Dirac Hamiltonian. The Hamiltonian for the interaction with the transverse field is written as $-cq\boldsymbol{\alpha} \cdot \mathbf{A}(\mathbf{r})$, where the three components of the vector $\boldsymbol{\alpha}$ are three Dirac matrices representing the velocity operator (**). The same technique we discussed in this subsection leads to the Breit interaction between particles 1 and 2 (***)

$$V_B = -\frac{q_1 q_2}{8\pi\varepsilon_0}\left\{\frac{\boldsymbol{\alpha}_1 \cdot \boldsymbol{\alpha}_2}{|\mathbf{r}_1 - \mathbf{r}_2|} + \frac{[\boldsymbol{\alpha}_1 \cdot (\mathbf{r}_1 - \mathbf{r}_2)][\boldsymbol{\alpha}_2 \cdot (\mathbf{r}_1 - \mathbf{r}_2)]}{|\mathbf{r}_1 - \mathbf{r}_2|^3}\right\}. \quad \text{(F.5)}$$

Likewise (F.4), V_B is an effective Hamiltonian describing the real interaction in an approximate way.

2. Van der Waals Interaction between Two Neutral Atoms

Consider two neutral atoms A and A' located at points $\mathbf{R}$ and $\mathbf{R}'$. We denote their distance D and $\mathbf{u}$ the unit vector of the line joining them:

$$D = |\mathbf{R} - \mathbf{R}'|; \quad \mathbf{u} = (\mathbf{R} - \mathbf{R}')/D. \quad \text{(F.6)}$$

Let us consider the effect of their interaction with the transverse field when they are both in their ground states $|a\rangle$ and $|a'\rangle$. We will use the electric dipole approximation. It is then useful to choose the point of view discussed in subsection 5 of the Appendix, where the Hamiltonian for the interaction with the transverse field is

$$H_I' = -\mathbf{d} \cdot \mathbf{E}_\perp(\mathbf{R}) - \mathbf{d}' \cdot \mathbf{E}_\perp(\mathbf{R}') \quad \text{(F.7)}$$

(*) See, for example, *Photons and Atoms—Introduction to Quantum Electrodynamics*, Chapter V, Section D.

(**) See, for example, *Photons and Atoms—Introduction to Quantum Electrodynamics*, Complement A_V.

(***) See, for example, Bethe and Salpeter, Section 38.

with **d** being the electric dipole moment of A, **d′** the electric dipole moment of A' and $\mathbf{E}_\perp$ the field operator given by formula (29) in the Appendix. Recall that, in this point of view, there is no longer any direct Coulomb interaction between A and A', and that $\mathbf{E}_\perp(\mathbf{R})$ represents the total electric field outside the systems of charges. The whole interaction between the two atoms is carried by the transverse field and appears to result from an exchange of virtual photons. In order to be able to identify the modification in the energy of the two atoms in their ground state, we consider the probability amplitude for the global system, initially in the state

$$|\varphi_i\rangle = |a, a'; 0\rangle \tag{F.8}$$

representing the two atoms in the photon vacuum to remain in this state after a time T.

The emission of a photon by one of the atoms in state $|a\rangle$ (resp. $|a'\rangle$) is necessarily accompanied by a transition to an excited state $|b\rangle$ (resp. $|b'\rangle$). Conservation of energy thus requires a subsequent absorption or emission, which allows the atom to return to the ground state. To the second order, we have only emission or absorption by the same atom (see diagram in Figure 32α), the other atom being simply a bystander. The first diagrams, in which both atoms are involved and in which they both return to their ground state in the final state, necessarily involve four interactions. Two examples are shown in Figure 36. The whole set of fourth-order diagrams is obtained by varying the temporal order of the different interactions.

The elementary process on which the Van der Waals interaction is based is thus, in the point of view adopted here [that is, by using the Hamiltonian (F.7)], the exchange of *pairs* of photons between the two atoms. Note that the different intermediate states are all nonresonant with the initial state (which is the ground state of the global system of atoms + photons). As for radiative corrections, the probability amplitude that the system will remain in the initial state (which is a discrete state of the total system, well-isolated from all others), allows us to calculate the shift ΔE of the energy of this initial state $|\varphi_i\rangle$ (see §B-1 in Chapter I). This shift depends on the distance D between A and A': this is the Van der Waals interaction. The general calculation of ΔE is too long to be discussed here (*), but we will identify below two simple limiting cases.

We first point out that the different transverse-field modes do not contribute uniformly to the interaction. There are few low-frequency modes (the density of modes varies as $k^2\, dk$) and the coupling matrix

(*) Cf. Power, Chapters VII and VIII.

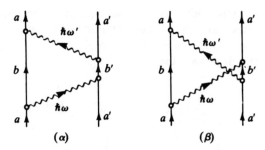

Figure 36. Examples of diagrams describing the Van der Waals interaction between two neutral atoms in the ground state by exchange of transverse photons. To interact with the field and return to the ground state, each atom must necessarily emit or absorb two photons, so that the interaction between the two atoms requires the exchange of at least two photons. Ten other diagrams may be obtained from diagrams α and β by changing the temporal order of the four vertices.

element varies as $\sqrt{\omega}$. Furthermore, the contributions of the modes having a large wave vector interfere destructively as a result of the rapid variation, from one mode to the next, of the phase difference between the factors $\exp(i\mathbf{k} \cdot \mathbf{R})$ and $\exp(i\mathbf{k} \cdot \mathbf{R}')$ which appear in the matrix elements of the interaction relative to each exchanged photon. Consequently, the significant contribution to the interaction results from the modes for which the wavelength is on the order of the distance D between atoms. We must distinguish two cases, depending on whether the energy $\hbar c/D$ of these photons is large or small as compared to the typical excitation energy $|E_a - E_b|$ of the atoms. In the first case, which corresponds to a distance between atoms smaller than $\lambda_{ab} = \hbar c / |E_a - E_b|$, the excitation energy of the atoms is negligible compared to the photon energy in the intermediate states. In the second case, the distance D is large compared to λ_{ab} and the excitation energy of the atoms predominates. In each of these extreme cases, the global process may be divided into two sequential processes.

a) SMALL DISTANCE: $D \ll \lambda_{ab}$

The energy of the virtual photons is high. They may thus exist for only an extremely short period of time. The two "photon lines" in Figure 36α are quasi-horizontal, as compared to the time interval separating these two lines and during which the two atoms are excited. The exchange of a photon between the two atoms may thus be considered as a quasi-instantaneous process, described by an effective Hamiltonian which simultaneously excites the two atoms. This process is analogous to the one discussed in subsection F-1 above, and the effective Hamiltonian that describes it has a structure identical to (F.3), except for the replacement of $q\mathbf{p}/m$ by $\mathbf{d}$, and of $\mathbf{A(R)}$ by $\mathbf{E}_\perp(\mathbf{R})$. It follows that

$$\delta V = -\sum_{\mathbf{k}\varepsilon} \frac{1}{2\varepsilon_0 L^3} (\varepsilon \cdot \mathbf{d})(\varepsilon \cdot \mathbf{d}') e^{i\mathbf{k}\cdot(\mathbf{R}-\mathbf{R}')} + \text{h.c.}$$

$$= -\frac{1}{\varepsilon_0} \sum_{ij} d_i \, d'_j \, \delta^\perp_{ij} (\mathbf{R} - \mathbf{R}') \tag{F.9}$$

where $\delta^\perp_{ij}$ is the transverse δ function (*) which, far from the origin, is

$$\delta^\perp_{ij} (\mathbf{R} - \mathbf{R}') = \frac{-\delta_{ij} + 3u_i u_j}{4\pi D^3}. \tag{F.10}$$

This effective Hamiltonian δV is simply the dipole-dipole interaction between two neutral atoms. It couples the state $|a, a'\rangle$ to the excited states $|b, b'\rangle$. Consequently, to the second order in δV, the energy shift of the state $|a, a'\rangle$ is given by

$$\Delta E = \sum_{bb'} \frac{\langle a, a'|\delta V|b, b'\rangle \langle b, b'|\delta V|a, a'\rangle}{E_a + E_{a'} - E_b - E_{b'}} \tag{F.11}$$

in which the D dependence stems from that of δV given by (F.9) and (F.10):

$$\Delta E = C_6/D^6. \tag{F.12}$$

The coefficient C_6 is a function of the dipole matrix elements for atoms A and A' and of the energy denominator appearing in (F.11). ΔE represents the usual Van der Waals interaction.

(*) The summation over the polarizations ε in the first line of (F.9) actually causes the function $\delta_{ij} - (k_i k_j/k^2)$ to appear [see formula (54) in Complement A_1]. Integration over $\mathbf{k}$ of the product of this function and $\exp i\mathbf{k} \cdot (\mathbf{R} - \mathbf{R}')$ then yields a Fourier transform which is simply the transverse δ function appearing in the second line of (F.9). For a discussion of the properties of $\delta^\perp_{ij}$, see *Photons and Atoms—Introduction to Quantum Electrodynamics*, Complement A_1.

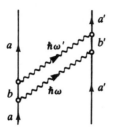

Figure 37. Example of a two-photon exchange process, which predominates at large distances. The energy of the virtually exchanged photons is very small compared with the excitation energy of the atoms, and the duration of the virtual excitation of each atom is very short compared to the time photons take to propagate from one atom to the other.

Remark

In the usual point of view (interaction Hamiltonian proportional to $\mathbf{A} \cdot \mathbf{p}$), there is an instantaneous Coulomb interaction between the two atoms, $V_{\text{Coul.}}^{AB}$. The effect of V_{Coul}^{AB} on the ground state of the two atoms is null to the first order and yields the Van der Waals interaction (F.12) to the second order.

b) LARGE DISTANCE $\lambda_{ab} \ll D$

The energy of the exchanged photons is small compared to the excitation energy of the atoms. The duration of the virtual excitation of each atom is short compared to the photon propagation time. Such a situation corresponds to processes such as those shown in Figure 37. In the case of very large distances, the overall process may be considered to be the quasi-instantaneous emission of a pair of photons by one of the atoms, followed by the absorption of this pair by the other atom after a certain time interval. The comparison with the case for short distances allows us to predict, without making any further calculations, the dependence of ΔE with regard to D. Let us consider the energy defects of the various intermediate states, that appear in the denominator of the expression for ΔE. During the processes of virtual excitation of each atom, through which the pair of photons is emitted or absorbed, the energy defect is on the order of $E_b - E_a$ or $E_{b'} - E_{a'}$, because $\hbar\omega$ and $\hbar\omega'$ are negligible as compared to these atomic excitation energies. During the photon propagation, it is equal to $\hbar\omega + \hbar\omega'$. In the processes considered in subsection a, the energy defects were successively $\hbar\omega$, $E_b + E_{b'} - E_a - E_{a'}$, $\hbar\omega'$ because in the first and third intermediate states, we could neglect the atomic excitation energies compared to $\hbar\omega$ and $\hbar\omega'$. The energy of the photons thus appears twice in the denominator instead of only once in the case corresponding to Figure 37. In the calculation of the Fourier integrals over $\mathbf{k}$ and $\mathbf{k}'$, multiplying the function to be integrated by k or k' causes an additional factor $1/D$ to appear in the result, for the simple reason of homogeneity. From that we can deduce that at a large distance

$$\Delta E \propto \frac{1}{D^7}. \tag{F.13}$$

The Van der Waals interaction decreases more rapidly than at a small distance (*). Recall that the transition from one regime to the other occurs for D close to λ_{ab}, the wavelength associated with the atomic transitions.

GENERAL REFERENCES

A general presentation of the different processes to lowest order for which they appear may be found in Heitler, Chapter V and in Power, Chapter VII. Also see the reference works cited in this chapter. They generally contain reviews on the phenomena discussed.

(*) This is the Casimir and Polder effect, *Phys. Rev.*, **73**, 360 (1948).

COMPLEMENT A$_{II}$

PHOTODETECTION SIGNALS AND CORRELATION FUNCTIONS

At the end of Section B, we saw how the dynamics of the absorption of a photon between two discrete atomic states depend on the state of the incident field. These dynamics also depends on the atomic transition, because, starting from a lower discrete atomic state, the subsequent evolution of the atom depends, on whether the upper level is discrete or belongs to a continuum.

In this complement, we present a synthetic treatment of these problems in terms of *correlation functions* of the two interacting systems, the atom and the field. Although limited to the lowest order of perturbation theory, such a treatment allows us to classify the different situations that appear by comparing the different characteristic times of the problem: duration of the interaction, correlation times characterizing the fluctuations of the atomic dipole and of the incident electric field, atomic relaxation time, etc. It is also clear that the quantitative analysis of the excitation processes of an atom placed in an incident field is essential if we want to extract from the excitation probabilities measured on the atom, that is from *photo-detection signals*, some information on the incident field. We show in this complement how it is possible to relate the photodetection signals to correlation functions of the incident radiation field. The expressions obtained in this way are the basis for the interpretation of several modern quantum optics experiments.

We begin (§1) by presenting a few simple models of atomic photodetectors and by specifying the type of measurements that can be made on them. We will then establish (§2), from a perturbative calculation of excitation amplitudes, a general expression relating the excitation probability of the atom to correlation functions of the atomic dipole and of the incident field. The results are then applied successively to the cases of broadband (§3) and narrow-band (§4) photodetectors. They allow us to show that it is then possible to measure either the instantaneous intensity of the field or its spectral density. Finally, we will analyze (§5) the correlations between the signals of two broadband photodetectors and show that they allow us to obtain higher-order correlation functions of the field.

1. Simple Models of Atomic Photodetectors

a) BROADBAND PHOTODETECTOR

Most broadband photodetectors are based on the photoelectric effect. Let us consider, for example, an atom A, having a ground state $|a\rangle$ and for which the ionization continuum begins at a distance E_I (E_I is the ionization energy) above the energy of $|a\rangle$ (Figure 1).

Let us assume that this atom, in state $|a\rangle$, interacts during a time Δt with a radiation field having all its frequencies above $E_I/\hbar$. We measure the probability that the atom has been photoionized after Δt, by detecting, for instance, the photoelectron produced by the ionization. (This detection process is assumed, for simplicity, to have a 100% efficiency.) The width of the spectral response of such a detector is measured by the width of the distribution of levels to which the ground state is efficiently coupled by the interaction Hamiltonian. The detector is called broadband if this width is large compared with the spectral width of the incoming radiation.

b) NARROW-BAND PHOTODETECTOR

Let us now assume that atom A has a discrete excited level $|b\rangle$ located at a distance $\hbar\omega_0$ above the ground state $|a\rangle$ (Figure 2). This atom in state $|a\rangle$ is irradiated for a time Δt by incident radiation having a spectrum of frequencies sufficiently concentrated about ω_0 so that no atomic transition other than $a \rightarrow b$ can be excited in any appreciable way. One measures, at the end of time Δt, the probability that the atom has been brought to the excited state b. For example, one can shine at this instant an intense radiation which can ionize an atom in state $|b\rangle$ with a 100% efficiency, but is unable to ionize it from $|a\rangle$, and detect the photoelectron produced by the ionization.

Later on we will see that the probability for the atom to be excited in $|b\rangle$ is related to the spectral density of incident radiation at frequency ω_0.

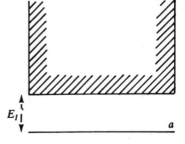

Figure 1. Ground-state and ionization continuum of an atom used as a broadband photodetector.

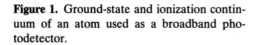

E_I

a

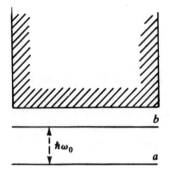

b

$\hbar\omega_0$

a

Figure 2. Ground state, discrete excited level, and ionization continuum of an atom used as a narrow-band photodetector.

Such a photodetector is thus narrow band. We will also discuss the influence of the spontaneous emission process which gives rise to the natural width Γ of the excited level $|b\rangle$ and the different regimes that may appear, depending on the relative values of Γ, $\Delta\omega_R$ (spectral width of incident radiation), Δt^{-1}, Ω_1 (Rabi frequency characterizing the interaction between the atom and the incident field).

Remark

The photodetector in Figure 1 may also be used as a narrow-band photodetector if a filtering device of the energy of the photoelectron is incorporated into the measuring device. As a result of energy conservation, a photoelectron having kinetic energy E_c, may actually only be produced by the components of the incident field at frequency ω such that $\hbar\omega = E_I + E_c$.

2. Excitation Probability and Correlation Functions

a) HAMILTONIAN. EVOLUTION OPERATOR

The Hamiltonian H of the total (photodetector atom + radiation) system is written

$$H = H_A + H_R + V. \tag{1}$$

H_A is the Hamiltonian of the atom (for which the center of mass is assumed to be fixed at **0**), H_R is the Hamiltonian of the radiation,

$$H_R = \sum_i \hbar\omega_i\left(a_i^+ a_i + \tfrac{1}{2}\right) \tag{2}$$

and V is the interaction Hamiltonian which, in the electric dipole point of view (see Appendix, §5), is equal to

$$V = -\mathbf{d} \cdot \mathbf{E}_\perp(0) \tag{3}$$

where $\mathbf{d}$ is the electric dipole moment of the atom and $\mathbf{E}_\perp(0)$ is the field operator (*), given by the expansion (89) in the Appendix and evaluated at the center of mass of the atom. To simplify these calculations, we ignore the vector character of $\mathbf{d}$ and $\mathbf{E}$ (we can assume, for example, that $\mathbf{d}$ and $\mathbf{E}$ are both parallel to $0z$), which allows (3) to be replaced by

$$V = -d E(0). \tag{4}$$

In the interaction representation with regard to $H_A + H_R$, the evolution operator $\tilde{U}(\Delta t, 0)$ between the instants 0 and Δt is written, to the first order in V [see expression (14) of Complement A_I].

$$\tilde{U}(\Delta t, 0) = \mathbb{1} + \frac{1}{i\hbar} \int_0^{\Delta t} dt\, \tilde{V}(t) \tag{5}$$

where

$$\tilde{V}(t) = e^{i(H_A + H_R)t/\hbar}\, V\, e^{-i(H_A + H_R)t/\hbar} = -\tilde{d}(t)\tilde{E}(0, t) \tag{6}$$

with

$$\tilde{d}(t) = e^{iH_A t/\hbar}\, d\, e^{-iH_A t/\hbar} \tag{7.a}$$

$$\tilde{E}(0, t) = e^{iH_R t/\hbar}\, E(0)\, e^{-iH_R t/\hbar}. \tag{7.b}$$

b) Calculation of the Probability That the Atom Has Left the Ground State after a Time Δt

At the initial instant $t = 0$, the atom is in the ground state $|a\rangle$ of H_A, and the incident radiation field is in the state $|\varphi_R\rangle$, so that the initial state of the total system is, in the interaction representation

$$|\tilde{\psi}(0)\rangle = |a\rangle \otimes |\varphi_R\rangle = |a, \varphi_R\rangle. \tag{8}$$

The atom is allowed to interact with the radiation for a time Δt. We calculate, in this subsection, the probability $P_{exc}(\Delta t)$ that, at the end of this

(*) To simplify notation, from now on, we will omit the index $\perp$ in $\mathbf{E}_\perp(0)$.

time interval, the atom has been brought to any excited state $|c\rangle$ (of the discrete or continuous spectrum of H_A).

Let $\{|\mu\rangle\}$ be an orthonormal basis of radiation states (for example, eigenstates of H_R). The probability that the total system has gone from the initial state (8) to the state $|c, \mu\rangle$ after a time Δt is, according to (5), (6), and (7), equal to

$$
\begin{aligned}
\left| \langle c, \mu | \tilde{U}(\Delta t, 0) | a, \varphi_R \rangle \right|^2 &= \\
&= \frac{1}{\hbar^2} \int_0^{\Delta t} dt' \int_0^{\Delta t} dt'' \langle \varphi_R | \tilde{E}(0, t') | \mu \rangle \langle \mu | \tilde{E}(0, t'') | \varphi_R \rangle \times \\
&\quad \times \langle a | \tilde{d}(t') | c \rangle \langle c | \tilde{d}(t'') | a \rangle.
\end{aligned} \tag{9}
$$

We have used the hermicity of d and $E(0)$ and the fact that $c \neq a$ [so that the first term of (5) does not contribute]. Because the final state of the radiation field is not observed, the excitation probability $P_{\text{exc}}(\Delta t)$ defined above is obtained by summing (9) over μ and over $c \neq a$

$$
P_{\text{exc}}(\Delta t) = \sum_\mu \sum_{c \neq a} \left| \langle c, \mu | \tilde{U}(\Delta t, 0) | a, \varphi_R \rangle \right|^2. \tag{10}
$$

Because the operator d is odd, we have

$$
\langle a | d | a \rangle = 0. \tag{11}
$$

The transition probability (9) is thus null for $c = a$ and the restriction $c \neq a$ is not necessary in (10). The summation over μ and c of expression (9) thus causes two closure relations to appear and results in

$$
P_{\text{exc}}(\Delta t) = \frac{1}{\hbar^2} \int_0^{\Delta t} dt' \int_0^{\Delta t} dt'' \, G_A^*(t', t'') G_R(t', t'') \tag{12}
$$

where

$$
G_A^*(t', t'') = \langle a | \tilde{d}(t') \tilde{d}(t'') | a \rangle \tag{13.a}
$$

$$
G_R(t', t'') = \langle \varphi_R | \tilde{E}(0, t') \tilde{E}(0, t'') | \varphi_R \rangle. \tag{13.b}
$$

The functions G_A or G_R are average values in the initial state of the atom or of the radiation of the product of two atomic or radiation operators taken, in the interaction representation, at different instants t' and t''. These "two-time averages" are also called *correlation functions*. The probability of excitation $P_{exc}(\Delta t)$ of the atom may thus be expressed, to lowest order in V, as a double integral of the product of two correlation functions of the two interacting systems, the correlation function of the dipole d and the correlation function of the field $E(0)$. Note that these correlation functions correspond to a free evolution, because the evolutions of $\bar{d}$ and $\bar{E}(0)$ are, according to (7), governed respectively by H_A and H_R.

Remarks

(i) The two-time averages (13) are not necessarily real. Their real and imaginary parts are related to quantities having precise physical meanings: the symmetric correlation function describing the dynamics of the fluctuations, and the linear susceptibility describing the linear response of the system to a perturbation (see Complement A_{IV}).

(ii) In the electric dipole point of view adopted here, the field operator $E(0)$, which appears in the interaction Hamiltonian (4), does not physically represent the transverse electric field in **0**, but rather the displacement, divided by ε_0 (see, for example, subsection 5-c of the Appendix, or *Photons and Atoms—Introduction to Quantum Electrodynamics*, Complement A_{IV}). The excitation probability (12) is thus actually related to the correlation function of the displacement in **0**. However, this displacement evolves freely [the evolution of $\bar{E}$ in (13.b) is due only to the Hamiltonian H_R] and the average value of the product of the two $\bar{E}$ is taken in the state $|\varphi_R\rangle$ that describes the free incident field in the absence of a photodetector. The free correlation function of the displacement in **0** in the state $|\varphi_R\rangle$ is thus the same as that of the free incident electric field in **0** in the absence of a photodetector. This is why we can consider the correlation function $G_R(t', t'')$ written in (13.b) as the correlation function of the incident electric field.

(iii) During the transition $|a, \varphi_R\rangle \rightarrow |c, \mu\rangle$, the energy of the atom varies by $E_c - E_a$. The average variation $\langle \Delta H_A \rangle$ of the atomic energy during time Δt is thus equal to

$$\langle \Delta H_A \rangle = \sum_{\mu} \sum_{c \neq a} (E_c - E_a) \left| \langle c, \mu | \bar{U}(\Delta t, 0) | a, \varphi_R \rangle \right|^2. \tag{14}$$

In addition, the substitution of (9) into (14) causes the quantity $(E_c - E_a)$ $\langle c | \bar{d}(t'') | a \rangle$ to appear. This quantity can be written also

$$(E_c - E_a)\langle c | \bar{d}(t'') | a \rangle = (E_c - E_a)\, e^{i(E_c - E_a)t''/\hbar} \langle c | d | a \rangle$$

$$= -i\hbar \frac{d}{dt''} \langle c | \bar{d}(t'') | a \rangle. \tag{15}$$

Finally, by using (15), we can, by following a procedure similar to that leading from (10) to (12), transform (14) into

$$\langle \Delta H_A \rangle = -\frac{i}{\hbar} \int_0^{\Delta t} dt' \int_0^{\Delta t} dt'' \, G_R(t', t'') \frac{d}{dt''} G_A^*(t', t'') \qquad (16)$$

This shows that physical quantities other than the excitation probability, such as the average variation of the atomic energy, may also be related to correlation functions of the two interacting systems (see Complement A$_{\text{IV}}$).

c) Atomic Dipole Correlation Function

Let us return to expression (13.a) for G_A^*. By using (7.a) and by inserting the closure relation over the eigenstates $|c\rangle$ of H_A between $\tilde{d}(t')$ and $\tilde{d}(t'')$, we obtain

$$G_A^*(t', t'') = \sum_c \langle a | e^{iH_A t'/\hbar} \, d \, e^{-iH_A t'/\hbar} | c \rangle \langle c | e^{iH_A t''/\hbar} \, d \, e^{-iH_A t''/\hbar} | a \rangle$$

$$= \sum_a |\langle a | d | c \rangle|^2 \, e^{-i\omega_{ca}(t'-t'')}. \qquad (17)$$

$G_A^*(t', t'')$ thus appears as a sum of exponentials evolving at the various Bohr frequencies $\omega_{ca} = (E_c - E_a)/\hbar$ of the transitions $c \leftrightarrow a$ starting from a, weighted by the squares of the moduli of the matrix elements of d between the two levels of these transitions.

It is possible to rewrite (17) in the form

$$G_A^*(t', t'') = G_A^*(t' - t'') = \int_{-\infty}^{+\infty} e^{-i\omega(t'-t'')} \mathcal{G}_A(\omega) \, d\omega \qquad (18)$$

where

$$\mathcal{G}_A(\omega) = \sum_c |\langle a | d | c \rangle|^2 \delta(\omega - \omega_{ca}). \qquad (19)$$

The Fourier transform $\mathcal{G}_A(\omega)$ of $G_A^*(t' - t'')$ is thus a superposition of delta functions centered on the various Bohr frequencies of the transitions starting from a, weighted by $|\langle a | d | c \rangle|^2$.

It is clear from (17) that $G_A^*(t', t'')$ depends only on $t' - t''$. This is due to the fact that the state $|a\rangle$, in which the correlation function is calculated, is an eigenstate of H_A, and thus a stationary state of the free atom, whose properties are invariant in a time translation.

d) FIELD CORRELATION FUNCTION

The operator $E(0)$ is, according to formula (89) of the Appendix, a linear superposition of annihilation and creation operators a_j and a_j^+ of the different field modes j. Because, according to (7.b), the evolution of $\tilde{E}(0, t)$ is due solely to H_R, each a_j is multiplied by $e^{-i\omega_j t}$ in $\tilde{E}(0, t)$, and each a_j^+ is multiplied by $e^{i\omega_j t}$, where ω_j is the frequency of the mode j. Let us now introduce the operator $\tilde{E}^{(+)}(0, t)$, which is the part of $\tilde{E}(0, t)$ that contains only exponentials in $\exp(-i\omega_j t)$, and the adjoint operator $\tilde{E}^{(-)}(0, t)$, which contains only exponentials in $\exp(i\omega_j t)$.

$$\tilde{E}^{(+)}(0, t) = i\sum_j \sqrt{\frac{\hbar\omega_j}{2\varepsilon_0 L^3}}\, a_j\, e^{-i\omega_j t} \qquad (20.a)$$

$$\tilde{E}^{(-)}(0, t) = -i\sum_j \sqrt{\frac{\hbar\omega_j}{2\varepsilon_0 L^3}}\, a_j^+\, e^{i\omega_j t}. \qquad (20.b)$$

$\tilde{E}^{(+)}(0, t)$ is called the positive-frequency component of the field, $\tilde{E}^{(-)}(0, t)$ the negative-frequency component. By replacing $\tilde{E}$ by $\tilde{E}^{(+)} + \tilde{E}^{(-)}$ in (13.b), we obtain

$$G_R(t', t'') = \langle \varphi_R | \tilde{E}^{(-)}(0, t') \tilde{E}^{(+)}(0, t'') | \varphi_R \rangle +$$
$$+ \langle \varphi_R | \tilde{E}^{(+)}(0, t') \tilde{E}^{(-)}(0, t'') | \varphi_R \rangle +$$
$$+ \langle \varphi_R | \tilde{E}^{(+)}(0, t') \tilde{E}^{(+)}(0, t'') | \varphi_R \rangle +$$
$$+ \langle \varphi_R | \tilde{E}^{(-)}(0, t') \tilde{E}^{(-)}(0, t'') | \varphi_R \rangle. \qquad (21)$$

Assume that the field is in a stationary state, that is, in a statistical mixture of eigenstates $|n_1, n_2 \cdots n_j \cdots \rangle$ of H_R (n_j being the number of photons in the mode j), with weights $p(n_1, n_2 \cdots n_j \cdots)$. It is then easy to see that the last two terms of (21) are zero, because they involve averages of the type $\langle a_j a_l \rangle$ or $\langle a_j^+ a_l^+ \rangle$ which are zero in each state $|n_1, n_2 \cdots n_j \cdots \rangle$ of the mixture. On the other hand, the first two terms in (21) are nonzero and depend only on $t' - t''$. For instance, the first term of (21), is equal to, taking (20) into account,

$$\langle \tilde{E}^{(-)}(0, t') \tilde{E}^{(+)}(0, t'') \rangle = G_R^N(t' - t'') =$$
$$= \frac{1}{2\varepsilon_0 L^3} \sum_j \hbar\omega_j \langle n_j \rangle\, e^{i\omega_j(t' - t'')} \qquad (22)$$

where

$$\langle n_j \rangle = \langle a_j^+ a_j \rangle = \sum_{n_1, n_2 \cdots n_j \cdots} n_j p(n_1, n_2 \cdots n_j \cdots) \qquad (23)$$

is the average number of photons in the mode j. The index N in G_R^N means that the operators are arranged in the normal order in the correlation function (22) (annihilation operators to the right of the creation operators). Relation (22) can also be written in the form

$$G_R^N(t' - t'') = \int_{-\infty}^{+\infty} \mathcal{G}_R^N(\omega) \, e^{i\omega(t'-t'')} \, d\omega \qquad (24.a)$$

where

$$\mathcal{G}_R^N(\omega) = \frac{1}{2\pi} \int_{-\infty}^{+\infty} G_R^N(t' - t'') \, e^{-i\omega(t'-t'')} \, d(t' - t'')$$

$$= \frac{1}{2\pi} \int_{-\infty}^{+\infty} \langle \tilde{E}^{(-)}(0, t') \tilde{E}^{(+)}(0, t'') \rangle \, e^{-i\omega(t'-t'')} \, d(t' - t'') \qquad (24.b)$$

is the Fourier transform of $G_R^N(t' - t'')$, and can be written, according to (22), in the form

$$\mathcal{G}_R^N(\omega) = \frac{1}{2\varepsilon_0 L^3} \sum_j \hbar \omega_j \langle n_j \rangle \delta(\omega - \omega_j). \qquad (24.c)$$

It is clear from (24.c) that $\mathcal{G}_R^N(\omega)$ characterizes the frequency distribution of the average energy of the field, that is, its spectral density. An analogous calculation yields for the second term of (21), which we denote $G_R^A(t' - t'')$ because its operators are arranged in antinormal order

$$G_R^A(t' - t'') = \int \mathcal{G}_R^A(\omega) \, e^{i\omega(t'-t'')} \, d\omega, \qquad (25.a)$$

where

$$\mathcal{G}_R^A(\omega) = \frac{1}{2\varepsilon_0 L^3} \sum_j \hbar \omega_j (\langle n_j \rangle + 1) \delta(\omega_j + \omega). \qquad (25.b)$$

Finally, let us assume that the incident radiation has a spectral density $\mathcal{G}_R^N(\omega)$, centered about a central frequency $\bar{\omega}$, and extends over an interval of width $\Delta\omega_R$. From Equation (24.a), it is possible to show that

$$G_R^N(t' - t'') = e^{i\bar{\omega}(t'-t'')} C_R^N(t' - t'') \qquad (26.a)$$

where $C_R^N(t' - t'')$ is an envelope function given by

$$C_R^N(t' - t'') = \int \mathscr{G}_R^N(\omega + \bar{\omega}) e^{i\omega(t'-t'')} d\omega. \qquad (26.b)$$

$C_R^N(t' - t'')$ is the Fourier transform of a function of width $\Delta\omega_R$, centered on $\omega = 0$, and thus tends to zero when $|t' - t''|$ becomes large compared to $1/\Delta\omega_R$, which may be considered to be the correlation time of the incident field.

To conclude this section, let us show that only the function $G_R^N(t' - t'')$ contributes significantly to the excitation probability (12). Consider the other correlation function of the field $G_R^A(t' - t'')$, given in (25), and compare it to the atomic correlation function $G_A^*(t' - t'')$ given in (17). Both of these correlation functions appear as a sum of exponentials with positive frequencies, varying, respectively, as $\exp[-i\omega_j(t' - t'')]$ and $\exp[-i\omega_{ca}(t' - t'')]$ (ω_{ca} is always positive because a is the ground state). The product of $G_A^*(t' - t'')$ by $G_R^A(t' - t'')$ is thus a function of $t' - t''$ evolving extremely rapidly, at optical frequencies. It follows that the integral over t' and t'' in (12) yields a negligible result if Δt is large compared to an optical period. On the other hand, $G_R^N(t' - t'')$ is, according to (24.a) and (24.c), a sum of exponentials with negative frequencies, as $\exp[i\omega_j(t' - t'')]$. The product of $G_R^N(t' - t'')$ and $G_A^*(t' - t'')$ may thus contain components that vary only slightly with $t' - t''$ [if the two spectral densities (19) and (24.c) overlap] and which will contribute significantly to the integral in (12). Therefore, in (12) we can neglect the contribution of $G_R^A(t' - t'')$ as compared to that of $G_R^N(t' - t'')$ and write

$$P_{exc}(\Delta t) \simeq \frac{1}{\hbar^2} \int_0^{\Delta t} dt' \int_0^{\Delta t} dt'' \, G_A^*(t' - t'') G_R^N(t' - t''). \qquad (27)$$

Such a result expresses that, for an atom leaving the ground state, the only process which can be resonant is the absorption process, during which the energy gained by the atom is lost by the field. The correlation function G_R^N, defined in (22), where the field operators are arranged in the normal order, characterizes the ability of the field to induce absorption processes, whereas the other correlation function (reverse order) G_R^A is involved in stimulated and spontaneous emission processes [associated, respectively, with terms $\langle n_j \rangle$ and 1 inside the parentheses in (25.b)].

3. Broadband Photodetection

a) CONDITION ON THE CORRELATION FUNCTIONS

Assume that all the incident radiation frequencies are greater than the ionization frequency $E_I/\hbar$ of the atom used as a photodetector. The excitation of discrete levels located below the ionization threshold may thus be ignored, because it can never be resonant, and only the contribution of the states of the ionization continuum can be retained in the expression (19) for $\mathscr{G}_A(\omega)$. The function $\mathscr{G}_A(\omega)$ characterizes the spectral sensitivity of the photodetector, that is, the way it reacts to the various incident frequencies. In order for the photodetector to be considered as a broadband photodetector, the width $\Delta\omega_A$ of $\mathscr{G}_A(\omega)$ must be very large compared to the width $\Delta\omega_R$ of $\mathscr{G}_R^N(\omega)$

$$\Delta\omega_A \gg \Delta\omega_R. \tag{28}$$

The response of the photodetector thus varies only slightly over the range of the incident radiation spectrum.

Condition (28) implies that the temporal width of $G_A^*(t' - t'')$, on the order of $1/\Delta\omega_A$ about $t' - t'' = 0$, is much smaller than that of the function $C_R^N(t' - t'')$ defined in (26.b) and which is on the order of $1/\Delta\omega_R$.

b) PHOTOIONIZATION RATE

Substitute (26.a) into (27). Because $C_R^N(t' - t'')$ varies much more slowly than $G_A^*(t' - t'')$ about $t' - t'' = 0$, we can replace $t' - t''$ by 0 in $C_R^N(t' - t'')$, which yields

$$C_R^N(0) = G_R^N(0) = \left\langle \tilde{E}^{(-)}(\mathbf{0}, t')\tilde{E}^{(+)}(\mathbf{0}, t') \right\rangle$$
$$= \left\langle E^{(-)}(\mathbf{0})E^{(+)}(\mathbf{0}) \right\rangle. \tag{29}$$

Because the field is in stationary state, the one-time average value $\left\langle \tilde{E}^{(-)}(\mathbf{0}, t')\tilde{E}^{(+)}(\mathbf{0}, t') \right\rangle$ does not actually depend on time. We thus get

$$P_{\text{exc}}(\Delta t) = \frac{1}{\hbar^2}\left\langle E^{(-)}(\mathbf{0})E^{(+)}(\mathbf{0}) \right\rangle \times$$

$$\times \int_0^{\Delta t} dt' \int_0^{\Delta t} dt'' \, G_A^*(t' - t'') \, e^{i\bar{\omega}(t' - t'')}. \tag{30}$$

The integration domain over t' and t'' in (30) is shown in Figure 3. Because of the exponential decrease of $G_A^*(t' - t'')$ with $t' - t''$, the

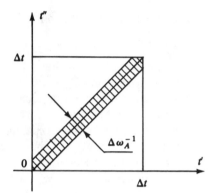

Figure 3. Domain of integration involved in the calculation of the excitation probability. The function to be integrated is significant only in the hatched zone.

integrand of (30) is different from 0 only in a strip of width $1/\Delta\omega_A$ about the principal diagonal (hatched portion of the figure).

Let us assume that

$$\Delta t \gg \Delta\omega_A^{-1}. \tag{31}$$

The hatched strip in Figure 3 is thus very narrow. Let us take as new variables t' and $\tau = t' - t''$. Condition (31) allows $-\infty$ and $+\infty$ to be taken as the limits of integration over τ, and we obtain

$$P_{exc}(\Delta t) = \frac{\Delta t}{\hbar^2} \langle E^{(-)}(0) E^{(+)}(0) \rangle \int_{-\infty}^{+\infty} G_A^*(\tau) e^{i\bar{\omega}\tau} d\tau. \tag{32}$$

Because $P_{exc}(\Delta t)$ is proportional to Δt, it is possible to define an excitation rate

$$w_{exc} = \frac{P_{exc}(\Delta t)}{\Delta t} = s \langle E^{(-)}(0) E^{(+)}(0) \rangle, \tag{33}$$

where the factor s, according to (32) and (18), is equal to

$$s = \frac{2\pi}{\hbar^2} \mathscr{G}_A(\bar{\omega}). \tag{34}$$

We have thus demonstrated that the ionization rate is proportional to the quantity $\langle E^{(-)}(0)E^{(+)}(0)\rangle$. This quantity can be considered as characterizing the total *intensity* of the field at 0, because, according to (24.a)

$$\langle E^{(-)}(0)E^{(+)}(0)\rangle = G_R^N(0) = \int \mathscr{G}_R^N(\omega)\,d\omega, \tag{35}$$

which shows that $\langle E^{(-)}(0)E^{(+)}(0)\rangle$ is the integral over all the frequencies of the spectral density of the field. The excitation rate w_{exc} is also proportional to a factor s which, according to (34), is related to the sensitivity of the photodetector at the average frequency $\bar{\omega}$ of the incident radiation.

Remarks

(i) The conditions for validity of result (33) are given in (28) and (31); to these conditions must be added

$$w_{\text{exc}}\Delta t \ll 1 \tag{36}$$

which expresses that the perturbative treatment of V is valid.

(ii) It is possible to generalize the calculation of this paragraph to the case in which the radiation is not in a stationary state. The same reasoning as that made at the end of subsection 2-d allows us to show that the last two terms of (21), which are now nonzero, do not significantly contribute to the integral in (12), because they vary too rapidly (in $\exp[\pm i\bar{\omega}(t' + t'')]$). Thus expression (27) remains valid. On the other hand, the average value (29) now depends on t', and for $P_{\text{exc}}(\Delta t)$ the following result is obtained:

$$P_{\text{exc}}(\Delta t) = s\int_0^{\Delta t} dt'\langle \bar{E}^{(-)}(0, t')\bar{E}^{(+)}(0, t')\rangle \tag{37}$$

where s is still given by (34). We thus find that the probability of photoionizing the atom between t' and $t' + dt'$ is equal to $w_I(0, t')\,dt'$, where the single counting rate

$$w_I(0, t') = s\langle \bar{E}^{(-)}(0, t')\bar{E}^{(+)}(0, t')\rangle \tag{38}$$

is proportional to the *instantaneous intensity* $\langle \bar{E}^{(-)}(0, t')\bar{E}^{(+)}(0, t')\rangle$.

4. Narrow-Band Photodetection

a) CONDITIONS ON THE INCIDENT RADIATION AND ON THE DETECTOR

Let us now assume that the incident radiation has a frequency spectrum such that it can resonantly excite only a single discrete excited state $|b\rangle$

located at a distance $\hbar\omega_0$ above a, with $E_b - E_a = \hbar\omega_0$. It is thus possible to retain only the contribution of this state in expression (19) of $\mathscr{G}_A(\omega)$ and to obtain

$$\mathscr{G}_A(\omega) = |d_{ab}|^2 \delta(\omega - \omega_0) \qquad (39.\text{a})$$

which yields for $G_A^*(t' - t'')$

$$G_A^*(t' - t'') = |d_{ab}|^2 e^{-i\omega_0(t' - t'')}. \qquad (39.\text{b})$$

Such a detector, which corresponds to the level diagram in Figure 2, and which, according to (39.a), responds to only a single frequency, is called a narrow-band detector. Later on (§c) we will see how it is possible to refine this model, by including in (39.b) the effect of the natural width of the excited level b.

b) EXCITATION BY A BROADBAND SPECTRUM

Assume that the spectral density $\mathscr{G}_R^N$ of the incident radiation has a nonzero width $\Delta\omega_R$. In the domain of integration over t' and t'' (see Figure 3), the product of the two functions $G_A^*(t' - t'')$ and $G_R^N(t' - t'')$ is nonzero only in a strip of width $\Delta\omega_R^{-1}$ (which is now determined by G_R^N, and not by G_A^* as was the case in subsection 3). If $\Delta t \gg \Delta\omega_R^{-1}$, this strip is extremely narrow, and a calculation analogous to that of subsection 3-b, allows, taking into account (39.b), the transformation of (27) to

$$P_{\text{exc}}(\Delta t) = \frac{\Delta t}{\hbar^2} |d_{ab}|^2 \times$$

$$\int_{-\infty}^{+\infty} \langle \tilde{E}^{(-)}(\mathbf{0}, t') \tilde{E}^{(+)}(\mathbf{0}, t'') \rangle e^{-i\omega_0(t' - t'')} \, d(t' - t''). \quad (40)$$

It is thus possible to define an excitation rate equal to [according to (24.b)]

$$w_{\text{exc}} = \frac{P_{\text{exc}}(\Delta t)}{\Delta t} = 2\pi \frac{|d_{ab}|^2}{\hbar^2} \mathscr{G}_R^N(\omega_0) \qquad (41)$$

A narrow-band photodetector, irradiated by broadband radiation, thus measures the spectral density of the incident radiation (namely, the Fourier transform of the correlation function of the field, arranged in the normal order) at the response frequency of the detector.

Finally, note that the foregoing perturbative treatment is valid only if $w_{\text{exc}}\Delta t \ll 1$. Moreover, Δt must be very large compared to $\Delta\omega_R^{-1}$, these two conditions on Δt are compatible only if $\Delta \omega_R \gg w_{exc}$, that is, if the incident radiation is not too intense.

Remark

When the incident radiation impinging on the detector is also monochromatic, expression (40) is no longer valid. Assume, for example, that in expression (24.c) for $\mathscr{S}_R^N(\omega)$, $\langle n_j \rangle$ is nonzero for only a single mode j. In addition, assume that the frequency ω_j of this mode coincides with the frequency ω_0 of the detector. We obtain

$$\mathscr{S}_R^N(\omega) = \langle E^{(-)}E^{(+)} \rangle \delta(\omega - \omega_0). \tag{42}$$

The coefficient of proportionality of the delta function is simply the intensity $\langle E^{(-)}E^{(+)} \rangle$. To see this, it is sufficient to integrate (42) over ω and to use (35). From (42), and taking into account (24.a), we deduce

$$G_R^N(t' - t'') = \langle E^{(-)}E^{(+)} \rangle \, e^{i\omega_0(t'-t'')}. \tag{43}$$

Then we substitute (39.b) and (43) into the expression (27) for $P_{exc}(\Delta t)$. Because the two exponentials of (39) and (43) cancel each other out, the integral over t' and t'' is immediate and yields

$$P_{exc}(\Delta t) = \frac{(\Delta t)^2}{\hbar^2} |d_{ab}|^2 \langle E^{(-)}E^{(+)} \rangle = \left(\frac{\Omega_1 \Delta t}{2} \right)^2 \tag{44}$$

where we have set

$$\left(\frac{\Omega_1}{2} \right)^2 = \frac{|d_{ab}|^2 \langle E^{(-)}E^{(+)} \rangle}{\hbar^2} = \frac{\langle V^2 \rangle}{\hbar^2}. \tag{45}$$

Thus there is no longer any transition rate. Actually, expression (44) is just the first term of the expansion in powers of V of the sinusoid $[\sin(\Omega_1 \Delta t/2)]^2 = [1 - \cos \Omega_1 \Delta t]/2$ that describes the reversible exchange of energy between the atom and the field at the *Rabi frequency* Ω_1 defined by (45).

c) INFLUENCE OF THE NATURAL WIDTH OF THE EXCITED ATOMIC LEVEL

Consider an isolated atom, without any incident radiation, in a linear superposition of states $|a\rangle$ and $|b\rangle$. The average value of d is thus nonzero and oscillates at the frequency ω_0. Such an oscillation cannot, however, continue indefinitely. Because the amplitude of finding the atom in the excited state decreases exponentially with a time constant $(\Gamma/2)^{-1}$, because of spontaneous emission (Γ is the natural width of the excited level b), the mean dipole is damped with the same time constant. One can show that the same damping occurs for the correlation function of the atomic dipole, so that an expression more precise than (39.b) for $G_A^*(t' - t'')$ is given by

$$G_A^*(t' - t'') = |d_{ab}|^2 \, e^{-i\omega_0(t' - t'')} \, e^{-\Gamma|t' - t''|/2}. \tag{46}$$

It follows that the Fourier transform $\mathscr{G}_A(\omega)$ of $G_A^*(t' - t'')$ is not an infinitely narrow delta function but rather a Lorentzian of width Γ. We will not demonstrate here the result (46) (*), we will instead study how such a photodetector, of width Γ, responds to a broad-line ($\Delta\omega_R \gg \Gamma$) or a narrow-line ($\Delta\omega_R \ll \Gamma$) excitation.

(i) *Broad-line excitation:* $\Delta\omega_R \gg \Gamma$. The situation is the same as that in subsection 4-b above. In expression (27), the function $G_R^N(t' - t'')$ decreases much more rapidly with $t' - t''$ than $G_A^*(t' - t'')$ so that, in the regions where $G_R^N(t' - t'')$ is nonzero, the decreasing exponential of (46) can be replaced by 1. The introduction of Γ thus has no effect on the excitation rate which is still given by (41).

(ii) *Narrow-line excitation:* $\Delta\omega_R \ll \Gamma$. Now it is $G_A^*(t' - t'')$ which decreases more rapidly with $t' - t''$ than $G_R^N(t' - t'')$. In the regions where $G_A^*(t' - t'')$ is nonzero, $G_R^N(t' - t'')$ can be considered as being purely monochromatic at frequency $\bar\omega = \omega_0$ (if it is assumed that the incident radiation spectrum is still centered on ω_0), and therefore we can use (43). If (46) and (43) are substituted into (27), and if $\Delta t \gg \Gamma^{-1}$, we then have

$$
\begin{aligned}
w_{\text{exc}} &= \frac{P_{\text{exc}}(\Delta t)}{\Delta t} = \frac{|d_{ab}|^2 \langle E^{(-)}E^{(+)} \rangle}{\hbar^2} \int_{-\infty}^{+\infty} e^{-(\Gamma/2)|\tau|} \, d\tau \\
&= \frac{4|d_{ab}|^2 \langle E^{(-)}E^{(+)} \rangle}{\hbar^2 \Gamma} = \frac{\Omega_1^2}{\Gamma}
\end{aligned}
\tag{47}
$$

using definition (45) for Ω_1.

It is thus possible, even with monochromatic excitation, that the excitation probability of the atom increases linearly with time, instead of quadratically, as would be the case with $\Gamma = 0$ [see (44)]. Thus we quantitatively justify the discussion in subsection B-4 of the chapter [see also subsection 3-b of Complement C_I, particularly relation (32) of this complement, which coincides with (47)].

(*) This result can be established using the quantum regression theorem (see Complements C_{IV} and A_V).

For the perturbative calculation to be valid, it is, of course, necessary that $w_{exc} \Delta t \ll 1$, that is, according to (47),

$$\Delta t \ll \frac{\Gamma}{\Omega_1^2}. \tag{48}$$

In addition, to establish (47), we have assumed

$$\Gamma^{-1} \ll \Delta t \tag{49}$$

(so that the strip in Figure 3 can be considered to be very narrow). These two equations are compatible only if

$$\Omega_1 \ll \Gamma \tag{50}$$

that is, if the incident intensity is low enough. In the opposite case, that is, if $\Omega_1 \gg \Gamma$, Δt must be much smaller for the perturbative calculation to be valid ($\Delta t \ll \Omega_1^{-1}$). The last exponential of (46) thus remains almost equal to 1 and we recover the result (44) derived with $\Gamma = 0$.

5. Double Photodetection Signals

a) CORRELATION BETWEEN TWO PHOTODETECTOR SIGNALS

Consider two atoms 1 and 2, having centers of mass assumed to be fixed at r_1 and r_2. These atoms are subjected to an incident radiation of spectral density such that the two atoms may be considered to be broadband photodetectors. Under such conditions, the two atoms may be ionized. We wish in this last subsection to calculate the probability $w_{II}(r_1, t_1; r_2, t_2) dt_1 dt_2$ that atom 1 will be ionized between t_1 and $t_1 + dt_1$ *and* atom 2 between t_2 and $t_2 + dt_2$. In general, such a probability is not simply equal to the product of $w_I(r_1, t_1) dt_1$ and $w_I(r_2, t_2) dt_2$ where w_I is the ionization rate of a single atom. The function w_{II} thus allows us to characterize the *correlations* between the ionizations produced by the same field at two different points and at two different instants.

b) SKETCH OF THE CALCULATION OF w_{II}

The procedure to be followed is quite similar to that in subsections 2 and 3 above. The Hamiltonian H of the problem is now

$$H = H_{A1} + H_{A2} + H_R - d_1 E(r_1) - d_2 E(r_2) \tag{51}$$

where H_{Ai} and d_i are the Hamiltonian and the dipole of atom i ($i = 1, 2$).

Consider the amplitude $\langle c_1, c_2, \mu | \bar{U}(\Delta t, 0) | a_1, a_2, \varphi_R \rangle$ that, between the instants 0 and Δt, each atom i goes from the ground state a_i to the excited state c_i, the field going from state $|\varphi_R\rangle$ to state $|\mu\rangle$. Because there are necessarily two interactions with the field, it is necessary to take the second-order term of the perturbative expansion of $\bar{U}(\Delta t, 0)$ [see formula (14) in Complement A$_1$]:

$$-\frac{1}{\hbar^2} \int_0^{\Delta t} dt'' \int_{t''>t'}^{\Delta t} dt' \langle c_1, c_2, \mu | \bar{V}(t'')\bar{V}(t') | a_1, a_2, \varphi_R \rangle. \tag{52}$$

We are not interested in processes in which the field interacts twice with the same atom. Because $\bar{V} = \bar{V}_1 + \bar{V}_2$, we can thus, in (52), replace $\bar{V}(t'')\bar{V}(t')$ by $\bar{V}_2(t'')\bar{V}_1(t') + \bar{V}_1(t'')\bar{V}_2(t')$, which yields

$$-\frac{1}{\hbar^2} \int_0^{\Delta t} dt'' \int_{t''>t'}^{\Delta t} dt' \langle c_2 | \bar{d}_2(t'') | a_2 \rangle \langle c_1 | \bar{d}_1(t') | a_1 \rangle \times$$
$$\times \langle \mu | \bar{E}^{(+)}(\mathbf{r}_2, t'') \bar{E}^{(+)}(\mathbf{r}_1, t') | \varphi_R \rangle + \text{similar terms } 1 \rightleftarrows 2. \tag{53}$$

The operators $\bar{E}$ have been replaced by $\bar{E}^{(+)}$, because the contribution of the operators $\bar{E}^{(-)}$, which are associated with emission processes, is negligible compared to that of $\bar{E}^{(+)}$ for amplitudes for which the atom starts from the ground state (see discussion at the end of subsection 2-d above). Because the two operators $\bar{E}^{(+)}$ of (53) commute (they contain only annihilation operators that commute between themselves), they may be put in reverse order (*). Thus invert t' and t'' in the second term of (53). The regrouping of the two terms yields a result equivalent to the first term without any restriction on the order of t' and t''.

To obtain the probability that the two atoms are excited between 0 and Δt, it is necessary to square the modulus of the first term of (53) (without restriction on the order of t' and t''), then sum over μ and over $c_1 \neq a_1$ and $c_2 \neq a_2$. Because the terms $c_1 = a_1$ and $c_2 = a_2$ are zero as a result of the fact that d_1 and d_2 are odd, they can be included to introduce closure relations that lead to the following expression of $P_{\text{exc}}(\Delta t)$

$$P_{\text{exc}}(\Delta t) = \frac{1}{\hbar^4} \int_0^{\Delta t} dt_1' \int_0^{\Delta t} dt_1'' \int_0^{\Delta t} dt_2' \int_0^{\Delta t} dt_2''$$
$$\times \langle a_1 | \bar{d}_1(t_2') \bar{d}_1(t_1') | a_1 \rangle \langle a_2 | \bar{d}_2(t_2'') \bar{d}_2(t_1'') | a_2 \rangle$$
$$\times \langle \varphi_R | \bar{E}^{(-)}(\mathbf{r}_1, t_2') \bar{E}^{(-)}(\mathbf{r}_2, t_2'') \bar{E}^{(+)}(\mathbf{r}_2, t_1'') \bar{E}^{(+)}(\mathbf{r}_1, t_1') | \varphi_R \rangle. \tag{54}$$

(*) Nevertheless, see the remark at the end of the paragraph which refers to a more precise calculation taking into account the source atoms emitting the radiation.

We should note once again the normal order of the field operators that appear in the field correlation function in (54).

Because the photodetectors are broadband, the two correlation functions of d_1 and d_2 decrease much more rapidly with $t_1' - t_2'$ and $t_1'' - t_2''$ than the correlation function of the field. A calculation analogous to the one in subsection 3 above thus allows (54) to be transformed into

$$P_{exc}(\Delta t) = s_1 s_2 \int_0^{\Delta t} dt' \int_0^{\Delta t} dt'' \times$$

$$\times \left\langle E^{(-)}(\mathbf{r}_1, t') E^{(-)}(\mathbf{r}_2, t'') E^{(+)}(\mathbf{r}_2, t'') E^{(+)}(\mathbf{r}_1, t') \right\rangle \quad (55)$$

where s_1 and s_2 are sensitivity factors given by expressions analogous to (34).

Finally, imagine that an aperture masks the field for atom 1 after t_1 and another one masks the field for atom 2 after t_2, with both t_1 and t_2 being less than Δt. The foregoing calculation remains valid on the condition that V_1 and V_2 be replaced by $V_1 \theta(t_1 - t)$ and $V_2 \theta(t_2 - t)$ where θ is the Heaviside function. The probability of having atom 1 being ionized between 0 and t_1 and atom 2 being ionized between 0 and t_2 is given by a formula analogous to (55), where $\int_0^{\Delta t} dt' \int_0^{\Delta t} dt''$ is replaced by $\int_0^{t_1} dt' \int_0^{t_2} dt''$. It is thus sufficient to take the derivative of this excitation probability with respect to t_1 and t_2 to obtain the elementary probability w_{II} defined in subsection 5-a above, which equals

$$w_{II}(\mathbf{r}_1, t_1; \mathbf{r}_2, t_2) =$$

$$= s_1 s_2 \left\langle \tilde{E}^{(-)}(\mathbf{r}_1, t_1) \tilde{E}^{(-)}(\mathbf{r}_2, t_2) \tilde{E}^{(+)}(\mathbf{r}_2, t_2) \tilde{E}^{(+)}(\mathbf{r}_1, t_1) \right\rangle. \quad (56)$$

Finally, note that since the operators $\tilde{E}^{(-)}$ and $\tilde{E}^{(+)}$ do not commute, w_{II} cannot be written in the form of an average value of the product of two intensities at $\mathbf{r}_1, t_1$ and $\mathbf{r}_2, t_2$

$$w_{II}(\mathbf{r}_1, t_1; \mathbf{r}_2, t_2) \neq s_1 s_2 \left\langle I(\mathbf{r}_1, t_1) I(\mathbf{r}_2, t_2) \right\rangle \quad (57)$$

where

$$I(\mathbf{r}_i, t_i) = \tilde{E}^{(-)}(\mathbf{r}_i, t_i) \tilde{E}^{(+)}(\mathbf{r}_i, t_i) \quad (i \simeq 1, 2).$$

This lack of commutation between $\tilde{E}^{(-)}$ and $\tilde{E}^{(+)}$ is at the origin of several quantum effects observable on double photodetection signals (*).

(*) See, for example, *Photons and Atoms—Introduction to Quantum Electrodynamics*, Complement A$_{III}$.

Remark

In the foregoing, we have assumed the incident radiation to be free. More precisely, in the absence of photodetector atoms, Hamiltonians (1) and (51) reduce to H_R. In certain cases, it can be worthwhile to include in the analysis the sources of the incident radiation, for example, if one wants to correlate the signals provided by the photodetectors with correlation functions of the dipole of the emitting atom (this is what we do for example in subsection D-1 of Chapter V). In this case, H_R must be replaced in (1) and (51) by

$$H_R + H_S + V_{SR} \tag{58}$$

where H_S is the Hamiltonian for the source atom and V_{SR} is the interaction between the radiation and the source atom. The calculation of the probabilities of excitation of photodetector atoms is quite similar to that done in this complement. However, we should now go to the interaction representation with respect to $H_A + H_R + H_S + V_{SR}$ and no longer with respect to $H_A + H_R$. The free field operators $\bar{E}^{(\pm)}(\mathbf{r}, t)$ of this complement must then be replaced by positive- and negative-frequency components $E^{(\pm)}(\mathbf{r}, t)$ of the field operator $E(\mathbf{r}, t)$, coupled to the source atom, and taken in the Heisenberg picture with respect to the Hamiltonian (58).

The single photodetection signals studied in subsections 3 and 4 above retain the same form, except that $\bar{E}^{(\pm)}(\mathbf{r}, t)$ is replaced by $E^{(\pm)}(\mathbf{r}, t)$. For example, the single counting rate provided by a broad band photodetector is proportional to $\langle E^{(-)}(\mathbf{r}, t)E^{(+)}(\mathbf{r}, t)\rangle$. A narrow-band photodetector measures the spectral density of the field at its frequency ω_0, that is, the Fourier transform of $\langle E^{(-)}(\mathbf{r}, t')E^{(+)}(\mathbf{r}, t'')\rangle$, evaluated at ω_0.

The calculation of the double counting rate requires a few more precautions. In expression (53), the operators $E^{(+)}(\mathbf{r}_1, t')$ and $E^{(+)}(\mathbf{r}_2, t'')$, which replace $\bar{E}^{(+)}(\mathbf{r}_1, t')$ and $\bar{E}^{(+)}(\mathbf{r}_2, t'')$, do not commute, and it is no longer possible to regroup the two terms of (53) into a single double integral. The two terms must be evaluated separately and respecting the temporal order $t'' > t'$ in the two products $E^{(+)}(\mathbf{r}_2, t'')E^{(+)}(\mathbf{r}_1, t')$ and $E^{(+)}(\mathbf{r}_1, t'')E^{(+)}(\mathbf{r}_2, t')$. However, only one of these two products contributes to the double counting rate $w_{II}(\mathbf{r}_1, t_1; \mathbf{r}_2, t_2)$, the first if $t_2 > t_1$, the second if $t_1 > t_2$. Finally, if $t_2 > t_1$, formula (56) remains valid, provided that $\bar{E}^{(\pm)}(\mathbf{r}, t)$ be replaced by $E^{(\pm)}(\mathbf{r}, t)$. If, on the other hand, $t_1 > t_2$, $\mathbf{r}_1, t_1$ and $\mathbf{r}_2, t_2$ must be interchanged. In conclusion, in the presence of sources, the higher-order correlation functions of the field which appear in the photodetection signals must be arranged, not only with the components of positive frequency to the right of components of negative frequency, but also with increasing time from the outside to the inside.

GENERAL REFERENCES

Glauber, Nussenzweig.

COMPLEMENT B$_{II}$

RADIATIVE CORRECTIONS IN THE PAULI–FIERZ REPRESENTATION

In Section E, we studied processes during which photons are emitted transiently, and then reabsorbed. These processes give rise to modifications of the effective properties of the emitting systems, atoms or particles, which are called "radiative corrections". Actually, one can consider a particle or an atom as permanently surrounded or "dressed" by a cloud of virtual photons which quantum-mechanically describe the state of the transverse field which is "tied" to the particle or the atom.

To understand such a tied transverse field, let us come back to classical theory. Consider, for example, a uniformly moving particle. In addition to the Coulomb field, there is also a magnetic field around the particle, which is linear in v/c, and a correction to the Coulomb field in v^2/c^2. These two latter fields are derived from the transverse vector potential. It follows that in quantum theory, the transverse field is not just the radiation emitted by the particles. Actually, this radiation vanishes for a uniformly moving particle. The transverse field also describes the part of the field dependent on the velocity of the particles, which is tied to them in a certain sense as long as this velocity does not change. It is this field which is described by the cloud of virtual photons.

It is therefore tempting to try to take as unperturbed states of the field + particles system new states that include the part of the transverse field that is bound to the particles. More precisely, the problem is to find a unitary transformation such that, in the new unperturbed states representing the particles in the presence of the "new vacuum", each particle is accompanied by the transverse field associated with its velocity. The transformation that Pauli and Fierz (*) introduced to study the emission of low-frequency radiation during an electronic collision (the so-called "infrared divergence" problem) realizes this goal on a part of the mode spectrum of the field. This transformation is actually simple only when it is limited to the long-wavelength modes.

In subsection 1, we determine the transverse field bound to a classical particle and establish the expression for the Pauli–Fierz transformation for a localized quantum particle. The effect of this transformation on the observables of the field and of the particle, and in particular on the Hamiltonian is analyzed in subsection 2. We then show (§3) that the new

(*) W. Pauli and M. Fierz, *Nuovo Cimento*, **15**, 167 (1938).

Hamiltonian explicitly contains the mass correction δm_α encountered in Section E of this chapter, and a correction to the potential energy that allows us to develop a physical understanding of the Lamb shift. Another important result is that the particles are coupled to the new transverse field only to second order in q_α and only if they are accelerated. Although the Pauli–Fierz transformation is interesting for discussing the radiative corrections in the long-wavelength approximation, it is not easily generalizable. Its limitations are discussed at the end of the complement.

1. The Pauli–Fierz Transformation

a) SIMPLIFYING ASSUMPTIONS

To simplify as much as possible the calculations and emphasize the physical ideas, we consider a single particle α, located near the origin by an external potential $U_e(\mathbf{r})$. Moreover, we take into account only the modes with a wave vector having a modulus less than a limit k_M such that the spatial variations of the corresponding fields are negligible over the extension a_0 of the wave functions of the particle in the potential $U_e(\mathbf{r})$. The long-wavelength approximation is thus applicable to all the modes considered, so that the interaction of the particle with the transverse field can be characterized by $\mathbf{A}(\mathbf{0})$ (the expansion of $\mathbf{A}$ is limited to the modes $|\mathbf{k}| < k_M$). The Coulomb-gauge Hamiltonian is written

$$H = \frac{1}{2m_\alpha}\left[\mathbf{p}_\alpha - q_\alpha \mathbf{A}(\mathbf{0})\right]^2 + \varepsilon_{\text{Coul}}^\alpha + q_\alpha U_e(\mathbf{r}_\alpha) +$$

$$+ \int_< d^3k \sum_\varepsilon \hbar\omega\left[a_\varepsilon^+(\mathbf{k})a_\varepsilon(\mathbf{k}) + \frac{1}{2}\right] \tag{1}$$

where the symbol $<$ indicates that the integral is limited to $|\mathbf{k}| < k_M$. The operators $a_\varepsilon(\mathbf{k})$ and $a_\varepsilon^+(\mathbf{k})$ characterize the *total* transverse field.

Our goal is to find a unitary transformation T that will allow us to separate the transverse field tied to the particle from the total field, leading to a new formulation of the electrodynamics in which the free states of a particle include this transverse field. In the new representation, the operators $a_\varepsilon(\mathbf{k})$ and $a_\varepsilon^+(\mathbf{k})$ must characterize the transverse field that is *not tied* to the particle. The first step of such a program thus consists of determining the transverse field tied to a particle.

b) TRANSVERSE FIELD TIED TO A CLASSICAL PARTICLE

Let us consider the particle α in uniform motion with a velocity $\mathbf{v}_\alpha$. The accompanying field is well known in classical electrodynamics. It derives, for instance, from the retarded potentials (*):

$$\Phi_P(\mathbf{r}, t) = \int d^3 r' \frac{\rho\left(\mathbf{r}', t - \dfrac{|\mathbf{r} - \mathbf{r}'|}{c}\right)}{4\pi\varepsilon_0 |\mathbf{r} - \mathbf{r}'|} \tag{2.a}$$

$$\mathbf{A}_P(\mathbf{r}, t) = \int d^3 r' \frac{\mathbf{j}\left(\mathbf{r}', t - \dfrac{|\mathbf{r} - \mathbf{r}'|}{c}\right)}{4\pi\varepsilon_0 c^2 |\mathbf{r} - \mathbf{r}'|} \tag{2.b}$$

$\rho(\mathbf{r}', t')$ and $\mathbf{j}(\mathbf{r}', t')$ being the charge and current densities associated with the particle α. For a motionless particle ($\mathbf{v}_\alpha = 0$), $\Phi_P(\mathbf{r})$ is the Coulomb potential and $\mathbf{A}_P$ is zero. For a moving particle, Equations (2) give the Lienard-Wiechert potentials, for which the first-order expansion in $\mathbf{v}_\alpha$ is written

$$\Phi_P(\mathbf{r}, t) = \frac{q_\alpha}{4\pi\varepsilon_0 |\mathbf{r} - \mathbf{r}_\alpha|} + \cdots \tag{3.a}$$

$$\mathbf{A}_P(\mathbf{r}, t) = \frac{q_\alpha \mathbf{v}_\alpha}{4\pi\varepsilon_0 c^2 |\mathbf{r} - \mathbf{r}_\alpha|} + \cdots \tag{3.b}$$

where $\mathbf{r}_\alpha$ is the position of the particle at the instant t. The first term of (3.b) is simply the vector potential ordinarily used in electrokinetics to describe the magnetic effects at the quasi-static limit. The Pauli–Fierz transformation neglects the higher-order terms in $\mathbf{v}_\alpha$ and takes into consideration only these first corrections to the Coulomb field, and assumes in addition that $\mathbf{v}_\alpha = \mathbf{p}_\alpha/m_\alpha$. The latter approximation is equivalent to neglecting the second term of the expression $[\mathbf{p}_\alpha - q_\alpha \mathbf{A}(\mathbf{r}_\alpha)]/m_\alpha$ of $\mathbf{v}_\alpha$, which would give a contribution in q_α^2. We thus calculate the field tied to the particle to the first order in $\mathbf{p}_\alpha$ and q_α. The potentials (3.a) and (3.b) do not satisfy the Coulomb gauge, because Φ_P differs from the Coulomb potential by second-order terms in v_α/c which do not explicitly appear in (3.a) and because $\mathbf{A}_P$ has a nonzero divergence. The potentials that describe the same field in the Coulomb gauge are, on the one hand, the Coulomb potential, and, on the other hand, the transverse part of $\mathbf{A}_P$ (because a gauge change does not modify the transverse component of the vector potential and because the vector potential is purely transverse in

(*) Jackson, Chapter 6, Landau and Lifchitz, *The Classical Theory of Fields*, §8-1 and 8-2; see also *Photons and Atoms—Introduction to Quantum Electrodynamics*, Exercise 4 of Complement C$_I$.

the Coulomb gauge—see Appendix, §1-b and §1-c). To obtain the transverse component of the vector potential (3.b), it is sufficient to write it in the form of a Fourier transform and to retain, for each value of $\mathbf{k}$, only the projection of the field onto the plane perpendicular to $\mathbf{k}$. We thus obtain

$$A_{P\perp}(\mathbf{r}) = \int \frac{d^3k}{(2\pi)^3} \sum_{\varepsilon} \frac{q_\alpha}{m_\alpha \varepsilon_0 c^2} \frac{\boldsymbol{\varepsilon} \cdot (\boldsymbol{\varepsilon} \cdot \mathbf{p}_\alpha)}{k^2} e^{i\mathbf{k}\cdot(\mathbf{r}-\mathbf{r}_\alpha)} \tag{4}$$

or

$$A_{P\perp}(\mathbf{r}) = \int d^3k \sum_{\varepsilon} \sqrt{\frac{\hbar}{2\varepsilon_0 \omega (2\pi)^3}} \left[\beta_\varepsilon(\mathbf{k}) \boldsymbol{\varepsilon} \, e^{i\mathbf{k}\cdot\mathbf{r}} + \text{c.c.} \right] \tag{5}$$

with

$$\beta_\varepsilon(\mathbf{k}) = \frac{q_\alpha}{m_\alpha \omega} \frac{1}{\sqrt{2\varepsilon_0 \hbar \omega (2\pi)^3}} \boldsymbol{\varepsilon} \cdot \mathbf{p}_\alpha \, e^{-i\mathbf{k}\cdot\mathbf{r}_\alpha}. \tag{6}$$

The quantity $\beta_\varepsilon(\mathbf{k})$ is thus the value of the normal variable describing how the mode $(\mathbf{k}, \varepsilon)$ is excited by the motion of the particle. Note that the state of the field + particle system characterized by $\mathbf{p}_\alpha$ and $\beta_\varepsilon(\mathbf{k})$ is stationary: In the absence of external perturbation, the uniform translation motion of the particle and of the associated field persists indefinitely.

Remark

Instead of starting from retarded potentials, then returning to the Coulomb gauge, we could have directly determined the expression for $A_{P\perp}$ by integrating the equations of motion of the normal variables of the field [equation (18.b) of the Appendix] in the presence of the current associated with the particle in uniform motion. This is done in Exercise 9. At the lowest order in q_α and $\mathbf{p}_\alpha$, we recover (6).

In the problem considered here, the particle moves in the potential $U_e(\mathbf{r})$ and its motion is not uniform. The potential $A_{P\perp}(\mathbf{r})$ does not thus describe the exact transverse field around the particle, even in the approximation under consideration. It is the field that should accompany the particle if its motion were uniform. It thus corresponds to the stationary state of the field + particle system tangent, at this instant, to its real motion. If we know only the values of the dynamic variables $\mathbf{r}_\alpha$ and $\mathbf{p}_\alpha$ of the particle at the instant t, the sum of the Coulomb field and of that derived from $\mathbf{A}_{P\perp}(\mathbf{r})$ represents the best approximation of the real field that can be constructed from these two data.

Since we will only consider here the modes for which the long-wavelength approximation is valid, it is convenient to make $\mathbf{k} \cdot \mathbf{r}_\alpha \simeq 0$ in (4) and (6). We thus get

$$\beta_\varepsilon(\mathbf{k}) = \frac{q_\alpha}{m_\alpha \omega} \frac{1}{\sqrt{2\varepsilon_0 \hbar \omega (2\pi)^3}} \, \boldsymbol{\varepsilon} \cdot \mathbf{p}_\alpha \tag{7}$$

and (*)

$$\mathbf{A}_P(\mathbf{r}) = \int_< \frac{d^3k}{(2\pi)^3} \sum_\varepsilon \frac{q_\alpha}{m_\alpha \varepsilon_0 c^2} \frac{\boldsymbol{\varepsilon}(\boldsymbol{\varepsilon} \cdot \mathbf{p}_a)}{k^2} e^{i\mathbf{k} \cdot \mathbf{r}} \tag{8}$$

c) DETERMINATION OF THE PAULI–FIERZ TRANSFORMATION

Now that we have identified the transverse field tied to the particle, we can return to quantum theory. We will use the Schrödinger representation in which the operators are time independent. In quantum theory, Equation (7) defines an operator $\beta_\varepsilon(\mathbf{k})$ relative to the particle [$\beta_\varepsilon(\mathbf{k})$ depends only on the dynamic variable $\mathbf{p}_\alpha$ of the particle which is represented by the operator $-i\hbar\boldsymbol{\nabla}_{\mathbf{r}_\alpha}$].

We seek a unitary transformation T which subtracts the transverse field tied to the particle from the total transverse field. To do this, it is sufficient that the transformation translate each operator $a_\varepsilon(\mathbf{k})$ of the quantity $\beta_\varepsilon(\mathbf{k})$ determined previously:

$$T a_\varepsilon(\mathbf{k}) T^+ = a_\varepsilon(\mathbf{k}) + \beta_\varepsilon(\mathbf{k}). \tag{9}$$

We will see in the following subsection that such a transformation allows us to subtract the transverse field tied to the particle from the total transverse field. First, we give the explicit form of T.

Because $\beta_\varepsilon(\mathbf{k})$ depends only on a single particle operator $\mathbf{p}_\alpha$ and because $\beta_\varepsilon(\mathbf{k})$ commutes with all the radiation operators, Equation (9) may be considered as defining a simple translation for the a and a^+. Such a translation operator is well known [see formula (66) and (67) of the Appendix]. It can be written

$$T = \exp\left\{ \int_< d^3k \sum_\varepsilon [\beta_\varepsilon^+(\mathbf{k}) a_\varepsilon(\mathbf{k}) - \beta_\varepsilon(\mathbf{k}) a_\varepsilon^+(\mathbf{k})] \right\}. \tag{10}$$

(*) To simplify notation, the index $\perp$ will henceforth be omitted in $\mathbf{A}_{P\perp}$.

Having thus determined the transformation, we will now determine the correspondence between physical variables and mathematical operators in the new point of view, as well as the new Hamiltonian.

2. The Observables in the New Picture

a) TRANSFORMATION OF THE TRANSVERSE FIELDS

In the Coulomb picture, noted (1), the mathematical operators

$$\mathbf{A}(\mathbf{r}) = \int_< d^3k \sum_\varepsilon \mathscr{A}_\omega \big[a_\varepsilon(\mathbf{k}) \boldsymbol{\varepsilon} \, e^{i\mathbf{k}\cdot\mathbf{r}} + a_\varepsilon^+(\mathbf{k}) \boldsymbol{\varepsilon} \, e^{-i\mathbf{k}\cdot\mathbf{r}} \big] \qquad (11.a)$$

$$\mathbf{E}_\perp(\mathbf{r}) = \int_< d^3k \sum_\varepsilon i \mathscr{E}_\omega \big[a_\varepsilon(\mathbf{k}) \boldsymbol{\varepsilon} \, e^{i\mathbf{k}\cdot\mathbf{r}} - a_\varepsilon^+(\mathbf{k}) \boldsymbol{\varepsilon} \, e^{-i\mathbf{k}\cdot\mathbf{r}} \big] \qquad (11.b)$$

represent, respectively, the total transverse vector potential and the transverse electric field (see Appendix, §1-f)

$$\mathbf{A}(\mathbf{r}) = \mathbf{A}^{(1)}(\mathbf{r}) \qquad (12.a)$$

$$\mathbf{E}_\perp(\mathbf{r}) = \mathbf{E}_\perp^{(1)}(\mathbf{r}). \qquad (12.b)$$

The operator representing the total transverse vector potential in the new picture, noted (2), is the transformed of $\mathbf{A}^{(1)}(\mathbf{r})$ by T and is written

$$\mathbf{A}^{(2)}(\mathbf{r}) = T\mathbf{A}^{(1)}(\mathbf{r})T^+ = T\mathbf{A}(\mathbf{r})T^+. \qquad (13)$$

By using expression (11.a) for $\mathbf{A}(\mathbf{r})$ and Equations (7) and (9), we thus obtain:

$$\mathbf{A}^{(2)}(\mathbf{r}) = \mathbf{A}(\mathbf{r}) + \mathbf{A}_P(\mathbf{r}) \qquad (14)$$

where $\mathbf{A}(\mathbf{r})$ is, again, the mathematical operator (11.a), and where $\mathbf{A}_P(\mathbf{r})$ is given by (8), $\mathbf{p}_\alpha$ being now considered as an operator. Because $\mathbf{A}_P(\mathbf{r})$ depends only on $\mathbf{p}_\alpha$, and because $\mathbf{p}_\alpha$ commutes with T, $\mathbf{A}_P(\mathbf{r})$ is invariant with respect to T and thus represents, in the two pictures, the same physical variable, that is, the transverse field tied to the particle:

$$\mathbf{A}_P(\mathbf{r}) = \mathbf{A}_P^{(1)}(\mathbf{r}) = \mathbf{A}_P^{(2)}(\mathbf{r}). \qquad (15)$$

The combination of (14) and (15) thus yields

$$\mathbf{A}(\mathbf{r}) = \mathbf{A}^{(2)}(\mathbf{r}) - \mathbf{A}_P^{(2)}(\mathbf{r}) = \delta\mathbf{A}^{(2)}(\mathbf{r}). \tag{16}$$

The operator $\mathbf{A}(\mathbf{r})$ thus describes, in the new picture, the difference between the total transverse field $\mathbf{A}^{(2)}(\mathbf{r})$ and the transverse field tied to the particle $\mathbf{A}_P^{(2)}(\mathbf{r})$, difference that we noted $\delta\mathbf{A}^{(2)}(\mathbf{r})$. The result obtained thus conforms to our initial motivation for introducing the transformation T.

Similarly, the transverse electric field is represented in the new picture by

$$\mathbf{E}_\perp^{(2)}(\mathbf{r}) = T\mathbf{E}_\perp^{(1)}(\mathbf{r})T^+ = T\mathbf{E}_\perp(\mathbf{r})T^+ \tag{17}$$

according to (12.b). By using expression (11.b) for $\mathbf{E}_\perp(\mathbf{r})$ and equation (9), we can verify that

$$T\mathbf{E}_\perp(\mathbf{r})T^+ = \mathbf{E}_\perp(\mathbf{r}) \tag{18}$$

so that the operator $\mathbf{E}_\perp$ still represents the transverse electric field in the new picture:

$$\mathbf{E}_\perp(\mathbf{r}) = \mathbf{E}_\perp^{(2)}(\mathbf{r}). \tag{19}$$

Such a result is not surprising. The deviation of the electric field with respect to the Coulomb field is of order v_α^2/c^2 and our calculation of $\beta_\varepsilon(\mathbf{k})$ takes into account only the terms linear in v_α.

b) TRANSFORMATION OF THE PARTICLE DYNAMICAL VARIABLES

We return to expression (10) for T. Using (7), it is possible to write (10) in the form:

$$T = \exp\left[\frac{i}{\hbar}\frac{q_\alpha}{m_\alpha}\mathbf{p}_\alpha \cdot \mathbf{Z}(\mathbf{0})\right] \tag{20}$$

where $\mathbf{Z}(\mathbf{r})$ is a quantum field (*) defined by

$$\mathbf{Z}(\mathbf{r}) = \int_< d^3k \sum_\varepsilon \sqrt{\frac{\hbar}{2\varepsilon_0\omega(2\pi)^3}}\left[\boldsymbol{\varepsilon}\frac{a_\varepsilon(\mathbf{k})}{i\omega}e^{i\mathbf{k}\cdot\mathbf{r}} - \boldsymbol{\varepsilon}\frac{a_\varepsilon^+(\mathbf{k})}{i\omega}e^{-i\mathbf{k}\cdot\mathbf{r}}\right]. \tag{21}$$

(*) A similar transformation can be introduced for a particle interacting with a classical external field. See W. C. Henneberger, *Phys. Rev. Lett.*, **21**, 838 (1968); or *Photons and Atoms—Introduction to Quantum Electrodynamics*, §B-4 of Chapter IV.

We now study how T transforms the position of the particle, which, in the Coulomb picture, is represented by the operator $\mathbf{r}_\alpha$ (this operator is defined as performing the *multiplication* by $\mathbf{r}_\alpha$):

$$\mathbf{r}_\alpha^{(1)} = \mathbf{r}_\alpha. \tag{22}$$

Because the particle operator $\mathbf{p}_\alpha$ commutes with the field operator $\mathbf{Z}(0)$, (20) is also a translation operator for $\mathbf{r}_\alpha$, and we get

$$\mathbf{r}_\alpha^{(2)} = T\mathbf{r}_\alpha^{(1)}T^+ = \mathbf{r}_\alpha + \frac{q_\alpha}{m_\alpha}\mathbf{Z}(0). \tag{23}$$

The last term of (23), which we write $\boldsymbol{\xi}_\alpha$, has the same form in the two pictures because $\mathbf{Z}(0)$ commutes with itself:

$$\boldsymbol{\xi}_\alpha = \frac{q_\alpha}{m_\alpha}\mathbf{Z}(0) = \boldsymbol{\xi}_\alpha^{(1)} = \boldsymbol{\xi}_\alpha^{(2)}. \tag{24}$$

The physical meaning of $\boldsymbol{\xi}_\alpha$ is obtained by calculating, to the first order in q_α, its second derivative with respect to time by using the Heisenberg equation. To lowest order in q_α, we can replace the total Hamiltonian by H_R to evaluate [in the representation (1)] the rate of change of $\mathbf{Z}(0)$. Taking the time derivative of $a_\varepsilon(\mathbf{k})$ is equivalent to multiplying $a_\varepsilon(\mathbf{k})$ by $-i\omega$. We deduce from this that $\ddot{\mathbf{Z}} = \mathbf{E}_\perp$ and consequently that

$$m_\alpha\ddot{\boldsymbol{\xi}}_\alpha = q_\alpha\mathbf{E}_\perp(0) + \text{higher-order terms in } q_\alpha. \tag{25}$$

The variable $\boldsymbol{\xi}_\alpha$ thus represents the motion the particle would have if it were subjected only to the transverse field (to first order in q_α): this is the linear response of the free particle to $\mathbf{E}_\perp(0)$. Substituting (24) into (23) thus yields

$$\mathbf{r}_\alpha = \mathbf{r}_\alpha^{(2)} - \boldsymbol{\xi}_\alpha^{(2)} \tag{26}$$

and shows that, in the picture (2), the operator $\mathbf{r}_\alpha$ represents the "average position" of the particle about which this particle accomplishes the vibrational motion $\boldsymbol{\xi}_\alpha^{(2)}$. We note this average position $\mathbf{r}_\alpha'^{(2)}$:

$$\mathbf{r}_\alpha = \mathbf{r}_\alpha'^{(2)}. \tag{27}$$

We now consider the other dynamical variable $\mathbf{p}_\alpha$. Because it commutes with T, the operator $\mathbf{p}_\alpha$ represents the same physical quantity in both pictures:

$$\mathbf{p} = \mathbf{p}_\alpha^{(1)} = \mathbf{p}_\alpha^{(2)}. \tag{28}$$

Later on we will see, once we have derived the expression for the Hamiltonian $H^{(2)}$, that $\mathbf{p}_\alpha^{(2)}$ is simply related to the average velocity $[\mathbf{r}_\alpha^{\prime(2)}, H^{(2)}]/i\hbar$.

It is interesting at this stage to compare Equations (16) and (26) giving the physical meaning for the operators $\mathbf{A}(\mathbf{r})$ and $\mathbf{r}_\alpha$ in the new picture. It then appears that the transformation T is equivalent to taking account in advance of the *linear response* of each of the systems (the particle or the transverse field) to the other one. Equation (16) actually means that the transverse field, produced at the first order in q_α by the particle, is subtracted from the total field, whereas (26) means that the displacement of the particle, produced to first order in q_α by the field, is subtracted from the instantaneous position. Because in the new representation, each operator of one system describes the deviation of the corresponding physical variable with respect to its linear response to the other system, we can easily guess that the new Hamiltonian must not contain any more interaction terms that are linear in q_α between the particle and the transverse field. This is what we will now verify by calculating $H^{(2)}$.

c) EXPRESSION FOR THE NEW HAMILTONIAN

The Hamiltonian $H^{(1)}$ in the Coulomb gauge is the operator H written in (1). By using (13), (14), (28), (23), and (9) we obtain

$$H^{(2)} = TH^{(1)}T^+ = THT^+ =$$

$$= \frac{1}{2m_\alpha}[\mathbf{p}_\alpha - q_\alpha\mathbf{A}(0) - q_\alpha\mathbf{A}_P(0)]^2 +$$

$$+ \varepsilon_{\mathrm{Coul}}^\alpha + q_\alpha U_e\left(\mathbf{r}_\alpha + \frac{q_\alpha}{m_\alpha}\mathbf{Z}(0)\right) +$$

$$+ \int_< d^3k \sum_\varepsilon \hbar\omega\left[(a_\varepsilon^+(\mathbf{k}) + \beta_\varepsilon^+(\mathbf{k}))(a_\varepsilon(\mathbf{k}) + \beta_\varepsilon(\mathbf{k})) + \frac{1}{2}\right]. \tag{29}$$

The first term of (29) contains $\mathbf{A}_P(0)$, which is written, taking into account (8),

$$\mathbf{A}_P(0) = \int_< \frac{d^3k}{(2\pi)^3} \sum_\varepsilon \frac{q_\alpha}{m_\alpha\varepsilon_0 c^2} \frac{\boldsymbol{\varepsilon}(\boldsymbol{\varepsilon} \cdot \mathbf{p}_\alpha)}{k^2}. \tag{30}$$

The summation over the transverse polarizations ε [see formula (54) in Complement A_1], and then the integral over the direction and the modulus of $\mathbf{k}$ present no difficulty and give

$$q_\alpha A_P(0) = \mathbf{p}_\alpha \frac{\delta m_\alpha}{m_\alpha} \tag{31}$$

where the mass δm_α, which is second order in q_α, equals

$$\delta m_\alpha = \frac{q_\alpha^2 k_M}{3\varepsilon_0 \pi^2 c^2}. \tag{32}$$

The first term of (29) is thus, taking into account (31), equal to

$$\frac{1}{2m_\alpha}\left[\mathbf{p}_\alpha\left(1 - \frac{\delta m_\alpha}{m_\alpha}\right) - q_\alpha A(0)\right]^2 = \frac{\mathbf{p}_\alpha^2}{2m_\alpha}\left(1 - \frac{\delta m_\alpha}{m_\alpha}\right)^2 -$$

$$- \frac{q_\alpha}{m_\alpha}\left(1 - \frac{\delta m_\alpha}{m_\alpha}\right)\mathbf{p}_\alpha \cdot A(0) + \frac{q_\alpha^2}{2m_\alpha}(A(0))^2. \tag{33.a}$$

Because, according to (32), δm_α is second order in q_α, the terms in $q_\alpha \delta m_\alpha$ and δm_α^2 of (33.a) are, respectively, third and fourth order in q_α. It is thus possible, up to the second order, to ignore them and to write the first term of (29) in the form (*):

$$\frac{\mathbf{p}_\alpha^2}{2m_\alpha} - \frac{\mathbf{p}_\alpha^2}{2m_\alpha}\frac{2\delta m_\alpha}{m_\alpha} - \frac{q_\alpha}{m_\alpha}\mathbf{p}_\alpha \cdot A(0) + \frac{q_\alpha^2}{2m_\alpha}(A(0))^2. \tag{33.b}$$

Let us now consider the last integral of (29). It gives rise to three types of terms. The first, in a^+a, has the form of a free field Hamiltonian. The second, linear in a and a^+, as well as in β and β^+, is simply, taking into account (7) and (11.a), $q_\alpha \mathbf{p}_\alpha \cdot A(0)/m_\alpha$. Finally, the third, which involves $\beta^+\beta$, is quadratic in $\mathbf{p}_\alpha$ and equals $(\mathbf{p}_\alpha^2/2m_\alpha)(\delta m_\alpha/m_\alpha)$. The third term of

(*) It is of course possible to retain all the terms of (33.a). If the mass m_α, which appears in the expression (20) for T, is slightly modified, an expression for $H^{(2)}$ can be obtained, which, while remaining compact, is exact to all orders in q_α [see Remark (i) at the end of the subsection g hereafter].

(29) is thus equal to

$$\int_< d^3k \sum_\varepsilon \hbar\omega \left[a_\varepsilon^+(\mathbf{k})a_\varepsilon(\mathbf{k}) + \frac{1}{2} \right] + \frac{q_\alpha}{m_\alpha} \mathbf{p}_\alpha \cdot \mathbf{A}(0) + \frac{\mathbf{p}_\alpha^2}{2m_\alpha} \frac{\delta m_\alpha}{m_\alpha}. \quad (33.c)$$

When (33.b) and (33.c) are added, the terms in $\mathbf{p}_\alpha \cdot \mathbf{A}(0)$ cancel each other exactly; the second term of (33.b) and the third term of (33.c) combine to give $-(\mathbf{p}_\alpha^2/2m_\alpha)(\delta m_\alpha/m_\alpha)$. Finally, to the second order in q_α for the kinetic energy term, $H^{(2)}$ is put in the form

$$H^{(2)} = H_\alpha' + H_R \quad (34.a)$$

where

$$H_\alpha' = \frac{\mathbf{p}_\alpha^2}{2m_\alpha} \left(1 - \frac{\delta m_\alpha}{m_\alpha} \right) +$$

$$+ q_\alpha U_e \left(\mathbf{r}_\alpha + \frac{q_\alpha}{m_\alpha} \mathbf{Z}(0) \right) + \varepsilon_{\text{Coul}}^\alpha + \frac{q_\alpha^2}{2m_\alpha} (\mathbf{A}(0))^2 \quad (34.b)$$

and where

$$H_R = \int_< d^3k \, \hbar\omega \sum_\varepsilon \left(a_\varepsilon^+(\mathbf{k})a_\varepsilon(\mathbf{k}) + \frac{1}{2} \right). \quad (34.c)$$

3. Physical Discussion

a) MASS CORRECTION

To interpret the first term of (34.b), we start by analyzing the physical meaning of $\mathbf{p}_\alpha$. In the Heisenberg picture, the equation of motion for $\mathbf{r}_\alpha$ is written

$$\dot{\mathbf{r}}_\alpha = \frac{1}{i\hbar} [\mathbf{r}_\alpha, H^{(2)}] = \frac{\mathbf{p}_\alpha}{m_\alpha} \left(1 - \frac{\delta m_\alpha}{m_\alpha} \right) \simeq \frac{\mathbf{p}_\alpha}{m_\alpha^*} \quad (35)$$

where

$$m_\alpha^* = m_\alpha + \delta m_\alpha \quad (36)$$

[the last equality of (35) is valid to the second order in q_α]. Taking into account (27) and (28), Equation (35) then gives

$$\mathbf{p}_\alpha^{(2)} = m_\alpha^* \dot{\mathbf{r}}_\alpha'^{(2)} \tag{37}$$

and expresses that $\mathbf{p}_\alpha^{(2)}$ is the momentum associated with the average position of the particle, the mass m_α being replaced by m_α^*. The first term of (34.b), which can be written $\mathbf{p}_\alpha^2/2m_\alpha^*$, thus represents the kinetic energy associated with the average motion of the particle.

The mass correction δm_α, given in (32), represents the contribution of long-wavelength modes to the electromagnetic mass of the particle. By returning to the terms of the Coulomb Hamiltonian that are at the origin of the correction $-(\mathbf{p}_\alpha^2/2m_\alpha)(\delta m_\alpha/m_\alpha)$ appearing in the first term of (34.b), it can be seen that this correction, which is globally negative, is the sum of two contributions: the energy of the transverse field tied to the particle [term in $\beta^+\beta$ of the last integral in (29)], which is positive, and the interaction energy of the particle with its own transverse field, which is negative and twice as large [term in $-2\mathbf{p}_\alpha \cdot \mathbf{A}_P(0)$ of the first term of (29)]. The kinetic energy correction in δm_α thus represents the field energy and the interaction energy of the transverse field tied to the particle. This energy correction was studied directly in the picture (1) in Chapter II [formula (E.4)].

Remark

Let us return to the calculation of $H^{(2)}$ presented in subsection 1-f above, and suppose that m_α is changed to $m_\alpha^* = m_\alpha + \delta m_\alpha$ in expression (20) for T

$$T' = \exp\left[\frac{i}{\hbar}\frac{q_\alpha}{m_\alpha^*}\mathbf{p}_\alpha \cdot \mathbf{Z}(0)\right]. \tag{20'}$$

To obtain the transform by T' of the kinetic-energy term of $H^{(1)}$, it is then sufficient to replace $\delta m_\alpha/m_\alpha$ by $\delta m_\alpha/m_\alpha^*$, inside the brackets of the first line of (33.a), which leads to replacing (33.a) by

$$\frac{1}{2m_\alpha}\left[\mathbf{p}_\alpha\left(1 - \frac{\delta m_\alpha}{m_\alpha^*}\right) - q_\alpha\mathbf{A}(0)\right]^2 =$$

$$= \frac{1}{2m_\alpha}\left[\mathbf{p}_\alpha\frac{m_\alpha}{m_\alpha^*} - q_\alpha\mathbf{A}(0)\right]^2$$

$$= \frac{\mathbf{p}_\alpha^2}{2m_\alpha^*}\frac{m_\alpha}{m_\alpha^*} - \frac{q_\alpha}{m_\alpha^*}\mathbf{p}_\alpha \cdot \mathbf{A}(0) + \frac{q_\alpha^2}{2m_\alpha}(\mathbf{A}(0))^2. \tag{33'.a}$$

Similarly, to obtain the transform of the term H_R in $H^{(1)}$, it is sufficient to replace m_α by m_α^* in (33.c), which gives

$$H_R + \frac{q_\alpha}{m_\alpha^*} \mathbf{p}_\alpha \cdot \mathbf{A}(0) + \frac{\mathbf{p}_\alpha^2}{2m_\alpha^*} \frac{\delta m_\alpha}{m_\alpha^*}. \tag{33'.c}$$

When we add (33'.a) and (33'.c), the terms in $\mathbf{p}_\alpha \cdot \mathbf{A}(0)$ still exactly cancel each other and the terms in $\mathbf{p}_\alpha^2/2m_\alpha^*$ combine to exactly give $\mathbf{p}_\alpha^2/2m_\alpha^*$. Finally, we obtain for $H^{(2)} = T'H^{(1)}T'^+$ a compact expression which generalizes (34) and remains valid at all orders in q_α

$$H^{(2)} = T'H^{(1)}T'^+$$

$$= \frac{\mathbf{p}_\alpha^2}{2m_\alpha^*} + q_\alpha U_e\left(\mathbf{r}_\alpha + \frac{q_\alpha}{m_\alpha^*} \mathbf{Z}(0)\right) +$$

$$+ \varepsilon_{\text{Coul}}^\alpha + H_R + \frac{q_\alpha^2}{2m_\alpha}(\mathbf{A}(0))^2. \tag{34'}$$

Expression (34') for $H^{(2)}$ is particularly useful in studying certain problems such as that of the "infrared catastrophe" (see Exercise 9). In fact, if the matrix elements of the second term of (34') are exactly calculable, we can study physical processes, such as the scattering of a charged particle by a potential, to first order with respect to the external potential, but to all orders with respect to the coupling with the transverse field.

b) New Interaction Hamiltonian between the Particle and the Transverse Field

It is only in the second term of (34.b), which can be written $q_\alpha U_e(\mathbf{r}_\alpha + \boldsymbol{\xi}_\alpha)$ taking into account (24), that both the variables of the particles $(\mathbf{r}_\alpha)$ and those of the transverse field $(\mathbf{Z}(0))$ appear. As we guessed at the end of subsection 2b, there is no longer any interaction term linear in q_α between the particle and the transverse field. The lowest-order term is obtained by expanding $q_\alpha U_e(\mathbf{r}_\alpha + \boldsymbol{\xi}_\alpha)$ in powers of $\boldsymbol{\xi}_\alpha$ and is written, according to (24),

$$q_\alpha \boldsymbol{\xi}_\alpha \cdot \nabla U_e(\mathbf{r}_\alpha) = q_\alpha \mathbf{Z}(0) \cdot \frac{q_\alpha}{m_\alpha} \nabla U_e(\mathbf{r}_\alpha). \tag{38}$$

It is indeed second order in q_α, but it is a one-photon term, because $\mathbf{Z}(0)$ is, according to (21), a linear superposition of the a_j and a_j^+. The term $q_\alpha \nabla U_e(\mathbf{r}_\alpha)/m_\alpha$, that appears in (38) is, except for its sign, the external electric force acting on the particle divided by m_α and evaluated at $\mathbf{r}_\alpha$, which is the average position of this particle. The interaction Hamiltonian (38) thus describes the coupling of the transverse field $\mathbf{Z}(0)$ with the acceleration of the average motion of the particle in the external field. Such a result recalls the fact that, classically, the field radiated by a charged particle is proportional to its acceleration. Because the new transverse field is essentially the radiation field, it is not surprising to find that it is the acceleration of the particle that is coupled to this field.

The following term in the expansion in powers of $\mathbf{Z}$ of the second term of (34.b) is written

$$\frac{1}{2} q_\alpha \frac{q_\alpha^2}{m_\alpha^2} \sum_{i,j=x,y,z} Z_i(0) Z_j(0) \left(\nabla_i \nabla_j U_e(\mathbf{r}_\alpha) \right). \tag{39}$$

This is a two-photon term because it is quadratic in $\mathbf{Z}$. It is third order in q_α.

The new picture is particularly convenient for studying the effect of vacuum fluctuations on the atomic energy levels. We now study the average value in the photon vacuum of the interaction Hamiltonians (38) and (39). The average value of the interaction Hamiltonian (38) is zero. On the other hand, the average of $Z_i Z_j$ is nonzero for $i = j$ and represents the average quadratic deviation of the vibrational motion of the particle under the action of vacuum fluctuations. The correction to the potential $U_e(\mathbf{r}_\alpha)$ resulting from the average in the vacuum of (39) thus represents the effect of the "averaging" of the potential U_e by the particle during its vibrational motion. For example, for the hydrogen atom, U_e is a Coulomb potential, so that

$$\sum_i \nabla_i^2 U_e(\mathbf{r}) = -\frac{q}{\varepsilon_0} \delta(\mathbf{r}) \tag{40}$$

where q is the proton charge. The effect of the averaging is then manifested by an effective potential:

$$\frac{q^4}{6\varepsilon_0 m^2} \langle \mathbf{Z}^2 \rangle_{\mathrm{vac}} \delta(\mathbf{r}) \tag{41}$$

which slightly increases the energies of the s states ($l = 0$). This is the physical origin (*) of the experimentally observed splitting between the state $2s_{1/2}$ and the state $2p_{1/2}$ (Lamb shift), which is discussed in

(*) T. A. Welton, *Phys. Rev.*, **74**, 1157 (1948).

subsection E-1-b of this chapter. In fact, we must also take into account
the effect of the coupling (38) with the transverse field at the second order:
a correction to (41) results, which essentially concerns the contribution of
the lowest-frequency modes (see Exercise 7). Note that we only evaluated
the contribution of the modes for which the long-wavelength approximation is justified.

Remarks

(i) Finally, we examine the last two terms of (34.b). $\varepsilon_{Coul}^{\alpha}$ is the Coulomb
self-energy of the particle. Using the reasoning that allows us to go from (24) to
(25), $\mathbf{A(0)}$ is, at the first order in q_α and except for the sign, the velocity of $\mathbf{Z(0)}$.
The last term of (34.b) may thus be written

$$\frac{1}{2}m_\alpha\left[\frac{q_\alpha\dot{\mathbf{Z}}(0)}{m_\alpha}\right]^2 = \frac{1}{2}m_\alpha\dot{\boldsymbol{\xi}}_\alpha^2. \tag{42}$$

It thus represents the kinetic energy of the vibrational motion of the particle in
the transverse field. Finally, the last term of (34.a) is the energy of the
transverse field in the new picture, that is, the energy of the transverse field not
tied to the particle.

(ii) The disappearance of the terms linear in q_α in the new interaction
Hamiltonian is an interesting property that can be used to characterize unitary
transformations more general than the Pauli–Fierz transformation. Suppose,
for example, that a unitary transformation that eliminates the interaction terms
linear in q_α and q_β is carried out on the Coulomb-gauge Hamiltonian of a
system of two particles coupled to the transverse field. The effects in $q_\alpha q_\beta$,
which were described in the Coulomb picture by an exchange of transverse
photons between the two particles (current-current interaction discussed in
Section F of Chapter II), must then explicitly appear in the new particle
Hamiltonian. Actually, in the new picture, any exchange of photons would lead
to higher-order effects in $q_\alpha q_\beta$. The same procedure may be used to derive
Hamiltonians describing magnetic interactions between two particles with spins
(see, for example, Exercise 11).

c) ADVANTAGES OF THE NEW REPRESENTATION

Finally, if we limit ourselves to lowest order in q_α and to the effect of
the long-wavelength modes, the Pauli–Fierz picture has several advantages.

First of all, the radiative corrections are manifested by simple modifications of the potential- and kinetic-energy terms of the particle, and the
physical meaning of these modifications is very clear.

Then, the interaction between the particle and the new transverse field, which is associated with "real photons", is much more physical in the sense that it directly involves the acceleration of the particle. The study of the emission of low-frequency radiation during a collision is simplified considerably (see Exercises 8 and 9).

Finally, the asymptotic states of a collision process have a much simpler form in the new picture. Actually, outside the zone of action of the potential U_e, the Hamiltonian (34) (or 34') is reduced to the sum of a free-particle Hamiltonian and a radiation-field Hamiltonian. The asymptotic states of scattering thus appear as the simple products of a particle plane wave multiplied by a radiation state. On the other hand, in the Coulomb picture, the interaction between the particle and the transverse field persists in the initial state and in the final state, because this interaction involves the velocity of the particle which, in contrast to what occurs with the acceleration, does not cancel out before the collision and after the collision. The incident or scattered particle is therefore, in the Coulomb picture, "dressed" by a cloud of "virtual" photons, and the construction of correct asymptotic states is thus more complex (see Complement B_{III}, §2).

All the foregoing advantages must not, however, make us forget the limitations of the Pauli–Fierz picture. First, there is the long-wavelength approximation. We could deal without it, but with more complex calculations by using unitary transformations that eliminate the interaction terms linear in q_α [see Remark (ii) of subsection 3-b], or which consist of replacing $\mathbf{Z}(\mathbf{0})$ by $\mathbf{Z}(\mathbf{r}_\alpha)$ in (20) (*). There are also more fundamental limitations that we will now address.

d) INADEQUACY OF THE CONCEPT OF A FIELD TIED TO A PARTICLE

The idea of separating the field tied to the electron from the free field represents a first version of the idea of "renormalization" envisaged by Kramers in 1938 (**). He attempted to construct a classical, then a quantum theory, in which the field tied to the electron and the rest of the electromagnetic field are constantly distinguishable. The contribution of the electromagnetic field to the mass of the particle would thus result only from the bound field. This is essentially what occurs at the lowest order: the correction to the kinetic energy appearing in the first term of (34.b) results from the transverse field tied to the particle and may be interpreted as resulting from a mass correction.

(*) See, for example, E. A. Power and T. Thirunamachandran, *Am. J. Phys.*, **46**, 370 (1978).
(**) H. A. Kramers, *Nuovo Cimento*, **15**, 108 (1938).

Actually, we can not go very far in this direction. The idea itself of a field tied to a particle must be questioned. It has meaning only for a classical particle in uniform motion. As soon as the particle accelerates or decelerates, the fields A_P and E_P, functions of r_α and p_α, change instantaneously in the whole space. Therefore such fields should not be considered, in relativistic theory, as physical fields. In the quantum framework, another difficulty appears: we cannot impose a uniform linear motion on the particle. We can only solve the coupled Heisenberg equations for the field and particle observables, for a particle initially in uniform translational motion in the presence of the vacuum. To first order in q_α, we recover the classical idea, because the calculated field results from the unperturbed motion of the particle. However, at higher orders, the motion of the particle is perturbed by the initial field, which, quantum mechanically, never vanishes, and its motion is no longer uniform. Finding a unitary transformation that exactly achieves the separation of the bound field is, in fact, impossible.

To become really efficient, the idea of "renormalization" should be considered in a completely different theoretical perspective. On the one hand, the particles must be described relativistically, and the calculations carried out in such a way as to involve only covariant variables. On the other hand, the idea itself of decomposing the electromagnetic field into two parts is abandoned. It is in calculating the probability amplitude of a process that we distinguish among the different contributions. Starting with a particle in a vacuum and turning on the coupling to the field, we identify, on a probability amplitude, the ensemble of interaction processes that appear, as being those describing the "dressing" of the particle by virtual photons, and which result from the interaction of the particle with the field produced by itself and with the vacuum fluctuations. The variation of the energy of the system produced by this coupling can be interpreted as a change in the mass of the particle. In the presence of other particles or an external field, we then distinguish in the probability amplitude the processes that are identical to the preceding ones and which will be expressed as a renormalization of the mass from the processes corresponding to an interaction with another particle or with the external field.

It is thus finally on the probability amplitudes of the processes and not on the field itself that we have to separate the contributions to the electromagnetic mass, and the contributions to the real production of radiation. This is the point of view adopted in the chapter.

CHAPTER III

Nonperturbative Calculation of Transition Amplitudes

In Chapter I, we introduced the concept of transition amplitude on which is based the quantum description of interaction processes between charged particles and photons. However, we limited ourselves to a *perturbative* calculation of these amplitudes, based on a splitting of the total Hamiltonian H of the system into an unperturbed part H_0 and a coupling V. For example, the calculation of transition amplitudes, presented in Complement A_I, was limited to the second order in V.

We also used a perturbative approach to analyze, in Chapter II, several physical effects that may be observed in atoms interacting with photons, each process being represented diagrammatically at the lowest order in V where it manifests itself. Nevertheless, throughout Chapter II, it appeared that a deeper understanding of certain physical phenomena requires going beyond perturbation theory and taking into account some effects of V to all orders. This is the case, for example, for the radiative decay of an excited atomic state, due to the spontaneous emission of a photon, or for resonant scattering. To analyze these phenomena, in Chapter II we used qualitative arguments based on a simplified model, introduced in Complement C_I, consisting of a discrete state coupled to a perfectly flat continuum. The aim of the present chapter is to treat these problems in a more general fashion and to present methods for calculating the evolution operator that are more powerful than those presented in Complement A_I, in that they lead to *nonperturbative* expressions for the transition amplitudes.

165

We begin in Section A by showing why, for the study of this type of problem, it is useful to introduce the *resolvent* $G(z) = 1/(z - H)$ of the Hamiltonian H, with z being a complex variable. The relation that exists between the resolvent $G(z)$ and the "unperturbed" resolvent $G_0(z) = 1/(z - H_0)$, relative to the unperturbed Hamiltonian H_0, is in fact an *algebraic* equation, much simpler to manipulate than the *integral* equation connecting the evolution operators $U(\tau)$ and $U_0(\tau)$ associated, respectively, with H and H_0. Once the matrix elements of $G(z)$ have been calculated, the corresponding matrix elements of $U(\tau)$ are deduced from them by a contour integral. The analytic properties of $G(z)$, which are involved in this contour integral, are analyzed in Complement A_{III}.

We then show in Section B how the algebraic equation connecting $G(z)$ and $G_0(z)$ may be used to obtain exact and compact expressions formally *resumming* the perturbation series and paying particular attention to some unperturbed states that play an important role in the physical processes under study. Two different approaches are used to introduce these *resummation* methods. One, which is extremely simple, relies on a *diagrammatic representation* of the perturbation series; the other, which is more rigorous, uses *projection operators*. The essential feature of the exact expressions thus obtained is that they are particularly suitable for introducing approximations. In this way, we can take into account the intermediate transitions involving certain unperturbed states to all orders in V, while still perturbatively treating the effect of intermediate transitions to the other eigenstates of H_0.

These *partial resummation* methods for the perturbation series are then applied in Section C to some quantum electrodynamics problems. We present a nonperturbative treatment of different phenomena associated with spontaneous emission (exponential decay and radiative shift of an excited atomic level, spectral distribution of emitted radiation, radiative cascades, effect of a coupling between two levels with different lifetimes). We also analyze the problem of multiphoton resonances and introduce the idea of generalized Rabi frequency.

Finally, we mention two other applications that are studied in the complements. Complement B_{III} shows how it is possible to use the resolvent $G(z)$ to study resonant scattering and to derive an exact expression for the scattering amplitude of a photon by an atom. Last, Complement C_{III} uses a simple model to study the corrections to the exponential decay for a discrete state coupled to a continuum of finite width. In particular, we show, by using simple graphic constructions, how increasing the coupling between the discrete state and the continuum change continuously the time evolution from a regime of exponential decay to a regime of Rabi oscillations.

A—EVOLUTION OPERATOR AND RESOLVENT

1. Integral Equation Satisfied by the Evolution Operator

As we previously indicated in subsection 2-a of Complement A_I, the evolution operator $U(t, t')$ associated with the Hamiltonian $H = H_0 + V$ is the solution to the differential equation

$$i\hbar \frac{d}{dt} U(t, t') = (H_0 + V)U(t, t') \qquad (A.1.a)$$

obeying the initial condition

$$U(t', t') = 1. \qquad (A.1.b)$$

It is easy to verify that the solution of Equations (A.1) may be written

$$U(t, t') = U_0(t, t') + \frac{1}{i\hbar} \int_{t'}^{t} dt_1 \, U_0(t, t_1) V U(t_1, t') \qquad (A.2)$$

where

$$U_0(t, t') = \exp\left[-iH_0(t - t')/\hbar\right] \qquad (A.3)$$

is the evolution operator associated with the unperturbed Hamiltonian H_0. The integral equation (A.2) satisfied by the evolution operator may be iterated to yield the perturbative expansion of $U(t, t')$ in powers of V. We obtain in this way Equation (13) of Complement A_I and the expansion for the transition amplitudes given by Equation (C.8) of Chapter I.

The constraint imposed that t_1 vary between t' and t prevents the integral in (A.2) from being a convolution product. To eliminate this constraint and to obtain a true convolution product that transforms into a simple product by Fourier transformation, we now introduce new operators, very closely related to $U(t, t')$.

2. Green's Functions—Propagators

Consider the operators K_+ and K_{0+} defined by

$$K_+(t, t') = U(t, t')\theta(t - t') \qquad (A.4)$$

$$K_{0+}(t, t') = U_0(t, t')\theta(t - t') \qquad (A.5)$$

where $\theta(x)$ is the Heaviside function, equal to 1 for $x > 0$, and equal to 0 for $x < 0$. By multiplying the two terms of (A.2) by $\theta(t - t')$ and by replacing $\int_{t'}^{t} dt_1$ by $\int_{-\infty}^{+\infty} dt_1 \, \theta(t - t_1)\theta(t_1 - t')$, it is easily verified that

$$K_+(t, t') = K_{0+}(t, t') + \frac{1}{i\hbar} \int_{-\infty}^{+\infty} dt_1 \, K_{0+}(t, t_1) V K_+(t_1, t'). \quad (A.6)$$

Equation (A.6), quite similar to (A.2), now involves a true convolution product.

Before taking the Fourier transform of (A.6), note that, taking into account (A.1.a) and the fact that $d\theta(x)/dx = \delta(x)$, $K_+(t, t')$ obeys the equation

$$\left(i\hbar \frac{d}{dt} - H \right) K_+(t, t') = \hbar\delta(t - t'). \quad (A.7)$$

The form of Equation (A.7) with a right-hand side proportional to $\delta(t - t')$, explains why the operator $K_+(t, t')$ is sometimes called the *"Green's function"*. In fact, it is a *retarded* Green's function because $K_+(t, t')$ is non-null only for $t > t'$. It is also possible to introduce advanced Green's functions

$$K_-(t, t') = -U(t, t')\theta(t' - t) \quad (A.8)$$

$$K_{0-}(t, t') = -U_0(t, t')\theta(t' - t). \quad (A.9)$$

K_- obeys the same evolution equation as K_+

$$\left(i\hbar \frac{d}{dt} - H \right) K_-(t, t') = i\hbar\delta(t - t') \quad (A.10)$$

but satisfies different boundary conditions.

We now introduce the Fourier transform of $K_+(t, t')$, which actually depends only on $\tau = t - t'$. Let us write

$$K_+(\tau) = -\frac{1}{2\pi i} \int_{-\infty}^{+\infty} dE \, e^{-iE\tau/\hbar} G_+(E) \quad (A.11)$$

the coefficient $-1/2\pi i$ being introduced to have the most simple possible form for G_+. By inverting (A.11), we obtain

$$G_+(E) = \frac{1}{i\hbar} \int_{-\infty}^{+\infty} d\tau \, e^{iE\tau/\hbar} K_+(\tau). \qquad (A.12)$$

Since $K_+(\tau) = e^{-iH\tau/\hbar} \theta(\tau)$, Equation (A.12) becomes

$$G_+(E) = \frac{1}{i\hbar} \int_0^\infty d\tau \, e^{i(E-H)\tau/\hbar}$$

$$= \lim_{\eta \to 0_+} \frac{1}{i\hbar} \int_0^\infty d\tau \, e^{i(E-H+i\eta)\tau/\hbar}$$

$$= \lim_{\eta \to 0_+} \frac{1}{E - H + i\eta}, \qquad (A.13)$$

where η is a positive real number that tends to zero. Similar calculations can be made for K_- and its Fourier transform $G_-(E)$ defined by

$$K_-(\tau) = -\frac{1}{2\pi i} \int_{-\infty}^{+\infty} dE \, e^{-iE\tau/\hbar} G_-(E). \qquad (A.14)$$

They lead to

$$G_-(E) = \lim_{\eta \to 0_+} \frac{1}{E - H - i\eta}. \qquad (A.15)$$

The operators $G_+(E)$ and $G_-(E)$, introduced in (A.13) and (A.15) are called advanced and retarded *propagators*. By Fourier transform, the integral appearing in (A.6) becomes a simple product and the integral equation (A.6) becomes an algebraic equation

$$G_+(E) = G_{0+}(E) + G_{0+}(E)VG_+(E) \qquad (A.16)$$

where $G_{0+}(E)$ is the retarded propagator associated with H_0. Note also that Equation (A.16) may be derived directly from the identity

$$\frac{1}{A} = \frac{1}{B} + \frac{1}{B}(B - A)\frac{1}{A} \qquad (A.17)$$

written with $A = E - H + i\eta$, $B = E - H_0 + i\eta$, and thus $B - A = H - H_0 = V$.

3. Resolvent of the Hamiltonian

The very simple form of $G_+(E)$ suggests the introduction of the operator $G(z)$, a function of the complex variable z, defined by

$$G(z) = \frac{1}{z - H}. \tag{A.18}$$

$G(z)$ is called the *resolvent* of the Hamiltonian H. The propagators $G_+(E)$ and $G_-(E)$ are simply the limits of $G(z)$ when z tends to the point E on the real axis, with a positive or negative value of its imaginary part

$$G_\pm(E) = \lim_{\eta \to 0_+} G(E \pm i\eta). \tag{A.19}$$

The evolution operator $U(\tau)$ is expressed by a simple contour integral of $G(z)$. Because $\theta(x) + \theta(-x) = 1$, Equations (A.4) and (A.8) lead to

$$U(\tau) = K_+(\tau) - K_-(\tau) \tag{A.20}$$

which yields, when (A.11) and (A.14) are taken into account

$$U(\tau) = \frac{1}{2\pi i} \int_{-\infty}^{+\infty} dE \, e^{-iE\tau/\hbar} [G_-(E) - G_+(E)]. \tag{A.21}$$

It is then sufficient to use (A.19) to obtain

$$U(\tau) = \frac{1}{2\pi i} \int_{C_+ + C_-} dz \, e^{-iz\tau/\hbar} G(z) \tag{A.22}$$

where $C_+ + C_-$ is the contour represented in Figure 1. C_+ and C_- are lines situated, respectively, immediately above and below the real axis and followed from right to left for C_+ and from left to right for C_-. For $\tau > 0$ ($\tau < 0$), the contribution of C_- (C_+) is zero.

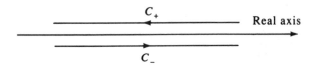

Figure 1. Integration contour appearing in the integral for $U(\tau)$ as a function of $G(z)$.

Because $U(\tau)$ can be deduced from $G(z)$ by a contour integral, it is clear that the analytic properties of the resolvent $G(z)$ play an important role in the determination of the properties of $U(\tau)$. We can show (see Complement A_{III}) that the matrix elements of $G(z)$ are analytic functions of z in the whole complex plane except for the real axis. The singularities, which are all on the real axis, consist of *poles* located at the discrete eigenvalues of the Hamiltonian H, and of *cuts* extending over the intervals corresponding to the continuous spectrum of H. A cut is characterized by the fact that the matrix elements of $G(z)$ do not tend to the same value, when z tends from below or from above toward a point on the real axis, located on the cut. The difference between these two values represents the difference between the two sides of the cut. Also note that it is possible, starting, for example, from the upper half-plane, to extend analytically the determination of $G(z)$ in the upper half-plane toward the lower half-plane. We thus explore the "second Riemann sheet". The continued function is no longer necessarily analytic outside the real axis and may have complex poles, which describe *unstable* states of the system (states having a complex energy and characterized by exponential damping). Later on we will have the occasion to return to this point.

Finally, let us give the perturbative expansion of $G(z)$ in powers of V. The identity (A.17), applied to $A = z - H$, $B = z - H_0$, yields

$$G(z) = G_0(z) + G_0(z)VG(z) \tag{A.23}$$

where $G_0(z)$ is the resolvent of H_0. This algebraic equation may be iterated to yield

$$G(z) = G_0(z) + G_0(z)VG_0(z) + G_0(z)VG_0(z)VG_0(z) + \cdots . \tag{A.24}$$

Let us take the matrix element of (A.24) between the eigenstates $\langle \varphi_k |$ and $| \varphi_l \rangle$ of H_0, with unperturbed energies E_k and E_l. By using the closure relation over the eigenstates $| \varphi_i \rangle$ of H_0, and by setting $G_{kl}(z) = \langle \varphi_k | G(z) | \varphi_l \rangle$, $V_{ij} = \langle \varphi_i | V | \varphi_j \rangle$, we obtain

$$G_{kl}(z) = \frac{1}{z - E_k} \delta_{kl} + \frac{1}{z - E_k} V_{kl} \frac{1}{z - E_l} +$$

$$+ \sum_i \frac{1}{z - E_k} V_{ki} \frac{1}{z - E_i} V_{il} \frac{1}{z - E_l} + \cdots . \tag{A.25}$$

The structure of (A.25) is quite simple and consists of products of matrix elements of V and of unperturbed energy denominators.

B—FORMAL RESUMMATION OF THE PERTURBATION SERIES

1. Diagrammatic Method Explained on a Simple Model

The simple form of the expansion (A.24) or (A.25) allows us to regroup the terms where an energy denominator $1/(z - E_b)$ involving a particular unperturbed state $|\varphi_b\rangle$ appears n times, then to formally sum the perturbation series. We get in this way compact and exact expressions for which interesting approximations can then be carried out.

Assume, for example, that $|\varphi_b\rangle$ is a discrete state of H_0, with energy E_b, well isolated from all the other discrete states (on the other hand, E_b may fall inside the continuous spectrum of H_0). The matrix element $G_{0b}(z) = \langle \varphi_b|G_0(z)|\varphi_b\rangle = 1/(z - E_b)$ of the unperturbed resolvent has a pole at $z = E_b$ and consequently varies very rapidly near $z = E_b$ along the contour $C_+ + C_-$ in Figure 1. Thus we expect similar variations for the matrix element of the perturbed resolvent $G_b(z) = \langle \varphi_b|G(z)|\varphi_b\rangle$, which tends to $G_{0b}(z)$ when V tends to zero. It is clear that near $z = E_b$, the energy denominators in $1/(z - E_b)$ of the expansion (A.25) of $G_b(z)$ play an essential role. Let us attempt to regroup the terms of this expansion which all contain the same number of denominators in $1/(z - E_b)$.

With $k = l = b$, the zero-order term in V of (A.25) is reduced to $1/(z - E_b)$, and thus contains this energy denominator once. Because $k = l = b$, the following terms of the expansion contain at least twice $1/(z - E_b)$. If we require them to contain $1/(z - E_b)$ only twice, all the other energy denominators must involve states $|\varphi_i\rangle$ other than $|\varphi_b\rangle$. The ensemble of these terms may thus be diagrammatically represented in the form

$$\text{(B.1)}$$

where a solid line represents a factor $1/(z - E_b)$, a dotted line a factor

$1/(z - E_i)$ with $i \neq b$, and a circle represents the matrix element of V between the two states associated with the lines located on both sides of the circle (the summation over the intermediate states $i, j \cdots$ is implicit). The sum (B.1) may be rewritten formally

$$\text{(B.2)}$$

where the square represents the sum

$$\text{(B.3)}$$

The explicit expression $R_b(z)$ for such a sum is equal to

$$R_b(z) = V_{bb} + \sum_{i \neq b} V_{bi} \frac{1}{z - E_i} V_{ib} +$$

$$+ \sum_{i \neq b} \sum_{j \neq b} V_{bi} \frac{1}{z - E_i} V_{ij} \frac{1}{z - E_j} V_{jb} + \cdots . \qquad \text{(B.4)}$$

The foregoing reasoning is easily generalized to terms that contain n denominators in $1/(z - E_b)$. For example, the ensemble of terms $n = 3$ is represented by

$$\text{(B.5)}$$

and contributes to $G_b(z)$ by

$$\frac{1}{(z - E_b)^3} [R_b(z)]^2. \tag{B.6}$$

Then it is sufficient to sum the contributions corresponding to different values of n to obtain

$$G_b(z) = \sum_{n=1}^{\infty} \frac{[R_b(z)]^{n-1}}{(z - E_b)^n} = \frac{1}{z - E_b} \sum_{n=0}^{\infty} \left[\frac{R_b(z)}{z - E_b}\right]^n \tag{B.7}$$

that is, also

$$G_b(z) = \frac{1}{z - E_b - R_b(z)}. \tag{B.8}$$

It is important to note that expression (B.8), which formally sums the perturbation series by giving a privileged role to the state $|\varphi_b\rangle$, is an *exact* expression. We have given here a diagrammatic demonstration of (B.8). We now present another demonstration of (B.8), which uses projection operators and which has the advantage of being applicable to situations in which it is interesting to single out not only one, but several discrete eigenstates of H_0.

2. Algebraic Method Using Projection Operators

a) PROJECTOR ONTO A SUBSPACE $\mathscr{E}_0$ OF THE SPACE OF STATES

Let $\mathscr{E}_0$ be a subspace of the space of states subtended by an ensemble of eigenvectors of $H_0 \{|\varphi_a\rangle, |\varphi_b\rangle \cdots |\varphi_l\rangle\}$, which play an important role in the physical process under study and that we wish to single out in the expansion of $G(z)$ in powers of V. If the states $|\varphi_a\rangle, |\varphi_b\rangle \cdots |\varphi_l\rangle$ are orthonormal, the projector onto the subspace $\mathscr{E}_0$ is written

$$P = |\varphi_a\rangle\langle\varphi_a| + |\varphi_b\rangle\langle\varphi_b| + \cdots + |\varphi_l\rangle\langle\varphi_l| \tag{B.9}$$

and satisfies the relations

$$P = P^+ \tag{B.10.a}$$

$$P^2 = P \tag{B.10.b}$$

characteristic of an orthogonal projector.

We use $\mathscr{S}_0$ to designate the supplementary subspace of $\mathscr{E}_0$ and use Q to designate the projector onto $\mathscr{S}_0$ which is written

$$Q = 1 - P \tag{B.11}$$

and satisfies relations similar to (B.10)

$$Q = Q^+ \tag{B.12.a}$$
$$Q^2 = Q. \tag{B.12.b}$$

Because the subspaces $\mathscr{E}_0$ and $\mathscr{S}_0$ are orthogonal, we also have

$$PQ = QP = 0. \tag{B.13}$$

Moreover, the fact that the states $|\varphi_a\rangle, |\varphi_b\rangle, \ldots, |\varphi_l\rangle$, are eigenstates of H_0 leads to

$$[P, H_0] = [Q, H_0] = 0 \tag{B.14}$$

and consequently, according to (B.13)

$$PH_0Q = QH_0P = 0. \tag{B.15}$$

We calculate now the projection $PG(z)P$ of $G(z)$ in the subspace $\mathscr{E}_0$. We will also calculate the other projections of $G(z)$: inside $\mathscr{S}_0$, $Q(G(z)Q$, and between $\mathscr{E}_0$ and $\mathscr{S}_0$, $PG(z)Q$ and $QG(z)P$.

b) CALCULATION OF THE PROJECTION OF THE RESOLVENT
   IN THE SUBSPACE $\mathscr{E}_0$

Let us begin with the equation defining $G(z)$

$$(z - H_0 - V)G(z) = 1 \tag{B.16}$$

that we multiply on the right by P, and on the left by P or Q. If we insert $P + Q = 1$ [see (B.11)] between $(z - H_0 - V)$ and $G(z)$ and use (B.15), we obtain the following two equations:

$$P(z - H)P[PG(z)P] - PVQ[QG(z)P] = P \tag{B.17}$$
$$- QVP[PG(z)P] + Q(z - H)Q[QG(z)P] = 0 \tag{B.18}$$

which are two operator equations connecting the two operators $PG(z)P$ and $QG(z)P$. To eliminate $QG(z)P$ from these two equations, it is sufficient to rewrite (B.18) in the form

$$QG(z)P = \frac{Q}{z - QH_0Q - QVQ} VPG(z)P \qquad (B.19)$$

and to then substitute (B.19) into (B.17) to obtain

$$P\left[z - H_0 - V - V\frac{Q}{z - QH_0Q - QVQ} V \right] PG(z)P = P. \quad (B.20)$$

We now introduce the operator $R(z)$ defined by

$$R(z) = V + V\frac{Q}{z - QH_0Q - QVQ} V. \qquad (B.21)$$

The operator $R(z)$ is called the *level-shift operator* for reasons that will be explained later. Its perturbative expansion in powers of V is written

$$R(z) = V + V\frac{Q}{z - H_0}V + V\frac{Q}{z - H_0}V\frac{Q}{z - H_0}V + \cdots . \quad (B.22)$$

We used (B.12.b) and (B.14) to write $Q[z - QH_0Q]^{-1} = Q[z - H_0]^{-1}$. The presence of projection operators Q in the expansion (B.22) results in the fact that the energy denominators appearing in (B.22) are all relative to eigenstates of H_0 other than those subtending $\mathscr{E}_0$. By substituting (B.21) into (B.20), we obtain for the projection of $G(z)$ in $\mathscr{E}_0$ the expression

$$PG(z)P = \frac{P}{z - PH_0P - PR(z)P} \qquad (B.23)$$

which generalizes Equation (B.8).

Let us first show that Equation (B.23) is actually equivalent to (B.8) when the subspace $\mathscr{E}_0$ contains only one eigenstate $|\varphi_b\rangle$ of H_0, in which case

$$P = |\varphi_b\rangle\langle\varphi_b| \qquad (B.24)$$

$$Q = \mathbb{1} - |\varphi_b\rangle\langle\varphi_b| = \sum_{i \neq b} |\varphi_i\rangle\langle\varphi_i|. \qquad (B.25)$$

The projections in $\mathscr{E}_0$ of $G(z)$, $R(z)$, H_0 that appear in (B.23), thus reduce to 1×1 matrices, that is, numbers equal to, respectively, $G_b(z) = \langle \varphi_b | G(z) | \varphi_b \rangle$, $R_b(z) = \langle \varphi_b | R(z) | \varphi_b \rangle$, and $\langle \varphi_b | H_0 | \varphi_b \rangle = E_b$. Equation (B.23) thus has the same form as (B.8). Moreover, by taking the matrix elements of the two sides of (B.22) between $\langle \varphi_b |$ and $| \varphi_b \rangle$ and by using the form (B.25) of Q, we verify that the quantity $R_b(z)$ appearing in (B.23) does indeed coincide with the expansion (B.4) derived above from a diagrammatic approach.

In Section C, we will also consider situations in which the subspace $\mathscr{E}_0$ contains two discrete eigenstates $| \varphi_b \rangle$ and $| \varphi_c \rangle$ of H_0, with eigenvalues E_b and E_c, in which case

$$P = | \varphi_b \rangle \langle \varphi_b | + | \varphi_c \rangle \langle \varphi_c | \qquad (B.26)$$

$$Q = \mathbb{1} - \sum_{i=b,c} | \varphi_i \rangle \langle \varphi_i |. \qquad (B.27)$$

The matrix representing $PG(z)P$ in the basis $\{| \varphi_b \rangle, | \varphi_c \rangle\}$ of $\mathscr{E}_0$ is thus a 2×2 matrix

$$\begin{pmatrix} G_{bb}(z) & G_{bc}(z) \\ G_{cb}(z) & G_{cc}(z) \end{pmatrix}. \qquad (B.28)$$

Equation (B.23) expresses that the 2×2 matrix written in (B.28) is the inverse of the matrix

$$\begin{pmatrix} z - E_b - R_{bb}(z) & -R_{bc}(z) \\ -R_{cb}(z) & z - E_c - R_{cc}(z) \end{pmatrix}. \qquad (B.29)$$

c) CALCULATION OF OTHER PROJECTIONS OF $G(z)$

Substituting (B.23) into (B.19) yields

$$QG(z)P = \frac{Q}{z - QH_0Q - QVQ} V \frac{P}{z - PH_0P - PR(z)P}. \qquad (B.30)$$

To obtain $PG(z)Q$, it is sufficient to take the adjoints of the two sides of (B.30) and to use the Hermiticity of P, Q, H_0, and V. By changing z^* to z in the equation thus obtained, we find

$$PG(z)Q = \frac{P}{z - PH_0P - PR(z)P} V \frac{Q}{z - QH_0Q - QVQ}. \qquad (B.31)$$

Finally, we calculate $QG(z)Q$. Multiply (B.16) on the right and on the left by Q. We get

$$Q(z - H)Q[QG(z)Q] - QVP[PG(z)Q] = Q. \qquad (B.32)$$

From (B.32) we deduce, taking into account (B.31)

$$QG(z)Q = \frac{Q}{z - QH_0Q - QVQ} +$$

$$+ \frac{Q}{z - QH_0Q - QVQ} V \frac{P}{z - PH_0P - PR(z)P} V \frac{Q}{z - QH_0Q - QVQ}.$$

$$(B.33)$$

d) INTERPRETATION OF THE LEVEL-SHIFT OPERATOR

The operator $PR(z)P$ appears in the denominator of expressions (B.23), (B.30), (B.31), and (B.33) giving the various projections of $G(z)$. Let us momentarily "forget" that $PR(z)P$ depends on z. The examination of (B.23) suggests that PRP be considered as a "Hamiltonian" in the subspace $\mathscr{E}_0$, being added to PH_0P and allowing us to determine the shifts of the perturbed levels relative to unperturbed levels. This is the reason for which the name level-shift operator is given to $R(z)$. This interpretation of PRP also allows us to simply understand the structure of the other expressions obtained previously. For example, reading expression (B.30) (from right to left) suggests that the system, starting from $\mathscr{E}_0$, evolves under the effect of the "Hamiltonian" $PH_0P + PRP$, then passes through the effect of the coupling V in the subspace $\mathscr{S}_0$, where it evolves under the effect of the Hamiltonian $QH_0Q + QVQ$.

Let us now describe in more detail the dependence on z of $PR(z)P$. In fact, in the contour integral (A.22) which allows us to calculate the matrix elements of $U(\tau)$ from those of $G(z)$, it is the values of $G(z)$ in the immediate neighborhood of the real axis that matter. It is thus important to calculate the matrix elements of $PR(z)P$ for $z = E \pm i\eta$, where E is a point on the real axis and η is an infinitely small and positive quantity. Let $|\varphi_b\rangle$ and $|\varphi_c\rangle$ be two eigenstates of H_0 belonging to $\mathscr{E}_0$. According to (B.21), we have

$$R_{bc}(E \pm i\eta) = V_{bc} + \langle\varphi_b|VQ\frac{1}{E - QHQ \pm i\eta}QV|\varphi_c\rangle. \qquad (B.34)$$

Because

$$\frac{1}{x \pm i\eta} = \frac{x}{x^2 + \eta^2} \mp \frac{i\eta}{x^2 + \eta^2} = \mathscr{P}\left(\frac{1}{x}\right) \mp i\pi\delta(x) \qquad (B.35)$$

we get

$$R_{bc}(E \pm i\eta) = V_{bc} + \hbar\left(\Delta_{bc}(E) \mp i\frac{\Gamma_{bc}(E)}{2}\right) \qquad \text{(B.36)}$$

where

$$\Delta_{bc}(E) = \frac{1}{\hbar}\mathscr{P}\langle\varphi_b|V\frac{Q}{E - QHQ}V|\varphi_c\rangle \qquad \text{(B.37)}$$

$$\Gamma_{bc}(E) = \frac{2\pi}{\hbar}\langle\varphi_b|VQ\delta(E - QHQ)QV|\varphi_c\rangle. \qquad \text{(B.38)}$$

Because operators V, H, P, and Q are Hermitian, it is clear from (B.37) and (B.38) that the matrix elements $\Delta_{bc}(E)$ and $\Gamma_{bc}(E)$ are Hermitian

$$\Delta_{bc}(E) = \Delta_{cb}^*(E)$$

$$\Gamma_{bc}(E) = \Gamma_{cb}^*(E). \qquad \text{(B.39)}$$

On the other hand, the presence of the factor i in (B.36) shows that, if $\Gamma_{bc}(E)$ is nonzero, the operator $PR(E \pm i\eta)P$ is represented in $\mathscr{E}_0$ by a non-Hermitian matrix. Even if it is legitimate to neglect the variation with E of $\Delta_{bc}(E)$ and $\Gamma_{bc}(E)$ about the unperturbed energy of the levels of $\mathscr{E}_0$, as we will see later on in several cases, $PR(E \pm i\eta)P$ is not a true Hamiltonian. The anti-Hermitian part of this operator (coming from the terms $\mp i\hbar\Gamma_{bc}$) describes dissipative phenomena, that is, expresses the fact that the system may leave $\mathscr{E}_0$ in an irreversible way.

Remark

When $\Gamma_{bc}(E)$ is nonzero, it also appears in (B.36) that $R_{bc}(E \pm i\eta)$ does not tend to the same value when z tends to the point E on the real axis from below or from above, which is characteristic of a cut. In fact, it can be shown by using calculations similar to those in Complement A$_{\text{III}}$ that $PR(z)P$ possesses cuts over all the intervals of the real axis corresponding to the continuous spectrum of QHQ. On the other hand, $PR(z)P$ diverges when z tends to a discrete eigenvalue E_α of QHQ, in which case $\Delta_{bc}(E)$ and $\Gamma_{bc}(E)$ are not defined in $E = E_\alpha$.

3. Introduction of Some Approximations

All the expressions derived in the preceding subsection are exact. We will now show how they are particularly well suited to approximations of a nonperturbative type by taking into account certain contributions of V to all orders and by singling out one or several unperturbed states.

a) PERTURBATIVE CALCULATION OF THE LEVEL-SHIFT OPERATOR. PARTIAL RESUMMATION OF THE PERTURBATION SERIES

Let us first consider the situation described in subsection B-1 in which H_0 has a discrete eigenstate $|\varphi_b\rangle$ well-isolated from all the other discrete eigenstates of H_0. As we have already indicated, it is near $z = E_b$ on the contour $C_+ + C_-$ of Figure 1 that $G_b(z)$ takes on the most important values. Let us now examine expression (B.4) near $z = E_b$. All the energy denominators involved in the expansion of $R_b(z)$ are large because the other discrete energies of H_0 are assumed to be far from E_b. Moreover, even if E_b falls within the continuous spectrum of H_0, the sums over the intermediate states associated with this continuous spectrum involve delta functions and principal parts which do not lead to any divergence (see the calculations in subsection C-1-a below). If V is sufficiently small compared with H_0, series (B.4) is rapidly convergent and it appears completely legitimate to approximate $R_b(z)$ by retaining only a finite number of terms in expansion (B.4) or (B.3). For example, we can retain only the first two terms of these expansions, which is equivalent to replacing (B.4) by

$$\hat{R}_b(z) = V_{bb} + \sum_{i \neq b} V_{bi} \frac{1}{z - E_i} V_{ib} \tag{B.40}$$

or (B.3) by

$$\Delta \;=\; \bigcirc \;+\; \begin{array}{c} \bigcirc \\ \vdots \\ \bigcirc \end{array} \tag{B.41}$$

The main interest of expression (B.8) is that a perturbative approximation for $R_b(z)$ does not correspond to a perturbative approximation for $G_b(z)$. In fact, the replacement in (B.8) of $R_b(z)$ by the approximate expression (B.40) is equivalent to approximating $G_b(z)$ by

$$G_b(z) \simeq \frac{1}{z - E_b - \hat{R}_b(z)} = \sum_{n=0}^{\infty} \frac{\left[\hat{R}_b(z)\right]^n}{(z - E_b)^{n+1}}. \tag{B.42}$$

It is clear from (B.42) that the approximate expression for $G_b(z)$, obtained according to approximation (B.40), contains arbitrarily high powers of V and thus does not correspond to a truncation of expansion (A.24). In other words, approximation (B.40) is equivalent to making a *partial resummation* of the perturbation series: the infinite sum

$$(B.43)$$

is replaced by another infinite sum

$$(B.44)$$

where all the denominators in $1/(z - E_b)$ are retained but only a single denominator in $1/(z - E_i)$ with $i \neq b$ is retained between two denominators in $1/(z - E_b)$. The foregoing considerations can easily be generalized to the case where H_0 has several discrete eigenstates forming an ensemble well isolated from the other eigenstates of H_0. If P is the projector onto the corresponding subspace $\mathscr{E}_0$, the approximation equivalent to (B.40) consists of keeping only the first two terms of expansion (B.22) and thus replacing, in (B.23), $PR(z)P$ by

$$P\hat{R}(z)P = PVP + PV\frac{Q}{z - H_0}VP. \qquad (B.45)$$

b) Approximation Consisting of Neglecting the Energy
     Dependence of the Level-Shift Operator

As in the preceding subsection, let us first assume that the subspace $\mathscr{E}_0$ contains only a single state $|\varphi_b\rangle$, and let us now consider expression (B.8) written for $z = E \pm i\eta$

$$G_b(E \pm i\eta) = \frac{1}{E \pm i\eta - E_b - R_b(E \pm i\eta)}. \tag{B.46}$$

Quite frequently it happens that the variations with E of $R_b(E \pm i\eta)$ are much slower than those of $G_b(E \pm i\eta)$, in particular near $E = E_b$, where the "trace" of the pole appearing in the unperturbed resolvent $G_{0b}(z) = 1/(z - E_b)$ is manifested by a resonant behavior of $G_b(E \pm i\eta)$. An approximation frequently carried out thus consists of neglecting the variation with E of $R_b(E \pm i\eta)$ and to substitute into (B.46) $R_b(E \pm i\eta)$ by

$$R_b(E \pm i\eta) \simeq R_b(E_b \pm i\eta). \tag{B.47}$$

We will return later on (§C-1-c) to the conditions of validity for such an approximation. Because $R_b(E_b \pm i\eta)$ is a complex quantity independent of E, it appears in (B.46) that such an approximation is equivalent to correcting the unperturbed energy E_b by a complex quantity, the real part of the correction representing an energy shift and the imaginary part representing a broadening.

In the general case, $\mathscr{E}_0$ has a dimension greater than 1. Assume that the states $|\varphi_a\rangle, |\varphi_b\rangle \cdots |\varphi_l\rangle$ which subtend $\mathscr{E}_0$ have the energies $E_a, E_b, \ldots, E_l$ grouped around E_0. The approximation equivalent to (B.47) thus consists of setting

$$PR(E \pm i\eta)P \simeq PR(E_0 \pm i\eta)P. \tag{B.48}$$

Remarks

(i) It is certainly possible to carry out the two previous approximations at the same time, that is, to replace $R_b(z)$ by a perturbative expression $\hat{R}_b(z)$, and then to neglect the variations with E of $\hat{R}_b(E \pm i\eta)$.

(ii) By replacing $R_b(E \pm i\eta)$ by $R_b(E_b \pm i\eta)$ in (B.46), we obtain an approximate expression for $G_b(E_b \pm i\eta)$ which no longer has poles near the other discrete eigenvalues of H_0. However, the exact expression for $G_b(z)$, $G_b(z) = \langle \varphi_b|(z - H)^{-1}|\varphi_b\rangle$ shows that $G_b(z)$ has poles at the discrete eigenvalues E_α of H, the residues corresponding to these poles being equal to $|\langle \varphi_b|\psi_\alpha\rangle|^2$ where $|\psi_\alpha\rangle$ is the eigenstate of H having an eigenvalue E_α. The residue $|\langle \varphi_b|\psi_\alpha\rangle|^2$ is very small if V is very small and if $|\psi_\alpha\rangle$ tends, when V tends to zero, to a discrete eigenstate of H_0 orthogonal to $|\varphi_b\rangle$. These are the poles, having very small residues, that the approximation (B.47) causes to disappear from $G_b(z)$.

C—STUDY OF A FEW EXAMPLES

We now apply the foregoing nonperturbative methods to a few quantum electrodynamic problems. An important example of the situation in which the subspace $\mathscr{E}_0$ contains only a single discrete state is that of the spontaneous emission of radiation from an excited discrete atomic state. In subsection 1 we will show how the study of the projection $PG(z)P$ of $G(z)$ onto $\mathscr{E}_0$ allows us to calculate the evolution of the excited state and to find the exponential decay of this excited state as well as its radiative shift. Continuing with this simple example, we will then show in subsection 2 how the calculation of $QG(z)P$ allows us to calculate the spectral distribution of spontaneously emitted photons and to analyze other physical phenomena such as radiative cascades. The last two subsections are devoted to the study of physical situations in which two discrete states $|\varphi_b\rangle$ and $|\varphi_c\rangle$ play an important role (subspace $\mathscr{E}_0$ of dimension 2). We begin by studying the effect of a coupling between two excited atomic levels having different lifetimes, an important example of such a situation being the quenching of the metastability of the $2s$ state of hydrogen by Stark coupling with the $2p$ state. Finally, we study light shifts and the generalized Rabi frequency associated with multiphoton resonance between two discrete atomic levels (*).

1. Evolution of an Excited Atomic State

a) Nonperturbative Calculation of the Probability Amplitude That the Atom Remains Excited

Let us consider an atom in a nondegenerate excited discrete state $|b\rangle$. For the sake of simplicity, we ignore its external degrees of freedom (global translation of the atom). Let

$$|\varphi_b\rangle = |b;0\rangle \tag{C.1}$$

be the state of the total atom + radiation system representing the excited state b in the vacuum of photons. The state $|\varphi_b\rangle$ is an eigenstate of the unperturbed Hamiltonian H_0, having an energy E_b which is assumed to be

---

(*) The physical examples considered in subsections 3 and 4 correspond to the general situation in which the two levels $|\varphi_b\rangle$ and $|\varphi_c\rangle$ have different widths. The case in which the two levels $|\varphi_b\rangle$ and $|\varphi_c\rangle$ are either stable or have the same width is encountered in experiments involving magnetic resonance in ground states or excited states. This situation results in simpler calculations (see Complement A_{VI}).

well isolated from all other discrete eigenvalues of H_0. We will calculate

$$G_b(z) = \langle b; 0|G(z)|b; 0\rangle \tag{C.2}$$

in order to subsequently deduce, from the contour integral (A.22), the probability amplitude

$$U_b(\tau) = \langle b; 0|U(\tau)|b; 0\rangle \tag{C.3}$$

that the atom, initially in state b, will remain in this state for a time τ.

According to formula (B.8), $G_b(z)$ is equal to

$$G_b(z) = \frac{1}{z - E_b - R_b(z)} \tag{C.4}$$

where $R_b(z)$ is given by an expression similar to (B.4). In quantum electrodynamics (see §C-1 in Chapter I), $V = H_{I1} + H_{I2}$ where H_{I1} and H_{I2} are first and second order with respect to the charges q_α (we ignore here the effects due to the spins, described by H_{I1}^s). As in subsection B-3-a, we replace $R_b(z)$ by its second-order expression in q_α which is written

$$\hat{R}_b(z) = \langle b; 0|H_{I2}|b; 0\rangle +$$

$$+ \sum_a \sum_{\mathbf{k}\varepsilon} \frac{\langle b; 0|H_{I1}|a; \mathbf{k}\varepsilon\rangle\langle a; \mathbf{k}\varepsilon|H_{I1}|b; 0\rangle}{z - E_a - \hbar\omega} \tag{C.5}$$

where $|a; \mathbf{k}\varepsilon\rangle$ represents the atom in the internal state a in the presence of a photon $\mathbf{k}\varepsilon$. In the long-wavelength approximation, the first term of (C.5) is just a constant independent of b, given by expression (E.5) in Chapter II and that we assume to be reintegrated into H_0. Expression (C.5) is thus reduced to the last term, for which the diagrammatic representation [analogous to (B.41)] is

$$\tag{C.6}$$

in which the dashed line of (B.41) has been replaced here by the combination of a straight line representing the atom in state a and a wavy line representing the photon **kε**. The replacement of $R_b(z)$ by $\hat{R}_b(z)$ in (C.4) thus results in summing in $G_b(z)$ the ensemble of terms

$$(C.7)$$

Thus we neglect all the processes in which several intermediate states (instead of a single one) appear between two states $|b; 0\rangle$ (represented by a simple atomic line b without a wavy photon line); for example

$$(C.8)$$

Let us now calculate the value of $\hat{R}_b(z)$ near the real axis. The last term of Equation (C.5) gives

$$\hat{R}_b(E \pm i\eta) = \sum_a \sum_{k\varepsilon} \frac{|\langle a; k\varepsilon|H_{I1}|b; 0\rangle|^2}{E \pm i\eta - E_a - \hbar\omega}. \qquad (C.9)$$

Taking into account (B.35), we get

$$\hat{R}_b(E \pm i\eta) = \hbar\hat{\Delta}_b(E) \mp i\hbar\frac{\hat{\Gamma}_b(E)}{2} \qquad (C.10)$$

$$\hat{\Delta}_b(E) = \frac{1}{\hbar}\mathscr{P}\sum_a\sum_{\mathbf{k}\varepsilon}\frac{|\langle a; \mathbf{k}\varepsilon|H_{I1}|b; 0\rangle|^2}{E - E_a - \hbar\omega} \qquad (C.11)$$

$$\hat{\Gamma}_b(E) = \frac{2\pi}{\hbar}\sum_a\sum_{\mathbf{k}\varepsilon}|\langle a; \mathbf{k}\varepsilon|H_{I1}|b; 0\rangle|^2\delta(E - E_a - \hbar\omega) \quad (C.12)$$

so that the approximate expression for $G_b(E \pm i\eta)$ is written

$$G_b(E \pm i\eta) \simeq \frac{1}{E \pm i\eta - E_b - \hbar\hat{\Delta}_b(E) \pm i(\hbar/2)\hat{\Gamma}_b(E)}. \qquad (C.13)$$

Before going further, we would like to emphasize some important properties of the functions $\hat{\Gamma}_b(E)$ and $\hat{\Delta}_b(E)$. First of all, from Equations (C.11) and (C.12), one gets the relation

$$\hat{\Delta}_b(E) = \frac{1}{2\pi}\mathscr{P}\int dE'\frac{\hat{\Gamma}_b(E')}{E - E'}. \qquad (C.14)$$

Relation (C.14) is a dispersion relation between two functions $\hat{\Delta}_b(E)$ and $\hat{\Gamma}_b(E)$ that are connected by a Hilbert transform. It also appears clearly in (C.12) that $\hat{\Gamma}_b(E)$ is positive and is different from zero only for $E > E_a$ (because $E_a + \hbar\omega$ varies between E_a and $+\infty$). In fact, $\hat{\Gamma}_b(E)\,dE$ characterizes the intensity of the coupling between the initial state $|b; 0\rangle$ and the ensemble of final states $|a; \mathbf{k}\varepsilon\rangle$ having an energy lying between E and $E + dE$. Because the density of states and the matrix elements of H_{I1} generally vary slowly with the energy E, it follows that $\hat{\Gamma}_b(E)$ and consequently $\hat{\Delta}_b(E)$ are functions that vary slowly with E. Finally, for $E = E_b$, $\hat{\Delta}_b(E)$ and $\hat{\Gamma}_b(E)$ take on values which we designate Δ_b and Γ_b, that are given by

$$\Delta_b = \hat{\Delta}_b(E_b) = \frac{1}{\hbar}\mathscr{P}\sum_a\sum_{\mathbf{k}\varepsilon}\frac{|\langle a; \mathbf{k}\varepsilon|H_{I1}|b; 0\rangle|^2}{E_b - E_a - \hbar\omega} \qquad (C.15)$$

$$\Gamma_b = \hat{\Gamma}_b(E_b) = \frac{2\pi}{\hbar}\sum_a\sum_{\mathbf{k}\varepsilon}|\langle a; \mathbf{k}\varepsilon|H_{I1}|b; 0\rangle|^2\delta(E_b - E_a - \hbar\omega) \quad (C.16)$$

and for which the physical interpretation is quite clear: $\hbar\Delta_b$ is the shift (to second order in q) of state b, due to the coupling with the radiation field. Γ_b is the rate of spontaneous emission of a photon from state b to a state a of lower energy.

b) RADIATIVE LIFETIME AND RADIATIVE LEVEL SHIFT

Let us now return to expression (C.13) for $G_b(E \pm i\hbar)$ and introduce the additional approximation consisting of replacing the functions $\hat{\Gamma}_b(E)$ and $\hat{\Delta}_b(E)$, which are functions that vary slowly with E, by the values they take near $E = E_b$. Thus we obtain

$$G_b(E \pm i\eta) \simeq \frac{1}{E - E_b - \hbar\Delta_b \pm i(\hbar/2)\Gamma_b}. \qquad \text{(C.17)}$$

The contour integral (A.22) of (C.17) then gives

$$U_b(\tau) = e^{-i(E_b + \hbar\Delta_b)\tau/\hbar} e^{-\Gamma_b|\tau|/2}. \qquad \text{(C.18)}$$

Expression (C.17) can also be written

$$G_b(E \pm i\eta) \simeq \frac{1}{E - \tilde{E}_b \pm i\hbar(\Gamma_b/2)} \qquad \text{(C.19.a)}$$

where

$$\tilde{E}_b = E_b + \hbar\Delta_b \qquad \text{(C.19.b)}$$

is the energy of state b corrected by the radiative level shift $\hbar\Delta_b$. The results presented in subsection A-1-c in Chapter II are thus demonstrated: The excited state b is shifted and the probability that the atom remains in the excited state decays exponentially with a lifetime $\tau_b = 1/\Gamma_b$. Actually, the various terms of the infinite partial sum (C.7) correspond by Fourier transform to the various terms of the expansion in powers of Δ_b and Γ_b of the exponential (C.18). If we had stopped the perturbation expansion at a certain order n, we would have obtained a polynomial in t, and not an exponential decay. Thus it appears quite clearly that a nonperturbative treatment is indispensable for describing the exponential decay of excited states.

c) Conditions of Validity for the Treatment of the Two
    Preceding Subsections

In the two preceding subsections, we have introduced two approximations in order to arrive at expression (C.18) for $U_b(\tau)$. First, in the exact
expression (C.4) for $G_b(z)$, we replaced $R_b(z)$ by the approximate expression $\hat{R}_b(z)$ given in (C.5). Then, we have ignored the variations with E of
$\hat{R}_b(E \pm i\eta)$ and replaced $\hat{R}_b(E \pm i\eta)$ by $\hat{R}_b(E_b \pm i\eta) = \hbar\Delta_b \mp i\hbar\Gamma_b/2$.

We begin by studying the second approximation. Let us examine the
result that we would have obtained for $U_b(\tau)$ if we had not ignored the
dependence on E of $\hat{R}_b(E \pm i\eta)$. Instead of (C.17), we now use (C.13).
According to (A.21), $U_b(\tau)$ is, to within a multiplicative factor, the Fourier
transform of $G_{b-}(E) - G_{b+}(E)$. More precisely, the relations (A.21),
(A.19), and (C.13) allow us to write

$$U_b(\tau) = \int_{-\infty}^{+\infty} dE \, \mathscr{U}_b(E) \, e^{-iE\tau/\hbar}, \qquad (C.20.a)$$

where

$$\mathscr{U}_b(E) = \frac{1}{2\pi i} [G_{b-}(E) - G_{b+}(E)]$$

$$= \frac{1}{2\pi i} \lim_{\eta \to 0_+} [G_b(E - i\eta) - G_b(E + i\eta)]$$

$$= \lim_{\eta \to 0_+} \frac{1}{\pi} \frac{(\hbar\hat{\Gamma}_b(E)/2) + \eta}{[E - E_b - \hbar\hat{\Delta}_b(E)]^2 + [(\hbar\hat{\Gamma}_b(E)/2) + \eta]^2} .$$

$$(C.20.b)$$

For the exponential decay of $U_b(\tau)$ described by (C.18) to be valid, the
variations of $\mathscr{U}_b(E)$ with E must be those of a Lorentzian. It is clear from
(C.20.b) that $\mathscr{U}_b(E)$ does not generally coincide with a Lorentzian (*). For
the Lorentzian approximation to be valid in (C.20.b), it is necessary that
$\hat{\Gamma}_b(E)$ and $\hat{\Delta}_b(E)$ change very slightly when E varies about $E_b + \hbar\Delta_b$ over
an interval of several $\hbar\Gamma_b$. The function $\mathscr{U}_b(E)$ is then quite close to the
Lorentzian leading to (C.18), over an interval of values of E equal to
several times the width of this Lorentzian. To express more precisely the

_____

(*) For this to be so, it is necessary that $\hat{\Gamma}_b(E)$ be independent of E, $\hat{\Delta}_b(E)$ being thus
null according to (C.14). Such a situation actually corresponds to the simple model of
continuum considered in Complement C_I.

foregoing condition, let us call $\hbar w_0$ the energy interval over which $\hat{\Gamma}_b(E)$ and $\hat{\Delta}_b(E)$ vary appreciably. For example, if b is the first excited state and a is the ground state, then $\hbar w_0$ is on the order of $E_b - E_a$. It is therefore necessary that

$$w_0 \gg \Gamma_b, \Delta_b. \tag{C.21}$$

When $|E - E_b|$ is not negligible compared to $\hbar w_0$, it is no longer possible to ignore in (C.20) the variations with E of $\hat{\Gamma}_b(E)$ and $\hat{\Delta}_b(E)$ which lead to a non-Lorentzian behavior of the wings of $\mathcal{U}_b(E)$. The corrections to exponential decay are discussed in Complement C_{III}, where a bell-shaped curve of width $\hbar w_0$ is used for $\hat{\Gamma}_b(E)$.

It remains for us to examine the other approximation used above which consists of replacing $R_b(z)$ by the perturbative expression written in (C.5). The infinitely small parameter that characterizes the perturbative expansion (B.4) of $R_b(z)$ is H_{I1}/H_0; that is on the order of magnitude of $[\Gamma/w_0]^{1/2}$ because $\Gamma \sim H_{I1}^2/w_0$. The inequality (C.21) results in the fact that, if the first terms (C.5) of the perturbative expansion of $R_b(z)$ are nonzero, the subsequent terms may be neglected.

Finally, note that the structure (C.20.b) of $\mathcal{U}_b(E)$ remains valid even if $R_b(E \pm i\eta)$ is not replaced by $\hat{R}_b(E \pm i\eta)$ (see Remark ii of subsection 3 in Complement A_{III}).

2. Spectral Distribution of Photons Spontaneously Emitted by an Excited Atom

a) RELEVANT MATRIX ELEMENT OF THE RESOLVENT OPERATOR

We will calculate here the matrix element of $G(z)$

$$\langle a; \mathbf{k}\boldsymbol{\varepsilon}|G(z)|b; 0\rangle \tag{C.22}$$

between the initial state $|b; 0\rangle$ (atom in the excited state b in the vacuum field) and the final state $|a; \mathbf{k}\boldsymbol{\varepsilon}\rangle$ (atom in the state a in the presence of a photon $\mathbf{k}\boldsymbol{\varepsilon}$). We assume that a is the ground state and b is the first excited state, with the transition $b \to a$ being allowed. Once (C.22) is calculated, the probability amplitude that after an elapsed time τ the atom will emit a photon $\mathbf{k}\boldsymbol{\varepsilon}$ and pass from b to a is obtained by using the contour integral (A.22). If τ is positive, only the contour C_+ contributes and

$$\langle a; \mathbf{k}\boldsymbol{\varepsilon}|U(\tau)|b; 0\rangle = \frac{1}{2\pi i} \int_{+\infty}^{-\infty} dE\, e^{-iE\tau/\hbar}\langle a; \mathbf{k}\boldsymbol{\varepsilon}|G(E + i\eta)|b; 0\rangle. \tag{C.23}$$

The first nonzero term of the expansion (A.24) allowing one to connect $|b; 0\rangle$ to $|a; \mathbf{k}\varepsilon\rangle$ is the term $G_0 H_{I1} G_0$, which yields

$$\langle a; \mathbf{k}\varepsilon | G(z) | b; 0 \rangle \simeq \frac{1}{z - E_a - \hbar\omega} \langle a; \mathbf{k}\varepsilon | H_{I1} | b; 0 \rangle \frac{1}{z - E_b} \quad (C.24)$$

which may be represented by the diagram

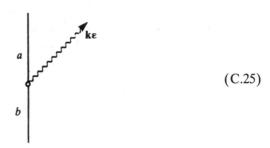

$$(C.25)$$

previously introduced in Chapter II (see Figure 1).

Let $\mathscr{E}_0$ be the subspace subtended by $|b; 0\rangle$. Because $|a; \mathbf{k}\varepsilon\rangle$ belongs to the supplementary subspace $\mathscr{S}_0$, the matrix element (C.22) is a matrix element of $QG(z)P$. The use of (B.30) thus allows us to obtain the following exact expression for (C.22)

$$\langle a; \mathbf{k}\varepsilon | G(z) | b; 0 \rangle = \langle a; \mathbf{k}\varepsilon | \frac{1}{z - QHQ} V | b; 0 \rangle \langle b; 0 | G(z) | b; 0 \rangle. \quad (C.26)$$

The last matrix element of (C.26) was discussed in the paragraph above and we obtained for this term an approximate expression (C.17), which is nonperturbative because it sums all the diagrams represented in (C.7). For $z = E + i\eta$, expression (C.26) becomes, taking into account (C.19.a)

$$\langle a; \mathbf{k}\varepsilon | G(E + i\eta) | b; 0 \rangle =$$

$$\langle a; \mathbf{k}\varepsilon | \frac{1}{E + i\eta - QHQ} V | b; 0 \rangle \frac{1}{E - \tilde{E}_b + i\hbar(\Gamma_b / 2)}. \quad (C.27)$$

We now consider the first matrix element of (C.27). To evaluate it, as in subsection B-1 above, we just consider the subensembles of diagrams resulting from the expansion of $G(E + i\eta)$ in powers of H_{I1}. We may thus attempt to sum all the diagrams in which the final state $|a; \mathbf{k}\varepsilon\rangle$ appears an arbitrary number of times, which results in an arbitrary number of denominators $1/(z - E_a - \hbar\omega)$ analogous to the one appearing in (C.24). On the other hand, between two such states $|a; \mathbf{k}\varepsilon\rangle$, we retain at most one "nonresonant" state that is different from $|b; 0\rangle$ or $|a; \mathbf{k}\varepsilon\rangle$. More precisely, we limit ourselves to the infinite sum of diagrams of the type (C.28)

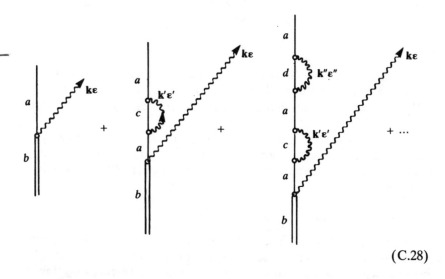

$$(C.28)$$

where the double line b represents the perturbed propagator of the initial state corresponding to the last term of (C.27) and symbolically representing the infinite sum (C.7). The diagrams represented in (C.28) suggests the following image: once the photon $\mathbf{k}\varepsilon$ has been emitted, the atom in state a virtually emits and reabsorbs photons continually. Such processes are responsible for the radiative level shift of the ground state a. Note that there is no instability for the state a because there is no lower-energy state to which the atom may radiatively decay from a. Because the photon $\mathbf{k}\varepsilon$ is a "spectator" in (C.28), we can easily convince ourselves by using the same approximations as those used to derive (C.17) that the summation of all the diagrams in (C.28) leads to replacing the first denominator of (C.24) (written for $z = E + i\eta$) by $1/(E + i\eta - \tilde{E}_a - \hbar\omega)$, where $\tilde{E}_a$ is the energy E_a corrected for the radiative level shift $\hbar\Delta_a$ of the state a. An approximation of the exact expression (C.26), which is better than (C.24), is thus finally

$$\langle a; \mathbf{k}\varepsilon | G(E + i\eta) | b; 0\rangle \simeq \frac{\langle a; \mathbf{k}\varepsilon | H_{I1} | b; 0\rangle}{\left(E + i\eta - \tilde{E}_a - \hbar\omega\right)\left[E - \tilde{E}_b + i\hbar(\Gamma_b/2)\right]}$$

$$(C.29)$$

and is represented by the diagram

$$(C.30)$$

where the two double lines represent the perturbed propagators of the initial state and of the final state. An approach more direct than that followed here, rigorously defining the final perturbed state, is presented in Complement B_{III}.

Before studying the spectral distribution of the emitted radiation, we will first generalize expression (C.29) to the case for which the radiative decay of the initial excited state causes several photons to appear in the final state.

b) Generalization to a Radiative Cascade

We will now consider the case in which the atom leaves a level c that is more excited than b and returns to the ground state a by emitting two photons $k_1 \varepsilon_1$ and $k_2 \varepsilon_2$ (see Figures 2α and 2β, representing the two possible temporal orders of emission of the two photons).

The presence of a level b located between c and a results in the fact that the amplitude of processes (2α) and (2β) may take on large values

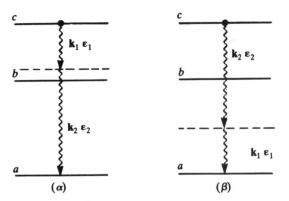

Figure 2. Passage from the excited state c to the ground state a by spontaneous emission of two photons $k_1\varepsilon_1$ and $k_2\varepsilon_2$. The two processes α and β differ in the order in which the two photons are emitted.

due to resonance in the intermediate state. This is what occurs when the frequency ω_1 of the photon $\mathbf{k}_1\boldsymbol{\varepsilon}_1$ for process (2α) or the frequency ω_2 of the photon $\mathbf{k}_2\boldsymbol{\varepsilon}_2$ for process (2β) are very close to the frequency ω_{cb} of the transition $c \rightarrow b$, the atom passing intermediately into state b. Such a situation corresponds to a radiative cascade $c \rightarrow b \rightarrow a$.

We assume in the following that the atomic frequencies ω_{cb} and ω_{ba} are noticeably different (the difference $|\omega_{cb} - \omega_{ba}|$ being very large compared to the widths Γ_c and Γ_b of c and b). In such a case, the two amplitudes associated with processes (2α) and (2β) can never both be large simultaneously (*). We will assume ω_1 to be near ω_{cb} and neglect process (2β). The photon $\mathbf{k}_1\boldsymbol{\varepsilon}_1$ is thus the one that is emitted first in the transition $c \rightarrow b$.

The matrix element

$$\langle a; \mathbf{k}_1\boldsymbol{\varepsilon}_1, \mathbf{k}_2\boldsymbol{\varepsilon}_2 | G(z) | c; 0 \rangle \tag{C.31}$$

allows us to calculate the transition amplitude

$$\langle a; \mathbf{k}_1\boldsymbol{\varepsilon}_1, \mathbf{k}_2\boldsymbol{\varepsilon}_2 | U(\tau) | c; 0 \rangle =$$

$$\lim_{\eta \to 0_+} \frac{1}{2\pi i} \int_{+\infty}^{-\infty} dE\, e^{-iE\tau/\hbar} \langle a; \mathbf{k}_1\boldsymbol{\varepsilon}_1, \mathbf{k}_2\boldsymbol{\varepsilon}_2 | G(E + i\eta) | c; 0 \rangle. \tag{C.32}$$

At the lowest order in H_{I1}, the third term of expansion (A.25) yields, for the contribution of process (2α) to amplitude (C.31), the result

$$\frac{\langle a; \mathbf{k}_1\boldsymbol{\varepsilon}_1, \mathbf{k}_2\boldsymbol{\varepsilon}_2 | H_{I1} | b; \mathbf{k}_1\boldsymbol{\varepsilon}_1 \rangle \langle b; \mathbf{k}_1\boldsymbol{\varepsilon}_1 | H_{I1} | c; 0 \rangle}{(z - \hbar\omega_1 - \hbar\omega_2 - E_a)(z - \hbar\omega_1 - E_b)(z - E_c)} \tag{C.33}$$

represented by the diagram

$$\tag{C.34}$$

As in the previous subsection, we may sum the other higher-order diagrams, of the same type as (C.35), which contain the "resonant states" $|c; 0\rangle$, $|b; k_1\varepsilon_1\rangle$, $|a; k_1\varepsilon_1, k_2\varepsilon_2\rangle$, as intermediate states, an arbitrary number of times, and contain no more than one nonresonant state between two such resonant states.

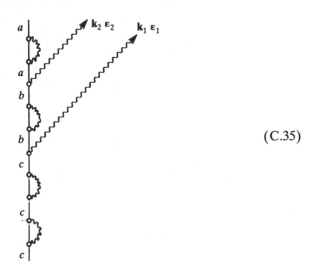

$$(C.35)$$

As above, such a summation is equivalent to replacing E_i by $\tilde{E}_i - i\hbar(\Gamma_i/2)$ in the denominator of (C.33) (written for $z = E + i\eta$) with $i = a, b, c$. We thus obtain

$$\langle a; k_1\varepsilon_1, k_2\varepsilon_2|H_{I1}|b; k_1, \varepsilon_1\rangle\langle b; k_1, \varepsilon_1|H_{I1}|c; 0\rangle \times$$

$$\times \left(\frac{1}{E + i\eta - \hbar\omega_1 - \hbar\omega_2 - \tilde{E}_a}\right)\left(\frac{1}{E + i\eta - \hbar\omega_1 - \tilde{E}_b + i\hbar(\Gamma_b/2)}\right) \times$$

$$\times \left(\frac{1}{E + i\eta - \tilde{E}_c + i\hbar(\Gamma_c/2)}\right). \tag{C.36}$$

corresponding to the diagram obtained by replacing the three single lines a, b, c in (C.34) by three double lines symbolizing the perturbed propagators [see the first diagram in (C.46)].

More elaborate methods, using projection operators over several orthogonal subspaces, may be developed to study the problem of radiative cascades. They confirm the results derived here by diagrammatic methods (*).

c) NATURAL WIDTH AND SHIFT OF THE EMITTED LINES

α) *Line Emitted from the First Excited Level*

We return to the spontaneous emission of a single photon from the first excited level b.

To obtain the spectral distribution of the emitted photon, we must calculate the emission amplitude (C.23) for a time τ sufficiently long so that the atom has certainly returned to state a ($\tau \gg \Gamma_b^{-1}$), then study the dependence on ω of the squared modulus of this amplitude.

We substitute (C.29) into (C.23) and calculate the integral over E by the residue method. For $\tau \gg \Gamma_b^{-1}$, only the pole at $\tilde{E}_a + \hbar\omega$ of (C.29) contributes, and we obtain

$$\langle a; \mathbf{k}\varepsilon|U(\tau)|b; 0\rangle = \frac{1}{\hbar} \frac{\langle a; \mathbf{k}\varepsilon|H_{I1}|b; 0\rangle}{\omega - \tilde{\omega}_{ba} + i(\Gamma_b/2)} \, e^{-i(\tilde{E}_a + \hbar\omega)\tau/\hbar} \quad \text{(C.37)}$$

where

$$\tilde{\omega}_{ba} = \left(\tilde{E}_b - \tilde{E}_a\right)/\hbar \tag{C.38}$$

represents the (angular) frequency of the transition connecting state b shifted by $\hbar\Delta_b$ to state a shifted by $\hbar\Delta_a$. We then square the modulus of (C.37). The dependence on ω of the numerator is much slower than that of the denominator and may be neglected. The distribution in frequency $I(\omega)$ of the emitted photons is thus proportional to

$$\frac{1}{\left(\omega - \tilde{\omega}_{ba}\right)^2 + (\Gamma_b/2)^2}. \tag{C.39}$$

It is given by a Lorentz curve of width at half-maximum Γ_b, called the *natural width*, centered on $\tilde{\omega}_{ba}$, which differs from ω_{ba} by the *shift* $\Delta_b - \Delta_a$.

β) *Lines Emitted in a Radiative Cascade*

To obtain the spectral distribution of the two photons emitted in the radiative cascade $c \to b \to a$ (see Figure 2α), we must now substitute

---

(*) See, for example, L. Mower, *Phys. Rev.* **142**, 799 (1966); A. S. Goldhaber and K. M. Watson, *Phys. Rev.* **160**, 1151 (1967); L. Mower, *Phys. Rev.* **165**, 145 (1968).

(C.36) into (C.32). For $\tau \gg \Gamma_c^{-1}, \Gamma_b^{-1}$, only the pole at $\tilde{E}_a + \hbar\omega_1 + \hbar\omega_2$ contributes to the integral and we obtain

$$\langle a; \mathbf{k}_1\mathbf{\varepsilon}_1, \mathbf{k}_2\mathbf{\varepsilon}_2 | U(\tau) | c; 0 \rangle = e^{-i(\tilde{E}_a + \hbar\omega_1 + \hbar\omega_2)\tau/\hbar} \times$$

$$\times \frac{\langle a; \mathbf{k}_1\mathbf{\varepsilon}_1, \mathbf{k}_2\mathbf{\varepsilon}_2 | H_{I1} | b; \mathbf{k}_1\mathbf{\varepsilon}_1 \rangle \langle b; \mathbf{k}_1\mathbf{\varepsilon}_1 | H_{I1} | c; 0 \rangle}{\left(\hbar\omega_1 + \hbar\omega_2 + \tilde{E}_a - \tilde{E}_c + i\hbar\dfrac{\Gamma_c}{2} \right) \left(\hbar\omega_2 + \tilde{E}_a - \tilde{E}_b + i\hbar\dfrac{\Gamma_b}{2} \right)}.$$

$$(C.40)$$

As previously, it is legitimate to neglect the variation with ω_1 and ω_2 of the numerator compared to that of the denominator, so that the combined distribution $I(\omega_1, \omega_2)$ of the frequencies of the emitted two photons is given by

$$I(\omega_1, \omega_2) \sim \frac{1}{\left[(\omega_1 + \omega_2 - \tilde{\omega}_{ca})^2 + \dfrac{\Gamma_c^2}{4} \right]\left[(\omega_2 - \tilde{\omega}_{ba})^2 + \dfrac{\Gamma_b^2}{4} \right]} \qquad (C.41)$$

where $\tilde{\omega}_{ba}$ is already given in (C.38) and where

$$\tilde{\omega}_{ca} = \left(\tilde{E}_c - \tilde{E}_a \right)/\hbar. \qquad (C.42)$$

Let us assume that we observe only the first photon ω_1. Its spectral distribution is obtained by integrating $I(\omega_1, \omega_2)$ over ω_2:

$$I(\omega_1) = \int d\omega_2 \, I(\omega_1, \omega_2). \qquad (C.43)$$

The integral over ω_2 of (C.41) may be easily calculated by the residue method. We get

$$I(\omega_1) \sim \frac{1}{(\omega_1 - \tilde{\omega}_{cb})^2 + [(\Gamma_b + \Gamma_c)/2]^2} \qquad (C.44)$$

where

$$\tilde{\omega}_{cb} = \left(\tilde{E}_c - \tilde{E}_b \right)/\hbar. \qquad (C.45)$$

The spectral distribution of photons emitted on the transition $c \to b$ is given by a Lorentzian centered at $\tilde{\omega}_{cb}$ and for which the width at half-maximum is the sum of the natural widths Γ_c and Γ_b associated with the initial and final levels of the transition.

If, on the other hand, we observe only the photon ω_2, we must integrate (C.41) over ω_1. We then find the same result as the one given in (C.39).

Remark

Strictly speaking, the amplitude of emission of the two photons is represented by a sum of two diagrams differing by the order in which the two photons are emitted.

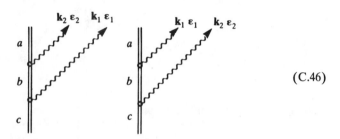

$$(C.46)$$

These two diagrams correspond to the two processes in Figure 2 and describe two different paths going from the same initial state to the same final state. The corresponding amplitudes must therefore interfere. In fact, if $|\tilde{\omega}_{cb} - \tilde{\omega}_{ba}| \gg \Gamma_c, \Gamma_b$, the two amplitudes may never be large simultaneously, so that the interference effects are negligible. The previous calculation is therefore valid. On the other hand, if the three levels a, b, and c are equidistant, then the two amplitudes associated with (C.46) may be comparable and their interference may no longer be ignored. An example of such a situation is analyzed in Exercise 15 which studies the spontaneous emission in cascade starting from the level $n = 2$ of a harmonic oscillator.

3. Indirect Coupling between a Discrete Level and a Continuum. Example of the Lamb Transition

a) INTRODUCING THE PROBLEM

We now consider the case in which a discrete level $|\varphi_c\rangle$ of H_0 is coupled to a continuum, not directly (as in the foregoing subsections) but indirectly, via another discrete level $|\varphi_b\rangle$ of H_0 having a nearby energy. This situation is encountered, for example, in the hydrogen atom, the metastable level $2s_{1/2}$ being very close in energy to the unstable level $2p_{1/2}$ (see Figure 3). We show in this subsection how a weak coupling

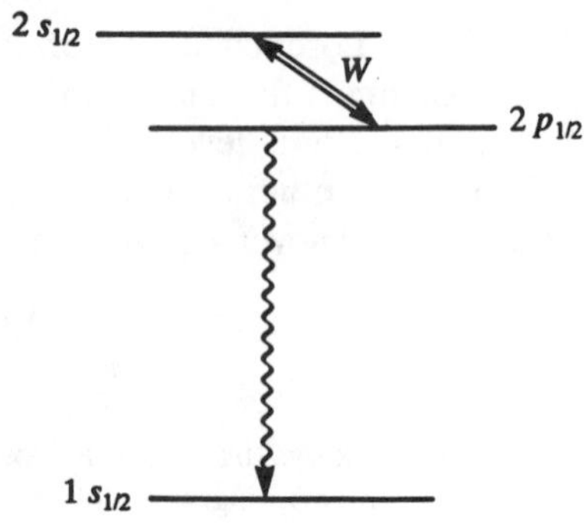

Figure 3. Diagram of the energy levels of the hydrogen atom considered in this subsection. Levels $2s_{1/2}$ and $2p_{1/2}$ are quasi-degenerate. Level $2p_{1/2}$ may decay to level $1s_{1/2}$ by emitting a photon. Level $2s_{1/2}$ is metastable in the absence of the coupling W.

between these two levels may lead to a significant reduction in the metastability of the level $2s_{1/2}$ (*).

The problem that we consider is thus that of a discrete level $|\varphi_c\rangle = |c; 0\rangle$ of $H_0 = H_P + H_R$ (atom in the excited state c without any photon), coupled by a Hamiltonian W to a nearby level $|\varphi_b\rangle = |b; 0\rangle$ (atom in the excited state b without any photon). This second state $|\varphi_b\rangle$ is itself coupled by H_{I1} to a continuum of states $|a; k\varepsilon\rangle$ (atom in the ground state a in the presence of a photon $k\varepsilon$). On the other hand, we assume that $|\varphi_c\rangle$ is not directly coupled to $|a; k\varepsilon\rangle$ by H_{I1} and is thus metastable with regard to spontaneous emission. A system initially prepared in state $|\varphi_c\rangle$ must therefore pass through the intermediate state $|\varphi_b\rangle$ before it is able to emit a photon. Thus it clearly appears that the evolution of the state $|\varphi_c\rangle$ cannot be studied independently of that of the state $|\varphi_b\rangle$. Thus we find ourselves in a situation in which the relevant subspace $\mathscr{E}_0$ is subtended by $|\varphi_b\rangle$ and $|\varphi_c\rangle$ and thus has a dimension equal to 2.

For the hydrogen example, we assume that the coupling W between the states $|\varphi_c\rangle$ and $|\varphi_b\rangle$ is due to an external static electric field $\mathbf{E}_0$ so that

$$W = -\mathbf{d} \cdot \mathbf{E}_0 \qquad (C.47)$$

where $\mathbf{d}$ is the electric dipole moment of the atom.

Remarks

(i) The theory developed below may be adapted easily to the case in which the coupling between $|\varphi_b\rangle$ and $|\varphi_c\rangle$ is due to a microwave electromagnetic field

(*) Actually, the $2s_{1/2}$ state may decay radiatively to state $1s_{1/2}$ by a spontaneous two-photon emission (see Chapter II, §D-1). Here we neglect such a process because the instability introduced by the coupling with the unstable state $2p_{1/2}$ is assumed to be more significant.

having a frequency close to $(E_c - E_b)/\hbar$. If we assume that initially there are N photons in this mode $\mathbf{k}_1\boldsymbol{\varepsilon}_1$ of the field and that the atom is in the state c, the coupling between $|c; N\mathbf{k}_1\boldsymbol{\varepsilon}_1\rangle$ and $|b; (N + 1)\mathbf{k}_1\boldsymbol{\varepsilon}_1\rangle$ may induce a transition between these two levels. Because the level $|b; (N + 1)\mathbf{k}_1\boldsymbol{\varepsilon}_1\rangle$ is unstable, as a result of its coupling to the continuum of states $|a; (N + 1)\mathbf{k}_1\boldsymbol{\varepsilon}_1, \mathbf{k}\boldsymbol{\varepsilon}\rangle$, the evolution of the system is described by equations similar to those obtained in the case of the Stark coupling (C.47). The relevant subspace $\mathscr{E}_0$ is then subtended by $|c; N\mathbf{k}_1\boldsymbol{\varepsilon}_1\rangle$ and $|b; (N + 1)\mathbf{k}_1\boldsymbol{\varepsilon}_1\rangle$.

(ii) We emphasize here the instability of the state $|\varphi_c\rangle$ induced by the coupling W. Another possible approach would consist of first determining the eigenstates $|\chi_1\rangle$ and $|\chi_2\rangle$ of the projection of $H_0 + W$ in the subspace $\mathscr{E}_0$ subtended by $|\varphi_b\rangle$ and $|\varphi_c\rangle$, then studying the effect of the radiative coupling between the states $|\chi_1\rangle$ and $|\chi_2\rangle$ thus obtained and the continuum $\{|a; \mathbf{k}\boldsymbol{\varepsilon}\rangle\}$. It would, however, be incorrect to think that the evolution of $|\chi_1\rangle$ and $|\chi_2\rangle$ may be deduced exclusively from the diagonal matrix elements $\langle\chi_1|R(E + i\eta)|\chi_1\rangle$ and $\langle\chi_2|R(E + i\eta)|\chi_2\rangle$. Because the states $|\chi_1\rangle$ and $|\chi_2\rangle$ are close in energy and coupled to the same continuum $|a; \mathbf{k}\boldsymbol{\varepsilon}\rangle$, the nondiagonal elements $\langle\chi_1|R(E + i\eta)|\chi_2\rangle$ and $\langle\chi_2|R(E + i\eta)|\chi_1\rangle$, must also be taken into account. Another physical example in which two degenerate levels are resonantly coupled to the same continuum is studied in Exercise 14, which deals with the problem of superradiance of a system of two atoms.

b) NONPERTURBATIVE CALCULATION OF THE TRANSITION AMPLITUDE

The system being in the state $|\varphi_c\rangle$ at $t = 0$, we calculate the probability amplitude $\langle\varphi_c|U(t, 0)|\varphi_c\rangle = U_{cc}(t)$ of finding it in the same state at a subsequent time t. To do this, first we consider $\langle\varphi_c|G(z)|\varphi_c\rangle = G_{cc}(z)$.

Because the evolutions of states $|\varphi_b\rangle$ and $|\varphi_c\rangle$ are coupled, we have to study the projection of $G(z)$ in the subspace $\mathscr{E}_0$ subtended by $|\varphi_b\rangle$ and $|\varphi_c\rangle$. It thus follows from subsection B-2-b that the matrix (B.28) representing $PG(z)P$ is the inverse of matrix (B.29). It is thus equal to

$$\begin{pmatrix} G_{bb}(z) & G_{bc}(z) \\ G_{cb}(z) & G_{cc}(z) \end{pmatrix} = \frac{1}{\mathscr{D}}\begin{pmatrix} z - E_c - R_{cc}(z) & R_{bc}(z) \\ R_{cb}(z) & z - E_b - R_{bb}(z) \end{pmatrix}$$

(C.48)

where

$$\mathscr{D} = [z - E_b - R_{bb}(z)][z - E_c - R_{cc}(z)] - R_{bc}(z)R_{cb}(z) \quad \text{(C.49)}$$

is the determinant of the matrix (C.29).

In the physical problem studied here, the Hamiltonian H is the sum of $H_0 = H_P + H_R$ and $V = H_{I1} + H_{I2} + W$. As in subsection C-1, we will carry out approximations on the exact expressions (C.48) and (C.49). We replace the matrix elements R_{ij} of R (with $i, j = b$ or c) by perturbative expressions $\hat{R}_{ij}$, calculated up to second order with regard to the charges q. Along the contour C_+ of Figure 1 (which has to be considered for $\tau > 0$), we also neglect the variation with E of $\hat{R}_{ij}(E + i\eta)$, by replacing $\hat{R}_{ij}(E + i\eta)$ by $\hat{R}_{ij}(E_0 + i\eta)$, where $E_0 = (E_b + E_c)/2$ is the mean energy of states $|\varphi_b\rangle$ and $|\varphi_c\rangle$.

Let us begin by calculating the nondiagonal element $\hat{R}_{bc}(E_0 + i\eta)$. The Hamiltonian H_{I2}, which does not act on particles in the long-wavelength approximation, cannot couple $|\varphi_b\rangle$ to $|\varphi_c\rangle$. Furthermore H_{I1} cannot couple $|\varphi_b\rangle$ and $|\varphi_c\rangle$ to the second order, because the atomic states $|b\rangle$ and $|c\rangle$ have opposite parities. Finally, the "crossed" coupling terms involving both H_{I1} and W are zero between the states $|\varphi_b\rangle$ and $|\varphi_c\rangle$ which both contain no photons. From that we deduce that, to the second order in q

$$\hat{R}_{bc}(E_0 + i\eta) = W_{bc} \qquad \text{(C.50.a)}$$

$$\hat{R}_{cb}(E_0 + i\eta) = W_{cb}. \qquad \text{(C.50.b)}$$

For the diagonal elements $\hat{R}_{bb}$ and $\hat{R}_{cc}$, we neglect the quadratic Stark effect due to the matrix elements of W between the state $|b\rangle$ (or $|c\rangle$) and the other distant levels. Because the crossed diagonal terms involving both H_{I1} and W are zero, and because $E_0 \simeq E_b \simeq E_c$, we obtain

$$\hat{R}_{bb}(E_0 + i\eta) = \langle b; 0|H_{I2}|b; 0\rangle + \sum_{\mathbf{k}\varepsilon} \frac{|\langle a; \mathbf{k}\varepsilon|H_{I1}|b; 0\rangle|^2}{E_b + i\eta - E_a - \hbar\omega}$$

$$= \hbar\Delta_b - i\frac{\hbar\Gamma_b}{2} \qquad \text{(C.50.c)}$$

and similarly

$$\hat{R}_{cc}(E_0 + i\eta) = \hbar\Delta_c. \qquad \text{(C.50.d)}$$

The absence of an imaginary term in $\hat{R}_{cc}(E_0 + i\eta)$ is due to the fact that the state $|c\rangle$ is metastable.

Finally, the approximations made above on $PR(E + i\eta)P$ are equivalent to considering that $PG(z)P$ is the resolvent of the "Hamiltonian"

$PH_0P + P\hat{R}(E_0 + i\eta)P$ that is represented by the matrix

$$\begin{pmatrix} \tilde{E}_b - i\hbar\dfrac{\Gamma_b}{2} & W_{bc} \\[2mm] W_{cb} & \tilde{E}_c \end{pmatrix} \qquad (C.51)$$

(where, as above, we have set $\tilde{E}_i = E_i + \hbar\Delta_i$ for $i = b, c$). The matrix (C.51) thus appears as a non-Hermitian effective Hamiltonian allowing us to study the evolution within the subspace $\mathscr{E}_0$.

Substituting (C.50) into (C.48) and (C.49) yields

$$G_{cc}(E + i\eta) = \frac{E - \tilde{E}_b + i\hbar\dfrac{\Gamma_b}{2}}{\left(E - \tilde{E}_b + i\hbar\dfrac{\Gamma_b}{2}\right)(E - \tilde{E}_c) - |W_{bc}|^2}. \qquad (C.52)$$

To obtain $U_{cc}(\tau)$, it is then sufficient to carry out the contour integral (A.22) by the residue method. We obtain in this way

$$U_{cc}(\tau) = \frac{\left(z_2 - \tilde{E}_b + i\hbar\dfrac{\Gamma_b}{2}\right)e^{-iz_2\tau/\hbar} - \left(z_1 - \tilde{E}_b + i\hbar\dfrac{\Gamma_b}{2}\right)e^{-iz_1\tau/\hbar}}{z_2 - z_1}$$

$$(C.53)$$

where z_1 and z_2 are the eigenvalues of the matrix (C.51).

c) WEAK COUPLING LIMIT. BETHE FORMULA

Let us consider first the case in which $|W_{bc}|$ is very small as compared with $\hbar\Gamma_b$ ($|W_{bc}| \ll \hbar\Gamma_b$). The eigenvalues of (C.51) are then respectively equal to

$$z_1 \simeq \tilde{E}_b - i\hbar\frac{\Gamma_b}{2} + \frac{|W_{bc}|^2}{\left(\tilde{E}_b - i\hbar\dfrac{\Gamma_b}{2}\right) - \tilde{E}_c} \qquad (C.54.a)$$

$$z_2 \simeq \tilde{E}_c - \frac{|W_{bc}|^2}{\left(\tilde{E}_b - i\hbar\dfrac{\Gamma_b}{2}\right) - \tilde{E}_c}. \qquad (C.54.b)$$

Now separate the real and imaginary parts of these eigenvalues. We then see the following quantities appear

$$\Delta' = -\tilde{\omega}_{bc}\frac{|W_{bc}|^2}{\hbar^2\left[\tilde{\omega}_{bc}^2 + \dfrac{\Gamma_b^2}{4}\right]} \tag{C.55}$$

$$\Gamma' = \Gamma_b\frac{|W_{bc}|^2}{\hbar^2\left(\tilde{\omega}_{bc}^2 + \dfrac{\Gamma_b^2}{4}\right)} \tag{C.56}$$

the meaning of which will be given later (recall that $\hbar\tilde{\omega}_{bc} = \tilde{E}_b - \tilde{E}_c$), and we obtain

$$z_1 \simeq \tilde{E}_b - \hbar\Delta' - i\hbar\frac{\Gamma_b}{2}\left(1 - \frac{\Gamma'}{\Gamma_b}\right)$$

$$\simeq \tilde{E}_b - \hbar\Delta' - i\hbar\frac{\Gamma_b}{2} \tag{C.57.a}$$

$$z_2 \simeq \tilde{E}_c + \hbar\Delta' - i\hbar\frac{\Gamma'}{2}. \tag{C.57.b}$$

In the case of weak coupling, Γ' and Δ' are very small compared with Γ_b and the eigenvalues z_1 and z_2 differ only slightly from the diagonal terms of the effective Hamiltonian (C.51), even when the energy difference $\hbar\tilde{\omega}_{bc}$ between levels b and c is zero. By substituting the values of z_1 and z_2 into the expression (C.53) for $U_{cc}(t)$, we find that the essential contribution comes from the pole z_2, so that

$$U_{cc}(t) \simeq e^{-i(\tilde{E}_c + \hbar\Delta')t/\hbar}\, e^{-\Gamma't/2}. \tag{C.58}$$

It follows from (C.58) that the coupling W has two consequences for level c: On the one hand, this level is shifted by the quantity $\hbar\Delta'$, on the other hand, it acquires a finite lifetime $1/\Gamma'$. As a result of its coupling with b, level c is "contaminated" by b and thus becomes unstable. We see in formula (C.56), also called the *Bethe formula*, that the decay rate Γ' varies with $\tilde{\omega}_{bc}$ as a Lorentzian curve centered at $\tilde{\omega}_{bc} = 0$. This probability is maximal when the energies of levels b and c are equal (including radiative shifts). On the other hand, the variation of Δ' as a function of $\tilde{\omega}_{bc}$ is that of a dispersion curve [see (C.55)]. Although they are coupled to each other, the energy levels continue to cross each other at $\tilde{\omega}_{bc} = 0$ when the energies $\tilde{E}_b$ and $\tilde{E}_c$ are scanned using an external parameter (*). The perturbation W thus does not lift the degeneracy of the energy levels when W is small compared to $\hbar\Gamma_b$. We find also that the eigenstates $|\varphi_1\rangle$ and $|\varphi_2\rangle$ of the matrix (C.51) differ only slightly from $|\varphi_b\rangle$ and $|\varphi_c\rangle$ (the corrections being at the maximum on the order of $|W_{bc}|/\hbar\Gamma_b$):

$$|\varphi_1\rangle \simeq |\varphi_b\rangle + \frac{W_{cb}}{\hbar\left(\tilde{\omega}_{bc} - i\dfrac{\Gamma_b}{2}\right)}|\varphi_c\rangle \qquad \text{(C.59.a)}$$

$$|\varphi_2\rangle \simeq |\varphi_c\rangle - \frac{W_{bc}}{\hbar\left(\tilde{\omega}_{bc} - i\dfrac{\Gamma_b}{2}\right)}|\varphi_b\rangle. \qquad \text{(C.59.b)}$$

Note that these two states are not orthogonal because the matrix (C.51) is not Hermitian.

d) Strong Coupling Limit. Rabi Oscillation

Let us now consider the other limiting situation where $|W_{bc}| \gg \hbar\Gamma_b$. In this case, the eigenvalues z_1 and z_2 of (C.51) are equal to

$$z_1 \simeq \tilde{E}_0 + \hbar\frac{\Omega}{2} - i\hbar\frac{\Gamma_1}{2} \qquad \text{(C.60.a)}$$

$$z_2 \simeq \tilde{E}_0 - \hbar\frac{\Omega}{2} - i\hbar\frac{\Gamma_2}{2} \qquad \text{(C.60.b)}$$

with

$$\tilde{E}_0 = \frac{\tilde{E}_b + \tilde{E}_c}{2} \qquad \text{(C.61)}$$

$$\Omega = \frac{1}{\hbar}\sqrt{4|W_{bc}|^2 + \hbar^2\tilde{\omega}_{bc}^2} \qquad \text{(C.62)}$$

$$\Gamma_1 = \frac{\Gamma_b}{2}\left(1 + \frac{\tilde{\omega}_{bc}}{\Omega}\right) \qquad \text{(C.63.a)}$$

(*) In the case of hydrogen, such a scanning is carried out by a Zeeman shift of the levels in a static magnetic field. See W. E. Lamb and R. C. Retherford, *Phys. Rev.* **79**, 549 (1950); **81**, 222 (1951); **86**, 1014 (1952).

$$\Gamma_2 = \frac{\Gamma_b}{2}\left(1 - \frac{\tilde{\omega}_{bc}}{\Omega}\right).$$ (C.63.b)

Substituting these values for z_1 and z_2 in (C.53) yields the transition amplitude $U_{cc}(t)$. At resonance ($\tilde{\omega}_{bc} = 0$), we find that

$$U_{cc}(t) = e^{-i\bar{E}_0 t/\hbar}\, e^{-\Gamma_b t/4}\cos\frac{|W_{bc}|t}{\hbar}.$$ (C.64)

The evolution of the system starting from the state $|\varphi_c\rangle$ thus appears as a damped Rabi oscillation. The frequency of this oscillation is proportional to the coupling matrix element $|W_{bc}|$ and the damping time of the oscillation is twice the lifetime of the level b. Under the effect of the strong coupling between the two states, the instability of the level b is found to be distributed equally between b and c. The eigenstates of the matrix (C.51) are in this case symmetric and antisymmetric linear combinations of $|\varphi_b\rangle$ and $|\varphi_c\rangle$ (if W_{bc} is real). Their energies are $\bar{E}_0 \pm |W_{bc}|$. In contrast to the previous case (weak coupling), the degeneracy of the two levels b and c is thus lifted by W.

If the system is detuned from resonance, then the mixing of the wave functions becomes less important and the eigenstates of (C.51) tend to $|\varphi_b\rangle$ and $|\varphi_c\rangle$ when $\hbar|\tilde{\omega}_{bc}| \gg |W_{bc}|$. The natural widths Γ_1 and Γ_2 of these levels [related to the imaginary parts of (C.60.a) and (C.60.b)] thus pass progressively from $\Gamma_b/2$ and $\Gamma_b/2$ to Γ_b and $\Gamma_c = 0$ [see (C.63.a) and (C.63.b)]. As for the energies [real parts of (C.60.a) and (C.60.b)], they equal $(\bar{E}_0 \pm \hbar\Omega)/2$ and form a "level anticrossing" when $\tilde{\omega}_{bc}$ varies (see

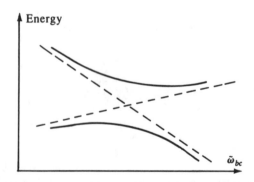

Figure 4. Level anticrossing. The energy levels in the absence of W are drawn with dashed lines and the perturbed energy levels are represented by solid lines.

Figure 4). The Rabi oscillation occurs at the frequency Ω given in (C.62) and its amplitude decreases.

Remark

In this entire subsection we have considered $|\varphi_c\rangle$ to be stable in the absence of coupling W. The results obtained may be extended easily to the case in which $|\varphi_c\rangle$ has a lifetime equal to Γ_c^{-1}. In this case, we must add an imaginary part $-i\hbar\Gamma_c/2$ to the diagonal term $\tilde{E}_c$ of the effective Hamiltonian (C.51). The level anticrossing situation thus corresponds to the inequality $|W_{bc}| \gg \hbar|\Gamma_b - \Gamma_c|$. In particular, when the two levels have the same lifetime, the coupling between $|\varphi_b|$ and $|\varphi_c\rangle$ always leads to a level anticrossing.

4. Indirect Coupling between Two Discrete States. Multiphoton Transitions

a) Physical Process and Subspace $\mathscr{E}_0$ of Relevant States

The goal of this last subsection is to show how the resolvent formalism may be applied to the study of multiphoton transitions. For example, let us consider the case of the resonant two-photon absorption between two excited discrete levels c and b (see Figure 5). The energy difference $E_b - E_c = \hbar\omega_{bc}$ between these levels is near $2\hbar\omega$ where ω is the frequency of the N incident photons $\mathbf{k}\boldsymbol{\varepsilon}$. We assume that the lower level of the transition c is metastable, whereas the higher level b decays by a radiative cascade, through the levels d, to the ground state a. Such a

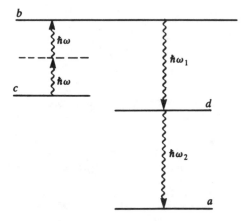

Figure 5. Energy-level diagram corresponding to the two-photon absorption process considered in this paragraph.

scheme corresponds, for example, to the case of the transition $2s \rightarrow 3s$ of the hydrogen atom, the radiative decay of the $3s$ level occurring almost completely via the $2p$ level.

Let $|\psi(t)\rangle$ be the state vector of the total atom + field system, corresponding to the initial state $|\psi(0)\rangle = |c; N\mathbf{k}\varepsilon\rangle$ (atom in the state c in the presence of N photons of the mode $\mathbf{k}\varepsilon$, all the other modes being empty). Among the eigenstates of $H_0 = H_P + H_R$ that will appear with a significant amplitude in the expansion of $|\psi(t)\rangle$, the state $|b; (N-2)\mathbf{k}\varepsilon\rangle$ plays a predominant role because the condition $\omega_{bc} \simeq 2\omega$ implies that this state is nearly degenerate with $|c; N\mathbf{k}\varepsilon\rangle$. Other states will also play an important role for long times, namely, those which result from the spontaneous emission of one photon from the excited state b ($|d; (N-2)\mathbf{k}\varepsilon, \mathbf{k}_1\varepsilon_1\rangle$) or of two photons ($|a; (N-2)\mathbf{k}\varepsilon, \mathbf{k}_1\varepsilon_1, \mathbf{k}_2\varepsilon_2\rangle$).

Note that these two latter states of the total system do not involve the atomic levels b and c. More generally, if it is possible to neglect all the spontaneous emission processes able to bring back the atom in levels b and c, these two atomic levels appear in the expansion of $|\psi(t)\rangle$ only through the states $|\varphi_b\rangle = |b; (N-2)\mathbf{k}\varepsilon\rangle$ and $|\varphi_c\rangle = |c; N\mathbf{k}\varepsilon\rangle$. The probability of finding the atom in the level b or c (regardless of the state of the field) is thus equal to that of finding the total system in state $|\varphi_b\rangle$ or $|\varphi_c\rangle$. The calculation of the evolution operator, and therefore that of the resolvent, may then be limited to the subspace $\mathscr{E}_0$ subtended by $|\varphi_b\rangle$ and $|\varphi_c\rangle$. Such a simplification would not be possible if the atomic state c were the ground state because it would be repopulated by spontaneous emission. Thus it would be necessary to take into account many other states of the total system to describe the atomic evolution (see §C-4-f).

b) Nonperturbative Calculation of the Transition Amplitude

We choose here to use the electric dipole point of view to describe the coupling between the atom and the field (*). The Hamiltonian H is then decomposed in $H_0 = H_P + H_R$ and $H_I' = -\mathbf{d} \cdot \mathbf{E}_\perp$.

To derive the evolution of the system, it is necessary, as in subsection C-3-b, to calculate beforehand the eigenvalues of the effective Hamiltonian $PH_0P + PR(E_0 + i\eta)P$ where E_0 is the mean energy of the levels $|\varphi_b\rangle$ and $|\varphi_c\rangle$. The matrix elements of $PR(E_0 + i\eta)P$, are calculated here at the lowest order in q_α where they are nonzero, namely, second order.

(*) One can actually show (see, for example, *Photons and Atoms—Introduction to Quantum Electrodynamics*, Exercise 2 in Complement E_{IV}) that the Hamiltonian $-\mathbf{d} \cdot \mathbf{E}_\perp$ is often more convenient to use than H_{I1} for the study of multiphoton transitions between discrete levels.

For the diagonal terms, we thus have

$$
\hat{R}_{bb}(E_0 + i\eta) \simeq \sum_i \sum_{\mathbf{k'\varepsilon'} \neq \mathbf{k\varepsilon}} \frac{\left| \langle b; (N-2)\mathbf{k\varepsilon} | H'_I | i; (N-2)\mathbf{k\varepsilon}, \mathbf{k'\varepsilon'} \rangle \right|^2}{E_b - E_i - \hbar\omega' + i\eta} +
$$

$$
+ \sum_i \frac{\left| \langle b; (N-2)\mathbf{k\varepsilon} | H'_I | i; (N-1)\mathbf{k\varepsilon} \rangle \right|^2}{E_b - E_i - \hbar\omega + i\eta} +
$$

$$
+ \sum_i \frac{\left| \langle b; (N-2)\mathbf{k\varepsilon} | H'_I | i; (N-3)\mathbf{k\varepsilon} \rangle \right|^2}{E_b - E_i + \hbar\omega + i\eta}. \tag{C.65}
$$

In the sum over $\mathbf{k'\varepsilon'}$ of the first term of (C.65), the photons $\mathbf{k\varepsilon}$ are spectators. Except for the contribution of the single mode $\mathbf{k\varepsilon}$ (which may be neglected) we find a sum analogous to (C.9), which thus describes the spontaneous radiative effects (lifetime, radiative level shift) previously discussed in subsection C-1 (*).

In the last two sums of (C.65), it is possible to eliminate the term $i\eta$ appearing in the denominator. In fact, in the example studied here, we exclude any single-photon resonance starting from b so that no energy denominator can ever vanish (**). The last two terms of (C.65) are therefore real. Let us call their sum $\hbar\Delta'_b$. The quantity $\hbar\Delta'_b$ actually describes the light shift of the level $|\varphi_b\rangle$ resulting from the interaction with the incident photons. By assuming that the number of photons in this mode $\mathbf{k\varepsilon}$ is very large ($N \gg 1$) and by using the expression (89) in the Appendix which gives the field operator $\mathbf{E}_\perp$, we can reexpress the last two terms of (C.65) in the form

$$
\hbar\Delta'_b \simeq \frac{N\hbar\omega}{2\varepsilon_0 L^3} \sum_i |\langle b | \mathbf{d} \cdot \boldsymbol{\varepsilon} | i \rangle|^2 \left(\frac{1}{E_b - E_i - \hbar\omega} + \frac{1}{E_b - E_i + \hbar\omega} \right). \tag{C.66}
$$

Or again, by using the intensity operator $I = \mathbf{E}_\perp^{(-)} \cdot \mathbf{E}_\perp^{(+)}$ introduced in

---

(*) Strictly speaking, the first sum in (C.65) does not coincide with (C.9). To obtain the correct radiative level shift $\hbar\Delta_b$, from the electric dipole Hamiltonian, it is necessary to add to this sum the contribution from the dipole self-energy term of the electric dipole Hamiltonian [see relation (76) in the Appendix and Exercise 7].

(**) In the case where there is a level i of the ionization continuum of the atom such that $E_i = E_b + \hbar\omega$, it would be necessary to retain the term in $i\eta$. The imaginary part of the last sum of (C.65) would then be associated with the lifetime of level b due to the photoionization process.

Complement A_{II}:

$$\hbar\Delta'_b = \langle I \rangle \sum_i |\langle b|\mathbf{d} \cdot \boldsymbol{\varepsilon}|i\rangle|^2 \left(\frac{1}{E_b - E_i - \hbar\omega} + \frac{1}{E_b - E_i + \hbar\omega} \right). \quad \text{(C.67)}$$

The light shift $\hbar\Delta'_c$ of the other level is given by a similar expression obtained by replacing b by c.

Finally, the diagonal terms of the effective Hamiltonian $PH_0P + P\hat{R}(E_0 + i\eta)P$ are written (with the notation $\tilde{E}_i = E_i + \hbar\Delta_i$ for $i = b, c$)

$$(N - 2)\hbar\omega + \tilde{E}_b + \hbar\Delta'_b - i\hbar\frac{\Gamma_b}{2} \quad \text{(C.68)}$$

for level b and

$$N\hbar\omega + \tilde{E}_c + \hbar\Delta'_c \quad \text{(C.69)}$$

for level c which was assumed to be metastable ($\Gamma_c = 0$).

Let us now consider the nondiagonal term $\tilde{R}_{bc}(E_0 + i\eta)$, which is equal to

$$\hat{R}_{bc}(E_0 + i\eta) =$$
$$\sum_i \frac{\langle b; (N - 2)\mathbf{k}\boldsymbol{\varepsilon}|H'_i|i; (N - 1)\mathbf{k}\boldsymbol{\varepsilon}\rangle\langle i; (N - 1)\mathbf{k}\boldsymbol{\varepsilon}|H'_i|c; N\mathbf{k}\boldsymbol{\varepsilon}\rangle}{E_c + \hbar\omega - E_i + i\eta}.$$
$$\text{(C.70)}$$

We have replaced E_0 by $N\hbar\omega + E_c$ in the energy denominator (we also could have taken $E_0 \simeq (N - 2)\hbar\omega + E_b$, which is only slightly different). Note that the term in $i\eta$ of the denominators may be eliminated because there is no level i located at an equal distance from levels b and c (no single-photon resonance). In contrast to the situation in subsection C-3, here there is no direct coupling between levels $|\varphi_b\rangle$ and $|\varphi_c\rangle$, but rather a second-order indirect coupling through the states $|i; (N - 1)\mathbf{k}\boldsymbol{\varepsilon}\rangle$. In the limit $N \gg 1$, we can rewrite $\hat{R}_{bc}(E_0)$ in the form

$$\hat{R}_{bc}(E_0) = \frac{N\hbar\omega}{2\varepsilon_0 L^3} \sum_i \frac{\langle b|\mathbf{d} \cdot \boldsymbol{\varepsilon}|i\rangle\langle i|\mathbf{d} \cdot \boldsymbol{\varepsilon}|c\rangle}{E_c + \hbar\omega - E_i}$$
$$= \langle I \rangle \sum_i \frac{\langle b|\mathbf{d} \cdot \boldsymbol{\varepsilon}|i\rangle\langle i|\mathbf{d} \cdot \boldsymbol{\varepsilon}|c\rangle}{E_c + \hbar\omega - E_i}. \quad \text{(C.71)}$$

Finally, the effective Hamiltonian that describes the evolution within $\mathscr{E}_0$ is written

$$(N-2)\hbar\omega + \begin{pmatrix} \tilde{E}_b + \hbar\Delta'_b - i\hbar\dfrac{\Gamma_b}{2} & \hat{R}_{bc}(E_0) \\[2mm] \hat{R}_{cb}(E_0) & 2\hbar\omega + \tilde{E}_c + \hbar\Delta'_c \end{pmatrix} \quad \text{(C.72)}$$

and has many similarities to the Hamiltonian (C.51) discussed in the preceding subsection. We may thus use the results of this subsection to determine the temporal evolution of the transition amplitudes $U_{cc}(t)$ and $U_{bc}(t)$. As above, we successively examine the case of weak coupling ($|\hat{R}_{bc}(E_0)| \ll \hbar\Gamma_b$) and that of strong coupling ($|\hat{R}_{bc}(E_0)| \gg \hbar\Gamma_b$).

c) Weak Coupling Case. Two-Photon Excitation Rate

Expression (C.58) shows that the population of state c decreases exponentially with a rate given by the imaginary part of the eigenvalue of the effective Hamiltonian (C.72) relative to the unperturbed state $|\varphi_c\rangle$. Let us write $-i\hbar\Gamma''/2$ the imaginary part of the eigenvalue of (C.72) closest to $N\hbar\omega + \tilde{E}_c + \hbar\Delta'_c$. Γ'' equals

$$\Gamma'' = \Gamma_b \frac{|\hat{R}_{bc}(E_0)|^2}{\hbar^2\left[(\tilde{\omega}_{bc} + \Delta'_b - \Delta'_c - 2\omega)^2 + (\Gamma_b^2/4)\right]} \quad \text{(C.73)}$$

and represents the two-photon excitation rate. Note that this transition probability is maximal when

$$\hbar\tilde{\omega}_{bc} + \hbar(\Delta'_b - \Delta'_c) = 2\hbar\omega \quad \text{(C.74)}$$

that is, when the energy of the two photons is resonant with the energy difference between levels b and c, corrected by spontaneous radiative level shifts (included in $\hbar\tilde{\omega}_{bc} = \tilde{E}_b - \tilde{E}_c$) and light shifts $\hbar\Delta'_b$ and $\hbar\Delta'_c$.

The resonance condition thus depends on light shifts: it follows that the center of the two-photon resonance shifts when the incident light intensity changes. The width at half-maximum is $\hbar\Gamma_b$. The variation of Γ'' with the intensity is quadratic because $\hat{R}_{bc}(E_0)$ varies as $\langle I \rangle$ [see (C.71)]. The two-photon transition probability is thus proportional to the square of the incident intensity.

Remark

In the case of weak coupling ($|\hat{R}_{bc}(E_0)| \ll \hbar\Gamma_b$), that is, when the two-photon transition is not "saturated", the light shifts are often small compared with the natural width $\hbar\Gamma_b$ of the levels studied. This allows one to use the two-photon transitions for high-resolution spectroscopy (see Chapter II, §D-2). Such a property results from the fact that the light shifts $\hbar\Delta'_b$ and $\hbar\Delta'_c$ vary as $\hat{R}_{bc}(E_0)$, proportionally to $\langle I \rangle$. There are thus many situations in which we can have $|\Delta'_b - \Delta'_c| \ll \Gamma_b$ while still having a non-negligible transition probability (*).

d) STRONG COUPLING LIMIT. TWO-PHOTON RABI OSCILLATION

When $|\hat{R}_{bc}(E_0)| \gg \hbar\Gamma_b$, the eigenvalues of the effective Hamiltonian (C.72) are given by expressions similar to (C.60.a) and (C.60.b) but where W_{bc} and $\tilde{\omega}_{bc}$ are replaced by $\hat{R}_{bc}(E_0)$ and $(\tilde{\omega}_{bc} + \Delta'_b - \Delta'_c - 2\omega)$. By analogy with the results obtained in subsection C-3-d, we can therefore predict that the evolution of the system will be a damped oscillation between levels b and c. For example, at resonance ($\tilde{\omega}_{bc} + \Delta'_b - \Delta'_c = 2\omega$), the probability of finding the system in the state $|\varphi_c\rangle$ is equal to

$$e^{-\Gamma_b t/2} \cos^2 \frac{|\hat{R}_{bc}(E_0)|}{\hbar} t. \tag{C.75}$$

The Rabi oscillation phenomenon thus also appears for a two-photon resonance. This oscillation occurs at the generalized Rabi frequency $2|\hat{R}_{bc}(E_0)|/\hbar$, which is, according to (C.71), proportional to the incident intensity.

e) HIGHER-ORDER MULTIPHOTON TRANSITIONS

The foregoing approach may easily be generalized to a multiphoton transition involving more than two photons.

We consider for example the three-photon resonant transition between the states $c = 2s$ and $b = 4f$ of the hydrogen atom. Near resonance, the states $|\varphi_c\rangle = |c; N\mathbf{k}\varepsilon\rangle$ and $|\varphi_b\rangle = |b; (N - 3)\mathbf{k}\varepsilon\rangle$ play a preponderant role. As above, the diagonal elements $R_{cc}(E_0 + i\eta)$ and $R_{bb}(E_0 + i\eta)$ (where E_0 is the mean of the energies of the states $|\varphi_b\rangle$ and $|\varphi_c\rangle$) appear to the second order in H'_I and represent the natural widths and spontaneous radiative level shifts of b and c, as well as the light shifts of these levels. On the other hand, $R_{bc}(E_0)$ now appears only to the third order in

(*) See, for example, Grynberg, Cagnac, and Biraben.

H'_I and equals, retaining only the first nonvanishing term,

$$\hat{R}_{bc}(E_0) = \left(\frac{N\hbar\omega}{2\varepsilon_0 L^3} \right)^{3/2} \sum_{i,j} \frac{\langle b | \mathbf{d} \cdot \boldsymbol{\varepsilon} | j \rangle \langle j | \mathbf{d} \cdot \boldsymbol{\varepsilon} | i \rangle \langle i | \mathbf{d} \cdot \boldsymbol{\varepsilon} | c \rangle}{(E_c + 2\hbar\omega - E_j)(E_c + \hbar\omega - E_i)}. \quad \text{(C.76)}$$

Depending on the relative values of $|\hat{R}_{bc}(E_0)|$ and $\hbar\Gamma_b$, we can define either a three-photon transition rate (if $|\hat{R}_{bc}(E_0)| \ll \hbar\Gamma_b$) equal to

$$\Gamma'' = \Gamma_b \frac{\left| \hat{R}_{bc}(E_0) \right|^2}{\hbar^2 \left[\left(\tilde{\omega}_{bc} + \Delta'_b - \Delta'_c - 3\omega \right)^2 + \frac{\Gamma_b^2}{4} \right]} \quad \text{(C.77)}$$

or a three-photon Rabi oscillation frequency (if $|\hat{R}_{bc}(E_0)| \gg \hbar\Gamma_b$) equal to $2|\hat{R}_{bc}(E_0)| / \hbar$.

Note that, when the transition is not saturated, the transition probability Γ'' is proportional to $|\hat{R}_{bc}(E_0)|^2$ which varies as the cube of the light intensity. Thus the light shifts play a more important role than they do in the case of two-photon resonance because the coupling matrix element $|\hat{R}_{bc}(E_0)|$ inducing the transition appears at an order of perturbation higher than Δ'_b and Δ'_c.

f) LIMITATIONS OF THE FOREGOING TREATMENT

In all the examples discussed in subsections C-3 and C-4, we were able to study the evolution of the system by limiting ourselves to a small number of selected levels. Such a situation is, nevertheless, far from being general. For example, let us consider one-photon resonant absorption from the ground state a. We call b the first excited level and assume that only a single mode of the field is populated, this mode containing N photons of frequency ω close to $(E_b - E_a)/\hbar$. To find the evolution of the system, we may be tempted to generalize the foregoing treatment by studying the projection of the resolvent onto the subspace subtended by $|a; N\mathbf{k}\boldsymbol{\varepsilon}\rangle$ and $|b; (N-1)\mathbf{k}\boldsymbol{\varepsilon}\rangle$. Nevertheless, it is clear that such an approach does not allow us to predict the probability of finding the atom in one of the levels a or b. Actually, the spontaneous emission process from the level b couples state $|b; (N-1)\mathbf{k}\boldsymbol{\varepsilon}\rangle$ to state $|a; (N-1)\mathbf{k}\boldsymbol{\varepsilon}, \mathbf{k}_1\boldsymbol{\varepsilon}_1\rangle$. A new absorption from this level leads to the state $|b; (N-2)\mathbf{k}\boldsymbol{\varepsilon}, \mathbf{k}_1\boldsymbol{\varepsilon}_1\rangle$, etc. It is thus impossible to limit ourselves only to the states $|a; N\mathbf{k}\boldsymbol{\varepsilon}\rangle$ and $|b; (N-1)\mathbf{k}\boldsymbol{\varepsilon}\rangle$ (except for times sufficiently short so that the probability of return by spontaneous emission to level a is negligible). For this type of problem, the use of the resolvent is not convenient because the subspace of selected states $\mathscr{E}_0$ has a dimension that is too large and it is necessary to use other methods (master equation, optical Bloch equations, dressed atom), which will be discussed in the following chapters.

GENERAL REFERENCES

Messiah (§§XVI-15, XXI-13, 14, 15), Goldberger and Watson (Chapter VIII), Roman (§4-5).

COMPLEMENT A$_{III}$

ANALYTIC PROPERTIES OF THE RESOLVENT

In Section A of this chapter, we introduced the resolvent

$$G(z) = \frac{1}{z - H} \tag{1}$$

of the Hamiltonian H of the system under study, z being a complex variable. We have shown that this operator has several interesting properties. In particular, its matrix elements between eigenstates of the unperturbed Hamiltonian H_0 are often more simple to calculate than those of the evolution operator $U(\tau)$. For instance, it is possible to formally sum the perturbation series in powers of $V = H - H_0$. Once the matrix element of $G(z)$ between the initial and final states of the process under study has been calculated, the corresponding matrix element of $U(\tau)$ can be deduced from it by a contour integral. It is thus important to specify the analytic properties of the matrix elements of $G(z)$ considered as functions of the complex variable z. This is the aim of this Complement.

We begin (§1) by showing that the matrix elements of $G(z)$ are analytic in the entire complex plane excluding the real axis. We then analyze (§2) the singularities of these matrix elements on the real axis and show that they are related to the discrete or continuous nature of the eigenvalues of H. Finally, we return to the problem of a discrete state of H_0 coupled to a continuum. We first show that such unstable states correspond to poles of the analytic continuation of the resolvent in the second Riemann sheet (§3). Exponential decays are associated with these poles. We then show (§4) that the contour integral also gives rise to nonexponential contributions.

1. Analyticity of the Resolvent outside the Real Axis

Let $|u\rangle$ be an arbitrary normalized state

$$\langle u|u\rangle = 1 \tag{2}$$

and

$$G_u(z) = \langle u|G(z)|u\rangle = \langle u|\frac{1}{z - H}|u\rangle \tag{3}$$

be the matrix element of $G(z)$ in this state.

To study the analytic properties of $G_u(z)$, it is convenient to introduce the eigenstates of H. Some are discrete eigenstates, $|\psi_i\rangle$ with eigenvalues E_i, the other eigenstates $|\psi_\mathbf{k}\rangle$ belonging to the continuous spectrum. If the states $|\psi_i\rangle$ and $|\psi_\mathbf{k}\rangle$ have the usual normalization, the closure relation over the eigenstates of H is written

$$\sum_i |\psi_i\rangle\langle\psi_i| + \int d^3k |\psi_\mathbf{k}\rangle\langle\psi_\mathbf{k}| = \mathbb{1}. \tag{4.a}$$

It is often convenient to have the energy E explicitly appearing as a quantum number labeling the states of the continuous spectrum, to which must be added other quantum numbers denoted by γ. The change of variables $\mathbf{k} \to E, \gamma$ then causes the density of states to appear in $d^3k = \rho(E, \gamma) dE\, d\gamma$ (see §3-c in Complement A_I) and (4.a) becomes

$$\sum_i |\psi_i\rangle\langle\psi_i| + \int\int dE\, d\gamma\, \rho(E, \gamma)|\psi(E, \gamma)\rangle\langle\psi(E, \gamma)| = \mathbb{1}. \tag{4.b}$$

The introduction of (4.b) between $1/(z - H)$ and $|u\rangle$ in (3) yields

$$G_u(z) = \sum_i \frac{|\langle u|\psi_i\rangle|^2}{z - E_i} +$$

$$+ \int\int dE'\, d\gamma'\, \rho(E', \gamma') \frac{|\langle u|\psi(E', \gamma')\rangle|^2}{z - E'}. \tag{5}$$

Let us now assume that z is a point in the complex plane located at a nonzero distance δ from the real axis. In (5), all the denominators $z - E_i$, $z - E'$ have a modulus larger than δ. Since all the numerators of (5) are positive, we can thus give an upper bound for $|G_u(z)|$

$$|G_u(z)| \leq \frac{1}{\delta}\left[\sum_i |\langle u|\psi_i\rangle|^2 + \right.$$

$$\left. + \int\int dE'\, d\gamma'\, \rho(E', \gamma')|\langle u|\psi(E', \gamma')\rangle|^2\right]. \tag{6}$$

The term between brackets in (6) is simply the square of the norm of $|u\rangle$, which, according to (2), equals 1, so that

$$|G_u(z)| \leq \frac{1}{\delta}. \tag{7}$$

A similar demonstration allows us to show that the derivative of $G_u(z)$, $G'_u(z)$ is also bounded.

Finally, $G_u(z)$ is, according to (5), an infinite sum of terms which are analytic functions of z outside the real axis. This sum is finite. Thus $G_u(z)$ exists outside the real axis, as well as its derivative. $G_u(z)$ is thus analytic outside the real axis.

Remark

The foregoing demonstration may be generalized to matrix elements of $G(z)$ between two different normalized states $|u\rangle$ and $|v\rangle$. Inequality (6) is then replaced by

$$|\langle u|G(z)|v\rangle| \le \frac{1}{\delta}\left[\sum_i |\langle u|\psi_i\rangle\langle\psi_i|v\rangle| + \right.$$

$$\left. + \int\int dE'\,d\gamma'\,\rho(E',\gamma')|\langle u|\psi(E',\gamma')\rangle\langle\psi(E',\gamma')|v\rangle|\right]. \quad (6')$$

By using for each of the products $|\langle u|\psi\rangle\langle\psi|v\rangle|$ the fact that $|\langle u|\psi\rangle\langle\psi|v\rangle| \le [|\langle u|\psi\rangle|^2 + |\langle v|\psi\rangle|^2]/2$, we obtain

$$|\langle u|G(z)|v\rangle| \le 1/\delta \quad (7')$$

and the foregoing conclusion remains valid.

2. Singularities on the Real Axis

First of all, at every point E of the real axis which is not an eigenvalue of H, $G_u(z)$ is analytic. In the foregoing demonstration, it suffices to replace δ by the distance between E and the eigenvalue of H closest to E.

When z tends to a discrete eigenvalue E_i of H, the term $|\langle u|\psi_i\rangle|^2/(z - E_i)$ of the sum (5) tends to infinity. The discrete eigenvalues of H are thus poles for $G_u(z)$ (except when $\langle u|\psi_i\rangle = 0$), having $|\langle u|\psi_i\rangle|^2$ as residues.

Let us now assume that z tends to an eigenvalue of E in the continuous spectrum, which is distinct from a discrete eigenvalue. Two situations are possible, depending on whether the imaginary part of z is positive or negative (Figure 1). To see this, let us calculate $\lim_{\eta \to 0_+} G_u(E \pm i\eta)$, which, according to (5), equals

$$\lim_{\eta \to 0_+} G_u(E \pm i\eta) = \lim_{\eta \to 0_+}\left[\sum_i \frac{|\langle u|\psi_i\rangle|^2}{E \pm i\eta - E_i} + \int dE' \frac{f_u(E')}{E \pm i\eta - E'}\right]$$

$$(8)$$

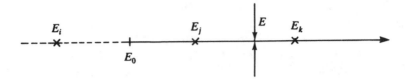

Figure 1. Discrete eigenvalues $(E_i, E_j, \dots)$ and continuous spectrum of H (starting from E_0).

where

$$f_u(E') = \int d\gamma' \, \rho(E', \gamma') |\langle u_1 \psi(E', \gamma') \rangle|^2. \tag{9}$$

The first term of the right-hand side of (8) simply tends, when $\eta \to 0_+$, to $\Sigma_i |\langle u|\psi_i \rangle|^2/(E - E_i)$, which is a real quantity. Using

$$\lim_{\eta \to 0_+} \frac{1}{x \pm i\eta} = \mathscr{P}\frac{1}{x} \mp i\pi\delta(x) \tag{10}$$

where $\mathscr{P}$ designates the principal part, one finds that the second term is equal to

$$\mathscr{P}\int dE' \, \frac{f_u(E')}{E - E'} \mp i\pi f_u(E). \tag{11}$$

The limits of $G_u(E \pm i\eta)$, when $\eta \to 0_+$, actually exist, but are different depending on whether z tends to E by positive or negative values of the imaginary part [if $f_u(E)$ is nonzero]. We say that the function $G_u(z)$ has a cut along the continuous spectrum. The difference, $-2i\pi f_u(E)$, between the values of $G_u(E + i\eta)$ and $G_u(E - i\eta)$, represents the gap between the two sides of the cut. The point at which the cut begins, E_0, is called the branch point. Because the density of states $\rho(E', \gamma)$ generally tends to zero when E' tends to E_0, the same is true for $f_u(E')$ according to (9). The gap between the two sides of the cut thus tends to zero when E approaches the branch point.

    Also note that the values taken by the function $G_u(z)$ on both sides of the cut are complex conjugates of each other, according to (11). More generally, the Hermiticity of H results in

$$G_u(z^*) = (G_u(z))^*. \tag{12}$$

The function $G_u(z)$ is analytic in the upper half-plane. We can make an analytic continuation of this function beyond the cut onto the lower half-plane. The value taken by the analytic continuation is different from the value taken at the same point by the initial determination of the function, because the cut precisely introduces a discontinuity into this determination. We say that $G_u(z)$ has been continued onto the second Riemann sheet, the complex plane constituting the first Riemann sheet. A similar continuation may be made from the lower half-plane to the upper half-plane.

In summary, the foregoing discussion has shown that the function $G_u(z)$ is analytic in the complex plane except for the discrete eigenvalues of H, which constitute poles, and for the continuous spectrum of H that makes up a cut. It is possible to analytically continue $G_u(z)$ beyond the cut onto the second Riemann sheet. The second determination $G_u^{II}(z)$ thus obtained may have poles. We will now see how these poles characterize the unstable states resulting, for example, from the coupling of a discrete state with a continuum.

3. Unstable States and Poles of the Analytic Continuation of the Resolvent

We return to the problem of spontaneous emission discussed in subsection C-1. The discrete state $|\varphi\rangle = |b; 0\rangle$, which is an eigenstate of the unperturbed Hamiltonian H_0, with eigenvalue E_b, represents the atom in the discrete excited state $|b\rangle$ in the photon vacuum $|0\rangle$. It is coupled by H_{I1} to the states $|a; \mathbf{k}\boldsymbol{\varepsilon}\rangle$ which represent the atom in the state a in the presence of a photon $\mathbf{k}\boldsymbol{\varepsilon}$. The states $|a; \mathbf{k}\boldsymbol{\varepsilon}\rangle$, of unperturbed energy $E_a + \hbar\omega$ form a continuum starting from E_a (for the sake of simplicity, we assume that there is no atomic state other than $|a\rangle$ with energy less than E_b).

The matrix element of the unperturbed resolvent in the state $|\varphi_b\rangle$

$$G_{0b}(z) = \langle\varphi_b|\frac{1}{z - H_0}|\varphi_b\rangle = \frac{1}{z - E_b} \tag{13}$$

has a pole at $z = E_b$. As for the matrix element of $G(z)$ in $|\varphi_b\rangle$, we have shown that it can be put in the form

$$G_b(z) = \frac{1}{z - E_b - R_b(z)} \tag{14}$$

and we have calculated, because H_{I1} is very small, an approximate value

$\hat{R}_b(z)$ for $R_b(z)$ given by (C.5), so that

$$G_b(z) \simeq \frac{1}{z - E_b - \hat{R}_b(z)}. \tag{15}$$

Near the real axis, $G_b(z)$ may be written

$$G_b(E \pm i\eta) \simeq \frac{1}{E \pm i\eta - E_b - \hbar\hat{\Delta}_b(E) \pm i\dfrac{\hbar}{2}\hat{\Gamma}_b(E)}. \tag{16}$$

The quantity $\hat{\Gamma}_b(E)$ is a positive function, which is zero for $E < E_a$, and which describes the intensity of the coupling of the discrete state $|\varphi_b\rangle$ with the shell of energy E in the continuum $\{|a;\mathbf{k}\varepsilon\rangle\}$ [see (C.12)]. The quantity $\hat{\Delta}_b(E)$ is simply related to $\hat{\Gamma}_b(E)$ by the dispersion relation (C.14). For $E = E_b$, $\hat{\Gamma}_b(E_b) = \Gamma_b$, and $\hbar\hat{\Delta}_b(E_b) = \hbar\Delta_b$ represent, respectively, the radiative decay rate of $|\varphi_b\rangle$ to the continuum and the radiative shift of the state $|\varphi_b\rangle$.

It is clear from (16) that $G_b(E + i\eta)$ and $G_b(E - i\eta)$ tend to well-defined limits when $\eta \to 0_+$, but these limits are different when $\hat{\Gamma}_b(E)$ is nonzero, because

$$G_b(E + i\eta) - G_b(E - i\eta) = \frac{-i\hbar\hat{\Gamma}_b(E)}{\left[E - E_b - \hbar\hat{\Delta}_b(E)\right]^2 + \left[\hbar\hat{\Gamma}_b(E)/2\right]^2}. \tag{17}$$

Thus, $G_b(z)$ possesses a cut on the real axis, starting from the branch point E_a [because $\hat{\Gamma}_b(E) = 0$ for $E < E_a$] and going to infinity.

The fact that the singularity changes from a pole for $G_{0b}(z)$ to a cut for $G_b(z)$ shows that H no longer has, as H_0 does, any discrete state of energy close to E_b. The coupling V has in a way "dissolved" the discrete state $|\varphi_b\rangle$ in the continuum. However, a trace of the discrete state remains, which is revealed by significant variations of $G_b(E + i\eta)$ and $G_b(E - i\eta)$ when E varies about $E_b + \hbar\Delta_b$. We will now show that these important variations of $G_b(E \pm i\eta)$ are in fact due to the existence of a pole in the analytic continuation of $G_b(z)$ near $z = E_b$.

Remarks

(i) The distance (17) between the two sides of the cut varies strongly with E near $E = E_b$. These variations, which are close to those of a Lorentzian centered at $E_b + \hbar\Delta_b$ and of width $\hbar\Gamma_b$, are just those of the density of the discrete state $|\varphi_b\rangle$ in the new continuum. In fact, if $|\psi(E, \gamma)\rangle$ designates the eigenstate of the total Hamiltonian H with eigenvalue E, the general expressions (8), (9), and (11) demonstrated above show that, for $|u\rangle = |\varphi_b\rangle$

$$G_b(E + i\eta) - G_b(E - i\eta) = -2i\pi \int d\gamma \, \rho(E, \gamma) |\langle \varphi_b | \psi(E, \gamma) \rangle|^2. \quad (18)$$

The integral of the right-hand side of (18) is just the density of the state $|\varphi_b\rangle$ in the new continuum. The comparison of (17) and (18) clearly shows that this density is significant near $E_b + \hbar\Delta_b$ and varies very rapidly with E over an interval of width $\hbar\Gamma_b$. Such a result generalizes the one obtained in Complement C_I on a very simple model.

(ii) It is possible to exactly calculate $R_b(z)$ and thus to obtain for $\Delta_b(E)$ and $\Gamma_b(E)$ more general expressions than the perturbative expressions $\hat{\Delta}_b(E)$ and $\hat{\Gamma}_b(E)$ given in (C.11) and (C.12). For this it is sufficient to use the general expressions (B.37) and (B.38) written in the case where the subspace $\mathscr{E}_0$ has only one dimension [we then have $\Delta_b(E) = \Delta_{bb}(E)$ and $\Gamma_b(E) = \Gamma_{bb}(E)$]. If the closure relation over the eigenstates of the operator QHQ is inserted between the two operators V of (B.37) and (B.38), we obtain exact expressions having the same structure as (C.11) and (C.12), $|a; k\varepsilon\rangle$, being replaced by an eigenstate of QHQ, and $E_a + \hbar\omega$ by the corresponding eigenvalue. It appears then that $\Gamma_b(E)$ is positive and nonzero for $E \geq \bar{E}_a$, where $\bar{E}_a$ is the energy of the ground state of QHQ, that is, the energy E_a of the ground state of H_0, corrected by the radiative shift of this state. The cut of $G_b(z)$ thus starts from $\bar{E}_a$ and not from E_a. Similarly, we can show that the dispersion relation (C.14) remains valid between the exact functions $\Delta_b(E)$ and $\Gamma_b(E)$.

Let us now return to (15). Near $z = E_b$, we can, according to (16), approximate $\hat{R}_b(z)$ by $\hbar[\Delta_b - i(\Gamma_b/2)]$ for Im $z > 0$, and by $\hbar[\Delta_b + i(\Gamma_b/2)]$ for Im $z < 0$. Near E_b and for Im $z > 0$,

$$G_b(z) \simeq \frac{1}{z - E_b - \hbar\Delta_b + i(\hbar\Gamma_b/2)}. \quad (19.a)$$

Near E_b and for Im $z < 0$,

$$G_b(z) \simeq \frac{1}{z - E_b - \hbar\Delta_b - i(\hbar\Gamma_b/2)}. \quad (19.b)$$

It is clear in (19) that $G_b(z)$ has no pole near E_b because the imaginary part of the denominator is never zero (Im z and $\pm \hbar\Gamma_b/2$ always have the same sign). Let us now make the analytic continuation of $G_b(z)$, given by (19.a), from the upper half-plane to the lower half-plane across the cut. Near E_b, the continuation of the function onto the second Riemann sheet,

$G_b^{\rm II}(z)$, has by continuity a form quite close to (19.a) but the domain of definition is now extended to Im $z < 0$. Near E_b and for Im $z < 0$,

$$G_b^{\rm II}(z) \simeq \frac{1}{z - E_b - \hbar\Delta_b + i(\hbar\Gamma_b/2)}. \tag{20}$$

Thus it is clear that $G_b^{\rm II}(z)$ has a pole at

$$z_0 \simeq E_b + \hbar\Delta_b - i(\hbar\Gamma_b/2). \tag{21}$$

Similarly we could make the analytic continuation (19.b) from the lower half-plane to the upper half-plane and show that the continued function has a pole in z_0^*.

We thus see that the pole of $G_{0b}(z)$, which was at $z = E_b$ on the real axis [see (13)], has not really disappeared in the change from H_0 to H. The effect of the coupling V was simply to slightly shift this pole onto the second Riemann sheet. In general, the unstable states, that is, the traces of the discrete states of H_0 coupled to continua, correspond to poles of the analytic continuation of the resolvent (such states are also called "resonances").

4. Contour Integral and Corrections to the Exponential Decay

Let us return to the contour integral, which allows us to connect $U_b(\tau)$ to $G_b(z)$ [see Figure 1 and expression (A.22) in this chapter]. If we assume that $\tau > 0$, the contribution of C_- is zero in (A.22) (we can close the contour to $z = -i\infty$ and there are no poles in the lower half-plane). The contour thus reduces to C_+, that is a line slightly above the real axis and followed from right to left, so that if $\tau > 0$

$$U_b(\tau) = \frac{1}{2\pi i} \int_{+\infty+i\eta}^{-\infty+i\eta} G_b(z)\, e^{-iz\tau/\hbar}\, dz. \tag{22}$$

To evaluate the integral by the residue method, it is necessary to close the contour toward the bottom [(Im z)τ must be negative]. Thus the contour must necessarily pass into the second Riemann sheet in its right-hand part, because $G_b(z)$ has a cut extending from E_a to infinity, whereas in its left-hand part, it remains in the first Riemann sheet. To close the contour, it is thus necessary to go back to turn around the branch point of the cut at $E = E_a$ (see Figure 2).

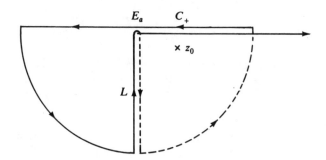

Figure 2. Closure toward the bottom of the contour integral C_+. The parts drawn with dashed lines are in the second Riemann sheet.

The contour thus encloses the pole located at z_0 in the second Riemann sheet, the existence of which has been demonstrated above [we shall assume that $G_b^{\mathrm{II}}(z)$ has no other pole within the contour]. The contribution of this pole to the contour integral is a damped exponential. The contribution of the semicircle is zero if its radius is sufficiently large. Finally, the contribution of the contour L formed by the contour turning around the branch point is written

$$\frac{1}{2\pi i} \int_{E_a - i\infty}^{E_a} \left[G_b(z) - G_b^{\mathrm{II}}(z) \right] e^{-iz\tau/\hbar} \, dz. \tag{23}$$

It represents corrections to the exponential decay associated with the pole at z_0. Expression (23) in particular allows us to understand why the decay of $U_b(\tau)$ at very long times $(\tau \gg \Gamma_b^{-1})$ is less rapid than an exponential decay. In fact, for very large τ, the exponential $\exp(-iz\tau/\hbar)$ damps the contribution of the pole much more than that of the contour L, because this contour has regions where Im z is as small as we like, whereas for the pole Im z remains equal to Im z_0. Actually, if $G_b(z) - G_b^{\mathrm{II}}(z)$ varies as $(z - E_a)^n$ near $z = E_a$, the integral of (23) which, for $\tau \gg \Gamma_b^{-1}$, involves only this neighborhood, varies as $1/\tau^{n+1}$ (for reasons of homogeneity).

GENERAL REFERENCE

Goldberger and Watson (Chapter VIII).

COMPLEMENT B_{III}

NONPERTURBATIVE EXPRESSIONS FOR THE SCATTERING AMPLITUDES OF A PHOTON BY AN ATOM

In Chapter II, we reviewed several processes in which photons are scattered by charged particles or by atoms. To describe these processes, we used scattering amplitudes calculated in lowest-order perturbation theory. As we have already seen for resonant scattering, such an approach is not always sufficient, in the sense that diagrams involving the resonant intermediate state can lead to divergences. It is therefore necessary to sum all these diagrams to obtain a finite expression.

An approach going beyond the lowest order is also necessary to correctly account for the interaction of an atom with a quantum electromagnetic field in the initial and final states of a scattering process. In fact, even if the incident photon is far from the atom, the atom interacts with the electromagnetic field by virtually emitting and reabsorbing photons. The atom-field interaction is thus always present and must be taken into account in the very definition of asymptotic scattering states.

These problems may be approached in different ways, in particular by using scattering states. We will present here another approach, which is related to the time evolution of the system and which uses the properties of the operator $G(z)$. In subsection 1, we will reconsider from such a point of view the second-order calculation of scattering amplitudes described in Complement A_I and we will complete it by the derivation of the Born expansion of the $\mathcal{T}$ matrix. We will also obtain a compact expression for this expansion using the resolvent and allowing us to treat the case of resonant scattering. Nevertheless, it appears that such an approach is not sufficient to treat the problem of the scattering of a quantum field, and that the asymptotic stationary states must be defined correctly. This problem is treated in subsection 2. Finally, in subsection 3, we derive an exact expression for the S-matrix for the scattering of a photon by an atom, taking into account all the radiative processes, in the intermediate state as well as in the initial and final states.

1. Transition Amplitudes between Unperturbed States

Throughout this subsection, we will consider that the initial and final states of the scattering are eigenstates of the unperturbed Hamiltonian

H_0. For the scattering of one photon, these states are

$$|\varphi_i\rangle = |a; \mathbf{k}\varepsilon\rangle \tag{1.a}$$

$$|\varphi_f\rangle = |a'; \mathbf{k}'\varepsilon'\rangle \tag{1.b}$$

and we have to evaluate the transition amplitude between the states $|\varphi_i\rangle$ and $|\varphi_f\rangle$.

At the limit where the interaction time tends to infinity, and if one eliminates the obvious phase factors due to the unperturbed evolution of the states $|\varphi_i\rangle$ and $|\varphi_f\rangle$, this amplitude is the element S_{fi} of the S-matrix:

$$S_{fi} = \langle \varphi_f | S | \varphi_i \rangle =$$

$$= \lim_{T \to \infty} \left[\exp(iE_f T/2\hbar) \langle \varphi_f | U(T/2, -T/2) | \varphi_i \rangle \exp(iE_i T/2\hbar) \right]. \tag{2}$$

In (2), $U(t, t')$ represents the evolution operator associated with the total Hamiltonian

$$H = H_0 + V \tag{3}$$

including the interaction V.

a) USING THE RESOLVENT

We will explicitly calculate (2) by using the properties of the resolvent $G(z)$. According to Equation (A.22) taken for $t > 0$,

$$S_{fi} = \lim_{T \to +\infty} \left\{ \exp \frac{iT}{2\hbar} (E_f + E_i) \left(\int_{C_+} dz \, \frac{e^{-izT/\hbar}}{2i\pi} \langle \varphi_f | G(z) | \varphi_i \rangle \right) \right\}. \tag{4}$$

We will not directly replace $G(z)$ by $1/(z - H)$ here. In fact, in order to obtain an expression having a well-defined limit when T tends to infinity, it is convenient to introduce $G_0(z)$ to the left and to the right of $G(z)$. The corresponding poles located at E_i and E_f thus give rise to oscillating terms compensating for those outside the integral. To do this, we use Equation (A.23) connecting G to G_0 in the form

$$G(z) = G_0(z) + G_0(z)VG(z) \tag{5.a}$$

then in the form

$$G(z) = G_0(z) + G(z)VG_0(z). \tag{5.b}$$

Replacing $G(z)$ by expression (5.b) in the second term of (5.a) thus yields

$$G(z) = G_0(z) + G_0(z)VG_0(z) + G_0(z)VG(z)VG_0(z) \qquad (6)$$

and consequently

$$\langle \varphi_f | G(z) | \varphi_i \rangle = \frac{\delta_{if}}{z - E_i} +$$

$$+ \frac{1}{(z - E_i)(z - E_f)} \{ V_{fi} + \langle \varphi_f | VG(z)V | \varphi_i \rangle \}. \qquad (7)$$

We assume for the time being that the matrix element $\langle \varphi_f | VG(z)V | \varphi_i \rangle$ has no real pole in the region $z = E_i$ or E_f. On the other hand, we do not exclude the existence of a cut corresponding to the continuous part of the spectrum of H. Complex poles may of course exist in the second Riemann sheet.

We must now substitute expression (7) into (4) and calculate the integral by the method of residues, closing the contour to $z = -i\infty$ along the path indicated in Figure 2 of Complement A_{III}. Let us first consider the contribution of the singularities of $\langle \varphi_f | VG(z)V | \varphi_i \rangle$. The complex poles of the second Riemann sheet yield terms of the type

$$\exp \frac{iT}{\hbar} \left\{ \frac{E_f + E_i}{2} - \left(E_b + \hbar \Delta_b - i\hbar \frac{\Gamma_b}{2} \right) \right\} \qquad (8)$$

which rapidly tend to zero when T tends to infinity. Similarly, as we saw in Complement A_{III}, the contribution of the contour L turning around the branch point also tends to zero when $T \to \infty$. Finally, possible discrete poles E_α, far from E_i and E_f, would yield contributions of the type

$$\frac{1}{(E_\alpha - E_i)(E_\alpha - E_f)} \text{Res}\{\langle \varphi_f | VG(E_\alpha)V | \varphi_i \rangle\} \times$$

$$\exp \frac{iT}{\hbar} \left(\frac{E_f + E_i}{2} - E_\alpha \right) \qquad (9)$$

According to the hypotheses made concerning the poles E_α, the denominators are large, and the coefficient of the exponential is therefore small. Moreover, when T tends to infinity, the exponential becomes an extremely rapidly oscillating function of E_i and E_f. Considered as a distribution, such a function is zero, because when multiplied by a slowly varying function of E_i and E_f and integrated over these variables, it yields an integral which tends to zero when $T \to \infty$. Finally, the singularities of $\langle \varphi_f | VG(z)V | \varphi_i \rangle$ do not contribute to the integral in (4).

b) TRANSITION MATRIX

The calculation of S_{fi} is thus reduced to that of the contribution of the poles of (7) located at $z = E_i$ and $z = E_f$. Let us write $\mathbb{T}_{fi}(E)$ the coefficient of $1/(z - E_i)(z - E_f)$ in (7), taken for $z = E + i\eta$:

$$\mathbb{T}_{fi}(E) = \langle \varphi_f | V + VG(E + i\eta)V | \varphi_i \rangle. \tag{10}$$

The contribution to (4) of the poles at E_i and E_f of (7) may then be written

$$S_{fi} = \lim_{T \to \infty} \left\{ \exp \frac{iT}{2\hbar}(E_f + E_i) \times \right.$$

$$\left. \times \left[\delta_{if}e^{-iE_iT/\hbar} + \mathbb{T}_{fi}(E_i)\frac{e^{-iE_iT/\hbar}}{E_i - E_f} + \mathbb{T}_{fi}(E_f)\frac{e^{-iE_fT/\hbar}}{E_f - E_i} \right] \right\}$$

$$= \delta_{fi} + \lim_{T \to \infty} \left\{ \frac{\mathbb{T}_{fi}(E_i) - \mathbb{T}_{fi}(E_f)}{E_i - E_f} \cos(E_f - E_i)\frac{T}{2\hbar} - \right.$$

$$\left. -i\left[\mathbb{T}_{fi}(E_i) + \mathbb{T}_{fi}(E_f) \right]\frac{\sin(E_f - E_i)(T/2\hbar)}{E_f - E_i} \right\}. \tag{11}$$

The coefficient of the cosine is finite for $E_i = E_f$. When T tends to infinity, the cosine is a rapidly oscillating function of $E_i - E_f$, of finite amplitude. We have seen previously that such a function, considered as a distribution, is zero. In the last term, we recognize the function $\pi\delta^{(T)}(E_f - E_i)$, so that at the limit $T \to \infty$, $E_i = E_f$ and S_{fi} reduces to the expression

$$S_{fi} = \delta_{fi} - 2i\pi\delta(E_f - E_i)\mathscr{T}_{fi} \tag{12}$$

where the transition matrix $\mathscr{T}_{fi}$ is given by expression (10) taken for $E = E_i$:

$$\mathscr{T}_{fi} = \langle \varphi_f | V + VG(E_i + i\eta)V | \varphi_i \rangle. \tag{13}$$

We have thus derived in a general way relations (B.13) and (B.15) of Chapter I, under the conditions of regularity assumed for $\langle \varphi_f | VG(z)V | \varphi_i \rangle$. If $G(E_i + i\eta)$ is replaced in (13) by its expansion in powers of V (A.24), the Born expansion of $\mathcal{T}_{fi}$ is obtained. The first terms of this expansion are the same as those previously encountered [relation (29) in Complement A$_1$].

c) APPLICATION TO RESONANT SCATTERING

For this application, the Hamiltonian H_0 is $H_0 = H_P + H_R$ and the coupling V is the interaction Hamiltonian $H_I = H_{I1} + H_{I2} + H_{I1}^S$ (see §3-b of the Appendix). We saw in Chapter II (§C-3) that three states play privileged roles in resonant scattering: the initial state $|a; \mathbf{k}\varepsilon\rangle$ coupled by the interaction H_I to the resonant excited state $|b; 0\rangle$, which is itself coupled to the continuum of final states $|a'; \mathbf{k}'\varepsilon'\rangle$ (see Figure 20 in Chapter II). Neglecting nonresonant processes concerning states $|a\rangle$ and $|a'\rangle$, we will attempt here to take into account all the intermediate transitions to the resonant state $|b; 0\rangle$ between the absorption of the initial photon and the emission of the final photon. For an electric dipole transition, the last two processes can only be achieved in a resonant fashion by H_{I1}, and the matrix $\mathcal{T}$ is written for this approximation:

$$\mathcal{T}_{fi}^{res} = \langle a'; \mathbf{k}'\varepsilon' | H_{I1} | b; 0 \rangle \langle b; 0 | G(E_a + \hbar\omega + i\eta) | b; 0 \rangle \langle b; 0 | H_{I1} | a; \mathbf{k}\varepsilon \rangle.$$

$$(14)$$

We saw in Chapter II that the approximation consisting of replacing G by G_0 in the intermediate state is equivalent to taking into account only process (15),

$$(15)$$

and leads to a divergence of $\mathcal{T}_{fi}^{res}$. A better approximation consists of replacing $\langle b; 0 | G(E_a + \hbar\omega + i\eta) | b; 0 \rangle$ in (14) by expression (C.17). This is equivalent to summing all the higher-order process represented in (16):

$$\text{(16)}$$

The expression for $\mathscr{T}_{fi}^{\text{res}}$ then becomes

$$\mathscr{T}_{fi}^{\text{res}} = \frac{\langle a'; \mathbf{k}'\varepsilon'|H_{I1}|b;0\rangle\langle b;0|H_{I1}|a;\mathbf{k}\varepsilon\rangle}{E_a + \hbar\omega - (E_b + \hbar\Delta_b) + i\hbar(\Gamma_b/2)} \tag{17}$$

and coincides, to within the radiative shift of the state b, with expression (C.5) in Chapter II. Using the resolvent has thus allowed us to derive a nonperturbative expression for the scattering amplitude at resonance.

d) Inadequacy of Such an Approach

In the calculation discussed above, only the higher-order radiative processes affecting the intermediate state were taken into account. Similar processes affecting the initial or final state were ignored. A consequence of this approximation can be seen in (17): The energy of the state a is not corrected by $\hbar\Delta_a$. Starting with the exact expression for $\langle\varphi_f|VG(z)V|\varphi_i\rangle$, we could consider the complete expansion of $G(z)$ and also sum the contributions of the processes in which the first and the last events are not the absorption of the initial photon and the emission of the final photon.

Such a process is shown, for example, in (18):

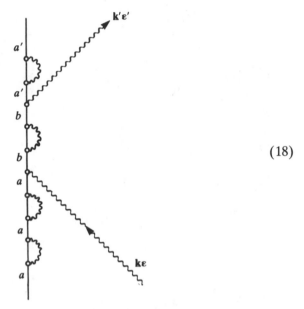

$$(18)$$

In doing so, we would take into account the processes that correct the energy of the initial and final states in the expression for $\mathcal{T}_{fi}$. However, the approach that leads from expression (7) to expression (13) is no longer justified. In fact, in $\langle \varphi_f | VG(z)V | \varphi_i \rangle$ real poles appear at energies $E_a + \hbar\Delta_a + \hbar\omega$ and $E'_a + \hbar\Delta_{a'} + \hbar\omega'$ close to E_i and E_f. Thus it is necessary to completely revise the derivation of the expression for S_{fi}.

The difficulty that we encounter here lies in the fact that the states $|\varphi_i\rangle$ and $|\varphi_f\rangle$ are not asymptotically stationary. In ordinary collisions, the interaction V between the particles may be neglected in the initial state and in the final state. The eigenstates of H_0 are then adequate for describing these states, in the sense that it is possible to build with these states quasi-monochromatic wave packets, having a finite extension and such that the particles are far from each other, so that V can be neglected. In the present case, H_I describes not only the interaction of the atom with the initial and final photons, but also the interaction of the atom with the whole transverse field, and the transverse field interacts permanently with the particles constituting the atom. This interaction is responsible for several physical effects (magnetic interaction between particles, retardation effects in the electromagnetic interaction, radiative corrections), and there is no justification for neglecting H_I in the initial state or in the final state. In other words, the correct asymptotic states of the scattering process must contain not only the free photons $\mathbf{k}\varepsilon$ or $\mathbf{k}'\varepsilon'$, but also the "virtual photon cloud" that describes the transverse field near the atom. We will thus reexamine the theory of scattering by introducing such states.

2. Introducing Exact Asymptotic States

a) THE ATOM IN THE ABSENCE OF FREE PHOTONS

The atomic states $|a\rangle, |a'\rangle, |b\rangle, \ldots$ are eigenstates of the particle Hamiltonian H_P given in Chapter I [formula (C.7.b)]. They are stationary only in the absence of coupling with the transverse field. The true Hamiltonian of the system of particles constituting the atom, interacting via the electromagnetic field, is the total Hamiltonian

$$H = H_p + H_R + H_I$$

$$= H_0 + H_I. \tag{19}$$

The only stable stationary state of the atom is thus the ground state of H, which we will call $|\psi_a\rangle$, of energy $\tilde{E}_a$

$$H|\psi_a\rangle = \tilde{E}_a|\psi_a\rangle. \tag{20}$$

The state $|\psi_a\rangle$ represents the atom in its ground state in the absence of free photons, but accompanied by the transverse field with which it is associated, as well as by vacuum fluctuations of the field throughout all space. The difference between $\tilde{E}_a$ and E_a represents the radiative shift of the ground state. We assume that the state $|\psi_a\rangle$ is normalized. Apart from $\tilde{E}_a$, the spectrum of H is a continuum extending from $\tilde{E}_a$ to $+\infty$.

Remarks

(i) The ground state of H may be degenerate, and can contain several sublevels $|\psi_a\rangle, |\psi_{a'}\rangle \cdots$. Certain levels close to the ground state can have a long radiative lifetime and can be considered as stable to a certain approximation. The spectrum of H is then made up of several continua starting from the energies $\tilde{E}_a, \tilde{E}_{a'} \cdots$.

(ii) The expression for $|\psi_a\rangle$ may be obtained in the form of a Wigner–Brillouin-type expansion. We denote by P the projector onto the eigensubspace of H_0, which contains the zero-order approximation $|\psi_a^0\rangle = |a; 0\rangle$ of $|\psi_a\rangle$ (and possibly the zero-order approximation for other sublevels $|\psi_{a'}\rangle$ of the ground state). The state vectors $|\psi_a\rangle$ and $|\psi_a^0\rangle$ are assumed to be normalized, and the constant representing the projection of $|\psi_a\rangle$ onto $|\psi_a^0\rangle$ is written $\sqrt{Z}$. Let us then project $|\psi_a\rangle$ onto the two subspaces defined by P and $Q = 1\!\!1 - P$:

$$|\psi_a\rangle = \sqrt{Z}\,|\psi_a^0\rangle + Q|\psi_a\rangle. \tag{21}$$

The expression for $Q|\psi_a\rangle$ is obtained as follows: Multiply (20) by Q on the left and insert $1\!\!1 = P + Q$ between H and $|\psi_a\rangle$. Because by definition $P|\psi_a\rangle$ is equal to $\sqrt{Z}\,|\psi_a^0\rangle$ and $PH_0Q = 0$, we have

$$Q|\psi_a\rangle = \frac{1}{\tilde{E}_a - QHQ}QH_IP|\psi_a^0\rangle\sqrt{Z}$$

$$= \frac{1}{\tilde{E}_a - QH_0Q - QH_IQ}QH_IP|\psi_a^0\rangle\sqrt{Z}. \tag{22}$$

The expansion in powers of H_I of the fraction in (22) and substitution into (21) gives the Wigner–Brillouin expansion of $|\psi_a\rangle$. The constant Z is determined by the normalization of $|\psi_a\rangle$.

The state $|\psi_a\rangle$ is not a zero-photon state. To see this, it is sufficient to verify that the operator $a_\varepsilon(\mathbf{k})$ does not give zero when it acts on $|\psi_a\rangle$ (see the calculation in the following). The photons present in state $|\psi_a\rangle$ are in fact the "virtual" photons that "dress" the atom in its ground state and describe the quantum transverse field associated with the atom.

Let us calculate more precisely $a_\varepsilon(\mathbf{k})|\psi_a\rangle$. The relation

$$Ha_\varepsilon(\mathbf{k}) - a_\varepsilon(\mathbf{k})H = -\hbar\omega a_\varepsilon(\mathbf{k}) + \left[H_I, a_\varepsilon(\mathbf{k})\right] \tag{23}$$

derived from $[H_0, a_\varepsilon(\mathbf{k})] = -\hbar\omega a_\varepsilon(\mathbf{k})$ leads to

$$\left(H - \tilde{E}_a + \hbar\omega\right)a_\varepsilon(\mathbf{k})|\psi_a\rangle = \left[H_I, a_\varepsilon(\mathbf{k})\right]|\psi_a\rangle. \tag{24}$$

Note that

$$V_\varepsilon^+(\mathbf{k}) = \left[a_\varepsilon(\mathbf{k}), H_I\right] = \frac{\partial}{\partial a_\varepsilon^+(\mathbf{k})}H_I \tag{25.a}$$

$$= -\sqrt{\frac{\hbar}{2\varepsilon_0\omega(2\pi)^3}}\,\sum_\alpha \boldsymbol{\varepsilon}\cdot\left(\mathbf{p}_\alpha - q_\alpha\mathbf{A}(\mathbf{r}_\alpha)\right)\frac{q_\alpha}{m_\alpha}\exp[-i\mathbf{k}\cdot\mathbf{r}_\alpha]. \tag{25.b}$$

Finally, we get

$$a_\varepsilon(\mathbf{k})|\psi_a\rangle = -\frac{1}{\hbar\omega + H - \tilde{E}_a}V_\varepsilon^+(\mathbf{k})|\psi_a\rangle. \tag{26}$$

The spectrum of H has a lower bound at $\tilde{E}_a$, so that the denominator equals at least $\hbar\omega$, whereas the numerator is first order in H_I. Therefore $a_\varepsilon(\mathbf{k})|\psi_a\rangle$ is first order with regard to the coupling between the particles and the transverse field.

b) THE ATOM IN THE PRESENCE OF A FREE PHOTON

Let us consider the state obtained by creating a photon $\mathbf{k}\varepsilon$ in the presence of the atom in its ground state

$$|\psi_a; \mathbf{k}\varepsilon\rangle = a_\varepsilon^+(\mathbf{k})|\psi_a\rangle. \tag{27}$$

These states $|\psi_a; \mathbf{k}\varepsilon\rangle$ are not exactly eigenstates of H, as was the state $|\psi_a\rangle$. Actually, we have

$$
\begin{aligned}
H|\psi_a; \mathbf{k}\varepsilon\rangle &= Ha_\varepsilon^+(\mathbf{k})|\psi_a\rangle \\
&= \left[H, a_\varepsilon^+(\mathbf{k})\right]|\psi_a\rangle + a_\varepsilon^+(\mathbf{k})H|\psi_a\rangle.
\end{aligned} \tag{28}
$$

The fact that $|\psi_a\rangle$ is an eigenstate of H allows us to write, taking into account the commutation relation $[H_0, a_\varepsilon^+(\mathbf{k})] = \hbar\omega a_\varepsilon^+(\mathbf{k})$ and Equation (25)

$$
\begin{aligned}
H|\psi_a; \mathbf{k}\varepsilon\rangle &= \left(\tilde{E}_a + \hbar\omega\right)|\psi_a; \mathbf{k}\varepsilon\rangle + \left[H_I, a_\varepsilon^+(\mathbf{k})\right]|\psi_a\rangle \\
&= \left(\tilde{E}_a + \hbar\omega\right)|\psi_a; \mathbf{k}\varepsilon\rangle + V_\varepsilon(\mathbf{k})|\psi_a\rangle.
\end{aligned} \tag{29}
$$

If the second term of the right-hand side of (29) were neglected, $|\psi_a; \mathbf{k}\varepsilon\rangle$ would be an eigenstate of H with eigenvalue $\tilde{E}_a + \hbar\omega$. We now show how it is possible to make such an approximation for wave packets constructed from $|\psi_a; \mathbf{k}\varepsilon\rangle$ and for which the photon is localized far from the atom. More precisely, let us consider the state vector $|\phi(t)\rangle$ obtained by combining the states $|\psi_a; \mathbf{k}\varepsilon\rangle$ with an amplitude $g_\varepsilon(\mathbf{k})$ and a time evolution factor $\exp[-i(\tilde{E}_a + \hbar\omega)t/\hbar]$:

$$|\phi(t)\rangle = \int d^3k \sum_\varepsilon g_\varepsilon(\mathbf{k})|\psi_a; \mathbf{k}\varepsilon\rangle \exp\left[-i\left(\tilde{E}_a + \hbar\omega\right)t/\hbar\right]. \tag{30}$$

Using (29) then leads without difficulty to

$$\left(i\hbar\frac{\partial}{\partial t} - H\right)|\phi(t)\rangle = -\int d^3k \sum_\varepsilon g_\varepsilon(\mathbf{k})\, e^{-i\omega t}\, V_\varepsilon(\mathbf{k})|\psi_a\rangle \exp\left(-i\tilde{E}_a t/\hbar\right). \tag{31}$$

We call $g(r, t)$ the transverse-plane wave packet constructed from $g_\varepsilon(k)/\sqrt{\omega}$.

$$g(\mathbf{r}, t) = \int \frac{d^3 k}{(2\pi)^{3/2} \sqrt{\omega}} \sum_\varepsilon g_\varepsilon(\mathbf{k}) \boldsymbol{\varepsilon} \exp i(\mathbf{k} \cdot \mathbf{r} - \omega t). \qquad (32)$$

Equation (31) is then written, taking into account (25):

$$\left(i\hbar \frac{\partial}{\partial t} - H\right) |\phi(t)\rangle = \sum_\alpha \sqrt{\frac{\hbar}{2\varepsilon_0 \omega}} \, q_\alpha \mathbf{v}_\alpha \cdot \mathbf{g}(\mathbf{r}_\alpha, t) |\psi_a\rangle \exp\left(-i\tilde{E}_a t/\hbar\right)$$

$$(33)$$

where $\mathbf{v}_\alpha = [\mathbf{p}_\alpha - q_\alpha \mathbf{A}(\mathbf{r}_\alpha)]/m_\alpha$ is the velocity operator of the particle α. Let us assume that the function $g_\alpha(k)$ is real, peaked around the mode $\mathbf{k}_0 \boldsymbol{\varepsilon}_0$, and of width Δk. The function $\mathbf{g}(r, t = 0)$ is then centered about the origin, the dimension of the wave packet being $1/\Delta k$. The time evolution of the wave packet is a propagation at the velocity c in the direction $\mathbf{k}_0$. After a time $|t| \gg 1/c\Delta k$, the wave packet has thus left the origin and $\mathbf{g}(0, t) \simeq 0$. Therefore, when $|t| \to \infty$, and to the extent that in the state $|\psi_a\rangle$ the coordinates $\mathbf{r}_\alpha$ of the particles are close to 0, we have

$$\mathbf{g}(\mathbf{r}_\alpha, t) |\psi_a\rangle \simeq 0 \qquad (34)$$

and consequently

$$\left(i\hbar \frac{\partial}{\partial t} - H\right) |\phi(t)\rangle \simeq 0. \qquad (35)$$

The state vector (30) is thus a solution of the Schrödinger equation for $|t| \to \infty$. The limit $|t| \to \infty$ simply indicates that the wave packet (32) has left the region where the atom is located. Next we go to the limit where Δk tends to zero. $|\phi(t)\rangle$ then essentially includes the state $|\psi_a; \mathbf{k}_0 \boldsymbol{\varepsilon}_0\rangle$ evolving with the energy $\tilde{E}_a + \hbar\omega_0$, and the wave packet has a spatial extension L_1 that tends to infinity. However, for times greater than L_1/c, the state $|\phi(t)\rangle$ is a solution of the Schrödinger equation. Thus, by giving this expression the meaning that we just defined by the double passage to the limit $\Delta k \to 0$ and $t \gg 1/c\Delta k$, we can consider the state $|\psi_a; \mathbf{k}_0 \boldsymbol{\varepsilon}_0\rangle$ to be asymptotically stationary, with energy $\tilde{E}_a + \hbar\omega_0$. Such a state describes for $t \to -\infty$ a quasi-monochromatic photon $|\psi_a; \mathbf{k}\boldsymbol{\varepsilon}\rangle$ incident on the atom, and for $t \to \infty$, the same photon leaving to infinity. The states $|\psi_a; \mathbf{k}\boldsymbol{\varepsilon}\rangle$ are thus well suited for describing the initial or the final state for the scattering of a photon by an atom.

3. Transition Amplitude between Exact Asymptotic States

a) NEW DEFINITION OF THE S-MATRIX

We just saw, that at the limit $|t| \to \infty$, the time evolution of asymptotic states is given by simple oscillating exponentials corresponding to the energies

$$\tilde{E}_{a;\mathbf{k}\varepsilon} = \tilde{E}_a + \hbar\omega \tag{36}$$

in which the exact energy $\tilde{E}_a$ of the state $|\psi_a\rangle$ appears. The generalization of the definition (2) of the S-matrix to the exact asymptotic states introduced above is thus

$$\tilde{S}_{fi} = \lim_{T \to \infty} \left[\exp(i\tilde{E}_f T/2\hbar) \langle \tilde{\varphi}_f | U(T/2, -T/2) | \tilde{\varphi}_i \rangle \exp(i\tilde{E}_i T/2\hbar) \right] \tag{37}$$

where

$$\tilde{E}_i = \tilde{E}_{a;\mathbf{k}\varepsilon} = \tilde{E}_a + \hbar\omega \tag{38.a}$$

$$\tilde{E}_f = \tilde{E}_{a';\mathbf{k}'\varepsilon'} = \tilde{E}_{a'} + \hbar\omega' \tag{38.b}$$

$$|\tilde{\varphi}_i\rangle = |\psi_a; \mathbf{k}\varepsilon\rangle \tag{38.c}$$

$$|\tilde{\varphi}_f\rangle = |\psi_{a'}; \mathbf{k}'\varepsilon'\rangle. \tag{38.d}$$

As in subsection 1, the calculation of the matrix element of U reduces to that of $G(z)$ between $|\tilde{\varphi}_i\rangle$ and $|\tilde{\varphi}_f\rangle$. Taking into account (38.c), (38.d), and (27), this matrix element is written

$$\langle \tilde{\varphi}_f | G(z) | \tilde{\varphi}_i \rangle = \langle \psi_{a'} | a_{\varepsilon'}(\mathbf{k}') G(z) a_\varepsilon^+(\mathbf{k}) | \psi_a \rangle. \tag{39}$$

Expression (4) must thus be replaced by

$$\tilde{S}_{fi} = \lim_{T \to \infty} \left\{ \exp \frac{iT}{2\hbar} \left(\tilde{E}_f + \tilde{E}_i \right) \times \right.$$

$$\left. \times \left[\int_{C_+} dz \, \frac{e^{-izT/\hbar}}{2i\pi} \langle \psi_{a'} | a_{\varepsilon'}(\mathbf{k}') G(z) a_\varepsilon^+(\mathbf{k}) | \psi_a \rangle \right] \right\}. \tag{40}$$

To calculate the matrix element appearing in (40), it is convenient to shift $G(z)$ to the left of $a_{\varepsilon'}(\mathbf{k}')$ or to the right of $a_\varepsilon^+(\mathbf{k})$, because the states $|\psi_a\rangle$ and $|\psi_{a'}\rangle$ are eigenstates of the operator H appearing in $G(z)$. Let us start with the commutator

$$\left[a_\varepsilon^+(\mathbf{k}), z - H\right] = \hbar\omega a_\varepsilon^+(\mathbf{k}) - \left[a_\varepsilon^+(\mathbf{k}), H_I\right] \quad (41)$$

which yields, taking into account (25)

$$a_\varepsilon^+(k)(z - \hbar\omega - H) = (z - H)a_\varepsilon^+(\mathbf{k}) + V_\varepsilon(\mathbf{k}). \quad (42)$$

It is then sufficient to multiply (42), on the left by $G(z)$, and on the right by $G(z - \hbar\omega)$, to obtain

$$G(z)a_\varepsilon^+(\mathbf{k}) = a_\varepsilon^+(\mathbf{k})G(z - \hbar\omega) + G(z)V_\varepsilon(\mathbf{k})G(z - \hbar\omega). \quad (43)$$

Finally, multiply (43) on the left by $a_{\varepsilon'}(\mathbf{k}')$. The term $a_{\varepsilon'}(\mathbf{k}')G(z)$ appears in the right-hand side and can be transformed once more by using the adjoint equation of (43) (z being replaced by z^* and $\mathbf{k}\varepsilon$ by $\mathbf{k}'\varepsilon'$)

$$a'_\varepsilon(\mathbf{k}')G(z) = G(z - \hbar\omega')a_{\varepsilon'}(\mathbf{k}') + G(z - \hbar\omega')V_{\varepsilon'}^+(\mathbf{k}')G(z). \quad (44)$$

Finally, we obtain

$$a_{\varepsilon'}(\mathbf{k}')G(z)a_\varepsilon^+(\mathbf{k}) = a_{\varepsilon'}(\mathbf{k}')a_\varepsilon^+(\mathbf{k})G(z - \hbar\omega) +$$

$$+ G(z - \hbar\omega')a_{\varepsilon'}(\mathbf{k}')V_\varepsilon(\mathbf{k})G(z - \hbar\omega) +$$

$$+ G(z - \hbar\omega')V_{\varepsilon'}^+(\mathbf{k}')G(z)V_\varepsilon(\mathbf{k})G(z - \hbar\omega) \quad (45)$$

and consequently, because $|\psi_a\rangle$ and $|\psi_{a'}\rangle$ are eigenstates of H with eigenvalues $\tilde{E}_a$ and $\tilde{E}_{a'}$

$$\langle\tilde{\varphi}_f|G(z)|\tilde{\varphi}_i\rangle = \langle\psi_{a'}|a_{\varepsilon'}(\mathbf{k}')a_\varepsilon^+(\mathbf{k})|\psi_a\rangle\frac{1}{z - \tilde{E}_a - \hbar\omega} +$$

$$+ \frac{1}{z - \tilde{E}_{a'} - \hbar\omega'}\langle\psi_{a'}|a_{\varepsilon'}(\mathbf{k}')V_\varepsilon(\mathbf{k})|\psi_a\rangle\frac{1}{z - \tilde{E}_a - \hbar\omega} +$$

$$+ \frac{1}{z - \tilde{E}_{a'} - \hbar\omega'}\langle\psi_{a'}|V_{\varepsilon'}^+(\mathbf{k}')G(z)V_\varepsilon(\mathbf{k})|\psi_a\rangle\frac{1}{z - \tilde{E}_a - \hbar\omega}. \quad (46)$$

In the first term of (46), $a_{\varepsilon'}(\mathbf{k}')$ may be brought to the right of the product of operators by a commutation relation:

$$\langle \tilde{\varphi}_f | G(z) | \tilde{\varphi}_i \rangle = \delta_{\varepsilon\varepsilon'} \delta(\mathbf{k} - \mathbf{k}') \frac{\delta_{aa'}}{z - \tilde{E}_i} +$$

$$+ \frac{1}{(z - \tilde{E}_i)} \langle \psi_{a'} | a_\varepsilon^+(\mathbf{k}) a_{\varepsilon'}(\mathbf{k}') | \psi_a \rangle +$$

$$+ \frac{1}{(z - \tilde{E}_i)(z - \tilde{E}_f)} \langle \psi_{a'} | a_{\varepsilon'}(\mathbf{k}') V_\varepsilon(\mathbf{k}) +$$

$$V_{\varepsilon'}^+(\mathbf{k}') G(z) V_\varepsilon(\mathbf{k}) | \psi_a \rangle. \tag{47}$$

Except for the second term, about which we will show that its contribution to (40) is negligible, expression (47) is quite similar to (7). We can use this similarity to apply to (47) the procedure of subsections 1-a, 1-b, and 1-c, after having shown that the initial assumption made after expression (7) is actually verified here, that is, that $\langle \psi_{a'} | V_{\varepsilon'}^+(\mathbf{k}') G(z) V_\varepsilon(\mathbf{k}) | \psi_a \rangle$ has no real poles in the region $z = \tilde{E}_i$ or $\tilde{E}_f$.

To do this, recall (see Complement A$_{III}$) that if $|u\rangle$ is a normalized state vector, $\langle u | G(z) | u \rangle$ has poles for the discrete eigenvalues of H and a cut on the real axis corresponding to the continuous spectrum. This property can be generalized easily to the matrix elements of $G(z)$ between two different states having a finite norm (see the remark in subsection 1 of Complement A$_{III}$). We first verify that $V_\varepsilon(\mathbf{k}) | \psi_a \rangle$ has indeed a finite norm. By using (25.b) and the expression for the velocity $\mathbf{v}_\alpha$ of the particle α, we obtain

$$\langle \psi_a | V_\varepsilon^+(\mathbf{k}) V_\varepsilon(\mathbf{k}) | \psi_a \rangle =$$

$$= \frac{\hbar}{2\varepsilon_0 \omega (2\pi)^3} \sum_\alpha \sum_{\alpha'} q_\alpha q_{\alpha'} \langle \psi_a | \boldsymbol{\varepsilon} \cdot \mathbf{v}_{\alpha'} e^{i\mathbf{k} \cdot (\mathbf{r}_\alpha - \mathbf{r}_{\alpha'})} \boldsymbol{\varepsilon} \cdot \mathbf{v}_\alpha | \psi_a \rangle. \tag{48}$$

The second member is the average, in the ground state of the system, of observables that are obviously finite (the velocity of the particles is bounded), so that the square of the norm of $V_\varepsilon(\mathbf{k}) | \psi_a \rangle$ is finite. It is the same for $V_{\varepsilon'}(\mathbf{k}') | \psi_{a'} \rangle$. Consequently, $\langle \psi_{a'} | V_{\varepsilon'}(\mathbf{k}') G(z) V_\varepsilon(\mathbf{k}) | \psi_a \rangle$ has no discrete poles other than $\tilde{E}_a$ and $\tilde{E}_{a'}$, which are far from $\tilde{E}_i = \tilde{E}_a + \hbar\omega$ and $\tilde{E}_f = \tilde{E}_{a'} + \hbar\omega$.

Therefore, using the same reasoning as that at the end of subsection 1-a allows us to ignore the contributions of the singularities of $G(z)$ to the contour integral (40), at the limit $T \to \infty$. The only nonzero contributions come from the poles at $\tilde{E}_i$ and $\tilde{E}_f$ of (47). A calculation similar to the one in subsection 1-b thus allows us to generalize [for the first and third terms of (47)] the results (12) and (13) in the form

$$\tilde{S}_{fi} = \delta_{\varepsilon\varepsilon'}\delta(\mathbf{k} - \mathbf{k}')\delta_{aa'} - 2i\pi\delta^{(T)}\left(\tilde{E}_i - \tilde{E}_f\right)\tilde{\mathcal{T}}_{fi} \qquad (49.\text{a})$$

with

$$\tilde{\mathcal{T}}_{fi} = \langle\psi_{a'}|a_{\varepsilon'}(\mathbf{k}')V_{\varepsilon}(\mathbf{k}) + V_{\varepsilon'}^+(\mathbf{k}')G\left(\tilde{E}_i + i\eta\right)V_{\varepsilon}(\mathbf{k})|\psi_a\rangle. \qquad (49.\text{b})$$

Let us finally calculate the contribution c_2 of the second term of (47) to the expression (40). Integration using the residue method immediately gives

$$c_2 = \lim_{T\to\infty}\left(\exp\frac{iT}{2\hbar}\left(\tilde{E}_f - \tilde{E}_i\right)\right) \times \langle\psi_{a'}|a_{\varepsilon}^+(\mathbf{k})a_{\varepsilon'}(\mathbf{k}')|\psi_a\rangle. \qquad (50)$$

When T tends to infinity, c_2 becomes a rapidly oscillating function of $\tilde{E}_i$ and $\tilde{E}_f$. Its amplitude is finite, and even small in so far as $a_{\varepsilon'}(\mathbf{k}')|\psi_a\rangle$ is first order in H_I (see end of subsection 2-a). Considered as a distribution, c_2 thus tends to zero.

Remark

Let us return to the comparison between (7) and (47). The preceding demonstration cannot be applied to $\langle\varphi_f|VG(z)V|\varphi_i\rangle$ because $|\varphi_i\rangle$ and $|\varphi_f\rangle$ are states of the continuous spectrum, which do not have a finite norm. In fact, using perturbative methods we have shown in subsection 1-d that $\langle\varphi_f|VG(z)V|\varphi_i\rangle$ has poles at $\tilde{E}_a + \hbar\omega$ and $\tilde{E}_{a'} + \hbar\omega'$.

b) New Expression for the Transition Matrix. Physical Discussion

Expression (49.b) for $\mathcal{T}_{fi}$ may still be transformed and put in a form very close to that of the perturbative expression describing, to second order in q, the same scattering process between uncoupled states $|a; k\varepsilon\rangle$ and $|a; k'\varepsilon'\rangle$ [formula (C.1) of Chapter II].

To do this, we will move $a_{\varepsilon'}(\mathbf{k}')$ to the right of $V_{\varepsilon}(\mathbf{k})$ in the first term of (49.b) and use result (26) obtained previously for $a_{\varepsilon'}(\mathbf{k}')|\psi_a\rangle$. We write $W_{\varepsilon'\varepsilon}(\mathbf{k}',\mathbf{k})$ the commutator of $a_{\varepsilon'}(\mathbf{k}')$ and $V_{\varepsilon}(\mathbf{k})$, which, according to (25), equals

$$W_{\varepsilon'\varepsilon}(\mathbf{k}',\mathbf{k}) = [a_{\varepsilon'}(\mathbf{k}'), V_{\varepsilon}(\mathbf{k})]$$
$$= [a_{\varepsilon'}(\mathbf{k}'), [H_I, a_{\varepsilon}^+(\mathbf{k})]]$$
$$= \frac{\hbar}{2\varepsilon_0(2\pi)^3\sqrt{\omega\omega'}} \sum_\alpha \boldsymbol{\varepsilon} \cdot \boldsymbol{\varepsilon}' \frac{q_\alpha^2}{m_\alpha} \exp[i(\mathbf{k} - \mathbf{k}') \cdot \mathbf{r}_\alpha]. \quad (51)$$

It is then possible to write the first term of (49.b) in the form

$$\langle \psi_{a'} | a_{\varepsilon'}(\mathbf{k}') V_\varepsilon(\mathbf{k}) | \psi_a \rangle = \langle \psi_{a'} | W_{\varepsilon'\varepsilon}(\mathbf{k}',\mathbf{k}) + V_\varepsilon(\mathbf{k}) a_{\varepsilon'}(\mathbf{k}') | \psi_a \rangle$$
$$= \langle \psi_{a'} | W_{\varepsilon'\varepsilon}(\mathbf{k}',\mathbf{k}) + V_\varepsilon(\mathbf{k}) G(\tilde{E}_a - \hbar\omega) V_{\varepsilon'}^+(\mathbf{k}') | \psi_a \rangle$$
$$(52)$$

where we have used (26) in going from the first line to the second. Finally, the new transition matrix $\mathcal{T}_{fi}$ can be written in the form

$$\tilde{\mathcal{T}}_{fi} = \langle \psi_{a'} | W_{\varepsilon'\varepsilon}(\mathbf{k}',\mathbf{k}) + V_\varepsilon(\mathbf{k}) G(\tilde{E}_a - \hbar\omega) V_{\varepsilon'}^+(\mathbf{k}') +$$
$$+ V_{\varepsilon'}^+(\mathbf{k}') G(\tilde{E}_a + \hbar\omega + i\eta) V_\varepsilon(\mathbf{k}) | \psi_a \rangle. \quad (53)$$

To facilitate the comparison of exact and perturbative expressions for $\mathcal{T}_{fi}$, it is convenient to rewrite formula (C.1) of Chapter II in a way which exhibits the operators $V_\varepsilon(\mathbf{k})$ and $V_{\varepsilon'}^+(\mathbf{k}')$ used here. To do so, it is sufficient to write

$$\langle b; 0 | H_{I1} | a; \mathbf{k}\varepsilon \rangle = \langle b; 0 | H_I | a; \mathbf{k}\varepsilon \rangle$$
$$= \langle b; 0 | [H_I, a_\varepsilon^+(\mathbf{k})] | a; 0 \rangle$$
$$= \langle b; 0 | V_\varepsilon(\mathbf{k}) | a; 0 \rangle \quad (54)$$

and

$$\langle a'; \mathbf{k}'\varepsilon' | H_{I2} | a; \mathbf{k}\varepsilon \rangle = \langle a'; 0 | a_{\varepsilon'}(\mathbf{k}') H_I a_\varepsilon^+(\mathbf{k}) | a; 0 \rangle$$
$$= \langle a'; 0 | [a_{\varepsilon'}(\mathbf{k}'), [H_i, a_\varepsilon^+(\mathbf{k})]] + [H_I, a_\varepsilon^+(\mathbf{k})] a_{\varepsilon'}(\mathbf{k}') +$$
$$+ a_{\varepsilon'}(\mathbf{k}') a_\varepsilon^+(\mathbf{k}) H_I | a; 0 \rangle$$
$$= \langle a'; 0 | W_{\varepsilon'\varepsilon}(\mathbf{k}',\mathbf{k}) | a; 0 \rangle + \delta_{\varepsilon\varepsilon'} \delta(\mathbf{k} - \mathbf{k}') \langle a'; 0 | H_I | a; 0 \rangle. \quad (55)$$

For $\mathbf{k}\varepsilon \neq \mathbf{k}'\varepsilon'$, the second member of (55) reduces to the first term and formula (C.1) of Chapter II is written in the form

$$\mathcal{T}_{fi} = \mathcal{T}_{fi}^{(\gamma)} + \mathcal{T}_{fi}^{(\beta)} + \mathcal{T}_{fi}^{(\alpha)} =$$
$$= \langle a'; 0 | W_{\varepsilon'\varepsilon}(\mathbf{k}', \mathbf{k}) + V_\varepsilon(\mathbf{k}) G_0(E_a - \hbar\omega) V_{\varepsilon'}^+(\mathbf{k}') +$$
$$+ V_{\varepsilon'}^+(\mathbf{k}') G_0(E_a + \hbar\omega + i\eta) V_\varepsilon(\mathbf{k}) | a; 0 \rangle \qquad (56)$$

which closely resembles the exact expression (53).

The comparison of (53) and (56) shows clearly that, to go from the approximate expression to the exact expression, $|a; 0\rangle$ must be replaced by $|\psi_a\rangle$, E_a by $\tilde{E}_a$, and G_0 by G. Thus the quantities relative to the bare ground state are replaced by those of the state "dressed" by the cloud of transverse photons and the free propagator G_0 is replaced by the perturbed propagator G.

A more exhaustive analysis of expression (53) lies beyond the scope of this Complement. We can, however, suggest the following ideas: a first approximation of the complete propagator of the atom is that of a "bare" atom in which the electrons have renormalized masses, and the interaction with the nucleus is modified at short distance. Thus, the use of the second-order formula (56), with renormalized parameters and complex energies for the excited states, actually constitutes an excellent approximation of (53). This result, which is in fact quite general, may be demonstrated within the framework of the renormalization theory. It justifies the current practice in all quantum optics and low-energy electrodynamics which consists of ignoring the radiative corrections, and of using the renormalized physical parameters directly in the formulas.

References

For the standard collision theory, see Goldberger and Watson. For the applications to field theory, this Complement was inspired by G. C. Wick, *Rev. Mod. Phys.*, **27**, 339 (1955), and by Kroll.

COMPLEMENT C_{III}

DISCRETE STATE COUPLED TO A FINITE-WIDTH CONTINUUM: FROM THE WEISSKOPF–WIGNER EXPONENTIAL DECAY TO THE RABI OSCILLATION

1. Introduction—Overview

The methods introduced in Section B of this chapter for partially summing the perturbation series correctly account for the exponential decay of a discrete level coupled to a continuum. We also discussed in subsection C-1-c the conditions of validity for such a result. In particular, the transition rate Γ_b from the discrete state $|\varphi_b\rangle$ to the continuum must be very small compared with a parameter w_0 characterizing the scale of variation with E of a function $\Gamma_b(E)$ that describes how the discrete state $|\varphi_b\rangle$ is coupled to the energy shell E in the continuum [see (C.12)]. In some cases, when $\Gamma_b(E)$ is a bell-shaped curve, $\hbar w_0$ is approximately the width of this curve and may therefore be considered as the *continuum width*.

Now imagine that we could progressively increase the coupling V, and thus $\Gamma_b(E)$, without modifying the width $\hbar w_0$ of $\Gamma_b(E)$. When Γ_b reaches values on the order of w_0, the deviations from the exponential decay become more and more important. At the limit $\Gamma_b \gg w_0$, it even seems that the width of the continuum could be neglected to a first approximation, so that the continuum could be considered as a discrete state. We therefore expect, for $\Gamma_b \gg w_0$, that the time evolution of the system tends toward a Rabi oscillation between two discrete levels.

The goal of this Complement is to present a model of a discrete state coupled to a continuum of finite width, sufficiently simple to allow one to understand by using *graphic constructions* how increasing the coupling V progressively changes the time evolution of the system (*). This model is presented in subsection 2 as well as a derivation of an *exact* expression for the Fourier transform $\mathscr{U}_b(E)$ of the amplitude $U_b(\tau)$ that the system, prepared at time 0 in the discrete state $|\varphi_b\rangle$, remains in this state after a time τ has elapsed. Contrary to what we did in subsection C-1 of this chapter, we no longer make the approximation consisting of replacing the expression $R_b(z)$ appearing in the denominator of (C.4) by an approximate expression $\hat{R}_b(z)$. The model is sufficiently simple to allow an *exact*

(*) C. Cohen-Tannoudji and P. Avan, in *Etats Atomiques et Moléculaires Couplés à un Continuum. Atomes et Molécules Hautement Excités* Editions du CNRS, Paris, 1977, p. 93.

calculation of $R_b(z)$, and therefore of $R_b(E \pm i\eta)$. After discussing the important physical parameters of the problem (§3), we then show (§4) how it is possible, for each value of E, to graphically construct $\mathscr{U}_b(E)$. We can then study the shape of $\mathscr{U}_b(E)$ in various regimes: weak coupling (§5); intermediate coupling (§6); and strong coupling (§7); and thus understand how $\mathscr{U}_b(E)$ changes progressively from a Lorentzian form (the Fourier transform of which is a damped exponential) to a set of two delta functions (for which the Fourier transform is a sinusoid).

2. Description of the Model

a) UNPERTURBED STATES

The unperturbed Hamiltonian H_0 is assumed to have a *single discrete eigenstate* $|\varphi_b\rangle$, of energy E_b, and a *single continuum* of states $|E, \beta\rangle$ with an energy E varying continuously from 0 to $+\infty$ (β is the ensemble of quantum numbers other than E required to characterize the states of the continuum).

$$H_0|\varphi_b\rangle = E_b|\varphi_b\rangle \tag{1.a}$$

$$H_0|E, \beta\rangle = E|E, \beta\rangle, \qquad 0 \le E < \infty. \tag{1.b}$$

b) ASSUMPTIONS CONCERNING THE COUPLING

The coupling is written λV where λ is a dimensionless parameter. If $\lambda \ll 1$, the coupling is weak; if $\lambda \gg 1$, the coupling is strong. V is assumed to have nonzero matrix elements only between the discrete state and the continuum. These matrix elements are written $v(E, \beta)$

$$\langle\varphi_b|V|\varphi_b\rangle = 0; \qquad \langle E, \beta|V|E', \beta'\rangle = 0 \tag{2.a}$$

$$\langle E, \beta|V|\varphi_b\rangle = v(E, \beta) \tag{2.b}$$

c) CALCULATION OF THE RESOLVENT AND OF THE PROPAGATORS

Equation (B.8) of the chapter becomes here

$$G_b(z) = \langle\varphi_b|\frac{1}{z - H_0 - \lambda V}|\varphi_b\rangle = \frac{1}{z - E_b - R_b(z)} \tag{3}$$

where $R_b(z)$ is given by an expression analogous to (B.4)

$$R_b(z) = \langle \varphi_b | \lambda V | \varphi_b \rangle + \sum_{i \neq b} \frac{\langle \varphi_b | \lambda V | \varphi_i \rangle \langle \varphi_i | \lambda V | \varphi_b \rangle}{z - E_i} +$$

$$+ \sum_{i,j \neq b} \frac{\langle \varphi_b | \lambda V | \varphi_j \rangle \langle \varphi_j | \lambda V | \varphi_i \rangle \langle \varphi_i | \lambda V | \varphi_b \rangle}{(z - E_j)(z - E_i)} + \cdots \qquad (4)$$

$|\varphi_i\rangle, |\varphi_j\rangle \cdots$ being eigenstates of H_0, other than $|\varphi_b\rangle$, and of energies $E_i, E_j \cdots$.

The assumptions made above concerning H_0 and V allow us to considerably simplify expression (4). First of all, the first term is zero according to (2.a). We now consider the third term. Because all the states $|\varphi_i\rangle$ and $|\varphi_j\rangle$ other than $|\varphi_b\rangle$ are necessarily the states $|E, \beta\rangle$ and $|E', \beta'\rangle$ of the continuum (see subsection 2-a above), the central matrix element of the third term of (4) reduces to $\langle E'\beta' | \lambda V | E, \beta \rangle$, which is zero according to (2.a). Analogous reasoning allows us to show that all the higher-order terms in V of the expansion (4) are zero. Finally, there only remains the second-order term where $|\varphi_i\rangle$ is replaced by $|E', \beta'\rangle$. $R_b(z)$ is then written, using (2.b):

$$R_b(z) = \lambda^2 \iint dE' \, d\beta' \, \rho(E', \beta') \frac{|v(E', \beta')|^2}{z - E'} \qquad (5)$$

where $\rho(E', \beta')$ is the density of states in the continuum. Thus, the simplifying assumptions on H_0 and V allow us to obtain an exact expression for $R_b(z)$. Moreover, $R_b(z)$ varies simply as λ^2.

The retarded and advanced propagators are written, using (A.19) and (3):

$$G_{b\pm}(E) = \lim_{\eta \to 0_+} G_b(E \pm i\eta) = \lim_{\eta \to 0_+} \frac{1}{E \pm i\eta - E_b - R_b(E \pm i\eta)}.$$

$$(6)$$

But, according to (5)

$$R_b(E \pm i\eta) = \lambda^2 \iint dE' \, d\beta' \, \rho(E', \beta') \frac{|v(E', \beta')|^2}{E - E' \pm i\eta}. \qquad (7)$$

Because

$$\lim_{\eta \to 0_+} \frac{1}{x \pm i\eta} = \mathscr{P}\frac{1}{x} \mp i\pi\delta(x) \tag{8}$$

Equation (6) gives

$$\lim_{\eta \to 0_+} R_b(E \pm i\eta) = \hbar\lambda^2 \left[\Delta_b(E) \mp i\frac{\Gamma_b(E)}{2} \right] \tag{9}$$

where

$$\Delta_b(E) = \mathscr{P}\frac{1}{\hbar} \int\int dE' \, d\beta' \, \rho(E', \beta') \frac{|v(E', \beta')|^2}{E - E'} \tag{10}$$

$$\Gamma_b(E) = \frac{2\pi}{\hbar} \int d\beta \, \rho(E, \beta)|v(E, \beta)|^2 \tag{11}$$

and finally

$$G_{b\pm}(E) = \lim_{\eta \to 0_+} \frac{1}{E - E_b - \hbar\lambda^2\Delta_b(E) \pm i\left[\hbar\lambda^2\dfrac{\Gamma_b(E)}{2} + \eta \right]} . \tag{12}$$

d) FOURIER TRANSFORM OF THE AMPLITUDE $U_b(\tau)$

According to (A.21), the Fourier transform $\mathscr{U}_b(E)$ of $U_b(\tau)$ is proportional to $G_{b-}(E) - G_{b+}(E)$. More precisely, Equations (A.21) and (12) give

$$U_b(\tau) = \int_{-\infty}^{+\infty} dE \, \mathscr{U}_b(E) \, e^{-iE\tau/\hbar} \tag{13.a}$$

where

$$\mathscr{U}_b(E) = \frac{1}{2\pi i}[G_{b-}(E) - G_{b+}(E)] =$$

$$= \lim_{\eta \to 0_+} \frac{1}{\pi} \frac{\hbar\lambda^2[\Gamma_b(E)/2] + \eta}{[E - E_b - \hbar\lambda^2\Delta_b(E)]^2 + [(\hbar\lambda^2\Gamma_b(E)/2) + \eta]^2} . \tag{13.b}$$

Finally, we have obtained equations for $U_b(\tau)$ and $\mathscr{U}_b(E)$ that are completely analogous to (C.20.a) and (C.20.b). Nevertheless, we must not forget that expressions (10) and (11) for $\Delta_b(E)$ and $\Gamma_b(E)$ are exact, within the framework of the simplified model that we have chosen, whereas expressions (C.11) and (C.12) result from a perturbative calculation of $R_b(E \pm i\eta)$, limited to the second order in V [see (C.5)]. When V increases beyond a certain value, expressions (C.11) and (C.12), and thus (C.20), are no longer valid, whereas (13.b) still remains exact. In the simple model chosen here, increasing the coupling simply amounts to increase the factor λ^2 multiplying $\Gamma_b(E)$ and $\Delta_b(E)$ in (13.b), while the functions $\Gamma_b(E)$ and $\Delta_b(E)$ still retain the same form.

Remark

We have derived formulas (13.a) and (13.b), which will serve as a basis for the rest of the complement, starting from the general formalism established in Section A of this chapter. The model chosen for H_0 and V in this complement is, however, so simple that Equations (13.a) and (13.b) can be obtained more directly. We expand the state vector $|\psi(t)\rangle$ in terms of the states $|\varphi_b\rangle$ and $|E, \beta\rangle$.

$$|\psi(t)\rangle = c_b(t)\,e^{-iE_b t/\hbar}|\varphi_b\rangle +$$

$$+ \int\int dE\,d\beta\,\rho(E,\beta)c(E,\beta,t)\,e^{-iEt/\hbar}|E,\beta\rangle. \tag{14}$$

The Schrödinger equation is then written

$$i\hbar\dot{c}_b(t) = \int\int dE\,d\beta\,\lambda v^*(E,\beta)\rho(E,\beta)c(E,\beta,t)\,e^{i(E_b-E)t/\hbar} \tag{15.a}$$

$$i\hbar\dot{c}(E,\beta,t) = \lambda v(E,\beta)c_b(t)\,e^{i(E-E_b)t/\hbar}. \tag{15.b}$$

By integrating (15.b), with the initial condition $c(E,\beta,0) = 0$, and by substituting the solution thus obtained into (15.a), we get

$$\dot{c}_b(t) = -\frac{\lambda^2}{2\pi\hbar}\int_{-\infty}^{+\infty}dE\int_0^t dt'\,\Gamma_b(E)\,e^{i(E_b-E)(t-t')}c_b(t') \tag{16.a}$$

where $\Gamma_b(E)$ is defined in (11); that is again

$$\dot{c}_b(t) = -\lambda^2\int_0^t d\tau\,\gamma_b(\tau)c_b(t-\tau) \tag{16.b}$$

where

$$\gamma_b(\tau) = \frac{1}{2\pi\hbar} \int_{-\infty}^{+\infty} dE\, \Gamma_b(E)\, e^{i(E_b - E)\tau/\hbar}. \tag{16.c}$$

The exact integro-differential Equation (16.b) can be used as a starting point for several treatments: a perturbative treatment [where $c_b(t - \tau)$ is replaced by $c_b(0) = 1$], a Weisskopf–Wigner treatment [where $c_b(t - \tau)$ is replaced by $c_b(t)$ and taken out from the integral]. We can also, without making any approximation, introduce Fourier–Laplace transforms of $c_b(t)$ and $\gamma_b(t)$ and transform (16.b) into an algebraic equation, from which it is possible to derive (13.b).

3. The Important Physical Parameters

Before discussing the graphical construction of $\mathcal{U}_b(E)$, we will specify the shape and the essential properties of the functions $\Gamma_b(E)$ and $\Delta_b(E)$ appearing in (13.b).

a) THE FUNCTION $\Gamma_b(E)$

According to its definition (11), $\Gamma_b(E)$ characterizes the coupling induced by V between the discrete state $|\varphi_b\rangle$ and the energy shell E in the continuum. It also clearly appears in (11) that

$$\Gamma_b(E) \geq 0. \tag{17}$$

Because the continuum starts from $E = 0$ [see (1.b)], the density of states $\rho(E, \beta)$ is zero for $E < 0$, so that

$$\Gamma_b(E) = 0 \qquad \text{for } E < 0. \tag{18}$$

When E increases, $\rho(E, \beta)$ is generally an increasing function of E, whereas $|v(E, \beta)|^2$ tends to 0 when $E \to \infty$. We assume here that $|v(E, \beta)|^2$ tends to 0 sufficiently rapidly so that

$$\Gamma_b(E) \to 0 \qquad \text{if } E \to \infty. \tag{19}$$

Finally, using all these results, we take for $\Gamma_b(E)$ a function for which the variations with E have the shape represented in Figure 1. The width at half-maximum of this function, $\hbar w_0$, may be considered as the width of the continuum.

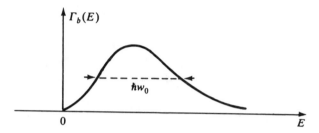

Figure 1. Shape of the function $\Gamma_b(E)$.

b) THE PARAMETER Ω_1 CHARACTERIZING THE COUPLING OF THE DISCRETE STATE WITH THE WHOLE CONTINUUM

Assume for a moment that all the states $|E, \beta\rangle$ and $|\varphi_b\rangle$ are degenerate in energy. The discrete state would then be coupled only to the state $V|\varphi_b\rangle$, that is, to the following normalized linear combination:

$$|\psi\rangle = \frac{\iint dE\, d\beta\, \rho(E, \beta)|E, \beta\rangle\langle E, \beta|V|\varphi_b\rangle}{\sqrt{\langle \varphi_b|VV|\varphi_b\rangle}} \qquad (20)$$

the coupling matrix element between $|\varphi_b\rangle$ and $|\psi\rangle$ being

$$\langle \psi|V|\varphi_b\rangle = \sqrt{\langle \varphi_b|VV|\varphi_b\rangle} =$$

$$= \sqrt{\iint dE\, d\beta\, \rho(E, \beta)|v(E, \beta)|^2}. \qquad (21)$$

The Rabi frequency Ω_1 of the oscillation of the system between the state $|\varphi_b\rangle$ and the state $|\psi\rangle$ would then be equal to (21) (divided by $\hbar$).

Actually, the states $|E, \beta\rangle$ are not degenerate with $|\varphi_b\rangle$ and have an energy spread over a range $\hbar w_0$. However, if the coupling λV is sufficiently strong, we expect that the system has enough time to make several oscillations between $|\varphi_b\rangle$ and the continuum before the effect of the energy spread $\hbar w_0$ of the states $|E, \beta\rangle$ is perceived. We are thus led to introduce the parameter

$$\Omega_1^2 = \frac{1}{\hbar^2} \iint dE\, d\beta\, \rho(E, \beta)|v(E, \beta)|^2 = \frac{1}{2\pi\hbar} \int_{-\infty}^{+\infty} dE\, \Gamma_b(E) \qquad (22)$$

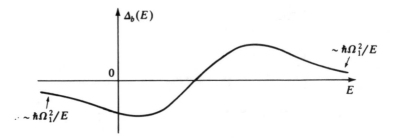

Figure 2. Shape of the function $\Delta_b(E)$.

[where we used (11) for the second equality]. Later we will see that $\lambda\Omega_1$ is actually the Rabi oscillation frequency appearing for $\lambda \gg 1$.

Finally, note that $\Gamma_b(E)$ must decrease sufficiently rapidly with E for the integral (22) to be convergent.

c) THE FUNCTION $\Delta_b(E)$

The relations (10) and (11) result in

$$\Delta_b(E) = \frac{1}{2\pi}\mathscr{P}\int dE' \frac{\Gamma_b(E')}{E - E'}. \tag{23}$$

The dispersion relation (23) allows us to predict the shape of $\Delta_b(E)$ from that of $\Gamma_b(E)$. If $\Gamma_b(E)$ resembles a bell-shaped curve (Figure 1), then $\Delta_b(E)$ resembles a dispersion curve (Figure 2).

When $|E| \gg \hbar w_0$, the denominator of (23) is equivalent to E because E' cannot, because of the numerator $\Gamma_b(E')$, be much greater than $\hbar w_0$. From this we deduce that if $|E| \gg \hbar w_0$,

$$\Delta_b(E) \simeq \frac{1}{2\pi E}\int dE' \,\Gamma_b(E') = \frac{\hbar\Omega_1^2}{E} \tag{24}$$

[we used (22) for the second equality]. The wings of the function $\Delta_b(E)$ thus decrease as $1/E$.

4. Graphical Discussion

a) CONSTRUCTION OF THE CURVE $\mathscr{U}_b(E)$

In Figure 3, three functions of E are represented: $\hbar\lambda^2\Gamma_b(E)$, $\hbar\lambda^2\Delta_b(E)$ and $E - E_b$ (straight line of slope 1, crossing the E axis at E_b). We now

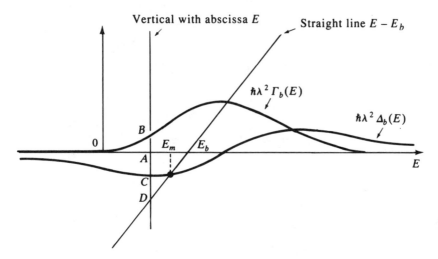

Figure 3. Graphical construction of $\mathcal{U}_b(E)$.

consider a vertical line with abscissa E, and denote by A, B, C, and D the intersections of this vertical line with, respectively, the E axis and the curves $\hbar\lambda^2\Gamma_b(E)$, $\hbar\lambda^2\Delta_b(E)$, $E - E_b$. We have

$$AB = \hbar\lambda^2\Gamma_b(E) \qquad CD = E - E_b - \hbar\lambda^2\Delta_b(E) \qquad (25)$$

so that the expression (13.b) for $\mathcal{U}_b(E)$ may be written

$$\mathcal{U}_b(E) = \lim_{\eta \to 0_+} \frac{1}{\pi} \frac{(AB/2) + \eta}{(CD)^2 + [(AB/2) + \eta]^2}. \qquad (26)$$

For each value of E, we can then measure AB and CD and determine $\mathcal{U}_b(E)$ by using (26).

Because all the quantities appearing in the denominator of (26) are positive, we expect to find a maximum of $\mathcal{U}_b(E)$ near the values of E such that the distance CD is zero. The abscissas E_m of the maxima of $\mathcal{U}_b(E)$ are then given, according to (25), by

$$E_m - E_b - \hbar\lambda^2\Delta_b(E_m) = 0. \qquad (27)$$

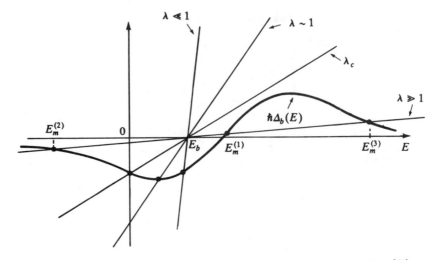

Figure 4. Graphical determination of the abscissas of the maxima of $\mathcal{U}_b(E)$.

b) GRAPHICAL DETERMINATION OF THE MAXIMA OF $\mathcal{U}_b(E)$.
CLASSIFICATION OF THE VARIOUS REGIMES

According to (27), the abscissas of the maxima of $\mathcal{U}_b(E)$ are obtained by finding the intersections of the curve $\hbar\Delta_b(E)$ with the straight line $(E - E_b)/\lambda^2$ passing through E_b, with the slope $1/\lambda^2$ (Figure 4).

For a weak coupling ($\lambda \ll 1$), $(E - E_b)/\lambda^2$ is an almost vertical line, and there is only a single point of intersection, having an abscissa very close to E_b: $E_m \simeq E_b$. A better approximation is obtained by replacing E_m by E_b in the last term of (27), which is very small. This gives

$$E_m \simeq E_b + \hbar\lambda^2\Delta_b(E_b). \tag{28}$$

When the coupling λ increases, Figure 4 shows that the abscissa of the intersection point of $\hbar\Delta_b(E)$ with $(E - E_b)/\lambda^2$ shifts to the left, until λ reaches a critical value λ_c, beyond which E_m becomes negative, that is, less than the beginning of the continuum starting at $E = 0$. (New intersections of the two curves may also appear.) The value of λ_c is obtained by setting $E_m = 0$ in (27):

$$\lambda_c^2 = -\frac{E_b}{\hbar\Delta_b(0)} = \frac{2\pi E_b}{\hbar\int_{-\infty}^{+\infty} dE\,\Gamma_b(E)/E}. \tag{29}$$

[We used (23) for the second equality.] Later on (§6-b) we will return to the physical meaning of the critical coupling.

For very strong couplings ($\lambda \gg 1$), the slope of the straight line $(E - E_b)/\lambda^2$ becomes very small, and in Figure 4 this line generally crosses the curve $\hbar\Delta_b(E)$ at three points, with abscissas $E_m^{(1)}$, $E_m^{(2)}$, and $E_m^{(3)}$. $E_m^{(1)}$ is quite close to the abscissa of the zero of $\Delta_b(E)$. The abscissas $E_m^{(2)}$ and $E_m^{(3)}$ correspond to points located in the wings of $\hbar\Delta_b(E)$, for which the asymptotic expression (24) for $\Delta_b(E)$ can be used. By substituting (24) into (27), we then obtain

$$E_m - E_b - \frac{\lambda^2\hbar^2\Omega_1^2}{E_m} = 0. \tag{30}$$

Because E_b can be neglected compared with E_m, Equation (30) gives

$$E_m^2 = \lambda^2\hbar^2\Omega_1^2 \tag{31}$$

that is also

$$E_m^{(2)} = -\lambda\hbar\Omega_1 \tag{32.a}$$

$$E_m^{(3)} = +\lambda\hbar\Omega_1. \tag{32.b}$$

5. Weak Coupling Limit

a) WEISSKOPF–WIGNER EXPONENTIAL DECAY

When $\lambda \ll 1$, the term $[E - E_b - \hbar\lambda^2\Delta_b(E)]^2$ of the denominator of (13.b) is very large compared with all the others, except in the neighborhood $E = E_b$, where it vanishes. In the very small terms $\hbar\lambda^2\Gamma_b(E)$ and $\hbar\lambda^2\Delta_b(E)$, it is thus legitimate to replace E by E_b because it is only near $E = E_b$ that these terms are non-negligible compared with $[E - E_b - \hbar\lambda^2\Delta_b(E)]^2$. We then have

$$\mathcal{U}_b(E) \simeq \frac{1}{\pi} \frac{\hbar\lambda^2\Gamma_b/2}{\left[E - E_b - \hbar\lambda^2\Delta_b\right]^2 + \left[\hbar\lambda^2\Gamma_b/2\right]^2} \tag{33}$$

where

$$\Gamma_b = \Gamma_b(E_b), \qquad \Delta_b = \Delta_b(E_b). \tag{34}$$

The Fourier transform of (33) is

$$U_b(\tau) = e^{-i(E_b + \hbar\lambda^2\Delta_b)\tau/\hbar} \, e^{-\lambda^2\Gamma_b|\tau|/2} \tag{35}$$

which is the well-known result derived by Weisskopf and Wigner. Γ_b is the transition rate from the discrete state $|\varphi_b\rangle$ to the continuum, calculated by the Fermi golden rule, and $\hbar\Delta_b$ is the shift of this discrete state due to the coupling with the continuum.

b) CORRECTIONS TO THE EXPONENTIAL DECAY

A better approximation than the one leading to (33) consists of still replacing $\Gamma_b(E)$ and $\Delta_b(E)$ by Γ_b and Δ_b in the denominator of (13.b), where the dependence on E is essentially determined by the term $[E - E_b - \hbar\lambda^2\Delta_b(E)]^2$, but retaining $\Gamma_b(E)$ in the numerator. We then get

$$\mathcal{U}_b(E) = \frac{1}{\pi} \frac{\hbar\lambda^2\Gamma_b(E)/2}{\left[E - E_b - \hbar\lambda^2\Delta_b\right]^2 + \left[\hbar\lambda^2\Gamma_b/2\right]^2}. \tag{36}$$

The fact that $\Gamma_b(E)$ is zero for $E < 0$ and tends to zero when $E \to \infty$ (see Figure 1) thus shows that the wings of $\mathcal{U}_b(E)$ tend to zero more rapidly than those of a Lorentzian. This results in corrections to the exponential decay (35), which we will now discuss.

First, we can consider the expression (36) to be the product of the Lorentzian (33) and the function $\Gamma_b(E)/\Gamma_b$, which is equal to 1 for $E = E_b$, and is negligible outside an interval of a few $\hbar w_0$. Therefore, the Fourier transform in (36) will be the convolution product of the exponential (35) and the Fourier transform of $\Gamma_b(E)/\Gamma_b$. Because the function $\Gamma_b(E)$ has a width $\hbar w_0$, its Fourier transform $\gamma_b(\tau)$ has a finite width in τ, which is on the order of $1/w_0$. Moreover, $\Gamma_b(E)$ is strictly zero for $E < 0$ and is not infinitely differentiable at $E = 0$. Thus its Fourier transform $\gamma_b(\tau)$ behaves like a power of $(1/\tau)$ at infinity. More precisely, if $\Gamma_b(E)$ starts off like E^n, then $\gamma_b(\tau)$ varies as $1/\tau^{n+1}$ at infinity. The convolution product of $\gamma_b(\tau)$ and the exponential (35) thus modifies the behavior of $U_b(\tau)$ at both short and long times.

At short times, the exponential (35) will be rounded and will no longer have a discontinuous derivative at $\tau = 0$. It turns out that we can directly determine the behavior at short times from (15). If $t \ll 1/w_0$, the exponentials $\exp[\pm i(E - E_b)t/\hbar]$ can be replaced by 1 (because $E - E_b$ is at most on the order of a few $\hbar w_0$). The integration of (15.b) between 0 and t

thus results in

$$c(E, \beta, t) \simeq t\lambda v(E, \beta)/i\hbar. \tag{37}$$

The conservation of the norm and (22) result in

$$|c_b(t)|^2 = 1 - \iint dE \, d\beta \, \rho(E, \beta)|c(E, \beta, t)|^2$$

$$= 1 - t^2\lambda^2\hbar^{-2} \iint dE \, d\beta \, \rho(E, \beta)|v(E, \beta)|^2$$

$$= 1 - \lambda^2\Omega_1^2 t^2. \tag{38}$$

It thus appears that for very short times ($t \ll 1/w_0$), $|c_b(t)|^2$ does not decrease linearly with t, as (35) would suggest, but rather quadratically.

For long times ($\tau \gg \Gamma_b^{-1}$), the convolution product of the functions (35) and $\gamma_b(\tau)$, behaves like the one decreasing the most slowly, thus like $1/\tau^{n+1}$.

6. Intermediate Coupling. Critical Coupling

When λ increases, the deviations of $\mathscr{U}_b(E)$ relative to a Lorentzian become more and more important. Large structures can appear in the tails of $\mathscr{U}_b(E)$, as well as new maxima. Before going any further, it is important to specify the shape of $\mathscr{U}_b(E)$ near one of the maxima.

a) POWER EXPANSION OF $\mathscr{U}_b(E)$ NEAR A MAXIMUM

Near a zero E_m of Equation (27), we can write

$$E - E_b - \hbar\lambda^2\Delta_b(E) =$$

$$= E_m - E_b - \hbar\lambda^2\Delta_b(E_m) + E - E_m - \hbar\lambda^2[\Delta_b(E) - \Delta_b(E_m)]$$

$$= (E - E_m)[1 - \hbar\lambda^2\Delta_b'(E_m)]. \tag{39}$$

We used (27) and expanded $\Delta_b(E) - \Delta_b(E_m)$ to first order in $(E - E_m)$. We write

$$\Gamma_b(E_m) = \Gamma_m \tag{40.a}$$

$$\Delta_b'(E_m) = \Delta_m' \tag{40.b}$$

and substitute (39) into the denominator of (13.b). We obtain for $\mathscr{U}_b(E)$

near E_m.

$$\mathscr{U}_b(E) = \frac{1}{1 - \hbar\lambda^2\Delta'_m} \lim_{\eta \to 0_+} \frac{1}{\pi} \frac{\hbar\gamma_m/2}{(E - E_m)^2 + (\hbar\gamma_m/2)^2} \qquad (41)$$

with

$$\hbar\gamma_m = \frac{2\eta + \hbar\lambda^2\Gamma_m}{1 - \hbar\lambda^2\Delta'_m}. \qquad (42)$$

Thus it appears that near a maximum with abscissa E_m, $\tilde{U}_b(E)$ has the shape of a Lorentzian, centered at $E = E_m$, of width $\hbar\gamma_m$ at half-maximum given by (42), and of weight $1/(1 - \hbar\lambda^2\Delta'_m)$. Of course, these results are valid only if $\Gamma_b(E)$ and $\Delta_b(E)$ vary slightly over the width $\hbar\gamma_m$ of this Lorentzian, that is, if

$$\gamma_m \ll w_0. \qquad (43)$$

b) Physical Meaning of the Critical Coupling

As soon as λ exceeds λ_c, it can be seen in Figure 4 that the line $(E - E_b)/\lambda^2$ intersects the curve $\hbar\Delta_b(E)$ at a point E_m having a negative abscissa. We then use the results of the preceding subsection to specify the shape of $\mathscr{U}_b(E)$ around this value E_m.

Because $\Gamma_b(E)$ is zero for negative E and because E_m is negative, $\Gamma_b(E_m) = \Gamma_m$ is zero and therefore, according to (42),

$$\hbar\gamma_m = \frac{2\eta}{1 - \hbar\lambda^2\Delta'_m} = 2\eta' \to 0. \qquad (44)$$

The approximation (41) of $\mathscr{U}_b(E)$ is then certainly valid because, γ_m being zero according to (44), condition (43) is fulfilled. Thus, around $E = E_m$,

$$\mathscr{U}_b(E) = \frac{1}{1 - \hbar\lambda^2\Delta'_m} \lim_{\eta \to 0_+} \frac{1}{\pi} \frac{\eta'}{(E - E_m)^2 + \eta'^2}$$

$$= \frac{1}{1 - \hbar\lambda^2\Delta'_m} \delta(E - E_m). \qquad (45)$$

We thus find that, for $\lambda > \lambda_c$, a delta function appears in $\mathscr{U}_b(E)$, centered at $E = E_m$ in the region $E < 0$. This delta function, according to (13.a), gives rise to an *undamped* oscillation in $U_b(\tau)$ of the form

$$\frac{1}{1 - \hbar\lambda^2\Delta'_m} e^{-iE_m\tau/\hbar}. \tag{46}$$

Physically, the coupling has shifted the initial discrete state $|\varphi_b\rangle$ to negative energies so much that this discrete state has passed below the continuum. This state has become stable because there are no more states of the continuum having the same energy toward which it can disintegrate.

7. Strong Coupling

For $\lambda \gg 1$, Figure 4 shows that there are three points of intersection between the straight line $(E - E_b)/\lambda^2$ and the curve $\hbar\Delta_b(E)$, having abscissas $E_m^{(1)}$, $E_m^{(2)}$, and $E_m^{(3)}$.

The one furthest to the left has a negative abscissa $E_m^{(2)}$. $\mathcal{U}_b(E)$ is thus near this point a delta function given by (45). As $\lambda \gg 1$, we can use the asymptotic value (32.a) for $E_m^{(2)}$, $E_m^{(2)} = -\lambda\hbar\Omega_1$, as well as the asymptotic expression (24) for $\Delta_b(E)$ to calculate $\Delta'_m = \Delta'_b(E_m)$. Thus $\Delta'_b(E) \simeq -\hbar\Omega_1^2/E^2$ and consequently

$$\Delta'_b(E_m) = \Delta'_b(-\lambda\hbar\Omega_1) = -\frac{\hbar\Omega_1^2}{\lambda^2\hbar^2\Omega_1^2} = -\frac{1}{\hbar\lambda^2} \tag{47}$$

so that

$$\frac{1}{1 - \hbar\lambda^2\Delta'_m} = \frac{1}{2}. \tag{48}$$

Thus the expression of $\mathcal{U}_b(E)$ contains, in $E = -\lambda\hbar\Omega_1$, a delta function of weight $\frac{1}{2}$.

Now consider the other intersection of the abscissa $E_m^{(3)} \simeq +\lambda\hbar\Omega_1$ (see 32.b). Near this value of E, $\mathcal{U}_b(E)$ behaves like a Lorentzian given by (41). A calculation, identical to the foregoing one, shows that the weight, $1/(1 - \hbar\lambda^2\Delta'_m)$, of this Lorentzian is also equal to $\frac{1}{2}$. Its width γ_m, according to (42), equals

$$\gamma_m = \frac{\lambda^2\Gamma_b(\lambda\hbar\Omega_1)}{1 - \hbar\lambda^2\Delta'_m} = \frac{1}{2}\lambda^2\Gamma_b(\lambda\hbar\Omega_1). \tag{49}$$

If $\Gamma_b(E)$ decreases more rapidly than $1/E^2$ when E tends to infinity, this

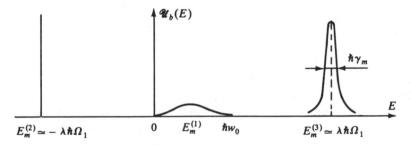

Figure 5. Shape of the variations of $\mathscr{U}_b(E)$ for $\lambda \gg 1$.

width γ_m tends to 0 when $\lambda \to \infty$. It tends to a constant if $\Gamma_b(E)$ varies as $1/E^2$. Finally, it diverges if $\Gamma_b(E)$ tends to zero less rapidly than $1/E^2$.

Finally, we examine the contribution of the third point of intersection at $E_m^{(1)}$, around the zero of $\Delta_b(E)$. We return to expression (13.b) for $\mathscr{U}_b(E)$. We see that, over an interval of a few $\hbar w_0$ on the right of $E = 0$, $E - E_b$ can be neglected compared to $\hbar \lambda^2 \Gamma_b(E)$ and $\hbar \lambda^2 \Delta_b(E)$, because $\lambda \gg 1$, so that $\mathscr{U}_b(E)$ is proportional to $1/\lambda^2$ in this interval. Thus it appears that in this same interval, $\mathscr{U}_b(E)$ behaves like a curve with a width on the order of $\hbar w_0$, whose weight, varying as $1/\lambda^2$, tends to zero when $\lambda \to \infty$. Actually the Lorentzian approximation (41) is no longer valid in this case, because $\Gamma_b(E)$ and $\Delta_b(E)$ vary a great deal in the interval $\hbar w_0$.

Finally, the variations of $\mathscr{U}_b(E)$ with E, for $\lambda \gg 1$, have the shape shown in Figure 5. As the weight of the central structure tends to zero when $\lambda \to \infty$, the behavior of $U_b(\tau)$ is essentially an oscillation of the form

$$\tfrac{1}{2} e^{i\lambda \hbar \Omega_1 \tau} + \tfrac{1}{2} e^{-i\lambda \hbar \Omega_1 \tau} = \cos \lambda \hbar \Omega_1 \tau \tag{50}$$

due to the beat note of the Fourier transforms of the two narrow curves in Figure 5, which each have a weight $\tfrac{1}{2}$ (for the time being, we neglect the width γ_m of the right-hand curve). We recover the Rabi oscillation at the frequency $\lambda \Omega_1$ when λ becomes very large. There are, however, corrections to this oscillation.

(i) For very short times ($\tau \ll w_0^{-1}$), small corrections appear, varying as $1/\lambda^2$, due to the broad central structure in Figure 5. They are damped in a time on the order of $1/w_0$.

(ii) The contribution of the right-hand curve in the figure is damped with a time constant on the order of $1/\gamma_m$, where γ_m is the width of this curve. The Rabi oscillation is thus damped with a time constant $1/\gamma_m$.

(iii) At very long times ($\tau \gg 1/\gamma_m$) only the contribution of the delta function, centered at $-\lambda\hbar\Omega_1$ persists. It is equal to $e^{i\lambda\Omega_1\tau}/2$ and is not damped.

Point (ii) clearly shows that it is impossible to obtain an undamped Rabi oscillation, as is the case for two truly discrete states. The coupling with a continuum introduces a fundamental irreversibility into the problem, which will never disappear, even if the couplings are strong.

To summarize, this very simple model shows the existence of a continuous transition between two extreme regimes, the Weisskopf–Wigner exponential decay, and the Rabi oscillation, while at the same time specifying the corrections that must be applied to these two extreme regimes.

CHAPTER IV

Radiation Considered as a Reservoir: Master Equation for the Particles

A—INTRODUCTION—OVERVIEW

In this chapter, we focus on the *evolution of the particles* and we derive equations describing such an evolution. A famous example of this approach is the one used by Einstein, who in 1917 introduced equations describing the effect of absorption, stimulated emission, and spontaneous emission processes between levels a and b of an atom immersed in the black-body radiation field (*)

$$\begin{cases} \dfrac{dN_b}{dt} = -A_{b \to a}N_b + u(\omega)(B_{a \to b}N_a - B_{b \to a}N_b) \\ \dfrac{dN_a}{dt} = A_{b \to a}N_b + u(\omega)(B_{b \to a}N_b - B_{a \to b}N_a). \end{cases} \tag{A.1}$$

In (A.1) N_a and N_b are the numbers of atoms in states a and b ($E_b > E_a$), $A_{b \to a}$ is the spontaneous emission rate from b to a, $B_{a \to b}$ (and $B_{b \to a}$) are the absorption (and stimulated emission) rates from a to b (and from b to a), and $u(\omega)$ is the energy density of the radiation field at the frequency $\omega = (E_b - E_a)/\hbar$. The purpose of this chapter is to justify and to generalize equations of this type starting from the basic equations of quantum electrodynamics describing the coupled evolution of the particles and the field.

(*) A. Einstein, *Phys. Z.* **18**, 121 (1917).

257

First note that the particles, forming a subsystem of a larger system, may be described only by a *density operator*. Indeed, even if the global system is in a pure state described by a state vector, the state of the particles is, in general, a *statistical mixture of states*. The density operator σ describing such a mixture is obtained by making a *partial trace* over the variables of the radiation field of the density operator ρ of the global system:

$$\sigma = \mathrm{Tr}_R \, \rho \qquad (A.2.a)$$

i.e., again, in terms of matrix elements

$$\langle a|\sigma|b \rangle = \sigma_{ab} = \sum_\mu \langle a,\mu|\rho|b,\mu \rangle = \sum_\mu \rho_{a\mu b\mu}. \qquad (A.2.b)$$

The Roman subscripts a and b describe the states of the particles, and the Greek subscripts μ describe the states of the radiation field (*). Thus, the state of the atom with two levels a and b introduced above is described by the *density matrix*:

$$\begin{pmatrix} \sigma_{bb} & \sigma_{ba} \\ \sigma_{ab} & \sigma_{aa} \end{pmatrix} \qquad (A.3)$$

In addition to the *populations* σ_{bb} and σ_{aa} of the two levels, proportional to the quantities N_b and N_a appearing in (A.1), there are the nondiagonal elements σ_{ab} and σ_{ba}, also called *coherences* between a and b, which are related to certain physical variables, evolving at the frequency $(E_b - E_a)/\hbar$, such as the electric dipole moment of the atom.

Before attempting to derive an evolution equation for σ, it might be useful to draw inspiration from other examples in classical physics where we are concerned with only one part of the global system.

Consider, for example, the *Brownian motion* of a heavy particle immersed in a gas or in a liquid of light particles, with which it undergoes constant collisions. As a result of the difference in mass, a very large number of collisions is required to make the velocity of the heavy particle vary appreciably. How can we describe the motion of the heavy particle? A first possibility is to introduce a *Langevin equation*, where the effect of the fluid on the particle is described by two types of force: a *friction force*, which describes the cumulative effect of the collisions and which damps

---

(*) It is clear that the information contained in σ is not as complete as the information contained in ρ. In particular, the reduced density operator σ describes neither the radiation field nor the correlations existing between the particles and the radiation field.

the velocity of the particle with a characteristic time T_R; a *Langevin force*, which describes the fluctuations of the instantaneous force about its average value, and varies with a characteristic time on the order of the collision time τ_c, which is much shorter than T_R. Another possibility is to derive an evolution equation for the statistical distribution function $f(\mathbf{r}, \mathbf{p})$ describing the position and the momentum of the heavy particle. In general, one studies the variation Δf of f over an interval of time Δt that is very short compared with T_R (so that the average velocity of the particle varies slightly during Δt), but very long compared with τ_c (so that many elementary collisions occur during this time interval). The equation giving $\Delta f / \Delta t$ is then simple. It is a *Fokker–Planck* equation, describing how the distribution function shifts and broadens under the influence of the collisions.

We retain from the foregoing example the following general ideas. First, the relevant particle $\mathscr{A}$ interacts with a system $\mathscr{R}$ having a very large number of degrees of freedom (the ensemble of the other particles of the gas or of the liquid). The "heat capacity" of $\mathscr{R}$ is therefore very large, and there is no macroscopic modification of the state of $\mathscr{R}$ under the influence of the coupling with $\mathscr{A}$. $\mathscr{R}$ can be considered to be a *reservoir*. Then, the evolution equation for $\mathscr{A}$ is simple, if there are *two distinct time scales* in the problem: a very short time τ_c characterizing the fluctuations of the perturbation exerted by $\mathscr{R}$ on $\mathscr{A}$, and a much longer time T_R characterizing the rate of variation of $\mathscr{A}$. If we consider only a *coarse-grained rate of variation*, averaged over a time Δt such that $\tau_c \ll \Delta t \ll T_R$, then simple *kinetic equations* can be obtained for the distribution functions of $\mathscr{A}$.

Is it possible to apply the foregoing ideas to the case where $\mathscr{A}$ is an ensemble of particles (that we assume here to form an atom or a molecule) and $\mathscr{R}$ is the radiation field? First, $\mathscr{R}$ actually has an infinite number of degrees of freedom, corresponding to the infinite number of modes of the electromagnetic field. If $\mathscr{A}$ is an atom (or a small number of atoms), it is legitimate to consider that the state of $\mathscr{R}$ changes only slightly as a result of its coupling with $\mathscr{A}$. What, then, are the conditions for the appearance of two distinct time scales? The dynamics of the field fluctuations (electric or magnetic) acting on the charged particles are described by the *correlation functions* of these fields. It can be shown (*) that, if the state of the field is the vacuum (which corresponds to the problem of spontaneous emission of photons by $\mathscr{A}$), these correlation functions $\langle E(t)E(t - \tau) + E(t - \tau)E(t)\rangle$ decrease very rapidly with τ. The correlation time τ_c of

(*) See, for example, *Photons and Atoms—Introduction to Quantum Electrodynamics*, III-C-3 and Complement C_{III}.

these vacuum fluctuations is very short, shorter than the period $2\pi/\omega_0$ of the relevant transition $b \to a$, which itself is much shorter than the lifetime $1/\Gamma$ of the level b, which characterizes the evolution of the atom. A similar result holds for an atom interacting with an incident wave whose spectral width $\Delta\omega$ is sufficiently large (*) and whose intensity is sufficiently low. Indeed, the correlation time of the incident field is on the order of $\tau_c = \Delta\omega^{-1}$, and the average time T_R after which an absorption or emission process occurs (evolution time for the atom) is inversely proportional to the light intensity I, so that, for appropriate values of $\Delta\omega$ and I, the condition $\tau_c \ll T_R$ is satisfied. This is the case in particular for black-body radiation, or for radiation emitted by ordinary sources (such as discharge lamps), which are neither completely monochromatic nor very intense.

Thus a large number of situations exist in which the radiation field can be considered as a reservoir exerting on the atom a perturbation that varies extremely rapidly on the atomic evolution time scale. In this Chapter, we show how the *master equation* giving the coarse-grained rate of variation of the density operator of $\mathscr{A}$ can be derived simply and in a *perturbative manner* when the coupling V between $\mathscr{A}$ and $\mathscr{R}$ has a weak effect during the correlation time τ_c of the fluctuations of $\mathscr{R}$. This situation is reminiscent of the weak-collision regime in Brownian motion. The condition expressing that the coupling has a weak effect during τ_c is known as the *motional narrowing* condition for reasons that will become clear later on. We begin (Section B) by deriving the master equation. We then provide a physical interpretation for the coefficients appearing in this equation (Section C), and we discuss the conditions of validity for the approximations used to derive the master equation (Section D). The results derived in these three parts are valid in general for any small system $\mathscr{A}$ coupled to a reservoir $\mathscr{R}$, provided the condition of motional narrowing is satisfied.

We return in Section E to the problem of a two-level atom coupled to the radiation field and we use the foregoing results to derive the evolution equations for the atomic density matrix under the influence of spontaneous emission, absorption, and stimulated emission processes. We are interested not only in internal degrees of freedom (populations of the two levels a and b and coherence between a and b), but also in the evolution of the external degrees of freedom (velocity of the center of mass) due to momentum exchanges between the atom and the photons.

Three complements continue the discussion of this chapter. Complement A_{IV} shows that the two-time averages appearing in the master

---

(*) We exclude here the case where the atom interacts with resonant monochromatic radiation (see Chapters V and VI).

equation can be related to two categories of statistical functions, the symmetric correlation functions describing the dynamics of the fluctuations of the observables of $\mathscr{A}$ and $\mathscr{R}$, and the linear susceptibilities describing the linear response of each system to an external perturbation. These functions can be used to show how the different physical effects described by the master equation can be simply interpreted by considering that the two interacting systems, $\mathscr{A}$ and $\mathscr{R}$, fluctuate and polarize each other. Complements B_{IV} and C_{IV} illustrate the different ideas introduced in this chapter by using the simple example of a harmonic oscillator coupled to a reservoir of harmonic oscillators. The evolution of the oscillator is studied in the Schrödinger representation (Complement B_{IV}), as well as in the Heisenberg representation (Complement C_{IV}). In particular, we show how the Heisenberg equations for the oscillator can be transformed and put in a form similar to that of the Langevin equation for Brownian motion. Such a calculation demonstrates the close connection that exists between fluctuations and dissipation.

B—DERIVATION OF THE MASTER EQUATION FOR A SMALL SYSTEM $\mathscr{A}$ INTERACTING WITH A RESERVOIR $\mathscr{R}$

1. Equation Describing the Evolution of the Small System in the Interaction Representation

Let

$$H = H_A + H_R + V \tag{B.1}$$

be the Hamiltonian of the global system $\mathscr{A} + \mathscr{R}$: H_A is the Hamiltonian of $\mathscr{A}$, H_R is the Hamiltonian of $\mathscr{R}$, and V is the interaction between $\mathscr{A}$ and $\mathscr{R}$. The density operator ρ of the global system $\mathscr{A} + \mathscr{R}$ obeys the evolution equation

$$\frac{d}{dt}\rho(t) = \frac{1}{i\hbar}[H, \rho(t)] \tag{B.2}$$

which becomes, in the interaction representation with respect to $H_A + H_R$:

$$\frac{d}{dt}\tilde{\rho}(t) = \frac{1}{i\hbar}[\tilde{V}(t), \tilde{\rho}(t)] \tag{B.3}$$

with

$$\tilde{\rho}(t) = e^{i(H_A + H_R)t/\hbar}\, \rho(t)\, e^{-i(H_A + H_R)t/\hbar} \tag{B.4.a}$$

$$\tilde{V}(t) = e^{i(H_A + H_R)t/\hbar}\, V\, e^{-i(H_A + H_R)t/\hbar}. \tag{B.4.b}$$

The advantage of the interaction representation is that, if V is sufficiently small, $\tilde{\rho}(t)$ evolves slowly and in particular no longer contains the unperturbed free evolution exponentials.

Integrating Equation (B.3) between t and $t + \Delta t$ yields

$$\tilde{\rho}(t + \Delta t) = \tilde{\rho}(t) + \frac{1}{i\hbar}\int_t^{t+\Delta t} dt'\,[\tilde{V}(t'), \tilde{\rho}(t')] \tag{B.5}$$

an equation that can be iterated to give

$$\Delta\tilde{\rho}(t) = \frac{1}{i\hbar}\int_t^{t+\Delta t} dt'\,[\tilde{V}(t'), \tilde{\rho}(t)] +$$

$$+ \left(\frac{1}{i\hbar}\right)^2 \int_t^{t+\Delta t} dt' \int_t^{t'} dt''\,[\tilde{V}(t'), [\tilde{V}(t''), \tilde{\rho}(t'')]] \tag{B.6}$$

where we have set

$$\Delta \tilde{\rho}(t) = \tilde{\rho}(t + \Delta t) - \tilde{\rho}(t). \tag{B.7}$$

Here we are interested in the evolution of the small system $\mathscr{A}$. Equation (A.2.a), which defines the reduced density operator σ of $\mathscr{A}$ from the density operator ρ of $\mathscr{A} + \mathscr{R}$ becomes, in the interaction representation

$$\tilde{\sigma}(t) = \mathrm{Tr}_R\, \tilde{\rho}(t). \tag{B.8}$$

By taking the trace with respect to $\mathscr{R}$ of Equation (B.6), we obtain

$$\Delta \tilde{\sigma}(t) = \frac{1}{i\hbar} \int_t^{t+\Delta t} dt' \, \mathrm{Tr}_R\big[\tilde{V}(t'), \tilde{\rho}(t)\big] +$$

$$+ \left(\frac{1}{i\hbar}\right)^2 \int_t^{t+\Delta t} dt' \int_t^{t'} dt'' \, \mathrm{Tr}_R\big[\tilde{V}(t'), \big[\tilde{V}(t''), \tilde{\rho}(t'')\big]\big]. \tag{B.9}$$

Up to this point, no approximation has been introduced and Equation (B.9) is exact. Before going further and introducing some approximations, we must now describe the assumptions concerning the reservoir.

2. Assumptions Concerning the Reservoir

a) STATE OF THE RESERVOIR

 Let

$$\tilde{\sigma}_R(t) = \mathrm{Tr}_A\, \tilde{\rho}(t) \tag{B.10}$$

be the density operator of $\mathscr{R}$ obtained by taking a partial trace over $\mathscr{A}$ of $\tilde{\rho}(t)$. Because $\mathscr{R}$ is a reservoir, the variation of $\tilde{\sigma}_R(t)$ due to the coupling with $\mathscr{A}$ is weak. To a first approximation, $\tilde{\sigma}_R(t)$ may be considered to be constant in the interaction representation (*):

$$\tilde{\sigma}_R(t) \simeq \tilde{\sigma}_R(0) = \sigma_R. \tag{B.11}$$

Moreover, we assume that the reservoir is in a stationary state, that is, that σ_R commutes with H_R:

$$[\sigma_R, H_R] = 0. \tag{B.12}$$

In other words, σ_R has no nondiagonal elements between eigenstates of H_R with different eigenvalues and can therefore be considered as a

---

(*) The coupling V, of course, causes weak correlations to appear between $\mathscr{A}$ and $\mathscr{R}$ that are essential for the evolution of $\mathscr{A}$ (see §D-4).

statistical mixture of eigenstates $|\mu\rangle$ of H_R

$$H_R|\mu\rangle = E_\mu|\mu\rangle \tag{B.13}$$

with weight p_μ

$$\sigma_R = \sum_\mu p_\mu|\mu\rangle\langle\mu|. \tag{B.14}$$

This is the case in particular when $\mathscr{R}$ is in thermodynamic equilibrium at temperature T, the p_μ being then equal to

$$p_\mu = Z^{-1} e^{-E_\mu/k_B T} \tag{B.15.a}$$

$$Z = \sum_\mu e^{-E_\mu/k_B T}. \tag{B.15.b}$$

b) ONE-TIME AND TWO-TIME AVERAGES FOR THE RESERVOIR OBSERVABLES

The interaction V between $\mathscr{A}$ and $\mathscr{R}$ will be taken as a product of an observable A of $\mathscr{A}$ and an observable R of $\mathscr{R}$.

$$V = -AR \tag{B.16}$$

which gives, in the interaction representation

$$\tilde{V}(t) = -\tilde{A}(t)\tilde{R}(t) \tag{B.17}$$

with

$$\tilde{A}(t) = e^{iH_A t/\hbar} A e^{-iH_A t/\hbar} \tag{B.18.a}$$

$$\tilde{R}(t) = e^{iH_R t/\hbar} R e^{-iH_R t/\hbar} \tag{B.18.b}$$

because the observables of $\mathscr{A}$ commute with those of $\mathscr{R}$.

Remark

The following calculations may be adapted easily to the more general case where V is a sum of products of operators A_p of $\mathscr{A}$ and R_p of $\mathscr{R}$, having the form $-\sum_p A_p R_p$.

We assume that the average value of R in the state σ_R of R is zero

$$\mathrm{Tr}[\sigma_R R] = \mathrm{Tr}\left[\sigma_R \tilde{R}(t)\right] = 0 \tag{B.19}$$

the first equation following from (B.18.b), (B.12), and the invariance of the trace of a product in a circular permutation. It follows that, for all t

$$\mathrm{Tr}_R\left[\sigma_R \tilde{V}(t)\right] = \tilde{A}(t)\mathrm{Tr}\left[\sigma_R \tilde{R}(t)\right] = 0. \tag{B.20}$$

The average value in σ_R of the coupling $\tilde{V}(t)$ is therefore zero. If this were not the case, it would suffice to reinsert $\mathrm{Tr}_R[\sigma_R V]$ into H_A and to take $V - (\mathrm{Tr}_R[\sigma_R V]) \otimes \mathbb{1}_R$ as the new interaction Hamiltonian, $\mathbb{1}_R$ being the unit operator in the state space of $\mathscr{R}$.

The average value of $\tilde{R}(t)$ in σ_R is a one-time average. We now consider the *two-time average*

$$g(t', t'') = \mathrm{Tr}\left[\sigma_R \tilde{R}(t')\tilde{R}(t'')\right] \tag{B.21}$$

equal to the average value in the state σ_R of a product of two observables $\tilde{R}(t')$ and $\tilde{R}(t'')$ taken at two different times t' and t''. In Complement A_{IV}, we analyze the physical meaning of the function $g(t', t'')$. In particular, we show that the real part of $g(t', t'')$ is a symmetric correlation function describing the *dynamics of the fluctuations of R* in the state σ_R, whereas the imaginary part of $g(t', t'')$ is related to a *linear susceptibility*. By using (B.12), (B.18), and the invariance of the trace of a product in a circular permutation, it is easy to show that $g(t', t'')$ depends only on $\tau = t' - t''$:

$$\mathrm{Tr}_R\left[\sigma_R \tilde{R}(t')\tilde{R}(t'')\right] = \mathrm{Tr}_R\left[\sigma_R\, e^{iH_R t'/\hbar}\, R\, e^{-iH_R(t'-t'')/\hbar}\, R\, e^{-iH_R t''/\hbar}\right]$$

$$= \mathrm{Tr}_R\left[\sigma_R \tilde{R}(\tau)\tilde{R}(0)\right] = g(\tau). \tag{B.22}$$

To evaluate $g(\tau)$ more precisely, we substitute expression (B.14) for σ_R into (B.22). This gives

$$g(\tau) = \mathrm{Tr} \sum_\mu \left\{ p_\mu |\mu\rangle\langle\mu| \tilde{R}(\tau)\tilde{R}(0) \right\}$$

$$= \sum_\mu p_\mu \langle\mu| \tilde{R}(\tau)\tilde{R}(0) |\mu\rangle$$

$$= \sum_\mu \sum_\nu p_\mu |R_{\mu\nu}|^2\, e^{i\omega_{\mu\nu}\tau} \tag{B.23}$$

where we have set

$$R_{\mu\nu} = \langle\mu|R|\nu\rangle \tag{B.24}$$

$$\omega_{\mu\nu} = \omega_\mu - \omega_\nu \tag{B.25.a}$$

$$\omega_\mu = E_\mu/\hbar. \tag{B.25.b}$$

Because p_μ and $|R_{\mu\nu}|^2$ are real, it is clear from (B.23) that

$$g(-\tau) = g(\tau)^*. \tag{B.26}$$

Expression (B.23) shows that $g(\tau)$ is a superposition of exponentials oscillating at the different Bohr frequencies $\omega_{\mu\nu}$ of $\mathscr{R}$. Because $\mathscr{R}$ is a reservoir, it has a very dense ensemble of energy levels and, consequently, a quasi-continuous spectrum of Bohr frequencies, so that the exponentials of (B.23) interfere destructively once τ becomes large enough. More precisely, we assume here that the function $g(\tau)$ tends rapidly to zero when τ increases, and we call τ_c the order of magnitude of the width in τ of $g(\tau)$.

Finally, the assumptions made about $\mathscr{R}$ are equivalent to assuming that $\mathscr{R}$ is in a stationary state and exerts on $\mathscr{A}$ a "force" fluctuating about a zero average value with a short correlation time τ_c.

3. Perturbative Calculation of the Coarse-Grained Rate of Variation of the Small System

We now return to the exact equation (B.9), and we derive from it a master equation for $\bar{\sigma}$ by introducing several approximations which will be discussed later on (Section D).

If V is sufficiently small, and if Δt is sufficiently short compared with the evolution time T_R of $\bar{\sigma}$, it seems legitimate to neglect the evolution of $\bar{\rho}$ between t and t'' in the last term of (B.9), which is already second-order in V and to replace $\bar{\rho}(t'')$ by $\bar{\rho}(t)$. Such an approximation is equivalent to an iteration of (B.5) in which only terms up to second order in V are retained.

After such an approximation, the right-hand side of (B.9) contains only $\bar{\rho}(t)$, which can still be written in the form

$$\bar{\rho}(t) = \mathrm{Tr}_R\,\bar{\rho}(t) \otimes \mathrm{Tr}_A\,\bar{\rho}(t) + \bar{\rho}_{\mathrm{correl}}(t) \tag{B.27}$$

where $\bar{\rho}_{\mathrm{correl}}(t)$, which is equal to the difference between $\bar{\rho}(t)$ and the product of the reduced density operators of $\mathscr{A}$ and $\mathscr{R}$, describes the correlations that exist between $\mathscr{A}$ and $\mathscr{R}$ at time t. In what follows, we will neglect the contribution of $\bar{\rho}_{\mathrm{correl}}$ to $\Delta\bar{\sigma}(t)$. Later on, in Section D, we will return to the conditions of validity for such an approximation, which assumes in particular that $\tau_c \ll \Delta t$. The general idea is that the initial correlations between $\mathscr{A}$ and $\mathscr{R}$ at time t disappear after a time τ_c and contribute little to the evolution of $\bar{\sigma}$ over the interval $[t, t + \Delta t]$, which is much longer than τ_c (*). Such an approximation is thus equivalent to writing, using (B.8), (B.10), and (B.11)

$$\tilde{\rho}(t) \simeq \tilde{\sigma}(t) \otimes \sigma_R. \tag{B.28}$$

We have thus introduced two approximations, one based on the condition $\Delta t \ll T_R$, and the other based on the condition $\Delta t \gg \tau_c$, which implies the existence of two very different time scales $T_R \gg \tau_c$:

$$\tau_c \ll \Delta t \ll T_R. \tag{B.29}$$

These two approximations allow Equation (B.9) to be written in a form relating the increase $\Delta\tilde{\sigma}$ of $\tilde{\sigma}$ between t and $t + \Delta t$ to $\tilde{\sigma}(t)$. Indeed, if in (B.9) we replace $\tilde{\rho}(t'')$ and $\tilde{\rho}(t)$ by (B.28) and divide both sides of the equation by Δt, we obtain [the first term of (B.9) is zero, according to (B.20)]

$$\frac{\Delta\tilde{\sigma}}{\Delta t} = -\frac{1}{\hbar^2}\frac{1}{\Delta t}\int_t^{t+\Delta t} dt' \int_t^{t'} dt'' \, \mathrm{Tr}_R\Big[\tilde{V}(t'), \big[\tilde{V}(t''), \tilde{\sigma}(t) \otimes \sigma_R\big]\Big]. \tag{B.30}$$

The rate of variation $\Delta\tilde{\sigma}/\Delta t$ is called the "coarse-grained" rate of variation because it can be considered to be the time average of the instantaneous rate $d\tilde{\sigma}/dt$ over an interval Δt. Indeed, $\Delta\tilde{\sigma}/\Delta t$ can be written

$$\frac{\Delta\tilde{\sigma}}{\Delta t} = \frac{\tilde{\sigma}(t + \Delta t) - \tilde{\sigma}(t)}{\Delta t} = \frac{1}{\Delta t}\int_t^{t+\Delta t} dt' \, \frac{d\tilde{\sigma}}{dt'}. \tag{B.31}$$

All the rapid variations of the instantaneous rate occurring on a time scale smaller than Δt are smoothed out in such an average. The fact that $\Delta\tilde{\sigma}/\Delta t$ depends only on the state $\tilde{\sigma}(t)$ of the system $\mathscr{A}$ at time t means that, examined with a time resolution that is not too high, the evolution of $\mathscr{A}$ depends only on the present and not on the past (Markov process).

Because, according to (B.17) and (B.18), $\tilde{V}(t')$ and $\tilde{V}(t'')$ are, like $\tilde{\sigma}(t) \otimes \sigma_R$, products of observables of $\mathscr{A}$ and of $\mathscr{R}$ commuting with each other, the trace over $\mathscr{R}$ of (B.30) concerns only products of the form $\sigma_R \tilde{R}(t')\tilde{R}(t'')$ or $\sigma_R \tilde{R}(t'')\tilde{R}(t')$. Thus it is clear that the integral of (B.30)

(*) New correlations between $\mathscr{A}$ and $\mathscr{R}$ appear between t and $t + \Delta t$, and these are the ones that cause $\tilde{\sigma}$ to evolve.

depends on the reservoir only through the two-time averages $g(\tau)$ or $g(-\tau)$ introduced above, with $\tau = t' - t''$. To take advantage of the fact that $g(\tau)$ decreases very rapidly with τ, it is convenient to change the variables of integration in (B.30), switching from the variables t' and t'' to the variables τ and t'. Figure 1 shows the domain of integration in t' and t'' of (B.30), which is the triangle $0AB$. The lines corresponding to equal values of τ are parallel to the first bisector $0B$, which corresponds to $\tau = 0$. For fixed τ, we can integrate over t' from $t + \tau$ to $t + \Delta t$, then integrate over τ from 0 to Δt, which gives

$$\int_t^{t+\Delta t} dt' \int_t^{t'} dt'' = \int_0^{\Delta t} d\tau \int_{t+\tau}^{t+\Delta t} dt'. \tag{B.32}$$

Because the two-time averages of the reservoir $g(\tau)$ and $g(-\tau)$ are negligible for $\tau \gg \tau_c$, the only region of the integration domain where the integrand is nonzero is a narrow band of width on the order of τ_c, near $0B$ (hatched region of Figure 1). As $\Delta t \gg \tau_c$, a negligible error is made if the upper bound of the integral in τ of (B.32) is extended to $+\infty$ and if the lower bound of the integral over t' is extended to t. Finally, after expanding the double commutator of (B.30), and using (B.17) and (B.22),

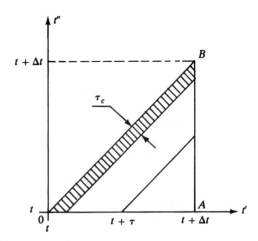

Figure 1. Domain of integration over t' and t''. As a result of the presence of the two-time averages $g(\tau)$ and $g(-\tau)$ of the reservoir, only a narrow strip of width τ_c near $0B$ contributes (hatched area).

we get

$$\frac{\Delta\tilde{\sigma}}{\Delta t} = -\frac{1}{\hbar^2}\int_0^\infty d\tau \frac{1}{\Delta t}\int_t^{t+\Delta t} dt' \times$$

$$\times\left\{g(\tau)\left[\tilde{A}(t')\tilde{A}(t'-\tau)\tilde{\sigma}(t) - \tilde{A}(t'-\tau)\tilde{\sigma}(t)\tilde{A}(t')\right] + \right.$$

$$\left. +g(-\tau)\left[\tilde{\sigma}(t)\tilde{A}(t'-\tau)\tilde{A}(t') - \tilde{A}(t')\tilde{\sigma}(t)\tilde{A}(t'-\tau)\right]\right\}. \quad \text{(B.33)}$$

To go further and carry out the integration over t', we now project the operator Equation (B.33) over a basis of states.

4. Master Equation in the Energy-State Basis

Let $|a\rangle$ be the eigenstates of H_A, having eigenvalues E_a:

$$H_A|a\rangle = E_a|a\rangle. \quad \text{(B.34)}$$

Equation (B.33) becomes, in the orthonormal basis $\{|a\rangle\}$

$$\frac{\Delta\tilde{\sigma}_{ab}}{\Delta t} = \sum_{cd}\gamma_{abcd}(t)\tilde{\sigma}_{cd}(t) \quad \text{(B.35)}$$

with

$$\gamma_{abcd}(t) = -\frac{1}{\hbar^2}\int_0^\infty d\tau \frac{1}{\Delta t}\int_t^{t+\Delta t} dt' \times$$

$$\times\left\{g(\tau)\left[\delta_{bd}\sum_n \tilde{A}_{an}(t')\tilde{A}_{nc}(t'-\tau) - \tilde{A}_{ac}(t'-\tau)\tilde{A}_{db}(t')\right] + \right.$$

$$\left. +g(-\tau)\left[\delta_{ac}\sum_n \tilde{A}_{dn}(t'-\tau)\tilde{A}_{nb}(t') - \tilde{A}_{ac}(t')\tilde{A}_{db}(t'-\tau)\right]\right\}. \quad \text{(B.36)}$$

The dependence on t' of the integrand of (B.36) can come only from the matrix elements $\tilde{A}_{an}(t')$, $A_{nc}(t'-\tau)\cdots$ which vary as $\exp i\omega_{an}t'$, $\exp i\omega_{nc}(t'-\tau)\cdots$, where the $\omega_{an}, \omega_{nc}\cdots$ are the Bohr frequencies of $\mathscr{A}$:

$$\omega_{an} = \omega_a - \omega_n \quad \text{(B.37.a)}$$

$$\omega_a = E_a/\hbar \qquad \omega_n = E_n/\hbar. \quad \text{(B.37.b)}$$

It is then simple to verify that all the terms of (B.36) inside brackets vary as $\exp i(\omega_{ab} - \omega_{cd})t'$. For example, the dependence on t' of the first term is (taking into account δ_{bd}, it follows that $\omega_b = \omega_d$)

$$\exp i(\omega_a - \omega_n + \omega_n - \omega_c)t' = \exp i(\omega_a - \omega_c)t' =$$
$$= \exp i(\omega_a - \omega_b + \omega_d - \omega_c)t' = \exp i(\omega_{ab} - \omega_{cd})t'. \quad (B.38)$$

The dependence of the second term is

$$\exp i(\omega_a - \omega_c + \omega_d - \omega_b)t' = \exp i(\omega_{ab} - \omega_{cd})t' \quad (B.39)$$

and analogous proofs can be made for the third and fourth terms. It follows that the integral over t' of (B.36) can be calculated easily and gives

$$\frac{1}{\Delta t}\int_t^{t+\Delta t} dt' \, e^{i(\omega_{ab}-\omega_{cd})t'} = e^{i(\omega_{ab}-\omega_{cd})t} f[(\omega_{ab} - \omega_{cd})\Delta t] \quad (B.40)$$

where

$$f(x) = e^{ix/2}\frac{\sin(x/2)}{(x/2)}. \quad (B.41)$$

If $|\omega_{ab} - \omega_{cd}| \ll 1/\Delta t$, the value of f in (B.40) is close to 1. By contrast, if $|\omega_{ab} - \omega_{cd}| \gg 1/\Delta t$, it is close to zero. It is thus legitimate to ignore the couplings between $\Delta\tilde{\sigma}_{ab}/\Delta t$ and $\tilde{\sigma}_{cd}$ if $|\omega_{ab} - \omega_{cd}| \gg 1/\Delta t$. Finally, if $|\omega_{ab} - \omega_{cd}| \sim 1/\Delta t$, we will see later on that the condition $T_R \gg \Delta t$ implies that the coupling between $\Delta\tilde{\sigma}_{ab}/\Delta t$ and $\tilde{\sigma}_{cd}$ has a weak effect. We will neglect this effect and retain only the terms coupling $\Delta\tilde{\sigma}_{ab}/\Delta t$ to $\tilde{\sigma}_{cd}$ with $|\omega_{ab} - \omega_{cd}| \ll 1/\Delta t$, terms called *secular* and for which $f = 1$. Finally, with this "secular approximation", the master equation (B.35) becomes

$$\frac{\Delta\tilde{\sigma}_{ab}}{\Delta t} = \sum_{c,d}^{(\text{sec})} e^{i(\omega_{ab}-\omega_{cd})t} \mathcal{R}_{abcd}\tilde{\sigma}_{cd}(t) \quad (B.42)$$

where $\sum_{c,d}^{(\text{sec})}$ indicates that the sum is restricted to states c, d such that $|\omega_{ab} - \omega_{cd}| \ll 1/\Delta t$, and where the $\mathcal{R}_{abcd}$ are coefficients independent of t and Δt and are given by the integral over τ of (B.36).

$$\mathcal{R}_{abcd} = -\frac{1}{\hbar^2}\int_0^\infty d\tau \times$$

$$\times \left\{ g(\tau)\left[\delta_{bd}\sum_n A_{an}A_{nc}\, e^{i\omega_{cn}\tau} - A_{ac}A_{db}\, e^{i\omega_{ca}\tau}\right] + \right.$$

$$\left. + g(-\tau)\left[\delta_{ac}\sum_n A_{dn}A_{nb}\, e^{i\omega_{nd}\tau} - A_{ac}A_{db}\, e^{i\omega_{bd}\tau}\right] \right\}. \quad (B.43)$$

Before calculating the coefficients $\mathcal{R}_{abcd}$ of the master equation and giving them a physical interpretation, we have to switch from the interaction representation to the Schrödinger representation, where the density operator of $\mathcal{A}$ is $\sigma(t)$. From the relation

$$\sigma_{ab}(t) = e^{-i\omega_{ab}t}\,\tilde{\sigma}_{ab}(t) \tag{B.44}$$

between the matrix elements $\sigma(t)$ and $\tilde{\sigma}(t)$ follows the relation

$$\frac{d\sigma_{ab}(t)}{dt} = -i\omega_{ab}\sigma_{ab}(t) + e^{-i\omega_{ab}t}\,\frac{d\tilde{\sigma}_{ab}(t)}{dt} \tag{B.45}$$

between the instantaneous rate of variation of σ_{ab} and $\tilde{\sigma}_{ab}$. We approximate the instantaneous rate $d\tilde{\sigma}_{ab}/dt$ appearing in (B.45) by using the coarse-grained rate $\Delta\tilde{\sigma}_{ab}/\Delta t$ calculated in (B.42). We then obtain, by using (B.44)

$$\frac{d}{dt}\sigma_{ab}(t) \simeq -i\omega_{ab}\sigma_{ab}(t) + \sum_{c,d}{}^{(\text{sec})}\mathcal{R}_{abcd}\sigma_{cd}(t). \tag{B.46}$$

The exponential appearing in (B.42) has disappeared. In the Schrödinger representation, the master equation, expanded over the basis of eigenstates of H_A, has the structure of a *linear differential system with time-independent coefficients*.

The first term of the right-hand side of (B.46) describes the free evolution of σ_{ab}, while the second term describes the effect of the interaction with $\mathcal{R}$. The coefficients $\mathcal{R}_{abcd}$ are thus on the order of $1/T_R$, where T_R is the evolution time of $\mathcal{A}$. If the matrix elements σ_{ab} and σ_{cd} have sufficiently different eigenfrequencies ω_{ab} and ω_{cd}, that is, if $|\omega_{ab} - \omega_{cd}| \gg 1/T_R$, the coupling $\mathcal{R}_{abcd}$ between them will have very weak effects (in the same way as in quantum mechanics where a coupling V_{ab} between two energy levels E_a and E_b has very weak effects if $|E_a - E_b| \gg V_{ab}$). Because $T_R \gg \Delta t$ [see (B.29)], the condition $|\omega_{ab} - \omega_{cd}| \sim 1/\Delta t$ indeed corresponds to $|\omega_{ab} - \omega_{cd}| \gg 1/T_R$. It is thus legitimate to neglect in (B.42) the coupling between σ_{ab} and σ_{cd} when $|\omega_{ab} - \omega_{cd}|$ is not very small compared with $1/\Delta t$.

C—PHYSICAL CONTENT OF THE MASTER EQUATION

1. Evolution of Populations

The populations σ_{aa} of the energy levels $|a\rangle$ of $\mathscr{A}$ all have the same free evolution frequency ($\omega_{aa} = 0$). The coupling terms that exist among them are thus all secular terms. Moreover, we assume that there is no coherence σ_{cd} with a very low free evolution frequency ($|\omega_{cd}| \ll 1/\Delta t$). Thus the populations are coupled only to populations and Equation (B.46) can be written

$$\frac{d\sigma_{aa}}{dt} = \sum_c \mathscr{R}_{aacc}\sigma_{cc}. \tag{C.1}$$

To calculate $\mathscr{R}_{aacc}$, we set $b = a$ and $d = c$ in (B.43). First assume that $c \neq a$. The two Kronecker symbols δ_{bd} and δ_{ac} are then zero. The two terms remaining inside the brackets in (B.43) are complex conjugates of each other [see (B.26)] and are regrouped to give

$$\mathscr{R}_{aacc}_{\substack{a \neq c}} = \frac{1}{\hbar^2}\int_{-\infty}^{+\infty} d\tau\, g(\tau)|A_{ac}|^2\, e^{i\omega_{ca}\tau}. \tag{C.2}$$

Replacing $g(\tau)$ by its expression (B.23) then gives

$$\mathscr{R}_{aacc}_{\substack{a \neq c}} = \frac{1}{\hbar^2}\sum_\mu p_\mu \sum_\nu \int_{-\infty}^{+\infty} d\tau\, e^{i(\omega_{\mu\nu}+\omega_{ca})\tau}|A_{ac}|^2|R_{\mu\nu}|^2. \tag{C.3}$$

The integral over τ is equal to $2\pi\delta(\omega_{\mu\nu} + \omega_{ca})$ and can be rewritten, using (B.25) and (B.37), as $2\pi\hbar\delta(E_\mu + E_c - E_\nu - E_a)$. Also, $|A_{ac}|^2|R_{\nu\mu}|^2$ is equal, according to (B.16), to $|\langle\nu, a|V|\mu, c\rangle|^2$. Finally, by setting

$$\mathscr{R}_{aacc}_{\substack{a \neq c}} = \Gamma_{c \to a} \tag{C.4}$$

we obtain for $\Gamma_{c \to a}$

$$\Gamma_{c \to a} = \frac{2\pi}{\hbar}\sum_\mu p_\mu \sum_\nu |\langle\nu, a|V|\mu, c\rangle|^2 \delta(E_\mu + E_c - E_\nu - E_a). \tag{C.5}$$

The physical interpretation of $\Gamma_{c \to a}$ is very clear. $\Gamma_{c \to a}$ is the probability per unit time for the system $\mathscr{A}$ to make a transition from level c to

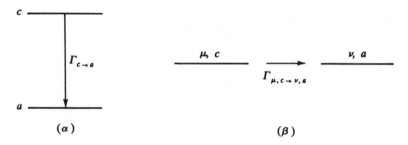

Figure 2. Transition rate between states c and a of $\mathscr{A}$ (α) and states μ, c and ν, a of $\mathscr{A} + \mathscr{R}$ (β). $\Gamma_{c \to a}$ is obtained from $\Gamma_{\mu, c \to \nu, a}$ by an average over μ and a sum over ν.

level a as a result of its coupling with $\mathscr{R}$. Indeed, such a transition corresponds, for the global system $\mathscr{A} + \mathscr{R}$, to a transition $|\mu, c\rangle \to |\nu, a\rangle$ (Figure 2). Equation (C.5) gives the rate for this transition (in agreement with the Fermi golden rule), *averaged over all the possible initial states* μ of the reservoir (weighted by p_μ) and *summed over all the final states* ν of the reservoir, with the delta function expressing the conservation of energy for the global system $\mathscr{A} + \mathscr{R}$.

$\mathscr{R}_{aaaa}$ remains to be evaluated. These are now the second and fourth terms inside the brackets of (B.43), which cancel the terms $n = a$ of the first and third sums. The remaining terms $n \neq a$ cause $\Gamma_{a \to n}$ to appear, and we obtain

$$\mathscr{R}_{aaaa} = - \sum_{n \neq a} \Gamma_{a \to n}. \tag{C.6}$$

Finally, the master equation for the populations (C.1) is written

$$\frac{d\sigma_{aa}}{dt} = -\sigma_{aa} \sum_{n \neq a} \Gamma_{a \to n} + \sum_{c \neq a} \sigma_{cc} \Gamma_{c \to a} \tag{C.7.a}$$

or as

$$\frac{d\sigma_{aa}}{dt} = \sum_{c \neq a} (\sigma_{cc} \Gamma_{c \to a} - \sigma_{aa} \Gamma_{a \to c}). \tag{C.7.b}$$

Equation (C.7.a) indicates that the population σ_{aa} of $\mathscr{A}$ decreases as a consequence of the transitions occurring from a to the other levels n, and increases as a result of transitions from other levels c to the level a. The form (C.7.b) exhibits the balance of transfers occurring for each pair of levels c and a and simply yields

$$\sum_a \frac{d}{dt}\sigma_{aa} = 0 \tag{C.8}$$

(conservation of the normalization of σ over time).

Remarks

(i) Very often, the steady-state solution of equation (C.7.b) corresponds to

$$\sigma_{aa}^{st}\Gamma_{a \to c} = \sigma_{cc}^{st}\Gamma_{c \to a}. \tag{C.9}$$

Condition (C.9), called the "detailed balance condition", shows that in the steady state, and for any pair of levels a and c, the number of transitions from a to c compensates for the number of transitions from c to a.

(ii) If the reservoir $\mathscr{R}$ is in thermodynamic equilibrium at temperature T, that is, if populations p_μ of the levels μ of $\mathscr{R}$ are given by (B.15), it can be shown by using (C.5) (see Exercise 16) that

$$e^{-E_a/k_BT}\,\Gamma_{a \to c} = e^{-E_c/k_BT}\,\Gamma_{c \to a}. \tag{C.10}$$

Comparing (C.9) with (C.10) then shows that the populations σ_{aa} of $\mathscr{A}$ tend, under the effect of the interaction with $\mathscr{R}$, to a steady state where they are proportional to $\exp(-E_a/k_BT)$. By interacting with a reservoir in thermodynamic equilibrium, the system $\mathscr{A}$ also reaches thermodynamic equilibrium at the same temperature T.

2. Evolution of Coherences

We now examine the evolution of the nondiagonal elements σ_{ab} of σ, and we first consider the case where the Bohr frequency ω_{ab} associated with the coherence σ_{ab} is nondegenerate, that is, the case where there are no other Bohr frequencies ω_{cd} differing from ω_{ab} by less than $1/\Delta t$. The only secular term is therefore the one coupling the coherence σ_{ab} to itself, and Equation (B.46) is written

$$\frac{d}{dt}\sigma_{ab} = -i\omega_{ab}\sigma_{ab} + \mathscr{R}_{abab}\sigma_{ab}. \tag{C.11}$$

To calculate $\mathscr{R}_{abab}$, set $c = a$, $d = b$ in (B.43). This gives

$$\mathscr{R}_{abab} = -\frac{1}{\hbar^2}\int_0^\infty d\tau \left\{ g(\tau)\left[\sum_n |A_{an}|^2 e^{i\omega_{an}\tau} - A_{aa}A_{bb}\right] + \right.$$

$$\left. + g(-\tau)\left[\sum_n |A_{bn}|^2 e^{-i\omega_{bn}\tau} - A_{aa}A_{bb}\right]\right\}. \tag{C.12}$$

We then replace $g(\tau)$ by (B.23), $g(-\tau)$ by $g(\tau)^*$, and evaluate the integrals from $\tau = 0$ to $\tau = \infty$ of the exponentials, leading to the appearance of principal parts and delta functions. We thus obtain

$$\mathscr{R}_{abab} = -\Gamma_{ab} - i\Delta_{ab} \tag{C.13}$$

where Δ_{ab} and Γ_{ab} are real quantities given by

$$\Delta_{ab} = \Delta_a - \Delta_b \tag{C.14}$$

with

$$\Delta_a = \frac{1}{\hbar}\mathscr{P}\sum_\mu p_\mu \sum_\nu \sum_n \frac{|\langle \nu, n|V|\mu, a\rangle|^2}{E_\mu + E_a - E_\nu - E_n} \tag{C.15}$$

and an analogous expression for Δ_b, and where

$$\Gamma_{ab} = \Gamma_{ab}^{\text{nonad.}} + \Gamma_{ab}^{\text{ad.}} \tag{C.16}$$

with

$$\Gamma_{ab}^{\text{nonad.}} = \frac{1}{2}\left(\sum_{n \neq a}\Gamma_{a \to n} + \sum_{n \neq b}\Gamma_{b \to n}\right) \tag{C.17}$$

and

$$\Gamma_{ab}^{\text{ad.}} = \frac{2\pi}{\hbar} \sum_{\mu} p_{\mu} \sum_{\nu} \delta(E_{\mu} - E_{\nu}) \times$$

$$\times \left(\frac{1}{2} |\langle \nu, a|V|\mu, a\rangle|^2 + \frac{1}{2} |\langle \mu, b|V|\nu, b\rangle|^2 - \right.$$

$$\left. - \text{Re}\langle \mu, a|V|\nu, a\rangle \langle \nu, b|V|\mu, b\rangle \right). \quad (C.18)$$

The quantity $\hbar\Delta_{ab}$ represents a *shift*, second order in V, of the frequency ω_{ab} due to the interaction between $\mathscr{A}$ and $\mathscr{R}$. Indeed, in (C.15), $\hbar\Delta_a$ is the shift of the state $|\mu, a\rangle$ of the global system $\mathscr{A} + \mathscr{R}$, weighted by the probability of occupation p_{μ} of the level μ of the reservoir and summed over μ. Thus $\hbar\Delta_a$ can be considered to be the average energy shift of the state $|a\rangle$ of $\mathscr{A}$.

The quantity Γ_{ab} represents the *damping* rate of the coherence σ_{ab} due to the interaction between $\mathscr{A}$ and $\mathscr{R}$. The first contribution to Γ_{ab}, shown in (C.17), comes from nonadiabatic effects, because expression (C.17) is the half sum of the rates with which the system leaves state a or state b. A process removing the system from state a or from state b actually perturbs the oscillation of any physical quantity sensitive to σ_{ab}. The second contribution, shown in (C.18), is called adiabatic because it is due to processes where $\mathscr{A}$ does not change state while interacting with $\mathscr{R}$, which goes from a state μ to a state ν having the same energy as μ.

Finally, we consider the case where the frequency ω_{ab} is degenerate. As in Equation (B.46), we must take into account the couplings between the coherence σ_{ab} and the other coherences σ_{cd} such that $|\omega_{cd} - \omega_{ab}| \ll 1/\Delta t$. The calculation of $\mathscr{R}_{abcd}$ from (B.43) (with $c \neq a$, $b \neq d$, $|\omega_{ab} - \omega_{cd}| \ll 1/\Delta t$) thus gives

$$\mathscr{R}_{abcd} = \frac{2\pi}{\hbar} \sum_{\mu} p_{\mu} \sum_{\nu} \langle \nu, a|V|\mu, c\rangle \langle \mu, d|V|\nu, b\rangle \times$$

$$\times \delta(E_{\mu} + E_c - E_{\nu} - E_a). \quad (C.19)$$

The couplings between different coherences having the same Bohr frequency are important for understanding the shift and the broadening of transitions of systems having equidistant levels, such as the harmonic oscillator or the dressed atom.

Remark

All the expressions derived in this part that relate physical parameters such as $\Gamma_{b \to a}, \Delta_a, \Gamma_{ab}^{\text{ad.}}, \Gamma_{ab}^{\text{nonad.}}, \mathcal{R}_{abcd} \cdots$ to matrix elements of V remain valid when V, instead of having the form $V = -AR$, is generalized as $V = -\sum_p A_p R_p$.

D—DISCUSSION OF THE APPROXIMATIONS

To specify the conditions of validity for the treatment presented in Section B above, we must first evaluate an order of magnitude of the evolution time T_R for the system $\mathscr{A}$.

1. Order of Magnitude of the Evolution Time for $\mathscr{A}$

We return to Equation (B.30) and try to evaluate an order of magnitude for the right-hand side. We saw above (end of subsection B-3) that the only part of the domain of integration in Figure 1 that contributes significantly to the integral is a strip of width τ_c near the first bisector $0B$. On this bisector, $t' = t''$, and the integrand of (B.30) is on the order of $\langle V^2 \rangle_R \tilde{\sigma}(t)$, where

$$\langle V^2 \rangle_R = \text{Tr}_R \, \tilde{\sigma}_R \tilde{V}(t')^2 = \text{Tr}_R \, \sigma_R V^2. \tag{D.1}$$

The order of magnitude of the right-hand side of (B.30) is thus obtained by multiplying (D.1) by $1/\hbar^2 \, \Delta t$ times the area of the cross-hatched portion of Figure 1, which is on the order of $\tau_c \, \Delta t$

$$\frac{\Delta \tilde{\sigma}}{\Delta t} \sim -\frac{\tau_c}{\hbar^2} \langle V^2 \rangle_R \tilde{\sigma}. \tag{D.2}$$

We denote v^2 the order of magnitude of $\langle V^2 \rangle_R$. The parameter v characterizes the intensity of the coupling between $\mathscr{A}$ and $\mathscr{R}$. Because the coefficient multiplying $\tilde{\sigma}$ on the right-hand side of (D.2) is on the order of the inverse of the evolution time T_R of $\mathscr{A}$, we obtain for $1/T_R$ the order of magnitude

$$\frac{1}{T_R} \sim \frac{v^2 \tau_c}{\hbar^2}. \tag{D.3}$$

2. Condition for Having Two Time Scales

Using (D.3), the condition $\tau_c \ll T_R$, on which the entire discussion in Section B above is based, can be written as

$$\frac{v \tau_c}{\hbar} \ll 1. \tag{D.4}$$

Equation (D.4) expresses that the evolution due to the coupling V between $\mathscr{A}$ and $\mathscr{R}$, characterized by the frequency $v/\hbar$, has a very weak influence during the correlation time τ_c (a situation similar to the weak-collision regime of Brownian motion where the velocity of the heavy particle changes only slightly during the collision time τ_c).

The parameter v introduced above also characterizes the dispersion of values of V. If the broadening of the spectral lines of $\mathscr{A}$ produced by the coupling with $\mathscr{R}$ were inhomogeneous, the width of the spectral lines would be on the order of $v/\hbar$. But, from (D.2), the width is approximately $1/T_R$, which is, according to (D.3) and (D.4)

$$\frac{v}{\hbar} \frac{v\tau_c}{\hbar} \ll \frac{v}{\hbar}. \tag{D.5}$$

Thus the fact that the interaction between $\mathscr{A}$ and $\mathscr{R}$ fluctuates rapidly, reduces the inhomogeneous width $v/\hbar$ by a factor $v\tau_c/\hbar \ll 1$, hence the name "motional narrowing" condition given to (D.4).

3. Validity Condition for the Perturbative Expansion

If Equation (B.5) is iterated beyond the second order [while replacing $\bar{\rho}(t)$ everywhere it appears by (B.28)], we obtain for the second member of (B.30), in addition to the double commutator already there, a triple commutator, a quadruple commutator, etc.

The same procedure as the one followed in subsection D-1 above may be used to evaluate the order of magnitude of the contributions of these terms that are higher order in V. For example, the triple commutator involves the integral over t_1, t_2, t_3 of a product of three operators $\tilde{V}(t_1)\tilde{V}(t_2)\tilde{V}(t_3)$ and of σ_R. Because the three times t_1, t_2, t_3 must be quite close to each other (to within τ_c), the significant volume of integration is on the order of $\tau_c^2 \, \Delta t$, so that the order of magnitude of the third-order term in V is

$$\frac{v^3}{\hbar^3}\tau_c^2 = \frac{v^2\tau_c}{\hbar^2}\frac{v\tau_c}{\hbar} \sim \frac{1}{T_R}\frac{v\tau_c}{\hbar}, \tag{D.6}$$

or $v\tau_c/\hbar$ times the second-order term. It thus appears that the small parameter $v\tau_c/\hbar$ also characterizes the perturbation expansion. If condition (D.4) is satisfied, it is legitimate to stop such an expansion at the second order in V.

4. Factorization of the Total Density Operator at Time t

We now examine the approximation consisting of neglecting $\tilde{\rho}_{\text{correl}}(t)$ in (B.27). Assume that the correlations [described by $\tilde{\rho}_{\text{correl}}(t)$] that exist between $\mathscr{A}$ and $\mathscr{R}$ at time t result from interactions that occurred between $\mathscr{A}$ and $\mathscr{R}$ prior to t. This is equivalent to assuming that at a given initial time t_0, prior to t, $\mathscr{A}$ and $\mathscr{R}$ are not correlated ($\mathscr{A}$ and $\mathscr{R}$, for example, start to interact at t_0). To lowest order in V, it is necessary to have at least one interaction $\tilde{V}$ prior to t to have $\tilde{\rho}_{\text{correl}}(t) \neq 0$. If we now consider that $\tilde{\rho}_{\text{correl}}(t)$ is nonzero, the contribution to $\Delta\tilde{\sigma}(t)$ of the first-order term in V of (B.9) is no longer zero because it is then no longer possible to use (B.20). Thus, to second order in V, a new contribution to $\Delta\tilde{\sigma}(t)$ appears that results from an interaction prior to t (which creates correlations between $\mathscr{A}$ and $\mathscr{R}$), and from another interaction in the interval $[t, t + \Delta t]$ (which produces a variation $\Delta\tilde{\sigma}$ of $\tilde{\sigma}$ from these correlations).

The order of magnitude of this contribution that was neglected is

$$\frac{\Delta\tilde{\sigma}}{\Delta t} \sim -\frac{1}{\hbar^2}\frac{1}{\Delta t}\int_{-\infty}^{t} dt'' \int_{t}^{t+\Delta t} dt' \langle \tilde{V}(t'')\tilde{V}(t')\rangle_R \tilde{\sigma} \qquad (D.7)$$

with t'' varying prior to t, and t' in the interval $[t, t + \Delta t]$. Because $\langle \tilde{V}(t')\tilde{V}(t'')\rangle_R$ is zero once $t' - t'' \gg \tau_c$, the domain of integration is reduced to two intervals of width τ_c on both sides of t, and the order of magnitude of the rate of variation associated with (D.7) is

$$\frac{v^2\tau_c^2}{\hbar^2 \Delta t} = \frac{v^2\tau_c}{\hbar^2}\frac{\tau_c}{\Delta t} = \frac{1}{T_R}\frac{\tau_c}{\Delta t}. \qquad (D.8)$$

Thus it appears, as we mentioned previously [see discussion before (B.28)], that the contribution of the last term of (B.27) is smaller than that of the first term by a factor $\tau_c/\Delta t$. Because we have taken $\Delta t \gg \tau_c$, it is completely legitimate to neglect such a contribution.

Thus the initial correlations between $\mathscr{A}$ and $\mathscr{R}$ at time t have an effect on the future of $\tilde{\sigma}$ only in the interval $[t, t + \tau_c]$, whereas new correlations are established permanently in the interval $[t, t + \Delta t]$ and cause $\tilde{\sigma}$ to evolve proportionally to Δt.

Remark

It is possible to imagine correlations between $\mathscr{A}$ and $\mathscr{R}$ at time t that can have a spectacular influence on the subsequent evolution of $\mathscr{A}$. Consider, for example, at time $t_0 = 0$, an atom initially excited in the state $|b\rangle$, with no incident photon. Radiative decay can occur between 0 and t. Then imagine that at time t, we apply to the state $|\psi(t)\rangle$ of the global system atom + radiation the time-reversal operator K. Because the Schrödinger equation has the time-reversal symmetry, we know that at time $2t$, the atom will have returned to the excited state (more precisely to the state $K|b\rangle$), in the radiation vacuum. It is therefore clear that the state $K|\psi(t)\rangle$ has very specific correlations because the evolution from this state does not at all have the irreversible behavior predicted by the master equation. Thus such an equation cannot always be valid, even if $\mathcal{R}$ is a reservoir. Nonetheless, we should note that correlations of the type contained in the state $K|\psi(t)\rangle$ are very peculiar and extremely difficult to achieve in an experiment. Therefore we exclude all situations of this type from our discussion.

5. Summary

Finally, the treatment presented in Section B is based on the motional narrowing condition (D.4). Under such a condition, two very distinct time scales T_R and τ_c can be defined, and a perturbative calculation can be performed for the variation $\Delta\tilde{\sigma}$ of $\tilde{\sigma}$ in the interval $[t, t + \Delta t]$ with $\tau_c \ll \Delta t \ll T_R$. It is then not necessary to take into account the initial correlations between $\mathcal{A}$ and $\mathcal{R}$ at time t, resulting from interactions between $\mathcal{A}$ and $\mathcal{R}$ prior to t. The fact that the coarse-grained rate of variation $\Delta\sigma/\Delta t$ is given (in the Schrödinger representation) by a differential linear system with constant coefficients implies that the previous approach is valid for all t. The master equation can therefore be used to predict the evolution of $\tilde{\sigma}$ over much longer times, on the order of a few T_R. A remarkable feature of this approach is that a perturbative study of the evolution of the system over a time interval that is short ($\Delta t \ll T_R$), but not too short ($\Delta t \gg \tau_c$), allows its behavior to be predicted over much longer times. In this sense, such an approach is nonperturbative and is similar to the procedure presented in Chapter III where a perturbative calculation of the level-shift operator allows nonperturbative expressions for the transition amplitudes to be obtained.

E—APPLICATION TO A TWO-LEVEL ATOM COUPLED
TO THE RADIATION FIELD

In this last part, the foregoing ideas are illustrated using a simple example. We derive and discuss the master equation describing the evolution of the density matrix of an atom with two levels a and b under the effect of spontaneous emission, absorption, and stimulated emission processes. First, the atom is assumed to be infinitely heavy and at rest, which allows us to study the evolution of just the internal degrees of freedom (§1). The translational degrees of freedom of the center of mass are then taken into account, and we will study the evolution of the atomic velocities resulting from momentum exchanges between the atom and the incident radiation field (§2).

1. Evolution of Internal Degrees of Freedom

The atomic electrons are assumed to evolve around a fixed point 0. In the electric dipole point of view, the interaction Hamiltonian between the atom and the radiation field is written [see formula (91) in the Appendix]

$$V = -\mathbf{d} \cdot \mathbf{E}_\perp (\mathbf{0}) = -i\mathbf{d} \cdot \sum_i \sqrt{\frac{\hbar \omega_i}{2\varepsilon_0 L^3}} \, \boldsymbol{\varepsilon}_i (a_i - a_i^+) \qquad (\text{E}.1)$$

where $\mathbf{d}$ is the electric dipole moment and where the field operator is evaluated at the point 0.

a) MASTER EQUATION DESCRIBING SPONTANEOUS EMISSION
FOR A TWO-LEVEL ATOM

We begin by assuming that the radiation field is in the vacuum state

$$\sigma_R = |0\rangle\langle 0|. \qquad (\text{E}.2)$$

We previously explained (Section A) why it is legitimate to consider the radiation field in such a state as a reservoir. The state (E.2) indeed satisfies conditions (B.12) and (B.20), taking into account expression (E.1) for V.

The spontaneous emission of a photon from the lower state a cannot conserve the total unperturbed energy. Equation (C.5) thus implies

$$\Gamma_{a \rightarrow b} = 0 \qquad (\text{E}.3)$$

By contrast, the same equation gives a nonzero result for $\Gamma_{b \to a}$:

$$\Gamma_{b \to a} = \frac{2\pi}{\hbar} \sum_{\mathbf{k}\varepsilon} |\langle a; \mathbf{k}\varepsilon|V|b; 0\rangle|^2 \delta(\hbar\omega - \hbar\omega_{ba}) \qquad \text{(E.4.a)}$$

which is just the rate Γ for the spontaneous emission of a photon, previously introduced in Chapter II and equal to the inverse of the radiative lifetime τ of level b

$$\Gamma_{b \to a} = \Gamma = \frac{1}{\tau}. \qquad \text{(E.4.b)}$$

Equations (C.7) for the populations are thus written here

$$\begin{cases} \dfrac{d}{dt}\sigma_{bb} = -\Gamma\sigma_{bb} \\[2mm] \dfrac{d}{dt}\sigma_{aa} = +\Gamma\sigma_{bb} \end{cases} \qquad \text{(E.5)}$$

and have the same form as the Einstein equations (A.1) [with $u(\omega) = 0$].
   For the evolution of the nondiagonal element of σ_{ba}, the general results of subsection C-2 give

$$\frac{d}{dt}\sigma_{ba} = -i(\omega_{ba} + \Delta_{ba})\sigma_{ba} - \frac{\Gamma}{2}\sigma_{ba} \qquad \text{(E.6)}$$

where $\hbar\Delta_{ba}$ is the difference in radiative shifts of the levels b and a

$$\Delta_{ba} = \Delta_b - \Delta_a \qquad \text{(E.7.a)}$$

$$\Delta_b = \frac{1}{\hbar}\mathscr{P}\sum_{\mathbf{k}\varepsilon} \frac{|\langle a; \mathbf{k}\varepsilon|V|b; 0\rangle|^2}{\hbar\omega_{ba} - \hbar\omega} \qquad \text{(E.7.b)}$$

$$\Delta_a = \frac{1}{\hbar}\mathscr{P}\sum_{\mathbf{k}\varepsilon} \frac{|\langle b; \mathbf{k}\varepsilon|V|a; 0\rangle|^2}{-\hbar\omega_{ba} - \hbar\omega} \qquad \text{(E.7.c)}$$

and where $\Gamma/2$, which is the half-sum of (E.3) and (E.4.b), is the nonadiabatic contribution to the damping of σ_{ab} [see (C.17)]. The adiabatic contribution (C.18) is zero because V has no diagonal elements in state $|a\rangle$ or state $|b\rangle$ ($\mathbf{d}$ is odd and $|a\rangle$ and $|b\rangle$ are assumed to have a well-defined parity).

Remark

We must not forget that the approximation consisting of considering only two atomic levels a and b causes several effects to be lost. First, even if a is the ground state and b is the first excited state, other states more excited than b can decay radiatively to a or b. Moreover, the contribution of other atomic levels c of the atom certainly cannot be ignored in nonresonant processes (virtual emission and reabsorption of a photon) which are the origin of the radiative shifts of a and b.

b) Additional Terms Describing the Absorption and Induced Emission of a Weak Broadband Radiation

We now assume that some photons are initially present. The density operator of the radiation is, according to (B.14), a statistical mixture of eigenstates $|n_1 \cdots n_i \cdots \rangle$ of H_R, representing n_1 photons in the mode $1, \cdots n_i$ photons in the mode $i \cdots$, with a weight $p(n_1 \cdots n_i \cdots)$

$$\sigma_R = \sum_{\{n_i\}} p(n_1 \cdots n_i \cdots)|n_1 \cdots n_i \cdots \rangle\langle n_1 \cdots n_i \cdots |. \quad \text{(E.8)}$$

Conditions (B.12) and (B.20) are still satisfied. We explained above (Section A), that if the spectral width of the radiation described by (E.8) is sufficiently broad, and if its intensity is sufficiently low, the condition $\tau_c \ll T_R$ is satisfied, which we assume here.

The transition rate from the lower level a to the higher level b, calculated from (C.5), is no longer zero when (E.2) is replaced by (E.8). We write it as Γ'. It represents the probability per unit time for the absorption of a photon from a and equals

$$\Gamma' = \Gamma_{a \to b} = \frac{2\pi}{\hbar} \sum_{\{n_i\}} p(n_1 \cdots n_i \cdots) \times$$

$$\times \sum_{\{n_i'\}} |\langle b; n_1' \cdots n_i' \cdots |V|a; n_1 \cdots n_i \cdots \rangle|^2 \delta(E_{\text{final}} - E_{\text{initial}}).$$

$$\text{(E.9)}$$

Because V is proportional to the field $\mathbf{E}_\perp$, and $\mathbf{E}_\perp$ is a linear combination of a_i and a_i^+ [see (E.1)], all the n' must be equal to n, except one, n_i' which must equal $n_i \pm 1$. In fact, because $E_b > E_a$, the conservation of energy results in the fact that only $n_i' = n_i - 1$ is possible, so that (E.9) is rewritten

$$\Gamma' = \frac{2\pi}{\hbar} \sum_{n_1 \cdots n_i \cdots} p(n_1 \cdots n_i \cdots) \times$$

$$\times \sum_i |\langle b; n_1 \cdots n_i - 1 \cdots |V| a; n_1 \cdots n_i \cdots \rangle|^2 \delta(\hbar\omega_i - \hbar\omega_{ba}).$$

$$(E.10)$$

Because $\langle n_i - 1 | a_i | n_i \rangle = \sqrt{n_i} \langle 0_i | a_i | 1_i \rangle$, Γ' is also equal to

$$\Gamma' = \frac{2\pi}{\hbar} \sum_i \left(\sum_{\{n_i\}} n_i p(n_1 \cdots n_i \cdots) \right) \times$$

$$\times |\langle b; 0 | V | a; 1_i \rangle|^2 \delta(\hbar\omega_i - \hbar\omega_{ba}) \qquad (E.11)$$

which causes the average number $\langle n_i \rangle$ of photons in the mode i to appear

$$\langle n_i \rangle = \sum_{\{n_i\}} n_i p(n_1 \cdots n_i \cdots). \qquad (E.12)$$

The radiation state thus appears in Γ' only through the average number of photons in each mode. Finally, the function $\delta(\hbar\omega_i - \hbar\omega_{ba})$ expressing the conservation of energy in (E.11) results in the fact that only the average numbers of photons in the resonant modes contribute. Γ' is thus proportional to the average intensity $I(\omega_{ba})$ of the incident radiation at the atomic frequency ω_{ba}.

A similar procedure can be followed for calculating $\Gamma_{b \to a}$. It is now the matrix elements of a_i^+ between $|b; n_1 \cdots n_i \cdots \rangle$ and $\langle a; n_1 \cdots n_i + 1 \cdots |$ that are involved and (E.11) must be replaced by

$$\Gamma_{b \to a} = \frac{2\pi}{\hbar} \sum_i (\langle n_i \rangle + 1) |\langle a; 1_i | V | b; 0 \rangle|^2 \delta(\hbar\omega_i - \hbar\omega_{ba}). \quad (E.13)$$

We used the normalization relation for $p(n_1 \cdots n_i \cdots)$. The contribution of the term 1 inside the parentheses ($\langle n_i \rangle + 1$) in (E.13) gives the previously calculated transition rate Γ associated with the spontaneous emission from b to a. As for the other term $\langle n_i \rangle$, it gives the transition probability Γ' given in (E.11) so that

$$\Gamma_{b \to a} = \Gamma + \Gamma'. \qquad (E.14)$$

Finally, the evolution equations (C.7) for populations become, in the presence of incident radiation,

$$
\begin{cases}
\dfrac{d}{dt}\sigma_{bb} = -\Gamma\sigma_{bb} + \Gamma'(\sigma_{aa} - \sigma_{bb}) \\[2mm]
\dfrac{d}{dt}\sigma_{aa} = +\Gamma\sigma_{bb} + \Gamma'(\sigma_{bb} - \sigma_{aa}).
\end{cases}
\tag{E.15}
$$

They actually have the general form of the Einstein equations (A.1) because Γ' is proportional to the light intensity at the frequency ω_{ba} and, for an atom with two nondegenerate levels a and b, it can be shown that $B_{a \to b} = B_{b \to a}$ in (A.1).

Remarks

(i) Until now, we have made no assumption concerning the angular distribution and the polarization of the incident radiation. If the incident radiation is isotropic and unpolarized, the average number of photons $\langle n_i \rangle$ in a mode i (i.e., a mode $\mathbf{k}_i \boldsymbol{\varepsilon}_i$) depends only on the frequency ω_i of this mode and not on $\boldsymbol{\kappa}_i = \mathbf{k}_i/k_i$ and $\boldsymbol{\varepsilon}_i$. The sum over i of (E.11) is in fact a sum over $\boldsymbol{\varepsilon}_i \perp \mathbf{k}_i$, followed by an angular integral over $\boldsymbol{\kappa}_i$ and an integral over $k_i = |\mathbf{k}_i|$. If $\langle n_i \rangle$ depends only on ω_i, the sum over $\boldsymbol{\varepsilon}_i$ and the integral over $\boldsymbol{\kappa}_i$ are the same in expressions (E.4.a) and (E.11) giving Γ and Γ'. The difference between Γ and Γ' appears only in the integral over ω_i. Because of the function $\delta(\hbar\omega_i - \hbar\omega_{ba})$ appearing in (E.4.a) and (E.11), this difference is simply a multiplying factor $\langle n(\omega_{ba}) \rangle$ for Γ'. For isotropic and unpolarized incident radiation, we have

$$
\Gamma' = \Gamma\langle n(\omega_{ba}) \rangle.
\tag{E.16}
$$

Γ' is thus equal to Γ multiplied by the average number of photons per resonant mode.

(ii) We now assume that the radiation is in thermodynamic equilibrium at temperature T. Such radiation is isotropic and unpolarized so that (E.16) is applicable. Moreover, in this case, we know the value of the probability $p(n_1 n_2 \cdots n_i \cdots)$ [see (B.15)] and, consequently, according to (E.12) the value of $\langle n(\omega_{ba}) \rangle$ which equals (*)

$$
\langle n(\omega_{ba}) \rangle = \frac{1}{e^{\hbar\omega_{ba}/k_B T} - 1}.
\tag{E.17}
$$

(*) See, for example, *Photons and Atoms—Introduction to Quantum Electrodynamics*, Exercise 4 in Complement D_{III}.

We then substitute (E.16) into the first equation (E.15), which gives, in the steady state ($d\sigma_{bb}/dt = 0$)

$$\frac{\sigma_{bb}}{\sigma_{aa}} = \frac{\Gamma'}{\Gamma + \Gamma'} = \frac{\langle n(\omega_{ba})\rangle}{1 + \langle n(\omega_{ba})\rangle}. \tag{E.18}$$

It is then sufficient to substitute (E.17) into (E.18) to obtain

$$\frac{\sigma_{bb}}{\sigma_{aa}} = e^{-\hbar\omega_{ba}/k_B T} = e^{-(E_b - E_a)/k_B T}. \tag{E.19}$$

Thus we have derived the Einstein equations from the first principles and showed that they imply that the atom reaches thermodynamic equilibrium.

(iii) The average number of photons per mode for isotropic radiation, $\langle n(\omega)\rangle$, is directly related to the radiation energy density at frequency ω, $u(\omega)$. To show this, we actually express in two different ways the radiation energy density in the frequency interval $[\omega, \omega + d\omega]$. On the one hand, this density is $u(\omega) d\omega$, and on the other hand, it is the average energy per mode $\langle n(\omega)\rangle\hbar\omega$ divided by the volume L^3, multiplied by the number of modes in the frequency interval $[\omega, \omega + d\omega]$, which, taking into account the two possible polarizations, equals $8\pi k^2 \, dk/(2\pi/L)^3$. Finally, we have

$$u(\omega) = \frac{\hbar\omega^3 \langle n(\omega)\rangle}{\pi^2 c^3}. \tag{E.20}$$

It remains to examine the new evolution equation for σ_{ba}, which is written

$$\frac{d}{dt}\sigma_{ba} = -i(\omega_{ba} + \Delta_{ba} + \Delta'_{ba})\sigma_{ba} - \frac{1}{2}(\Gamma + 2\Gamma')\sigma_{ba}. \tag{E.21}$$

The new terms, indicated by a prime superscript, describe, on the one hand, an additional damping (in Γ') of σ_{ba}, associated with a shortening of the lifetime of levels a and b as a result of the absorption and stimulated emission processes and, on the other hand, an additional shift Δ'_{ba} of the line $b \leftrightarrow a$

$$\Delta'_{ba} = \Delta'_b - \Delta'_a \tag{E.22}$$

associated with the light shifts of levels a and b produced by the incident radiation. For example, Δ'_a is given by

$$\hbar \Delta'_a = \mathscr{P} \sum_{n_1 \cdots n_i \cdots} p(n_1 \cdots n_i \cdots) \times$$

$$\times \sum_i \frac{|\langle b; n_1 \cdots n_i - 1 \cdots |V|a; n_1 \cdots n_i \cdots \rangle|^2}{\hbar \omega_i - \hbar \omega_{ba}}. \quad \text{(E.23.a)}$$

A calculation analogous to that carried out on (E.10) yields

$$\Delta'_a \sim \mathscr{P} \int \frac{I(\omega)\,d\omega}{\omega - \omega_{ba}} \quad \text{(E.23.b)}$$

and shows that Δ'_a may be significant if the excitation profile $I(\omega)$ is quasi-resonant, while remaining asymmetric with respect to ω_{ba}. We have thus precisely justified the qualitative considerations of subsection E-2 in Chapter II concerning the shift and broadening of the levels of an atom produced by the incident light.

Remarks

(i) The foregoing discussion can be generalized to the case in which levels a and b each contain several Zeeman sublevels. The master equation can then be used to quantitatively describe the population transfers between Zeeman sublevels of a by absorption-spontaneous emission or absorption-stimulated emission cycles ("optical pumping") as well as the evolution of the various coherences between Zeeman sublevels (*).

(ii) In (E.23), we considered only processes in which one photon is virtually absorbed from a and then reemitted. Actually, a photon can also be virtually emitted from a, and then reabsorbed, which leads to an expression analogous to (E.23.a), where the quantity appearing in Σ_i is replaced by

$$\frac{|\langle b; n_1 \cdots n_i + 1 \cdots |V|a; n_1 \cdots n_i \cdots \rangle|^2}{-\hbar \omega_i - \hbar \omega_{ba}}. \quad \text{(E.24)}$$

The term independent of n_i in (E.24) gives the radiative shift Δ_a of state a [see (E.7.c)]. As for the term proportional to n_i, it gives a contribution similar to (E.23), except for the replacement of $\omega - \omega_{ba}$ by $-\omega - \omega_{ba}$ in the denominator (which demonstrates that this last correction to Δ'_a is very small when the

(*) See, for example, Cohen-Tannoudji.

excitation is quasi-resonant). A more correct expression for Δ'_a is thus

$$\Delta'_a \sim \mathscr{P} \int d\omega\, I(\omega) \left[\frac{1}{\omega - \omega_{ba}} + \frac{1}{-\omega - \omega_{ba}} \right] = 2\omega_{ba} \mathscr{P} \int \frac{I(\omega)\, d\omega}{\omega^2 - \omega_{ba}^2}. \quad (E.25)$$

For a real atom, we should also take into account the virtual excitation of all other transitions $a \to c$ starting from a.

2. Evolution of Atomic Velocities

The goal of this final section is to obtain a master equation describing the evolution of the atomic velocity distribution function due to momentum exchanges between the atom and the radiation field.

The atom is still represented by a two-level system a and b, but we now take into account the motion of its center of mass, with position $\mathbf{R}$ and momentum $\mathbf{P}$. The eigenfunctions of the Hamiltonian $\mathbf{P}^2/2M$ of the center of mass (where M is the mass of the atom) are plane waves on which periodic boundary conditions are imposed in a cube-shaped box of side L (as for the modes of the radiation field). These eigenfunctions are written

$$\varphi_{\mathbf{p}}(\mathbf{R}) = \frac{1}{\sqrt{L^3}} \exp(i\mathbf{p} \cdot \mathbf{R}/\hbar). \quad (E.26)$$

The atom interacts with a homogeneous isotropic unpolarized radiation field for which the spectral width is sufficiently large and the intensity sufficiently weak so that it is possible to write a master equation for the atom.

We begin (§a) by generalizing equations (E.15) and by deriving equations describing the coupled evolution of internal and translational degrees of freedom. We then show how it is possible, under certain conditions, to eliminate the internal variables and to obtain a Fokker–Planck equation for the atomic velocity distribution function (§b). Finally, we discuss the physical meaning of this equation (§c and §d).

a) TAKING INTO ACCOUNT THE TRANSLATIONAL DEGREES OF FREEDOM IN THE MASTER EQUATION

The atomic density matrix elements are now identified by internal (a or b) and external ($\mathbf{p}$) quantum numbers. We are interested in the populations $\pi_{b,\mathbf{p}} = \sigma_{b\mathbf{p},b\mathbf{p}}$ and $\pi_{a,\mathbf{p}} = \sigma_{a\mathbf{p},a\mathbf{p}}$ that represent the probability of the atom's being in state b or a with total momentum $\mathbf{p}$.

In the electric dipole representation (see §5 of the Appendix), the Hamiltonian V_R for the coupling between the atom, assumed to be neutral, and the radiation field is equal to

$$V_R = -\mathbf{d} \cdot \mathbf{E}_\perp (\mathbf{R}) =$$

$$= -i\mathbf{d} \cdot \sum_i \sqrt{\frac{\hbar\omega_i}{2\varepsilon_0 L^3}} \left[\boldsymbol{\varepsilon}_i a_i \, e^{i\mathbf{k}_i \cdot \mathbf{R}} - \boldsymbol{\varepsilon}_i a_i^+ \, e^{-i\mathbf{k}_i \cdot \mathbf{R}} \right] \quad \text{(E.27)}$$

where $\mathbf{d}$ is the electric dipole moment of the atom and where the field operator $\mathbf{E}_\perp (\mathbf{R})$ is evaluated at the center of mass. The equations generalizing (E.15) are thus written

$$\dot{\pi}_{b,\mathbf{p}} = \sum_{\mathbf{p}'} \Gamma_{a\mathbf{p}' \to b\mathbf{p}} \pi_{a,\mathbf{p}'} - \sum_{\mathbf{p}'} \Gamma_{b\mathbf{p} \to a\mathbf{p}'} \pi_{b,\mathbf{p}} \quad \text{(E.28.a)}$$

$$\dot{\pi}_{a,\mathbf{p}} = \sum_{\mathbf{p}'} \Gamma_{b\mathbf{p}' \to a\mathbf{p}} \pi_{b,\mathbf{p}'} - \sum_{\mathbf{p}'} \Gamma_{a\mathbf{p} \to b\mathbf{p}'} \pi_{a,\mathbf{p}} \quad \text{(E.28.b)}$$

where the coefficients $\Gamma_{a\mathbf{p}' \to b\mathbf{p}}$ and $\Gamma_{b\mathbf{p}' \to a\mathbf{p}}$ are deduced from (C.5) and (E.10). Thus, by setting $\omega_{ba} = \omega_0$, we have

$$\Gamma_{a\mathbf{p}' \to b\mathbf{p}} = \frac{2\pi}{\hbar} \sum_{\{n_i\}} p(\{n_i\}) \times$$

$$\times \sum_i |\langle b, \mathbf{p}; n_1 \cdots n_i - 1 \cdots |V_R| a, \mathbf{p}'; n_1 \cdots n_i \cdots \rangle|^2 \times$$

$$\times \delta \left[\hbar\omega_0 - \hbar\omega_i + \left(\frac{\mathbf{p}^2}{2M} - \frac{\mathbf{p}'^2}{2M} \right) \right].$$

$$\text{(E.29)}$$

From (E.27) it is possible to split the matrix element of V_R into a part depending on the center of mass variables and a part depending on the internal variables and radiation field variables. We thus obtain, using the same notation as in subsection E-1:

$$\langle b, \mathbf{p}; n_1 \cdots n_i - 1 \cdots |V_R| a, \mathbf{p}'; n_1 \cdots n_i \cdots \rangle =$$

$$= \sqrt{n_i} \, \langle \mathbf{p} | e^{i\mathbf{k}_i \cdot \mathbf{R}} | \mathbf{p}' \rangle \langle b; 0 |V| a; 1_i \rangle. \quad \text{(E.30)}$$

From Equation (E.26) we get

$$\langle \mathbf{p}| \exp(i\mathbf{k}_i \cdot \mathbf{R})|\mathbf{p}'\rangle = \delta_{\mathbf{p}-\hbar\mathbf{k}_i,\,\mathbf{p}'} \qquad (\text{E.31})$$

which allows us to rewrite $\Gamma_{a\mathbf{p}' \to b\mathbf{p}}$ in the form

$$\Gamma_{a\mathbf{p}' \to b\mathbf{p}} = \frac{2\pi}{\hbar} \sum_i \langle n_i\rangle |\langle b; 0|V|a; 1_i\rangle|^2 \times$$

$$\times \, \delta_{\mathbf{p}-\hbar\mathbf{k}_i,\,\mathbf{p}'}\delta(\hbar\omega_0 - \hbar\omega_i + \hbar\xi_D - \hbar\xi_R) \qquad (\text{E.32})$$

where ξ_D and ξ_R are the frequency shifts due, respectively, to the Doppler effect and to the recoil effect

$$\xi_D = \mathbf{k}_i \cdot \mathbf{p}/M \qquad (\text{E.33})$$

$$\xi_R = \hbar\mathbf{k}_i^2/2M. \qquad (\text{E.34})$$

A similar calculation gives

$$\Gamma_{b\mathbf{p}' \to a\mathbf{p}} = \frac{2\pi}{\hbar} \sum_i (\langle n_i\rangle + 1)|\langle b,0|V|a,1_i\rangle|^2 \times$$

$$\times \, \delta_{\mathbf{p}+\hbar\mathbf{k}_i,\,\mathbf{p}'}\delta(\hbar\omega_0 - \hbar\omega_i + \hbar\xi_D + \hbar\xi_R) \qquad (\text{E.35})$$

where ξ_D and ξ_R are still given by (E.33) and (E.34).

Remarks

(i) By writing Equations (E.28), we have implicitly assumed that the diagonal term $\sigma_{b\mathbf{p},\,b\mathbf{p}}$ is not coupled to any nondiagonal term $\sigma_{a\mathbf{p}',\,a\mathbf{p}''}$. The absence of such couplings does not result from a secular approximation similar to the one in subsection B.4. In fact, $\mathbf{p}'$ and $\mathbf{p}''$ can have the same modulus and different directions, so that the free evolution frequency of the coherence $\sigma_{a\mathbf{p}',\,a\mathbf{p}''}$ can be arbitrarily small. Actually, the term coupling $\sigma_{b\mathbf{p},\,b\mathbf{p}}$ to $\sigma_{a\mathbf{p}',\,a\mathbf{p}''}$ is, according to (C.19), equal to

$$\frac{2\pi}{\hbar} \sum_{\{n_i\}} p(\{n_i\}) \sum_i \langle b, \mathbf{p}; n_1 \cdots n_i - 1 \cdots |V_\mathbf{R}|a, \mathbf{p}'; n_1 \cdots n_i \cdots\rangle \times$$

$$\times \langle a, \mathbf{p}''; n_1 \cdots n_i \cdots |V_\mathbf{R}|b, \mathbf{p}; n_1 \cdots n_i - 1 \cdots\rangle \times$$

$$\times \delta\left[\hbar\omega_0 - \hbar\omega_i + \left(\frac{\mathbf{p}^2}{2M} - \frac{\mathbf{p}'^2}{2M}\right)\right]. \qquad (\text{E.36})$$

We then separate, in the matrix elements of V_R, the part dependent on the center-of-mass variables as we previously did in (E.30). The first matrix element is proportional to $\delta_{\mathbf{p}-\hbar\mathbf{k}_i,\mathbf{p}'}$ and the second is proportional to $\delta_{\mathbf{p}-\hbar\mathbf{k}_i,\mathbf{p}''}$. It is then clear that we must have $\mathbf{p}' = \mathbf{p}''$. It follows that, for the master equation considered here, there is no coupling between the diagonal and the nondiagonal terms of the density matrix σ.

(ii) A more precise study including the second-order Doppler effect is sometimes necessary. In this case (E.33) must be replaced by

$$\xi_D = \frac{\mathbf{k}_i \cdot \mathbf{p}}{M} - \frac{\omega_0}{2}\frac{\mathbf{p}^2}{M^2c^2}. \tag{E.37}$$

We now substitute (E.32) and (E.35) into Equations (E.28). The sums over $\mathbf{p}'$ reduce to a single term, because of the delta functions expressing momentum conservation:

$$\dot{\pi}_{b,\mathbf{p}} = \sum_i \frac{2\pi}{\hbar}|\langle b;0|V|a;1_i\rangle|^2 \times$$

$$\times \delta(\hbar\omega_0 - \hbar\omega_i + \hbar\xi_D - \hbar\xi_R) \times$$

$$\times \{\langle n_i\rangle \pi_{a,\mathbf{p}-\hbar\mathbf{k}_i} - (\langle n_i\rangle + 1)\pi_{b,\mathbf{p}})\} \tag{E.38.a}$$

$$\dot{\pi}_{a,\mathbf{p}} = \sum_i \frac{2\pi}{\hbar}|\langle b;0|V|a;1_i\rangle|^2 \times$$

$$\times \delta(\hbar\omega_0 - \hbar\omega_i + \hbar\xi_D + \hbar\xi_R) \times$$

$$\times \{(\langle n_i\rangle + 1)\pi_{b,\mathbf{p}+\hbar\mathbf{k}_i} - \langle n_i\rangle \pi_{a,\mathbf{p}}\}. \tag{E.38.b}$$

To simplify the calculations, we ignore the dependence of $\sum_{\varepsilon_i}|\langle b;0|V|a;1_i\rangle|^2$ with the direction $\boldsymbol{\kappa}_i = \mathbf{k}_i/k_i$ of $\mathbf{k}_i$, which is equivalent to taking an isotropic radiation pattern (see, however, the remark following). This quantity thus depends only on ω_i [it varies linearly with $\omega_i = ck_i$, according to (E.1)] and can be expressed as a function of Γ

$$\Gamma = \frac{2\pi}{\hbar}\sum_i|\langle b;0|V|a;1_i\rangle|^2\delta(\hbar\omega_i - \hbar\omega_0)$$

$$= \frac{L^3}{\pi\hbar^2c^3}\omega_0^3\sum_{\varepsilon_i}\frac{|\langle b;0|V|a;1_i\rangle|^2}{\omega_i}. \tag{E.39}$$

By taking the limit $L \to \infty$ and by calling $\pi_j(\mathbf{p})$ (with $j = a, b$) the population density of level j in momentum space, we finally obtain for (E.38):

$$\dot{\pi}_b(\mathbf{p}) = \Gamma \int_0^{+\infty} \frac{\omega^3}{\omega_0^3} \, d\omega \int \frac{d\Omega}{4\pi} \delta(\omega_0 - \omega + \xi_D - \xi_R) \times$$

$$\times \{\langle n(\omega)\rangle \pi_a(\mathbf{p} - \hbar\mathbf{k}) - [\langle n(\omega)\rangle + 1]\pi_b(\mathbf{p})\} \quad \text{(E.40.a)}$$

$$\dot{\pi}_a(\mathbf{p}) = \Gamma \int_0^{+\infty} \frac{\omega^3}{\omega_0^3} \, d\omega \int \frac{d\Omega}{4\pi} \delta(\omega_0 - \omega + \xi_D + \xi_R) \times$$

$$\times \{[\langle n(\omega)\rangle + 1]\pi_b(\mathbf{p} - \hbar\mathbf{k}) - \langle n(\omega)\rangle \pi_a(\mathbf{p})\} \quad \text{(E.40.b)}$$

where $d\Omega$ is the elementary solid angle about $\boldsymbol{\kappa} = \mathbf{k}/k$ and where, for each direction $\boldsymbol{\kappa}$, the modulus of $\mathbf{k}$ is determined by the argument of the delta function, i.e., by the Doppler and recoil shifts.

The comparison of Equations (E.15) and (E.40) clearly shows what are the new effects connected with the atomic motion. First, when the atom goes from one level to another by absorption or emission of a photon, its momentum increases or decreases by a quantity $\hbar\mathbf{k}$ equal to the momentum of the absorbed or emitted photon. Second, the rates of absorption and stimulated emission involve the average numbers of photons having a frequency which is no longer ω_0, but is now corrected by the Doppler and recoil shifts.

Remark

If the radiation pattern is not isotropic, it is sufficient to replace $d\Omega/4\pi$ in (E.40) by $I(\boldsymbol{\kappa})\,d\Omega/4\pi$, where $I(\boldsymbol{\kappa})$ is the normalized radiation intensity ($\int d\Omega \, I(\boldsymbol{\kappa})/4\pi = 1$), which has the important property of being even [$I(\boldsymbol{\kappa}) = I(-\boldsymbol{\kappa})$]. The friction coefficient γ and the diffusion coefficient D introduced further on are then tensors and no longer scalars.

b) FOKKER–PLANCK EQUATION FOR THE ATOMIC VELOCITY
DISTRIBUTION FUNCTION

The probability density that the momentum of the atom is equal to $\mathbf{p}$, regardless of its internal state, is given by

$$\pi(\mathbf{p}) = \pi_a(\mathbf{p}) + \pi_b(\mathbf{p}). \quad \text{(E.41)}$$

In the electric dipole representation, the momentum $\mathbf{p}$ is identical to $M\mathbf{v}$ [see Equation (78) of the Appendix], so that the function $\pi(\mathbf{p})$ is also the distribution of atomic velocity $\mathbf{v}$. We now attempt to derive an evolution equation for $\pi(\mathbf{p})$ from Equations (E.40).

α) *The Small Parameters of the Problem*

In what follows, we assume that $\hbar k$ is small compared with the width Δp of the functions $\pi_a(\mathbf{p}), \pi_b(\mathbf{p})$ and that ξ_D and ξ_R are small compared with the width $\Delta\omega$ of the spectral distribution of $\omega^3\langle n(\omega)\rangle$. We thus introduce two small parameters

$$\eta_1 = \frac{\hbar k_0}{\Delta p} \tag{E.42.a}$$

$$\eta_2 = \frac{\xi_D}{\Delta\omega} \sim \frac{k_0 \Delta p}{M\Delta\omega} \tag{E.42.b}$$

from which it is possible to express $\xi_R/\Delta\omega$

$$\frac{\xi_R}{\Delta\omega} \sim \frac{\hbar k_0^2}{2M\Delta\omega} \sim \eta_1\eta_2. \tag{E.42.c}$$

The rate of variation $\dot{\pi}(\mathbf{p})$ of $\pi(\mathbf{p})$ is obtained by adding the two equations (E.40). It appears that $\dot{\pi}(\mathbf{p})$ is canceled out if we set $\xi_D = \xi_R = 0$ in the argument of the delta function and $\hbar k = 0$ in $\pi_a(\mathbf{p} - \hbar k)$ and $\pi_b(\mathbf{p} + \hbar k)$. To order 0, in η_1 and η_2, we thus have $\dot{\pi}(\mathbf{p}) = 0$.

Remark

The comparison of the second-order Doppler effect $\xi_D^{(2)}$, introduced in (E.37), with $\Delta\omega$ shows that

$$\frac{\xi_D^{(2)}}{\Delta\omega} \sim \left(\frac{k_0\Delta p}{M\Delta\omega}\right)^2 \frac{\Delta\omega}{\omega_0} \sim \eta_2^2 \frac{\Delta\omega}{\omega_0} \tag{E.43}$$

In situations where $\Delta\omega \lesssim \omega_0$, the second-order Doppler effect introduces small corrections of order η_2^2 at most.

β) *Perturbative Expansion of the Evolution Equation for $\pi(\mathbf{p})$*

Because $\dot{\pi}(\mathbf{p})$ is canceled out to order 0 in η_1 and η_2, it is convenient to expand, on the one hand $\pi_j(\mathbf{p} \pm \hbar k)$ (with $j = a, b$) in powers of $\eta_1 = \hbar k/\Delta p$, and on the other hand, the delta function in powers of ξ_D and ξ_R (which is equivalent after integration over ω to expanding the factor multiplying the delta function in powers of $\xi_D/\Delta\omega$ and $\xi_R/\Delta\omega$):

$$\pi_j(\mathbf{p} \pm \hbar\mathbf{k}) = \pi_j(\mathbf{p}) \pm \hbar\mathbf{k} \cdot \nabla\pi_j + \frac{1}{2}\sum_{l,m}k_l k_m\frac{\partial^2\pi_j}{\partial k_l\partial k_m} + \cdots \quad (\text{E.44})$$

with $l, m = x, y, z$:

$$\delta(\omega_0 - \omega + \xi_D \pm \xi_R) = \delta(\omega_0 - \omega) + (\xi_D \pm \xi_R)\delta'(\omega_0 - \omega) +$$

$$+ \tfrac{1}{2}(\xi_D \pm \xi_R)^2\delta''(\omega_0 - \omega) + \cdots. \quad (\text{E.45})$$

The expansion (E.45) can be ordered as a function of the small parameters introduced in the preceding paragraph so as to cause the successive appearance of terms of order zero, $\eta_2, \eta_1\eta_2, \eta_2^2 \cdots$, etc.

$$\delta(\omega_0 - \omega + \xi_D \pm \xi_R) = \delta(\omega_0 - \omega) +$$

$$+ \xi_D\delta'(\omega_0 - \omega) \pm \xi_R\delta'(\omega_0 - \omega) + \tfrac{1}{2}\xi_D^2\delta''(\omega_0 - \omega) + \cdots. \quad (\text{E.46})$$

In the equation giving $\dot{\pi}(\mathbf{p})$ obtained by summing the two equations (E.40), the contribution of first-order terms in η_1 or η_2 is zero because of the angular integral

$$\int\frac{d\Omega}{4\pi}k_l = 0 \quad (\text{E.47})$$

(where $l = x, y, z$). The terms in $\eta_1\eta_2$ come either from products of the second terms of developments (E.44) and (E.46) or from the third term of (E.46). By using

$$\int\frac{d\Omega}{4\pi}k_l k_m = \frac{k^2}{3}\delta_{lm} \quad (\text{E.48})$$

we find that the contribution of the terms in $\eta_1\eta_2$ to $\dot{\pi}$ is written

$$-\Gamma\frac{d\langle n(\omega_0)\rangle}{d\omega_0}\frac{\hbar k_0^2}{3M}\nabla \cdot \{\mathbf{p}[\pi_a(\mathbf{p}) - \pi_b(\mathbf{p})]\} +$$

$$+ \Gamma\frac{5\hbar\omega_0}{3Mc^2}\nabla \cdot \{\mathbf{p}[(\langle n(\omega_0)\rangle + 1)\pi_b(\mathbf{p}) - \langle n(\omega_0)\rangle\pi_a(\mathbf{p})]\}. \quad (\text{E.49})$$

The terms in η_1^2 come from the second-order terms of the expansion

(E.44) of π_j. Their contribution to $\dot{\pi}$ is equal to

$$\Gamma \frac{\hbar^2 k_0^2}{6} \left[\langle n(\omega_0) \rangle \Delta \pi(\mathbf{p}) + \Delta \pi_b(\mathbf{p}) \right]. \qquad (E.50)$$

Finally, the terms in η_2^2 cancel each other in (E.40.a) and (E.40.b) and disappear from the equation for $\dot{\pi}$.

Thus, up to order 2 in η_1 and η_2, $\dot{\pi}(\mathbf{p})$ reduces to the sum of (E.49) and (E.50). It clearly appears that $\dot{\pi}(\mathbf{p})$ does not depend only on $\pi(\mathbf{p})$ [and the derivatives of $\pi(\mathbf{p})$ relative to $\mathbf{p}$]. The presence of $\pi_a(\mathbf{p})$ and $\pi_b(\mathbf{p})$ in (E.49) and (E.50) is a manifestation of the coupling that exists between internal and external variables.

Remark

In the order of perturbation considered here, $\dot{\pi}$ remains equal to the sum of (E.49) and (E.50) even when the second-order Doppler effect is included in ξ_D. This results from the fact that the linear term in ξ_D, obtained by expanding the delta function, disappears from the sum of (E.40.a) and (E.40.b) when we set $\mathbf{k} = 0$ in $\pi_j(\mathbf{p} \pm \hbar \mathbf{k})$.

γ) *Adiabatic Elimination of Internal Variables*

The fact that $\dot{\pi}(\mathbf{p})$ is smaller than $\dot{\pi}_a(\mathbf{p})$ or $\dot{\pi}_b(\mathbf{p})$ by a factor on the order of $\eta_1 \eta_2$ or η_1^2 means that the characteristic evolution time of the external variables is much longer than that of the internal variables. We will use this difference between the two time scales to eliminate the internal variables from the equation giving $\dot{\pi}(\mathbf{p})$.

Because the terms (E.49) and (E.50), whose sum gives $\dot{\pi}(\mathbf{p})$, are already second order in η_1 and η_2, we can limit ourselves to the order 0 in η_1 and η_2 to study the evolution of the populations $\pi_a(\mathbf{p})$ and $\pi_b(\mathbf{p})$ that appear in these terms. Thus, to order 0 in η_1 and η_2, Equations (E.40) become

$$\dot{\pi}_b(\mathbf{p}) = \Gamma' \pi_a(\mathbf{p}) - (\Gamma + \Gamma') \pi_b(\mathbf{p}) \qquad (E.51.a)$$

$$\dot{\pi}_a(\mathbf{p}) = -\Gamma' \pi_a(\mathbf{p}) + (\Gamma + \Gamma') \pi_b(\mathbf{p}) \qquad (E.51.b)$$

where

$$\Gamma' = \Gamma \langle n(\omega_0) \rangle \qquad (E.51.c)$$

because the radiation is isotropic [see (E.16)]. Therefore, it follows that, to order 0 in η_1 and η_2, $\pi_a(\mathbf{p})$ and $\pi_b(\mathbf{p})$ tend with a very short time constant equal to $1/(\Gamma + 2\Gamma')$, to the values

$$\pi_a(\mathbf{p}) = \frac{\Gamma + \Gamma'}{\Gamma + 2\Gamma'} \pi(\mathbf{p}) = \frac{1 + \langle n(\omega_0) \rangle}{1 + 2\langle n(\omega_0) \rangle} \pi(\mathbf{p}) \qquad \text{(E.52.a)}$$

$$\pi_b(\mathbf{p}) = \frac{\Gamma'}{\Gamma + 2\Gamma'} \pi(\mathbf{p}) = \frac{n\langle(\omega_0)\rangle}{1 + 2\langle n(\omega_0) \rangle} \pi(\mathbf{p}). \qquad \text{(E.52.b)}$$

We can therefore consider that $\pi_a(\mathbf{p})$ and $\pi_b(\mathbf{p})$ adapt themselves quasi-instantaneously to the much slower variations of $\pi(\mathbf{p})$ and at each instant take the values given in (E.52). Substituting these equations into (E.49) and (E.50) thus allows us to "adiabatically eliminate" π_a and π_b in favor of π. The contribution of the second line of (E.49) is canceled and we obtain the following equation for $\dot{\pi}$:

$$\frac{\partial}{\partial t} \pi(\mathbf{p}, t) = \gamma \nabla \cdot [\mathbf{p}\pi(\mathbf{p}, t)] + \frac{D}{3} \Delta \pi(\mathbf{p}, t) \qquad \text{(E.53)}$$

where the coefficients γ and D are given by

$$\gamma = -\frac{\hbar k_0^2}{3M} \Gamma \frac{\Gamma}{\Gamma + 2\Gamma'} \frac{d\langle n(\omega_0) \rangle}{d\omega_0} \qquad \text{(E.54)}$$

$$D = \hbar^2 k_0^2 \frac{\Gamma'(\Gamma + \Gamma')}{\Gamma + 2\Gamma'}. \qquad \text{(E.55)}$$

The partial differential equation (E.53) is a Fokker–Planck equation which is very similar to the one encountered in the study of Brownian motion.

c) Evolutions of the Momentum Mean Value and Variance

To physically interpret the coefficient γ in (E.53), we calculate the rate of variation of the momentum mean value

$$\langle \mathbf{p} \rangle = \int d^3p \, \mathbf{p}\pi(\mathbf{p}, t) \qquad \text{(E.56)}$$

by multiplying both sides of (E.53) by $\mathbf{p}$ and integrating over $\mathbf{p}$. Integrating the two terms of the right-hand side by parts and assuming that $\pi(\mathbf{p})$ tends to zero sufficiently rapidly when $|\mathbf{p}| \to \infty$, we obtain

$$\frac{d}{dt}\langle\mathbf{p}\rangle = -\gamma\langle\mathbf{p}\rangle. \tag{E.57}$$

We assume in what follows that $d\langle n(\omega_0)\rangle/d\omega_0$ is negative, i.e., that the mean number of photons per mode is a decreasing function of ω near ω_0. The coefficient γ defined in (E.54) is then positive and Equation (E.57) shows that it is the damping rate of the average momentum value (friction coefficient).

The same type of calculation results in

$$\frac{d}{dt}\langle\mathbf{p}^2\rangle = -2\gamma\langle\mathbf{p}^2\rangle + 2D \tag{E.58}$$

which shows that the variance σ_p^2 of $\mathbf{p}$

$$\sigma_p^2 = \langle\mathbf{p}^2\rangle - \langle\mathbf{p}\rangle^2 \tag{E.59}$$

obeys, taking into account (E.57) and (E.58), the following equation:

$$\frac{d}{dt}\sigma_p^2 = -2\gamma\sigma_p^2 + 2D. \tag{E.60}$$

The coefficient $2D$ thus characterizes the rate of increase of the variance of $\mathbf{p}$ which is also damped at a rate 2γ. The coefficient D is thus a diffusion coefficient for the momentum (see §1-b in Complement C_{IV}).

We now attempt to qualitatively understand why the atomic velocity is damped if $d\langle n(\omega_0)\rangle/d\omega_0$ is negative. To do that, we consider a simplified one-dimensional model where the momenta of the atom and the photons are parallel or antiparallel to Ox. If the atom goes to the right ($v > 0$), the photons that propagate in the opposite direction must have a frequency lower than ω_0, $\omega_0 - \omega_0 v/c$, in order to interact in a resonant manner with the atom, whereas the photons propagating in the same direction must have a higher frequency, $\omega_0 + \omega_0 v/c$ (Doppler effect). If $d\langle n(\omega_0)\rangle/d\omega_0$ is negative, the atom thus "sees" more resonant photons arriving in the direction opposite to its motion than in the same direction as its motion. The momentum that it will absorb will thus be directed preferentially in the direction opposite to its motion and the atom will be slowed down.

The foregoing reasoning can be made quantitative. Consider the mean momentum gained by an atom of momentum $\mathbf{p}$ in absorption and stimulated emission processes. The atom has a relative probability $\hat{\pi}_a(\mathbf{p}) =$

$\pi_a(\mathbf{p})/\pi(\mathbf{p})$ of being in state a, so that during the time dt, it absorbs $\Gamma\langle n(\omega_0 + \mathbf{k} \cdot \mathbf{v})\rangle\hat{\pi}_a(\mathbf{p})\, dt\, d\Omega/4\pi$ photons whose wave vector $\mathbf{k}$ points into the solid angle $d\Omega$. The number of photons emitted through stimulated emission is given by a similar expression where $\hat{\pi}_a$ is replaced by $\hat{\pi}_b$. The rate of variation of the atomic momentum due to these processes is thus, to first order in $\mathbf{k} \cdot \mathbf{v}/\Delta\omega$,

$$\frac{d\mathbf{p}}{dt} = \hbar\mathbf{k}\Gamma\left[\langle n(\omega_0)\rangle + \frac{\mathbf{k} \cdot \mathbf{p}}{M}\frac{d\langle n(\omega_0)\rangle}{d\omega_0}\right][\hat{\pi}_a(\mathbf{p}) - \hat{\pi}_b(\mathbf{p})]\frac{d\Omega}{4\pi}. \quad \text{(E.61)}$$

The angular integration over $\mathbf{\kappa} = \mathbf{k}/k$ thus gives, taking into account (E.47) and (E.48),

$$\frac{d\mathbf{p}}{dt} = \frac{\hbar k_0^2}{3M}\frac{d\langle n(\omega_0)\rangle}{d\omega_0}\Gamma[\hat{\pi}_a(\mathbf{p}) - \hat{\pi}_b(\mathbf{p})]\mathbf{p}. \quad \text{(E.62)}$$

Spontaneous emission does not contribute to $d\mathbf{p}/dt$ because, in the reference frame of the atom, it occurs with equal probabilities in opposite directions and therefore does not cause the mean velocity of the atom to vary. Finally, by substituting into (E.62) the adiabatic expressions (E.52) for π_a and π_b, and by averaging over the initial momentum $\mathbf{p}$ of the atom, we recover equation (E.57) with the expression (E.54) for γ.

To understand the expression (E.55) for the diffusion coefficient D, note that for each elementary process of absorption, stimulated emission or spontaneous emission, the momentum of the atom varies by a quantity whose modulus equals $\hbar k_0$, and whose direction varies randomly from one process to another. In momentum space, the momentum of the atom thus makes a *random walk* with a step $\hbar k_0$. Let dn be the mean total number of elementary processes during dt for an atom of momentum $\mathbf{p}$

$$dn = [\Gamma'\hat{\pi}_a(\mathbf{p}) + (\Gamma + \Gamma')\hat{\pi}_b(\mathbf{p})]\, dt. \quad \text{(E.63)}$$

According to the well-known properties of a random walk, the increase in $\mathbf{p}^2$ over a period dt equals

$$d\mathbf{p}^2 = \hbar^2 k_0^2\, dn = \hbar^2 k_0^2[\Gamma'\hat{\pi}_a(\mathbf{p}) + (\Gamma + \Gamma')\hat{\pi}_b(\mathbf{p})]\, dt. \quad \text{(E.64)}$$

It is then sufficient to substitute the adiabatic expressions (E.52) for π_a and π_b into (E.64), and then to average over the initial momentum $\mathbf{p}$ of the atom, to find that $\mathbf{p}^2$ increases as $2Dt$, where D is given by (E.55).

d) STEADY-STATE DISTRIBUTION. THERMODYNAMIC EQUILIBRIUM

The Fokker–Planck equation (E.53) can be written in the form of a continuity equation

$$\frac{\partial}{\partial t}\pi(\mathbf{p}, t) + \nabla \cdot \mathbf{J}(\mathbf{p}, t) = 0 \tag{E.65}$$

where the "current" $\mathbf{J}(\mathbf{p}, t)$ is given by

$$\mathbf{J}(\mathbf{p}, t) = -\gamma\mathbf{p}\pi(\mathbf{p}, t) - \frac{D}{3}\nabla\pi(\mathbf{p}, t). \tag{E.66}$$

It is then easy to verify that the Fokker–Planck equation has a steady-state solution $(\partial\pi/\partial t = 0)$, given by the solution to the equation $\mathbf{J} = \mathbf{0}$, which is written

$$\pi_{st}(\mathbf{p}) = \mathcal{N}\exp\left[-\frac{3\gamma}{2D}\mathbf{p}^2\right] \tag{E.67}$$

where $\mathcal{N}$ is a normalization constant. In the steady state, the distribution function of momenta or velocities is thus Gaussian.

Finally, we consider the specific case where the radiation field interacting with the atom is in thermodynamic equilibrium. We must then use expression (E.17) for $\langle n(\omega_0)\rangle$. Using (E.16), we obtain for the coefficients γ and D given in (E.54) and (E.55)

$$\gamma = \frac{\Gamma}{6\sinh(\hbar\omega_0/k_BT)}\frac{\hbar\omega_0}{k_BT}\frac{\hbar\omega_0}{Mc^2} \tag{E.68.a}$$

$$D = \frac{\Gamma}{2\sinh(\hbar\omega_0/k_BT)}\frac{\hbar^2\omega_0^2}{c^2}. \tag{E.68.b}$$

Substituting (E.68) into (E.67) shows that

$$\pi_{st}(\mathbf{p}) = \mathcal{N}\exp\left[-\frac{\mathbf{p}^2}{2Mk_BT}\right] \tag{E.69}$$

which is just a thermodynamic equilibrium distribution at temperature T. We thus generalize the results of subsection E-1 above and show that the external degrees of freedom, as well as the internal degrees of freedom, reach thermodynamic equilibrium when the atoms interact with a radiation field which is itself in thermodynamic equilibrium.

GENERAL REFERENCES

Abragam (Chapter VIII), Louisell (Chapter 6), Cohen-Tannoudji (§4), Agarwal (Chapter 6). Also see the original article by R. K. Wangsness and F. Bloch, *Phys. Rev.*, **89**, 728 (1953).

The discussion of the evolution of atomic velocities in subsection E-2 was derived for the most part from the thesis of J. Dalibard, Université Pierre et Marie Curie, Paris (1986).

COMPLEMENT A_{IV}

FLUCTUATIONS AND LINEAR RESPONSE
APPLICATION TO RADIATIVE PROCESSES

We showed in this chapter that, when the radiation field can be considered as a reservoir, the evolution of the density operator of the particles is described by a master equation. In this equation, the reservoir is involved only through a single function $g(\tau)$, which is a two-time average of an observable R of the reservoir.

The purpose of this complement is first to relate the real and imaginary parts of $g(\tau)$ to two statistical functions of the reservoir, a symmetric correlation function and a linear susceptibility function. We show in subsection 1 how these statistical functions can be used to characterize the way the small system is affected by its interaction with the reservoir. From this analysis, it is possible to derive a very general physical interpretation, according to which each of the two-coupled systems, the small system and the reservoir, fluctuates and polarizes the other. More precisely, on the one hand, there are processes in which the small system evolves according to its own dynamics, polarizes the reservoir, which in turn modifies the motion of the system; on the other hand, there are phenomena in which it is the fluctuations of the reservoir that polarize the small system and modify its properties.

We then apply (§2) these general results to the case of an atom interacting with a homogeneous and isotropic radiation field. Black-body radiation and the vacuum field are two examples of this type of situation. We evaluate the statistical functions associated with the atom and with the radiation field, and we interpret the atomic level shifts, as well as the energy exchanges between the atom and the radiation field as resulting from fluctuations of the radiation field and from the radiation reaction.

1. Statistical Functions and Physical Interpretation of the Master Equation

We start with expression (B.22) for the function $g(\tau)$, from which the coefficients (B.43) of the master equation are defined:

$$g(\tau) = T_R\left[\sigma_R \tilde{R}(\tau)\tilde{R}(0)\right]. \tag{1}$$

This function $g(\tau)$ is not real, even though R is Hermitian, because, in general, $\tilde{R}(\tau)$ and $\tilde{R}(0)$ do not commute. To separate the real and imaginary parts of $g(\tau)$, we write

$$g(\tau) = \frac{1}{2}\langle \tilde{R}(\tau)\tilde{R}(0) + \tilde{R}(0)\tilde{R}(\tau)\rangle_R + \frac{i}{2}\langle [\tilde{R}(\tau),\tilde{R}(0)]/i\rangle_R \quad (2)$$

where $\langle\ \rangle_R$ indicates the average over the reservoir in the state defined by σ_R. The first term in (2) is a symmetric correlation function, and the second is related to a linear susceptibility of the reservoir. We analyze the properties of these two functions before returning to the interpretation of the master equation.

a) Symmetric Correlation Function

Let $C_R(\tau)$ be the symmetric correlation function of the observable R

$$C_R(\tau) = \frac{1}{2}\langle \tilde{R}(\tau)\tilde{R}(0) + \tilde{R}(0)\tilde{R}(\tau)\rangle_R. \quad (3)$$

This function is real and tends to the ordinary correlation function in the classical limit. It gives a physical description of the dynamics of the fluctuations of the observable R in the state σ_R. If R were a random classical function of frequency ω_0,

$$R(t) = R_0 \exp[-i(\omega_0 t + \varphi)] + \text{c.c.} \quad (4)$$

and if the amplitude R_0 and the phase φ were independent random variables, φ being uniformly distributed between 0 and 2π, then the correlation function of R would equal

$$C_R(\tau) = \langle R_0^2\rangle e^{-i\omega_0\tau} + \text{c.c.} \quad (5)$$

The explicit expression for the quantum correlation function defined by (3) is given by the real part of expression (B.23) for $g(\tau)$

$$C_R(\tau) = \sum_\mu p_\mu \sum_\nu |R_{\mu\nu}|^2 \cos(\omega_{\mu\nu}\tau). \quad (6)$$

It appears in the form of a sum of expressions of the same type as (5) corresponding to the different Bohr frequencies involved in the motion of R. The function $C_R(\tau)$ is even in τ.

We also use the Fourier transform $\hat{C}_R(\omega)$ of $C_R(\tau)$ defined by

$$C_R(\tau) = \frac{1}{2\pi} \int d\omega \, \hat{C}_R(\omega) \, e^{-i\omega\tau} \tag{7}$$

and which equals

$$\hat{C}_R(\omega) = \sum_\mu p_\mu \sum_\nu \pi |R_{\mu\nu}|^2 \left[\delta(\omega + \omega_{\mu\nu}) + \delta(\omega - \omega_{\mu\nu}) \right]. \tag{8}$$

The function $\hat{C}_R(\omega)$ is also real and even.

Remark

The function $C_R(\tau)$ is the autocorrelation function of the variable R. The correlation function for two different (real) variables R_p and R_q is defined by

$$C_{pq}(\tau) = \frac{1}{2} \left\langle \tilde{R}_p(0) \tilde{R}_q(\tau) + \tilde{R}_q(\tau) \tilde{R}_p(0) \right\rangle_R. \tag{9}$$

The previously described properties are generalized as follows:

$$C_{pq}(\tau) = C_{pq}^*(\tau); \qquad C_{pq}(\tau) = C_{qp}(-\tau) \tag{10.a}$$

$$\hat{C}_{pq}^*(\omega) = \hat{C}_{pq}(-\omega); \qquad \hat{C}_{pq}(\omega) = \hat{C}_{qp}(-\omega). \tag{10.b}$$

b) LINEAR SUSCEPTIBILITY

The symmetric correlation function (3) describes the fluctuations of the observable R of the reservoir. Another statistical function, the linear susceptibility, allows its linear response to an external perturbation to be characterized.

First recall some simple results for a classical damped harmonic oscillator. When such an oscillator is subjected to a periodic force

$$f(t) = F e^{-i\omega t} + \text{c.c.} \tag{11}$$

it has a forced oscillation motion of complex amplitude X:

$$x(t) = X e^{-i\omega t} + \text{c.c.} \tag{12}$$

The amplitude X is proportional to F

$$X = \hat{\chi}(\omega) F. \tag{13}$$

The coefficient of proportionality $\hat{\chi}(\omega)$ is the linear susceptibility of the oscillator at the frequency ω. This is a function of the parameters characterizing the oscillator. Separation of the real and imaginary parts of $\hat{\chi}(\omega)$

$$\hat{\chi}(\omega) = \hat{\chi}'(\omega) + i\hat{\chi}''(\omega) \tag{14}$$

results in the appearance of the components $\hat{\chi}'F$ and $\hat{\chi}''F$ of the motion of x, respectively, in phase and in quadrature with the force $f(t)$. More generally, if $f(t)$ is a sum of oscillating exponentials:

$$f(t) = \int_{-\infty}^{+\infty} \frac{d\omega}{2\pi} F(\omega) e^{-i\omega t} \tag{15}$$

the response of the oscillator is the sum of the responses associated with each frequency ω:

$$x(t) = \int_{-\infty}^{+\infty} \frac{d\omega}{2\pi} \hat{\chi}(\omega) F(\omega) e^{-i\omega t}. \tag{16}$$

By inverting the Fourier transform (15) and introducing the Fourier transform of the function $\hat{\chi}(\omega)$

$$\chi(\tau) = \int_{-\infty}^{+\infty} \frac{d\omega}{2\pi} \hat{\chi}(\omega) e^{-i\omega\tau} \tag{17}$$

expression (16) is written in the form

$$x(t) = \int_{-\infty}^{+\infty} \chi(t - t') f(t') \, dt'. \tag{18}$$

We now return to the reservoir $\mathscr{R}$, which is a quantum system. If it is subjected to a perturbation

$$V(t) = -R\lambda(t) \tag{19}$$

where $\lambda(t)$ is a classical function, its state will become slightly different from the initial steady state σ_R. From perturbation theory, we can determine the motion induced by this perturbation on the mean value of an observable, for example, R, whose mean value in the initial steady state is assumed to be zero for the sake of simplicity. To first order in λ (*), an

---

(*) See *Photons and Atoms—Introduction to Quantum Electrodynamics*, Exercise 6 in Complement E$_{IV}$.

expression similar to (18) is obtained for the mean value of R at time t,

$$\langle R \rangle_t = \int_{-\infty}^{+\infty} \chi_R(\tau) \lambda(t - \tau) \, d\tau \tag{20}$$

where the linear susceptibility $\chi_R(\tau)$ of the reservoir is given by

$$\chi_R(\tau) = \frac{i}{\hbar} \theta(\tau) \left\langle \left[\tilde{R}(0), \tilde{R}(-\tau) \right] \right\rangle_R. \tag{21}$$

The mean value is taken as it was previously in the state σ_R of $\mathcal{R}$ and $\theta(\tau)$ equals 1 for $\tau > 0$ and 0 for $\tau < 0$. The comparison of (2) and (21) yields the following relation between $\chi_R(\tau)$ and the imaginary part of $g(\tau)$:

$$\chi_R(\tau) = \frac{2}{\hbar} \theta(\tau) \operatorname{Im} g(-\tau). \tag{22}$$

The explicit expression for $\chi_R(\tau)$ is thus directly deduced from (B.23):

$$\chi_R(\tau) = -\frac{2}{\hbar} \sum_\mu P_\mu \sum_\nu |R_{\mu\nu}|^2 \theta(\tau) \sin \omega_{\mu\nu}\tau, \tag{23}$$

as well as its Fourier transform, defined as in (7):

$$\hat{\chi}_R(\omega) = \frac{-1}{\hbar} \sum_\mu P_\mu \sum_\nu |R_{\mu\nu}|^2 \left[\mathcal{P} \frac{1}{\omega_{\mu\nu} + \omega} + \mathcal{P} \frac{1}{\omega_{\mu\nu} - \omega} \right.$$
$$\left. - i\pi\delta(\omega_{\mu\nu} + \omega) + i\pi\delta(\omega_{\mu\nu} - \omega) \right]. \tag{24}$$

In contrast to $\hat{C}_R(\omega)$, the function $\hat{\chi}_R(\omega)$ is not real, and it is advantageous to separate its real and imaginary parts which characterize the response in phase and in quadrature at the frequency ω:

$$\hat{\chi}_R(\omega) = \hat{\chi}_R'(\omega) + i\hat{\chi}_R''(\omega) \tag{25}$$

$$\hat{\chi}_R'(\omega) = -\frac{1}{\hbar} \sum_\mu P_\mu \sum_\nu |R_{\mu\nu}|^2 \left[\mathcal{P} \frac{1}{\omega_{\mu\nu} + \omega} + \mathcal{P} \frac{1}{\omega_{\mu\nu} - \omega} \right] \tag{26.a}$$

$$\hat{\chi}_R''(\omega) = \frac{\pi}{\hbar} \sum_\mu P_\mu \sum_\nu |R_{\mu\nu}|^2 \left[\delta(\omega_{\mu\nu} + \omega) - \delta(\omega_{\mu\nu} - \omega) \right]. \tag{26.b}$$

Note that $\hat{\chi}_R'(\omega)$ is even in ω, and that $\hat{\chi}_R''(\omega)$ is odd.

Remark

The generalization to the case of a coupling $V = -\Sigma_q R_q \lambda_q$ for formulas (20) and (21) is written

$$\langle R_p \rangle_t = \sum_q \int_{-\infty}^{+\infty} X_{pq}(\tau) \lambda_q(t - \tau) \, d\tau \tag{27}$$

$$X_{pq}(\tau) = \frac{i}{\hbar} \theta(\tau) \langle [\tilde{R}_p(0), \tilde{R}_q(-\tau)] \rangle_R. \tag{28}$$

If $\hat{\chi}'_{pq}$ and $\hat{\chi}''_{pq}$ are the reactive and dissipative parts defined by separating the principal parts and the delta functions as in (26.a) and (26.b) [these are no longer the real and imaginary parts of $\hat{\chi}_{pq}(\omega)$], their symmetry properties are

$$\hat{\chi}_{pq}(\omega) = \hat{\chi}^*_{pq}(-\omega) \tag{29}$$

$$\hat{\chi}'_{pq}(\omega) = \left[\hat{\chi}'_{qp}(\omega) \right]^* \qquad \hat{\chi}''_{pq}(\omega) = \left[\hat{\chi}''_{qp}(\omega) \right]^*. \tag{30}$$

c) POLARIZATION ENERGY AND DISSIPATION

The linear susceptibility is also involved in the energy exchanges between the system and the external medium near equilibrium. To see this, we return to the example of the classical damped oscillator introduced at the beginning of the preceding subsection and we show that $\hat{\chi}'(\omega)$ and $\hat{\chi}''(\omega)$ appear, respectively, in the expressions giving the polarization energy of the oscillator and the energy dissipated by the damping processes. The results will suggest a physical interpretation for the quantum expressions giving the level shifts of the small system $\mathscr{A}$, due to its interaction with the reservoir (§1-d), and the energy exchange rates between $\mathscr{A}$ and $\mathscr{R}$ (§1-e).

In the steady-state regime, the work per unit time carried out by the external force (11), averaged over a period T, is equal to:

$$\dot{\mathscr{Q}} = \frac{1}{T} \int_0^T dt f(t) \dot{x}(t)$$

$$= \frac{1}{T} \int_0^T dt [-i\omega X e^{-i\omega t} + \text{c.c.}][F e^{-i\omega t} + \text{c.c.}]$$

$$= (-i\omega XF^* + \text{c.c.})$$

$$= \omega \hat{\chi}''(\omega)(2FF^*). \tag{31}$$

This power is dissipated by the damping process of the oscillator. For this reason $\hat{\chi}''(\omega)$ is called the dissipative part of the susceptibility. It appears clearly in the first line of (31) that only the part f_v of the force $f(t)$ in phase with the velocity $\dot{x}$ contributes to this process.

The other part of the force f_x, in phase with x, does no work in the steady state; it does work only when the oscillation amplitude varies. If this variation is slow, the corresponding work is stored in a reversible form in the oscillator. It is equal to

$$W = \int^t f_x(t')\dot{x}(t')\,dt'. \tag{32.a}$$

The quasi-stationary part of W equals

$$\overline{W} = \int^t dt' \left[F_x^* \dot{X} + \text{c.c.} \right] \tag{32.b}$$

where $\dot{X}$ is the slow variation of the oscillation amplitude and where

$$F_x = \frac{\hat{\chi}'(\omega)}{|\hat{\chi}(\omega)|^2} X \tag{33}$$

is the part of the force in phase with X. In the presence of the driving force, a variation in the oscillation amplitude implies also a variation in the interaction energy

$$W_I = -f(t)x \tag{34}$$

which equals, on average over a period (only f_x contributes),

$$\overline{W}_I = -(F_x^* X + \text{c.c.}). \tag{35}$$

Therefore, in the presence of the driving force, and for slow variations of the parameters, the quantity $U = \overline{W} + \overline{W}_I$ plays the role of potential energy. When Equation (32.b) for $\overline{W}$ is integrated by parts and combined with (35), one finds that the contribution from $\overline{W}_I$ is cancelled and that U

is given by

$$U = -\int^t dt'\left[\dot{F}_x^* X + \text{c.c.}\right]$$

$$= -\hat{\chi}'(\omega)\int^t dt'\left[\dot{X}^* X + \text{c.c.}\right]/|\hat{\chi}(\omega)|^2$$

$$= -\hat{\chi}'(\omega)|X|^2/|\hat{\chi}(\omega)|^2 = -\frac{1}{2}\hat{\chi}'(\omega)(2FF^*). \qquad (36)$$

U represents the reversible variation of the oscillator energy due to its "polarization" in the presence of the force $f(t)$. It is this reactive part $\hat{\chi}'(\omega)$ of the polarizability that is associated with this physical effect. We will call U the polarization energy, by analogy with the effect of a static electric field on a dielectric medium.

d) PHYSICAL INTERPRETATION OF THE LEVEL SHIFTS

The interaction of the small system $\mathscr{A}$ with the reservoir $\mathscr{R}$ among other effects, gives rise to a shift of the energy levels of $\mathscr{A}$ given by formula (C.15) in the chapter. In this expression, the matrix element $\langle \mu, a|V|\nu, n\rangle$ can be factored into one part relative to $\mathscr{A}$ and another part relative to $\mathscr{R}$:

$$\Delta_a = \frac{1}{\hbar^2}\sum_\mu p_\mu \sum_\nu |\langle \mu|R|\nu\rangle|^2 \times$$

$$\times\left(\sum_n |\langle a|A|n\rangle|^2 \mathscr{P}\frac{1}{\omega_{\mu\nu} + \omega_{an}}\right). \qquad (37)$$

To make the statistical functions discussed above appear in this expression, we express the fraction $1/(\omega_{\mu\nu} + \omega_{an})$ as a product of functions relative to each of the two systems, each of these functions having a given parity:

$$\mathscr{P}\frac{1}{\omega_{\mu\nu} + \omega_{an}} = \frac{1}{4}\int d\omega \times$$

$$\times\left\{\left(\mathscr{P}\frac{1}{\omega_{\mu\nu} + \omega} + \mathscr{P}\frac{1}{\omega_{\mu\nu} - \omega}\right)(\delta(\omega + \omega_{an}) + \delta(\omega - \omega_{an})) + \right.$$

$$\left. + \left(\mathscr{P}\frac{1}{\omega_{an} + \omega} + \mathscr{P}\frac{1}{\omega_{an} - \omega}\right)(\delta(\omega + \omega_{\mu\nu}) + \delta(\omega - \omega_{\mu\nu}))\right\}. \qquad (38)$$

When (38) is substituted into (37), $\hat{\chi}'_R(\omega)$, $\hat{\chi}'_{Aa}(\omega)$, $\hat{C}_R(\omega)$, and $\hat{C}_{Aa}(\omega)$, appear, where $\hat{C}_{Aa}(\omega)$ and $\hat{\chi}_{Aa}(\omega)$ are the symmetric correlation function and the linear susceptibility of the observable A of the system $\mathscr{A}$ in the state $|a\rangle$, given by expressions analogous to (8) and (24) and where only $p_a = 1$ is nonzero. The shift Δ_a appears in the form of a sum of two terms:

$$\Delta_a = \Delta_a^{\text{fr}} + \Delta_a^{rr} \tag{39}$$

where

$$\hbar\Delta_a^{\text{fr}} = -\frac{1}{2}\int_{-\infty}^{+\infty}\frac{d\omega}{2\pi}\hat{\chi}'_{Aa}(\omega)\hat{C}_R(\omega) \tag{40}$$

$$\hbar\Delta_a^{rr} = -\frac{1}{2}\int_{-\infty}^{+\infty}\frac{d\omega}{2\pi}\hat{\chi}'_R(\omega)\hat{C}_{Aa}(\omega). \tag{41}$$

The interpretation of these expressions is simple if they are compared to formula (36): $\hat{C}_R(\omega)\,d\omega/2\pi$ represents the spectral density defined in (8) for the fluctuations of the observable R of the reservoir in the frequency band $d\omega$. The quantity $-(\frac{1}{2})\hat{\chi}'_{Aa}(\omega)\hat{C}_R(\omega)\,d\omega/2\pi$ is the polarization energy of the system $\mathscr{A}$ in the state a perturbed by these fluctuations. Therefore, $\hbar\Delta_a^{\text{fr}}$ represents the polarization energy of the small system in the state a induced by the fluctuations of the variable R of the reservoir (the superscript fr indicates fluctuations of the reservoir).

In (41), the roles of $\mathscr{A}$ and $\mathscr{R}$ are reversed. Δ_a^{rr} represents the polarization energy of the reservoir by the small system. The motion of the observable A in the state $|a\rangle$, characterized by $\hat{C}_{Aa}$, perturbs the equilibrium of the reservoir, and the interaction of A with the polarization thus created in $\mathscr{R}$, proportional to $\hat{\chi}'_R$, gives rise to the energy shift $\hbar\Delta_a^{rr}$. This shift therefore represents the effect of the "reservoir reaction" on the small system (which explains the superscript rr). As before, only the reactive part $\hat{\chi}'_R$ of the polarizability of the reservoir contributes to the polarization energy.

e) PHYSICAL INTERPRETATION OF THE ENERGY EXCHANGES

We now consider energy exchanges between the system $\mathscr{A}$ and the reservoir $\mathscr{R}$. To do this, we start with the rate of variation of the mean atomic energy for the atom initially in the state a:

$$\frac{d}{dt}\langle H_A\rangle_a = \sum_b (E_b - E_a)\Gamma_{a\to b} \tag{42}$$

where $\Gamma_{a \to b}$ is the transition rate from a to b, produced by the interaction with the reservoir and given by (C.5).

The replacement of $\Gamma_{a \to b}$ by its expression (C.5) gives a general explicit expression for $d\langle H_A \rangle_a / dt$:

$$\frac{d}{dt} \langle H_A \rangle_a = \sum_b \hbar \omega_{ba} \left\{ \sum_\mu \frac{2\pi}{\hbar} P_\mu \sum_\nu |\langle \mu | R | \nu \rangle|^2 |\langle a | A | b \rangle|^2 \delta(\hbar \omega_{\mu\nu} + \hbar \omega_{ab}) \right\}$$

$$= \frac{2\pi}{\hbar} \sum_\mu P_\mu \sum_\nu |\langle \mu | R | \nu \rangle|^2 \left(\sum_b \omega_{ba} |\langle a | A | b \rangle|^2 \delta(\omega_{\mu\nu} + \omega_{ab}) \right).$$

$$(43)$$

We proceed as in subsection 1.d and write

$$\omega_{ba} \delta(\omega_{\mu\nu} + \omega_{ab}) = \frac{1}{4} \int d\omega \, \omega \times$$

$$\times \left\{ \left[\delta(\omega_{\mu\nu} + \omega) + \delta(\omega_{\mu\nu} - \omega) \right] \left[\delta(\omega + \omega_{ab}) - \delta(\omega - \omega_{ab}) \right] + \right.$$

$$\left. + \left[\delta(\omega_{\mu\nu} - \omega) - \delta(\omega_{\mu\nu} + \omega) \right] \left[\delta(\omega + \omega_{ab}) + \delta(\omega - \omega_{ab}) \right] \right\}. \quad (44)$$

When (44) is substituted into (43), the statistical functions for $\mathscr{A}$ and $\mathscr{R}$ appear, and $d\langle H_A \rangle_a / dt$ appears in the form of a sum of two terms

$$\frac{d}{dt} \langle H_A \rangle_a = \dot{\mathscr{Q}}^{\text{fr}} + \dot{\mathscr{Q}}^{\text{rr}} \qquad (45)$$

where

$$\dot{\mathscr{Q}}^{\text{fr}} = \int \frac{d\omega}{2\pi} \omega \hat{C}_R(\omega) \hat{\chi}''_{Aa}(\omega) \qquad (46.a)$$

$$\dot{\mathscr{Q}}^{\text{rr}} = -\int \frac{d\omega}{2\pi} \omega \hat{\chi}''_R(\omega) \hat{C}_{Aa}(\omega). \qquad (46.b)$$

The two terms (46.a) and (46.b) have an especially clear physical meaning when they are compared with expression (31) describing the absorption of energy by a system. The first involves the symmetric correlation function for the reservoir and the linear susceptibility of the atom. It thus describes the absorption of energy by the small system, when this system is perturbed by the reservoir fluctuations. The second term involves the symmetric correlation function of the small system and the linear susceptibility of the reservoir. It describes the damping by the reservoir of the atomic motion. Because $\dot{\mathcal{E}}^{rr}$ is the energy gained by the atom, $-\dot{\mathcal{E}}^{rr}$ is in fact the energy absorbed by the reservoir, according to the interpretation of expression (31).

2. Applications to Radiative Processes

We now assume that $\mathcal{A}$ is an atom fixed at the origin **0** of the coordinate system and that $\mathcal{R}$ is a homogeneous and isotropic broadband radiation field, defined by the probabilities $p(n_1, n_2, \cdots n_i \cdots)$ as in Section E of the chapter. The average number of photons in the mode i, $\langle n_i \rangle$, is defined by (E.12). It depends only on ω_i and is written $\langle n(\omega_i) \rangle$. This average number of photons per mode at the frequency ω_i is related to the energy density of the radiation field by (E.20). For the sake of simplicity, we take a simple model for the atom which consists of a single electron moving in a potential with spherical symmetry about the origin, inside a volume having small dimensions compared with the wavelengths of the incident radiation field. Thus, we can make the long-wavelength approximation for all the modes considered, whose frequency is less than a cutoff frequency ω_M.

Let **r** and **p** be operators for the position and momentum of the electron, q and m its charge and its mass, and **A(0)** the vector potential at **0**. With the aforementioned approximations, the coupling between the atom and the radiation field is reduced to (*)

$$V = V_x + V_y + V_z \tag{47.a}$$

$$V_x = -\left(\frac{q}{m}p_x\right)A_x(0). \tag{47.b}$$

The coupling does not have the simple form $V = -AR$, but is instead a sum of such couplings. However, because the system is isotropic, we can easily show that the rate of variation of the density matrix σ_A is the sum of the rates corresponding to V_x, V_y, V_z. Thus, it is sufficient to calculate the statistical functions relative to $R = A_x(0)$ for the field, and $A = qp_x/m$ for the atom and then to sum over x, y, z.

(*) In the long-wavelength approximation, the Hamiltonian H_{I2} does not act on the electron variables.

a) Calculation of the Statistical Functions

We begin with the calculations for the field. Expressions (8) for $\hat{C}_R(\omega)$ and (24) for $\hat{\chi}_R(\omega)$ have the same form:

$$S_{\pm}(\omega) = \sum_{\mu} p_{\mu} \sum_{\nu} |R_{\mu\nu}|^2 f^{\pm}(\omega_{\mu\nu}, \omega) \tag{48}$$

where $f^{\pm}(\omega_{\mu\nu}, \omega)$ is a function of given parity $\pm$ with regard to $\omega_{\mu\nu}$: $+$ for $\hat{C}$, and $-$ for $\hat{\chi}$. Using a more precise notation for (48), we obtain

$$S_{\pm}(\omega) = \sum_{\{n_i\}} p(n_1, \ldots, n_i, \ldots) \times$$

$$\times \sum_{j} \left\{ \left| \langle \ldots, n_j, \ldots | A_x(0) | \ldots, n_j + 1, \ldots \rangle \right|^2 f^{\pm}(-\omega_j, \omega) + \right.$$

$$\left. + \left| \langle \ldots, n_j, \ldots | A_x(0) | \ldots, n_j - 1, \ldots \rangle \right|^2 f^{\pm}(\omega_j, \omega) \right\}. \tag{49}$$

We used the fact that $A_x(0)$ is linear in a_j and a_j^+, and changes n_j by ± 1. The matrix elements of (49) equal:

$$\left| \langle \ldots, n_j, \ldots | A_x(0) | \ldots, n_j \pm 1, \ldots \rangle \right|^2 = \frac{\hbar}{2\varepsilon_0 \omega_j L^3} \varepsilon_{jx}^2 \left(n_j + \frac{1}{2} \pm \frac{1}{2} \right). \tag{50}$$

The sum over the distributions of photons $\{n_i\}$ weighted by $p(\ldots, n_i, \ldots)$ causes $\langle n_i \rangle$ to appear, so that

$$S_{\pm}(\omega) = \sum_{j} \frac{\hbar}{2\varepsilon_0 \omega_j L^3} \varepsilon_{jx}^2 \left[\langle n_j + 1 \rangle f^{\pm}(-\omega_j, \omega) + \langle n_j \rangle f^{\pm}(\omega_j, \omega) \right]$$

$$= \sum_{j} \frac{\hbar}{2\varepsilon_0 \omega_j L^3} \varepsilon_{jx}^2 \left[\pm \langle n_j + 1 \rangle + \langle n_j \rangle \right] f^{\pm}(\omega_j, \omega). \tag{51}$$

We replace the discrete sum over the modes by a sum over the polarizations, calculated by using formula (54) from Complement A_I, and an integral over **k** (for which the angular part can be calculated explicitly because $\langle n_j \rangle$ is assumed to depend only on ω_j). This yields

$$S_{\pm}(\omega) = \frac{1}{6\pi^2\varepsilon_0 c^3} \int_0^{\omega_M} d\omega' \, \hbar\omega' \big[\pm\langle n(\omega') + 1\rangle + \langle n(\omega')\rangle\big] f^{\pm}(\omega', \omega).$$

(52)

This result applied successively to $\hat{C}_R^{xx}$, $\hat{\chi}_R'^{xx}$, $\hat{\chi}_R''^{xx}$ gives

$$\hat{C}_R^{xx}(\omega) = \int_0^{\omega_M} \frac{d\omega'}{3\pi\varepsilon_0 c^3} \hbar\omega' \langle n(\omega') + 1/2\rangle \big[\delta(\omega' + \omega) + \delta(\omega' - \omega)\big]$$

$$= \frac{1}{3\pi\varepsilon_0 c^3} \hbar|\omega|\langle n(|\omega|) + 1/2\rangle$$

(53)

$$\hat{\chi}_R'^{xx}(\omega) = \frac{1}{6\pi^2\varepsilon_0 c^3} \int_0^{\omega_M} d\omega' \, \omega' \left[\mathscr{P}\frac{1}{\omega' + \omega} + \mathscr{P}\frac{1}{\omega' - \omega}\right]$$

(54)

$$\hat{\chi}_R''^{xx}(\omega) = -\frac{1}{6\pi\varepsilon_0 c^3} \int_0^{\omega_M} d\omega' \, \omega' \big[\delta(\omega' + \omega) - \delta(\omega' - \omega)\big]$$

$$= \frac{1}{6\pi\varepsilon_0 c^3} \omega.$$

(55)

Before commenting on these expressions in the following subsection, we give the expressions $\hat{C}_{Aa}^{xx}(\omega)$ and $\hat{\chi}_{Aa}^{xx}(\omega)$ of the correlation function for the atomic variable qp_x/m, and the corresponding susceptibility, when the atom is in the state $|a\rangle$. The equations equivalent to (8), (26.a), and (26.b) are

$$\hat{C}_{Aa}^{xx}(\omega) = \sum_b \frac{q^2}{m^2}|\langle a|p_x|b\rangle|^2\pi\big[\delta(\omega_{ab} + \omega) + \delta(\omega_{ab} - \omega)\big]$$

(56)

$$\hat{\chi}_{Aa}'^{xx}(\omega) = \sum_b \frac{-q^2}{\hbar m^2}|\langle a|p_x|b\rangle|^2\left[\mathscr{P}\frac{1}{\omega_{ab} + \omega} + \mathscr{P}\frac{1}{\omega_{ab} - \omega}\right]$$

(57)

$$\hat{\chi}_{Aa}''^{xx}(\omega) = \sum_b \frac{q^2}{\hbar m^2}|\langle a|p_x|b\rangle|^2\pi\big[\delta(\omega_{ab} + \omega) - \delta(\omega_{ab} - \omega)\big].$$

(58)

b) PHYSICAL DISCUSSION

The spectral density of the field fluctuations appears as the sum of two terms:

$$\hat{C}_R^{xx}(\omega) = \frac{1}{3\pi\varepsilon_0 c^3}\left[\hbar|\omega|\langle n(|\omega|)\rangle + \frac{1}{2}\hbar|\omega|\right]. \tag{59}$$

The first involves the average number of photons per mode. By using (E.20), it can be put in the form $\pi u(\omega)/3\varepsilon_0\omega^2$, thus showing that it is proportional to the energy density of the radiation at the frequency ω, a completely classical quantity. The second term adds $1/2$ to $\langle n(\omega)\rangle$ for all the modes. It corresponds to the contribution of vacuum fluctuations, whose zero-point energy equals $\hbar\omega/2$. Therefore, the spectral density of vacuum fluctuations is identical to that of a random isotropic classical electromagnetic field whose frequency distribution corresponds to a half quantum per mode.

The situation is completely different for the susceptibility of the field, which is independent of $\langle n(\omega)\rangle$. This property is related to the linearity of Maxwell equations: The field created by a given source is independent of the existing field. In particular, this results in the fact that the susceptibility of the field is the same for a microscopic source as for a classical macroscopic source. Indeed, (54) and (55) do not contain $\hbar$. The susceptibility of the field is a classical variable.

Finally, note that, for $\omega > 0$,

$$\hat{C}_R^{xx}(\omega) = 2\hbar\langle n(\omega) + 1/2\rangle\hat{\chi}_R^{\prime\prime xx}(\omega). \tag{60}$$

At thermodynamic equilibrium, according to (E.17) we have

$$2\langle n(\omega) + 1/2\rangle = \left[\tanh\frac{\hbar\omega}{2k_BT}\right]^{-1} \tag{61}$$

and

$$\hat{C}_R^{xx}(\omega) = \hbar\left[\tanh\frac{\hbar\omega}{2k_BT}\right]^{-1}\hat{\chi}_R^{\prime\prime xx}(\omega). \tag{62}$$

In this particular case, this relation is an expression for the fluctuation-dissipation theorem, relating the spectral density of the fluctuations to the dissipative part of the susceptibility.

c) Level Shifts due to the Fluctuations of the Radiation Field

To calculate the effect on the atomic energy level $|a\rangle$ of the fluctuations of the radiation field, we explicitly calculate expression (40) for the case of the coupling of the atom with the component $A_x(0)$ of the vector potential. By using expressions (57) and (53) for $\hat{\chi}'^{xx}_{Aa}$ and $\hat{C}^{xx}_R$, we obtain

$$\hbar\Delta^{fr}_{ax} = -\frac{1}{2}\int\frac{d\omega}{2\pi} \times$$

$$\times\left\{-\frac{1}{\hbar}\sum_b\frac{q^2}{m^2}|\langle a|p_x|b\rangle|^2\left[\mathscr{P}\frac{1}{\omega_{ab}+\omega}+\mathscr{P}\frac{1}{\omega_{ab}-\omega}\right]\right\} \times$$

$$\times\left\{\frac{1}{3\pi\varepsilon_0 c^3}\int_0^{\omega_M}d\omega'\,\hbar\omega'\langle n(\omega')+1/2\rangle[\delta(\omega'+\omega)+\delta(\omega'-\omega)]\right\}.$$

$$(63)$$

In the integral over ω, the contributions of $\delta(\omega'+\omega)$ and $\delta(\omega'-\omega)$ are equal, because the rest of the integrand is even in ω. We must then add to (63) analogous contributions coming from $A_y(0)$ and $A_z(0)$, which gives

$$\hbar\Delta^{fr}_a = \frac{q^2}{6\pi^2\varepsilon_0 m^2 c^3}\sum_b|\langle a|\mathbf{p}|b\rangle|^2 \times$$

$$\times\int_0^{\omega_M}d\omega'\,\omega'\langle n(\omega')+1/2\rangle\left[\mathscr{P}\frac{1}{\omega_{ab}+\omega'}+\mathscr{P}\frac{1}{\omega_{ab}-\omega'}\right]. \quad (64)$$

The energy-level shift due to the fluctuations of the field appears as the sum of the effect Δ^{fr}_a of the incident photons, proportional to $\langle n(\omega)\rangle$, and of the effect of the fluctuations of the vacuum Δ^{fv}_a corresponding to the "1/2 per mode" term in (64).

First we expand Δ^{fv}_a. By using the following results

$$\int_0^{\omega_M}\omega'\,d\omega'\,\mathscr{P}\frac{1}{\omega'\pm\omega_0} = \omega_M \mp \omega_0\ln\frac{\omega_M}{\omega_0} + 0\left(\frac{\omega_0}{\omega_M}\right) \quad (65.a)$$

$$\int_0^{\omega_M}\omega'\,d\omega'\left[\mathscr{P}\frac{1}{\omega_0+\omega'}+\mathscr{P}\frac{1}{\omega_0-\omega'}\right] = -2\omega_0\ln\frac{\omega_M}{\omega_0} \quad (65.b)$$

we obtain

$$\hbar\Delta_a^{\text{fv}} = \frac{q^2}{6\pi^2\varepsilon_0 m^2 c^3} \sum_b |\langle a|\mathbf{p}|b\rangle|^2 (-\omega_{ab}) \ln \frac{\omega_M}{|\omega_{ab}|}. \tag{66}$$

As shown in Exercise 7, this expression can be transformed into

$$\hbar\Delta_a^{\text{fv}} = \frac{\alpha}{3\pi}\left(\frac{\hbar}{mc}\right)^2 \left(\ln\frac{\omega_M}{cK_a}\right)\langle a|\frac{q^2}{\varepsilon_0}\delta(\mathbf{r})|a\rangle \tag{67}$$

(α = fine structure constant, $\hbar cK_a$ = mean atomic excitation energy). This shift affects only the s states: this is the Lamb shift discussed previously in the treatment' of radiative corrections (Chapter II, §E.1.b, Complement B$_{\text{II}}$ and Exercise 7). The preceding calculation clearly shows its physical origin: the vacuum fluctuations cause the electron to undergo a vibrational motion during which it averages the Coulomb potential over a distance of approximately $\sqrt{\alpha}\,\hbar/mc$. The average potential differs from the value of the potential at the average position only inside the source of the Coulomb potential, hence the function $\delta(\mathbf{r})$. Note that the coefficient $\alpha(\hbar/mc)^2$ is proportional to $\hbar$: the Lamb shift is a quantum effect, just like the vacuum fluctuations that give rise to it. Nevertheless, once the existence of these fluctuations is acknowledged, their effect is the same as the one which would be produced by a random classical field characterized by a spectral density corresponding to an energy $\hbar\omega/2$ per mode.

We return now to the part $\Delta_a'^{\text{fr}}$ of Δ_a^{fr}, proportional to $\langle n(\omega')\rangle$ [see (64)]. It corresponds to the stimulated radiative corrections that we described in subsection E.2 of Chapter II. For the sake of simplicity, assume that $\langle n(\omega')\rangle$ is appreciable only near a frequency ω_L close enough to ω_{ab} so that only a particular level contributes to the sum over all the levels b of (64). The corresponding shift of the level a equals

$$\hbar\Delta_a'^{\text{fr}} \simeq \frac{q^2}{6\pi^2\varepsilon_0 m^2 c^3}|\langle a|\mathbf{p}|b\rangle|^2 \frac{2\omega_{ab}\omega_L}{\omega_{ab}^2 - \omega_L^2}\int d\omega' \langle n(\omega')\rangle. \tag{68}$$

This is the light shift due to the radiation field. Note that this quantity can also be expressed as a function of the mean quadratic value of the electric radiation field. In fact, $\langle n(\omega)\rangle\,d\omega$ is related to the energy density in the same frequency band $u(\omega)\,d\omega$ by formula (E.20), and the integral over ω'

of $u(\omega')$ is simply the energy density of the radiation

$$\int d\omega'\, u(\omega') = \frac{\varepsilon_0}{2}\langle E_\perp^2 + c^2 B^2 \rangle = \varepsilon_0 \langle E_\perp^2 \rangle. \tag{69}$$

(For a free and stationary radiation field $\langle E_\perp^2 \rangle = c^2 \langle B^2 \rangle$.) Using the relation $\langle a|\mathbf{p}|b\rangle = \mathrm{Im}\, \omega_{ab}\langle a|\mathbf{r}|b\rangle$, and the approximation $|\omega_{ab}| \simeq \omega_L$, we then have

$$\hbar\Delta_a^{'\mathrm{fr}} = \frac{\omega_{ab}}{\hbar(\omega_{ab}^2 - \omega_L^2)} q^2 \frac{|\langle a|\mathbf{r}|b\rangle|^2}{3} \langle E_\perp^2 \rangle. \tag{70}$$

This expression, which causes the polarization energy of the atom by the electric field to appear, justifies the name "dynamic Stark effect" also given to $\Delta_a^{'\mathrm{fr}}$.

d) Level Shifts due to Radiation Reaction

Expression (41) for Δ_a^{rr} involves the "susceptibility" $\hat{\chi}_R'$ of the field, which, according to (54), does not depend on $\langle n(\omega)\rangle$. The phenomenon of radiation reaction is therefore independent of the field incident on the atom.

By using expressions (54) for $\hat{\chi}_R'^{xx}$ relative to $A_x(0)$ and (56) for $\hat{C}_{Aa}^{xx}$ relative to p_x, then by summing over the analogous expressions with y and z, we obtain

$$\hbar\Delta_a^{\mathrm{rr}} = -\frac{1}{2}\int \frac{d\omega}{2\pi} \times$$

$$\times \left\{ \frac{1}{6\pi^2\varepsilon_0 c^3} \int_0^{\omega_M} d\omega'\, \omega' \left[\mathscr{P}\frac{1}{\omega' + \omega} + \mathscr{P}\frac{1}{\omega' - \omega} \right] \right\} \times$$

$$\times \left\{ \sum_b \frac{q^2}{m^2} |\langle a|\mathbf{p}|b\rangle|^2 \pi [\delta(\omega_{ab} + \omega) + \delta(\omega_{ab} - \omega)] \right\}$$

$$= -\frac{q^2}{12\pi^2\varepsilon_0 m^2 c^3} \sum_b \int_0^{\omega_M} d\omega'\, \omega' |\langle a|\mathbf{p}|b\rangle|^2 \times$$

$$\times \left[\mathscr{P}\frac{1}{\omega' + \omega_{ab}} + \mathscr{P}\frac{1}{\omega' - \omega_{ab}} \right]. \tag{71}$$

The integral over ω' up to the limit ω_M thus yields, taking into account (65.a):

$$
\begin{aligned}
\hbar \Delta_a^{\text{rr}} &= \frac{-q^2}{6\pi^2 \varepsilon_0 m^2 c^3} \sum_b |\langle a|\mathbf{p}|b\rangle|^2 \omega_M \\
&= -\frac{4}{3} \frac{\delta m}{m} \langle a|\frac{\mathbf{p}^2}{2m}|a\rangle
\end{aligned}
\tag{72}
$$

where δm is the correction to the mass of the particle associated with the Coulomb field and given by $\delta m c^2 = \varepsilon_{\text{Coul}}$. Hence, the expression (72) is the correction to the kinetic energy of the particle associated with this change in mass. The factor $\frac{4}{3}$ is due to the noncovariant cutoff procedure we used to eliminate the effect of the high-frequency modes. The same factor also appears in classical theory.

Note that $\varepsilon_{\text{Coul}}$ describes the energy associated with the interaction of the particle with its own Coulomb field, in the same way that Δ_a^{rr} describes the effect on the particle of its interaction with its own transverse field. These two effects are the same as those occurring in classical electromagnetism, as demonstrated by the absence of $\hbar$ in the expression for δm. They are independent of the atomic potential and are therefore the same for a free particle. Another consequence of the increase in the mass of the particle is the decrease in its cyclotron frequency in a magnetic field. One can thus qualitatively explain why the Landé factor of the electron is greater than 2 (see Exercises 10 and 12).

e) ENERGY EXCHANGES BETWEEN THE ATOM AND THE RADIATION

We will now consider successively the rates $\dot{\mathscr{E}}^{\text{fr}}$ and $\dot{\mathscr{E}}^{\text{rr}}$ associated, respectively, with radiation fluctuations and radiation reaction.

With expressions (46.a), (59), and (58), and after summation of the effects relative to the three components x, y, z, $\dot{\mathscr{E}}^{\text{fr}}$ is put in the form:

$$
\begin{aligned}
\dot{\mathscr{E}}^{\text{fr}} = \frac{-q^2}{3\pi \varepsilon_0 \hbar m^2 c^3} \sum_b |\langle a|\mathbf{p}|b\rangle|^2 \times \\
\times \omega_{ab}\left[\hbar|\omega_{ab}|\langle n(|\omega_{ab}|)\rangle + \frac{1}{2}\hbar|\omega_{ab}|\right].
\end{aligned}
\tag{73}
$$

It is advantageous to write (73) in a form that exhibits the spontaneous emission rate Γ_{ab}^{sp} relative to the transition ab (which occurs of course from the highest level toward the lowest level). By expanding (E.4.a), we find

$$\Gamma_{ab}^{sp} = \frac{q^2 |\langle a|\mathbf{p}|b\rangle|^2 |\omega_{ab}|}{3\pi\varepsilon_0 \hbar m^2 c^3} \tag{74}$$

so that

$$\dot{\mathcal{E}}^{fr} = \sum_b (E_b - E_a) \Gamma_{ab}^{sp} \left[\langle n(|\omega_{ab}|)\rangle + \tfrac{1}{2} \right]. \tag{75}$$

Just like a random classical perturbation, the fluctuations of the radiation field transfer the population of the state a toward other higher or lower levels b. As in subsection 2-b, the incident radiation field contributes to this process proportionately to $\langle n(|\omega_{ab}|)\rangle$ per mode, and vacuum fluctuations contribute proportionately to $\frac{1}{2}$. Note also that the latter yield a transition rate equal to only half of Γ_{ab}^{sp}.

The quantity $\dot{\mathcal{E}}^{rr}$ is calculated in the same way from (46.b), (55), and (56). We find:

$$\dot{\mathcal{E}}^{rr} = \frac{-1}{6\pi\varepsilon_0 c^3} \sum_b \frac{q^2}{m^2} |\langle a|\mathbf{p}|b\rangle|^2 \omega_{ab}^2. \tag{76}$$

This expression can be put into a particularly simple form, because $\langle a|\mathbf{p}|b\rangle\omega_{ab}$ is directly related to the matrix element of the acceleration $\ddot{r}$ of the electron in the atomic potential:

$$\begin{aligned}
\frac{1}{m}\langle a|\mathbf{p}|b\rangle\omega_{ab} &= \frac{1}{m\hbar}\langle a|[H_A, \mathbf{p}]|b\rangle \\
&= -\frac{i}{m}\langle a|\dot{\mathbf{p}}|b\rangle \\
&= -i\langle a|\ddot{\mathbf{r}}|b\rangle.
\end{aligned} \tag{77}$$

We can then use the closure relation over b to obtain

$$\dot{\mathcal{E}}^{rr} = -\frac{2}{3}\frac{q^2}{4\pi\varepsilon_0 c^3}\langle a|(\ddot{\mathbf{r}})^2|a\rangle. \tag{78}$$

The energy lost by the atom per unit time due to the radiation reaction is equal to the average in the state $|a\rangle$ of the power radiated classically by the particle, with this power being proportional to the square of the particle acceleration. $\dot{\mathcal{E}}^{rr}$ is, moreover, always negative and corresponds to an energy loss, even for the ground state of the atom. Thus, we find here some results that are well known in classical electrodynamics: the Lorentz damping force, and the fall of the electron onto the nucleus. Of course, it is not possible to consider only the effect of the radiation reaction, because even in the absence of radiation, vacuum fluctuations exist and play a role, as we saw in the preceding paragraph. To compare their respective roles, it is interesting to put (76) in the same form as (75). We obtain

$$\dot{\mathscr{E}}^{\text{rr}} = \sum_b - |E_b - E_a| \frac{\Gamma_{ab}^{\text{sp}}}{2} \qquad (79)$$

whereas

$$\dot{\mathscr{E}}^{\text{fv}} = \sum_b (E_b - E_a) \frac{\Gamma_{ab}^{\text{sp}}}{2}. \qquad (80)$$

If $|a\rangle$ is the ground state, $|E_b - E_a| = (E_a - E_b)$ and $\dot{\mathscr{E}}^{\text{rr}} + \dot{\mathscr{E}}^{\text{fv}} = 0$. The atomic ground state is stable, because vacuum fluctuations exactly cancel out the effect of the radiation reaction. The quantized character of the field, of which vacuum fluctuations are a manifestation, is therefore essential for preserving the stability of the ground state of the atom. This result is very general (*): The coupling of a quantized system to a classical dynamic system would lead to incoherences connected in the present case to the fact that the uncertainty relations would be violated for the electron if it were to lose energy from the ground state. If $|a\rangle$ is an excited level, the two energy variation rates cancel out for transitions toward levels b above a (the atom cannot gain energy) and are added together for transitions toward the lower levels b. Thus, we can say that vacuum fluctuations and the radiation reaction contribute equally to the total spontaneous emission transition rate Γ_{ab}^{sp} (**).

REFERENCES

A general discussion of correlation functions and susceptibilities can be found in Martin, Les Houches 1967. See also Landau and Lifchitz, *Statistical Physics*, Chapter XII.

For the particular application considered in this complement, more detailed discussions can be found in J. Dalibard, J. Dupont-Roc, and C. Cohen-Tannoudji, *J. Phys.* (*Paris*), **43**, 1617 (1982); **45**, 637 (1984), and in the references therein cited.

(*) I. R. Senitzky *Phys. Rev. Lett.*, **20**, 1062 (1968).

(**) This is why heuristic calculations taking into account only one of these two effects lead to rates which are too small by a factor $\frac{1}{2}$.

COMPLEMENT B_{IV}

MASTER EQUATION FOR A DAMPED HARMONIC OSCILLATOR

1. The Physical System

In Section E of the chapter, we worked out the master equation for a two-level atom interacting with the radiation field. In this complement, we consider the case where the small system $\mathscr{A}$ is a one-dimensional harmonic oscillator of frequency ω_0 whose Hamiltonian is

$$H_A = \hbar\omega_0\left(b^+b + \tfrac{1}{2}\right) \tag{1}$$

where b^+ and b are the creation and annihilation operators of this oscillator. As in the example in Section E, the reservoir $\mathscr{R}$ consists of an infinite number of one-dimensional harmonic oscillators i, of frequency ω_i with creation and annihilation operators a_i^+ and a_i, so that the Hamiltonian H_R for $\mathscr{R}$ is written

$$H_R = \sum_i \hbar\omega_i\left(a_i^+a_i + \tfrac{1}{2}\right). \tag{2}$$

Finally, we take a very simple coupling Hamiltonian between $\mathscr{A}$ and $\mathscr{R}$ of the form

$$V = \sum_i \left(g_i b^+ a_i + g_i^* b a_i^+ \right) \tag{3}$$

where g_i is the coupling constant between $\mathscr{A}$ and the oscillator i of $\mathscr{R}$. Each term in (3) describes processes in which $\mathscr{A}$ gains (or loses), one energy quantum $\hbar\omega_0$, whereas one oscillator i of $\mathscr{R}$ loses (or gains) one quantum $\hbar\omega_i$. For the sake of simplicity, we have not included terms of the type ba_i or $b^+a_i^+$ in V because the antiresonant processes to which they give rise cannot contribute to the damping of $\mathscr{A}$.

Such a model can describe several different physical situations. For example, $\mathscr{A}$ can be a matter harmonic oscillator, i.e., a charge elastically bound to the origin (and constrained to move on one axis, so as to be similar to a one-dimensional oscillator), and the oscillators i of $\mathscr{R}$ can be the different modes of the radiation field. The master equation for $\mathscr{A}$ thus describes how the motion of the elastically-bound charge is damped by the spontaneous emission, absorption and stimulated emission of radiation.

We can also consider $\mathscr{A}$ to be a mode of an electromagnetic cavity, the oscillators i being matter oscillators contained in the walls of the cavity. The master equation for $\mathscr{A}$ then describes the damping of this eigenmode of the cavity due to losses inside the walls.

Aside from its intrinsic interest in connection with the two preceding physical situations, the example discussed in this complement has the great advantage of leading to coupled Heisenberg equations for the operators b and a_i which are *linear* with respect to these operators. It is then very easy to derive from these equations an evolution equation for b that closely resembles the Langevin equation used to describe Brownian motion in classical theory. This is done in Complement C_{IV}, which also establishes the relationship that exists between fluctuations and dissipation, in classical theory as well as in quantum theory.

To determine the form of the master equation for the harmonic oscillator, it is obvious that we can use the results from Section C of the chapter, and explicitly calculate the coefficients of the master equation in this particular case. Here we take a different approach, and starting from (B.30), directly derive an operator equation for the motion of σ, without making reference to any particular basis (§2). We then use this equation to derive the evolution equations for the populations of the energy levels of H_A, and for the average values of certain observables (§3). Finally, by projecting this equation onto a coherent state basis, we derive the evolution equation for the distribution of quasi-probability $P_N(\beta, \beta^*)$, which has the form of a Fokker–Planck equation (§4).

2. Operator Form of the Master Equation

We rewrite equation (B.30) with the change of variables (B.32) and use the fact that $\tau_c \ll \Delta t$ to extend to $+\infty$ the upper bound of the integral over $\tau = t' - t''$ and extend to t the lower bound of the integral over t' (see discussion at the end of subsection B-3). We then have

$$\frac{\Delta \bar{\sigma}}{\Delta t} = -\frac{1}{\hbar^2} \frac{1}{\Delta t} \int_0^\infty d(t' - t'') \times$$

$$\times \int_t^{t+\Delta t} dt'\, \mathrm{Tr}_R \left[\bar{V}(t'), \left[\bar{V}(t''), \bar{\sigma}(t) \otimes \sigma_R \right] \right]. \tag{4}$$

We write V in the form

$$V = -\left[b^+ R^{(+)} + b R^{(-)} \right] \tag{5}$$

where

$$R^{(+)} = -\sum_i g_i a_i \qquad R^{(-)} = -\sum_i g_i^* a_i^+. \tag{6}$$

Assume that σ_R is diagonal in the basis $\{|n_1, n_2 \cdots n_i \cdots \rangle\}$ of the eigenstates of H_R [σ_R is still given by formula (E.8)]. Of the two operators $\tilde{V}(t')$ and $\tilde{V}(t'')$ of (4), one must contain a_i, the other containing a_i^+, so that the trace over R of their product multiplied by σ_R gives a nonzero result in (4). Such a restriction results in only two possibilities, the first consisting of taking $\tilde{V}(t') = -\tilde{b}^+(t')\tilde{R}^{(+)}(t')$ and $\tilde{V}(t'') = -\tilde{b}(t'')\tilde{R}^{(-)}(t'')$, while the second follows from the first by exchanging t' and t''. Moreover, we can check that the two terms of (4) corresponding to these two possibilities are Hermitian conjugates of each other. In the interaction representation, the operators $\tilde{b}(t), \tilde{a}_i(t) \cdots$ have a very simple time dependence, $\tilde{b}(t) = b\,e^{-i\omega_0 t}$, $\tilde{a}_i(t) = a_i\,e^{-i\omega_i t} \cdots$ as a result of the simplicity of the commutators $[b, H_A], [a_i, H_R] \cdots$.

Finally, by using the invariance of a trace in a circular permutation and the fact that the integrand of (4) depends only on $t' - t''$, we obtain

$$\frac{\Delta\bar{\sigma}}{\Delta t} = -\frac{1}{\hbar^2} \times$$

$$\times \left\{ (b^+ b\bar{\sigma} - b\bar{\sigma}b^+) \int_0^\infty \langle \tilde{R}^{(+)}(t')\tilde{R}^{(-)}(t'') \rangle_R\, e^{i\omega_0(t'-t'')}\, \mathrm{d}(t'-t'') + \right.$$

$$\left. + (\bar{\sigma}bb^+ - b^+\bar{\sigma}b) \int_0^\infty \langle \tilde{R}^{(-)}(t'')\tilde{R}^{(+)}(t') \rangle_R\, e^{i\omega_0(t'-t'')}\, \mathrm{d}(t'-t'') \right\}$$

$$+ \text{ h.c.} \tag{7}$$

We now calculate the two-time averages that appear in (7). By using (6) and (E.8), we obtain

$$\langle \tilde{R}^{(+)}(t')\tilde{R}^{(-)}(t'') \rangle_R = \sum_i |g_i|^2 \langle a_i a_i^+ \rangle_R\, e^{-i\omega_i(t'-t'')}$$

$$= \sum_i |g_i|^2 (\langle n_i \rangle + 1)\, e^{-i\omega_i(t'-t'')} \tag{8}$$

where $\langle n_i \rangle$ is the average number of excitation quanta of the oscillator i, which is still given by (E.12). An analogous calculation gives

$$\langle \tilde{R}^{(-)}(t'')\tilde{R}^{(+)}(t') \rangle_R = \sum_i |g_i|^2 \langle n_i \rangle\, e^{-i\omega_i(t'-t'')}. \tag{9}$$

The first integral of (7) thus equals

$$\frac{1}{\hbar^2} \int_0^\infty d\tau \sum_i |g_i|^2 (\langle n_i \rangle + 1) e^{-i(\omega_i - \omega_0)\tau}$$

$$= \frac{1}{\hbar^2} \sum_i |g_i|^2 (\langle n_i \rangle + 1) \left[i\mathscr{P} \frac{1}{\omega_0 - \omega_i} + \pi \delta(\omega_0 - \omega_i) \right]$$

$$= \frac{\Gamma + \Gamma'}{2} + i(\Delta + \Delta') \tag{10}$$

where the quantities $\Gamma, \Gamma', \Delta, \Delta'$, which will be interpreted physically later on, are given by

$$\Gamma = \frac{2\pi}{\hbar} \sum_i |g_i|^2 \delta(\hbar\omega_0 - \hbar\omega_i) \tag{11.a}$$

$$\Gamma' = \frac{2\pi}{\hbar} \sum_i |g_i|^2 \langle n_i \rangle \delta(\hbar\omega_0 - \hbar\omega_i) \tag{11.b}$$

$$\hbar\Delta = \mathscr{P} \sum_i \frac{|g_i|^2}{\hbar\omega_0 - \hbar\omega_i} \tag{12.a}$$

$$\hbar\Delta' = \mathscr{P} \sum_i \frac{|g_i|^2 \langle n_i \rangle}{\hbar\omega_0 - \hbar\omega_i}. \tag{12.b}$$

For the second integral of (7) a similar calculation yields

$$\frac{1}{\hbar^2} \int_0^\infty d\tau \sum_i |g_i|^2 \langle n_i \rangle e^{-i(\omega_i - \omega_0)\tau} = \frac{\Gamma'}{2} + i\Delta'. \tag{13}$$

Finally, by substituting (10) and (13) into (7), by using $[b, b^+] = 1$, and by returning to the Schrödinger representation, we obtain the following operator form for the master equation

$$\frac{d\sigma}{dt} = -\frac{\Gamma}{2} [\sigma, b^+ b]_+ - \Gamma' [\sigma, b^+ b]_+ - \Gamma' \sigma -$$

$$- i(\omega_0 + \Delta)[b^+ b, \sigma] +$$

$$+ \Gamma b\sigma b^+ + \Gamma' (b^+ \sigma b + b\sigma b^+). \tag{14}$$

In (14), $[A, B]_+$ represents the anticommutator $AB + BA$. Also note the absence of terms involving Δ'.

Remark

As in Section E, some simplifications appear if the average number of quanta $\langle n_i \rangle$ of the oscillator i depends only on the energy of this oscillator, and not on the parameters other than the energy that are used to characterize the oscillators. The same procedure as that leading to (E.16) then yields

$$\Gamma' = \Gamma \langle n(\omega_0) \rangle \tag{15}$$

where $\langle n(\omega_0) \rangle$ is the average number of quanta in the oscillators of the reservoir having the same frequency ω_0 as the oscillator $\mathscr{A}$. If, moreover, $\mathscr{R}$ is in thermodynamic equilibrium, $\langle n(\omega_0) \rangle$ is equal to $[\exp(\hbar\omega_0/k_B T) - 1]^{-1}$.

3. Master Equation in the Basis of the Eigenstates of H_A

a) EVOLUTION OF THE POPULATIONS

Let σ_{nn} be the population of the energy level $|n\rangle$ of $\mathscr{A}$. Equation (14) taken in average value in the state $|n\rangle$ gives

$$\frac{d}{dt}\sigma_{nn} = -n\Gamma\sigma_{nn} + (n + 1)\Gamma\sigma_{n+1\,n+1} +$$

$$+ (n + 1)\Gamma'(\sigma_{n+1\,n+1} - \sigma_{nn}) + n\Gamma'(\sigma_{n-1\,n-1} - \sigma_{nn}). \tag{16}$$

The interpretation of Γ and Γ' follows immediately from this equation. If we assume that $\mathscr{A}$ is an atomic oscillator and $\mathscr{R}$ the radiation field, then Γ is associated with the spontaneous emission process, and Γ' with the absorption and stimulated emission processes. More precisely, $n\Gamma$ is the spontaneous emission rate from the level $|n\rangle$ to the lower level $|n - 1\rangle$: The level $|n\rangle$ empties by spontaneous emission to the level $|n - 1\rangle$ at a rate $n\Gamma$ and is filled from the level $|n + 1\rangle$ at a rate $(n + 1)\Gamma$ (wavy lines in Figure 1). The factors n and $n + 1$ are simply related to the squares of the moduli of the matrix elements of b and b^+ appearing in V [see (3)] between $|n\rangle$ and $|n - 1\rangle$ or between $|n + 1\rangle$ and $|n\rangle$. Similarly, the absorption processes empty the level $|n\rangle$ toward the level $|n + 1\rangle$ at a rate $(n + 1)\Gamma'$ and fill it from level $|n - 1\rangle$ at a rate $n\Gamma'$ (rising arrows in Figure 1), the stimulated emission processes occurring in the opposite direction at the same rate (falling arrows).

The parameter Δ appearing in (14) is related to "spontaneous" radiative shifts occurring in the absence of any excitation of $\mathscr{R}$ [$\hbar\Delta$ does not depend on $\langle n_i \rangle$—see (12.a)]. Due to the structure of V, the ground

Figure 1. Transfer rate between energy levels of the harmonic oscillator associated with spontaneous emission processes (wavy lines) and absorption and stimulated emission processes (straight lines).

state $|0\rangle$ of the oscillator in the presence of the reservoir in the state $|0_1 0_2 \cdots 0_i\rangle$ is not coupled to any other state. The spontaneous radiative shift $\hbar \Delta_0$ of $|0\rangle$ is therefore zero. By contrast, if the oscillator is in the state $|1\rangle$, the coupling V given in (3) allows transitions in which $\mathscr{A}$ falls from $|1\rangle$ to $|0\rangle$, whereas an oscillator i of $\mathscr{R}$ goes from $|0_i\rangle$ to $|1_i\rangle$. The shift $\hbar \Delta$ is simply the radiative shift associated with such a process (summed over all the oscillators i). The same reasoning can be applied to an oscillator initially in the state $|n\rangle$, which then goes from $|n\rangle$ to $|n - 1\rangle$, whereas an oscillator i goes from $|0_i\rangle$ to $|1_i\rangle$. Because $|\langle n - 1|b|n\rangle|^2 = n|\langle 0|b|1\rangle|^2$, the spontaneous radiative shift of the state $|n\rangle$ equals $n\hbar \Delta$. Finally, the difference between the shifts $\hbar \Delta_n$ and $\hbar \Delta_{n-1}$ of two adjacent levels is independent of n and equals $\hbar \Delta$, which shows that the perturbed levels of the oscillator remain equidistant with an apparent frequency $\omega_0 + \Delta$, as expressed by the fourth term in (14).

   Finally, we consider the last parameter Δ'. Because it depends on the $\langle n_i \rangle$ [see (12.b)], it represents a radiative shift due to the incoming radiation. We calculate such a shift for the level $|1\rangle$. A stimulated emission process causing the oscillator $\mathscr{A}$ to fall from $|1\rangle$ to $|0\rangle$ and an additional quantum $\hbar \omega_i$ to appear in oscillator i shifts level $|1\rangle$ by the quantity $\hbar \Delta'$ given in (12.b). However, we must not forget the absorption processes that result in a jump from $|1\rangle$ to $|2\rangle$ for $\mathscr{A}$ and the disappearance of a quantum $\hbar \omega_i$ for $\mathscr{R}$. Such a process shifts the level $|1\rangle$ by $-2\hbar \Delta'$, the negative sign resulting from the fact that the energy defect in the intermediate state now equals $\hbar(\omega_i - \omega_0)$, instead of $\hbar(\omega_0 - \omega_i)$ and the factor 2 coming from $|\langle 2|b^+|1\rangle|^2$. Finally, the global shift of the level $|1\rangle$ equals $+\hbar \Delta' - 2\hbar \Delta' = -\hbar \Delta'$. The same results holds for the level $|n\rangle$, which is shifted by $n\hbar \Delta' - (n + 1)\hbar \Delta' = -\hbar \Delta'$. Thus, it is clear that the absorption and stimulated emission processes shift all the levels of $\mathscr{A}$ by the same amount and therefore do not change the frequency of the oscillator. This is the reason why Δ' does not appear in (14).

b) EVOLUTION OF A FEW AVERAGE VALUES

First we calculate $d\langle b\rangle/dt$, which allows us to study how the mean position and the mean velocity of $\mathscr{A}$ are damped. Because $\langle b\rangle = \mathrm{Tr}(\sigma b)$

$$\frac{d}{dt}\langle b\rangle = \frac{d}{dt}\mathrm{Tr}(\sigma b) = \mathrm{Tr}\left(\frac{d\sigma}{dt}b\right). \qquad (17)$$

We then substitute (14) into the last term of (17). By using the invariance of a trace in a circular permutation and the commutator $[b, b^+] = 1$, we easily obtain

$$\frac{d}{dt}\langle b\rangle = -\left[i(\omega_0 + \Delta) + \frac{\Gamma}{2}\right]\langle b\rangle. \qquad (18)$$

Under the influence of the coupling of $\mathscr{A}$ with $\mathscr{R}$, the oscillation of $\langle b\rangle$ thus undergoes a frequency shift Δ and is damped with a rate $\Gamma/2$.

The result (18) calls for two comments. First, the damping rate of the oscillator does not depend on its initial excitation, contrary to what Figure 1 might suggest, by showing transition rates between levels of $\mathscr{A}$ that increase with the energy of these levels. Second, the damping rate depends only on Γ and not on Γ'. To attempt to understand these two points, we write the evolution equation of the coherence $\sigma_{n+1\,n}$ involved in $\langle b\rangle = \sum_n\sqrt{n+1}\,\sigma_{n+1\,n}$. Projecting (14) over $\langle n + 1|$ and $|n\rangle$ yields

$$\frac{d}{dt}\sigma_{n+1\,n} = -\frac{1}{2}(2n + 1)\Gamma\sigma_{n+1\,n} -$$

$$-2(n + 1)\Gamma'\sigma_{n+1\,n} - i(\omega_0 + \Delta)\sigma_{n+1\,n} +$$

$$+\sqrt{(n + 1)(n + 2)}\,(\Gamma + \Gamma')\sigma_{n+2\,n+1} + \sqrt{n(n + 1)}\,\Gamma'\sigma_{n\,n-1}. \qquad (19)$$

The first two terms of (19) essentially describe a destruction of the coherence between $|n + 1\rangle$ and $|n\rangle$ due to processes that empty the levels n and $n + 1$. The global coefficient of $\sigma_{n+1\,n}$ is indeed, according to (C.17), the half-sum of all the transition rates that leave from $|n\rangle$ and $|n + 1\rangle$ and which all increase with n. By contrast, the last two terms of (19) describe coherence transfers, supplying $\sigma_{n+1\,n}$ from $\sigma_{n+2\,n+1}$ by spontaneous and stimulated emission, and from $\sigma_{n\,n-1}$ by absorption. These transfer terms exist only because the levels of $\mathscr{A}$ are equidistant: the coherences $\sigma_{n+1\,n}$, $\sigma_{n+2\,n+1}$, and $\sigma_{n\,n-1}$ evolve at the same frequency (see the end of subsection C-2). These coherence transfers result in the fact that $\langle b \rangle$ is damped at a rate independent of Γ' and much lower than the coefficient of $\sigma_{n+1\,n}$ in (19) (*).

Finally, we calculate the evolution of $\langle b^+b \rangle$

$$\frac{d}{dt}\langle b^+b \rangle = \mathrm{Tr}\left(\frac{d\sigma}{dt}b^+b\right). \tag{20}$$

The same calculation method as that leading to (18) yields

$$\frac{d}{dt}\langle b^+b \rangle = -\Gamma\langle b^+b \rangle + \Gamma'. \tag{21}$$

The spontaneous emission processes thus damp $\langle b^+b \rangle$ with a rate Γ, whereas the absorption and stimulated emission processes give rise to the source term Γ'. In the steady state

$$\langle b^+b \rangle_{\mathrm{st}} = \frac{\Gamma'}{\Gamma}. \tag{22}$$

If the reservoir $\mathscr{R}$ is in thermodynamic equilibrium, we can use (15) to show that

$$\langle b^+b \rangle_{\mathrm{st}} = \langle n(\omega_0) \rangle = \frac{1}{e^{\hbar\omega_0/k_BT} - 1} \tag{23}$$

which is indeed the value corresponding to the oscillator $\mathscr{A}$ at thermodynamic equilibrium.

4. Master Equation in a Coherent State Basis

In classic statistical mechanics, the state of the oscillator $\mathscr{A}$ is defined by a given probability distribution in phase space. In quantum mechanics, the uncertainty relations between x and p prevent us from defining a state corresponding to a point in phase space. We can, however, replace these points by minimum uncertainty states, such as coherent states, and expand σ in the form of a sum of projectors over such states. The evolution equation of the quasi-probability describing the expansion of σ

(*) Such a result derived from the master equation is reminiscent of the result in Exercise 15. The equidistance between the energy levels of the harmonic oscillator gives rise to interferences between the different amplitudes associated with a radiative cascade and results in the fact that the spectral distribution of the emitted radiation is independent of the initial excitation of the oscillator.

over such a basis has the simple form of a Fokker–Planck equation which can be integrated exactly.

a) BRIEF REVIEW OF COHERENT STATES AND THE REPRESENTATION P_N OF THE DENSITY OPERATOR (*)

The coherent states of the oscillator $\mathscr{A}$ are the eigenstates of the annihilation operator b, having eigenvalue β

$$b|\beta\rangle = \beta|\beta\rangle \tag{24.a}$$

$$\langle\beta|b^+ = \beta^*\langle\beta|. \tag{24.b}$$

They can also be written in the form

$$|\beta\rangle = e^{-\beta\beta^*/2} e^{\beta b^+}|0\rangle \tag{25}$$

or equivalently

$$|\beta\rangle = e^{-\beta\beta^*/2} \sum_{n=0}^{\infty} \frac{\beta^n}{\sqrt{n!}}|n\rangle. \tag{26}$$

A representation P_N (also known as a "Glauber representation") exists for the density operator σ, if it can be written in the form

$$\sigma = \int d^2\beta \, P_N(\beta, \beta^*)|\beta\rangle\langle\beta| \tag{27}$$

where $d^2\beta = d\,\mathrm{Re}\,\beta\,d\,\mathrm{Im}\,\beta$. The advantage of the function $P_N(\beta, \beta^*)$ is that it closely resembles, in certain ways, a classical probability distribution. The hermiticity of σ ($\sigma = \sigma^+$) and its normalization ($\mathrm{Tr}\,\sigma = 1$) result in the fact that $P_N(\beta, \beta^*)$ is a real and normalized function. Also, it is easy to show that the mean value in the state σ of any function of b and b^+ arranged in the normal order (i.e., with all the b^+ to the left of the b) is thus quite simply expressed as a function of $P_N(\beta, \beta^*)$ (the index N in P_N refers to normal order):

$$\mathrm{Tr}\,\sigma(b^+)^l(b)^m = \int d^2\beta(\beta^*)^l(\beta)^m P_N(\beta, \beta^*). \tag{28}$$

The function $P_N(\beta, \beta^*)$ can, however, take negative values and relations as simple as (28) are no longer valid for orders other than the normal order. For this reason $P_N(\beta, \beta^*)$ is actually a quasi-probability distribution and not a probability distribution.

(*) For more details, see, for example, *Photons and Atoms—Introduction to Quantum Electrodynamics*, §C-4 in Chapter III and Exercises 5 and 6 of Complement D_{III}.

Remark

Other quasi-probability distributions are also used. For example, the distribution $P_A(\beta, \beta^*)$ gives expressions as simple as (28) for the mean values of products of operators $b^m(b^+)^l$ arranged in antinormal order. Likewise, the Wigner distribution $W(\beta_1, \beta_2)$ is adapted to the calculation of mean values of symmetrized products of Hermitian operators $b_1 = (b + b^+)/\sqrt{2}$ and $b_2 = i(b^+ - b)/\sqrt{2}$. It can be shown (*) that P_A and W can be expressed as a function of P_N:

$$P_A(\beta, \beta^*) = = \frac{1}{\pi} \int d^2\gamma \, P_N(\gamma, \gamma^*) \exp(-|\beta - \gamma|^2), \tag{29}$$

$$W(\beta_1, \beta_2) = \frac{2}{\pi} \int d^2\gamma \, P_N(\gamma, \gamma^*) \exp(-2|\gamma - \beta_1 - i\beta_2|^2). \tag{30}$$

b) Evolution Equation for $P_N(\beta, \beta^*, t)$

When expression (27) is put into the master Equation (14), terms of the form $b|\beta\rangle\langle\beta|$ and $|\beta\rangle\langle\beta|b^+$ appear, which, according to (24), equal $\beta|\beta\rangle\langle\beta|$ and $\beta^*|\beta\rangle\langle\beta|$, respectively. Other terms, of the form $b^+|\beta\rangle\langle\beta|$ and $|\beta\rangle\langle\beta|b$ also appear, and we now explain how to calculate them. We consider β and β^* to be independent variables (instead of Re β and Im β) and take the derivative with respect to β or β^* of the expression

$$|\beta\rangle\langle\beta| = e^{-\beta\beta^*} e^{\beta b^+} |0\rangle\langle 0| e^{\beta^* b} \tag{31}$$

which follows from (25). We then have

$$\begin{cases} \dfrac{\partial}{\partial\beta} |\beta\rangle\langle\beta| = -\beta^*|\beta\rangle\langle\beta| + b^+|\beta\rangle\langle\beta| \\[2mm] \dfrac{\partial}{\partial\beta^*} |\beta\rangle\langle\beta| = -\beta|\beta\rangle\langle\beta| + |\beta\rangle\langle\beta|b \end{cases} \tag{32}$$

which yields

$$b^+|\beta\rangle\langle\beta| = \left(\beta^* + \frac{\partial}{\partial\beta}\right)|\beta\rangle\langle\beta|$$

$$|\beta\rangle\langle\beta|b = \left(\beta + \frac{\partial}{\partial\beta^*}\right)|\beta\rangle\langle\beta|. \tag{33}$$

(*) See, for example, H. Haken, *Rev. Mod. Phys.* **47**, 67 (1975), in particular page 116, and the references therein cited.

It is possible to transform each of the terms obtained after substituting (27) into (14). Consider, for example, the terms in $b^+b\sigma$ of (14). They can be written, using (24) and (33)

$$b^+b\sigma = \int d^2\beta \, P_N(\beta, \beta^*) b^+ b |\beta\rangle\langle\beta|$$

$$= \int d^2\beta \, P_N(\beta, \beta^*) \beta b^+ |\beta\rangle\langle\beta|$$

$$= \int d^2\beta \, \beta P_N(\beta, \beta^*) \left(\beta^* + \frac{\partial}{\partial\beta}\right) |\beta\rangle\langle\beta|. \tag{34}$$

Integrating the term in $\partial/\partial\beta$ by parts and assuming that $P_N(\beta, \beta^*)$ vanishes at infinity, we then obtain

$$b^+b\sigma = \int d^2\beta \left\{ \beta^*\beta P_N(\beta, \beta^*) - \frac{\partial}{\partial\beta}[\beta P_N(\beta, \beta^*)] \right\} |\beta\rangle\langle\beta|. \tag{35}$$

Analogous calculations can be made for all the other terms of (14). By setting the coefficient of $|\beta\rangle\langle\beta|$ equal to zero in the integral over $d^2\beta$, we obtain

$$\frac{\partial}{\partial t} P_N(\beta, \beta^*, t) = \left(\frac{\Gamma}{2} + i\tilde{\omega}_0\right) \frac{\partial}{\partial\beta}[\beta P_N(\beta, \beta^*, t)] +$$

$$+ \left(\frac{\Gamma}{2} - i\tilde{\omega}_0\right) \frac{\partial}{\partial\beta^*}[\beta^* P_N(\beta, \beta^*, t)] +$$

$$+ \Gamma' \frac{\partial^2}{\partial\beta \, \partial\beta^*} P_N(\beta, \beta^*, t) \tag{36}$$

where we have set $\tilde{\omega}_0 = \omega_0 + \Delta$.

c) PHYSICAL DISCUSSION

The first two terms of (36), in $\partial/\partial\beta$ and $\partial/\partial\beta^*$, are drift terms describing how the quasi-probability distribution drifts over time. The last term, in $\partial^2/\partial\beta \, \partial\beta^*$, is a diffusion term describing the broadening of the distribution.

Equations of the same type as (36) are known as Fokker–Planck equations and are used in a very general way to give an approximate description of the evolution of certain probability distributions [see, for example, Equation (E.53)]. It should be noted that, for the simple model described here, the Fokker–Planck equation was derived without any approximation from the master equation.

One can check by substitution that the function

$$\frac{1}{\pi \langle n(\omega_0) \rangle (1 - e^{-\Gamma t})} \times$$

$$\times \exp - \frac{\left[\beta - \beta_0\, e^{-(\Gamma/2 + i\bar{\omega}_0)t} \right]\left[\beta^* - \beta_0^*\, e^{-(\Gamma/2 - i\bar{\omega}_0)t} \right]}{\langle n(\omega_0) \rangle (1 - e^{-\Gamma t})} \tag{37}$$

which reduces to $\delta(\beta - \beta_0)\delta(\beta^* - \beta_0^*)$ for $t = 0$, is a solution of (36) and is therefore the Green's function for this equation. It is a Gaussian centered at a point in the complex plane that executes a spiral motion having angular velocity $\bar{\omega}_0$ and a radius that decreases exponentially with time. The width of the Gaussian increases with time, from the value 0 for $t = 0$ to the value $\sqrt{\langle n(\omega_0) \rangle}$ for $t = \infty$. For $t \to \infty$, the expression (37) gives the representation P_N of the equilibrium state

$$P_N^{\text{eq}}(\beta, \beta^*) = \frac{1}{\pi \langle n(\omega_0) \rangle} \exp\left[-\frac{\beta^* \beta}{\langle n(\omega_0) \rangle} \right] \tag{38}$$

$\langle n(\omega_0) \rangle$ being given by (E.17) if the reservoir is in thermodynamic equilibrium.

Remark

Note a remarkable property of $P_N(\beta, \beta^*)$: If $\langle n(\omega_0) \rangle = 0$, i.e., if the reservoir is in the ground state $|0\rangle$, then the Green's function (37) remains a delta function of zero width regardless of t. Such a result expresses that, if a matter oscillator is initially in a coherent state (for which the representation P_N is a delta function), without any incident photons, it continues to remain in a coherent state when it decays by spontaneous emission.

REFERENCES

W. H. Louisell and J. H. Marburger, *IEEE J. Quantum Electron.*, **QE-3**, 348 (1967), and references therein. Louisell, Chapter VI. Glauber, §§8 and 9.

COMPLEMENT C_{IV}

QUANTUM LANGEVIN EQUATIONS FOR A SIMPLE PHYSICAL SYSTEM

The master equation derived in this chapter describes, in the Schrödinger representation, how a small system $\mathscr{A}$ evolves under the influence of its interaction with a large reservoir $\mathscr{R}$. The purpose of this Complement is to approach the same problem in the *Heisenberg* representation. To simplify the calculations as much as possible, we limit ourselves to the simple model, introduced in Complement B_{IV}, of a harmonic oscillator $\mathscr{A}$ coupled to a reservoir $\mathscr{R}$ of harmonic oscillators. Our purpose is twofold: first, to show that the Heisenberg equations for the observables of $\mathscr{A}$ can be put in a form similar to that of the Langevin equations for classical Brownian motion; then to use these *Heisenberg–Langevin* equations to analyze the relations that exist between *fluctuations and dissipation* and to calculate *correlation functions* for observables of $\mathscr{A}$.

We begin (§1) by reviewing the essentials of the classical theory of Brownian motion and by analyzing the Langevin equation used to describe such motion. We then derive (§2) the quantum Heisenberg–Langevin equations relative to the simple model of Complement B_{IV} and discuss for these equations the fluctuation-dissipation theorem as well as the quantum regression theorem relative to the correlation functions for $\mathscr{A}$.

1. Review of the Classical Theory of Brownian Motion

a) Langevin Equation

A large particle having mass M and momentum **p** is immersed in a homogeneous fluid of lighter particles with which it undergoes continuous collisions that make its motion erratic. The equation introduced by Langevin to phenomenologically describe such motion is written (for a component p of **p**):

$$\frac{d}{dt}p(t) = -\gamma p(t) + F(t). \tag{1}$$

The total force acting on the particle is separated into two parts. The first, $-\gamma p(t)$, describes the cumulative effect of collisions, which is to damp the momentum of the particle with a friction coefficient γ. The second is a fluctuating force $F(t)$, called the Langevin force, which describes the fluctuations of the instantaneous force around its average value. In Equation (1), $F(t)$ is considered to be an external force, independent of $p(t)$, described by a random stationary function of t, satisfying the following properties:

(i) The average value of $F(t)$ is zero:

$$\overline{F(t)} = 0. \tag{2.a}$$

(ii) The correlation function for the Langevin force, which describes the dynamics of the fluctuations of $F(t)$, has the form

$$\overline{F(t)F(t')} = 2Dg(t - t') \tag{2.b}$$

where D is a coefficient (that will be interpreted later on) and $g(t - t')$ is an even function of $t - t'$, of width τ_c, whose integral over $t - t'$ equals 1. The properties of parity and of invariance by translation in time of g result from the fact that $\overline{F(t)F(t')}$ is a classical stationary autocorrelation function. The correlation time τ_c of $F(t)$ is on the order of the collision time between the Brownian particle and a particle in the fluid. It is much shorter than the damping time $T_R = \gamma^{-1}$ of $\overline{p(t)}$, because several collisions are necessary to appreciably change the momentum of the heavy particle

$$\tau_c \ll T_R = \gamma^{-1}. \tag{3}$$

With regard to the functions of $t - t'$ varying slowly on the scale of τ_c, $g(t - t')$ thus appears as a delta function, so that the correlation function (2.b) can be approximated by

$$\overline{F(t)F(t')} \simeq 2D\delta(t - t'). \tag{4}$$

b) INTERPRETATION OF THE COEFFICIENT D. CONNECTION
   BETWEEN FLUCTUATIONS AND DISSIPATION

We now study the evolution of $p(t)$ starting from a well-defined initial value $p(t_0) = p_0$. Integrating (1) gives

$$p(t) = p_0 e^{-\gamma(t - t_0)} + \int_{t_0}^{t} dt' \, F(t') e^{-\gamma(t - t')}. \tag{5}$$

Using (2.a), we then obtain for $\overline{p(t)}$

$$\overline{p(t)} = p_0 \, e^{-\gamma(t-t_0)}. \tag{6}$$

The average momentum therefore is damped with a rate γ. We now calculate the evolution of the variance $\sigma_p^2(t)$ of $p(t)$:

$$\sigma_p^2(t) = \overline{[p(t)]^2} - [\overline{p(t)}]^2 = \overline{[p(t) - \overline{p(t)}]^2}. \tag{7}$$

Using (5) and (6)

$$p(t) - \overline{p(t)} = \int_{t_0}^{t} dt' \, F(t') \, e^{-\gamma(t-t')} \tag{8}$$

so that

$$\sigma_p^2(t) = \int_{t_0}^{t} dt' \int_{t_0}^{t} dt'' \, \overline{F(t')F(t'')} \, e^{-\gamma(t-t')} e^{-\gamma(t-t'')}. \tag{9}$$

The two exponentials of (9) vary slowly on the scale of τ_c. Using (4) then gives (for $t - t_0 \gg \tau_c$)

$$\sigma_p^2(t) = 2D \int_{t_0}^{t} dt' \, e^{-2\gamma(t-t')} = \frac{D}{\gamma} [1 - e^{-2\gamma(t-t_0)}]. \tag{10}$$

Equation (10) shows that $\sigma_p^2(t)$, i.e., the dispersion of the possible values of p, begins by increasing linearly with $t - t_0$:

$$\sigma_p^2(t) \simeq 2D(t - t_0) \qquad \text{for } \tau_c \ll t - t_0 \ll \gamma^{-1}. \tag{11}$$

D then appears as a *momentum diffusion coefficient*.

At long times ($t - t_0 \gg \gamma^{-1}$), Equation (10) shows that $\sigma_p^2(t)$ tends to an equilibrium value equal to D/γ, whereas $\overline{p(t)}$ tends to zero according to (6). It follows from (7) that the equilibrium value of $\overline{p^2}$ is

$$\overline{p^2} = \frac{D}{\gamma}. \tag{12}$$

If the fluid is in thermodynamic equilibrium, this equilibrium value must satisfy

$$\frac{\overline{p^2}}{2M} = \frac{1}{2}k_BT.$$ (13)

Substituting (12) into (13) and solving for D then gives the Einstein equation

$$D = Mk_BT\gamma$$ (14)

establishing a relationship between the diffusion coefficient D characterizing the *fluctuations* of the Langevin force acting on p and the friction coefficient γ characterizing the *dissipation* force acting on this same variable p. Recall that, according to (11), D also appears as the diffusion coefficient of p.

Remark

Let

$$\delta p(t) = p(t_0 + \delta t) - p(t_0) = p(t_0 + \delta t) - p_0$$ (15)

be the increase of $p(t)$ between t_0 and $t_0 + \delta t$ with $\tau_c \ll \delta t \ll \gamma^{-1}$. Equations (6) and (11) then give

$$\overline{\delta p(t)} = p_0 e^{-\gamma\delta t} - p_0 \simeq -p_0\gamma\,\delta t$$ (16)

$$\overline{\left[\delta p(t) - \overline{\delta p(t)}\right]^2} = \overline{\left[\delta p(t)\right]^2} - \left[\overline{\delta p(t)}\right]^2 = 2D\,\delta t.$$ (17)

Because $[\overline{\delta p(t)}]^2$ varies as $p_0^2\gamma^2\delta t^2$ according to (16) and is thus negligible as compared to $D\,\delta t$, we can write (17) in the form

$$\overline{\left[\delta p(t)\right]^2} = 2D\,\delta t.$$ (18)

Equations (16) and (18) give the first- and second-order moments of the increase δp of p over a short time.

c) A Few Correlation Functions

We make t_0 tend to $-\infty$ in (5). The random function $p(t)$ has then lost any memory of the initial conditions and can be written at time t':

$$p(t') = \int_{-\infty}^{t'} dt''\, F(t'')\, e^{-\gamma(t'-t'')}.$$ (19)

α) Correlation between the Momentum and the Langevin Force

Multiply both sides of (19) by $F(t)$ and take the average value. This gives

$$\overline{p(t')F(t)} = \int_{-\infty}^{t'} dt'' \, \overline{F(t'')F(t)} \, e^{-\gamma(t'-t'')}. \tag{20}$$

If t is in the future of t' and if $t - t' \gg \tau_c$, we have in (20) $t - t'' \gg \tau_c$ (because $t'' < t'$), so that $\overline{F(t'')F(t)}$ is zero:

$$\overline{p(t')F(t)} = 0 \qquad \text{if } t - t' \gg \tau_c. \tag{21}$$

Equation (21) shows that $p(t')$, which depends on the Langevin force in the past of t', cannot be correlated with the Langevin force $F(t)$ at a time t in the future of t' if $t - t' \gg \tau_c$.

If t is in the past of t' and if $t' - t \gg \tau_c$, we can use (4) and Equation (20) then gives:

$$\overline{p(t')F(t)} = 2D \, e^{-\gamma(t'-t)} \qquad \text{if } t' - t \gg \tau_c. \tag{22}$$

Finally, for t close to $t'(|t - t'| \lesssim \tau_c)$, $\overline{p(t')F(t)}$ varies rapidly between $2D$ and 0 over an interval of width τ_c, and takes a value equal to D for $t = t'$ [this results from Equation (20) written for $t' = t$ and from the fact that $\overline{F(t'')F(t)}$ is even].

All these results are summarized in Figure 1, which shows the variations with t of $\overline{p(t')F(t)}$. The important point, that we will use later on,

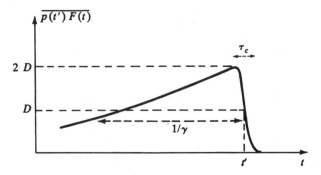

Figure 1. Variations with t of $\overline{p(t')F(t)}$.

is that $p(t')$ is decorrelated from $F(t)$ if $t \geq t'$, except in a small interval of width τ_c near $t = t'$, where $\overline{p(t')F(t)}$ remains finite and on the order of D.

β) Momentum Autocorrelation Function

Multiply both sides of (1) by $p(t')$ and take the average value. This gives

$$\frac{d}{dt} \overline{p(t)p(t')} = -\gamma\overline{p(t)p(t')} + \overline{F(t)p(t')} \tag{23}$$

with the initial condition $\overline{p(t = t')p(t')} = \overline{p^2}$. We will now show that the second term of the right-hand side of (23) can be ignored for $t \geq t'$. Replace t by t'' in (23) and integrate over t'' from t' to t. The contribution of $\overline{F(t'')p(t')}$ to $\overline{p(t)p(t')}$ is equal to $\int_{t'}^{t} dt'' \ \overline{F(t'')p(t')}$ and remains bounded by $D\tau_c$ according to Figure 1. Using (3), this contribution is very small compared with $D\gamma^{-1}$, which is simply the initial value $\overline{p^2}$ of $\overline{p(t)p(t')}$ according to (12). One can thus ignore it and consider that, for $t \geq t'$, $\overline{p(t)p(t')}$ obeys

$$t \geq t' \qquad \frac{d}{dt} \overline{p(t)p(t')} = -\gamma\overline{p(t)p(t')}. \tag{24}$$

It thus appears that $\overline{p(t)p(t')}$ decreases exponentially with $t - t'$ starting from an initial value given by (12). Because $\overline{p(t)p(t')}$ is an autocorrelation function and is thus even in $t - t'$, we finally have

$$\overline{p(t)p(t')} \simeq \overline{p^2} e^{-\gamma|t-t'|}. \tag{25}$$

Equation (25) shows that the correlations between $p(t)$ and $p(t')$ decrease, i.e., "regress" in the same way as the average value $\overline{p(t)}$, which obeys the equation

$$\frac{d}{dt} \overline{p(t)} = -\gamma\overline{p(t)} \tag{26}$$

analogous to (24).

We could have obtained expression (25) directly from (19) and (4). The advantage of the derivation presented here is that it can be generalized in quantum theory, leading to the "quantum regression" theorem (see §2-g).

Remark

The effect of the correction introduced by the last term in (23) is to round off the sharp point of (25) in $t = t'$ and to make $\overline{p(t)p(t')}$ start with a horizontal tangent at this point. To derive such a result, we consider the Fourier transforms $\tilde{p}(\omega)$ and $\tilde{F}(\omega)$ of $p(t)$ and $F(t)$. Equation (1) leads to

$$| \tilde{p}(\omega)|^2 = \frac{|\tilde{F}(\omega)|^2}{\gamma^2 + \omega^2}. \tag{27}$$

We can then use the Wiener–Khinchine theorem, which states that the Fourier transforms of $|\tilde{p}(\omega)|^2$ and $|\tilde{F}(\omega)|^2$ are (except for multiplicative coefficients) the autocorrelation functions $\overline{p(t)p(t')}$ and $\overline{F(t)F(t')}$. According to (27), $\overline{p(t)p(t')}$ is thus the convolution product of the Fourier transform of $(\gamma^2 + \omega^2)^{-1}$, i.e., $e^{-\gamma|t-t'|}$, by the Fourier transform of $|\tilde{F}(\omega)|^2$, i.e., $g(t - t')$. Because the width τ_c of $g(t - t')$ is much smaller than that, γ^{-1}, of $\exp(-\gamma|t - t'|)$, we again find that $\overline{p(t)p(t')}$ is nearly equal to $\exp(-\gamma|t - t'|)$, apart from small corrections in τ_c/T_R which round off the sharp point at $t = t'$ over an interval of width τ_c.

2. Heisenberg–Langevin Equations for a Damped Harmonic Oscillator

We now return to the model, introduced in Complement B_{IV}, of a harmonic oscillator $\mathscr{A}$ (of frequency ω_0, with creation and annihilation operators b^+ and b) coupled to a reservoir of oscillators i (of frequency ω_i, with creation and annihilation operators a_i^+ and a_i). The total Hamiltonian H can be expressed as

$$H = H_A + H_R + V$$

$$= \hbar\omega_0\left(b^+b + \frac{1}{2}\right) + \sum_i \hbar\omega_i\left(a_i^+a_i + \tfrac{1}{2}\right) + \sum_i \left(g_i b^+ a_i + g_i^* b a_i^+\right) \tag{28}$$

where g_i is a coupling constant.

a) COUPLED HEISENBERG EQUATIONS

The Heisenberg equation for $b(t)$ is written

$$i\hbar\frac{d}{dt}b(t) = [b(t), H] = \hbar\omega_0 b(t) + \sum_i g_i a_i(t). \tag{29}$$

It depends on the operators $a_i(t)$ of the reservoir whose evolution equation has a similar form:

$$i\hbar\frac{d}{dt}a_i(t) = [a_i(t), H] = \hbar\omega_i a_i(t) + g_i^* b(t). \tag{30}$$

The evolutions of the annihilation operators $b(t)$ and $a_i(t)$ are thus coupled to each other by the linear equations (29) and (30). Note that the simplicity of these equations results from the bilinear form in b or b^+ and a_i or a_i^+ that we have chosen for the interaction V and from the fact that the commutators $[b, b^+]$ and $[a_i, a_i^+]$ simply equal 1. In Complement A_V we will consider the case where the oscillator $\mathscr{A}$ is replaced by a two-level system. The last term of the equation analogous to (29) is then nonlinear because, in this case, it depends not only on the $a_i(t)$, but also on the operators of $\mathscr{A}$.

It will be convenient for what follows to set:

$$b(t) = \hat{b}(t) e^{-i\omega_0 t} \tag{31.a}$$

$$a_i(t) = \hat{a}_i(t) e^{-i\omega_i t} \tag{31.b}$$

so that $\hat{b}$ and $\hat{a}_i$ evolve only under the influence of V. Equations (29) and (30) then become

$$i\hbar\frac{d}{dt}\hat{b}(t) = \sum_i g_i \hat{a}_i(t) e^{i(\omega_0-\omega_i)t} \tag{32}$$

$$i\hbar\frac{d}{dt}\hat{a}_i(t) = g_i^* \hat{b}(t) e^{i(\omega_i-\omega_0)t}. \tag{33}$$

b) THE QUANTUM LANGEVIN EQUATION AND QUANTUM LANGEVIN FORCES

We integrate Equation (33) to get

$$\hat{a}_i(t) = \hat{a}_i(t_0) - \frac{i}{\hbar}g_i^*\int_{t_0}^t dt'\, \hat{b}(t')\, e^{i(\omega_i-\omega_0)t'} \tag{34}$$

and insert the result obtained into (32). This gives

$$\frac{d}{dt}\hat{b}(t) = -\int_0^{t-t_0} d\tau\, \kappa(\tau)\hat{b}(t-\tau) + \hat{F}(t) \tag{35}$$

where

$$\kappa(\tau) = \frac{1}{\hbar^2} \sum_i |g_i|^2 \, e^{i(\omega_0 - \omega_i)\tau} \tag{36}$$

$$\hat{F}(t) = -\frac{i}{\hbar} \sum_i g_i \hat{a}_i(t_0) \, e^{i(\omega_0 - \omega_i)t}. \tag{37}$$

First consider the function $\kappa(t)$ given in (36). Since $\mathscr{R}$ is a reservoir, the frequencies ω_i of the oscillators cover a very wide range. Besides, $|g_i|^2$ generally varies slowly with ω_i. It follows that the set of oscillating exponentials of (36) interfere destructively when τ increases starting from 0, so that $\kappa(\tau)$ becomes negligible as soon as $\tau \gg \tau_c$, where τ_c is the correlation time of the reservoir. In addition, $\hat{b}(t - \tau)$ varies much more slowly with τ, over a time scale $T_R \gg \tau_c$, where T_R is the damping time of $\mathscr{A}$. We can then neglect the variation with τ of $\hat{b}(t - \tau)$ as compared to that of $\kappa(\tau)$ in the integral of (35), and replace $\hat{b}(t - \tau)$ by $\hat{b}(t)$ which may then be removed from the integral. Assuming $t - t_0 \gg \tau_c$, we get in this way

$$\int_0^\infty \kappa(\tau) \, d\tau = \frac{1}{\hbar^2} \lim_{\eta \to 0_+} \sum_i |g_i|^2 \int_0^\infty e^{i(\omega_0 - \omega_i + i\eta)\tau} \, d\tau$$

$$= \frac{1}{\hbar^2} \sum_i |g_i|^2 \left\{ \pi\delta(\omega_0 - \omega_i) + i\mathscr{P} \frac{1}{\omega_0 - \omega_i} \right\}$$

$$= \frac{\Gamma}{2} + i\Delta \tag{38}$$

where Γ and $\hbar\Delta$ are the parameters that were introduced in expressions (11.a) and (12.a) in Complement B_{IV} and which describe respectively the spontaneous emission rate and the spontaneous radiative shift of the oscillator. Equation (35) can then be rewritten:

$$\frac{d}{dt}\hat{b}(t) = -\left(\frac{\Gamma}{2} + i\Delta\right)\hat{b}(t) + \hat{F}(t). \tag{39}$$

To show that Equation (39) can be considered as a Langevin equation, we must now study the properties of the operator $\hat{F}(t)$. According to (37), $\hat{F}(t)$ is an operator of the reservoir. At the initial time t_0, where both the Schrödinger and Heisenberg representations coincide, we assume that the density operator of the global system is factored in the form $\sigma_A \otimes \sigma_R$

where σ_R is given by the statistical mixture of the states $|n_1 n_2 \cdots n_i \cdots \rangle$ with weights $p(n_1 n_2 \cdots n_i \cdots)$ [see formula (E.8)]. Because $a_i(t_0)$ has no diagonal elements in the state $|n_i\rangle$, it follows that

$$\langle \hat{F}(t) \rangle = \text{Tr}\, \sigma_A \sigma_R F(t) = 0. \tag{40}$$

The average value of Equation (39) gives

$$\frac{d}{dt}\langle \hat{b}(t) \rangle = -\left(\frac{\Gamma}{2} + i\Delta\right)\langle \hat{b}(t) \rangle \tag{41}$$

which coincides with Equation (18) of Complement B_{IV} [when we return from $\hat{b}(t)$ to $b(t)$ by using (31.a)]. We now calculate the correlation functions of $\hat{F}(t)$ and $\hat{F}^+(t)$. Because only products such as $a_i^+(t_0)a_i(t_0)$ and $a_i(t_0)a_i^+(t_0)$ have nonzero average values (respectively equal to $\langle n_i\rangle$ and $\langle n_i\rangle + 1$) in the state (E.8) of the reservoir, we get

$$\langle \hat{F}(t')\hat{F}(t)\rangle = \langle \hat{F}^+(t')\hat{F}^+(t)\rangle = 0 \tag{42}$$

$$\langle \hat{F}^+(t')\hat{F}(t)\rangle = \sum_i \frac{1}{\hbar^2}|g_i|^2\langle n_i\rangle\, e^{i(\omega_0 - \omega_i)(t - t')} \tag{43}$$

$$\langle \hat{F}(t)\hat{F}^+(t')\rangle = \sum_i \frac{1}{\hbar^2}|g_i|^2(\langle n_i\rangle + 1)\, e^{i(\omega_0 - \omega_i)(t - t')}. \tag{44}$$

As above, the oscillating exponentials appearing in (43) and (44) interfere destructively as soon as $|t - t'| \gg \tau_c$. The two-time averages (43) and (44) are thus very narrow functions of $\tau = t - t'$. We call $2D_N$ and $2D_A$ the integrals over τ of these functions between $-\infty$ and $+\infty$:

$$2D_N = \int_{-\infty}^{+\infty} d\tau \langle \hat{F}^+(t - \tau)F(t)\rangle \tag{45}$$

$$2D_A = \int_{-\infty}^{+\infty} d\tau \langle \hat{F}(t)\hat{F}^+(t - \tau)\rangle \tag{46}$$

(the subscripts N and A indicate the normal or antinormal order of $\hat{F}^+$ and $\hat{F}$). Using (43), (44), and formulas (11.a) and (11.b) from Complement

B_{IV}, these integrals equal

$$2D_N = \frac{1}{\hbar^2} \int_{-\infty}^{+\infty} d\tau \sum_i |g_i|^2 \langle n_i \rangle \, e^{i(\omega_0 - \omega_i)\tau} = \Gamma' \tag{47}$$

$$2D_A = \frac{1}{\hbar^2} \int_{-\infty}^{+\infty} d\tau \sum_i |g_i|^2 (\langle n_i \rangle + 1) \, e^{i(\omega_0 - \omega_i)\tau} = \Gamma' + \Gamma. \tag{48}$$

We can thus rewrite (43) and (44) in the form

$$\langle \hat{F}^+(t)\hat{F}(t') \rangle = 2D_N g_N(t - t') \tag{49}$$

$$\langle \hat{F}(t)\hat{F}^+(t') \rangle = 2D_A g_A(t - t') \tag{50}$$

where $g_N(\tau)$ and $g_A(\tau)$ are two normalized functions of τ, of width τ_c (*).

Finally, if we assume that $\langle n_i \rangle$ depends only on ω_i, we can use relation (15) from Complement B_{IV}, with $\Gamma' = \Gamma \langle n(\omega_0) \rangle$ [where $\langle n(\omega_0) \rangle$ is the average number of quanta of the reservoir modes having the same frequency ω_0 as $\mathscr{A}$], to rewrite (47) and (48) in the form

$$2D_N = \Gamma \langle n(\omega_0) \rangle \tag{51}$$

$$2D_A = \Gamma (1 + \langle n(\omega_0) \rangle). \tag{52}$$

The operators $\hat{F}$ and $\hat{F}^+$ can therefore be considered to be Langevin forces fluctuating very rapidly around a zero average value. However, it should be noted that these operators do not commute with each other, as shown by the difference between (51) and (52).

c) CONNECTION BETWEEN FLUCTUATIONS AND DISSIPATION

Equations (51) and (52) establish a quantitative relationship between the fluctuations of $\hat{F}$ and $\hat{F}^+$, characterized by D_N and D_A, and the damping Γ characterizing the dissipation associated with the motions of b and b^+. If the reservoir is in thermodynamic equilibrium, $\langle n(\omega_0) \rangle$ is equal to $[\exp(\hbar\omega_0/k_B T) - 1]^{-1}$ and expression (51), for example, becomes

$$2D_N = \frac{\Gamma}{e^{\hbar\omega_0/k_B T} - 1}, \tag{53}$$

(*) Functions g_N and g_A do not have a well-defined parity in τ. With regard to the functions of τ varying over time scales $T_R \gg \tau_c$, they can be considered to be a sum of functions $\delta, \delta', \delta'' \cdots$. At the lowest order in τ_c/T_R, they can be replaced by a $\delta(\tau)$ function.

an equation that can be considered as an expression for a quantum fluctuation-dissipation theorem. In contrast to what we did in subsection 1-b above, relation (53) is derived from Heisenberg equations of motion rather than from a phenomenological equation. It is also valid whatever the relative values of $\hbar\omega_0$ and $k_B T$. In particular, if $\hbar\omega_0 \ll k_B T$, Equation (53) becomes

$$2D_N = \frac{\Gamma k_B T}{\hbar\omega_0} \qquad (54)$$

and, as in (14), establishes a relationship between D and $\Gamma k_B T$.

The relation (53) between D_N and Γ was derived from (51), i.e., from the explicit calculation of the two-time average $\langle \hat{F}^+(t - \tau)\hat{F}(t)\rangle$ of the Langevin forces appearing in expression (45) for $2D_N$. It is also possible to relate the coefficient D_N to the rate of increase of the "crossed" variance of b^+ and b defined as

$$\mathscr{V}_N(t) = \langle \hat{b}^+(t)\hat{b}(t)\rangle - \langle \hat{b}^+(t)\rangle\langle \hat{b}(t)\rangle \qquad (55)$$

with the coefficient D_A being related to the rate of increase of the variance

$$\mathscr{V}_A(t) = \langle \hat{b}(t)\hat{b}^+(t)\rangle - \langle \hat{b}(t)\rangle\langle \hat{b}^+(t)\rangle \qquad (56)$$

corresponding to the antinormal order of b and b^+. Such a result, which will be derived in subsection 2-e below, in a certain sense generalizes the relation (11) demonstrated above for the classical Langevin equation, and shows that the rate of increase of the variance of p is proportional to D.

Before calculating $d\mathscr{V}_N/dt$ and $d\mathscr{V}_A/dt$, we begin by deriving some useful results concerning the two-time averages $\langle \hat{F}^+(t)b(t')\rangle$ or $\langle \hat{b}^+(t')\hat{F}(t)\rangle$.

d) Mixed Two-Time Averages Involving Langevin Forces and
   Operators of $\mathscr{A}$

We integrate Equation (39) between t_0 and t'

$$\hat{b}(t') = \hat{b}(t_0)\, e^{-(\Gamma/2 + i\Delta)(t' - t_0)} + \int_{t_0}^{t'} dt''\, F(t'')\, e^{-(\Gamma/2 + i\Delta)(t' - t'')}. \qquad (57)$$

For $t_0 \to -\infty$, the first term of (57) is negligible and we then obtain, by multiplying both sides of (57) by $\hat{F}^+(t)$:

$$\left\langle \hat{F}^+(t)\hat{b}(t') \right\rangle = \int_{t_0}^{t'} dt'' \left\langle \hat{F}^+(t)\hat{F}(t'') \right\rangle e^{-(\Gamma/2 + i\Delta)(t'-t'')}. \qquad (58)$$

The same reasoning as that used in subsection 1-c-α then gives, for the variations of $\langle \hat{F}^+(t)\hat{b}(t') \rangle$ with t, results analogous to those shown in Figure 1. For $t > t'$, $\hat{F}^+(t)$ is not correlated with $\hat{b}(t')$, except in an interval of width τ_c near $t = t'$. When t increases and crosses the value t', $\langle \hat{F}^+(t)\hat{b}(t') \rangle$ goes from the value $2D_N$ to 0 over an interval of width τ_c. In particular, the value for $t = t'$ is given by

$$\left\langle \hat{F}^+(t)\hat{b}(t) \right\rangle = \int_{t_0}^{t} dt'' \left\langle \hat{F}^+(t)\hat{F}(t'') \right\rangle e^{-(\Gamma/2 + i\Delta)(t-t'')}$$

$$\simeq \int_{t_0}^{t} dt'' \left\langle \hat{F}^+(t)\hat{F}(t'') \right\rangle \qquad (59)$$

because $\langle \hat{F}^+(t)\hat{F}(t'') \rangle$ varies much more rapidly with $t - t''$ than the exponential. In the same way, we find

$$\left\langle \hat{b}^+(t)\hat{F}(t) \right\rangle \simeq \int_{t_0}^{t} dt'' \left\langle \hat{F}^+(t'')\hat{F}(t) \right\rangle. \qquad (60)$$

The change of variable $\tau = t - t''$ and the stationary character of the two-time averages of $\hat{F}^+$ and $\hat{F}$ allow us to recombine (59) and (60) in the form

$$\left\langle \hat{b}^+(t)\hat{F}(t) \right\rangle + \left\langle \hat{F}^+(t)\hat{b}(t) \right\rangle = \int_{-(t-t_0)}^{+(t-t_0)} d\tau \left\langle \hat{F}^+(t-\tau)\hat{F}(t) \right\rangle = 2D_N. \qquad (61)$$

We used (45) and assumed that $t - t_0 \gg \tau_c$.

e) RATE OF VARIATION OF THE VARIANCES $\mathcal{V}_N$ AND $\mathcal{V}_A$

The evolution of the average value $\langle \hat{b}(t) \rangle$ was determined by the relaxation Equation (41). The evolution Equation (39) of the operator itself results in the appearance of the same rate of variation, which we call the "drift" term, written $\mathscr{D}(\hat{b}(t))$:

$$\mathscr{D}(\hat{b}(t)) = -\left(\frac{\Gamma}{2} + i\Delta\right)\hat{b}(t). \tag{62}$$

The rate of variation of $\hat{b}(t)$ is then simply the sum of the drift term and the Langevin force

$$\frac{d}{dt}\hat{b}(t) = \mathscr{D}(\hat{b}(t)) + F(t). \tag{63}$$

Similarly

$$\frac{d}{dt}\hat{b}^+(t) = \mathscr{D}(\hat{b}^+(t)) + F^+(t) \tag{64}$$

with

$$\mathscr{D}(\hat{b}^+(t)) = \left(-\frac{\Gamma}{2} + i\Delta\right)\hat{b}^+(t). \tag{65}$$

We now consider the rate of variation of the operators $\hat{b}^+\hat{b}$ and $\hat{b}\hat{b}^+$ appearing in the variances $\mathscr{V}_N$ and $\mathscr{V}_A$ defined in (55) and (56). For $\hat{b}^+\hat{b}$, we first obtain

$$\frac{d}{dt}(\hat{b}^+(t)\hat{b}(t)) = \left(\frac{d}{dt}\hat{b}^+(t)\right)\hat{b}(t) + \hat{b}^+(t)\left(\frac{d}{dt}\hat{b}(t)\right) =$$

$$= \mathscr{D}(\hat{b}^+(t))\hat{b}(t) + \hat{b}^+(t)\mathscr{D}(\hat{b}(t)) + \hat{F}^+(t)\hat{b}(t) + \hat{b}^+(t)\hat{F}(t). \tag{66}$$

Taking the average value of both sides of (66) and using (61) (*), we get

$$\frac{d}{dt}\langle\hat{b}^+(t)\hat{b}(t)\rangle = \langle\mathscr{D}(\hat{b}^+(t))\hat{b}(t)\rangle + \langle\hat{b}^+(t)\mathscr{D}(\hat{b}(t))\rangle + 2D_N \tag{67}$$

(*) The fact that the last two terms of (66) have, according to (61), a nonzero average value prevents us from considering their sum to be a Langevin force in the equation of motion of the operator $\hat{b}^+(t)\hat{b}(t)$.

i.e., also, using (62) and (65):

$$\frac{d}{dt}\langle \hat{b}^+(t)\hat{b}(t)\rangle = -\Gamma\langle \hat{b}^+(t)\hat{b}(t)\rangle + 2D_N. \tag{68}$$

According to (63) and (64), we have

$$\frac{d}{dt}\langle \hat{b}^+(t)\rangle\langle \hat{b}(t)\rangle = \langle \mathscr{D}(\hat{b}^+(t))\rangle\langle \hat{b}(t)\rangle + \langle \hat{b}(t)\rangle\langle \mathscr{D}(\hat{b}^+(t))\rangle$$

$$= -\Gamma\langle \hat{b}^+(t)\rangle\langle \hat{b}(t)\rangle \tag{69}$$

so that, by subtraction, Equations (68) and (69) yield, using (55)

$$\frac{d}{dt}\mathscr{V}_N(t) = -\Gamma\mathscr{V}_N(t) + 2D_N. \tag{70}$$

Similar calculations yield

$$\frac{d}{dt}\langle \hat{b}(t)\hat{b}^+(t)\rangle = -\Gamma\langle \hat{b}(t)\hat{b}^+(t)\rangle + 2D_A \tag{71}$$

and consequently

$$\frac{d}{dt}\mathscr{V}_A(t) = -\Gamma\mathscr{V}_A(t) + 2D_A \tag{72}$$

The variances thus tend to increase linearly with t under the influence of the Langevin forces, whose effect is represented by the constant terms $2D_N$ and $2D_A$ of (70) and (72). These terms can thus indeed be considered as diffusion coefficients. The damping Γ limits this diffusion motion and the equilibrium values of (70) and (72) again give the relations among the coefficients D, Γ, and the temperature [see (53)].

Remark

The fact that D_N and D_A are different results from the quantum nature of $F(t)$ and $F^+(t)$ and is essential for preserving the commutation relation between b and b^+. By subtracting (68) from (71), we find that a necessary condition for

$\langle[\hat{b}(t), \hat{b}^+(t)]\rangle$ remaining equal to 1 is that

$$D_A - D_N = \frac{\Gamma}{2} \tag{73}$$

a condition that is indeed satisfied by (47) and (48).

f) GENERALIZATION OF EINSTEIN'S RELATION

We return to Equation (67). Because the operator $\hat{b}^+(t)\hat{b}(t)$ is an operator of the small system $\mathcal{A}$, we can associate with it a drift velocity $\mathcal{D}(\hat{b}^+\hat{b})$ whose average value is by definition the rate of variation of $\langle b^+b \rangle$.

$$\frac{d}{dt}\langle \hat{b}^+(t)\hat{b}(t)\rangle = \langle \mathcal{D}(\hat{b}^+(t)b(t))\rangle \tag{74}$$

In principle the master equation for $\mathcal{A}$ allows us to calculate $d\langle \hat{b}^+\hat{b}\rangle/dt$ and thus to derive $\mathcal{D}(\hat{b}^+\hat{b})$ from it. Indeed, Equation (21) of Complement B$_{\text{IV}}$ gives

$$\mathcal{D}(\hat{b}^+(t)\hat{b}(t)) = -\Gamma\hat{b}^+(t)\hat{b}(t) + \Gamma'. \tag{75}$$

Equation (74) allows us to rewrite (67) in the form

$$2D_N = \langle \mathcal{D}(\hat{b}^+\hat{b}) - \mathcal{D}(\hat{b}^+)\hat{b} - \hat{b}^+\mathcal{D}(\hat{b})\rangle \tag{76}$$

all the operators of the right-hand side of (76) being taken at the same time t. Equation (76) can be considered to be a generalization of the Einstein relation (14) between D and Γ. It expresses the diffusion coefficient D_N relative to $\hat{b}^+\hat{b}$ as a function of the drift velocities that describe the damping of $\hat{b}^+\hat{b}$, $\hat{b}^+$, and $\hat{b}$ and which can all be calculated starting from the master equation.

It is also clear from (76) that the diffusion coefficient D_N is associated only with the *non-Hamiltonian* part of the drift velocities. If we indeed had, for any operator G_A of $\mathcal{A}$

$$\mathcal{D}(G_A) = \frac{i}{\hbar}[H, G_A], \tag{77}$$

then $2D_N$ would be zero because

$$\left[H, \hat{b}^+\hat{b}\right] - [H, \hat{b}^+]\hat{b} - \hat{b}^+[H, \hat{b}] = 0. \tag{78}$$

It is for this reason that the radiative shifts of the levels of $\mathscr{A}$, which can be described by an effective Hamiltonian, do not appear in the expression for diffusion coefficients.

In the particular case under consideration here, it is sufficient to substitute (75), (62), and (65) into (76) to immediately obtain expression (47) for D_N.

g) Calculation of Two-Time Averages for Operators of $\mathscr{A}$. Quantum Regression Theorem

We take the Hermitian conjugate of Equation (39), multiply both sides on the right by $b(t')$ and take the average value. This gives

$$\frac{d}{dt}\left\langle \hat{b}^+(t)\hat{b}(t')\right\rangle = -\left(\frac{\Gamma}{2} - i\Delta\right)\left\langle \hat{b}^+(t)\hat{b}(t')\right\rangle + \left\langle \hat{F}^+(t)\hat{b}(t')\right\rangle. \tag{79}$$

Assume that $t \geq t'$. According to what we saw in subsection 2-d concerning the two-time averages $\langle \hat{F}^+(t)\hat{b}(t')\rangle$, the last term of (79) is zero except in a small interval of width τ_c, near $t = t'$, where it is on the order of D_N. It is thus totally legitimate to neglect such a term for $t \gg t'$, the error made on $\langle \hat{b}^+(t)\hat{b}(t')\rangle$ being on the order of $D_N\tau_c$, i.e., according to (51), on the order of $\Gamma'\tau_c \ll 1$. The two-time averages $\langle \hat{b}^+(t)\hat{b}(t')\rangle$ can therefore, to a very good approximation, be considered as obeying the equation

$$t \geq t' \qquad \frac{d}{dt}\left\langle \hat{b}^+(t)\hat{b}(t')\right\rangle = -\left(\frac{\Gamma}{2} - i\Delta\right)\left\langle \hat{b}^+(t)\hat{b}(t')\right\rangle \tag{80}$$

exactly analogous to the Hermitian conjugate of Equation (41):

$$t \geq t' \qquad \frac{d}{dt}\left\langle \hat{b}^+(t)\right\rangle = -\left(\frac{\Gamma}{2} - i\Delta\right)\left\langle \hat{b}^+(t)\right\rangle \tag{81}$$

giving the evolution of the one-time averages.

The foregoing result constitutes the quantum regression theorem which allows us to calculate the evolution of two-time averages using equations having the same structure as those giving the evolution of one-time averages, which are themselves obtained from the master equation.

Finally, using the simple model of an oscillator coupled to a reservoir of oscillators, we were able to extend most of the results obtained from the Langevin description of Brownian motion to a quantum system. We will see in Complement A_V that it is also possible to generalize these results to the case of a two-level atom coupled to a reservoir of oscillators (in this case, the modes of the radiation field), and to thus obtain the Bloch–Langevin equations. Such a generalization is nontrivial because a two-level atom is nonlinear, which causes nonlinear terms to appear in the coupled evolution equations, and makes the manipulation of these equations more difficult than in the linear case studied here.

REFERENCES

For classical Brownian motion theory, see Van Kampen.

For quantum Langevin equations of a system coupled to a reservoir, see Lax or else M. Lax, *Phys. Rev.*, **145**, 110 (1966), and the references therein cited.

For the specific problem of the harmonic oscillator, see I. R. Senitzky, *Phys. Rev.*, **119**, 670 (1960); Louisell, Chapter 7; Sargent, Scully, and Lamb Chapter 19.

CHAPTER V

Optical Bloch Equations

After having described in the previous chapter the evolution of an atomic system placed in the vacuum of the radiation field or in the presence of broadband radiation, we will now consider situations where the atom interacts with a *monochromatic* incident radiation field whose frequency is very close to an atomic eigenfrequency. Our goal is to understand, on the one hand, the temporal evolution of the atomic system, and, on the other hand, the essential characteristics of the light that the atom reemits in the presence of the radiation field ("fluorescence").

The treatment presented here is not perturbative with respect to the incident field. Indeed, we would like to study the case where the incident light intensity is sufficiently high to "saturate" the atomic transition (the corresponding Rabi frequency is then large compared with the natural width of the levels). Such a situation is common in experiments where laser sources are used. It is therefore no longer possible to calculate only the linear response of the atom or to study scattering processes involving a single incident photon, as we did in Section C of Chapter II or in Complement B_{III}.

The physical problem studied in this chapter cannot be approached simply by the nonperturbative methods used in Chapter III, which are based on the calculation of the resolvent $G(z)$ of the Hamiltonian in a subspace $\mathscr{E}_0$ of privileged states. Indeed, when the incident radiation contains several photons and when it excites an atomic transition starting from the ground state a, a very large number of states of the atom + field system must be taken into account (see §C-4-f in Chapter III). These states correspond to the many "fluorescence cycles" that the atom can undergo starting from the ground state a (by passing from a to b by absorbing an incident photon $\mathbf{k}\boldsymbol{\varepsilon}$, then by returning from b to a by spontaneously emitting a photon into another mode $\mathbf{k}'\boldsymbol{\varepsilon}'$). The subspace $\mathcal{E}_0$ of privileged states thus has a dimension that is too high for simple calculations to be made with the projection of $G(z)$ onto this subspace.

Rather than studying the evolution of the atom + field global system, we can instead more modestly attempt to derive an evolution equation for the atomic system alone. Nevertheless, it is clear that the methods used in Chapter IV are not directly applicable to the problem studied here, because the radiation field can no longer be considered as a reservoir. The correlation time of the monochromatic incident radiation is actually infinitely long. It is thus impossible to derive equations analogous to Equations (E.15) and (E.21) in Chapter IV that allow us to describe the physical phenomena in terms of transition rates between levels, corresponding to absorption, stimulated emission, and spontaneous emission processes.

Moreover, if it were possible to take into account only the interaction of the atom with the incident radiation, considered as an external monochromatic field, then we could certainly write the Schrödinger equation describing the evolution of the atomic state vector under the influence of this sinusoidal perturbation. However, it is absolutely necessary to also consider the interaction of the atom with the empty modes of the field. It is, in fact, these interactions and the corresponding relaxation processes that allow the atom to reach a steady state, resulting from the competition between the excitation by the incident wave and the damping due to spontaneous emission. Moreover, the light that the atom reemits and that we wish to analyze is simply the light that appears in these initially empty modes. The optical Bloch equations, which we will consider in this Chapter, express precisely that the rate of change of the atomic density matrix is a sum of two terms describing, respectively, the contribution of the coupling with the incident wave and the contribution of the coupling with the empty modes. We begin in Section A by deriving these optical Bloch equations for a two-level atom and by specifying the different possible ways to write these equations. We will then analyze, in Section B, the physical meaning of the optical Bloch equations by emphasizing what distinguishes them from other types of evolution equations. Finally, we will review several physical problems that these equations allow us to treat. Section C is devoted to a study of the temporal evolution of the internal and external degrees of freedom of the atom. Section D then describes various properties (total intensity, spectral distribution, etc.) of the fluorescence emitted by the atom. Such an analysis is based on the study of correlation functions for the atomic dipole and uses the quantum regression theorem that is derived in Complement A_V starting from the coupled Heisenberg equations for the atom and the field.

A—OPTICAL BLOCH EQUATIONS FOR A TWO-LEVEL ATOM

To simplify the calculations as much as possible, we will consider a single atom, at rest at the coordinate origin 0, with only two discrete nondegenerate states, the ground state a and the first excited state b, located at a distance $\hbar\omega_0$ above a and having a natural width Γ.

1. Description of the Incident Field

The incident radiation is treated as an external field, and is thus described by a given classical function of time. The particles interact, on the one hand, with this external field, and on the other hand, with the quantum radiation field which we assume to be initially in the vacuum state $|0\rangle$. The Hamiltonian for the atom + field system is then written (within the long-wavelength approximation and in the electric dipole picture)

$$H = H_A + H_R - \mathbf{d} \cdot \left[\mathbf{E}_e(\mathbf{0}, t) + \mathbf{E}_\perp(\mathbf{0}) \right] \qquad (A.1)$$

where H_A is the Hamiltonian of the atom, H_R is the Hamiltonian of the quantum radiation field, $\mathbf{d}$ is the electric dipole moment of the atom, $\mathbf{E}_e$ is the external field associated with the incident radiation, and $\mathbf{E}_\perp(\mathbf{0})$ is the quantum radiation field.

Moreover, we assume an incident monochromatic field, with frequency ω_L and amplitude $\mathscr{E}_0$:

$$\mathbf{E}_e(\mathbf{0}, t) = \mathscr{E}_0 \cos \omega_L t. \qquad (A.2)$$

Remarks

(i) The description of the incident radiation by an external field is not an approximation if the quantum field is initially in a coherent state. This result is shown in Exercise 17.

(ii) We also neglect any modification of the incident radiation due to its interaction with the atom. More precisely, we consider a single atom interacting with "macroscopic" incident radiation such as laser light. We are then justified in neglecting the absorption or amplification of such radiation by the atom. Our treatment therefore excludes situations where an ensemble of atoms interacts with a field originating from the atoms themselves. Such a situation is encountered, for example, for atoms inside a cavity, or for optically dense media. In this case, we must solve coupled evolution equations for the atoms and the field that are sometimes known as Bloch–Maxwell equations (*).

(*) See, for example, Sargent, Scully, and Lamb, Chapter VIII; Allen and Eberly, Chapter IV.

2. Approximation of Independent Rates of Variation

If the external field (A.2) were zero, we would know how to write the evolution equation for the atomic density matrix σ. Indeed, when the initial state of the total field is the vacuum, the radiation can be treated as a reservoir. The evolution equation for σ is the master equation describing the influence of spontaneous emission on a two-level atom [see Equations (E.5) and (E.6) in Chapter IV].

Similarly, if we could neglect the effect of the coupling with the quantum field $\mathbf{E}_\perp(0)$ in (A.1), the evolution of σ would simply be given by the Schrödinger equation

$$i\hbar\dot{\sigma} = [H_A - \mathbf{d}\cdot\boldsymbol{\mathscr{E}}_0\cos\omega_L t, \sigma]. \tag{A.3}$$

The approximation of independent rates of variation consists of independently adding the rates of variation of σ associated with the two previous couplings $-\mathbf{d}\cdot\boldsymbol{\mathscr{E}}_0\cos\omega_L t$ and $-\mathbf{d}\cdot\mathbf{E}_\perp(0)$, and calculated as if each coupling acted *alone*. By expanding Equation (A.3) in the basis $\{|a\rangle, |b\rangle\}$ and by adding the damping terms of Equations (E.5) and (E.6) from Chapter IV that describe the effect of spontaneous emission, we get

$$\dot{\sigma}_{bb} = i\Omega_1\cos\omega_L t(\sigma_{ba} - \sigma_{ab}) - \Gamma\sigma_{bb} \tag{A.4.a}$$

$$\dot{\sigma}_{aa} = -i\Omega_1\cos\omega_L t(\sigma_{ba} - \sigma_{ab}) + \Gamma\sigma_{bb} \tag{A.4.b}$$

$$\dot{\sigma}_{ab} = i\omega_0\sigma_{ab} - i\Omega_1\cos\omega_L t(\sigma_{bb} - \sigma_{aa}) - \frac{\Gamma}{2}\sigma_{ab} \tag{A.4.c}$$

$$\dot{\sigma}_{ba} = -i\omega_0\sigma_{ba} + i\Omega_1\cos\omega_L t(\sigma_{bb} - \sigma_{aa}) - \frac{\Gamma}{2}\sigma_{ba}. \tag{A.4.d}$$

We have assumed that the radiative shifts of levels a and b have been included in the atomic frequency ω_0 and we have set

$$\hbar\Omega_1 = -\mathbf{d}_{ab}\cdot\boldsymbol{\mathscr{E}}_0 \tag{A.5}$$

where

$$\mathbf{d}_{ab} = \langle a|\mathbf{d}|b\rangle = \langle b|\mathbf{d}|a\rangle \tag{A.6}$$

is the matrix element of $\mathbf{d}$, which is assumed to be real. The frequency Ω_1 is called the *Rabi frequency*. It characterizes the strength of the coupling between the atom and the incident wave. Each of the two rates that we have added in (A.4) conserves the normalization of σ, i.e., satisfies $\dot{\sigma}_{aa} + \dot{\sigma}_{bb} = 0$.

Because we describe the effect of spontaneous emission by using the same terms as those derived in the absence of radiation, we neglect the modifications of spontaneous emission connected with the presence of the incident radiation. Such an approximation is valid if the effect of the coupling with this radiation can be neglected during the correlation time τ_c of the vacuum fluctuations that are responsible for spontaneous emission. One can show (as we do in Complement A_V and in Chapter VI) that this is indeed the case if the Rabi frequency Ω_1 is very small compared with the frequency ω_0 of the transition $a \leftrightarrow b$

$$\Omega_1 \ll \omega_0. \tag{A.7}$$

Note that such a condition already underlies the approximation which takes into account only two of the atomic levels. If it were not satisfied, we would need to also take into account the coupling between the incident wave and all the other atomic transitions. To be able to limit ourselves to two levels, we must also, in addition to (A.7), assume that

$$|\omega_L - \omega_0| \ll \omega_0 \tag{A.8}$$

a condition expressing the fact that the incident radiation, with frequency ω_L, is quasi-resonant with the transition $a \leftrightarrow b$.

Remark

The only relaxation process considered here is spontaneous emission whose correlation time is shorter than the optical period ω_0^{-1}. It would be possible to include in Equations (A.4) some terms describing the effect of other relaxation processes, such as, for example, collisions (see Exercise 18). The approximation consisting of independently adding the rates of variation is valid only if

$$\Omega_1 \ll \tau_c^{-1} \tag{A.7$'$}$$

$$|\omega_L - \omega_0| \ll \tau_c^{-1} \tag{A.8$'$}$$

where τ_c is the correlation time of these other relaxation processes. In the case of collisional relaxation, τ_c is on the order of the collision time and the conditions (A.7$'$) and (A.8$'$) define the "impact limit" (see Complement B_{VI}).

3. Rotating-Wave Approximation

a) Elimination of Antiresonant Terms

The dipole **d**, which is purely nondiagonal in the basis $\{|a\rangle, |b\rangle\}$, may be written, using (A.6)

$$\mathbf{d} = \mathbf{d}_{ab}(|b\rangle\langle a| + |a\rangle\langle b|) = \mathbf{d}_+ + \mathbf{d}_- \tag{A.9}$$

with

$$\mathbf{d}_{\pm} = \mathbf{d}_{ab}\mathcal{S}_{\pm} \qquad \text{(A.10.a)}$$

$$\mathcal{S}_{+} = |b\rangle\langle a| \qquad \text{(A.10.b)}$$

$$\mathcal{S}_{-} = |a\rangle\langle b|. \qquad \text{(A.10.c)}$$

The operators $\mathcal{S}_{+}$ and $\mathcal{S}_{-}$ are, respectively, the raising and lowering operators from a to b and from b to a. We rewrite the interaction Hamiltonian, $-\mathbf{d} \cdot \mathbf{\mathcal{E}}_0 \cos \omega_L t$, as a function of $\mathcal{S}_{+}, \mathcal{S}_{-}$ and of the exponentials $\exp(-i\omega_L t)$ and $\exp(+i\omega_L t)$ which come from $\cos \omega_L t$ and which are associated, respectively, with the absorption and the emission of a photon. We then have, using (A.5)

$$-\mathbf{d} \cdot \mathbf{\mathcal{E}}_0 \cos \omega_L t = \tfrac{1}{2}\hbar\Omega_1\left[\mathcal{S}_{+}\, e^{-i\omega_L t} + \mathcal{S}_{-}\, e^{i\omega_L t} + \mathcal{S}_{-}\, e^{-i\omega_L t} + \mathcal{S}_{+}\, e^{i\omega_L t}\right].$$
$$\text{(A.11)}$$

The first two terms inside the brackets describe processes where the atom rises from a to b by absorbing a photon or falls from b to a by emitting a photon. These processes are resonant near $\omega = \omega_0$ and are much more important than the nonresonant processes associated with the last two terms in (A.11) (the atom falls from b to a by absorbing a photon or rises from a to b by emitting a photon). Therefore, in the rest of this chapter, we will neglect (*) the last two terms of (A.11). The corresponding approximation is called the rotating-wave approximation, for reasons that will become apparent later on (§A.4).

b) TIME-INDEPENDENT FORM OF THE OPTICAL BLOCH EQUATIONS

If we keep only the first two terms of (A.11), Equations (A.4) are modified: $\sigma_{ba} \cos \omega_L t$ and $\sigma_{ab} \cos \omega_L t$ are replaced by $\sigma_{ba}[\exp(i\omega_L t)]/2$ and $\sigma_{ab}[\exp(-i\omega_L t)]/2$, respectively, in the right-hand side of (A.4.a) and (A.4.b); $\cos \omega_L t$ is replaced by $(\exp i\omega_L t)/2$ in (A.4.c) and by $[\exp(-i\omega_L t)]/2$ in (A.4.d). It is then possible to suppress any time dependence in the coefficients of the equations by introducing new variables

$$\hat{\sigma}_{ba} = \sigma_{ba}\, e^{i\omega_L t}$$

$$\hat{\sigma}_{ab} = \sigma_{ab}\, e^{-i\omega_L t} \qquad \text{(A.12)}$$

$$\hat{\sigma}_{aa} = \sigma_{aa} \qquad \hat{\sigma}_{bb} = \sigma_{bb}$$

(*) We then lose certain physical effects, such as the Bloch–Siegert shift of the resonance $\omega_0 = \omega$, to which we will return later on in Complement A_{VI}.

which leads to the following equations

$$\frac{d}{dt}\hat{\sigma}_{bb} = i\frac{\Omega_1}{2}(\hat{\sigma}_{ba} - \hat{\sigma}_{ab}) - \Gamma\hat{\sigma}_{bb} \tag{A.13.a}$$

$$\frac{d}{dt}\hat{\sigma}_{aa} = -i\frac{\Omega_1}{2}(\hat{\sigma}_{ba} - \hat{\sigma}_{ab}) + \Gamma\hat{\sigma}_{bb} \tag{A.13.b}$$

$$\frac{d}{dt}\hat{\sigma}_{ab} = -i\delta_L\hat{\sigma}_{ab} - i\frac{\Omega_1}{2}(\hat{\sigma}_{bb} - \hat{\sigma}_{aa}) - \frac{\Gamma}{2}\hat{\sigma}_{ab} \tag{A.13.c}$$

$$\frac{d}{dt}\hat{\sigma}_{ba} = i\delta_L\hat{\sigma}_{ba} + i\frac{\Omega_1}{2}(\hat{\sigma}_{bb} - \hat{\sigma}_{aa}) - \frac{\Gamma}{2}\hat{\sigma}_{ba} \tag{A.13.d}$$

where

$$\delta_L = \omega_L - \omega_0 \tag{A.14}$$

is the detuning between the frequency ω_L of the incident wave and the atomic frequency ω_0. Note that we still have $d(\hat{\sigma}_{aa} + \hat{\sigma}_{bb})/dt = 0$.

c) OTHER FORMS OF THE OPTICAL BLOCH EQUATIONS

It will be useful for what follows to rewrite Equations (A.13) in terms of average values of operators. We introduce the three operators

$$S_+ = e^{-i\omega_L t}\mathscr{S}_+ = e^{-i\omega_L t}|b\rangle\langle a| \tag{A.15.a}$$

$$S_- = e^{i\omega_L t}\mathscr{S}_- = e^{i\omega_L t}|a\rangle\langle b| \tag{A.15.b}$$

$$S_Z = \tfrac{1}{2}(|b\rangle\langle b| - |a\rangle\langle a|) \tag{A.15.c}$$

whose average values are

$$\langle S_+\rangle = \mathrm{Tr}(\sigma\mathscr{S}_+ e^{-i\omega_L t}) = \sigma_{ab}e^{-i\omega_L t} = \hat{\sigma}_{ab} \tag{A.16.a}$$

$$\langle S_-\rangle = \mathrm{Tr}(\sigma\mathscr{S}_- e^{i\omega_L t}) = \sigma_{ba}e^{i\omega_L t} = \hat{\sigma}_{ba} \tag{A.16.b}$$

$$\langle S_Z\rangle = \mathrm{Tr}[\sigma\tfrac{1}{2}(|b\rangle\langle b| - |a\rangle\langle a|)] = \tfrac{1}{2}(\hat{\sigma}_{bb} - \hat{\sigma}_{aa}). \tag{A.16.c}$$

In conjunction with the normalization condition

$$\sigma_{aa} + \sigma_{bb} = \hat{\sigma}_{aa} + \hat{\sigma}_{bb} = 1 \tag{A.17}$$

these equations allow us to rewrite (A.13) in the form

$$\langle \dot{S}_+ \rangle = -\left(i\delta_L + \frac{\Gamma}{2} \right)\langle S_+ \rangle - i\Omega_1 \langle S_Z \rangle \qquad (A.18.a)$$

$$\langle \dot{S}_- \rangle = -\left(-i\delta_L + \frac{\Gamma}{2} \right)\langle S_- \rangle + i\Omega_1 \langle S_Z \rangle \qquad (A.18.b)$$

$$\langle \dot{S}_Z \rangle = \frac{i\Omega_1}{2}[\langle S_- \rangle - \langle S_+ \rangle] - \Gamma\left(\langle S_Z \rangle + \frac{1}{2} \right). \qquad (A.18.c)$$

Finally, we introduce other variables

$$u = \frac{1}{2}\left(\hat{\sigma}_{ab} + \hat{\sigma}_{ba} \right) \qquad (A.19.a)$$

$$v = \frac{1}{2i}\left(\hat{\sigma}_{ab} - \hat{\sigma}_{ba} \right) \qquad (A.19.b)$$

$$w = \frac{1}{2}\left(\hat{\sigma}_{bb} - \hat{\sigma}_{aa} \right) \qquad (A.19.c)$$

which are very often used in this type of problem (u, v, and w are the three components of the "Bloch vector"). When Equations (A.13) are rewritten as a function of u, v, and w, they become

$$\dot{u} = \delta_L v - \frac{\Gamma}{2}u \qquad (A.20.a)$$

$$\dot{v} = -\delta_L u - \Omega_1 w - \frac{\Gamma}{2}v \qquad (A.20.b)$$

$$\dot{w} = \Omega_1 v - \Gamma w - \frac{\Gamma}{2} \qquad (A.20.c)$$

w represents half the difference between the populations of the two levels b and a. To interpret u and v, we calculate the average value of **d**

$$\langle \mathbf{d} \rangle = \text{Tr}(\sigma \mathbf{d}) = \mathbf{d}_{ab}(\sigma_{ab} + \sigma_{ba})$$
$$= \mathbf{d}_{ab}\left(\hat{\sigma}_{ab}\, e^{i\omega_L t} + \hat{\sigma}_{ba}\, e^{-i\omega_L t} \right)$$
$$= 2\mathbf{d}_{ab}(u \cos \omega_L t - v \sin \omega_L t). \qquad (A.21)$$

The comparison of (A.21) and (A.2) shows that u and v are, respectively, proportional to the components of $\langle \mathbf{d} \rangle$ in phase and in quadrature with the incident field [because $-\sin \omega_L t = \cos(\omega_L t + \pi/2)$].

4. Geometric Representation in Terms of a Fictitious Spin $\frac{1}{2}$

Every two-level system is formally equivalent to a fictitious spin $\frac{1}{2}$. We associate the states $|-\rangle$ and $|+\rangle$ of such a spin with states $|a\rangle$ and $|b\rangle$.

$$|a\rangle \leftrightarrow |-\rangle \qquad |b\rangle \leftrightarrow |+\rangle \qquad (A.22)$$

and we introduce the (dimensionless) spin operators $\mathscr{S}_x$, $\mathscr{S}_y$, and $\mathscr{S}_z$, represented in the basis $\{|+\rangle, |-\rangle\}$ by the matrices

$$(\mathscr{S}_x) = \tfrac{1}{2}\begin{pmatrix} 0 & 1 \\ 1 & 0 \end{pmatrix} \qquad (\mathscr{S}_y) = \tfrac{1}{2}\begin{pmatrix} 0 & -i \\ i & 0 \end{pmatrix} \qquad (\mathscr{S}_z) = \tfrac{1}{2}\begin{pmatrix} 1 & 0 \\ 0 & -1 \end{pmatrix}$$

$$(A.23)$$

which are simply, except for the factor $\frac{1}{2}$, the Pauli matrices. Each operator of the two-level system is represented in the basis $\{|b\rangle, |a\rangle\}$ by a 2×2 matrix which can always be expanded over the three matrices in (A.23) and the unit matrix ($\mathbb{1}$). Thus

$$(H_A) = \hbar\omega_0\begin{pmatrix} 1 & 0 \\ 0 & 0 \end{pmatrix} = \frac{\hbar\omega_0}{2}(\mathbb{1}) + \hbar\omega_0(\mathscr{S}_z) \quad (A.24.a)$$

$$(-\mathbf{d}\cdot\mathbf{\mathscr{E}}_0\cos\omega_L t) = \hbar\Omega_1\cos\omega_L t\begin{pmatrix} 0 & 1 \\ 1 & 0 \end{pmatrix} = 2\hbar\Omega_1\cos\omega_L t(\mathscr{S}_x).$$

$$(A.24.b)$$

The Hamiltonians H_A and $-\mathbf{d}\cdot\mathbf{\mathscr{E}}_0\cos\omega_L t$ may therefore be considered as Hamiltonians describing the interaction of the fictitious spin with the magnetic fields $\mathbf{B}_0$ and $2\mathbf{B}_1\cos\omega_L t$, which are, respectively, parallel to $0z$ and $0x$, and have amplitudes such that the Larmor spin precession frequencies around these two fields are ω_0 and $2\Omega_1\cos\omega_L t$ (Figure 1a). The field $2\mathbf{B}_1\cos\omega_L t$ parallel to $0x$ can be decomposed into two fields, having the same amplitude B_1, rotating in the plane $x0y$ at the frequency ω_L in the clockwise and counterclockwise directions. If $\omega_L = \omega_0$, the counterclockwise component accompanies the spin in its Larmor precession around $\mathbf{B}_0$ and thus can act efficiently on it, whereas the other component rotates much too rapidly relative to the spin (at the frequency $-2\omega_L$) to have any appreciable effect. The approximation of the rotating wave consists precisely of retaining only the component rotating in the same direction as the spin, hence the name for this approximation.

We now put ourselves in the $0XYZ$ reference frame rotating about $0z$ at the frequency ω_L. In this reference frame, the rotating component that we have retained from the field $2\mathbf{B}_1\cos\omega_L t$ becomes a *time-independent*

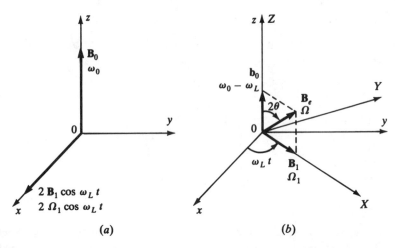

Figure 1. Magnetic fields acting on the fictitious spin. (*a*) In the laboratory reference frame $0xyz$. (*b*) In the reference frame $0XYZ$ rotating around $0z$ at the frequency ω_L (the component of the field rotating in the opposite direction is neglected).

field $\mathbf{B}_1$ parallel to $0X$ and the field along $0Z$ is reduced from $\mathbf{B}_0$ to $\mathbf{b}_0$ because the Larmor spin precession about $0Z$ is reduced from ω_0 to $\omega_0 - \omega_L$ (Figure 1*b*). In the rotating reference frame, the spin therefore "sees" the resultant $\mathbf{B}_e$ of $\mathbf{b}_0$ and $\mathbf{B}_1$, called the effective field, about which it precesses at the frequency

$$\Omega = \sqrt{\Omega_1^2 + \delta_L^2} \qquad (A.25)$$

At resonance ($\delta_L = \omega_L - \omega_0 = 0$), $\mathbf{b}_0$ is zero and the spin precesses about $\mathbf{B}_1$ at the Rabi frequency Ω_1. We recover the Rabi oscillation of the system between the two levels a and b.

We now show that the variables u, v, and w introduced in (A.19) are in fact average values of the spin components in the rotating frame. Note that the correspondence (A.22) and formulas (A.23) result in the usual relations $\mathscr{S}_\pm = \mathscr{S}_x \pm i\mathscr{S}_y$. Therefore, definition (A.19.a) for u and relations (A.16) allow us to write u in the form

$$\begin{aligned}
u &= \tfrac{1}{2}\,\mathrm{Tr}\left\{\sigma\left[\left(\mathscr{S}_x + i\mathscr{S}_y\right)e^{-i\omega_L t} + \left(\mathscr{S}_x - i\mathscr{S}_y\right)e^{i\omega_L t}\right]\right\} \\
&= \langle\mathscr{S}_x\rangle\cos\omega_L t + \langle\mathscr{S}_y\rangle\sin\omega_L t \\
&= \langle\mathscr{S}\rangle \cdot \mathbf{e}_X
\end{aligned} \qquad (A.26)$$

where $\mathbf{e}_X$ is the unit vector of the axis $0X$ of the rotating frame. Similar relations can be derived for v and w, so that

$$u = \langle \mathscr{S}_X \rangle \qquad v = \langle \mathscr{S}_Y \rangle \qquad w = \langle \mathscr{S}_Z \rangle. \qquad (A.27)$$

The fact that v has a $+\pi/2$ phase shift relative to u is then geometrically obvious. With these new notations, Equations (A.20) become

$$\frac{d}{dt} \langle \mathscr{S}_X \rangle = \delta_L \langle \mathscr{S}_Y \rangle - \frac{\Gamma}{2} \langle \mathscr{S}_X \rangle \qquad (A.28.a)$$

$$\frac{d}{dt} \langle \mathscr{S}_Y \rangle = -\delta_L \langle \mathscr{S}_X \rangle - \Omega_1 \langle \mathscr{S}_Z \rangle - \frac{\Gamma}{2} \langle \mathscr{S}_Y \rangle \qquad (A.28.b)$$

$$\frac{d}{dt} \langle \mathscr{S}_Z \rangle = \Omega_1 \langle \mathscr{S}_Y \rangle - \Gamma \langle \mathscr{S}_Z \rangle - \frac{\Gamma}{2} \qquad (A.28.c)$$

and closely resemble the Bloch equations for magnetic resonance (with relaxation times T_1 and T_2 which equal here, for spontaneous emission, $1/\Gamma$ and $2/\Gamma$).

B—PHYSICAL DISCUSSION—DIFFERENCES WITH OTHER EVOLUTION EQUATIONS

1. Differences with Relaxation Equations. Couplings between Populations and Coherences

As opposed to the relaxation equations examined in Chapter IV, the Bloch equations (A.13) do not separate into two distinct groups where either only the populations $\hat{\sigma}_{aa}$ and $\hat{\sigma}_{bb}$ of the two levels appear, or only the "coherences" $\hat{\sigma}_{ab}$ and $\hat{\sigma}_{ba}$ (nondiagonal density matrix elements) appear. It is therefore not possible to directly interpret them in terms of transition rates between levels a and b.

The terms in Ω_1 of Equations (A.13) couple populations and coherences. This originates from the coherent nature of the external field. The phase difference between the oscillation of the average electric dipole moment $\langle \mathbf{d} \rangle$ and the driving field $\mathscr{E}_0 \cos \omega_L t$ is crucial for determining whether the atom will absorb energy ($\hat{\sigma}_{bb} - \hat{\sigma}_{aa}$ increases) or lose energy ($\hat{\sigma}_{bb} - \hat{\sigma}_{aa}$ decreases). The rate of variation of $\hat{\sigma}_{bb} - \hat{\sigma}_{aa}$ thus necessarily depends on $\langle \mathbf{d} \rangle$, i.e., according to (A.21), on $\hat{\sigma}_{ba}$ and $\hat{\sigma}_{ab}$ [more precisely on the component in quadrature $v = (\hat{\sigma}_{ab} - \hat{\sigma}_{ba})/2i$ which is involved in the work done by the field on the dipole].

Remark

There are situations where the Bloch equations can be transformed into relaxation equations involving populations only. If the coherences evolve much more rapidly than the populations (as a result of an additional relaxation process which damps them more efficiently than the population differences), they can "adiabatically follow" the evolution of the populations, i.e., at each time reach the steady-state condition corresponding to the value of the populations assumed to be "frozen" at this time. It is then possible to rewrite at each time the coherences as a function of the populations at the same time and to thus obtain evolution equations involving populations only. An example of a physical problem where the coherences can be adiabatically eliminated is analyzed in Exercise 18. We consider in this exercise a two-level atom undergoing collisions which introduce a collisional broadening much larger than Γ.

2. Differences with Hamiltonian Evolution Equations

In Chapter III we saw that the instability of the excited state b, due to spontaneous emission, may often be described by the addition of an imaginary part, $-i\hbar\Gamma/2$, to the energy $E_b = \hbar\omega_0$ of this state (we take $E_a = 0$). It may then be asked whether it would be possible to write

Equations (A.4) in a Hamiltonian form, by replacing the atomic Hamiltonian H_A in Equation (A.3) by the non-Hermitian Hamiltonian:

$$(H'_A) = \hbar \begin{pmatrix} \omega_0 - i\dfrac{\Gamma}{2} & 0 \\ 0 & 0 \end{pmatrix} \tag{B.1}$$

and by also replacing the commutator $H_A\sigma - \sigma H_A$ by $H'_A\sigma - \sigma H'^{+}_A$, so as
to preserve the Hermiticity of $d\sigma/dt$, and therefore of σ. Such a procedure results in the correct equations of motion (A.4.a), (A.4.c), and (A.4.d)
for $\dot{\sigma}_{bb}$, $\dot{\sigma}_{ab}$, and $\dot{\sigma}_{ba}$. However, the equation of motion for $\dot{\sigma}_{aa}$ obtained in
this way does not contain the transfer term from b to a, $\Gamma\sigma_{bb}$, appearing
in the right-hand side of the correct equation (A.4.b).

It is therefore not possible to find an atomic Hamiltonian, not even a
generalized one including complex energies, that allows us to write the
optical Bloch equations as a Schrödinger equation. In particular, this
shows that the atomic system must be described by a density operator, and
that it is impossible to describe the effect of spontaneous emission in
terms of a rate of variation of an atomic state vector. This is, of course,
related to the fact that the optical Bloch equations describe the evolution
of a subsystem (the atom), which is part of a larger set (the atom +
quantum radiation).

3. Differences with Heisenberg–Langevin Equations

Written in the form (A.18), the optical Bloch equations appear to
describe the evolution of average values of the atomic operators S_+, S_-,
and S_z. One might ask whether it would be correct to write the same
equations for the operators themselves instead of the average values.

A simple counterexample allows us to show that such a procedure
would be incorrect. Rewritten in the particular case $\Omega_1 = 0$ and without
averaging, Equations (A.18.a) and (A.18.b) would indeed result in both S_+
and S_- decreasing exponentially to zero, which is incompatible with the
operator identity $S_+S_- + S_-S_+ = |b\rangle\langle b| + |a\rangle\langle a| = \mathbb{1}$, which results
from (A.15) and which must be satisfied at all times.

In fact, the Heisenberg equations for S_+, S_-, and S_z may be put in a
form similar to (A.18), but, in addition "Langevin forces" F_+, F_-, and F_z
appear in the right-hand side, representing to some extent the fluctuating
part of the force exerted on the atom by the quantum radiation, while the
cumulative effect of this force is represented by the friction terms in Γ
(see Complement A_V). More precisely, if the Bloch equations (A.18) for

the *average values* $\langle S_q \rangle$ (with $q = +, -, z$) are written in the condensed form:

$$\langle \dot{S}_q \rangle = \sum_{q' = +, -, z} \mathscr{B}_{qq'} \langle S_{q'} \rangle + \lambda_q \tag{B.2}$$

the Heisenberg–Langevin equations for the operators S_q,

$$\dot{S}_q = \sum_{q' = +, -, z} \mathscr{B}_{qq'} S_{q'} + \lambda_q + F_q \tag{B.3}$$

differ from (B.2) by the term F_q, which is an operator acting on the degrees of freedom of both the atom and the radiation field. The average value of F_q is zero, so that the Heisenberg–Langevin equations (B.3) indeed give, in average value, the Bloch equations (B.2). The presence of F_q is nevertheless required in the operator equation (B.3) to ensure that the various commutation relations among S_+, S_-, and S_z are preserved over time. Finally, note that F_q fluctuates very rapidly, in the sense that the two-time averages $\langle F_q(t)F_{q'}(t + \tau) \rangle$ decrease very rapidly with τ, on a time scale τ_c (correlation time of vacuum fluctuations) much shorter than the relaxation time Γ^{-1} associated with spontaneous emission or the Rabi period Ω_1^{-1}.

C—FIRST APPLICATION—EVOLUTION OF ATOMIC AVERAGE VALUES

Studying the solutions of optical Bloch equations allows us to analyze the time evolution of the quantities u, v, and w, which are average values of "internal" *atomic observables*, such as the mean dipole or the population difference between states b and a. This is what we do in subsection C.1. It is also possible to express the average force exerted by a driving field on the oscillating atomic dipole as a function of u and v. The optical Bloch equations thus allow us to also analyze the mean *radiative forces* exerted by a light beam on an atom. These forces, which act on the "external" or translational degrees of freedom of the atom, are discussed in subsection C.2.

1. Internal Degrees of Freedom

a) TRANSIENT REGIME

The optical Bloch equations (A.20) or (B.2) are linear differential equations with constant coefficients. The solutions to these equations are thus superpositions of exponentials $\exp(-r_\lambda t)$. We will not give here the general expression for the eigenvalues $-r_\lambda$ of the Bloch matrix $(\mathscr{B}_{qq'})$, but we will instead study some limiting cases.

In the limit $\Omega_1 \to 0$ and for $\omega_L = \omega_0$ ($\delta_L = 0$), we see in (A.18) that two r_λ appear that are equal to $\Gamma/2$, and the third one to Γ. At very low intensities, and at resonance, the transient regime is thus purely damped (without oscillation).

By contrast, for $\Omega_1 \gg \Gamma$ and δ_L still zero, the transient response must reflect the Rabi oscillation at the frequency Ω_1. This is clearly seen from the representation of the problem in terms of the fictitious spin $\frac{1}{2}$ (see Figure 1*b*). In the rotating reference frame $0XYZ$, the field $\mathbf{b}_0$ is zero and the spin precesses at the frequency Ω_1 around $\mathbf{B}_1$ which is aligned on $0X$. The component of the spin in the plane $Y0Z$ thus rotates very rapidly at the frequency Ω_1 by going successively over the axes $0Y$ and $0Z$, where the damping rates are equal to $\Gamma/2$ and Γ, respectively. Therefore, we expect that the average damping rate (over a period $2\pi/\Omega_1$) is the half-sum of $\Gamma/2$ and Γ, i.e., $3\Gamma/4$. Two of the exponentials of the transient regime must therefore be $\exp(\pm i\Omega_1 t)\exp(-3\Gamma t/4)$. The third corresponds to the motion of the spin along $0X$, which is purely damped with a rate $\Gamma/2$. The calculation of r_1, r_2, and r_3 in fact confirms that if

$\Omega_1 \gg \Gamma$ and if $\delta_L = 0$

$$r_1 = i\Omega_1 + \frac{3\Gamma}{4} \qquad r_2 = -i\Omega_1 + \frac{3\Gamma}{4} \qquad r_3 = \frac{\Gamma}{2}. \qquad \text{(C.1)}$$

Finally, we consider the case $|\delta_L| \gg |\Omega_1|, \Gamma$. The effective field $\mathbf{B}_e$ of Figure 1*b* is then practically aligned on $0Z$. The motion of the spin in the plane XOY is, in this case, a precession at the frequency δ_L, damped with a rate $\Gamma/2$, whereas the motion on $0Z$ is purely damped with a rate Γ. It is found that if $|\delta_L| \gg \Omega_1, \Gamma$

$$r_1 = i\delta_L + \frac{\Gamma}{2} \qquad r_2 = -i\delta_L + \frac{\Gamma}{2} \qquad r_3 = \Gamma. \qquad \text{(C.2)}$$

b) STEADY-STATE REGIME

The steady-state solution of the Bloch equations (A.20) is

$$u_{\text{st}} = \frac{\Omega_1}{2} \frac{\delta_L}{\delta_L^2 + (\Gamma^2/4) + (\Omega_1^2/2)} \qquad \text{(C.3.a)}$$

$$v_{\text{st}} = \frac{\Omega_1}{2} \frac{\Gamma/2}{\delta_L^2 + (\Gamma^2/4) + (\Omega_1^2/2)} \qquad \text{(C.3.b)}$$

$$w_{\text{st}} + \frac{1}{2} = \sigma_{bb}^{\text{st}} = \frac{\Omega_1^2}{4} \frac{1}{\delta_L^2 + (\Gamma^2/4) + (\Omega_1^2/2)}. \qquad \text{(C.3.c)}$$

We have given the value for $w_{\text{st}} + \frac{1}{2}$ rather than for w_{st}, because it represents the steady-state population of the upper state b [see (A.17) and (A.19.c)].

In Equations (C.3) the component in quadrature of the dipole (v_{st}) and the population of the upper state vary with the detuning $\delta_L = \omega_L - \omega_0$ as an absorption curve, centered at $\delta_L = 0$, with a half-width $[(\Gamma^2/4) + (\Omega_1^2/2)]^{1/2}$, whereas the component in phase (u_{st}) varies as a dispersion curve. When, δ_L being fixed, Ω_1 increases, u_{st} and v_{st} start to increase linearly with Ω_1, then go through a maximum and tend to zero for very large Ω_1. At very high intensities, the mean dipole is therefore zero. The population σ_{bb}^{st} starts to increase quadratically with Ω_1, and then tends to a limit value equal to $\frac{1}{2}$ when Ω_1 tends to infinity. An intense excitation thus equalizes the populations of the two levels. It is said that the transition is "saturated". The solution (C.3) is usually expressed as a function of the *saturation parameter*

$$s = \frac{\Omega_1^2/2}{\delta_L^2 + (\Gamma^2/4)} \tag{C.4}$$

which represents the degree of saturation of the transition. We then have

$$u_{st} = \frac{\delta_L}{\Omega_1} \frac{s}{1+s} \qquad v_{st} = \frac{\Gamma}{2\Omega_1} \frac{s}{1+s} \qquad \sigma_{bb}^{st} = \frac{1}{2} \frac{s}{1+s}. \tag{C.5}$$

Finally, note that, because of the presence of Ω_1^2 in the denominator of Equations (C.3), these expressions are not perturbative with regard to the driving field. (The expansion in series of the fractions causes arbitrarily large powers of Ω_1 to appear.)

c) ENERGY BALANCE. MEAN NUMBER OF INCIDENT PHOTONS ABSORBED PER UNIT TIME

Between t and $t + dt$, the atomic electron moves from $\mathbf{r}$ to $\mathbf{r} + d\mathbf{r}$ and the driving field $\mathscr{E}_0 \cos \omega_L t$ carries out work on it:

$$dW = q\mathscr{E}_0 \cos \omega_L t \cdot d\mathbf{r}. \tag{C.6}$$

The average power absorbed by the atom thus equals, taking into account the fact that $q\langle \mathbf{r} \rangle = \langle \mathbf{d} \rangle$:

$$\left\langle \frac{dW}{dt} \right\rangle = \mathscr{E}_0 \cos \omega_L t \cdot \langle \dot{\mathbf{d}} \rangle. \tag{C.7}$$

We then insert into (C.7) expression (A.21) giving $\langle \mathbf{d} \rangle$ as a function of u and v, and average (C.7) over an optical period. We then have

$$\overline{\left\langle \frac{dW}{dt} \right\rangle} = -2\mathbf{d}_{ab} \cdot \mathscr{E}_0 \omega_L \left[\overline{\cos^2 \omega_L t}\, v + \overline{\sin \omega_L t \cos \omega_L t}\, u \right]$$

$$= \hbar\Omega_1 \omega_L v \tag{C.8}$$

where we have used the definition (A.5) for Ω_1. It is thus clear that the mean absorbed power is related only to the quadrature component v of the mean atomic dipole. Finally, by dividing (C.8) by the energy $\hbar\omega_L$ of each incident photon, we simply obtain the mean number of photons absorbed per unit time by the atom.

$$\left\langle \frac{dN}{dt} \right\rangle = \Omega_1 v. \tag{C.9}$$

Equation (C.9) thus suggest a simple interpretation of the third Bloch equation (A.20.c). By replacing w by $(\sigma_{bb} - \sigma_{aa})/2 = \sigma_{bb} - (\tfrac{1}{2})$ in this equation, we obtain

$$\dot{\sigma}_{bb} = \left\langle \frac{dN}{dt} \right\rangle - \Gamma \sigma_{bb}. \tag{C.10}$$

Any disappearance of an incident photon corresponds to a transition of the atom from a to b. This is expressed by the first term of (C.10). The second term describes a departure from b due to spontaneous emission. In steady state, $\dot{\sigma}_{bb}$ is zero. The number of photons absorbed per unit time is therefore equal to the number of photons emitted spontaneously per unit time

$$\left\langle \frac{dN}{dt} \right\rangle_{st} = \Gamma \sigma_{bb}^{st}. \tag{C.11}$$

2. External Degrees of Freedom. Mean Radiative Forces

Until now, we have considered the atom as being infinitely heavy and at rest at the coordinate system origin. If its translational degrees of freedom are to be taken into consideration, the Hamiltonian (A.1) must be replaced by

$$H = \frac{\mathbf{P}^2}{2M} + H_A + H_R - \mathbf{d} \cdot \left[\mathbf{E}_e(\mathbf{R}, t) + \mathbf{E}_\perp(\mathbf{R}) \right] \tag{C.12}$$

where $\mathbf{P}$ and $\mathbf{R}$ are the momentum and the position of the center of mass of the atom, and M is the total mass. The first term of (C.12) represents the translational kinetic energy of the atom. The external field and the radiation field are now evaluated at center of mass $\mathbf{R}$ of the atom.

a) Equation of Motion of the Center of the Atomic Wave Packet

The Heisenberg equations for $\mathbf{R}$ and $\mathbf{P}$ are written

$$\dot{\mathbf{R}} = \frac{\partial H}{\partial \mathbf{P}} = \frac{\mathbf{P}}{M} \tag{C.13.a}$$

$$\dot{\mathbf{P}} = M\ddot{\mathbf{R}} = -\frac{\partial H}{\partial \mathbf{R}} = \sum_{j=x,y,z} d_j \nabla_{\mathbf{R}} \left[E_{ej}(\mathbf{R}, t) + E_{\perp j}(\mathbf{R}) \right]. \tag{C.13.b}$$

The average value of Equation (C.13.b), taken over the atomic wave function, gives (Ehrenfest equation):

$$M\langle\ddot{\mathbf{R}}\rangle = \sum_j \langle d_j \nabla_{\mathbf{R}}[E_{ej}(\mathbf{R}, t) + E_{\perp j}(\mathbf{R})]\rangle. \tag{C.14}$$

Let $\mathbf{r}_G = \langle\mathbf{R}\rangle$ be the center of the atom wave packet. The left-hand side of (C.14) is simply $M\ddot{\mathbf{r}}_G$. To evaluate the right-hand side, we will introduce two approximations.

i) Small Atomic Wave Packet Limit

Because of the large value of M, the de Broglie wavelength of the atom, $\lambda_{DB} = h/Mv$, is in general much smaller than the optical wavelength λ, which characterizes the scale of spatial variations of the driving field. It is therefore possible to construct atomic wave packets having very small dimensions compared with the optical wavelength. For such wave packets, it is completely legitimate to replace the operator $\mathbf{R}$ in the right-hand side of (C.14) by its mean value $\langle\mathbf{R}\rangle = \mathbf{r}_G$. It can be shown that the last term of (C.14), which represents the contribution of the gradient of the quantum radiation field at $\mathbf{r}_G$, is zero (*). Equation (C.14) is then written

$$M\ddot{\mathbf{r}}_G = \sum_{j=x,y,z} \langle d_j\rangle\nabla E_{ej}(\mathbf{r}_G, t). \tag{C.15}$$

The right-hand side can be interpreted as the force that governs the motion of the center $\mathbf{r}_G$ of the atomic wave packet. This force is expressed as a function of the driving field evaluated at this point.

ii) Existence of Two Distinct Time Scales for the Evolution of Internal and External Degrees of Freedom

As we saw above, the internal degrees of freedom of the atom evolve appreciably over time scales on the order of $T_{int} = \Gamma^{-1}$ (or Ω_1^{-1} if $\Omega_1^{-1} \gg \Gamma$). We will consider here only atoms of very slow velocity v, which, during T_{int}, travel over a distance $vT_{int} = v\Gamma^{-1}$ very small compared with the scale of the variation of the light wave (on the order of λ). Moreover, we will see later on that, under the influence of radiative forces, the velocity itself evolves over time scales on the order of $T_{ext} = \hbar/E_{rec}$ where $E_{rec} = \hbar^2 k^2/2M$ is the recoil energy of the atom when it absorbs a photon $k\varepsilon$. For most of the allowed transitions, $\hbar\Gamma \gg E_{rec}$, which results in the fact that $T_{int} \ll T_{ext}$ (for example, for the yellow lines of sodium, $\hbar\Gamma \simeq 400E_{rec}$).

(*) See the remark at the end of Section 2 of Complement A$_V$.

The large difference between T_{int} and T_{ext} results in the fact that, if $\mathbf{r}_G = \mathbf{0}$ initially, the mean dipole $\langle \mathbf{d} \rangle$ has the time to reach the steady-state regime calculated in subsection C-1-b before $\mathbf{r}_G$ has changed appreciably under the influence of the mean radiative force written in the right-hand side of (C.15). In what follows in this part, we will consider the mean radiative force exerted on an atom initially at rest at $\mathbf{0}$. To calculate such a force, we can thus replace $\langle d_j \rangle$ in (C.15) by its steady-state value calculated above.

Remarks

(i) We can, of course, also consider the mean radiative force acting on an atom having a velocity $\mathbf{v}$. Here again, the condition $T_{int} \ll T_{ext}$ allows us to neglect the variation of $\mathbf{v}$ during the time needed for $\langle d_j \rangle$ to reach a steady-state regime. The Bloch equations used for calculating $\langle d_j \rangle$ must then take into account the fact that the driving field "seen" by an atom in motion is not the same as for an atom at rest. The interest of such velocity-dependent radiative forces is that, in certain cases, they can damp the velocity of the atom and therefore constitute an efficient means of cooling.

(ii) In this entire section, we are interested only in the mean radiative force exerted on the atom. The fluctuations of this force around its mean value are responsible for a diffusion of the atomic momentum and a heating of the translational degrees of freedom. These physical phenomena may be analyzed by using Bloch–Langevin equations analogous to those introduced in Complement A_V (*).

b) The Two Types of Forces for an Atom Initially at Rest

Near the origin $\mathbf{0}$ where the atom is located, the driving field is written

$$\mathbf{E}_e(\mathbf{r}, t) = \mathbf{e}\mathcal{E}_0(\mathbf{r}) \cos[\omega_L t + \phi(\mathbf{r})]. \qquad (C.16)$$

Its amplitude $\mathcal{E}_0(\mathbf{r})$ and phase $\phi(\mathbf{r})$ vary in space. By contrast, we assume for the sake of simplicity that the polarization vector $\mathbf{e}$ does not depend on $\mathbf{r}$ (**). The time origin can always be chosen so that the phase $\phi(\mathbf{r})$ is zero at $\mathbf{r} = \mathbf{0}$.

$$\phi(\mathbf{0}) = 0. \qquad (C.17)$$

We thus obtain the field (A.2) introduced previously.

(*) See, for example, J. P. Gordon and A. Ashkin, *Phys. Rev. A*, **21**, 1606 (1980).
(**) Note, however, that there exist efficient cooling mechanisms associated with polarization gradients and multilevel atoms. See, for example, C. Cohen-Tannoudji and W. D. Phillips, *Physics Today*, October 1990, p. 33, and references therein.

Using (C.16) and (C.17), the field gradient appearing in (C.15) is written

$$\nabla E_{ej} = e_j[\cos \omega_L t \, \nabla \mathscr{E}_0 - \sin \omega_L t \, \mathscr{E}_0 \nabla \phi] \qquad (C.18)$$

where ∇E_{ej}, $\nabla \mathscr{E}_0$, $\nabla \phi$, and $\mathscr{E}_0$ are evaluated at $\mathbf{r} = 0$. We saw in the preceding subsection that it is possible to take, for the mean dipole $\langle \mathbf{d} \rangle$, expression (A.21) obtained above where u and v are replaced by their steady-state values.

$$\langle d_j \rangle = 2(\mathbf{d}_{ab})_j[u_{\text{st}} \cos \omega_L t - v_{\text{st}} \sin \omega_L t]. \qquad (C.19)$$

We then substitute (C.18) and (C.19) into (C.15) and take the average over an optical period. The expression we obtain for the mean radiative force $\mathscr{F}$ acting on the atom is

$$\mathscr{F} = \sum_j \overline{\langle d_j \rangle \nabla E_{ej}} = (\mathbf{e} \cdot \mathbf{d}_{ab})[u_{\text{st}} \nabla \mathscr{E}_0 + v_{\text{st}} \mathscr{E}_0 \nabla \phi] \qquad (C.20)$$

which causes two types of forces to appear: a force that we call reactive, proportional to the amplitude gradient and to the in-phase component of the dipole

$$\mathscr{F}_{\text{react}} = (\mathbf{e} \cdot \mathbf{d}_{ab}) u_{\text{st}} \nabla \mathscr{E}_0 \qquad (C.21)$$

and a force that we call dissipative, proportional to the phase gradient and to the quadrature component of the dipole

$$\mathscr{F}_{\text{dissip}} = (\mathbf{e} \cdot \mathbf{d}_{ab}) v_{\text{st}} \mathscr{E}_0 \nabla \phi. \qquad (C.22)$$

It will be convenient for what follows to reexpress these two forces as a function of the Rabi frequency Ω_1 which is written, according to (A.5) and (C.16):

$$\Omega_1 = -\mathbf{d}_{ab} \cdot \mathbf{e} \, \mathscr{E}_0 / \hbar. \qquad (C.23)$$

We then obtain

$$\mathscr{F}_{\text{react}} = -\hbar \Omega_1 u_{\text{st}} \boldsymbol{\alpha} \qquad (C.24.a)$$

with

$$\boldsymbol{\alpha} = \frac{\nabla \Omega_1}{\Omega_1} \qquad (C.24.b)$$

and

$$\mathscr{F}_{\text{dissip}} = -\hbar \Omega_1 v_{\text{st}} \boldsymbol{\beta} \qquad (C.25.a)$$

with

$$\boldsymbol{\beta} = \boldsymbol{\nabla}\phi. \tag{C.25.b}$$

c) DISSIPATIVE FORCE. RADIATION PRESSURE

The simplest example of a light wave having a phase gradient is the plane wave with wave vector $\mathbf{k}_L$

$$\mathbf{E}_e(\mathbf{r}, t) = \mathbf{e}\mathcal{E}_0 \cos(\omega_L t - \mathbf{k}_L \cdot \mathbf{r}) \tag{C.26}$$

which has a constant amplitude and a phase $\phi(\mathbf{r}) = -\mathbf{k}_L \cdot \mathbf{r}$ so that

$$\boldsymbol{\beta} = \boldsymbol{\nabla}\phi = -\mathbf{k}_L. \tag{C.27}$$

For such a wave, the reactive force is zero (because $\boldsymbol{\nabla}\mathcal{E}_0 = 0$) and the dissipative force is written, using (C.25) and (C.27):

$$\mathcal{F}_{\text{dissip}} = \Omega_1 v_{\text{st}} \hbar \mathbf{k}_L. \tag{C.28}$$

The equality (C.9) then allows us to transform (C.28) into

$$\mathcal{F}_{\text{dissip}} = \left\langle \frac{dN}{dt} \right\rangle_{\text{st}} \hbar \mathbf{k}_L \tag{C.29}$$

where $\langle dN/dt \rangle_{\text{st}}$ is the mean number of incident photons disappearing per unit time in the steady state regime. The physical interpretation of Equation (C.29) is very clear. Each incident photon carries a momentum $\hbar \mathbf{k}_L$ which is gained by the atom during the absorption of such a photon. If the atom returns to the ground state by stimulated emission of a photon, it loses this momentum which is then regained by the incident beam. By contrast, if the atom returns to the ground state by spontaneous emission, the loss of momentum is zero on average, because spontaneous emission occurs with equal probabilities in two opposite directions. Moreover, a fluorescence cycle, i.e., an absorption-spontaneous emission cycle, definitely causes a photon to disappear from the incident beam. Thus we can understand why the average momentum gained per unit time by the atom, i.e., the mean force acting on it, is $\hbar \mathbf{k}_L$ times the mean number of incident photons that disappear per unit time. For this reason, the force (C.29) is often called the "radiation pressure force" or the "resonant scattering force". Equation (C.11) also allows us to write (C.29) in the form

$$\mathcal{F}_{\text{dissip}} = \Gamma \sigma_{bb}^{\text{st}} \hbar \mathbf{k}_L \tag{C.30}$$

and shows that the dissipative force is also equal to $\hbar \mathbf{k}_L$ times the number of photons spontaneously emitted per unit time. When σ_{bb}^{st} is replaced by its value (C.3.c) we get

$$\mathcal{F}_{\text{dissip}} = \hbar k_L \frac{\Gamma}{2} \frac{\Omega_1^2/2}{\left(\omega_L - \omega_0\right)^2 + \left(\Gamma^2/4\right) + \left(\Omega_1^2/2\right)}. \qquad (C.31)$$

The dissipative force varies with the detuning $\delta_L = \omega_L - \omega_0$ as a Lorentzian centered at $\omega_L = \omega_0$, of full width at half-maximum $[\Gamma^2 + 2\Omega_1^2]^{1/2}$. At low intensity, the force is proportional to Ω_1^2, and thus to the intensity. At high intensity, it tends toward a limit

$$\mathcal{F}_{\text{sat}} = \hbar k_L \frac{\Gamma}{2} \qquad (C.32)$$

independent of the intensity.

Remarks

(i) We give an order of magnitude for $\mathcal{F}_{\text{sat}}$, or rather for the acceleration

$$\gamma = \frac{\mathcal{F}_{\text{sat}}}{M} \qquad (C.33)$$

that such a force can communicate to the atom. For the sodium atom excited on the yellow lines, we find $|\gamma| = 10^6$ m/s^2, or approximately 10^5 times the acceleration due to gravity.

(ii) The foregoing results allow us to evaluate an order of magnitude of the time T_{ext} characterizing the temporal variations of the external degrees of freedom. Consider an atom initially at rest, illuminated by a resonant plane wave. Under the influence of the corresponding radiation pressure, its velocity will increase from 0 to γt during a time t, γ being the modulus of (C.33). The time T_{ext} is the time after which the velocity has reached a value v_m such that the atom gets out of resonance with the incident wave because of the Doppler effect. It is then no longer affected by this wave. The time T_{ext} and the velocity v_m are thus given by the equations

$$\gamma T_{\text{ext}} \simeq \frac{\hbar k_L}{M} \Gamma T_{\text{ext}} = v_m \qquad (C.34.a)$$

$$k_L v_m \simeq \Gamma. \qquad (C.34.b)$$

Eliminating v_m one gets

$$T_{\text{ext}} \sim \frac{M}{\hbar k_L^2} \sim \frac{\hbar}{E_{\text{rec}}} \qquad (C.35)$$

which justifies the value $\hbar/E_{\text{rec}}$ taken above as the order of magnitude of T_{ext}.

d) REACTIVE FORCE. DIPOLE FORCE

A plane wave has no amplitude gradient. To have $\nabla \mathcal{E}_0$, and consequently $\mathcal{F}_{\text{react}}$, be nonzero, it is necessary to superimpose several plane waves with different wave vectors.

As an example, we consider the simple case of a standing wave resulting from the superposition of two plane waves of opposite wave vectors $+\mathbf{k}_L$ and $-\mathbf{k}_L$. If $\mathbf{k}_L$ is parallel to $0z$ and $\mathbf{e}$ is parallel to $0x$, the incident wave (C.16) is written

$$\mathbf{E}_e(\mathbf{r}, t) = \mathbf{e}_x \mathcal{E}_0 \cos k_L z \cos \omega_L t \qquad (\text{C.36})$$

and it has an amplitude that varies sinusoidally in space with a period $\lambda = 2\pi/k_L$. By contrast, the phase of the wave is constant. The reactive force, which, according to (C.21), depends on the component u of the dipole in phase with the driving field, does not involve any exchange of energy between the atom and the driving field. According to (C.8), these exchanges depend only on the quadrature component v. The absence of a global exchange of energy between the atom and the field does not, however, prevent a *redistribution* of energy among the different plane waves making up the driving wave. For example, an atom in a standing wave can indeed absorb a photon in the wave $+\mathbf{k}_L$ and then emit in a stimulated way a photon into the *other* wave $-\mathbf{k}_L$. At the end of this cycle of absorption into one wave and then stimulated emission into another wave, the global energy of the field has not changed. However, a photon has passed from wave $+\mathbf{k}_L$ to wave $-\mathbf{k}_L$, which causes the momentum of the field to vary by $-2\hbar\mathbf{k}_L$, and therefore the momentum of the atom to vary by $+2\hbar\mathbf{k}_L$.

To illustrate the foregoing argument, we represent in the complex plane (see Figure 2) the fields E_1 and E_2 of the two plane waves $\mathbf{k}_L$ and $-\mathbf{k}_L$ at a point where we assume, for example, that these fields are in quadrature (for more generality, we take E_1 and E_2 with different amplitudes). Consider the reactive response of the dipole to the total field E. This response u is in phase with E (or has a π phase shift depending on the sign of the detuning $\omega_L - \omega_0$). Let u_1 and u_2 be the projections of u on E_1 and E_2. The component u_1 in phase with E_1 does not absorb any energy from E_1. The same is true for u_2 and E_2. By contrast, u_2 has a $+\pi/2$ phase shift with respect to E_1, whereas u_1 has a $-\pi/2$ phase shift with respect to E_2. It follows that if wave E_1 loses energy by interacting with u_2, wave E_2 gains energy by interacting with u_1. Moreover, because $|E_1||u_2| = |E_2||u_1|$, then the energy lost by one wave is gained by the other. The foregoing scheme thus explains, on the one hand, the existence

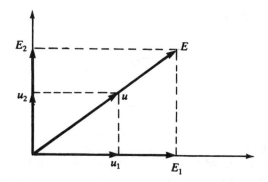

Figure 2. Representation in the complex plane of a field E resulting from the superposition of two fields E_1 and E_2 in quadrature. u, u_1, and u_2 are the dipoles in phase respectively with E, E_1, and E_2.

of a redistribution of energy between the two waves, and on the other hand, the *coherent nature* of this redistribution because its direction $(1 \to 2$ or $2 \to 1)$ depends only on the relative phases of the two waves at the point where the atom is located. Finally, note that, depending on whether u has a zero or π phase shift with respect to E, the direction of the redistribution is different. The same scheme also allows us to predict the direction of the redistribution, and thus the sign of the reactive force, depending on the sign of the detuning $\omega_L - \omega_0$.

To obtain the general expression for the reactive force, we substitute the expression (C.3.a) for u_{st} into (C.24.a). We then have

$$\mathscr{F}_{\text{react}} = -\frac{\hbar(\omega_L - \omega_0)}{4} \frac{\nabla(\Omega_1^2)}{(\omega_L - \omega_0)^2 + \frac{\Gamma^2}{4} + \frac{\Omega_1^2}{2}}. \qquad \text{(C.37)}$$

From this we find that $\mathscr{F}_{\text{react}}$, which varies with the detuning $\omega_L - \omega_0$ as a dispersion curve, changes sign with $\omega_L - \omega_0$. For $\omega_L < \omega_0$ (red detuning) the force attracts the atom toward regions of high intensity. For $\omega_L > \omega_0$ (blue detuning) the force repels the atoms away from the high intensity regions.

For each value of the intensity, and thus of Ω_1^2, the value of the detuning $\delta_L = \omega_L - \omega_0$ that maximizes $|\mathscr{F}_{\text{react}}|$ is different. We find that this value for $|\delta_L|$ is on the order of $|\Omega_1|$, so that the maximal value of

$|\mathcal{F}_{\text{react}}|$ is on the order of

$$|\mathcal{F}_{\text{react}}| \sim \frac{\hbar|\nabla\Omega_1^2|}{\Omega_1} \sim \hbar|\nabla\Omega_1|. \tag{C.38}$$

By contrast with the dissipative force, the reactive force increases without bound when the intensity of the wave increases. Because $|\nabla\Omega_1|$ is at most on the order of $k_L\Omega_1$, where $k_L = 2\pi/\lambda$, the maximum value of $|\mathcal{F}_{\text{react}}|$ is about $\hbar k_L\Omega_1$, which corresponds to momentum exchanges of $\hbar k_L$ occurring at a rate Ω_1, as should be the case for a force involving absorption-stimulated emission cycles. Such a result can be compared with the maximal value (C.32) of the dissipative force, which is on the order of $\hbar k_L$ times the rate of spontaneous emission Γ.

Finally, note that the reactive force (C.37) derives from a potential

$$\mathcal{F}_{\text{react}} = -\nabla U \tag{C.39}$$

where

$$U = \frac{\hbar(\omega_L - \omega_0)}{2} \ln\left[1 + \frac{\Omega_1^2/2}{(\omega_L - \omega_0)^2 + (\Gamma^2/4)}\right]. \tag{C.40}$$

For a red detuning, the regions of maximum intensity, such as the focal zone of a laser beam, thus appear as potential wells for the atom. Such wells can be used to trap neutral atoms.

The reactive force $|\mathcal{F}_{\text{react}}|$ is also called the dipole force. We will see later on in Chapter VI that it can be given another physical interpretation, in terms of energy levels for the atom dressed by the incident photons. This picture will also allow us to understand the mechanism of fluctuations of this force about its mean value. These fluctuations considerably limit the stability of the optical traps based on this force.

D—PROPERTIES OF THE LIGHT EMITTED BY THE ATOM

In this last part, we show how the optical Bloch equations, which are atomic evolution equations, also allow us to analyze the properties of the fluorescence emitted by the atom. To do this, we use the fact that the field radiated by the atom onto the photodetector observing the emitted light is proportional to the atomic dipole. The photodetection signals, which, according to the results in Complement A_{II}, are proportional to the correlation functions of the field arriving at the detector, can therefore be expressed in terms of one- or two-time averages of the emitting dipole moment (§1). An important example of a light signal proportional to a one-time average is the total intensity emitted by the atom. Using this simple example, we will show how it is possible to distinguish in the emitted light the contribution of the mean dipole from that of the dipole fluctuations around its mean value (§2). Most of the other signals are proportional to two-time averages. The quantum regression theorem, derived in Complement A_V, allows us to study the evolution of two-time averages of the atomic dipole, by using the optical Bloch equations. We will apply it here to the study of the spectral distribution of the emitted light (§3).

1. Photodetection Signals. One- and Two-Time Averages of the Emitting Dipole Moment

a) CONNECTION BETWEEN THE RADIATED FIELD AND THE EMITTING DIPOLE MOMENT

We consider the field, at the position r_D of the detector and at time t, radiated by the atom located at $\mathbf{0}$. This field is proportional to the dipole $\mathbf{d}$ of the atom (*) at time $t - (r_D/c)$:

$$E(\mathbf{r}_D, t) = \eta d\left(t - \frac{r_D}{c}\right) \qquad (D.1)$$

where η is a proportionality coefficient. To simplify notation, we have ignored the vector nature of E and d. Equation (D.1) is valid between operators in the Heisenberg representation. We are interested in the evolution of operators having frequencies close to $\pm\omega_L$ (or $\pm\omega_0$). We call $E^{(+)}$ and $E^{(-)}$ the positive and negative frequency components of the

(*) Strictly speaking, E is proportional to $\ddot{d}$ (see *Photons and Atoms—Introduction to Quantum Electrodynamics*, Exercise 6 in Complement E_{IV}). However, we are interested here in the motion of d having frequencies close to ω_L so that $\ddot{d} \simeq -\omega_L^2 d$.

operator $E(\mathbf{r}_D, t)$. Similarly, according to (A.9), (A.10), and (A.15), the dipole d can be split into positive and negative frequency components $S_- \exp(-i\omega_L t)$ and $S_+ \exp(+i\omega_L t)$. Starting with (D.1), it is possible to write

$$E^{(\pm)}(\mathbf{r}_D, t) = \eta \, e^{\mp i\omega_L(t - r_D/c)} \, S_\mp\left(t - \frac{r_D}{c}\right). \qquad (D.2)$$

Remark

Strictly speaking, expression (D.1) does not represent the *total* field at the detector position, even if, as we assume here, this detector is located outside the incident laser beam, so that $E_e(\mathbf{r}_D, t) = 0$. Integrating the equations of motion for the operators a and a^+ of the field in the presence of sources actually causes two contributions to appear [see, for example, Equations (23)–(26) in Complement A_V]: one that depends on the emitting dipole moment and gives rise to the radiated field or *source field*, and the other, which is independent of the dipole and represents the *quantum vacuum field*. It can be shown that the latter does not contribute to the photodetection signals [see Remarks (ii) and (iii) at the end of the following subsection]. We will thus ignore it here.

b) EXPRESSION OF PHOTODETECTION SIGNALS

In Complement A_{II}, we introduced two correlation functions for the field E arriving at the detector

$$C_1(t, \tau) = \left\langle E^{(-)}(\mathbf{r}_D, t + \tau) E^{(+)}(\mathbf{r}_D, t) \right\rangle \qquad (D.3)$$

$$C_2(t, \tau) = \left\langle E^{(-)}(\mathbf{r}_D, t) E^{(-)}(\mathbf{r}_D, t + \tau) E^{(+)}(\mathbf{r}_D, t + \tau) E^{(+)}(\mathbf{r}_D, t) \right\rangle. \qquad (D.4)$$

Several photodetection signals may be expressed in terms of these correlation functions. For instance,

$$C_1(t, 0) = \left\langle E^{(-)}(\mathbf{r}_D, t) E^{(+)}(\mathbf{r}_D, t) \right\rangle = \left\langle I(t) \right\rangle \qquad (D.5)$$

is a one-time average which is equal to the *total average intensity* at time t, as it is measured by a broadband photodetector. The Fourier transform of $C_1(t, \tau)$ with respect to τ

$$\mathscr{I}(\omega) = \frac{1}{2\pi} \int_{-\infty}^{+\infty} d\tau \, e^{-i\omega\tau} \left\langle E^{(-)}(\mathbf{r}_D, t + \tau) E^{(+)}(\mathbf{r}_D, t) \right\rangle \qquad (D.6)$$

gives the *spectral density* of the radiation, as measured by a narrow-band photodetector. Finally, $C_2(t, \tau)$ represents a *photon correlation signal*, proportional to the probability density of detecting one photoelectron at time t *and* another at time $t + \tau$.

In the two following subsections, we will study the total intensity (D.5) as well as the spectral distribution (D.6) of the emitted light, the photon correlation signals (D.4) being discussed in subsection E-3 of Chapter VI. Using (D.2) in (D.5) and (D.6) then gives

$$\langle I(t) \rangle = \eta^2 \langle S_+(t - (r_D/c)) S_-(t - (r_D/c)) \rangle \qquad (D.7)$$

$$\mathscr{I}(\omega) = \frac{\eta^2}{2\pi} \int_{-\infty}^{+\infty} d\tau \, e^{i(\omega_L - \omega)\tau} \times$$
$$\times \langle S_+[t + \tau - (r_D/c)] S_-[t - (r_D/c)] \rangle. \qquad (D.8)$$

Remarks

(i) Because we take into consideration here the source field E, this field is not free and the two operators $E^{(+)}$ in (D.4) do not commute with each other; nor do the two $E^{(-)}$. Recall that the interpretation of (D.4) as a photon correlation signal is valid only if τ is positive (see the remark at the end of Complement A_{II}).

(ii) We return now to the contribution of the vacuum field to the photodetection signals. Equation (D.2) must be replaced by

$$E^{(\pm)}(\mathbf{r}_D, t) = E_0^{(\pm)}(\mathbf{r}_D, t) + \eta \, e^{\mp i\omega_L(t - (r_D/c))} S_\mp[t - (r_D/c)] \quad (D.9)$$

where E_0 is the quantum vacuum field. Several terms depending on $E_0^{(\pm)}$ then appear in expressions for C_1 and C_2 when (D.9) is inserted into (D.3) and (D.4). However, it is possible to show that all these new terms equal zero. Consider, for example, the function C_1. The state of the quantum field is the vacuum, so that

$$E_0^{(+)}(\mathbf{r}_D, t)|0\rangle = 0 \qquad (D.10.a)$$

$$\langle 0|E_0^{(-)}(\mathbf{r}_D, t) = 0. \qquad (D.10.b)$$

Because the $E_0^{(\pm)}$ appear in the normal order in (D.3), (D.10) then shows that all the new terms appearing in C_1 are zero.

(iii) The foregoing demonstration does not apply to C_2. Some terms having the form

$$\langle 0|S_+(t - (r_D/c)) S_+(t + \tau - (r_D/c)) \times$$
$$\times E_0^{(+)}(\mathbf{r}_D, t + \tau) S_-(t - (r_D/c))|0\rangle \qquad (D.11)$$

remain, because the presence of S_- to the right of $E_0^{(+)}$ prevents $E_0^{(+)}$ from acting on $|0\rangle$. The fact that τ is positive in C_2 [see Remark (i)] nevertheless allows us to show that $E_0^{(+)}(\mathbf{r}_D, t + \tau)$ commutes with $S_-[t - (r_D/c)]$ and can thus act on $|0\rangle$ to yield a zero result. We will give here only a simple physical argument to explain this result (*). The quantum field $E_0^{(+)}(\mathbf{r}_D, t + \tau)$ does not commute with the quantum field $E_0^{(-)}(\mathbf{r}', t')$ if $\mathbf{r}', t'$ is on the light cone of $\mathbf{r}_D, t + \tau$. Such a light cone intersects the time axis for the emitting atom located at $\mathbf{0}$ at time $t + \tau - (r_D/c)$ which is in the future of the time $t - (r_D/c)$ appearing in the operator S_- of (D.11), because τ is positive. However, the dipole $S_-(t - (r_D/c))$ depends only on the quantum field $E_0^{(+)}$ at its position that acted on it prior to $t - (r_D/c)$. It then commutes with $E_0^{(+)}(\mathbf{r}_D, t + \tau)$ for $\tau > 0$.

2. Total Intensity of the Emitted Light

a) PROPORTIONALITY TO THE POPULATION OF THE ATOMIC EXCITED STATE

Because the two operators S_+ and S_- appearing in expression (D.7) for $\langle I(t) \rangle$ are taken at the same time, we can use the relation $S_+ S_- = |b\rangle\langle b|$, which gives

$$\langle I(t) \rangle = \eta^2 \sigma_{bb}(t - (r_D/c)). \tag{D.12}$$

Such a result expresses physically that the light energy arriving at the detector at time t is proportional to the population σ_{bb} of the atomic excited state at time $t - (r_D/c)$. It is the basis for optical detection methods which consist of using the signal given by a broadband photodetector to follow the evolution of atomic populations σ_{bb} and $\sigma_{aa} = 1 - \sigma_{bb}$, in the transient regime as well as in the steady state. In the latter case, the study of the variations of σ_{bb} with $\omega_L - \omega_0$ allows the phenomenon of optical resonance to be detected.

b) COHERENT SCATTERING AND INCOHERENT SCATTERING

We rewrite the operators $S_\pm$ of (D.7) in the form

$$S_\pm(t - (r_D/c)) = \langle S_\pm(t - (r_D/c)) \rangle + \delta S_\pm(t - (r_D/c)) \tag{D.13}$$

where $\langle S_\pm \rangle$ is the average value of $S_\pm$ and where

$$\delta S_\pm(t - (r_D/c)) = S_\pm(t - (r_D/c)) - \langle S_\pm(t - (r_D/c)) \rangle \tag{D.14}$$

(*) For a more detailed demonstration, see, for example, B. R. Mollow, *J. Phys. A*, **8**, L130 (1975).

is the difference between $S_\pm$ and its average value, satisfying

$$\langle \delta S_\pm(t - (r_D/c)) \rangle = 0. \tag{D.15}$$

Such a separation allows us to distinguish between two components in the light radiated by the "instantaneous" dipole $S_\pm$. First, the radiation of the average dipole $\langle S_\pm \rangle$ which is the radiation of a classical oscillating dipole with a phase that is well defined relative to the incident laser field. The light radiated by $\langle S_\pm \rangle$ can then interfere with the incident field. It is for this reason that the radiation of $\langle S_\pm \rangle$ is frequently associated with a *coherent scattering* process. The other component $\delta S_\pm$ of $S_\pm$ radiates a field which does not have a phase that is well defined relative to the incident field because this radiation comes from the fluctuating part of the atomic dipole. The corresponding scattering process is called *incoherent*.

c) RESPECTIVE CONTRIBUTIONS OF COHERENT AND INCOHERENT SCATTERING TO THE TOTAL INTENSITY EMITTED IN STEADY STATE

In the steady state, $\langle S_+(t) \rangle$ and $\langle S_+(t)S_-(t) \rangle$ do not depend on t. By inserting (D.13) into (D.7), we obtain

$$\langle I \rangle = \eta^2 \langle S_+ \rangle \langle S_- \rangle + \eta^2 \langle \delta S_+ \delta S_- \rangle. \tag{D.16}$$

The first term of (D.16) represents the contribution to $\langle I(t) \rangle$ of the mean dipole, whereas the second term represents the contribution of the fluctuations of the dipole. We call the two contributions $\langle I_{coh} \rangle$ and $\langle I_{incoh} \rangle$.

To calculate $\langle I_{coh} \rangle$ and $\langle I_{incoh} \rangle$, it is sufficient to use the steady-state solution to the optical Bloch equations given in subsection C-1-b above. We then obtain, using (A.15), (A.16), (A.19), and (C.5):

$$\frac{1}{\eta^2} \langle I_{coh} \rangle = |\langle S_+ \rangle|^2 = |u_{st} + iv_{st}|^2 = \frac{1}{2} \frac{s}{(1 + s)^2} \tag{D.17}$$

where s is the saturation parameter defined in (C.4)

$$\frac{1}{\eta^2} \langle I_{incoh} \rangle = \langle S_+ S_- \rangle - |\langle S_+ \rangle|^2$$

$$= \sigma_{bb}^{st} - |u_{st} + iv_{st}|^2$$

$$= \frac{1}{2} \frac{s^2}{(1 + s)^2}. \tag{D.18}$$

Figure 3 shows the variations of $\langle I_{coh} \rangle$ and $\langle I_{incoh} \rangle$ with s (which is

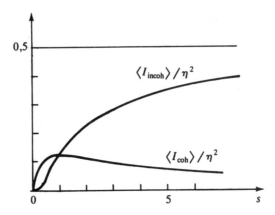

Figure 3. Variations of $\langle I_{\mathrm{coh}} \rangle$ and $\langle I_{\mathrm{incoh}} \rangle$ with the saturation parameter s defined in (C.4).

proportional to the incident laser intensity I_L). For low values of I_L ($s \ll 1$), $\langle I_{\mathrm{coh}} \rangle$ is proportional to s, and thus to I_L, whereas $\langle I_{\mathrm{incoh}} \rangle$ is proportional to s^2, and thus to I_L^2. The scattering process is then essentially coherent and we can define a scattering cross section, because the total radiated intensity is proportional to I_L, and thus to the incident flux. At high intensities ($s \gg 1$), $\langle I_{\mathrm{coh}} \rangle$ tends to zero. Indeed, as a consequence of the saturation of the atomic transition, the mean dipole $\langle S_+ \rangle$ tends to zero. By contrast, $\langle I_{\mathrm{incoh}} \rangle$ is important and becomes almost independent of I_L. Such a result expresses the fact that the atom spends half its time in the higher state b. Therefore, it cannot emit more than $\Gamma/2$ photons per unit time, regardless of the incident intensity. Finally, it can be seen that at high intensities, the scattering is essentially incoherent.

3. Spectral Distribution of the Emitted Light in Steady State

a) RESPECTIVE CONTRIBUTIONS OF COHERENT AND INCOHERENT
    SCATTERING. ELASTIC AND INELASTIC SPECTRA

In steady state, $\langle S_+ \rangle$ is time independent. Substituting (D.13) into (D.8) yields

$$\mathscr{I}(\omega) = \mathscr{I}_{\mathrm{coh}}(\omega) + \mathscr{I}_{\mathrm{incoh}}(\omega) \tag{D.19}$$

where

$$\mathscr{I}_{\text{coh}}(\omega) = \frac{\eta^2}{2\pi} \int_{-\infty}^{+\infty} d\tau \, e^{i(\omega_L - \omega)\tau} \left| \langle S_+ \rangle \right|^2 \tag{D.20}$$

$$\mathscr{I}_{\text{incoh}}(\omega) = \frac{\eta^2}{2\pi} \int_{-\infty}^{+\infty} d\tau \, e^{i(\omega_L - \omega)\tau} \langle \delta S_+(t + \tau) \delta S_-(t) \rangle. \tag{D.21}$$

Integrating (D.20) gives, using (D.17)

$$\mathscr{I}_{\text{coh}}(\omega) = \langle I_{\text{coh}} \rangle \delta(\omega - \omega_L). \tag{D.22}$$

The light emitted in the steady state by the mean dipole is therefore monochromatic, with frequency ω_L. Such a result is indeed in accordance with the physical picture of a mean dipole oscillating in the forced regime at the frequency ω_L under the influence of laser excitation, and thus radiating a field having the same frequency. In the low-intensity limit, the corresponding scattering process is indeed described by the lowest-order elastic scattering diagram, shown in Figure 29α of Chapter II. The spectrum (D.22) is thus an elastic scattering spectrum.

The contribution (D.21) of the incoherent scattering to $\mathscr{I}(\omega)$ is simply the spectral density of the dipole fluctuations. We will show later on that such a density is not monochromatic. It is for this reason that $\mathscr{I}_{\text{incoh}}(\omega)$ is an inelastic scattering spectrum.

b) Outline of the Calculation of the Inelastic Spectrum

In the steady state, the two-time average appearing in (D.21) does not depend on t and is equal to $\langle \delta S_+(\tau) \delta S_-(0) \rangle$. The fact that $S_+ = (S_-)^+$ and the invariance by time translation leads to

$$\langle \delta S_+(\tau) \delta S_-(0) \rangle = \langle \delta S_+(0) \delta S_-(\tau) \rangle^* = \langle \delta S_+(-\tau) \delta S_-(0) \rangle^* \tag{D.23}$$

which allows us to rewrite (D.21) in the form

$$\mathscr{I}_{\text{incoh}}(\omega) = \frac{\eta^2}{2\pi} 2 \,\text{Re} \int_0^{\infty} d\tau \, e^{i(\omega_L - \omega)\tau} \langle \delta S_+(\tau) \delta S_-(0) \rangle \tag{D.24}$$

which involves only the positive values of τ.

We now write the Bloch–Langevin equations satisfied by the three operators $\delta S_q(\tau)$ (with $q = +, -, z$). To do this, subtract from (B.3) Equation (B.2), which is satisfied by the average values $\langle S_q \rangle$. The source terms λ_q disappear, and we then have

$$\delta\dot{S}_q(\tau) = \sum_{q'}\mathcal{B}_{qq'}\delta S_{q'}(\tau) + F_q(\tau) \tag{D.25}$$

where $\mathcal{B}_{qq'}$ is the Bloch matrix and $F_q(\tau)$ is the Langevin force acting on $S_q(\tau)$. Because τ is positive in (D.24), we can then use the quantum regression theorem, derived in Complement A_V, which states that, for $\tau > 0$, the three two-time averages $\langle\delta S_q(\tau)\delta S_-(0)\rangle$ obey the same equations as the three one-time averages $\langle\delta S_q(\tau)\rangle$; that is,

$$\text{if } \tau > 0, \quad \frac{d}{d\tau}\langle\delta S_q(\tau)\delta S_-(0)\rangle = \sum_{q'}\mathcal{B}_{qq'}\langle\delta S_{q'}(\tau)\delta S_-(0)\rangle. \tag{D.26}$$

It thus appears that $\langle\delta S_q(\tau)\delta S_-(0)\rangle$ is a superposition of three exponentials $\exp(-r_\lambda\tau)$, where the $-r_\lambda$ $(\lambda = 1, 2, 3)$ are the three eigenvalues of the Bloch matrix $\mathcal{B}_{qq'}$.

In the following subsection, we study only the few limiting cases previously considered in subsection C-1-a above in connection with the transient evolution of one-time averages. A more complete calculation can be made, using, for example, a Fourier–Laplace transformation of Equations (D.26) (see the references at the end of the chapter).

c) INELASTIC SPECTRUM IN A FEW LIMITING CASES

First assume that $\omega_L = \omega_0$ and $\Omega_1 \ll \Gamma$ (low-intensity resonant excitation). We saw above (see §C-1-a) that two r_λ are equal to $\Gamma/2$, whereas the third one equals Γ. The inelastic spectrum is then, according to (D.24), composed of two lines having a total width at half-maximum of 2Γ and Γ, both centered at ω_L.

We now consider the limit $|\omega_L - \omega_0| \gg \Omega_1, \Gamma$ which is the limit of a nonresonant excitation. According to formula (C.2) given above, the spectrum $\mathcal{S}_{incoh}(\omega)$ is made up of three lines: a central line at ω_L, of width 2Γ (associated with r_3) and two sidebands centered at $\omega_L + (\omega_L - \omega_0) = 2\omega_L - \omega_0$ and $\omega_L - (\omega_L - \omega_0) = \omega_0$, of width Γ (associated, respectively, with r_1 and r_2). The two sidebands are associated with the second-order nonlinear scattering process represented in Figure 29β of Chapter II (see also Figure 30 of Chapter II). The central line of width 2Γ can be interpreted as being associated with a *inverse* scattering process starting from the excited state b (spontaneous emission of a photon ω from b, then absorption of a laser photon ω_L and return to state b), the atom being initially prepared in the state b by the three-photon process shown in Figure 28 of Chapter II.

Finally, at high intensities and for a resonant excitation ($\Omega_1 \gg \Gamma$, $\omega_L = \omega_0$), we can use result (C.1), which leads to the fact that the inelastic spectrum is made up of a central line at ω_L of width Γ (associated with r_3) and two sidebands centered at $\omega_L + \Omega_1$ and $\omega_L - \Omega_1$ of width $3\Gamma/2$ (associated with r_1 and r_2) (*). This is the fluorescence triplet for which it is not possible to give a perturbative interpretation in terms of scattering processes involving several photons. We shall reexamine this problem in Chapter VI by using the dressed atom approach and we will also study the other optical detection signal:

$$C_2(t, \tau) = \eta^4 \langle S_+(t - (r_D/c)) S_+(t + \tau - (r_D/c)) \times$$

$$\times S_-(t + \tau - (r_D/c)) S_-(t - (r_D/c)) \rangle \quad \text{(D.27)}$$

which, for $\tau > 0$, represents a photon-correlation signal.

GENERAL REFERENCES

Bloch equations in nuclear magnetic resonance: Abragam.

Optical Bloch equations: Allen and Eberly (Chapter 2), Cohen-Tannoudji. The transient solutions of the Bloch equations can be found in H. C. Torrey, *Phys. Rev.*, **76**, 1059 (1949).

Average radiative forces: The study of these forces using the Heisenberg equations and the Ehrenfest theorem can be found in R. J. Cook, *Phys. Rev. A*, **20**, 224 (1979).

For review articles on radiative forces and their applications, see A. Ashkin, Science, **210**, 1081 (1980); R. J. Cook, *Comments At. Mol. Phys.*, **10**, 267 (1981); V. S. Letokhov and V. G. Minogin, *Phys. Rep.*, **73**, 1 (1981); S. Stenholm, *Rev. Mod. Phys.* **58**, 699 (1986).

Calculation of fluorescence signals starting from optical Bloch equations: B. R. Mollow, *Phys. Rev.*, **188**, 1969 (1969). See also review articles by Mollow and Cohen-Tannoudji.

(*) For experimental observations of this spectrum, see F. Schuda, C. R. Stroud, and M. Hercher, *J. Phys. B*, **7**, L 198 (1974); R. E. Grove, F. Y. Wu, and S. Ezekiel, *Phys. Rev. A*, **15**, 227 (1977); J. D. Cresser, J. Hager, G. Leuchs, M. Rateike, and H. Walther, in *Dissipative Systems in Quantum Optics*, Vol. 27 of *Topics in Current Physics*, edited by R. Bonifacio (Springer-Verlag, Berlin, 1982).

COMPLEMENT A_V

BLOCH–LANGEVIN EQUATIONS AND
QUANTUM REGRESSION THEOREM

In Section D of the chapter, we showed how the characteristics of the light emitted by an atom excited by a laser beam are related to the two-time averages of the emitting atomic dipole. The purpose of this complement is to show that the dynamics of these two-time averages are determined by equations that are completely analogous to the optical Bloch equations. The results obtained in Complement C_{IV} for a harmonic oscillator can then be generalized to a two-level atom. Note, however, that such a generalization is not obvious because of the nonlinearities associated with a two-level system.

We begin by writing the Heisenberg equations for the atomic dipole and for the quantum radiation field in the presence of the driving field, which is assumed to be monochromatic and to have a frequency close to the atomic eigenfrequency (§1). Then, by formally solving the equations of motion for the field, we decompose the total quantum field into two parts: the free field and the field produced by the dipole (§2). When this expression for the total field is substituted into the Heisenberg equation for the atomic dipole, this equation then takes the form of a Langevin equation. The rate of variation of the atomic observables indeed appears as the sum of an average rate (similar to that of the Bloch equations) and a Langevin force, which is zero on average and fluctuates rapidly (§3). Using these Bloch–Langevin equations, we finally derive the quantum regression theorem, which states that the quantum correlations regress as the average values.

1. Coupled Heisenberg Equations for the Atom and the Field

We consider exactly the same system as the one described in the chapter. One atom, represented by a two-level system, is fixed at the coordinate-system origin **0**. It interacts with the driving field $E_e(0, t)$ of frequency ω_L given in (A.2) and with the quantum radiation field $E(0)$, assumed to be initially in the vacuum state $|0\rangle$.

a) HAMILTONIAN AND OPERATOR BASIS FOR THE SYSTEM

Within the long-wavelength approximation, and in the electric dipole representation, the Hamiltonian of the system is the one previously used in the chapter and is given by (A.1). We write it in the form

$$H = H_A + H_R + H_{Ie} + H_I. \tag{1}$$

H_A is the atomic Hamiltonian which we assume to include the dipole self energy $\varepsilon_{\rm dip}$ [see formula (76) in the Appendix]. The atom is considered as a two-level system, whose internal energy is measured from the ground state $|a\rangle$ so that H_A is simply

$$H_A = \hbar\omega_0|b\rangle\langle b|. \tag{2}$$

H_R is the Hamiltonian of the quantum radiation field

$$H_R = \int d^3k \sum_{\varepsilon \perp k} \hbar\omega\left[a_\varepsilon^+(\mathbf{k})a_\varepsilon(\mathbf{k}) + \tfrac{1}{2}\right]. \tag{3}$$

Within the long-wavelength approximation, the sum over the modes $\mathbf{k}\varepsilon$ is limited to $|\mathbf{k}| < k_M$, with k_M satisfying the long-wavelength condition.

The electric dipole moment operator $\mathbf{d}$ is assumed to have a real matrix element between $|a\rangle$ and $|b\rangle$, directed along $0z$, and written $d_{ab}\mathbf{e}_z$:

$$\mathbf{d} = (|b\rangle\langle a| + |a\rangle\langle b|)d_{ab}\mathbf{e}_z. \tag{4}$$

The Hamiltonian for the interaction with the driving field H_{Ie} is written:

$$\begin{aligned} H_{Ie} &= -\mathbf{d} \cdot \mathbf{E}_e(0, t) \\ &= \hbar\Omega_1(|b\rangle\langle a| + |a\rangle\langle b|) \cos \omega_L t \end{aligned} \tag{5}$$

where Ω_1 is the Rabi frequency defined by (A.5). Finally, H_I is the interaction Hamiltonian with the quantum field:

$$\begin{aligned} H_I &= -\mathbf{d} \cdot \mathbf{E}_\perp(0) \\ &= -d_{ab}E_{\perp z}(0)(|b\rangle\langle a| + |a\rangle\langle b|). \end{aligned} \tag{6}$$

Recall that the field $\mathbf{E}_\perp(\mathbf{r})$ is defined by expression (89) in the Appendix

$$\mathbf{E}_\perp(\mathbf{r}) = i\int d^3k \sum_{\varepsilon \perp k} \mathscr{E}_\omega\left(a_\varepsilon(\mathbf{k})\boldsymbol{\varepsilon}\, e^{i\mathbf{k}\cdot\mathbf{r}} - a_\varepsilon^+(\mathbf{k})\boldsymbol{\varepsilon}\, e^{-i\mathbf{k}\cdot\mathbf{r}}\right) \tag{7}$$

in which the sum over $\mathbf{k}$ is limited to $|k| < k_M$.

The equivalence of the two-level system with a spin $\frac{1}{2}$, described in Subsection A-4 of the chapter, allows us to express any atomic observable as a linear combination of the four operators

$$\{\mathscr{S}_+, \mathscr{S}_z, \mathscr{S}_-, \mathbb{1}\} \tag{8}$$

with

$$\mathscr{S}_+ = |b\rangle\langle a| \tag{9.a}$$
$$\mathscr{S}_- = |a\rangle\langle b| \tag{9.b}$$
$$\mathscr{S}_z = \tfrac{1}{2}[|b\rangle\langle b| - |a\rangle\langle a|] \tag{9.c}$$

$\mathbb{1}$ being the unit operator. Therefore

$$H_A = \hbar\omega_0[\mathscr{S}_z + 1/2] \tag{10.a}$$
$$H_{Ie} + H_I = [\hbar\Omega_1 \cos \omega_L t - d_{ab}E_{\perp z}(0)](\mathscr{S}_+ + \mathscr{S}_-). \tag{10.b}$$

Similarly, the observables of the field are expressed as a function of the operators

$$\{\ldots, a_\varepsilon(\mathbf{k}), a_\varepsilon^+(\mathbf{k}), \ldots\} \tag{11}$$

so that it is sufficient to know the evolution of the operators (8) and (11) to determine the evolution of the total system.

b) EVOLUTION EQUATIONS FOR THE ATOMIC AND FIELD OBSERVABLES

In the Heisenberg representation, the atomic operators are time dependent, and their commutation relations at a given time t,

$$[\mathscr{S}_+(t), \mathscr{S}_-(t)] = 2\mathscr{S}_z(t) \tag{12.a}$$
$$[\mathscr{S}_z(t), \mathscr{S}_\pm(t)] = \pm\mathscr{S}_\pm(t) \tag{12.b}$$

allow us to explicitly write the Heisenberg equation

$$\dot{A}(t) = \frac{1}{i\hbar}[A(t), H(t)] \tag{13}$$

for each of the three operators $\mathscr{S}_+$, $\mathscr{S}_-$, and $\mathscr{S}_z$. We then have

$$\frac{d}{dt}\mathscr{S}_+(t) = \frac{1}{i\hbar}\left[\mathscr{S}_+(t), \hbar\omega_0\left(\mathscr{S}_z(t) + \frac{1}{2}\right)\right] +$$

$$+ \frac{1}{i\hbar}[\mathscr{S}_+(t), \mathscr{S}_+(t) + \mathscr{S}_-(t)](\hbar\Omega_1 \cos \omega_L t - d_{ab}E_{\perp z}(0, t))$$

$$= i\omega_0\mathscr{S}_+(t) - 2i\mathscr{S}_z(t)\left(\Omega_1 \cos \omega_L t - \frac{d_{ab}}{\hbar}E_{\perp z}(0, t)\right). \tag{14}$$

The adjoint of (14) gives the equation for $\mathcal{S}_-(t)$

$$\frac{d}{dt}\mathcal{S}_-(t) = -i\omega_0\mathcal{S}_-(t) + 2i\mathcal{S}_z(t)\left(\Omega_1\cos\omega_L t - \frac{d_{ab}}{\hbar}E_{\perp z}(0,t)\right).$$

$$(15)$$

Similarly, we obtain for $\mathcal{S}_z(t)$

$$\frac{d}{dt}\mathcal{S}_z(t) = -i(\mathcal{S}_+(t) - \mathcal{S}_-(t))\left(\Omega_1\cos\omega_L t - \frac{d_{ab}}{\hbar}E_{\perp z}(0,t)\right). \quad (16)$$

To derive these equations, we used the fact that the field operator $E_{\perp z}(0,t)$ commutes with the atomic operators taken at the same time.

The evolution of the field is entirely determined by that of the operators $a_\varepsilon(\mathbf{k})$ and their adjoints. The operator $a_\varepsilon(\mathbf{k})$ commutes with the atomic operators taken at the same time, and satisfies the usual commutation relations for the creation and annihilation operators. The Heisenberg equation for the operator $a_\varepsilon(\mathbf{k})$ is written

$$\dot{a}_\varepsilon(\mathbf{k},t) = \frac{1}{i\hbar}[a_\varepsilon(\mathbf{k},t),H(t)]$$

$$= \frac{1}{i\hbar}[a_\varepsilon(\mathbf{k},t),H_R(t) - \mathbf{d}\cdot\mathbf{E}_\perp(0,t)]$$

$$= -i\omega a_\varepsilon(\mathbf{k},t) + \frac{1}{\hbar}\mathcal{E}_\omega\boldsymbol{\varepsilon}\cdot\mathbf{d}(t). \quad (17)$$

c) Rotating-Wave Approximation. Change of Variables

As in the chapter, we are interested only in quasi-resonant radiative processes and therefore we make the rotating-wave approximation. In the Heisenberg picture, such an approximation consists of neglecting in the right-hand side of the equations of motion (14)–(17) the coupling terms between the atom and the field having eigenfrequencies that are very different from those of the operator appearing in the left-hand side. For example, in Equation (14), we retain only the component $(\exp i\omega_L t)/2$ of $\cos\omega_L t$ and the terms in $a_\varepsilon^+(\mathbf{k})$ of $E_{\perp z}(0,t)$. It is also convenient to introduce operators whose rates of variation are slow relative to the optical frequencies. For the atom, we therefore use the operators S_+, S_z, and S_-, defined by (A.15). For the field, we use the operators

$$a_\varepsilon(\mathbf{k},t) = \hat{a}_\varepsilon(\mathbf{k},t)\,e^{-i\omega t}.$$

Equations (14)–(17) then become

$$\dot{S}_+(t) = -i\delta_L S_+(t) - i\Omega_1 S_z(t) + 2i\frac{d_{ab}}{\hbar} E_z^{(c)}(0,t)\, e^{-i\omega_L t}\, S_z(t) \quad (18.a)$$

$$\dot{S}_-(t) = i\delta_L S_-(t) + i\Omega_1 S_z(t) - 2i\frac{d_{ab}}{\hbar} E_z^{(a)}(0,t)\, e^{i\omega_L t}\, S_z(t) \quad (18.b)$$

$$\dot{S}_z(t) = i\frac{\Omega_1}{2}(S_-(t) - S_+(t)) + i\frac{d_{ab}}{\hbar} E_z^{(a)}(0,t)\, e^{i\omega_L t}\, S_+(t) -$$

$$- i\frac{d_{ab}}{\hbar} E_z^{(c)}(0,t)\, e^{-i\omega_L t}\, S_-(t) \quad (18.c)$$

$$\dot{\hat{a}}_\varepsilon(\mathbf{k},t) = \frac{1}{\hbar}\mathcal{E}_\omega \varepsilon_z d_{ab} S_-(t)\, e^{i(\omega-\omega_L)t} \quad (19)$$

where δ_L was defined in (A.14)

$$\delta_L = \omega_L - \omega_0 \quad (20)$$

and where $E_z^{(a)}$ and $E_z^{(c)}$ are the parts of $E_{\perp z}$ containing, respectively, the annihilation and creation operators of the different modes. By using the operators $\hat{a}$ and $\hat{a}^+$, we have

$$E_z^{(a)}(0,t) = \int d^3k \sum_\varepsilon i\varepsilon_z \mathcal{E}_\omega \hat{a}_\varepsilon(\mathbf{k},t)\, e^{-i\omega t} \quad (21)$$

$$E_z^{(c)}(0,t) = \left(E_z^{(a)}(0,t)\right)^+. \quad (22)$$

Note that, in the absence of interaction between the atom and the empty modes of the radiation field, Equations (18) form a system of linear equations with constant coefficients whose eigenfrequencies are those of the spin in the rotating reference frame, i.e., $\pm\sqrt{\Omega_1^2 + \delta_L^2}$ and 0, whereas $\hat{a}_\varepsilon(\mathbf{k})$ is constant. According to the assumptions we made, these frequencies are small compared with ω_L and ω_0.

d) Comparison with the Harmonic Oscillator Case

To derive the Heisenberg–Langevin equation for the atom, in the following section we will use a procedure quite similar to the one followed in Complement C_{IV}, where the atom was represented by a harmonic oscillator. First we compare the Heisenberg equations of the two systems.

Equation (19) is very similar to Equation (33) in C_{IV} giving the evolution of $\hat{a}_i$. It is formally integrated in the same way

$$\hat{a}_\varepsilon(\mathbf{k}, t) = \hat{a}_\varepsilon(\mathbf{k}, t_0) + \frac{\varepsilon_z \mathscr{E}_\omega d_{ab}}{\hbar} \int_0^{t-t_0} d\tau \, S_-(t - \tau) \, e^{i(\omega - \omega_L)(t - \tau)}. \quad (23)$$

When they are multiplied by $\exp(-i\omega t)$ and substituted into the quantum expression for the fields, the two terms in the right-hand side of (23) give, respectively, the initial field having evolved freely from t_0 to t, and the field created by the atomic dipole between these two times. Thus, the two terms in (23) represent what are usually called the "free field" and the "source field" (see references at the end of the chapter).

The three equations (18) are equivalent to the equation of motion (32) for $\hat{b}$ in C_{IV}, and the adjoint equation for $\hat{b}^+$. One can first notice a difference between Equations (18) and (32): the *external* field introduces couplings between the evolutions of S_+, S_z, and S_-, whereas $\hat{b}$ and $\hat{b}^+$ evolve independently in the absence of coupling with the reservoir. However, this difference is not essential, because we could make linear combinations of the three equations (18) so as to display the eigenmodes for the evolution of the atomic observables in presence of the external field. A second, more important, difference relates to the form of the coupling terms with the reservoir, represented here by the quantum field in the vacuum state. Whereas these coupling terms depended only on reservoir operators, thus playing the role of source terms in the equation of motion for $\hat{b}$, they appear here as *products* of an atomic operator and a field operator. This property is related to the fact that the response of a two-level system to a field is *nonlinear* (the response to the sum of two fields is not the sum of the responses to each of them). It makes the transition to the Heisenberg–Langevin equation more difficult, because we must pay attention to the *order* in which the two operators of the product are written. The atomic operators $S_q(t)$ ($q = +, z, -$) and those of the field, $E_z^{(a)}(0, t)$ and $E_z^{(c)}(0, t)$, commute when they are taken at the same time. This is not true for the free field and the source field taken separately. Indeed, the free field depends on $\hat{a}_\varepsilon(\mathbf{k}, t_0)$, whereas the source field depends on $S_+(t - \tau)$ [first and second terms of (23)]. These two operators do not necessarily commute with $S_q(t)$. Therefore, the coupling terms with the free field and the source field appear in different forms, depending on the order selected for writing the product of $S_q(t)$ and $E_\perp(0, t)$. We will use this property in the following subsection to simplify the calculations as much as possible.

2. Derivation of the Heisenberg–Langevin Equations

In Equations (18), the fields $E_z^{(a)}(0, t)$ and $E_z^{(c)}(0, t)$ are now split into a free field part and a source-field part

$$E_z^{(a)}(0, t) = E_{0z}^{(a)}(0, t) + E_{sz}^{(a)}(0, t) \tag{24.a}$$

$$E_z^{(c)}(0, t) = E_{0z}^{(c)}(0, t) + E_{sz}^{(c)}(0, t) \tag{24.b}$$

where

$$E_{0z}^{(a)}(0, t) = \int d^3k \sum_\varepsilon i\varepsilon_z \mathcal{E}_\omega \hat{a}_\varepsilon(\mathbf{k}, t_0) e^{-i\omega t} \tag{25}$$

$$E_{sz}^{(a)}(0, t) = \int d^3k \sum_\varepsilon \frac{i}{\hbar} \varepsilon_z^2 \mathcal{E}_\omega^2 d_{ab} e^{-i\omega_L t} \times$$

$$\times \int_0^{t-t_0} d\tau\, S_-(t - \tau) e^{-i(\omega - \omega_L)\tau} \tag{26}$$

and

$$E_{0z}^{(c)} = \left(E_{0z}^{(a)} \right)^+ \qquad E_{sz}^{(c)} = \left(E_{sz}^{(a)} \right)^+. \tag{27}$$

a) CHOICE OF THE NORMAL ORDER

Later on we will take the average in the vacuum state of equations (18), or of similar equations derived from them and giving the evolution of products of atomic operators. It is then convenient to write, in these products, the operators $\hat{a}_\varepsilon(\mathbf{k}, t_0)$ on the right and the operators $\hat{a}_\varepsilon^+(\mathbf{k}, t_0)$ on the left insofar as the relations

$$\hat{a}_\varepsilon(\mathbf{k}, t_0)|0\rangle = 0 \tag{28.a}$$

$$\langle 0|\hat{a}_\varepsilon^+(\mathbf{k}, t_0) = 0 \tag{28.b}$$

can be used for all the modes $\mathbf{k}, \varepsilon$. Therefore, in the second terms of Equations (18), we place the operators $E_z^{(a)}(0, t)$ on the right of operators $S_q(t)$, and the operators $E_z^{(c)}(0, t)$ on the left (which is possible, because these operators commute). The contributions of $E_{0z}^{(a)}$ and $E_{0z}^{(c)}$ to these terms will therefore be zero on average. Only the source fields $E_{sz}^{(a)}$ and $E_{sz}^{(c)}$ will contribute to the averages. This way of ordering the operators corresponds to what is known as the normal order. With this choice, we can, for example, explicitly write Equation (18.a) by using (24.b), (27), and (26):

$$\dot{S}_+(t) = -i\delta_L S_+(t) - i\Omega_1 S_z(t) + \frac{2id_{ab}}{\hbar} E_{0z}^{(c)}(0,t) e^{-i\omega_L t} S_z(t) +$$

$$+ 2\frac{d_{ab}^2}{\hbar^2} \int d^3k \sum_\varepsilon \varepsilon_z^2 \mathcal{E}_\omega^2 \int_0^{t-t_0} d\tau \, e^{i(\omega - \omega_L)\tau} S_+(t-\tau) S_z(t). \quad (29)$$

The penultimate term comes from $E_{0z}^{(c)}$. It is zero on average in the vacuum and we will see later on that this term can be considered as the Langevin force $F_+(t)$ acting on S_+. The last term represents the contribution of the source field that we will now evaluate. Before doing so, note that the equation for $S_-(t)$ is obtained by taking the Hermitian conjugate of (29), whereas the equation for $S_z(t)$ is written

$$\dot{S}_z(t) = \frac{i\Omega_1}{2}(S_-(t) - S_+(t)) + \frac{id_{ab}}{\hbar}\left[e^{i\omega_L t} S_+(t) E_{0z}^{(a)}(0,t) - \text{h.c.}\right] -$$

$$- \frac{d_{ab}^2}{\hbar^2} \int d^3k \sum_\varepsilon \varepsilon_z^2 \mathcal{E}_\omega^2 \int_0^{t-t_0} d\tau \left[e^{-i(\omega - \omega_L)\tau} S_+(t) S_-(t-\tau) + \text{h.c.}\right].$$

$$(30)$$

b) CONTRIBUTION OF THE SOURCE FIELD

The last terms in (29) and (30) appear in the form of an integral over τ of $S_q(t-\tau)$ multiplied by the function

$$f(\tau) = \int d^3k \sum_\varepsilon \varepsilon_z^2 \mathcal{E}_\omega^2 \, e^{-i(\omega - \omega_L)\tau} \quad (31)$$

or by $f^*(\tau)$. The operators $S_q(t-\tau)$ are slowly varying functions of τ on the time scales ω_0^{-1} and ω_L^{-1}. By contrast, the integral over $\mathbf{k}$ in (31) leads to summing over a very broad spectrum of frequencies (between 0 and ck_M), so that the function $f(\tau)$ is very narrow around $\tau = 0$, with a width on the order of $1/ck_M$. Over such a short time interval, it is reasonable to approximate the evolution of $S_q(t-\tau)$ by its unperturbed evolution. We can choose for such an evolution, either the one that occurs in the presence of the external field at frequencies $\pm\sqrt{\Omega_1^2 + \delta_L^2}$ and 0, or the evolution in the absence of any interaction with the field which occurs at frequencies $\pm\delta_L$ and 0. These two choices lead to identical results to within terms in Ω_1/ω_0 or in δ_L/ω_0, which are of the same order of magnitude as those neglected in the rotating-wave approximation. We take the second option here, which gives expressions directly comparable to those used in the chapter for the Bloch equations. Thus, in (29) and (30), we replace $S_\pm(t - \tau)$ by

$$S_\pm(t - \tau) = S_\pm(t)\, e^{\pm i\delta_{L}\tau}. \tag{32}$$

Thus in the last term of (29), appear on the one hand the operator $S_+(t)S_z(t)$ which, according to Pauli matrix algebra, is equal to $-S_+(t)/2$, and on the other hand, the integral

$$2\frac{d_{ab}^2}{\hbar^2} \int d^3k \sum_\varepsilon \varepsilon_z^2 \mathscr{E}_\omega^2 \int_0^{t-t_0} d\tau \, e^{i(\omega-\omega_0)\tau} =$$

$$= 2\int d^3k \sum_\varepsilon \frac{d_{ab}^2}{\hbar^2} \varepsilon_z^2 \mathscr{E}_\omega^2 \left[-i\mathscr{P}\frac{1}{(\omega_0 - \omega)} + \pi\delta(\omega - \omega_0) \right]$$

$$= 2\left[-i\Delta + \frac{\Gamma}{2} \right] \tag{33}$$

where Γ is the spontaneous emission rate from the state b [formula (E.4.a) in Chapter IV] and $\hbar\Delta$ is the radiative shift of the state b [formula (E.7.b) also from Chapter IV; within the rotating wave approximation, the radiative shift of state a is zero]. The contribution of the source field to the right-hand side of (29) is thus finally written

$$\left(-\frac{\Gamma}{2} + i\Delta \right) S_+(t). \tag{34}$$

By proceeding in the same fashion, we find that the contribution of the source field to (30) is

$$-\frac{d_{ab}^2}{\hbar^2} \int d^3k \sum_\varepsilon \varepsilon_z^2 \mathscr{E}_\omega^2 \times$$

$$\times \int_0^{t-t_0} d\tau \left[e^{-i(\omega-\omega_0)\tau} S_+(t)S_-(t) + e^{i(\omega-\omega_0)\tau} S_+(t)S_-(t) \right] =$$

$$= -\Gamma S_+(t)S_-(t) = -\Gamma\left(S_z(t) + \frac{1}{2} \right). \tag{35}$$

c) SUMMARY. PHYSICAL DISCUSSION

Using (34) and (35), the equations of motion for $S_+(t)$, $S_z(t)$, and $S_-(t)$ are written

$$\dot{S}_+(t) = -\left(i\tilde{\delta}_L + \frac{\Gamma}{2}\right)S_+(t) - i\Omega_1 S_z(t) + F_+(t) \tag{36.a}$$

$$\dot{S}_z(t) = -i\frac{\Omega_1}{2}S_+(t) - \Gamma S_z(t) + i\frac{\Omega_1}{2}S_-(t) - \frac{\Gamma}{2} + F_z(t) \tag{36.b}$$

$$\dot{S}_-(t) = i\Omega_1 S_z(t) - \left(-i\tilde{\delta}_L + \frac{\Gamma}{2}\right)S_-(t) + F_-(t) \tag{36.c}$$

where

$$\tilde{\delta}_L = \omega_L - \omega_0 - \Delta \tag{36.d}$$

is the detuning between the laser frequency ω_L and the atomic frequency corrected by the radiative shift Δ, and where the Langevin forces are given by

$$F_+(t) = \frac{2id_{ab}}{\hbar} E_{0z}^{(c)}(\mathbf{0}, t)\, e^{-i\omega_L t} S_z(t) \tag{37.a}$$

$$F_z(t) = i\frac{d_{ab}}{\hbar}\left[e^{i\omega_L t} S_+(t) E_{0z}^{(a)}(\mathbf{0}, t) - \text{h.c.}\right] \tag{37.b}$$

$$F_-(t) = -\frac{2id_{ab}}{\hbar}\, e^{i\omega_L t} S_z(t) E_{0z}^{(a)}(\mathbf{0}, t). \tag{37.c}$$

In what follows, we take the average values of the operators in the Heisenberg representation in a state of the global system having the form $|\Psi_A; 0\rangle$, representing the atom in the state $|\Psi_A\rangle$ in the presence of the quantum field in the vacuum state. The average values of the Langevin forces (37) in such a state are zero as a consequence of relations (28). It remains to be shown that their correlation time is short, which we will prove in the following subsection.

Equations (36) are the Bloch–Langevin equations. They differ from the Bloch equations (A.18) only by the Langevin forces. If we take the average value in the vacuum state, they give exactly the Bloch equations, so that the derivation of Equations (36) given in this complement constitutes a justification for the approximation of independent rates of variation, introduced in the chapter under the assumptions $|\delta_L|, \Omega_1 \ll \omega_L, \omega_0$.

Finally, we return to the physical interpretation of the calculations presented in this subsection. These calculations might lead us to believe that the damping and the radiative shifts are due solely to the interaction of the atom with its own field (radiation reaction), whereas the fluctuations of the free field are involved only in the Langevin forces. However, a choice other than the normal order would have led to different conclusions, although the global effect on the evolution of $\langle S_q \rangle$ remains, of course, the same.

In fact, to be able to give a physical interpretation to each of the rates of variation, the one due to the radiation reaction as well as the one due to the vacuum fluctuations, one must consider only Hermitian variables of the system and require that their respective rates of variation be separately Hermitian. One can then show (*) that these conditions are satisfied only if the products $S_q(t)E_{\perp z}(0, t)$ with $(q = x, y, z)$ are made symmetric. The resulting conclusions concerning the contribution of the radiation reaction and the contribution of the vacuum fluctuations are then in complete agreement with those given in Complement A_{IV}.

Remark

It can be seen in (23) that the term relative to the source field depends only on ω and ε_z. It is thus even in **k**. As a result, the source field is even in **r**. In particular, its gradient at the origin is zero. The force exerted on the atom by its own field is therefore zero, a result we have admitted in subsection (C-2-a) of the chapter. The same result holds also for the average force due to the vacuum field (25). To see this, it is sufficient, for example, to choose the normal order in expression (C.14).

3. Properties of Langevin Forces

We showed in subsection 2 above that the three operators $S_q(t)$ $(q = +, z, -)$ obey the Bloch–Langevin equations (36) which can be written in the form

$$\frac{d}{dt}S_q(t) = \sum_{q'}\mathscr{B}_{qq'}S_{q'}(t) + \lambda_q + F_q(t) \tag{38}$$

where $\mathscr{B}_{qq'}$ and λ_q are constant coefficients identical to those appearing in the optical Bloch equations (A.18), and where the $F_q(t)$ are the forces

(*) J. Dalibard, J. Dupont-Roc, and C. Cohen-Tannoudji, *J. Phys.* (Paris), **43**, 1617 (1982); **45**, 637 (1984).

given in (37) having an average value equal to zero

$$\langle F_q(t) \rangle = 0. \tag{39}$$

These forces may be considered as Langevin forces, insofar as their correlation functions are characterized by a correlation time τ_c that is very short compared to Γ^{-1}. To prove this we will show that

$$\mathcal{G}_{qq'}(t, t') = \langle F_q^+(t) F_{q'}(t') \rangle = 2 D_{qq'} g(t - t') \tag{40}$$

where $D_{qq'}$ is a diffusion coefficient, on the order of Γ, and where $g(t - t')$ is a function of $t - t'$ of width $\tau_c = 1/ck_M$ and having an integral equal to 1 (this function is not necessarily even in $t - t'$ as a result of the quantum nature of the forces F_q).

In the case of the harmonic oscillator studied in Complement C_{IV}, the Langevin forces depend only on the field operators, and calculating their correlation functions is very easy. Expressions (37) show that the Langevin forces of the two-level system are products of free field operators and atomic operators. Calculating their correlation functions therefore requires prior knowledge of the commutation relations between atomic operators and free field operators (*).

a) COMMUTATION RELATIONS BETWEEN THE ATOMIC DIPOLE MOMENT AND THE FREE FIELD

We first write explicitly the commutator $[S_q(t_2), E_{0z}^{(a)}(\mathbf{0}, t_1)]$ by using (25):

$$\left[S_q(t_2), E_{0z}^{(a)}(\mathbf{0}, t_1) \right] = \int d^3k \sum_\varepsilon i\varepsilon_z \mathcal{E}_\omega \left[S_q(t_2), \hat{a}_\varepsilon(\mathbf{k}, t_0) \right] e^{-i\omega t_1}. \tag{41}$$

Using (23), we can reexpress $\hat{a}_\varepsilon(\mathbf{k}, t_0)$ as a function of $\hat{a}_\varepsilon(\mathbf{k}, t_2)$ [which commutes with $S_q(t_2)$ because it is taken at the same time] and of the source field. We then have

$$\left[S_q(t_2), E_{0z}^{(a)}(\mathbf{0}, t_1) \right] = - \int d^3k \sum_\varepsilon \frac{i}{\hbar} \varepsilon_z^2 \mathcal{E}_\omega^2 d_{ab} \times$$

$$\times \int_0^{t_2 - t_0} d\tau \left[S_q(t_2), S_-(t_2 - \tau) \right] e^{-i\omega t_1} e^{i(\omega - \omega_L)(t_2 - \tau)}. \tag{42}$$

The sum over $\mathbf{k}$ and ε causes the function f, introduced in (31), to appear

(*) See also B. R. Mollow, *J. Phys. A*, **8**, L130 (1975).

with the argument $\tau - (t_2 - t_1)$. This function f is very narrow, with width $1/ck_M$ around 0, and the integral over τ extends only to positive values of τ because $t_2 - t_0$ is positive.

First assume that $t_2 < t_1$, more precisely that $t_2 - t_1 \ll -1/ck_M$. The integral of $f[\tau - (t_2 - t_1)]$ is then zero because the interval of integration $[0, t_2 - t_0]$ does not contain the interval of width $1/ck_M$ about $t_2 - t_1$ in which f is appreciable. It follows that

$$\left[S_q(t_2), E_{0z}^{(a)}(\mathbf{0}, t_1) \right] \simeq 0 \qquad \text{if } t_2 \ll t_1 - (1/ck_M). \qquad (43)$$

Such a result is physically satisfying. Indeed, the operator $S_q(t_2)$ depends only on the vacuum field $\mathbf{E}_0^{(\pm)}(\mathbf{0}, t')$ which acted on the atomic dipole at times t' located in the past of t_2 $(t' < t_2)$. Therefore, this operator commutes with the vacuum field $\mathbf{E}_0^{(\pm)}(\mathbf{0}, t_1)$ in the future of t_2 $(t_1 > t_2)$ (*).

Assume now that $|t_2 - t_1| \leq 1/ck_M$. The integral over τ in (42) is no longer zero and can be calculated by using the approximation (32), which is written here

$$S_-(t_2 - \tau) \simeq S_-(t_1) e^{-i\delta_L(\tau - t_2 + t_1)} \qquad (44)$$

and we then have

$$\left[S_q(t_2), E_{0z}^{(a)}(\mathbf{0}, t_1) \right] \simeq - \left[S_q(t_2), S_-(t_1) \right] e^{-i\omega_L t_1} \times$$

$$\times \int d^3k \sum_\varepsilon \frac{d_{ab}^2}{\hbar} \varepsilon_z^2 \mathscr{E}_\omega^2 \left[\mathscr{P} \frac{1}{\omega - \omega_0} + i\pi\delta(\omega - \omega_0) \right] e^{i(\omega - \omega_0)(t_2 - t_1)}. \qquad (45)$$

This is a function of t_1 and t_2, on the order of $\Gamma\hbar/d_{ab}$. By taking the Hermitian conjugate of (43) and (45), we obtain analogous relations for $E_{0z}^{(c)}(\mathbf{0}, t)$.

b) CALCULATION OF THE CORRELATION FUNCTIONS OF LANGEVIN FORCES

For example, we calculate the correlation function $\mathscr{G}_{++} = \langle F_+^+(t)F_+(t') \rangle$ where F_+ is defined in (37.a)

$$\mathscr{G}_{++}(t, t') = \frac{4d_{ab}^2}{\hbar^2} \langle S_z(t) E_{0z}^{(a)}(\mathbf{0}, t) e^{i\omega_L(t - t')} E_{0z}^{(c)}(\mathbf{0}, t') S_z(t') \rangle. \qquad (46)$$

(*) Recall that the commutator $[E_{0i}^{(+)}(\mathbf{r}_1, t_1), E_{0j}^{(-)}(\mathbf{r}_2, t_2)]$ is nonzero only if the two events $(\mathbf{r}_1, t_1)$ and $(\mathbf{r}_2, t_2)$ are located on the same light cone.

We reverse the order of the two fields. Equation (25) then gives

$$e^{i\omega_L(t-t')}\left[E_{0z}^{(a)}(\mathbf{0},t),E_{0z}^{(c)}(\mathbf{0},t')\right]=$$

$$=\int d^3k\ \sum_{\varepsilon}\varepsilon_z^2\mathcal{E}_\omega^2\ e^{-i(\omega-\omega_L)(t-t')}=f(t-t')\qquad(47)$$

where f is the function previously introduced in (31). This function $f(t-t')$ is centered at $t-t'=0$ and has a width on the order of $1/ck_M$. Expression (46) is thus written

$$\mathcal{G}_{++}(t,t')=\frac{4d_{ab}^2}{\hbar^2}\langle S_z(t)S_z(t')\rangle f(t-t')\ +$$

$$+4\frac{d_{ab}^2}{\hbar^2}\langle S_z(t)E_{0z}^{(c)}(\mathbf{0},t')E_{0z}^{(a)}(\mathbf{0},t)S_z(t')\rangle\ e^{i\omega_L(t-t')}.\quad(48)$$

In the first term, $S_z(t')$ is slowly varying on the scale of the width $1/ck_M$ of the function $f(t-t')$. Thus we can replace t' by t to arrive at $S_z^2(t)$ which is equal to $\frac{1}{4}$. Note that the integral over t from $-\infty$ to $+\infty$ of this first term is simply, using (31), the sum over all the modes of $(d_{ab}^2/\hbar^2)\varepsilon_z^2\mathcal{E}_\omega^2 2\pi\delta(\omega-\omega_L)$. According to (33), this sum equals Γ. Therefore, this first term can be simply written $\Gamma g(t-t')$, where $g(\tau)$, proportional to $f(\tau)$, is a function of τ having a width $1/ck_M$ and an integral equal to 1. The second term of (48) is negligible: indeed, if $|t-t'|\gg 1/ck_M$, we can use (43) and its analog for $E_{0z}^{(c)}$, either to make $S_z(t)$ commute with $E_{0z}^{(c)}(\mathbf{0},t')$, or $E_{0z}^{(a)}(\mathbf{0},t)$ with $S_z(t')$, depending on the sign of $t-t'$. One of these fields then acts directly on the state $|\psi_A;0\rangle$ which is involved in the average value of (46) and gives a zero result; if $|t-t'|\le 1/ck_M$, the foregoing commutators remain finite, on the order of $\Gamma\hbar/d_{ab}$ according to (45), so that the second term is on the order of $(d_{ab}^2/\hbar^2)(\Gamma\hbar/d_{ab})^2=\Gamma^2$. In the same interval of width $1/ck_M$, the first term is equal to $\Gamma(ck_M)\gg\Gamma^2$. It is therefore justified to neglect the second term of (48) relative to the first one and finally we have:

$$\mathcal{G}_{++}(t,t')=\Gamma g(t-t')\qquad(49.a)$$

which gives, according to (40),

$$2D_{++}=\Gamma\qquad(49.b)$$

Similar calculations give for the other diffusion coefficients

$$D_{--} = D_{-+} = D_{+-} = D_{z-} = D_{-z} = 0 \qquad (50.\text{a})$$

$$2D_{zz} = \Gamma(\langle S_z \rangle + \tfrac{1}{2}) \qquad (50.\text{b})$$

$$2D_{z+} = \Gamma \langle S_+ \rangle \qquad (50.\text{c})$$

$$2D_{+z} = \Gamma \langle S_- \rangle. \qquad (50.\text{d})$$

c) QUANTUM REGRESSION THEOREM

This theorem is a consequence of properties (39) and (40) of the Langevin forces, and of the existence of the two very distinct time scales, Γ^{-1} and $\tau_c = 1/ck_M$, with $\tau_c \ll \Gamma^{-1}$. To derive this theorem, we will follow a procedure quite similar to the one used in Complement C$_{IV}$. We give here only an outline of the proof. As in subsection 2-d in Complement C$_{IV}$, we begin by showing

$$\langle F_q(t) S_{q'}(t') \rangle = 0 \qquad \text{if } t - t' \gg \tau_c. \qquad (51)$$

For this it is sufficient to formally integrate the system of linear equations (38) to show that $S_{q'}(t')$ depends linearly on the Langevin forces $F_{q''}(t'')$ in the past of t' and therefore, as a consequence of (40), cannot be correlated with the Langevin force $F_q(t)$ for t sufficiently distant in the future of t'.

We then multiply the two sides of (38) by $S_{q'}(t')$ (after having replaced the index of summation q' by q'') and we take the average value. This gives

$$\frac{d}{dt} \langle S_q(t) S_{q'}(t') \rangle = \sum_{q'} \mathscr{B}_{qq''} \langle S_{q''}(t) S_{q'}(t') \rangle +$$

$$+ \lambda_q \langle S_{q'}(t') \rangle + \langle F_q(t) S_{q'}(t') \rangle. \qquad (52)$$

For $t - t' \gg \tau_c$, the last term in (52) is negligible, using (51). For $t - t' \leq \tau_c$, we can show that it remains finite and on the order of $D_{qq'}$, that is, on the order of Γ, so that its contribution to the solution of (52) for $t \geq t'$ is on the order of $\Gamma \tau_c \ll 1$. Finally, we can ignore the last term of (52) for $t \geq t'$. This shows that the two-time averages $\langle S_q(t) S_{q'}(t') \rangle$ obey, for $t > t'$, the equations

$$\text{if } t \geq t' \qquad \frac{d}{dt} \langle S_q(t) S_{q'}(t') \rangle = \sum_{q''} \mathscr{B}_{qq''} \langle S_{q''}(t) S_{q'}(t') \rangle + \lambda_q \langle S_{q'}(t') \rangle$$

$$(53)$$

having the same structure as the equations describing the evolution of the one-time averages $\langle S_q(t) \rangle$

$$\frac{d}{dt}\langle S_q(t) \rangle = \sum_{q''}\mathcal{B}_{qq''}\langle S_{q''}(t) \rangle + \lambda_q. \tag{54}$$

It is such a result that allowed us, in Section D of the chapter, to use optical Bloch equations to study the dynamics of the fluctuations of the atomic dipole.

d) GENERALIZED EINSTEIN RELATIONS

To end this complement, we show how the diffusion coefficients $D_{qq'}$ introduced in (40) can be calculated directly from optical Bloch equations, without having to use the expressions (37) for the Langevin forces, as we did in subsection 3-b above. To do this, we follow a procedure similar to the one used in subsections 2-e and 2-f of Complement C_{IV}.

We return to (38) and call "drift velocity of S_q" the quantity

$$\mathcal{D}(S_q) = \sum_{q''}\mathcal{B}_{qq''}S_{q''} + \lambda_q \tag{55}$$

so that the rate of variation of S_q given by (38) is just the sum of the drift velocity and the Langevin force:

$$\frac{d}{dt}S_q(t) = \mathcal{D}(S_q(t)) + F_q(t). \tag{56}$$

Note that since $F_q(t)$ is zero on average, only the drift velocity remains for $\langle S_q(t) \rangle$:

$$\frac{d}{dt}\langle S_q(t) \rangle = \langle \mathcal{D}(S_q(t)) \rangle. \tag{57}$$

We now consider the operator $S_q^+(t)S_{q'}(t)$, which is an atomic operator at time t. Pauli matrix algebra allows us to express it as a function of the $S_{q''}(t)$ and to calculate the average drift velocity

$$\frac{d}{dt}\langle S_q^+(t)S_{q'}(t) \rangle = \langle \mathcal{D}(S_q^+(t)S_{q'}(t)) \rangle \tag{58}$$

as a function of the coefficients $\mathcal{B}$ and λ appearing in the optical Bloch equations (54).

We can also write directly, using (56):

$$\frac{d}{dt}\langle S_q^+(t)S_{q'}(t)\rangle = \langle \dot{S}_q^+(t)S_{q'}(t)\rangle + \langle S_q^+(t)\dot{S}_{q'}(t)\rangle$$
$$= \langle \mathscr{D}(S_q^+(t))S_{q'}(t)\rangle + \langle S_q^+(t)\mathscr{D}(S_{q'}(t))\rangle +$$
$$+ \langle F_q^+(t)S_{q'}(t) + S_q^+(t)F_{q'}(t)\rangle. \qquad (59)$$

To calculate the last term in (59), note that, according to (51), the forces $F_q(t)$ are not correlated with the operators $S_{q'}(t - \Delta t)$ if $\Delta t \gg 1/ck_M$. Thus we integrate Equation (56) between $t - \Delta t$ and t, so that we can express $S_{q'}(t)$ as a function of $S_{q'}(t - \Delta t)$ in the product $\langle F_q^+(t)S_{q'}(t)\rangle$. We choose Δt to be relatively small as compared to Γ^{-1} so that a linear approximation of the variation of $S_q(t)$ between $t - \Delta t$ and t is justified.

$$S_{q'}(t) = S_{q'}(t - \Delta t) + \Delta t\, \mathscr{D}(S_{q'}(t - \Delta t)) + \int_{t-\Delta t}^{t} F_{q'}(t')\,dt'. \quad (60)$$

When we substitute this expression into the last term of (59) and use definition (54) for $\mathscr{D}(S_{q'})$, average values of the type $\langle F_q^+(t)S_{q''}(t - \Delta t)\rangle$ and $\lambda_{q'}\langle F_q^+(t)\rangle$ appear, which are zero. What remains is the contribution of the last term in (60)

$$\langle F_q^+(t)S_{q'}(t) + S_q^+(t)F_{q'}(t)\rangle =$$
$$= \int_{t-\Delta t}^{t} dt'\langle F_q^+(t)F_{q'}(t') + F_q^+(t')F_{q'}(t)\rangle$$
$$= 2D_{qq'}\int_{t-\Delta t}^{t} dt'[g(t - t') + g(t' - t)]$$
$$= 2D_{qq'}\int_{-\Delta t}^{+\Delta t} d\tau\, g(\tau) = 2D_{qq'}. \qquad (61)$$

To derive (61), we used (40) and the fact that the integral of the function g over an interval that is large as compared to its width is equal to 1.

Combining (58), (59), and (61) gives the generalized Einstein relation

$$2D_{qq'} = \langle \mathscr{D}(S_q^+ S_{q'}) - \mathscr{D}(S_q^+)S_{q'} - S_q^+\mathscr{D}(S_{q'})\rangle \qquad (62)$$

where it is understood that all the quantities are taken at time t.

Like the similar relation (76) derived in Complement C_{IV}, Equation (62) is the expression of a connection between fluctuations (characterized by the diffusion coefficient $D_{qq'}$) and dissipation (characterized by damping coefficients appearing in the drift velocities). It is also clear from (62) that the diffusion coefficient is associated with the non-Hamiltonian part of the drift velocities.

Finally, we can verify in the case of D_{++} that the result given by (62) coincides with that found above. To do this, we must use the relations

$$S_+^+ S_+ = S_- S_+ = -S_z + \frac{1}{2} \tag{63.a}$$

$$S_z S_\pm = \pm \frac{1}{2} S_\pm \tag{63.b}$$

$$\mathscr{D}(S_+^+ S_+) = \mathscr{D}\left(-S_z + \frac{1}{2}\right) = \Gamma S_z + i\frac{\Omega_1}{2}(S_+ - S_-) + \frac{\Gamma}{2} \tag{63.c}$$

$$\mathscr{D}(S_+) = -\left(i\tilde{\delta}_L + \frac{\Gamma}{2}\right)S_+ - i\Omega_1 S_z. \tag{63.d}$$

We thus obtain $2D_{++} = \Gamma$, in agreement with (49.b).

REFERENCES

For the derivation of the Heisenberg equations for the coupled atom and field system, and for the separation of the field into the vacuum field and the source field, see J. R. Ackerhalt and J. H. Eberly, *Phys. Rev. D*, **10**, 3350 (1974); R. Saunders, R. K. Bullough, and F. Ahmad, *J. Phys. A*, **8**, 759 (1975); H. J. Kimble and L. Mandel, *Phys. Rev. A*, **13**, 2123 (1976); Allen and Eberly (Chapter 7).

See also the references in Complement C_{IV}.

CHAPTER VI

The Dressed Atom Approach

A—INTRODUCTION: THE DRESSED ATOM

This last chapter, just as the preceding one, is devoted to the study of an atomic system interacting with a monochromatic resonant or quasi-resonant laser field. However, instead of treating the laser field as a classical external field, we now consider it as a single-mode quantum field. The motivation for such a quantum treatment is twofold. First, because the Hamiltonian of the atom + field global system is time independent, it is possible to introduce true energy levels and to study the manifestations of the atom-laser interaction on the positions and the wave functions of these levels. Second, with a quantized laser field, it is possible to clearly identify the elementary processes of absorption and stimulated emission of laser photons, as well as the spontaneous emission processes for fluorescence photons. We can then shed new light on several important physical phenomena, such as the spectral distribution of the light emitted or absorbed by the atom, the photon correlations, or the fluctuations of dipole forces.

Figure 1 shows the different subsystems considered in this chapter, along with their respective Hamiltonians and the various couplings. An atom A, with Hamiltonian H_A, is first coupled by the Hamiltonian V_{AL} to a particular mode of the field, which initially contains photons and which we call the "laser mode" L, with Hamiltonian H_L. Atom A is also coupled by V_{AR} to the "reservoir" R of the initially empty modes of the field, with Hamiltonian H_R. The global Hamiltonian of the $A + L + R$

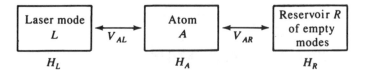

Figure 1. Various subsystems considered in this chapter, with respective Hamiltonians H_A, H_L, and H_R. The Hamiltonians V_{AL} and V_{AR} describe the interactions between A and L on the one hand, and between A and R on the other hand.

system is then

$$H = H_A + H_L + H_R + V_{AL} + V_{AR}. \qquad (A.1)$$

In Section B below, we begin by neglecting V_{AR} and by considering the system $A + L$ formed by the coupled-system atom + laser mode. The Hamiltonian of $A + L$ is

$$H_{AL} = H_A + H_L + V_{AL}. \qquad (A.2)$$

Such a system, which we call the "atom dressed by laser photons", has an infinite ladder of *discrete* energy levels. The positions and the wave functions of these energy levels are studied in detail and we show how the processes of absorption and stimulated emission of laser photons appear in the basis of uncoupled states (eigenstates of $H_A + H_L$) and in the basis of dressed states (eigenstates of H_{AL}).

We then introduce the coupling V_{AR} in Section C. Considering the system $A + L$ as spontaneously emitting photons into the reservoir R of empty modes allows us to introduce the picture of the radiative cascade of the dressed atom. In this picture, the sequence of photons emitted by the atom under the influence of laser excitation appears as a sequence of photons emitted in a cascade by the dressed atom descending along the ladder of its energy levels. Several physical phenomena are reviewed and interpreted from this point of view.

To provide quantitative support for the pictures given in Section C, we study in Section D the master equation which describes the spontaneous emission of the dressed atom. Using the results from Chapter IV, we derive the evolution equation for the dressed atom density operator σ_{AL} under the influence of its coupling with the reservoir R. The projection of this equation onto the dressed state basis is analyzed in detail in the secular limit where it is possible to neglect the couplings between populations of dressed levels and coherences between dressed levels. We also show how the optical Bloch equations are contained in these more general equations.

In Section E, the foregoing results are then applied to several concrete physical problems: widths and weights of the various components of the fluorescence triplet; absorption spectrum recorded on a weak laser beam, probing either the transition excited by the first laser beam or another transition sharing a common level with the first one; photon correlation signals; average value and fluctuations of dipole forces.

Finally, two complements continue the discussions of this chapter in other directions. Complement A_{VI} reviews several applications of the dressed atom method in the area of magnetic resonance. Spontaneous emission is then negligible, but the intensity of the field can be large enough for the Rabi frequency to be comparable to the eigenfrequency of the transition. Complement B_{VI} shows how it is possible to apply the dressed atom approach to the problem of the collisional redistribution of radiation.

B—ENERGY LEVELS OF THE DRESSED ATOM

1. Model of the Laser Beam

To simplify the quantum description of the laser field as much as possible, we consider a cavity *without losses*, having only one mode, with frequency ω_L, which contains photons. This mode is chosen in such a way that its field is similar to that of the laser beam. Such a model thus associates with the laser beam a free mode of a cavity, the laser mode L. The atom is assumed to be placed inside the cavity (Figure 2b). The fact that it can absorb and emit photons in a stimulated fashion in this mode L represents the only possible cause for the variation in the number of laser photons.

In the real experiment, the atom is not generally in a cavity, but instead interacts in free space with an incident laser wave (Figure 2a). The model corresponding to Figure 2b can be applied to the real experiment only if several conditions are fulfilled.

(i) The field "seen" by the atom in the cavity must have the same value and the same local spatial variations as in the real experiment.

(ii) The cavity must be sufficiently large so that the spontaneous emission of the atom is not modified.

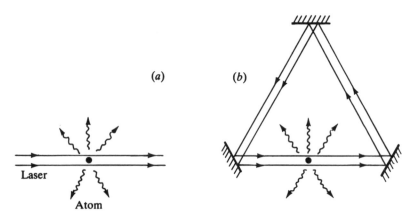

Figure 2. (a) Real experiment. The atom interacts with an incident laser wave. (b) Model used here. The atom interacts with the field of a lossless cavity having only one mode containing photons.

Condition (i) shows that the exact value of the average number $\langle N \rangle$ of photons in the laser mode is not important. Only the ratio $\langle N \rangle / V$ (where V is the volume of the cavity) is significant, because it is related to the *energy density*. Condition (ii) requires that V must be very large. Indeed, we will make $\langle N \rangle$ and V tend to infinity while keeping $\langle N \rangle / V$ constant.

We also assume that the width ΔN of the distribution of the number of laser photons is very large (for example, if the laser mode is in a coherent state, $\Delta N = \sqrt{\langle N \rangle}$). Another important parameter is the number of photons emitted spontaneously by the atom during the interaction time T (corresponding to the actual experiment). This number is at most on the order of $\Gamma T / 2$, and is also equal to the number of photons that disappear from the laser mode, because each fluorescence cycle (absorption followed by spontaneous emission) transfers a photon from the laser mode to initially empty modes. At the limit $\langle N \rangle \rightarrow \infty$, $V \rightarrow \infty$, we can also take $\Delta N \rightarrow \infty$, and thus $\Delta N \gg \Gamma T$, which allows us to neglect the number of absorbed laser photons compared with ΔN and to consider that, during the entire duration T of the experiment, the atom is subjected to the same laser intensity.

Remark

It is, of course, possible to also consider experiments on atoms actually located in a cavity having sufficiently small dimensions so that the spontaneous emission of these atoms is modified. We must then explicitly consider the modes of this real cavity. We will not deal here with these "cavity quantum electrodynamics" effects (see the references given at the end of subsection A-1-c in Chapter II).

In what follows, we thus assume

$$\langle N \rangle, \Delta N \gg 1, \Gamma T. \tag{B.1}$$

We also assume that ΔN, while being very large compared with 1, is very small compared with $\langle N \rangle$

$$\langle N \rangle \gg \Delta N \gg 1 \tag{B.2}$$

which is equivalent to taking, for the distribution of the photon number in the laser mode, a relatively narrow distribution around a large average value.

2. Uncoupled States of the Atom + Laser Photons System

We begin by neglecting the interaction between the atom and the laser photons. The Hamiltonian of the global system is then $H_L + H_A$, where

$$H_L = \hbar\omega_L\left(a^+a + \tfrac{1}{2}\right) \tag{B.3}$$

is the Hamiltonian of the laser mode, a^+ and a being the creation and annihilation operators, and

$$H_A = \hbar\omega_0|b\rangle\langle b| \tag{B.4}$$

is the atomic Hamiltonian, having eigenstates $|b\rangle$ and $|a\rangle$ with energies $\hbar\omega_0$ and 0.

The eigenstates of $H_A + H_L$ are then labeled by two quantum numbers: the number N of laser photons, and the atomic quantum number b or a. The states $|a, N + 1\rangle$ and $|b, N\rangle$ are close to each other near resonance, i.e., when

$$|\delta_L| \ll \omega_0 \tag{B.5}$$

where

$$\delta_L = \omega_L - \omega_0 \tag{B.6}$$

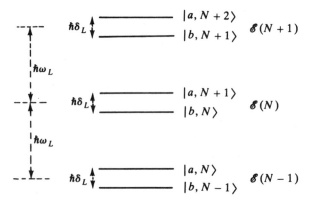

Figure 3. Manifolds $\mathscr{E}(N + 1)$, $\mathscr{E}(N)$, and $\mathscr{E}(N - 1)$ of uncoupled states of the atom + laser photons system. The energy difference $\hbar\delta_L$ between the two levels of a given manifold is very small compared with the gap $\hbar\omega_L$ between two adjacent manifolds.

is the detuning between the laser frequency ω_L and the atomic frequency ω_0. The energy difference between these two levels equals $\hbar\delta_L$, the level $|a, N + 1\rangle$ being above $|b, N\rangle$ if δ_L is positive. The manifold formed by these two levels is written

$$\mathscr{E}(N) = \{|a, N + 1\rangle, \quad |b, N\rangle\}. \tag{B.7}$$

Figure 3 shows such a manifold, located at a distance $\hbar\omega_L$ below $\mathscr{E}(N + 1)$, formed by $|a, N + 2\rangle$ and $|b, N + 1\rangle$ and at a distance $\hbar\omega_L$ above $\mathscr{E}(N - 1)$, formed by $|a, N\rangle$ and $|b, N - 1\rangle$.

3. Atom-Laser Photons Coupling

a) INTERACTION HAMILTONIAN

In the electric dipole representation (see Appendix, §5), the interaction Hamiltonian V_{AL} between the atom and the laser mode is written $-\mathbf{d} \cdot \mathbf{E}_\perp(\mathbf{R})$, where $\mathbf{d}$ is the atomic dipole and $\mathbf{E}_\perp(\mathbf{R})$ is the laser field operator, evaluated at the position $\mathbf{R}$ of the atom. Here we treat $\mathbf{R}$ classically. With an appropriate choice for the coordinate origin, the field $\mathbf{E}_\perp(\mathbf{R})$ is equal to

$$\mathbf{E}_\perp(\mathbf{R}) = \sqrt{\frac{\hbar\omega_L}{2\varepsilon_0 V}} \, \boldsymbol{\varepsilon}_L(a + a^+) \tag{B.8}$$

where $\boldsymbol{\varepsilon}_L$ is the polarization of the laser mode. Because

$$\mathbf{d} = \mathbf{d}_{ab}(\mathscr{S}_+ + \mathscr{S}_-) \tag{B.9}$$

where $\mathbf{d}_{ab} = \langle a|\mathbf{d}|b\rangle = \langle b|\mathbf{d}|a\rangle$ is assumed to be real and where

$$\mathscr{S}_+ = |b\rangle\langle a| \quad \mathscr{S}_- = |a\rangle\langle b| \tag{B.10}$$

V_{AL} is finally written

$$V_{AL} = -\mathbf{d} \cdot \mathbf{E}_\perp(\mathbf{R}) = g(\mathscr{S}_+ + \mathscr{S}_-)(a + a^+) \tag{B.11}$$

where g is a coupling constant equal to

$$g = -\boldsymbol{\varepsilon}_L \cdot \mathbf{d}_{ab}\sqrt{\frac{\hbar\omega_L}{2\varepsilon_0 V}}. \tag{B.12}$$

b) Resonant and Nonresonant Couplings

The interaction Hamiltonian V_{AL} couples the two states of each manifold $\mathscr{E}(N)$ to each other. The corresponding matrix element is written

$$v_N = \langle b, N|V_{AL}|a, N + 1\rangle = g\sqrt{N + 1}. \tag{B.13}$$

Physically, such a coupling expresses the fact that the atom in the state $|a\rangle$ can absorb a laser photon and go to state $|b\rangle$.

The state $|a, N + 1\rangle$ is also coupled by V_{AL} to the state $|b, N + 2\rangle$, which belongs to $\mathscr{E}(N + 2)$. Similarly, the state $|b, N\rangle$ is coupled to $|a, N - 1\rangle$, which belongs to $\mathscr{E}(N - 2)$. In addition to the resonant couplings within the same manifold, V_{AL} thus also introduces nonresonant couplings between states belonging to different manifolds, separated by $\pm 2\hbar\omega_L$. We neglect here these nonresonant couplings. We will see in Complement A_{VI} (§4-b) that, among other effects, they give rise to the Bloch–Siegert shift of magnetic resonance lines.

c) Local Periodicity of the Energy Diagram

The relative variation of the coupling v_N over the range ΔN of the distribution of the number of laser photons is written, using (B.13)

$$\frac{\Delta v_N}{v_N} = \frac{\Delta\sqrt{N + 1}}{\sqrt{N + 1}} \simeq \frac{1}{2}\frac{\Delta N}{\langle N\rangle}. \tag{B.14}$$

Using (B.2), such a variation can be neglected, which is equivalent to approximating v_N by

$$v_N \simeq g\sqrt{\langle N\rangle}. \tag{B.15}$$

Taking a coupling v_N independent of N leads to the fact that the energy-level diagram resulting from the diagonalization of $H_A + H_L + V_{AL}$ can be considered to be periodic over a range ΔN of values of N around $\langle N\rangle$.

d) Introduction of the Rabi Frequency

We assume here that the laser mode is in a coherent state $|\alpha\exp(-i\omega_L t)\rangle$ with α being real. Equation (B.8) then gives in average value, and by using the fact that $\langle N \rangle = \alpha^2$

$$\langle \alpha e^{-i\omega_L t}|\mathbf{E}_\perp(\mathbf{R})|\alpha e^{-i\omega_L t}\rangle = \mathcal{E}_0 \cos \omega_L t \qquad \text{(B.16.a)}$$

$$\mathcal{E}_0 = 2\varepsilon_L \sqrt{\frac{\hbar\omega_L}{2\varepsilon_0 V}} \sqrt{\langle N \rangle}. \qquad \text{(B.16.b)}$$

We know that it is then correct (see Exercise 17) to treat the laser field as a classical external field given by (B.16.a). This is what was done in Chapter V. We then set:

$$\hbar\Omega_1 = -\mathbf{d}_{ab} \cdot \mathcal{E}_0 \qquad \text{(B.17)}$$

where Ω_1 is the Rabi frequency that we assume to be positive. By substituting (B.16.b) into (B.17) and by using (B.12), we can write (B.15) in the form

$$v_N = \hbar\Omega_1/2 \qquad \text{(B.18)}$$

which expresses the resonant coupling associated with V_{AL} as a function of the Rabi frequency Ω_1 introduced in Chapter V.

4. Dressed States

a) ENERGY LEVELS AND WAVE FUNCTIONS

When we take into account the coupling v_N between the two states $|a, N+1\rangle$ and $|b, N\rangle$ of $\mathcal{E}(N)$, we obtain two perturbed states $|1(N)\rangle$ and $|2(N)\rangle$ that we call dressed states. For N varying in the interval ΔN around $\langle N \rangle$, we can ignore the variation of v_N with N and use (B.18). We then find that the two dressed states are separated by an interval

$$\hbar\Omega = \hbar\sqrt{\delta_L^2 + \Omega_1^2} \qquad \text{(B.19)}$$

and are symmetrically located with respect to the unperturbed levels (Figure 4).

By convention, the state $|1(N)\rangle$ will be the one with the greatest energy, so that the eigenstates $|1(N)\rangle$ and $|2(N)\rangle$ are written

$$|1(N)\rangle = \sin\theta|a, N+1\rangle + \cos\theta|b, N\rangle \qquad \text{(B.20.a)}$$

$$|2(N)\rangle = \cos\theta|a, N+1\rangle - \sin\theta|b, N\rangle \qquad \text{(B.20.b)}$$

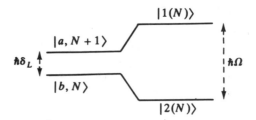

Figure 4. Uncoupled levels (left-hand part) and perturbed levels (right-hand part) of the manifold $\mathscr{E}(N)$.

where the angle θ is defined by

$$\tan 2\theta = -\frac{\Omega_1}{\delta_L} \qquad 0 \le 2\theta < \pi. \tag{B.21}$$

b) ENERGY DIAGRAM VERSUS $\hbar\omega_L$

Figure 5 shows the variations with $\hbar\omega_L$ of the energies of the dressed states $|1(N)\rangle$ and $|2(N)\rangle$. The uncoupled state $|a, N\rangle$ is chosen as the energy origin. The dashed lines represent the energies of the uncoupled states $|a, N + 1\rangle$ (straight line with slope 1 passing through the origin) and $|b, N\rangle$ (horizontal line with ordinate $\hbar\omega_0$). They intersect at $\hbar\omega_L = \hbar\omega_0$. The energies of the dressed levels $|1(N)\rangle$ and $|2(N)\rangle$ form the two branches of a hyperbola having the above-mentioned lines as asymptotes. The minimum distance between the two branches of the hyperbola occurs when $\hbar\omega_L = \hbar\omega_0$ and is equal to $\hbar\Omega_1$.

By contrast with the uncoupled states which cross each other for $\hbar\omega_L = \hbar\omega_0$, the dressed levels $|1(N)\rangle$ and $|2(N)\rangle$ repel each other and form an "anticrossing" (*). The advantage of such a graphic representation, as well as the others that will be given later on in this chapter, is that it gives an overall view on several physical phenomena. For instance, when the detuning $\delta_L = \omega_L - \omega_0$ is made to vary from positive values to negative values, it can be seen in Figure 5 that the state $|1(N)\rangle$ passes continuously from the uncoupled state $|a, N + 1\rangle$ to the uncoupled state $|b, N\rangle$, while, for $\delta_L = 0$, it is a linear superposition of these two states

(*) Recall that we have not yet considered spontaneous emission. When we take this phenomenon into account, the perturbed levels continue to cross each other if Ω_1 is not sufficiently large compared with Γ (see §C-3-c in Chapter III).

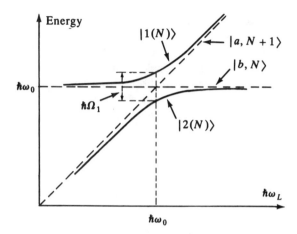

Figure 5. Energies of the dressed levels $|1(N)\rangle$ and $|2(N)\rangle$ (solid lines) versus $\hbar\omega_L$. The dashed lines represent the energies of the uncoupled states $|a, N + 1\rangle$ and $|b, N\rangle$.

with equal weights. The distances between the perturbed levels and their asymptotes for $|\delta_L| \gg \Omega_1$ represent the light shifts (or a.c. Stark shifts) of atomic states a and b due to the coupling with the laser. It can be seen, for example, in Figure 5, that the light shift of the state a is positive for $\delta_L > 0$ and negative for $\delta_L < 0$, with the situation for the state b being the reverse.

5. Physical Effects Associated with Absorption and Induced Emission

As we have already seen above, the couplings described by v_N correspond to processes of absorption ($|a, N + 1\rangle \rightarrow |b, N\rangle$) or stimulated emission ($|b, N\rangle \rightarrow |a, N + 1\rangle$). The effects of these processes can be considered from a dynamic or a static point of view.

We begin with the dynamic point of view. Assume that the system is in the state $|a, N + 1\rangle$ at time $t = 0$. What is the probability $\mathscr{P}(t)$ of finding it in the state $|b, N\rangle$ at time t? This problem is well known in quantum mechanics. The probability $\mathscr{P}(t)$ is a sinusoidal function of time, oscillating at the Bohr frequency Ω associated with the two perturbed levels $|1(N)\rangle$ and $|2(N)\rangle$. The oscillation is a full oscillation when the two uncoupled states $|a, N + 1\rangle$ and $|b, N\rangle$ both contain equal proportions of the coupled states $|1(N)\rangle$ and $|2(N)\rangle$. We then find the "Rabi nutation" phenomenon, which, at resonance ($\delta_L = 0$), occurs at the Rabi frequency Ω_1.

From the static point of view, the effect of the coupling is manifested by the fact that the new stationary states of the global system have energies and wave functions that differ from those of the uncoupled states. We will see later on that these differences give rise to several important phenomena, such as the modification of the emission or absorption spectra of the atom when it interacts with the laser beam.

C—RESONANCE FLUORESCENCE INTERPRETED AS A
RADIATIVE CASCADE OF THE DRESSED ATOM

After having discussed the energy levels of the dressed atom Hamiltonian H_{AL}, we now introduce the coupling of this system with the empty modes of the field (right-hand portion of Figure 1). Such a coupling causes photons to appear in modes that were initially empty. These are the *fluorescence* photons that an atom emits when it is excited by a resonant laser wave, and which appear here as photons spontaneously emitted by the dressed atom. The purpose of this section is to analyze such a picture in more detail and to show that resonance fluorescence can indeed be described in terms of a radiative cascade of the dressed atom descending along its energy-level ladder as a result of successive spontaneous emission processes. The emphasis here will be put on simple physical arguments. A more quantitative study, based on the master equation for describing the spontaneous emission of the dressed atom, will be presented in Sections D and E. After a brief review of the different time scales of the problem (§1), we describe the radiative cascade of the dressed atom and the different physical phenomena associated with it. Two different bases are used: the uncoupled basis (§2) adapted to the temporal aspect of the processes, and the dressed state basis (§3) adapted to the energy aspect.

1. The Relevant Time Scales

In Chapter IV, we introduced two characteristic time scales for a relaxation process such as spontaneous emission: the correlation time τ_c, which is very short (τ_c is less than one optical period $1/\omega_0$ because its inverse characterizes the range of variation of the spectral density of vacuum fluctuations which varies as ω^3); the radiative lifetime $\tau_R = \Gamma^{-1}$, which is much longer and is the average time after which an initially excited atom emits a photon. When an atom is excited, one cannot predict with certainty the time at which the photon will be emitted. Only the emission rate Γ is known. When emission occurs, the process itself lasts a time on the order of τ_c.

Now consider the processes of absorption and stimulated emission. According to subsection B.5, the characteristic time scale associated with them is the Rabi nutation period, which equals Ω_1^{-1} at resonance and Ω^{-1} in the general case [see (B.19)]. In this entire chapter (just as in the previous chapter), we will limit ourselves to the case

$$\tau_c \ll \Omega^{-1}. \tag{C.1}$$

Condition (C.1), which is equivalent to $\omega_0 \gg \Omega_1, |\delta_L|$, expresses the fact that the duration τ_c of an elementary spontaneous emission process is too short for the interaction with the laser photons to manifest itself during this time τ_c: The laser photons remain "spectators." By contrast, between two spontaneous processes (time scale Γ^{-1}), the atom and the laser photons have time to get coupled.

The physical picture that we will give below for resonance fluorescence depends, of course, on the experiment being considered. If we are interested in the temporal aspect of the phenomena, by measuring, for example, the various emission times of the photons emitted by the atom using a very wide-band photodetector, it is clear that the description of the evolution of the system will be simpler in the uncoupled basis, because, for each elementary spontaneous emission process, the laser photons remain spectators. If, by contrast, we attempt to measure the energy of the emitted photons by means of a narrow-band photodetector of bandwidth $\delta\nu$, the measurement time (on the order of $\delta\nu^{-1}$) is sufficiently long to allow the Bohr frequencies of the coupled atom + laser photons system to be measured. The dressed state basis is then more interesting.

2. Radiative Cascade in the Uncoupled Basis

a) Time Evolution of the System

In Figure 6, we have represented three manifolds $\mathscr{E}(N)$, $\mathscr{E}(N-1)$, and $\mathscr{E}(N-2)$ of uncoupled levels of the atom + laser photons system (the levels containing b are to the right of the levels containing a). The couplings induced by V_{AL} between levels of the same manifold describe absorption or stimulated emission processes (horizontal arrows pointing, respectively, to the right and to the left). Because the laser photons remain spectators during an elementary spontaneous emission process in which the atom goes from b to a, the corresponding transitions connect states b and a with the same value of N (wavy lines).

The time evolution can then be described in the following way. Assume that the system is initially in $\mathscr{E}(N)$, where it makes a Rabi oscillation between $|a, N+1\rangle$ and $|b, N\rangle$. At a certain time, a "quantum jump" (of duration τ_c) occurs from $|b, N\rangle$ to $|a, N\rangle$ with the spontaneous emission of a photon. A new Rabi oscillation then begins between $|a, N\rangle$ and $|b, N-1\rangle$, until the next quantum jump $|b, N-1\rangle \rightarrow |a, N-1\rangle$, and so on.

b) Photon Antibunching

The foregoing discussion shows that, immediately after a spontaneous emission process, the atom is found in the lower state a, with the global

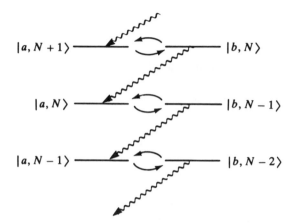

Figure 6. Radiative cascade in the uncoupled basis. The horizontal arrows correspond to absorption and stimulated emission processes, and the wavy arrows correspond to spontaneous emission processes.

system being, for example, in the state $|a, N\rangle$ of $\mathcal{E}(N - 1)$. No spontaneous emission occurs from $|a, N\rangle$ (see Figure 6) and a certain amount of time is needed before the atom goes into the state $|b, N - 1\rangle$ (by absorption of a laser photon) and then becomes able to spontaneously emit another photon again.

Two spontaneous emissions of photons by the same atom are thus separated by a time interval τ whose probability distribution tends to 0 when $\tau \to 0$. Such an effect is called *photon antibunching* and has been observed experimentally (*). It is a typical quantum effect.

c) TIME INTERVALS BETWEEN TWO SUCCESSIVE SPONTANEOUS EMISSIONS

Assume that a photon is emitted at time $t = 0$, for example, during a transition $|b, N\rangle \to |a, N\rangle$. Just after $t = 0$, the system is therefore in the state $|a, N\rangle$. Because the next photon is necessarily emitted from $|b, N - 1\rangle$, the probability that the *following* emission will occur between τ and $\tau + d\tau$ is equal to $K(\tau) d\tau$ where $K(\tau)$ is Γ times the probability of being in $|b, N - 1\rangle$ at time τ, knowing that the global system started from $|a, N\rangle$ at $t = 0$. To calculate $K(\tau)$, we can then limit ourselves to the two

(*) H. J. Kimble, M. Dagenais, and L. Mandel, *Phys. Rev. Lett.*, **39**, 691 (1977); M. Dagenais and L. Mandel, *Phys. Rev. A*, **18**, 2217 (1978); J. D. Cresser, J. Hager, G. Leuchs, M. Rateike, and H. Walther in *Dissipative Systems in Quantum Optics*, Vol. 27 of *Topics in Current Physics*, edited by R. Bonifacio (Springer-Verlag, Berlin, 1982).

states of the manifold $\mathscr{E}(N-1)$. The problem is then reduced to the evolution of two coupled states, one being the stable state $|a, N\rangle$, and the other being the unstable state $|b, N-1\rangle$. We previously studied such a problem in subsection C.3 of Chapter III and showed that the evolution within the manifold of the two states is described by a non-Hermitian effective Hamiltonian. One must actually solve the system of equations

$$i\dot{c}_a = \frac{\Omega_1}{2}c_b$$

$$i\dot{c}_b = -\delta_L c_b - i\frac{\Gamma}{2}c_b + \frac{\Omega_1}{2}c_a \tag{C.2}$$

with the initial condition $c_a(0) = 1$ and $c_b(0) = 0$, and deduce $K(\tau) = \Gamma|c_b(\tau)|^2$. Such a calculation is straightforward and leads to, for a resonant excitation ($\delta_L = 0$):

$$K(\tau) = \Gamma\frac{\Omega_1^2}{\lambda^2}\left(\sin^2\frac{\lambda\tau}{2}\right)e^{-\Gamma\tau/2} \tag{C.3.a}$$

with

$$\lambda^2 = \Omega_1^2 - \frac{\Gamma^2}{4}. \tag{C.3.b}$$

(We assume that $\lambda^2 > 0$, i.e., $\Omega_1 \geq \Gamma/2$. If not, we must change the sign of λ^2 and replace sin by sinh.)

The function $K(\tau)$ characterizes the distribution of intervals τ between two *successive* emissions. We now consider the sequence of random times $t_1, t_2 \cdots t_n \cdots$ at which the atom has spontaneously emitted photons during its radiative cascade. The time t_n of emission of the nth photon is correlated with the time t_{n-1} of emission of the previous photon. This is the antibunching effect mentioned above. By contrast, the intervals $\tau_n = t_n - t_{n-1}$ and $\tau_{n-1} = t_{n-1} - t_{n-2}$ are independent random variables, because, after emission of a photon at time t_{n-1}, the system starts again from a state of the type $|a, N\rangle$ independently of everything that happened before t_{n-1}. Knowing the distribution $K(\tau)$ of the intervals between two successive emissions is thus sufficient for the characterization of the statistical properties of the random process formed by the sequence of emission times. If the emitted photons are observed with a photodetector and if the efficiency η of the detector is known, it is easy to imagine that it is possible to calculate, using η and $K(\tau)$, all the characteristics of the number of photons detected during a time T (photon counting signals) (*).

In subsection E.3, we will calculate the photon-correlation signal from the master equation describing the evolution of the dressed atom. We will then justify in detail the introduction of the function $K(\tau)$ and all the physical pictures we have just presented.

3. Radiative Cascade in the Dressed State Basis

a) ALLOWED TRANSITIONS BETWEEN DRESSED STATES

To find the allowed spontaneous transitions between dressed states, we must determine the pairs of dressed states between which the atomic dipole **d** has a nonzero matrix element. We saw in subsection C.2 that, in the uncoupled basis, **d** can only connect the two levels $|b, N\rangle$ and $|a, N\rangle$ belonging to two adjacent manifolds $\mathscr{E}(N)$ and $\mathscr{E}(N - 1)$ (arrow with dashed lines in Figure 7). The two levels $|1(N)\rangle$ and $|2(N)\rangle$ of $\mathscr{E}(N)$ are both "contaminated" by $|b, N\rangle$ (see B.20). Similarly, the two levels $|1(N - 1)\rangle$ and $|2(N - 1)\rangle$ of $\mathscr{E}(N - 1)$ are both contaminated by $|a, N\rangle$. One thus concludes that the four transitions connecting the two dressed states of two adjacent manifolds are all allowed (wavy arrows in Figure 7). By contrast, because the two states $|1(N)\rangle$ and $|2(N)\rangle$ are not contaminated simultaneously by $|a, N\rangle$ and $|b, N\rangle$, there is no allowed transition between the two dressed states $|1(N)\rangle$ and $|2(N)\rangle$ of the same manifold. Similarly, we would show that there are no allowed transitions between $\mathscr{E}(N)$ and $\mathscr{E}(N - p)$ if $p > 1$.

Quantitatively, the four transition matrix elements between $|i(N)\rangle$ and $|j(N - 1)\rangle$ are proportional to the matrix elements $\langle i(N)|\mathscr{S}_+|j(N - 1)\rangle$. For N varying in the interval ΔN about $\langle N\rangle$, we can use expression (B.20) for the states $|i(N)\rangle$. The matrix elements between $\langle i(N)|$ and $|j(N - 1)\rangle$ of the operator $\mathscr{S}_+$ defined in (B.10) are then independent of N, and are written $(\mathscr{S}_+)_{ij}$. They equal

$$\langle 1(N)|\mathscr{S}_+|1(N - 1)\rangle = (\mathscr{S}_+)_{11} = \sin\theta\cos\theta \qquad \text{(C.4.a)}$$

$$\langle 2(N)|\mathscr{S}_+|2(N - 1)\rangle = (\mathscr{S}_+)_{22} = -\sin\theta\cos\theta \qquad \text{(C.4.b)}$$

$$\langle 1(N)|\mathscr{S}_+|2(N - 1)\rangle = (\mathscr{S}_+)_{12} = \cos^2\theta \qquad \text{(C.4.c)}$$

$$\langle 2(N)|\mathscr{S}_+|1(N - 1)\rangle = (\mathscr{S}_+)_{21} = -\sin^2\theta. \qquad \text{(C.4.d)}$$

(*) See, for example, S. Reynaud, *Ann. Phys. Fr.*, **8**, 315 (1983).

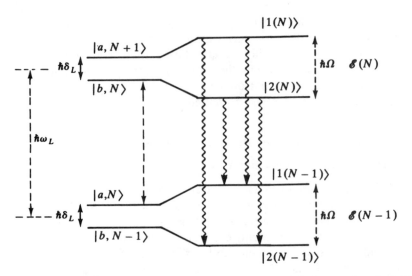

Figure 7. Allowed spontaneous transitions between uncoupled states (on the left) and dressed states (on the right).

b) FLUORESCENCE TRIPLET

The central frequencies of the lines spontaneously emitted in the allowed transitions in Figure 7 equal:

$$\text{transition } |1(N)\rangle \rightarrow |2(N-1)\rangle : \text{frequency } \omega_L + \Omega$$

$$\text{transition } |2(N)\rangle \rightarrow |1(N-1)\rangle : \text{frequency } \omega_L - \Omega \qquad \text{(C.5)}$$

$$\text{transition } |i(N)\rangle \rightarrow |i(N-1)\rangle : \text{frequency } \omega_L \qquad (i = 1,2).$$

The dressed atom approach thus provides a very simple interpretation of the triplet structure of the fluorescence spectrum, formed by a central line at ω_L and two sidebands at $\omega_L \pm \Omega$.

To determine the widths of these three lines, it is not sufficient to calculate the natural width of each dressed state, which is equal to the spontaneous emission rate of a photon from this level (given by the Fermi golden rule). Indeed, because the dressed atom energy diagram is periodic, coherence transfers can occur (as for the harmonic oscillator; see Complement B_{IV}, §3.b) between pairs of states corresponding to the same Bohr frequency. Later on, we will see how using the master equation for the dressed atom solves this problem. We will also obtain simple expressions for the weights of the lines as a function of the populations of the levels and of the transition rates.

c) Time Correlations between Frequency Filtered Fluorescence Photons

Assume that $\Omega \gg \Gamma$. The three lines of the triplet, having a width on the order of Γ as we will see below, are then well separated. By placing in front of the broadband detector that observes the fluorescence light a filter with a frequency bandwidth $\delta\omega$ such that

$$\Gamma \ll \delta\omega \ll \Omega \qquad (C.6)$$

and whose central frequency coincides with ω_L, $\omega_L + \Omega$, or $\omega_L - \Omega$, one can then determine from which line of the triplet the detected photon was emitted while still maintaining a temporal resolution better than Γ^{-1}.

To interpret the signals observed under such conditions, the picture of the radiative cascade between dressed levels is particularly convenient. For example, in the cascade represented in Figure 8, the system starting from $|1(N)\rangle$ emits a photon $\omega_L + \Omega$, which brings it to $|2(N-1)\rangle$, then a photon $\omega_L - \Omega$, which brings it to $|1(N-2)\rangle$, then a photon ω_L, which brings it to $|1(N-3)\rangle$, and so on.

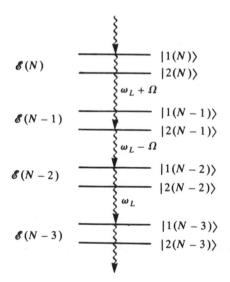

Figure 8. Radiative cascade in the dressed state basis.

The foregoing picture allows us to simply understand the temporal correlations between photons emitted in the various lines of the triplet. For example, after the emission of a "large" photon ($\omega_L + \Omega$), the dressed atom is in a type-2 state from which it can emit only a "small" photon ($\omega_L - \Omega$) or a "medium" photon (ω_L). Between two emissions of large photons ($\omega_L + \Omega$), there must necessarily be the emission of a small photon ($\omega_L - \Omega$), which brings the dressed atom from a type-2 state to a type-1 state. The correlations between photons $\omega_L + \Omega$ and $\omega_L - \Omega$ have been experimentally observed (*).

Remark

We return to the photon antibunching effect observable with a broadband detector (see §C.2.b above). How will it appear in the dressed state basis? Immediately after a photon is detected, we know that the system is in a state that contains only a, for example, $|a, N\rangle$. However, the state $|a, N\rangle$ is a linear superposition of the two dressed states $|1(N-1)\rangle$ and $|2(N-1)\rangle$ which does not radiate. The subsequent evolution causes the phase difference between $|1(N-1)\rangle$ and $|2(N-1)\rangle$ to vary at the rate Ω and again makes possible the emission of light. In other words, the Rabi oscillations observed on the population of $|b, N\rangle$, and therefore on the emission rate, appear in the dressed state basis as a quantum beat signal, the initial coherence between the two levels of $\mathscr{E}(N-1)$ being suddenly prepared by the detection of the first photon. By placing filters (of the type described in this subsection) in front of the detector, one suppresses these beats in the same way as, in a quantum beat experiment, any attempt to determine from which state the photon is emitted suppresses the interference which is at the origin of the beat.

(*) A. Aspect, G. Roger, S. Reynaud, J. Dalibard, and C. Cohen-Tannoudji, *Phys. Rev. Lett.*, **45**, 617 (1980).

D—MASTER EQUATION FOR THE DRESSED ATOM

In subsection E-a-1 of Chapter IV, we derived the master equation describing the spontaneous emission of an atom with two levels a and b. For the discussion here in Section D, it will be useful to rewrite this equation in the form

$$\frac{d}{dt}\sigma_A = -\frac{i}{\hbar}[H_A, \sigma_A] - \frac{\Gamma}{2}(\mathcal{S}_+\mathcal{S}_-\sigma_A + \sigma_A\mathcal{S}_+\mathcal{S}_-) + \Gamma\mathcal{S}_-\sigma_A\mathcal{S}_+. \quad (D.1)$$

In (D.1), σ_A is the atomic density operator, H_A is the atomic Hamiltonian (B.4), and $\mathcal{S}_+$ and $\mathcal{S}_-$ are the operators given above in (B.10). The projection of (D.1) on the basis $\{|a\rangle, |b\rangle\}$ gives

$$\dot{\sigma}_{bb} = -\Gamma\sigma_{bb}$$

$$\dot{\sigma}_{aa} = \Gamma\sigma_{bb} \qquad\qquad (D.2)$$

$$\dot{\sigma}_{ab} = i\omega_0\sigma_{ab} - \frac{\Gamma}{2}\sigma_{ab}$$

which are indeed the Equations (E.5) and (E.6) of Chapter IV (we assume that the radiative shifts have been included in ω_0).

In this Section D, we generalize Equation (D.1) for the atom + laser mode system; i.e., for the dressed atom with Hamiltonian $H_{AL} = H_A + H_L + V_{AL}$. We begin (§1) by deriving the general form of the master equation within the approximation of independent rates of variation. We then project this equation over the dressed state basis and physically interpret the results obtained at the secular limit $\Omega \gg \Gamma$ (§2). Finally, we show how a quasi-steady state can be defined for the radiative cascade of the dressed atom (§3).

1. General Form of the Master Equation

a) Approximation of Independent Rates of Variation

The physical discussion in subsection C-1 above shows that, if the condition $\tau_c \ll \Omega^{-1}$ is fulfilled, the laser photons remain "spectators" during an elementary spontaneous emission process, because there is not sufficient time for the Rabi oscillation with frequency Ω to occur during the correlation time τ_c of such a process. By contrast, between two spontaneous emission processes, there is sufficient time for the coupling V_{AL} between the atom and the laser photons to manifest itself. The approximation of independent rates of variation consists of independently adding the rates of variation of the density operator σ_{AL} of the dressed atom, due, respectively, to the processes of absorption and induced emission on the one hand, and spontaneous emission on the other hand, and calculated as if each type of process were acting alone. We thus generalize Equation (D.1) in the form

$$\frac{d}{dt}\sigma_{AL} = -\frac{i}{\hbar}[H_{AL},\sigma_{AL}]$$
$$-\frac{\Gamma}{2}(\mathscr{S}_+\mathscr{S}_-\sigma_{AL} + \sigma_{AL}\mathscr{S}_+\mathscr{S}_-) + \Gamma\mathscr{S}_-\sigma_{AL}\mathscr{S}_+. \quad \text{(D.3)}$$

It should be noted that H_{AL} (which contains V_{AL}) has replaced H_A in the first term (*). Later on (see the Remark in subsection D-2-b and Complement B_{VI}) we will return to the corrections that can be made to the approximation of independent rates of variation.

The last two terms of (D.3) describe the effect of spontaneous emission. By projecting over the uncoupled basis of eigenstates of $H_A + H_L$, they give

$$\langle b, N|\dot{\sigma}_{AL}|b, N'\rangle = -\Gamma\langle b, N|\sigma_{AL}|b, N'\rangle$$
$$\langle a, N|\dot{\sigma}_{AL}|a, N'\rangle = \Gamma\langle b, N|\sigma_{AL}|b, N'\rangle \quad \text{(D.4)}$$
$$\langle b, N|\dot{\sigma}_{AL}|a, N'\rangle = -\frac{\Gamma}{2}\langle b, N|\sigma_{AL}|a, N'\rangle.$$

We recover the same equations as for σ_A alone [see (D.2)], the quantum numbers N and N' of the laser photons being unchanged during the process.

To simplify notation, we will now write σ instead of σ_{AL} for the density operator of the dressed atom.

b) COMPARISON WITH OPTICAL BLOCH EQUATIONS

We introduce

$$\langle S_+\rangle = \sum_N \langle a, N|\sigma|b, N\rangle$$
$$\langle S_-\rangle = \sum_N \langle b, N|\sigma|a, N\rangle \quad \text{(D.5)}$$
$$\langle S_z\rangle = \tfrac{1}{2}\sum_N [\langle b, N|\sigma|b, N\rangle - \langle a, N|\sigma|a, N\rangle].$$

(*) The operators $\mathscr{S}_\pm$ of (D.3) must also be understood as the tensor products of operators (B.10) by the unit operator of the laser mode.

The equations of motion for $\langle S_+ \rangle$, $\langle S_- \rangle$, and $\langle S_z \rangle$ can be obtained from (D.3). One can easily check that these equations are identical to the optical Bloch equations (A.18) in Chapter V. For the spontaneous emission terms, such a result is not surprising, considering the similarity between (D.2) and (D.4). For the absorption and stimulated emission terms [contribution of the first term of (D.3)], it is sufficient to note that the projection of H_{AL} onto a manifold $\mathscr{E}(N)$ of the dressed atom is represented in the basis $\{|b, N\rangle, |a, N + 1\rangle\}$ by the following matrix, except for a constant term.

$$\frac{\hbar}{2} \begin{pmatrix} -\delta_L & \Omega_1 \\ \Omega_1 & \delta_L \end{pmatrix}. \tag{D.6}$$

In Chapter V we introduced a geometric representation for the Rabi oscillation (associated with absorption and stimulated emission processes) in terms of a fictitious spin coupled to magnetic fields (see subsection A-4 in Chapter V). It is easy to see that the matrix (D.6) coincides with the matrix that represents, in the rotating reference frame $0XYZ$, the interaction Hamiltonian of the fictitious spin with the effective field $\mathbf{B}_e$ (see Figure 1 in Chapter V). Thus, there is a strong analogy between the dressed states $|1(N)\rangle$ and $|2(N)\rangle$ of a manifold $\mathscr{E}(N)$ of the dressed atom and the eigenstates of the fictitious spin along the direction of the effective field $\mathbf{B}_e$. The angle 2θ introduced in (B.21) corresponds to the angle between the two fields $\mathbf{B}_e$ and $\mathbf{b}_0$ in Figure 1 of Chapter V.

We have therefore shown that the optical Bloch equations in Chapter V are contained in the more general equations (D.3). However, it should not be forgotten that the "reduction" operation associated with (D.5) (sum over N) "condenses" all the manifolds $\mathscr{E}(N)$ into a single manifold, and thus causes the loss of the picture of the radiative cascade $\mathscr{E}(N) \rightarrow \mathscr{E}(N - 1) \rightarrow \mathscr{E}(N - 2) \cdots$ and all the resulting simple physical interpretations (see Section C above).

2. Master Equation in the Dressed State Basis in the Secular Limit

a) ADVANTAGES OF THE COUPLED BASIS IN THE SECULAR LIMIT

The operator equation (D.3) can be projected over any basis. Because the dressed state basis diagonalizes H_{AL}, such a projection is particularly simple for the first term in (D.3). However, in this basis, the last two terms of (D.3) give equations that are not as simple as (D.4). In particular, couplings between dressed state populations and coherences between two dressed states appear.

However, the situation is simplified if the splitting $\hbar\Omega$ between the two states of a manifold is large compared to $\hbar\Gamma$. We can then ignore the "nonsecular" couplings between populations and coherences and formulate a simple physical picture in terms of spontaneous transfer rates between dressed states. In the rest of Section D we will assume that

$$\Omega \gg \Gamma. \tag{D.7}$$

b) EVOLUTION OF POPULATIONS

We project Equation (D.3) over $|i(N)\rangle$ on the right, and over $\langle i(N)|$ on the left (with $i = 1$ or 2). When the nonsecular couplings between populations and coherences are ignored, we obtain for the population

$$\pi_{i(N)} = \langle i(N)|\sigma|i(N)\rangle \tag{D.8}$$

of the dressed state $|i(N)\rangle$ the evolution equation

$$\dot{\pi}_{i(N)} = -\left(\sum_{j=1,2}\Gamma_{i \to j}\right)\pi_{i(N)} + \sum_{l=1,2}\Gamma_{l \to i}\pi_{l(N+1)} \tag{D.9}$$

where the transition rates

$$\Gamma_{i \to j} = \Gamma|\langle i(N)|\mathcal{S}_+|j(N-1)\rangle|^2 = \Gamma|(\mathcal{S}_+)_{ij}|^2 \tag{D.10}$$

equal, using (C.4)

$$\Gamma_{1 \to 1} = \Gamma_{2 \to 2} = \Gamma \cos^2\theta \sin^2\theta$$

$$\Gamma_{2 \to 1} = \Gamma \sin^4\theta \tag{D.11}$$

$$\Gamma_{1 \to 2} = \Gamma \cos^4\theta.$$

The interpretation of Equation (D.9), written for $i = 1$,

$$\dot{\pi}_{1(N)} = -\pi_{1(N)}(\Gamma_{1 \to 1} + \Gamma_{1 \to 2}) + \pi_{2(N+1)}\Gamma_{2 \to 1} + \pi_{1(N+1)}\Gamma_{1 \to 1} \tag{D.12}$$

is very clear (see Figure 9a). The state $|1(N)\rangle$ decays by spontaneous emission to the levels $|1(N-1)\rangle$ (with a rate $\Gamma_{1 \to 1}$) and $|2(N-1)\rangle$ (with a rate $\Gamma_{1 \to 2}$) and is populated from the levels $|1(N+1)\rangle$ (with a rate $\Gamma_{1 \to 1}$) and $|2(N+1)\rangle$ (with a rate $\Gamma_{2 \to 1}$).

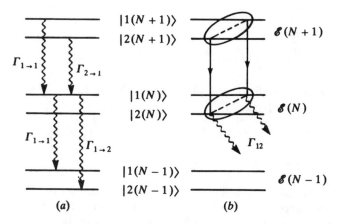

Figure 9. (*a*) Transfer of populations responsible for the variation of the population of the state $|1(N)\rangle$. (*b*) Transfer of coherences from $\mathscr{E}(N+1)$ to $\mathscr{E}(N)$.

Remark

Instead of deriving (D.9) from (D.3), we could use the general equations (C.7) and (C.5) of Chapter IV. The expressions (D.11) for the transition rates would only be found if the spectral density of vacuum fluctuations (which vary as ω^3) can be considered as constant from one component of the fluorescence triplet to another, i.e., if

$$\omega_L^3 \simeq (\omega_L - \Omega)^3 \simeq (\omega_L + \Omega)^3 \simeq \omega_0^3. \tag{D.13}$$

Condition (D.13) is simply the validity condition (C.1) for the approximation of the independent rates of variation (because $\tau_c \lesssim \omega_0^{-1}$). It then appears that by directly using the general results from Chapter IV, one could correct the foregoing approximation and obtain a master equation for the spontaneous emission of the dressed atom, which is more general than the one derived from (D.1) and (D.3).

c) EVOLUTION OF COHERENCES—TRANSFER OF COHERENCES

The projection of (D.3) on the right over $|2(N)\rangle$ and on the left over $\langle 1(N)|$ gives

$$\frac{d}{dt}\langle 1(N)|\sigma|2(N)\rangle = -i\Omega\langle 1(N)|\sigma|2(N)\rangle -$$

$$- \Gamma_{12}\langle 1(N)|\sigma|2(N)\rangle + K_{12}\langle 1(N+1)|\sigma|2(N+1)\rangle \tag{D.14}$$

where

$$\Gamma_{12} = \frac{1}{2}[\Gamma_{1\to1} + \Gamma_{1\to2} + \Gamma_{2\to1} + \Gamma_{2\to2}] = \frac{\Gamma}{2} \qquad (D.15)$$

is the half sum of all the transition rates starting from the states $|1(N)\rangle$ and $|2(N)\rangle$ and where

$$K_{12} = \Gamma\langle 1(N)|\mathscr{S}_-|1(N+1)\rangle\langle 2(N+1)|\mathscr{S}_+|2(N)\rangle$$
$$= \Gamma(\mathscr{S}_+)^*_{11}(\mathscr{S}_+)_{22} = -\Gamma\sin^2\theta\cos^2\theta. \qquad (D.16)$$

The first term of the right-hand side of (D.14) describes the free evolution of the coherence $\langle 1(N)|\sigma|2(N)\rangle$ at the frequency Ω. The second term describes the damping due to radiative transitions of the states involved to lower states of $\mathscr{E}(N-1)$. Finally, the last term describes the radiative transfer of coherence from $\mathscr{E}(N+1)$ to $\mathscr{E}(N)$ (see Figure 9b). As for the harmonic oscillator (see Complement B_{IV}, §3-b), it is due to the periodicity of the energy diagram which gives rise to coherences evolving at the same frequency.

Remark

It is also possible to consider coherences between states belonging to different manifolds (with $|N - N'| \ll \Delta N$). For example, for the evolution equation of $\langle 1(N)|\sigma|2(N')\rangle$ we obtain the expression

$$\frac{d}{dt}\langle 1(N)|\sigma|2(N')\rangle = -i[\Omega + (N - N')\omega_L]\langle 1(N)|\sigma|2(N')\rangle -$$
$$- \Gamma'_{12}\langle 1(N)|\sigma|2(N')\rangle + K'_{12}\langle 1(N+1)|\sigma|2(N'+1)\rangle$$
$$\qquad (D.17)$$

where Γ'_{12}, which is the half-sum of the transition rates from $|1(N)\rangle$ to $|1(N-1)\rangle$ and $|2(N-1)\rangle$ and from $|2(N')\rangle$ to $|1(N'-1)\rangle$ and $|2(N'-1)\rangle$, is equal to Γ_{12} and where

$$K'_{12} = \Gamma\langle 1(N)|\mathscr{S}_-|1(N+1)\rangle\langle 2(N'+1)|\mathscr{S}_+|2(N')\rangle$$
$$= \Gamma(\mathscr{S}_+)^*_{11}(\mathscr{S}_+)_{22} = K_{12}. \qquad (D.18)$$

From this we deduce that, except for the free evolution term, the coherences $\langle 1(N)|\sigma|2(N')\rangle$ and $\langle 1(N)|\sigma|2(N)\rangle$ obey the same evolution equations. A similar procedure would show that the coherences $\langle i(N)|\sigma|i(N')\rangle$ with $i = 1, 2$ obey, except for the free evolution term, the same equations as the populations $\langle i(N)|\sigma|i(N)\rangle = \pi_{i(N)}$.

d) REDUCED POPULATIONS AND REDUCED COHERENCES

The evolution equations for the "reduced" variables

$$\pi_i = \sum_N \pi_{i(N)} \tag{D.19.a}$$

$$\sigma_{12} = \sum_N \langle 1(N)|\sigma|2(N)\rangle \tag{D.19.b}$$

are easily derived from Equations (D.9) and (D.14).
For the reduced populations π_1 and π_2, we obtain

$$\dot{\pi}_1 = -\pi_1\Gamma_{1\rightarrow2} + \pi_2\Gamma_{2\rightarrow1}$$

$$\dot{\pi}_2 = -\pi_2\Gamma_{2\rightarrow1} + \pi_1\Gamma_{1\rightarrow2}. \tag{D.20}$$

The steady-state solution of these equations

$$\pi_1^{st} = \frac{\Gamma_{2\rightarrow1}}{\Gamma_{1\rightarrow2} + \Gamma_{2\rightarrow1}} = \frac{\sin^4\theta}{\cos^4\theta + \sin^4\theta} \tag{D.21.a}$$

$$\pi_2^{st} = \frac{\Gamma_{1\rightarrow2}}{\Gamma_{1\rightarrow2} + \Gamma_{2\rightarrow1}} = \frac{\cos^4\theta}{\sin^4\theta + \cos^4\theta} \tag{D.21.b}$$

obeys the equation

$$\pi_1^{st}\Gamma_{1\rightarrow2} = \pi_2^{st}\Gamma_{2\rightarrow1} \tag{D.22}$$

called the "detailed balance condition" and also satisfies the normalization condition

$$\pi_1^{st} + \pi_2^{st} = 1. \tag{D.23}$$

The transient regime is given by

$$\pi_i(\tau) = \pi_i^{st} + \left[\pi_i(0) - \pi_i^{st}\right]e^{-\tau/\tau_{pop}} \tag{D.24}$$

where the time constant τ_{pop} is the inverse of

$$\Gamma_{pop} = \Gamma_{1\rightarrow2} + \Gamma_{2\rightarrow1} = \Gamma(\cos^4\theta + \sin^4\theta). \tag{D.25}$$

The reduced coherence σ_{12} obeys the evolution equation

$$\dot{\sigma}_{12} = -(i\Omega + \Gamma_{coh})\sigma_{12} \qquad (D.26)$$

where the damping rate Γ_{coh} equals

$$\Gamma_{coh} = \Gamma_{12} - K_{12} = \Gamma\left(\tfrac{1}{2} + \cos^2\theta\sin^2\theta\right) \qquad (D.27)$$

[we used (D.15) and (D.16)]. Note the importance of the transfer term $(-K_{12})$, which results in the fact that Γ_{coh} is not simply equal to $\Gamma_{12} = \Gamma/2$.

Remark

We saw above (§D-1-b) that the reduction operation (D.5) in the uncoupled basis yields the usual optical Bloch equations. Here in (D.19) we carry out a similar reduction in the dressed state basis. It can be shown that the equations obtained ((D.20) and (D.26)) are identical to the optical Bloch equations, when the following two transformations are used.

The Bloch equations must be written, not in the basis $\{| \pm \rangle\}$ of eigenstates of S_z, but instead in the basis of eigenstates of the fictitious spin along the effective field $\mathbf{B}_e$ (Figure 1 in Chapter V). Such a basis change is equivalent to diagonalizing the Hamiltonian part of the optical Bloch equations.

In this new basis, the relaxation terms due to spontaneous emission are less simple than in the basis $\{| \pm \rangle\}$. To obtain (D.20) and (D.26) we must retain only the secular relaxation terms (in the new basis).

As above, one should not forget that the reduction (D.19) causes the picture of the radiative cascade to be lost in the dressed state basis (see Figure 8).

More generally, one can consider the reduced coherences

$$\sigma_{ij}^{(q)} = \sum_{N}\langle i(N-q)|\sigma|j(N)\rangle \qquad (D.28)$$

involving coherences between states belonging to different manifolds, separated by $q\hbar\omega_L$. Equations (D.17), (D.18), and (D.27) then result in

$$\dot{\sigma}_{12}^{(q)} = [i(q\omega_L - \Omega) - \Gamma_{coh}]\sigma_{12}^{(q)}. \qquad (D.29)$$

Except for the free evolution term, the reduced coherences $\sigma_{12}^{(q)}$ and σ_{12} defined in (D.19.b) obey the same evolution equations. Similarly, one would show that $\sigma_{21}^{(q)}$ obeys the evolution equation

$$\dot{\sigma}_{21}^{(q)} = [i(q\omega_L + \Omega) - \Gamma_{coh}]\sigma_{21}^{(q)} \qquad (D.30)$$

and that the two reduced coherences $\sigma_{ii}^{(q)}$ with $i = 1, 2$ obey the coupled equations

$$\dot{\sigma}_{11}^{(q)} = iq\omega_L\sigma_{11}^{(q)} - \sigma_{11}^{(q)}\Gamma_{1\rightarrow 2} + \sigma_{22}^{(q)}\Gamma_{2\rightarrow 1}$$

$$\dot{\sigma}_{22}^{(q)} = iq\omega_L\sigma_{22}^{(q)} - \sigma_{22}^{(q)}\Gamma_{2\rightarrow 1} + \sigma_{11}^{(q)}\Gamma_{1\rightarrow 2} \tag{D.31}$$

which are indeed analogous to the evolution equations (D.20) for reduced populations.

3. Quasi-Steady State for the Radiative Cascade

a) Initial Density Matrix

At the initial time $t = 0$, the atom in the state $|a\rangle$ is placed in the presence of the laser mode in the state σ_L, so that the density operator of the global system is

$$\sigma(0) = |a\rangle\langle a| \otimes \sigma_L. \tag{D.32}$$

Assume that the state σ_L is a coherent state $|\alpha\rangle\langle\alpha|$ with

$$|\alpha\rangle = \sum_{N=0}^{\infty} c_N|N\rangle = \sum_{N=0}^{\infty} e^{-\alpha^2/2}\frac{\alpha^N}{\sqrt{N!}}|N\rangle \tag{D.33}$$

(we have assumed that α is real). The initial distribution $p_0(N)$ of the number of laser photons,

$$p_0(N) = \langle N|\sigma_L|N\rangle = c_N^2 = e^{-\alpha^2}\frac{\alpha^{2N}}{N!} \tag{D.34}$$

is then centered at $\langle N\rangle = \alpha^2$ and has a width $\Delta N = \sqrt{\langle N\rangle}$ satisfying (B.2). Using (D.32), (D.34), and (B.20), one can calculate all the initial values of the matrix elements of σ. For example,

$$\langle 1(N)|\sigma(0)|2(N')\rangle = \sin\theta\cos\theta c_{N+1}c_{N'+1}$$

$$= \sin\theta\cos\theta\sqrt{p_0(N+1)p_0(N'+1)}. \tag{D.35}$$

If we consider the coherences between adjacent manifolds ($|N - N'| \ll \Delta N$), the square root in (D.35) can be approximated by $p_0(N)$, so that $\langle 1(N)|\sigma(0)|2(N')\rangle$ is close to $\langle 1(N)|\sigma(0)|2(N)\rangle$. More generally, for small $|N - N'|$, we have

$$\langle i(N)|\sigma(0)|j(N')\rangle \simeq \langle i(N)|\sigma(0)|j(N)\rangle. \qquad (D.36)$$

b) Transient Regime and Quasi-Steady State

Let $\sigma_N(t)$ be the projection of $\sigma(t)$ onto the manifold $\mathscr{E}(N)$ of the two states $|1(N)\rangle$ and $|2(N)\rangle$. According to the foregoing, the evolution of $\sigma_N(t)$ should reflect the existence of two phenomena. In the first place, the transient evolution of the reduced density matrix introduced in subsection D-2-d should appear; the reduced populations and coherences must tend to their steady-state values π_1^{st}, π_2^{st} [given in (D.21)] and $\sigma_{12}^{st} = 0$ with the time constants Γ_{pop}^{-1} and Γ_{coh}^{-1} on the order of Γ^{-1}. Second, because laser photons are permanently absorbed, the distribution $p(N)$ of the number of laser photons must shift downward. During a time T, the order of magnitude of this shift is ΓT.

As soon as $\Gamma T \gg 1$, the reduced density matrix has achieved its steady state. Because $\langle N \rangle$ and ΔN are very large compared with ΓT [condition (B.1)], we can ignore the downward shift of the distribution $p_0(N)$ and consider only a quasi-steady state for the cascade in which

$$\pi_{i(N)} \simeq \pi_i^{st} p_0(N) \qquad (D.37)$$

$$\langle 1(N)|\sigma|2(N)\rangle \simeq \sigma_{12}^{st} p_0(N) = 0. \qquad (D.38)$$

Remarks

(i) Such a quasi-steady state can be considered as resulting from a dynamic equilibrium. When N varies by ± 1, $p_0(N)$ varies only slightly and (D.37) gives $\pi_{i(N+1)} \simeq \pi_{i(N)}$. The cancellation of $\dot{\pi}_{1(N)}$ in the quasi-steady state is first due to the fact that the population which arrives from $|1(N + 1)\rangle$ is compensated for by the population which leaves $|1(N)\rangle$ toward $|1(N - 1)\rangle$ (see Figure 9a). Then, the detailed balance condition (D.22) for π_1^{st} and π_2^{st} shows, using (D.37), that the population which arrives in $|1(N)\rangle$ from $|2(N + 1)\rangle$ is compensated for by the population which leaves $|1(N)\rangle$ toward $|2(N - 1)\rangle$. The state $|1(N)\rangle$ is therefore traversed by population fluxes which cancel each other.

(ii) We have only considered here the quasi-steady state of $\langle i(N)|\sigma|j(N)\rangle$. To obtain the quasi-steady state of $\langle i(N)|\sigma|j(N')\rangle$, it is sufficient to note that the two matrix elements obey the same evolution equations, except for the free evolution term (see Remark, §D-2-c), and also that their initial values are identical [see (D.36)]. In the quasi-steady state, their values thus only differ by the free evolution exponential. In particular

$$\langle i(N)|\sigma|i(N')\rangle = e^{-i(N-N')\omega_L t} \pi_i^{st} p_0(N) \qquad (D.39)$$

$$\langle 1(N)|\sigma|2(N')\rangle = e^{-i(N-N')\omega_L t} \sigma_{12}^{st} p_0(N) = 0. \qquad (D.40)$$

E—DISCUSSION OF A FEW APPLICATIONS

1. Widths and Weights of the Various Components of the Fluorescence Triplet

As a first application of the dressed atom approach, we will determine in this subsection the widths and the weights of the three components of the fluorescence triplet emitted by the atom. To do this, we start with the expression for the steady-state spectrum [see formulas (D.8) and (A.15) in Chapter V]:

$$\mathcal{I}(\omega) = \frac{\Gamma}{2\pi} \int_{-\infty}^{+\infty} \langle \mathcal{S}_+(\tau)\mathcal{S}_-(0) \rangle \, e^{-i\omega\tau} \, d\tau$$

$$= \frac{\Gamma}{\pi} \, \mathrm{Re} \int_0^{\infty} \langle \mathcal{S}_+(\tau)\mathcal{S}_-(0) \rangle \, e^{-i\omega\tau} \, d\tau. \tag{E.1}$$

Here the average is taken in the quasi-steady state defined above (§D-3). The normalization of $\mathcal{I}(\omega)$ is selected so that the total intensity

$$I = \int d\omega \, \mathcal{I}(\omega) = \Gamma \langle \mathcal{S}_+(0)\mathcal{S}_-(0) \rangle = \Gamma \sigma_{bb} \tag{E.2}$$

corresponds to the number of photons emitted per second.

To evaluate the two-time average which appears in the second line of (E.1), we use the quantum regression theorem which says that, for $\tau > 0$, $\langle \mathcal{S}_+(\tau)\mathcal{S}_-(0) \rangle$ evolves as $\langle \mathcal{S}_+(\tau) \rangle$. We must then begin by discussing the evolution of $\langle \mathcal{S}_+(\tau) \rangle$, i.e., the evolution of the mean dipole moment.

a) EVOLUTION OF THE MEAN DIPOLE MOMENT

In the dressed state basis, the operator $\mathcal{S}_+$ is written, with the notations introduced in (C.4),

$$\mathcal{S}_+ = \sum_{i,j,N} (\mathcal{S}_+)_{ij} |i(N)\rangle\langle j(N-1)| \tag{E.3}$$

that is, also

$$\mathcal{S}_+ = \mathcal{S}_+^{12} + \mathcal{S}_+^{21} + \sum_{i=1,2} \mathcal{S}_+^{ii} \tag{E.4}$$

with

$$\mathcal{S}_+^{ij} = (\mathcal{S}_+)_{ij} \sum_N |i(N)\rangle\langle j(N-1)|. \tag{E.5}$$

The advantage of using (E.4) is that the three terms on the right-hand side contribute to the mean dipole moment with different eigenfrequencies.

Consider first the average value of $\mathcal{S}_+^{12}$, which is equal to

$$\langle \mathcal{S}_+^{12}\rangle = (\mathcal{S}_+)_{12} \sum_N \langle 2(N-1)|\sigma|1(N)\rangle = (\mathcal{S}_+)_{12}\sigma_{21}^{(1)} \tag{E.6}$$

where $\sigma_{21}^{(1)}$ is a reduced coherence of the same type as those introduced above in (D.28), and whose evolution equation according to (D.30), is given by

$$\dot{\sigma}_{21}^{(1)} = \left[i(\omega_L + \Omega) - \Gamma_{coh}\right]\sigma_{21}^{(1)}. \tag{E.7}$$

Using (E.7) in (E.6) then shows that

$$\frac{d}{d\tau}\langle \mathcal{S}_+^{12}(\tau)\rangle = \left[i(\omega_L + \Omega) - \Gamma_{coh}\right]\langle \mathcal{S}_+^{12}(\tau)\rangle. \tag{E.8}$$

The component $\langle \mathcal{S}_+^{12}(\tau)\rangle$ of the mean dipole moment thus evolves at the frequency $\omega_L + \Omega$ and is damped with a rate equal to Γ_{coh}. An analogous calculation for $\langle \mathcal{S}_+^{21}(\tau)\rangle$ yields

$$\frac{d}{d\tau}\langle \mathcal{S}_+^{21}(\tau)\rangle = \left[i(\omega_L - \Omega) - \Gamma_{coh}\right]\langle \mathcal{S}_+^{21}(\tau)\rangle. \tag{E.9}$$

We now consider the average value of the operators $\mathcal{S}_+^{ii}$ appearing in the last term of (E.4)

$$\langle \mathcal{S}_+^{ii}\rangle = (\mathcal{S}_+)_{ii} \sum_N \langle i(N-1)|\sigma|i(N)\rangle = (\mathcal{S}_+)_{ii}\sigma_{ii}^{(1)}. \tag{E.10}$$

By using the evolution equations (D.31) for the reduced coherences $\sigma_{ii}^{(1)}$ and the fact that, according to (C.4.a) and (C.4.b), $(\mathcal{S}_+)_{11} = -(\mathcal{S}_+)_{22}$, we obtain for the average values $\langle \mathcal{S}_+^{11}(\tau)\rangle$ and $\langle \mathcal{S}_+^{22}(\tau)\rangle$ the coupled evolution equations

$$\frac{d}{dt}\langle \mathcal{S}_+^{11}(\tau)\rangle = (i\omega_L - \Gamma_{1\to 2})\langle \mathcal{S}_+^{11}(\tau)\rangle - \Gamma_{2\to 1}\langle \mathcal{S}_+^{22}(\tau)\rangle$$

$$\frac{d}{dt}\langle \mathcal{S}_+^{22}(\tau)\rangle = -\Gamma_{1\to 2}\langle \mathcal{S}_+^{11}(\tau)\rangle + (i\omega_L - \Gamma_{2\to 1})\langle \mathcal{S}_+^{22}(\tau)\rangle. \tag{E.11}$$

The 2×2 matrix associated with such a linear system has eigenvalues equal to $i\omega_L$ and $i\omega_L - \Gamma_{pop}$ where Γ_{pop} is given in (D.25). It follows that, after a transient regime damped with a rate Γ_{pop}, the average values $\langle \mathscr{S}_+^{ii}(\tau) \rangle$ tend to a steady state, or more precisely to a forced oscillation regime given, according to (E.10) and (D.39), by

$$\langle \mathscr{S}_+^{ii}(\tau) \rangle_{st} = (\mathscr{S}_+)_{ii} \pi_i^{st} e^{i\omega_L t}. \tag{E.12}$$

b) WIDTHS AND WEIGHTS OF THE SIDEBANDS

We now consider the contribution of each of the terms of (E.4) to the spectrum (E.1). The first term, $\mathscr{S}_+^{12}(\tau)$, gives rise to the two-time average $\langle \mathscr{S}_+^{12}(\tau) \mathscr{S}_-(0) \rangle$. The quantum regression theorem indicates that, for $\tau > 0$, this two-time average obeys a quantum evolution equation identical to the equation for the one-time average $\langle \mathscr{S}_+^{12}(\tau) \rangle$. Using (E.8) then gives, for $\tau > 0$,

$$\frac{d}{dt}\langle \mathscr{S}_+^{12}(\tau) \mathscr{S}_-(0) \rangle = [i(\omega_L + \Omega) - \Gamma_{coh}]\langle \mathscr{S}_+^{12}(\tau) \mathscr{S}_-(0) \rangle \tag{E.13}$$

an equation whose solution is

$$\langle \mathscr{S}_+^{12}(\tau) \mathscr{S}_-(0) \rangle = \langle \mathscr{S}_+^{12} \mathscr{S}_- \rangle \exp[i(\omega_L + \Omega) - \Gamma_{coh}]\tau. \tag{E.14}$$

Substituting (E.14) into (E.1) gives

$$\Gamma\langle \mathscr{S}_+^{12} \mathscr{S}_- \rangle \frac{1}{\pi} \frac{\Gamma_{coh}}{(\omega - \omega_L - \Omega)^2 + \Gamma_{coh}^2} \tag{E.15}$$

i.e., a Lorentzian line centered at $\omega_L + \Omega$, with a full width at half maximum $2\Gamma_{coh}$ and having a weight (*) $\Gamma\langle \mathscr{S}_+^{12} \mathscr{S}_- \rangle$, which, according to (E.5), is equal to

$$\Gamma\langle \mathscr{S}_+^{12} \mathscr{S}_- \rangle = \Gamma(\mathscr{S}_+)_{12} \, \text{Tr} \sum_N (|1(N)\rangle\langle 2(N-1)|\mathscr{S}_- \sigma)$$

$$= \Gamma(\mathscr{S}_+)_{12} \sum_{i,N} (\mathscr{S}_-)_{2i}\langle i(N)|\sigma|1(N)\rangle. \tag{E.16}$$

According to (D.38), $\langle 2(N)|\sigma|1(N)\rangle$ is zero in the quasi-steady state. We

---

(*) Using (E.2) and (E.15), the weight of a line of the triplet is the number of photons emitted per second in this line.

thus obtain, using (D.37) and (D.10),

$$\Gamma\langle \mathscr{S}_+^{12}\mathscr{S}_-\rangle = \Gamma|(\mathscr{S}_+)_{12}|^2 \pi_1^{\text{st}} \sum_N P_0(N) = \pi_1^{\text{st}}\Gamma_{1\to 2}. \qquad (E.17)$$

The physical interpretation of such a result is very clear. The line at $\omega_L + \Omega$ is emitted in a transition $|1(N)\rangle \to |2(N - 1)\rangle$ [see (C.5)]. The weight of the line is the product of the population of the states $|1(N)\rangle$ from which the line starts and the rate $\Gamma_{1\to 2}$ of the transitions $|1(N)\rangle \to |2(N - 1)\rangle$.

A completely analogous calculation gives a Lorentzian centered at $\omega_L - \Omega$ for the other sideband at $\omega_L - \Omega$, with a full width at half maximum $2\Gamma_{\text{coh}}$ and a weight

$$\Gamma\langle \mathscr{S}_+^{21}\mathscr{S}_-\rangle = \pi_2^{\text{st}}\Gamma_{2\to 1}. \qquad (E.18)$$

This weight is equal to the weight of the other sideband given by (E.17), according to the detailed balance condition (D.22). Because the two sidebands have the same weight and the same width, the fluorescence spectrum is symmetric about ω_L.

c) STRUCTURE OF THE CENTRAL LINE

The last term in (E.4) yields the sum of the two averages $\langle \mathscr{S}_+^{ii}(\tau)\mathscr{S}_-(0)\rangle$, with $i = 1, 2$. According to the quantum regression theorem, for $\tau > 0$, both these two-time averages obey coupled evolution equations

$$\frac{d}{d\tau}\langle \mathscr{S}_+^{ii}(\tau)\mathscr{S}_-(0)\rangle = \sum_j \beta_{ij}\langle \mathscr{S}_+^{jj}(\tau)\mathscr{S}_-(0)\rangle \qquad (E.19)$$

involving the same coefficients β_{ij} as the evolution equations (E.11) concerning the one-time averages $\langle \mathscr{S}_+^{ii}(\tau)\rangle$. Because the eigenvalues of the matrix β_{ij} are equal to $i\omega_L$ and $i\omega_L - \Gamma_{\text{pop}}$, we deduce that the contribution of the last term in (E.4) has the general form

$$\langle [\mathscr{S}_+^{11}(\tau) + \mathscr{S}_+^{22}(\tau)]\mathscr{S}_-(0)\rangle = A\exp[i\omega_L - \Gamma_{\text{pop}}]\tau + B\exp i\omega_L\tau. \qquad (E.20)$$

When the last term of (E.20) is inserted into (E.1), it gives rise to the coherent (or elastic) component of the fluorescence spectrum

$$\mathscr{I}_{\text{coh}}(\omega) = \Gamma B \delta(\omega - \omega_L) \tag{E.21}$$

with weight ΓB. The first term of (E.20) corresponds to the central inelastic component of the fluorescence triplet, centered at ω_L, with weight ΓA and full width at half-maximum $2\Gamma_{\text{pop}}$.

To calculate A and B, first we set $\tau = 0$ in (E.20). This gives

$$A + B = \langle \mathscr{S}_+^{11} \mathscr{S}_- \rangle + \langle \mathscr{S}_+^{22} \mathscr{S}_- \rangle. \tag{E.22}$$

A calculation analogous to the one leading from (E.16) to (E.17) then yields for the total weight $\Gamma(A + B)$ of the two lines centered at ω_L the result

$$\Gamma(A + B) = \pi_1^{\text{st}} \Gamma_{1 \to 1} + \pi_2^{\text{st}} \Gamma_{2 \to 2}. \tag{E.23}$$

The interpretation of (E.23) is analogous to the one given above for (E.16). Finally, we take $\tau \gg \Gamma_{\text{pop}}^{-1}$ in (E.20). The first term of the right-hand side of (E.20) is then zero and the two operators of the left-hand side are decorrelated so that

$$\begin{aligned}
B \exp(i\omega_L \tau) &= \langle \mathscr{S}_+^{11}(\tau) + \mathscr{S}_+^{22}(\tau) \rangle_{\text{st}} \langle \mathscr{S}_+(0) \rangle_{\text{st}}^* \\
&= \left[(\mathscr{S}_+)_{11} \pi_1^{\text{st}} + (\mathscr{S}_+)_{22} \pi_2^{\text{st}} \right] e^{i\omega_L \tau} \langle \mathscr{S}_+(0) \rangle_{\text{st}}^* \tag{E.24}
\end{aligned}$$

according to (E.12). However, according to (E.8) and (E.9), only the last term in (E.4) contributes to $\langle \mathscr{S}_+(0) \rangle_{\text{st}}^*$, which is thus equal to the complex conjugate of the term in brackets of (E.24). Using the fact that $(\mathscr{S}_+)_{22} = -(\mathscr{S}_+)_{11}$, we find the weight ΓB of the coherent line

$$I_{\text{coh}} = \Gamma B = \Gamma |(\mathscr{S}_+)_{11}|^2 (\pi_1^{\text{st}} - \pi_2^{\text{st}})^2 = \Gamma_{1 \to 1} (\pi_1^{\text{st}} - \pi_2^{\text{st}})^2. \tag{E.25}$$

Finally, the dressed atom method allowed us to obtain analytical expressions for all the characteristics (widths and weights) of the three lines of the fluorescence spectrum, within the secular limit $\Omega \gg \Gamma$ where these lines are well separated. By using expressions (D.25) and (D.27) for Γ_{pop} and Γ_{coh}, as well as the definition (B.21) of θ, we can recover all the results obtained in subsection D-3-c in Chapter V for the widths of the lines in the two limits $\Omega_1 \gg \Gamma$, $\omega_L = \omega_0$ on the one hand, and $|\delta_L| \gg \Omega_1$, Γ on the other hand. In addition, we also obtain here very physical expressions for the weights of the lines [see (E.17), (E.18), and (E.23)] in terms of populations of dressed states and spontaneous emission rates.

2. Absorption Spectrum of a Weak Probe Beam

a) Physical Problem

In addition to the laser beam at ω_L, which is in general intense and perturbs the atom on the transition $a \leftrightarrow b$ of frequency ω_0, we now consider a second laser beam, of very low intensity, whose frequency ω is close, either to ω_0 (Figure 10α), or to the frequency ω_0' of another transition $b \leftrightarrow c$ sharing a common level (b) with the foregoing transition (Figure 10β).

The absorption of the probe beam is measured by means of a detector. We next discuss the modification of the absorption spectrum of the probe beam due to the presence of the beam of frequency ω_L.

b) Case Where the Two Lasers Are Coupled to the Same Transition

Because the second laser beam is of low intensity, we neglect any perturbation introduced by this beam on the positions, widths, and quasi-steady-state populations of the states of the atom "dressed by the photons ω_L". Figure 11 shows the manifolds $\mathscr{E}(N)$ and $\mathscr{E}(N-1)$ of this dressed atom. The circles represent the populations $\pi_1^{st} p_0(N)$ and $\pi_2^{st} p_0(N)$ of the type 1 and type 2 states in the quasi-steady state. We have assumed $\delta_L < 0$ ($\omega_L < \omega_0$), so that the state $|1(N)\rangle$, which is more contaminated by $|b, N\rangle$ than $|2(N)\rangle$, is more unstable and thus less populated ($\pi_1^{st} < \pi_2^{st}$).

The transitions "probed" by the second laser beam correspond to pairs of dressed states between which the atomic dipole **d** has a nonzero matrix

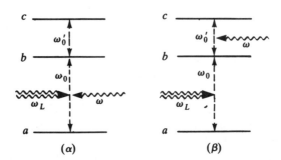

Figure 10. The second laser beam, of frequency ω, probes either the transition $a \leftrightarrow b$ previously perturbed by the first laser of frequency ω_L (α), or the transition $b \leftrightarrow c$ (β).

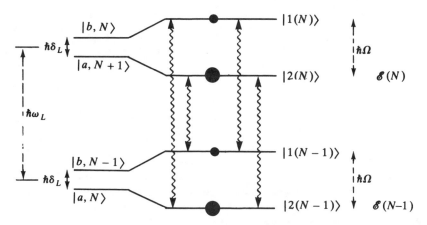

Figure 11. Dressed levels and quasi-steady-state populations of the atom + photons ω_L system. The wavy arrows indicate the transitions probed by the second laser beam.

element. These are therefore the same transitions as those involved in the fluorescence triplet, and they are represented by the wavy arrows in Figure 11. Without giving the details of the calculations (which are indeed quite similar to those in subsection E-1 above), one immediately understands that the transition $|1(N)\rangle \leftrightarrow |2(N-1)\rangle$ (of frequency $\omega_L + \Omega$) is absorbing because, the lower level being more populated than the upper level, the absorption processes $|2(N-1)\rangle \rightarrow |1(N)\rangle$ overcome the stimulated emission processes $|1(N)\rangle \rightarrow |2(N-1)\rangle$. These conclusions are reversed for the transition $|2(N)\rangle \leftrightarrow |1(N-1)\rangle$ (with frequency $\omega_L - \Omega$), which is amplifying because the upper level is more populated than the lower level. Finally, the two transitions $|i(N)\rangle \rightarrow |i(N-1)\rangle$ with $i = 1, 2$ (with frequency ω_L) do not result in any amplification or any absorption because they connect equally populated levels (*).

The dressed atom approach thus allows us to easily predict that for $\omega_L < \omega_0$ and at the secular limit ($\Omega \gg \Gamma$), the absorption spectrum of the probe beam over the transition $a \leftrightarrow b$ is composed of an absorption

(*) These results are valid only within the framework of the secular approximation. The nonsecular terms give rise to amplification or attenuation processes for $|\omega_L - \omega| \sim \Gamma$. Their importance is nevertheless lower by a factor Γ/Ω_1. For more details, see B. R. Mollow, *Phys. Rev. A*, **5**, 2217 (1972); G. Grynberg, E. Le Bihan, and M. Pinard, *J. Phys.* (*Paris*), **47**, 1321 (1986); M. T. Gruneisen, K. R. MacDonald, and R. W. Boyd, *J. Opt. Soc. Am. Ser. B*, **5**, 123 (1988).

line, centered at $\omega_L + \Omega$, of width $2\Gamma_{\text{coh}}$ (damping rate of the reduced coherences σ_{12}), and with weight $\Gamma_{1\to 2}$ $(\pi_2^{\text{st}} - \pi_1^{\text{st}})$, and of an amplifying line, centered at $\omega_L - \Omega$, with width $2\Gamma_{\text{coh}}$ and weight $\Gamma_{2\to 1}$ $(\pi_2^{\text{st}} - \pi_1^{\text{st}})$ (*).

Remarks

(i) The foregoing results can be interpreted perturbatively in the limit $\omega_0 - \omega_L \gg \Omega_1, \Gamma$. The states $|1(N)\rangle$ and $|2(N-1)\rangle$ then differ only slightly from $|b, N\rangle$ and $|a, N\rangle$, so that the absorption line $|2(N-1)\rangle \to |1(N)\rangle$, whose frequency is very close to ω_0, corresponds to the ordinary absorption $a \to b$ [at order zero in $\Omega_1/(\omega_0 - \omega_L)$]. By contrast, the transition $|2(N)\rangle \to |1(N-1)\rangle$, whose frequency is very close to $2\omega_L - \omega_0$, involves the contamination of $|2(N)\rangle$ by $|b, N\rangle$ and the contamination of $|1(N-1)\rangle$ by $|a, N\rangle$. This transition is therefore at least second order in $\Omega_1/(\omega_0 - \omega_L)$. The amplification mechanism is related to a three-photon process: The atom goes from a to b by the absorption of two laser photons ω_L and the stimulated emission of a photon $2\omega_L - \omega_0$ (see Figure 28 in Chapter II).

(ii) The results discussed above are obviously only valid for atoms at rest. When we consider moving atoms, new effects appear because, for each velocity group, the dressing and probe beams have their frequencies shifted by the Doppler effect. The case where these two beams have the same frequency and propagate in opposite directions (geometry of saturated absorption spectroscopy) is discussed in Exercise 20 in the limit $ku \gg \Omega_1 \gg \Gamma$ (ku being the Doppler width of the transition). It is shown in this exercise that, even when $\omega = \omega_L = \omega_0$, the medium is not totally transparent for the probe beam.

c) PROBING ON A TRANSITION TO A THIRD LEVEL.
THE AUTLER–TOWNES EFFECT

We now consider the situation in Figure 10β and assume that ω_0 and ω_0' are sufficiently different so that we can ignore any perturbation produced by the intense laser with frequency $\omega_L \sim \omega_0$ on the state c (the laser ω_L is too far from resonance for the transition $b \leftrightarrow c$).

In the left-hand part of Figure 12, we have represented the uncoupled states $|a, N+1\rangle$, $|b, N\rangle$, and $|c, N\rangle$. Because ω_L is close to ω_0, the states $|a, N+1\rangle$ and $|b, N\rangle$ are close and separated by $\hbar|\delta_L|$ (we assume $\omega_L < \omega_0$ so that $|b, N\rangle$ is above $|a, N+1\rangle$). The distance between $|b, N\rangle$ and $|c, N\rangle$ is $\hbar\omega_0'$. Under the influence of the coupling V_{AL}, the states $|a, N+1\rangle$ and $|b, N\rangle$ give rise to the two dressed states $|1(N)\rangle$ and $|2(N)\rangle$ separated by $\hbar\Omega$: By contrast, $|c, N\rangle$, which is not close to the

(*) These effects were observed experimentally. See, for example, F. Y. Wu, S. Ezekiel, M. Ducloy, and B. R. Mollow, *Phys. Rev. Lett.*, **38**, 1077 (1977).

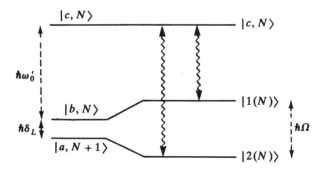

Figure 12. Unperturbed levels (left-hand part) and dressed levels (right-hand part) involved in the interpretation of the Autler–Townes doublet (wavy arrows).

levels $|b, N \pm 1\rangle$ to which it is coupled, remains unchanged (*) (right-hand side of Figure 12).

The transitions, with frequencies close to ω'_0, which are probed by the second laser beam, are those that reduce to the $|c, N\rangle \leftrightarrow |b, N\rangle$ transition in the limit $\Omega_1 = 0$. Because the two dressed states $|1(N)\rangle$ and $|2(N)\rangle$ are both contaminated by $|b, N\rangle$, we see that, for $\Omega_1 \neq 0$, two transitions appear: $|c, N\rangle \leftrightarrow |1(N)\rangle$ with frequency $\omega'_0 - (\Omega + \delta_L)/2$ and $|c, N\rangle \leftrightarrow |2(N)\rangle$ with frequency $\omega'_0 + (\Omega - \delta_L)/2$ represented by the wavy arrows in Figure 12.

Thus, the dressed atom approach allows us to easily predict that the excitation by an intense laser of a transition $a \leftrightarrow b$ splits into two components the absorption line for a laser probe beam tuned to another transition $b \leftrightarrow c$ sharing a common level with $a \leftrightarrow b$. The corresponding doublet is called the Autler–Townes doublet (**). If Ω_1 is fixed and if $\omega_L - \omega_0$ is varied, the splitting between the two components of the doublet changes and reaches a minimum equal to Ω_1 for $\omega_L = \omega_0$. If the level $|c, N\rangle$ is unpopulated in the absence of the probe beam, the weight of each component $|i(N)\rangle \leftrightarrow |c, N\rangle$ of the Autler–Townes doublet is proportional to the product of the steady-state population π_i^{st} of the level $|i(N)\rangle$ [given in (D.21)] and the transition rate from $|i(N)\rangle$ to $|c, N\rangle$ proportional to $|\langle i(N)|b, N\rangle|^2$. The width of the line $|i(N)\rangle \leftrightarrow |c, N\rangle$ is

---

(*) These nonresonant couplings actually introduce a slight light shift of the state $|c, N\rangle$ that we neglect here.

(**) S. H. Autler and C. H. Townes, *Phys. Rev.*, **100**, 703 (1955). See also C. H. Townes, and A. L. Schawlow, *Microwave Spectroscopy*, Dover, New York, 1975, §10-9.

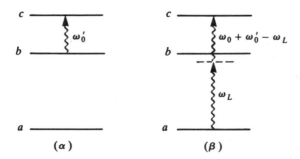

Figure 13. Perturbative interpretation of the Autler–Townes doublet at the limit $\omega_0 - \omega_L \gg \Omega_1, \delta_L$. α—Direct process between b and c involving the absorption of one probe photon. β—Two-photon process (one laser photon and one probe photon) between a and c.

the sum of the width of the level c and the width of the level $|i(N)\rangle$ equal to $\sum_j \Gamma_{i \to j}$.

Remark

As in subsection E-2-b above, it is possible to give a perturbative interpretation of the two lines at the limit $\omega_0 - \omega_L \gg \Omega_1, \Gamma$. The state $|1(N)\rangle$ then differs only slightly from $|b, N\rangle$ and the line $|1(N)\rangle \to |c, N\rangle$, having a frequency close to ω_0', corresponds to the ordinary $b \to c$ absorption (Figure 13α). The transition $|2(N)\rangle \to |c, N\rangle$ around $\omega_0' + \omega_0 - \omega_L$ involves the contamination of $|2(N)\rangle$ by $|b, N\rangle$. This process is thus at least first order in $\Omega_1/(\omega_0 - \omega_L)$. It corresponds to the two-photon absorption process in Figure 13β where the atom goes from a to c by absorbing one laser photon ω_L and one probe laser photon.

3. Photon Correlations

In Section D of Chapter V, we introduced the two-time average

$$\langle \mathscr{S}_+(t)\mathscr{S}_+(t+\tau)\mathscr{S}_-(t+\tau)\mathscr{S}_-(t)\rangle \qquad (\text{E.26})$$

which is involved in the probability density of detecting, in the light emitted by the atom, two photons separated by the interval τ [see formula (D.27) in Chapter V, which, in the steady-state regime, depends only on τ]. We calculate here such a photon-correlation signal and show that the result obtained confirms the picture of the radiative cascade for the dressed atom given above (§C-2).

a) CALCULATION OF THE PHOTON-CORRELATION SIGNAL

In (E.26), the operators are considered in the Heisenberg representation. For example,

$$\mathscr{S}_+(t + \tau) = e^{iH(t+\tau-t_0)/\hbar}\, \mathscr{S}_+\, e^{-iH(t+\tau-t_0)/\hbar} \qquad (E.27)$$

where H is the Hamiltonian (A.1) of the atom + laser mode + empty modes global system and t_0 is an initial time. The states of the global system will be described by using the uncoupled basis $|s, N; \{n_i\}\rangle$, where $s = a$ or b, and where the n_i represent the numbers of photons in modes other than the laser mode. In particular, $|\{0\}\rangle$ represents the vacuum for all these modes.

Let $\sigma_{ALR}(t_0)$ be the density operator of the global system (for the sake of clarity, in this subsection we return to more detailed notation, of the type σ_{ALR} for the global system, and σ_{AL} for the atom + laser mode system...). Using the invariance of a trace by circular permutation, we write (E.26) as

$$\mathscr{C}_2(\tau) = \text{Tr}\left[e^{iH(t-t_0)/\hbar}\, \mathscr{S}_+\mathscr{S}_+(\tau)\mathscr{S}_-(\tau)\mathscr{S}_-\, e^{-iH(t-t_0)/\hbar}\, \sigma_{ALR}(t_0)\right]$$

$$= \text{Tr}\left[\mathscr{S}_+\mathscr{S}_-\, e^{-iH\tau/\hbar}\, \mathscr{S}_-\sigma_{ARL}(t)\mathscr{S}_+\, e^{iH\tau/\hbar}\right] \qquad (E.28)$$

where the relation

$$e^{-iH(t-t_0)/\hbar}\, \sigma_{ALR}(t_0)\, e^{iH(t-t_0)/\hbar} = \sigma_{ALR}(t) \qquad (E.29)$$

was used. Using (B.10) and the fact that all the operators of (E.28) act in the state space of the global system enables us to write the operator $\mathscr{S}_+\mathscr{S}_-$ as

$$\mathscr{S}_+\mathscr{S}_- = \sum_{N, \{n_i\}} |b, N; \{n_i\}\rangle\langle b, N; \{n_i\}| \qquad (E.30)$$

which finally gives, for the trace appearing in (E.28)

$$\mathscr{C}_2(\tau) = \sum_{N, \{n_i\}} \langle b, N; \{n_i\}| e^{-iH\tau/\hbar}\, \mathscr{S}_-\sigma_{ARL}(t)\mathscr{S}_+\, e^{iH\tau/\hbar} |b, N; \{n_i\}\rangle.$$

$$(E.31)$$

Up to this point, we have not made any approximation. We now neglect the correlations between the dressed atom (system $A + L$) and the reservoir R of empty modes, and approximate the density matrix of the global system by

$$\sigma_{ALR}(t) \simeq \sigma_{AL}(t) \otimes |\{0\}\rangle\langle\{0\}| =$$

$$= \sum_{\substack{s, N \\ s', N'}} |s, N; \{0\}\rangle\langle s, N|\sigma_{AL}(t)|s', N'\rangle\langle s', N'; \{0\}| \quad \text{(E.32)}$$

where $\sigma_{AL}(t)$ is the density matrix of the dressed atom. Such an approximation was previously discussed in detail in subsection D-4 of Chapter IV. It is justified by the fact that the correlations which exist at the time t between $A + L$ and the reservoir R disappear after a time on the order of the correlation time τ_c of R and are therefore not important if we restrict ourselves in (E.26) to times $\tau \gg \tau_c$. It is actually the same type of approximation that was used to derive the quantum regression theorem.

The foregoing approximation and expression (B.10) for $\mathcal{S}_{\pm}$ allow us to write the operator $\mathcal{S}_- \sigma_{ALR}(t)\mathcal{S}_+$ appearing in (E.31) in the form

$$\mathcal{S}_- \sigma_{ALR}(t)\mathcal{S}_+ = \sum_{N', N''} |a, N'; \{0\}\rangle\langle b, N'|\sigma_{AL}(t)|b, N''\rangle\langle a, N''; \{0\}|.$$

$$\text{(E.33)}$$

When this expression is substituted into (E.31), the product of the two following probability amplitudes appears:

$$\langle b, N; \{n_i\}| \exp(-iH\tau/\hbar)|a, N'; \{0\}\rangle$$

and

$$\langle b, N; \{n_i\}| \exp(-iH\tau/\hbar)|a, N''; \{0\}\rangle^*.$$

The first amplitude is non-negligible only if $N' - 1 - N = \Sigma_i n_i$. The state $|a, N'\rangle$ indeed belongs to the manifolds $\mathscr{E}(N' - 1)$ of the dressed atom, whereas the state $|b, N\rangle$ belongs to the manifold $\mathscr{E}(N)$, and the difference between the integers $N' - 1$ and N identifying these two manifolds must correspond to the number of spontaneously emitted photons. Similarly, the other amplitude is non-negligible only if $N'' - 1 - N = \Sigma_i n_i$. From this we deduce that the only non-negligible terms coming from the double sum over N' and N'' of (E.33) are the terms $N' = N''$. This, finally, gives

$$
\mathscr{C}_2(\tau) = \sum_{N, N'} \left(\sum_{\{n_i\}} |\langle b, N; \{n_i\}| e^{-iH\tau/\hbar} |a, N'; \{0\}\rangle|^2 \right) \times
$$

$$
\times \langle b, N'|\sigma_{AL}(t)|b, N'\rangle. \tag{E.34}
$$

b) PHYSICAL DISCUSSION

We now interpret the general term of the sum appearing on the right-hand side of (E.34). The second factor is simply the population $\pi_{b, N'}(t)$ of the state $|b, N'\rangle$ at the time t. For $t - t_0 \gg \Gamma^{-1}$, this population tends to the quasi-steady population $\pi_{b, N'}$. The first factor physically represents the probability $\mathscr{P}(b, N; \tau|a, N'; 0)$ that the dressed atom will be in the state $|b, N\rangle$ at the time τ, knowing that the system departed at time 0 from the state $|a, N'\rangle$. Because the evolution during this time τ is due either to absorption and stimulated emission processes that cause the system to oscillate between the states $|a, N'\rangle$ and $|b, N' - 1\rangle$, or to spontaneous emission processes that cause it to fall to lower manifolds (see Figure 6), $\mathscr{P}(b, N; \tau|a, N'; 0)$ is nonzero only if $N = N' - 1$, $N' - 2, N' - 3 \dots$. Finally, we have shown that the photon-correlation signal (E.26) can be written

$$
\sum_{N'} \sum_{p=1,2,3\dots} \mathscr{P}(b, N' - p; \tau|a, N'; 0)\pi_{b, N'}. \tag{E.35}
$$

The physical interpretation of (E.35) is quite clear. To emit the first photon that is detected at time t, the atom must be in the excited state b. It is for this reason that the probability $\pi_{b, N'}$ of finding the atom in the state b, N' appears in (E.35). Immediately after this emission, the atom is in state a, the laser photons remaining spectators: This is the quantum jump $b, N' \to a, N'$ discussed in subsection C-2-a. To emit the second photon, which will be detected at time τ after the first one, the atom must again go from a to b during this time τ. The second term of (E.35) gives the probability of such a passage, which is itself decomposed into several contributions $|a, N'\rangle \to |b, N' - p\rangle$ with $p = 1, 2 \dots$. For $p = 1$, the second photon detected is emitted from $|b, N' - 1\rangle$, which belongs to the same manifold as $|a, N'\rangle$. The term $p = 1$ of (E.35) therefore corresponds to the case where the second photon detected at $t + \tau$ is the *first* fluorescence photon emitted after the first photon detected at t. The term $p = 2$, which corresponds to an emission from $|b, N' - 2\rangle$, corresponds to the case where the second photon detected at $t + \tau$ is the *second* fluorescence photon emitted after the first photon detected at time t, the first fluorescence photon (emitted from $|b, N' - 1\rangle$) being not detected, and so on. Thus we see how the photon correlation signal can be interpreted by using the picture of the radiative cascade introduced above.

Finally, we show that it is possible to transform expression (E.35). In (E.35), the first factor depends only on p and not on N' (for N' varying within the width ΔN of the distribution of the number of laser photons). The sum over p of this term represents the probability $\mathscr{P}(b; \tau | a; 0)$ of being in b at time τ starting from a at time 0. The sum over N' of the second factor of (E.35) represents the probability π_b of being in the state b. It is therefore possible to write (E.35) in the form

$$\mathscr{P}(b; \tau | a; 0)\pi_b. \tag{E.36}$$

The result (E.36) could have been obtained from the optical Bloch equations and the quantum regression theorem (*). The advantage of the dressed atom approach is that it leads to expressions like (E.35) in which the contributions of the different steps of the radiative cascade appear explicitly. In particular, the term $p = 1$ involves the function $\mathscr{P}(b, N' - 1; \tau | a, N'; 0)$, which is directly related to the function $K(\tau)$ introduced above (§C-2-c) and giving the distribution of intervals between two *successive* emissions. We have

$$K(\tau) = \Gamma\mathscr{P}(b, N' - 1; \tau | a, N'; 0). \tag{E.37}$$

Such a function is not easy to calculate from the optical Bloch equations, which instead lead to the function

$$J(\tau) = \Gamma\mathscr{P}(b; \tau | a; 0) \tag{E.38}$$

giving the distribution of time intervals between *any* two emissions (not necessarily successive).

Remark

Compare the two functions $J(\tau)$ and $K(\tau)$ for a resonant and saturating excitation ($\delta_L = 0$, $\Omega_1 \gg \Gamma$). Using the optical Bloch equations, we find

$$J(\tau) \simeq \frac{\Gamma}{2}\left[1 - e^{-3\Gamma\tau/4} \cos \Omega_1\tau\right] \tag{E.39.a}$$

(*) See, for example, Cohen-Tannoudji, p. 98.

whereas the expression (C.3) for $K(\tau)$ gives

$$K(\tau) \simeq \frac{\Gamma}{2}[1 - \cos \Omega_1 \tau]\, e^{-\Gamma\tau/2}. \tag{E.39.b}$$

For $\tau \ll \Gamma^{-1}$, spontaneous emission can be ignored. The two functions are then equal and reflect the resonant Rabi oscillation between $|a, N'\rangle$ and $|b, N' - 1\rangle$. By contrast, for $\tau \gg \Gamma^{-1}$, $K(\tau)$ tends to zero (the first emission after the first detection has certainly occurred) whereas $J(\tau)$ tends to $\Gamma/2$, which is simply $\Gamma \pi_b^{st}$. Indeed, $\mathscr{P}(b; \tau|a; 0)$ has then lost any memory of the initial conditions at $\tau = 0$ and tends to the steady-state population of b.

c) GENERALIZATION TO A THREE-LEVEL SYSTEM: INTERMITTENT FLUORESCENCE

Another advantage of the dressed atom approach, for the discussion of photon correlations, is the possibility of generalizing it to situations more complex than those considered here. Consider, for example, the case where two atomic transitions $a \leftrightarrow b$ and $a \leftrightarrow b'$ starting both from the ground state a and, having eigenfrequencies ω_0 and ω_0', are simultaneously excited by two laser beams L and L' with frequencies ω_L and ω_L', respectively, close to ω_0 and ω_0' (Figure 14). The natural widths of the excited levels b and b' are Γ and Γ', respectively.

The function $K(\tau)$ giving the distribution of the intervals τ between two *successive* emissions (whether these emissions take place on the transition $b \to a$ or $b' \to a$) can be calculated by a method that is completely analogous to the one presented in subsection C-2-c. For this we consider the atom dressed by the two laser modes L and L'. The uncoupled levels of the system $A + L + L'$, labeled by the atomic quantum numbers a, b or b' and the numbers N and N' of laser photons ω_L

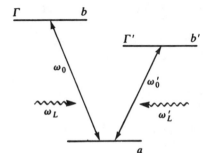

Figure 14. Configuration of levels allowing intermittent fluorescence to be observed.

and ω'_L, are grouped now into three-dimensional manifolds $\mathscr{E}(N, N')$:

$$\mathscr{E}(N, N') = \{|a, N + 1, N' + 1\rangle, |b, N, N' + 1\rangle, |b', N + 1, N'\rangle\}. \tag{E.40}$$

Immediately after the emission of a photon, the atom is projected into the state $|a\rangle$ and the global system is found, for example, in the state $|a, N + 1, N' + 1\rangle$ of $\mathscr{E}(N, N')$. The interactions between A and L and between A and L', characterized by the Rabi frequencies Ω_1 and Ω'_1, couple such a state to the two other states $|b, N, N' + 1\rangle$ and $|b', N + 1, N'\rangle$ of $\mathscr{E}(N, N')$ from which the atom can spontaneously emit a photon with rates Γ and Γ'. As long as the atom has not spontaneously emitted a photon, the system remains within the manifold $\mathscr{E}(N, N')$ and its evolution can then be described by a non-Hermitian effective Hamiltonian (see Chapter III, §C-3). More precisely, if

$$|\psi(t)\rangle = c_a(t)|a, N + 1, N' + 1\rangle + c_b(t)|b, N, N' + 1\rangle +$$
$$+ c_{b'}(t)|b', N + 1, N'\rangle \tag{E.41}$$

is the projection onto $\mathscr{E}(N, N')$ of the state vector of the system, the coefficients c_a, c_b, and $c_{b'}$ obey the following coupled equations analogous to (C.2)

$$i\dot{c}_a = (\Omega_1/2)c_b + (\Omega'_1/2)c_{b'}$$
$$i\dot{c}_b = -\delta_L c_b - i(\Gamma/2)c_b + (\Omega_1/2)c_a \tag{E.42}$$
$$i\dot{c}_{b'} = -\delta'_L c_{b'} - i(\Gamma'/2)c_{b'} + (\Omega'_1/2)c_a$$

where $\delta_L = \omega_L - \omega_0$ and $\delta'_L = \omega'_L - \omega'_0$ are the frequency detunings of the two laser excitations.

The function $K(\tau)$ giving the distribution of time intervals τ between two successive emissions is equal to the probability of emission at the time τ of a photon by the dressed atom in the state (E.41), with the initial condition $|\psi(0)\rangle = |a, N + 1, N' + 1\rangle$. It is therefore equal to

$$K(\tau) = \Gamma|c_b(\tau)|^2 + \Gamma'|c_{b'}(\tau)|^2 \tag{E.43}$$

where $c_b(\tau)$ and $c_{b'}(\tau)$ are the solutions of Equations (E.42) satisfying the initial conditions $c_a(0) = 1$, $c_b(0) = c_{b'}(0) = 0$.

The solutions $c_b(\tau)$ and $c_{b'}(\tau)$ of Equations (E.42) are the superpositions of exponentials $\exp(-\lambda_\alpha \tau)$ where the $-\lambda_\alpha$ are the characteristic roots of the system (E.42). A particularly interesting situation occurs when $\Gamma \gg \Gamma'$ (the level b' being, for example, a metastable level with a lifetime that is much longer than that of b). For sufficiently low values of Ω_1', we then find that one of the roots $-\lambda_\alpha$ gives a damping time [on the order of $(\Gamma')^{-1}$] much longer than the two others. As a result, $K(\tau)$ then has a component which decreases very slowly with τ. Such a result indicates that there is a nonzero probability that two successive emissions will be separated by a very long time that is much longer than the average time (on the order of Γ^{-1}) separating two successive emissions when only the transition $b \leftrightarrow a$ is excited by a resonant and intense laser. Adding the laser excitation at ω_L' on the transition $a \leftrightarrow b'$ can thus stop the fluorescence of the atom for a long period of time and cause "black periods" to appear in this fluorescence.

The dressed atom method thus gives a simple (*) approach to this phenomenon of "intermittent fluorescence" whose qualitative interpretation is the following. Any absorption of a laser photon ω_L' by the atom puts the atom into the level b' where it can escape, for a time on the order of $(\Gamma')^{-1}$, from the action of the laser at ω_L. The advantage of such a method, called the "shelving method" (**), is to allow the absorption of a single photon ω_L' (the one which puts the atom on the shelf b') to be detected by the absence of a very large number of fluorescence photons on the transition $a \leftrightarrow b$. Another interest is the possibility that the "quantum jumps" that the atom makes between a and b' under the influence of the laser excitation at ω_L' (***) be detected by the stopping or reestablishment of the fluorescence. Several experiments have allowed such quantum jumps to be observed on a single trapped ion (****).

Remark

Another interesting situation arises when an atom possesses two lower levels a and a', which are both coupled to a single excited state b (see Figure 15).

When the two transitions $a \leftrightarrow b$ and $a' \leftrightarrow b$ are simultaneously excited by two laser beams with frequencies ω_L and ω_L' such that the Raman resonance

---

(*) See, for example, C. Cohen-Tannoudji and J. Dalibard, *Europhys. Lett.*, **1**, 441 (1986).

(**) Such a method was proposed by H. G. Dehmelt, *Bull. Am. Phys. Soc.*, **20**, 60 (1975).

(***) The connection between quantum jumps and intermittent fluorescence was suggested by R. J. Cook and H. J. Kimble, *Phys. Rev. Lett.*, **54**, 1023 (1985).

(****) W. Nagourney, J. Sandberg, and H. Dehmelt, *Phys. Rev. Lett.*, **56**, 2797 (1986); The. Sauter, W. Neuhauser, R. Blatt, and P. E. Toschek, *Phys. Rev. Lett.*, **57**, 1696 (1986); J. C. Bergquist, R. G. Hulet, W. M. Itano, and D. J. Wineland, *Phys. Rev. Lett.*, **57**, 1699 (1986).

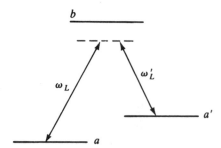

Figure 15. Configuration of levels leading to the dark resonance phenomenon.

condition $\hbar(\omega_L - \omega'_L) = E_{a'} - E_a$ is satisfied, one observes a complete disappearance of the fluorescence (black resonances) (*). In the dressed atom
approach, such a phenomenon is due to the existence within each manifold
$\mathcal{E}(N, N')$ of a linear combination of $|a', N, N' + 1\rangle$ and $|a, N + 1, N'\rangle$ which
is not coupled to $|b, N, N'\rangle$ by the atom-laser interaction and which is stable
with regard to spontaneous emission. The atoms get trapped into this state
(trapping state) (**).

4. Dipole Forces

To end Section E, we now study the dipole forces related to the
amplitude gradients of the laser field (see §C-2-d in Chapter V). First we
introduce energy diagrams that give the energies of the dressed levels as a
function of the position of the atom in the laser field (§a). Then we show
how these energy diagrams allow us to simply understand the average
value (§b) and the fluctuations (§c) of these dipole forces.

a) Energy Levels of the Dressed Atom in a Spatially
   Inhomogeneous Laser Wave

If the amplitude of the wave associated with the laser mode varies in
space, so does the Rabi frequency Ω_1, which depends on the point **r** where
the atom is located and which we write $\Omega_1(\mathbf{r})$ (***).

(*) The first observation of this phenomenon was reported in G. Alzetta, A. Gozzini, L.
Moi, and G. Orriols, *Nuovo Cimento*, **36B**, 5 (1976).

(**) See, for example, P. M. Radmore and P. L. Knight, *J. Phys. B*, **15**, 561 (1982); J.
Dalibard, S. Reynaud, and C. Cohen-Tannoudji, in *Interaction of Radiation with Matter*, *A
Volume in Honor of Adriano Gozzini*, Scuola Normale Superiore, Pisa, 1987, p. 29.

(***) The eigenmodes used to quantify the field are not plane waves here, but rather
modes reproducing the spatial dependence of the laser field.

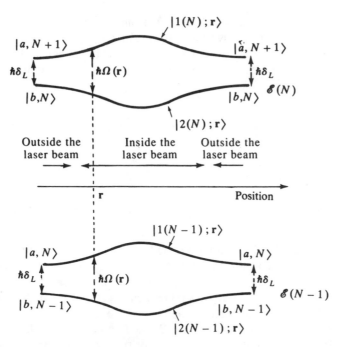

Figure 16. Variations, as a function of the position **r** of the atom, of the energies of a few dressed levels.

Figure 16 shows the variations, as a function of the position **r** of the atom, of the energies of the dressed levels belonging to the manifolds $\mathscr{E}(N)$ and $\mathscr{E}(N-1)$. Outside the laser beam, $\Omega_1(\mathbf{r})$ tends to zero and these levels tend to the uncoupled states $|a, N+1\rangle$ and $|b, N\rangle$ for $\mathscr{E}(N)$, $|a, N\rangle$ and $|b, N-1\rangle$ for $\mathscr{E}(N-1)$, which are separated by $\hbar\delta_L$ (the figure represents the case $\delta_L > 0$). Within the laser beam $\Omega_1(\mathbf{r})$ is nonzero and the splitting

$$\hbar\Omega(\mathbf{r}) = \hbar\sqrt{\Omega_1^2(\mathbf{r}) + \delta_L^2} \tag{E.44}$$

between the two dressed levels of the same manifolds increases with $\Omega_1^2(\mathbf{r})$.

We begin by neglecting spontaneous emission. Assume that the system is in the state $|1(N)\rangle$ or $|2(N)\rangle$ and that the motion of the atom is described by a wave packet having very small dimensions (compared to the wavelength of the laser field) and moving sufficiently slowly so that one can neglect any nonadiabatic transition from one level to another. The system will then adiabatically follow the level $|1(N)\rangle$ or $|2(N)\rangle$ in which it is found initially, and the energy curves in Figure 16 appear in this case as potential energy curves $V_{1N}(\mathbf{r})$ and $V_{2N}(\mathbf{r})$. The dressed atom therefore experiences a force

$$\mathbf{F}_1 = -\nabla V_{1N}(\mathbf{r}) = -\frac{\hbar}{2}\nabla\Omega(\mathbf{r}) \qquad \text{(E.45.a)}$$

if it is on a type-1 level and a force

$$\mathbf{F}_2 = -\nabla V_{2N}(\mathbf{r}) = +\frac{\hbar}{2}\nabla\Omega(\mathbf{r}) = -\mathbf{F}_1 \qquad \text{(E.45.b)}$$

if it is on a type-2 level. We are then led to the picture of a force acting on the dressed atom that depends on the internal state, either type 1 or type 2, in which it is found. This is, in a certain sense, an optical Stern–Gerlach effect very similar to the usual Stern–Gerlach effect that occurs for a spin $\frac{1}{2}$ in an inhomogeneous magnetic field.

Spontaneous emission causes the dressed atom to cascade along its energy diagram. The radiative transitions of the atom can change the type of state (1 or 2) in which it is found (for example, the transition $|1(N)\rangle \to |2(N-1)\rangle$ is a transition from state 1 to state 2), which causes the sign of the force acting on it to change abruptly. We thus arrive at the physical picture of a light force acting on the atom and whose sign changes randomly over time.

b) Interpretation of the Mean Dipole Force

In steady state, the mean dipole force appears as the mean of the forces $\mathbf{F}_1$ and $\mathbf{F}_2 = -\mathbf{F}_1$ given in (E.45), weighted by the proportions of times spent in the type-1 and -2 levels, which are simply the steady-state reduced populations π_1^{st} and π_2^{st} introduced in subsection D-2-d above

$$\langle \mathbf{F}_{\text{dip}} \rangle = \mathbf{F}_1 \pi_1^{\text{st}} + \mathbf{F}_2 \pi_2^{\text{st}} = -\frac{\hbar}{2}\nabla\Omega(\mathbf{r})\left[\pi_1^{\text{st}} - \pi_2^{\text{st}}\right]. \qquad \text{(E.46)}$$

Using the values given above at the secular limit for $\Omega(\mathbf{r})$, π_1^{st} and π_2^{st} [see (E.44), (D.21), and the definition (B.21) for θ], one can show that expression (E.46) indeed coincides with the value obtained in subsection C-2-d in Chapter V from the optical Bloch equations (see the references at the end of the chapter).

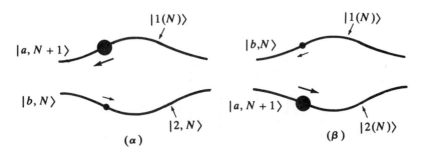

Figure 17. Dressed levels and steady-state reduced populations for a detuning $\delta_L > 0 \; (\alpha)$ and for a detuning $\delta_L < 0 \; (\beta)$.

We confine ourselves here to interpreting the sign of $\langle F_{\text{dip}} \rangle$ (see Figure 17). For $\omega_L > \omega_0 \; (\delta_L > 0)$ (Figure 17α), it is the level $|1(N)\rangle$ that tends to $|a, N + 1\rangle$ outside the laser beam. This level is thus less contaminated by $|b, N\rangle$ than the other dressed level $|2(N)\rangle$. Consequently, it is more stable and therefore more populated: $\pi_1^{\text{st}} > \pi_2^{\text{st}}$. The dressed atom thus spends more time in the type-1 levels where it is pushed outside the laser beam. The mean dipole force is thus expelling the atom outside the regions of high intensity. These conclusions are reversed for $\omega_L < \omega_0$ $(\delta_L < 0)$, where it is level $|2(N)\rangle$ that tends to $|a, N + 1\rangle$ and which is thus more populated: $\pi_2^{\text{st}} > \pi_1^{\text{st}}$ (see Figure 17β). For $\omega_L < \omega_0$, the mean dipole force thus attracts the atom to regions of high intensity. Finally, for $\omega_L = \omega_0 \; (\delta_L = 0)$, the two dressed levels are equally contaminated by $|a, N + 1\rangle$, $\pi_1^{\text{st}} = \pi_2^{\text{st}}$ and $\langle \mathbf{F}_{\text{dip}} \rangle = \mathbf{0}$.

c) FLUCTUATIONS OF THE DIPOLE FORCE

Figure 18 shows the evolution over time of the instantaneous force acting on the atom. This force jumps at random times between the two possible values $\mathbf{F}_1$ and $\mathbf{F}_2 = -\mathbf{F}_1$ [see (E.45)]. The time intervals τ_1 and τ_2 spent in each type-1 or -2 dressed level between two successive quantum jumps are random variables, whose mean values are on the order of Γ^{-1} for a quasi-resonant excitation.

The fluctuating force in Figure 18 produces a diffusion of the atomic momentum characterized by the diffusion coefficient

$$D_{\text{dip}} = \int_0^\infty d\tau \left[\langle F_{\text{dip}}(t) F_{\text{dip}}(t + \tau) \rangle - \langle F_{\text{dip}} \rangle^2 \right]. \qquad (E.47)$$

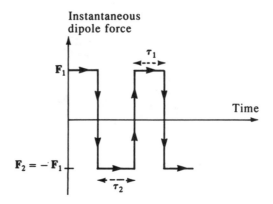

Figure 18. Time variation of the instantaneous dipole force acting on the atom.

The order of magnitude of this coefficient is given by the product of the value of the term in brackets of (E.47) at $\tau = 0$, which is equal to F_1^2, and the correlation time of the fluctuating force in Figure 18, which is on the order of $\langle \tau_1 \rangle \sim \langle \tau_2 \rangle \sim \Gamma^{-1}$. At resonance, $F_1 \sim -\hbar \nabla \Omega_1$, so that

$$D_{\text{dip}} \sim \frac{\hbar^2 (\nabla \Omega_1)^2}{\Gamma} . \tag{E.48}$$

We can see in (E.48) that the diffusion coefficient associated with the dipole forces increases very rapidly with the laser intensity, as Ω_1^2, whereas the maximal depth of the potential well associated with such a force increases only as Ω_1 (see subsection C-2-d in Chapter V). We can then understand why the heating associated with D_{dip} makes it difficult to trap neutral atoms with dipole forces (*).

GENERAL REFERENCES

C. Cohen-Tannoudji and S. Reynaud, in *Multiphoton Processes* edited by J. Eberly and P. Lambropoulos, Wiley, New York, 1978, p. 103; S. Reynaud, Thèse Paris 1981, *Ann. Phys. Fr.* **8**, 315 (1983); **8**, 371 (1983).

(*) A more precise calculation of the integral (E.47) and a more quantitative discussion of the fluctuations of radiative forces are presented in the reference given at the end of the chapter.

Application to fluorescence and absorption spectra: C. Cohen Tannoudji and S. Reynaud, *J. Phys. B*, **10**, 345 (1977). For three-level systems, see also C. Cohen-Tannoudji and S. Reynaud, *J. Phys. B*, **10**, 365 (1977); **10**, 2311 (1977).

Application to photon correlations: C. Cohen-Tannoudji and S. Reynaud, *Philos. Trans. R. Soc. London Ser. A*, **293**, 233 (1979).

Application to dipole forces: J. Dalibard and C. Cohen-Tannoudji, *J. Opt. Soc. Am. B*, **2**, 1707 (1985).

Other applications: The dressed atom approach has also been applied to other problems, such as the four-wave mixing: G. Grynberg, M. Pinard, and P. Verkerk, *Opt. Commun.*, **50**, 261 (1984); *J. Phys. (Paris)*, **47**, 617 (1986).

For applications in the areas of radio frequencies and collisions, see the references in Complements A_{VI} and B_{VI}.

Other theoretical approaches related to the dressed atom approach: Two-level atom in a real single-mode cavity: E. T. Jaynes and F. W. Cummings, *Proc. IEEE*, **51**, 89 (1963).

Use of the Floquet theorem and quasi-periodic states in a semi-classical approach: J. H. Shirley, *Phys. Rev.* **138**, B979 (1965); Ya. B. Zel'dovich, *Sov. Phys. JETP*, **24**, 1006 (1967); Shih-I Chu, *Adv. At. Mol. Phys.*, **21**, 197 (1985), and references therein.

COMPLEMENT A_{VI}

THE DRESSED ATOM IN THE RADIO-FREQUENCY DOMAIN

In this chapter, we introduced the dressed atom approach to describe an atom interacting with a resonant or quasi-resonant laser beam. The same approach can be used to describe the effect of transitions (real or virtual) induced by a radio-frequency field between fine, hyperfine, or Zeeman sublevels of an atom.

When the radio-frequency domain is compared to the optical domain, two clear differences appear. Spontaneous emission is negligible and the number of levels to be considered is finite. This simplifies the evolution equations and situations where the fields are very intense, i.e., when the Rabi frequency is on the same order as or greater than the transition frequency can then be handled easily (*).

The purpose of this complement is to present the particularly simple case of a two-level system interacting with a radio-frequency field. The splitting $\hbar\omega_{ab}$ between the two unperturbed atomic sublevels is varied by means of an external parameter (for example, a static magnetic field). The energy diagram of the dressed atom therefore presents an infinite number of energy levels which exhibit "crossings" or "anticrossings" when plotted versus $\hbar\omega_{ab}$. We begin (§1) by reviewing the conditions for which resonances can be associated with level crossings or anticrossings. We then present (§2) the system considered, a spin $\frac{1}{2}$ interacting with a static magnetic field and a radio-frequency (rf) field. Two polarizations of the field are considered. For circular polarization (§3), exact calculations can be made. This example allows us to give a full quantum description of the magnetic resonance phenomenon, which appears when the Larmor frequency ω_0 of the spin is equal to the frequency ω of the rf field. Other resonant effects, which can be observed near $\omega_0 = 0$ or $\omega_0 = 2\omega$ are also analyzed, for arbitrarily high intensities of the radio-frequency field. The study of the linear polarization case (§4) also allows several examples of higher-order effects to be discussed and calculated (multiphoton transi-

---

(*) Certainly, such situations can also be considered in the optical domain. However, at such high intensities, the two levels connected by an optical transition are in general strongly coupled to several other excited levels, in particular to ionization continua. The calculation of the corresponding nonlinear phenomena is thus much more complex. By contrast, in the radio-frequency domain, the splitting between the sublevels is considerably smaller than the optical distances between them and other excited levels.

tions, radiative shift of resonances, modification of the magnetic properties of the atom).

Obviously, all these effects could be studied by using a classical description of the rf field and Bloch equations. The dressed atom approach allows a more global analysis to be made. It has the advantage of correlating all the observable resonance phenomena with the properties of the energy diagram of a time-independent Hamiltonian.

1. Resonance Associated with a Level Crossing or Anticrossing

Consider a system described by an unperturbed Hamiltonian H_0. Let $|\varphi_a\rangle$ and $|\varphi_b\rangle$ be two of its eigenstates, with energies E_a and E_b close to each other. When a perturbation V is added, the energies and the wave functions of these states are perturbed in a way that essentially depends on the energy splitting between the levels and the properties of V. In this subsection, we introduce the notions of level anticrossing and level crossing and we show that resonances of different types can be associated with them.

a) ANTICROSSING FOR A TWO-LEVEL SYSTEM

First we consider the simple case where V couples only the states $|\varphi_a\rangle$ and $|\varphi_b\rangle$ to each other

$$V_{aa} = V_{bb} = 0 \qquad V_{ab} = \hbar\Omega_1/2 \tag{1}$$

(assumed to be real) and we write

$$\omega_{ab} = (E_a - E_b)/\hbar. \tag{2}$$

The energies of the two perturbed levels are

$$E_1 = \frac{E_a + E_b}{2} + \frac{\hbar\Omega}{2} \tag{3.a}$$

$$E_2 = \frac{E_a + E_b}{2} - \frac{\hbar\Omega}{2} \tag{3.b}$$

with

$$\Omega = \sqrt{\omega_{ab}^2 + \Omega_1^2}. \tag{4}$$

As functions of ω_{ab}, the energies E_1 and E_2 vary as two branches of a

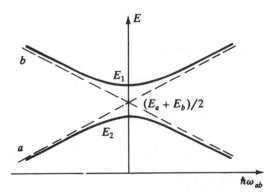

Figure 1. Level anticrossing. Variation with $\hbar\omega_{ab}$ of the unperturbed energies E_a and E_b (taken to be equal, respectively, to $\hbar\omega_{ab}/2$ and $-\hbar\omega_{ab}/2$ and represented by dashed lines), and the perturbed energies (represented by solid lines).

hyperbola having E_a and E_b as asymptotes (Figure 1). They form an anticrossing.

The perturbed states $|\chi_1\rangle$ and $|\chi_2\rangle$ are given by

$$|\chi_1\rangle = \sin\theta|\varphi_a\rangle + \cos\theta|\varphi_b\rangle \tag{5.a}$$

$$|\chi_2\rangle = \cos\theta|\varphi_a\rangle - \sin\theta|\varphi_b\rangle \tag{5.b}$$

with

$$\tan 2\theta = -\frac{\Omega_1}{\omega_{ab}}, \qquad 0 \le 2\theta < \pi. \tag{6}$$

In the middle of the anticrossing ($E_a = E_b$), the eigenstates are superpositions of $|\varphi_a\rangle$ and $|\varphi_b\rangle$ with equal amplitudes.

The time evolution of the system, and in particular the transition amplitude from state $|\varphi_a\rangle$ to state $|\varphi_b\rangle$, can be derived by different methods. In anticipation of more complex situations, we use the resolvent method here. The matrix element $G_{ba}(z)$, which according to formula (C.48) in Chapter III, is equal to

$$G_{ba}(z) = \frac{V_{ba}}{(z - E_a)(z - E_b) - |V_{ab}|^2} \tag{7}$$

has E_1 and E_2 as poles. The transition amplitude $U_{ba}(t)$, for $t > 0$, is found from the usual integral [see (A.22) in Chapter III],

$$U_{ba}(t) = \frac{1}{2i\pi} \int_{C_+} dz\, e^{-izt/\hbar}\, G_{ba}(z)$$

$$= V_{ba}\left[\frac{1}{E_1 - E_2} \exp(-iE_1 t/\hbar) + \frac{1}{E_2 - E_1} \exp(-iE_2 t/\hbar)\right]$$

$$= \frac{-i\Omega_1}{\Omega} \exp\left[-i(E_1 + E_2)t/2\hbar\right] \sin\frac{\Omega t}{2}. \tag{8}$$

The transition probability $P_{ba}(t) = |U_{ba}(t)|^2$ then equals, using (4),

$$P_{ba}(t) = \frac{\Omega_1^2}{\omega_{ab}^2 + \Omega_1^2} \sin^2\left[\sqrt{\omega_{ab}^2 + \Omega_1^2}\,\frac{t}{2}\right]. \tag{9}$$

It displays Rabi oscillations whose amplitude equals 1 at the center of the anticrossing ($\omega_{ab} = 0$). The amplitude of the oscillations decreases as ω_{ab} moves away from the center, while their frequency increases.

In several experiments, the Rabi oscillation cannot be observed indefinitely due to the finite lifetime of the system. For example, if the levels a and b are sublevels of an excited atomic state with a lifetime $1/\Gamma$, spontaneous emission causes the system to decay after a random time interval t whose distribution is given by

$$\mathscr{P}(t) = \Gamma \exp(-\Gamma t). \tag{10}$$

For ground-state sublevels, the lifetime of these levels is infinite, but the system can move and leave the observation region, or else the Rabi oscillation can be interrupted by a relaxation process resulting from the interaction of the system with its environment.

In most cases, it is necessary to continuously renew the system by means of an adequate preparation process (optical excitation, filling of a cell by an atomic beam...). We assume here that the relevant systems are continuously prepared in the state $|\varphi_a\rangle$ and that they evolve under the influence of the Hamiltonian $H_0 + V$ during a random time t whose distribution is given by (10), with the observation being made on the average population of the state $|\varphi_b\rangle$. This last quantity is simply the average $\bar{P}_{ba}$ of $P_{ba}(t)$ with weight (10):

$$\bar{P}_{ba} = \int_0^\infty dt \, \Gamma \exp(-\Gamma t) P_{ba}(t) \tag{11.a}$$

or also, using (9),

$$\bar{P}_{ba} = \frac{1}{2} \frac{\Omega_1^2}{\Omega^2} \left[1 - \frac{1}{2} \left(\frac{\Gamma}{\Gamma - i\Omega} + \frac{\Gamma}{\Gamma + i\Omega} \right) \right]$$

$$= \frac{1}{2} \frac{\Omega_1^2}{\Gamma^2 + \Omega_1^2 + \omega_{ab}^2} \, . \tag{11.b}$$

Thus, under the conditions described previously for preparation and detection, the anticrossing is manifested by a population transfer from $|\varphi_a\rangle$ to $|\varphi_b\rangle$, resonant in the center of the anticrossing ($\omega_{ab} = 0$). This resonant transfer is proportional to Ω_1^2 for $\Omega_1 \ll \Gamma$, and tends to a limit equal to $\frac{1}{2}$ for $\Gamma \ll \Omega_1$. The half-width of the resonance (variation of $\bar{P}_{ba}$ with ω_{ab}) is $[\Gamma^2 + \Omega_1^2]^{1/2}$. The fact that the width of the resonance increases with the intensity Ω_1^2 of the coupling is sometimes called "radiative broadening" or "power broadening".

b) Higher-Order Anticrossing

It may happen that the perturbation V does not directly couple the state $|\varphi_a\rangle$ to the state $|\varphi_b\rangle$ ($V_{ab} = 0$), but that instead this coupling exists in higher orders through one or several levels $|\varphi_c\rangle$, whose energies are far from E_a and E_b. More precisely, we assume

$$|V_{ac}| \ll |E_a - E_c| . \tag{12}$$

To calculate the transition amplitude $U_{ba}(t)$, we then consider the projection of the resolvent onto the subspace subtended by $|\varphi_a\rangle$ and $|\varphi_b\rangle$. If P is the projector onto this subspace, the expression for $PG(z)P$ is given by Equation (B.23) in Chapter III:

$$PG(z)P = \frac{P}{z - H_0 - PR(z)P} . \tag{13}$$

To the extent that the energy levels E_a and E_b are close to each other and far from other levels $|\varphi_c\rangle$, the level shift operator $R(z)$ can be approximated by its value for $z = E_0 = (E_a + E_b)/2$. As we saw in subsection B-3 in Chapter III, this is equivalent to describing the time evolution in the subspace $\mathscr{E}_0$ by using the effective Hamiltonian

$$H_{\text{eff}} = \begin{pmatrix} E_a + R_{aa}(E_0) & R_{ab}(E_0) \\ R_{ba}(E_0) & E_b + R_{bb}(E_0) \end{pmatrix}. \tag{14}$$

Taking into account condition (12), it is sufficient in the expansion of $R(E_0)$ [see formula (B.22) in Chapter III] to retain only the first nonzero term. Let p be the order of the first nonzero term of the nondiagonal element $R_{ab}(E_0)$. As for the two diagonal elements, it should be pointed out that, for all finite values of p, there exists at least one nonzero element V_{ac} and another nonzero element $V_{bc'}$. The second-order terms of the expansion of $R_{aa}(E_0)$ and $R_{bb}(E_0)$ are thus necessarily nonzero, and these are the preponderant terms, because we assume that $V_{aa} = V_{bb} = 0$.

With these approximations, we are brought back to the problem of the preceding subsection with the substitution

$$E_i \rightarrow \tilde{E}_i = E_i + R_{ii}(E_0) \qquad (i = a, b) \tag{15.a}$$

$$V_{ab} \rightarrow R_{ab}(E_0). \tag{15.b}$$

The levels $|\varphi_a\rangle$ and $|\varphi_b\rangle$ are first *shifted* by the quantities $R_{aa}(E_0)$ and $R_{bb}(E_0)$, so that their crossing point $\tilde{E}_a = \tilde{E}_b$ is shifted by

$$\hbar\delta\omega_{ab} = -[R_{aa}(E_0) - R_{bb}(E_0)] \tag{16}$$

(see Figure 2). Moreover, these levels are coupled by $R_{ab}(E_0)$ and form an anticrossing. The minimum distance between the two levels E_1 and E_2 is $R_{ab}(E_0)$, which is therefore of order p in V. This is an anticrossing of order p. It should be noted that, if $p > 2$, the level shift, which is second order, is much more important than the gap of the anticrossing.

The calculation of $U_{ba}(t)$ from the expression

$$G_{ba}(z) = \frac{R_{ba}(E_0)}{\left(z - \tilde{E}_a\right)\left(z - \tilde{E}_b\right) - |R_{ab}(E_0)|^2} \tag{17}$$

allows us to associate the same physical phenomena with the higher-order anticrossings as with the first order anticrossing: Rabi oscillation for $P_{ba}(t)$, resonance for $\bar{P}_{ba}$. It is sufficient to make the substitutions (15) in (9) and (11). The only new phenomena are, on the one hand, the shift of the center of resonance due to the replacement of E_a and E_b by $\tilde{E}_a$ and

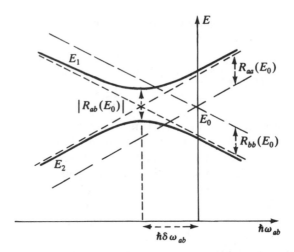

Figure 2. Higher-order anticrossing. As in Figure 1, the long-dashed straight lines represent the energies of unperturbed levels $|\varphi_a\rangle$ and $|\varphi_b\rangle$. The short-dashed straight lines represent the levels shifted by the quantities R_{aa} and R_{bb}. The solid lines represent the perturbed levels which anticross.

$\tilde{E}_b$, and on the other hand, the radiative broadening of the resonance which varies as $|R_{ab}|$, thus as $|V|^p$, and no longer as $\Omega_1 \sim |V|$.

c) LEVEL CROSSING. COHERENCE RESONANCE

It may happen that $R_{ba}(z)$ is zero to all orders if, for symmetry reasons for example, the states $|\varphi_a\rangle$ and $|\varphi_b\rangle$ belong to two subspaces which are not connected by the perturbation V. The energy levels then continue to cross while being shifted. Their energies E_a and $\tilde{E}_b$ are solutions to the implicit equations

$$\tilde{E}_a = E_a + R_{aa}(\tilde{E}_a) \qquad \tilde{E}_b = E_b + R_{bb}(\tilde{E}_b). \qquad (18)$$

The corresponding eigenstates are written $|\tilde{\varphi}_a\rangle$ and $|\tilde{\varphi}_b\rangle$. The time evolution respects the symmetry just as the Hamiltonian does, so

$$\langle \varphi_b | U(t) | \varphi_a \rangle = \langle \tilde{\varphi}_b | U(t) | \tilde{\varphi}_a \rangle = 0. \qquad (19)$$

No transitions are induced by V between the level $|\varphi_a\rangle$ and the level $|\varphi_b\rangle$.

Despite the absence of transition, we can observe a resonance at the shifted crossing point defined by

$$\tilde{E}_a = \tilde{E}_b \tag{20}$$

if the systems are prepared in a superposition of states $|\tilde{\varphi}_a\rangle$ and $|\tilde{\varphi}_b\rangle$:

$$|\psi_i\rangle = \lambda_a |\tilde{\varphi}_a\rangle + \lambda_b |\tilde{\varphi}_b\rangle + \cdots \tag{21}$$

and detected in a state $|\psi_f\rangle$, which is itself also a superposition of $|\tilde{\varphi}_a\rangle$ and $|\tilde{\varphi}_b\rangle$

$$|\psi_f\rangle = \mu_a |\tilde{\varphi}_a\rangle + \mu_b |\tilde{\varphi}_b\rangle + \cdots \tag{22}$$

(the series of dots indicates that $|\psi_i\rangle$ and $|\psi_f\rangle$ can have nonzero projections onto other states). This resonance is related to the interference, in the transition amplitude from $|\psi_i\rangle$ to $|\psi_f\rangle$, between two terms associated with two "paths" going, respectively, through $|\tilde{\varphi}_a\rangle$ or $|\tilde{\varphi}_b\rangle$. More precisely, the contribution of these two terms to the transition amplitude is written

$$\langle \psi_f | U(t) | \psi_i \rangle = \langle \psi_f | \tilde{\varphi}_a \rangle \exp\left(-i\tilde{E}_a t/\hbar\right) \langle \tilde{\varphi}_a | \psi_i \rangle +$$

$$+ \langle \psi_f | \tilde{\varphi}_b \rangle \exp\left(-i\tilde{E}_b t/\hbar\right) \langle \tilde{\varphi}_b | \psi_i \rangle$$

$$= \mu_a^* \lambda_a \exp\left(-i\tilde{E}_a t/\hbar\right) + \mu_b^* \lambda_b \exp\left(-i\tilde{E}_b t/\hbar\right). \tag{23}$$

In the transition probability $P_{fi}(t)$, the interference between these two terms in (23) gives rise to a modulation at the frequency $\tilde{\omega}_{ba} = (\tilde{E}_b - \tilde{E}_a)/\hbar$:

$$P_{fi}(t) = |\mu_a|^2 |\lambda_a|^2 + |\mu_b|^2 |\lambda_b|^2 +$$

$$+ \left[\mu_a^* \mu_b \lambda_a \lambda_b^* \exp(-i\tilde{\omega}_{ab} t) + \text{c.c.}\right] \tag{24}$$

which manifests itself in the averaged probability $\bar{P}_{fi}$ defined in (11.a) by a resonant term at $\tilde{\omega}_{ab} = 0$:

$$\bar{P}_{fi} = |\mu_a|^2 |\lambda_a|^2 + |\mu_b|^2 |\lambda_b|^2 +$$

$$+ \left[\mu_a^* \mu_b \lambda_a \lambda_b^* \frac{\Gamma}{\Gamma + i\tilde{\omega}_{ab}} + \text{c.c.}\right]. \tag{25}$$

It should be noted that the resonance is observable only if the four numbers λ_a, λ_b, μ_a, and μ_b are simultaneously nonzero. It is then necessary to prepare and detect coherent superpositions of the levels that cross. This type of resonance is thus called coherence resonance (*).

Remark

It may happen that two levels $|\varphi_a\rangle$ and $|\varphi_b\rangle$ are coupled indirectly by $R_{ab}(E)$, but that this coupling tends to zero when the crossing point is approached. One can then show that, despite their coupling, the perturbed levels continue to cross. The resonances observed near such a crossing are quite similar to coherence resonances, but they can appear even if the system is prepared in the state $|\varphi_a\rangle$ and detected in the state $|\varphi_b\rangle$ (**).

2. Spin $\frac{1}{2}$ Dressed by Radio-Frequency Photons

a) DESCRIPTION OF THE SYSTEM

The fine, hyperfine, or Zeeman structures of atoms and molecules involve a finite number of sublevels, frequently very close to each other in comparison to electronic excitation energies. We consider here the most simple case where there are only two sublevels. A model of such a system is provided by a paramagnetic atom assumed to be fixed at the origin, having a spin $\frac{1}{2}$ and interacting with a static magnetic field $\mathbf{B}_0$. The magnetic moment $\boldsymbol{\mu}$ associated with the spin $\mathbf{S}$ is given by

$$\boldsymbol{\mu} = \gamma \hbar \mathbf{S} \tag{26}$$

where γ is the gyromagnetic ratio, and where S_x, S_y, and S_z are the three Pauli matrices, multiplied by $\frac{1}{2}$. The angular precession velocity of the spin around the direction $0z$ of the magnetic field $\mathbf{B}_0$ is

$$\omega_0 = -\gamma B_0. \tag{27}$$

Such a spin is also assumed to interact with a monochromatic radio-frequency field. Because the spatial variations of the radio-frequency field are negligible on the atomic scale, the atom is sensitive only to the value $\mathbf{B}_1(\mathbf{0}, t)$ of this field at the point $\mathbf{0}$ where the atom is located. A mode $\mathbf{k}\varepsilon$ of the free field can then be used to simulate the action of $\mathbf{B}_1(\mathbf{0}, t)$ on the

(*) A concrete example of level-crossing resonance (Hanle or Franken effect) is discussed in Exercise 6.

(**) For more details, see G. Grynberg, J. Dupont-Roc, S. Haroche, and C. Cohen-Tannoudji, *J. Phys. (Paris)*, **34**, 523 (1973); **34**, 537 (1973).

atom, provided that its frequency ω, its polarization $\boldsymbol{\varepsilon}$, and its quantum state $|\psi(t)\rangle$ are such that the average value of the magnetic field operator $\mathbf{B}(0)$ is equal to the experimental value $\mathbf{B}_1(0, t)$. We choose here for $|\psi(t)\rangle$ a coherent state $|\alpha(t)\rangle$ (with $\alpha \gg 1$) satisfying

$$\langle \alpha(t) | \mathbf{B}(0) | \alpha(t) \rangle = \mathbf{B}_1(0, t). \tag{28}$$

The advantage of this description of the radio-frequency field is obviously that, instead of a time-dependent perturbation problem, we can now consider the problem of the evolution of an isolated system (dressed atom), for which there are conserved variables, such as energy, angular momentum, etc., which greatly facilitate the calculations and the physical interpretation of the results.

In the absence of interaction between the atom and the radio-frequency field, the Hamiltonian H_0 of the system is the sum of the atomic Hamiltonian H_A and the Hamiltonian H_R of the mode of the radio-frequency field. The influence of other modes of the radio-frequency field is neglected, as is the zero-point energy of the mode:

$$H_0 = H_A + H_R \tag{29}$$

$$H_A = -\boldsymbol{\mu} \cdot \mathbf{B}_0 = \hbar\omega_0 S_z \tag{30}$$

$$H_R = \hbar\omega a^+ a. \tag{31}$$

If $|+\rangle$ and $|-\rangle$ are the two eigenstates of S_z, are $|N\rangle$ is the state with N photons of the mode being considered, the eigenstates of H_0 are $|\pm, N\rangle$, with energies $E_{\pm, N}$

$$H_0 |\pm, N\rangle = E_{\pm, N} |\pm, N\rangle \tag{32}$$

$$E_{\pm, N} = \pm(\hbar\omega_0/2) + N\hbar\omega. \tag{33}$$

As a function of the splitting $\hbar\omega_0$ between the two atomic sublevels, the energy diagram appears as an infinite number of straight lines of slope $\pm \frac{1}{2}$ and with ordinates $N\hbar\omega$ for $\omega_0 = 0$ (Figure 3). In what follows, we will consider levels such as $N \simeq \langle N \rangle \gg 1$.

b) INTERACTION HAMILTONIAN BETWEEN THE ATOM AND THE RADIO-FREQUENCY FIELD

The interaction Hamiltonian between the magnetic moment carried by the spin and the magnetic field of the mode with polarization $\boldsymbol{\varepsilon}$ is

$$V = -\boldsymbol{\mu} \cdot \mathbf{B}(0). \tag{34}$$

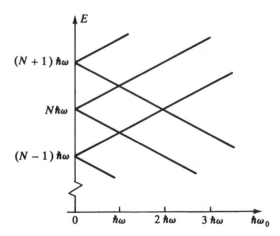

Figure 3. Unperturbed energy levels of the spin $\frac{1}{2}$ + radio-frequency photons system.

The expression for $\mathbf{B}(0)$ results from formula (30) in the Appendix:

$$\mathbf{B}(0) = \sqrt{\frac{\hbar\omega}{2\varepsilon_0 c^2 L^3}} \left[a \frac{i(\mathbf{k} \times \boldsymbol{\varepsilon})}{k} + a^+ \frac{(-i)(\mathbf{k} \times \boldsymbol{\varepsilon}^*)}{k} \right]. \qquad (35)$$

Using (26) and (35), expression (34) for V can be put in a simple form:

$$V = \lambda[(\mathbf{e} \cdot \mathbf{S})a + (\mathbf{e}^* \cdot \mathbf{S})a^+] \qquad (36)$$

where $\mathbf{e}$ is the polarization vector of the rf magnetic field and where λ plays the role of a coupling constant having the dimension of an energy.

Consider now more precisely the matrix elements of V. They are nonzero only between the states $|\pm, N\rangle$ whose values of N differ by one unit:

$$\Delta N = \pm 1. \qquad (37)$$

The selection rules for the eigenvalues m_s of S_z depend on the polarization $\mathbf{e}$ of the rf magnetic field. Two particular cases are considered.

i) *Circular Polarization*

The circular polarizations σ_+ and σ_- relative to the static magnetic field $\mathbf{B}_0$ are obtained by selecting for the polarization $\mathbf{e}$:

$$\mathbf{e}_\pm = \frac{1}{\sqrt{2}}(\mathbf{e}_x \pm i\mathbf{e}_y). \tag{38}$$

The coupling is then written (with $S_\pm = S_x + iS_y$)

$$V_{\sigma+} = \frac{\lambda}{\sqrt{2}}(aS_+ + a^+S_-) \tag{39.a}$$

and

$$V_{\sigma-} = \frac{\lambda}{\sqrt{2}}(aS_- + a^+S_+). \tag{39.b}$$

It is easy to check that $V_{\sigma+}$ conserves the quantum number $m_s + N$. This selection rule is a physical consequence of the conservation of the z component of the total angular momentum. Its interpretation is very simple: Each photon of the mode carries a unit of angular momentum along $0z$, so that $(m_s + N)\hbar$ represents the total angular momentum of the system, which is conserved. Similarly, $V_{\sigma-}$ conserves $m_s - N$, each photon having an angular momentum $-\hbar$.

ii) *Linear Polarization*

A polarization of the radio-frequency field that is linear and perpendicular to $0z$ is obtained, for example, for $\mathbf{e} = \mathbf{e}_x$. We call such a polarization σ:

$$V_\sigma = \lambda S_x(a + a^+). \tag{40}$$

Note that V_σ changes N by ± 1 and m_s by ± 1, and thus $m_s + N$ by 0 or by ± 2. This shows that V_σ conserves the quantum number

$$\eta(m_s, N) = (-)^{m_s + N - 1/2}. \tag{41}$$

The latter takes the values $+1$ or -1, depending on the levels.

Remark

In formulas (39.a), (39.b), and (40), the operators a^+ and a are creation and annihilation operators for photons of the corresponding modes [σ_+ for (39.a), σ_- for (39.b), and σ for (40)]. To simplify notation, they are written with the same letters.

c) PREPARATION AND DETECTION

Different methods can be used to prepare paramagnetic atoms in a given spin state: The deflection of atoms in an atomic beam by magnetic field gradients depends on their spin state and can be used for ground or metastable states. Optical pumping can be used under the same conditions, or for atoms contained in a cell. The states prepared in this way are not necessarily eigenstates $|\pm\rangle$ of H_0, but may also be coherent superpositions of these eigenstates. This is what happens, for example, if transverse optical pumping (i.e., perpendicular to the static field $\mathbf{B}_0$) prepares the spins in an eigenstate of S_x or S_y. Optical excitation with circularly polarized light allows one to prepare excited electronic states with angular momentum $\frac{1}{2}$ in a given Zeeman sublevel or a superposition of these sublevels. Finally, Boltzmann equilibrium at low temperatures in a high magnetic field is characterized by unequal populations of the two sublevels $|+\rangle$ and $|-\rangle$. Pulsed radio-frequency techniques may then allow superposition states to be prepared.

To be specific, we assume in what follows that the preparation scheme provides atoms in the state

$$|\varphi_a\rangle = |-\rangle \tag{42}$$

or, for a coherent excitation, in the state

$$|\psi_i\rangle = [|+\rangle - |-\rangle]/\sqrt{2} \tag{43}$$

(the eigenstate of S_x with eigenvalue $-\frac{1}{2}$). The initial state of the total system is thus the tensor product of these states and the state of the radio-frequency field at the time of preparation, which is assumed to be either a state $|N\rangle$ or a coherent state $|\alpha(t)\rangle$. Magnetic deflection and optical methods (for example, measuring the absorption of a circularly polarized light beam) can also be used for detection. The signals obtained are proportional to the population of a given state that we assume to be

$$|\varphi_b\rangle = |+\rangle \tag{44}$$

or, for coherent detection,

$$|\psi_f\rangle = [|+\rangle + |-\rangle]/\sqrt{2}. \tag{45}$$

For the dressed atom, the populations of the states $|+, N\rangle$ or $|\psi_f, N\rangle = |\psi_f\rangle \otimes |N\rangle$ must be summed over N.

Remark

We assume implicitly that the preparation and detection processes are not affected by the interaction with the radio-frequency field (or that they take place outside the region where the radiofrequency field acts). Such a condition implies that the typical correlation time for the preparation and detection processes are shorter than the inverse of the pertinent Bohr frequencies of the system, or that the entry (and exit) times of the atoms within the interaction region are sufficiently short so that the sudden approximation can be applied.

3. The Simple Case of Circularly Polarized Photons

This case is particularly simple, because the energy diagram and the resonance signals can be determined analytically.

a) ENERGY DIAGRAM

The coupling $V_{\sigma+}$ has nonzero matrix elements only between eigenstates of H_0 characterized by the same value of $m_s + N$. The diagonalization of H can thus be carried out separately inside each of these subspaces. Because m_s can only take the values $\pm \frac{1}{2}$, $m_s + N$ takes only half-integer values, and a two-dimensional subspace is associated with each of these values. Therefore, the subspace associated with the value $N + \frac{1}{2}$ is subtended by the two states:

$$|\varphi_a\rangle = |-, N + 1\rangle \qquad |\varphi_b\rangle = |+, N\rangle \qquad (46)$$

whose unperturbed energies

$$E_a = (N + 1)\hbar\omega - (\hbar\omega_0/2) \qquad E_b = N\hbar\omega + (\hbar\omega_0/2) \qquad (47)$$

are separated by the Bohr frequency $\omega_{ab} = \omega - \omega_0$. The matrix elements of $V_{\sigma+}$ in this subspace are given by

$$\langle \varphi_a | V_{\sigma+} | \varphi_a \rangle = \langle \varphi_b | V_{\sigma+} | \varphi_b \rangle = 0 \qquad (48.a)$$

$$\langle \varphi_b | V_{\sigma+} | \varphi_a \rangle = \frac{\lambda}{\sqrt{2}} \sqrt{N + 1}. \qquad (48.b)$$

Because the dispersion of the values of N around $\langle N \rangle$ is very low in relative value, we can approximate the matrix element (48.b) by

$$\langle \varphi_b | V_{\sigma+} | \varphi_a \rangle \simeq \hbar\Omega_1/2 \qquad (49)$$

where

$$\hbar\Omega_1 = \lambda\sqrt{2\langle N \rangle}. \tag{50}$$

The perturbed states resulting from the states $|+, N\rangle$ and $|-, N + 1\rangle$ thus form a simple anticrossing of the type discussed in subsection 1-a. The energy levels $E_1(N)$ and $E_2(N)$ are two hyperbolas centered on $\omega_0 = \omega$, whose equations are given by (3.a) and (3.b) with the values (47) for E_a and E_b:

$$\frac{1}{\hbar}E_1(N) = \left(N + \frac{1}{2}\right)\omega + \sqrt{\left(\frac{\omega - \omega_0}{2}\right)^2 + \left(\frac{\Omega_1}{2}\right)^2} \tag{51.a}$$

$$\frac{1}{\hbar}E_2(N) = \left(N + \frac{1}{2}\right)\omega - \sqrt{\left(\frac{\omega - \omega_0}{2}\right)^2 + \left(\frac{\Omega_1}{2}\right)^2}. \tag{51.b}$$

The corresponding eigenstates $|\chi_1(N)\rangle$ and $|\chi_2(N)\rangle$ are expressed as a function of the unperturbed states (46) by the formulas (5).

$$|\chi_1(N)\rangle = \sin\theta|-, N + 1\rangle + \cos\theta|+, N\rangle \tag{52.a}$$

$$|\chi_2(N)\rangle = \cos\theta|-, N + 1\rangle - \sin\theta|+, N\rangle \tag{52.b}$$

$$\tan 2\theta = -\frac{\Omega_1}{\omega - \omega_0} \qquad 0 \le 2\theta < \pi. \tag{53}$$

The energy levels are shown in Figure 4. They form a set of hyperbolas having the unperturbed levels from Figure 3 as asymptotes and for which the gap between the two branches is equal to $\hbar\Omega_1$.

We now consider the other level crossings in the unperturbed energy-level diagram. In the presence of the coupling V, these level crossings do not transform into anticrossings. Indeed, the levels that cross belong to subspaces corresponding to different values of $N + m_s$, and they are thus not coupled by V to any order.

b) MAGNETIC RESONANCE INTERPRETED AS A LEVEL-ANTICROSSING RESONANCE OF THE DRESSED ATOM

According to the discussion in subsection 1-a, a population transfer between the unperturbed states $|\varphi_a\rangle = |-, N + 1\rangle$ and $|\varphi_b\rangle = |+, N\rangle$ is associated with each of the anticrossings in Figure 4. This transition between these two states of the dressed atom is interpreted as a spin flip

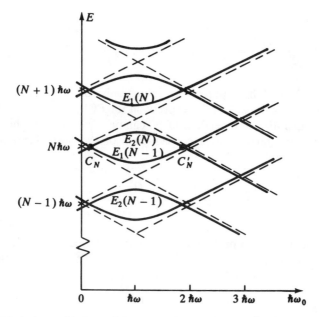

Figure 4. Variations with $\hbar\omega_0$ of the energy levels of a spin $\frac{1}{2}$ dressed by circularly polarized radio-frequency photons.

from the state $|-\rangle$ to the state $|+\rangle$ with the absorption of a radio-frequency photon. This transfer is resonant at $\omega_{ab} = 0$, or $\omega = \omega_0$. If atoms are prepared continuously in the state $|-\rangle$ in the presence of $N + 1$ photons, the population of the state $|+, N\rangle$ is given by formula (11.b):

$$\bar{P}(+, N; -, N + 1) = \frac{1}{2} \frac{\Omega_1^2}{\Gamma^2 + \Omega_1^2 + (\omega - \omega_0)^2}. \tag{54}$$

In this expression, we have assumed $N \simeq \langle N \rangle$, and have used the approximation (49). Actually, because the preparation is not made with a well-defined value of N and the observation concerns only the spin state, it is necessary to take the average over the different initial values of N, with the weight $|\langle N|\alpha\rangle|^2$, and for each value of N, to sum over all the possible final values of N' of the number of photons. Because the state $|-, N + 1\rangle$ is coupled only to the state $|+, N\rangle$, the sum over N' contains only the term

(54). We then have

$$\bar{P}_{+-} = \sum_N |\langle N|\alpha\rangle|^2 \bar{P}(+,N;-,N+1).$$ (55)

Because $|\langle N|\alpha\rangle|^2$ has appreciable values only for $N \approx \langle N\rangle = |\alpha|^2$, the approximation (49) made in the expression for $\bar{P}(+,N;-,N+1)$ is valid. This latter quantity, independent of N, can be removed from the sum over N, which then equals 1 as a consequence of the normalization of $|\alpha\rangle$. Finally, we have

$$\bar{P}_{+-} = \frac{1}{2} \frac{\Omega_1^2}{\Gamma^2 + \Omega_1^2 + (\omega - \omega_0)^2}.$$ (56)

The average transition probability $\bar{P}_{+-}$ varies with $\omega_0 - \omega$ as a resonance curve, centered at $\omega_0 = \omega$, having a width $2[\Gamma^2 + \Omega_1^2]^{1/2}$ and whose height tends to $\frac{1}{2}$ when Ω_1/Γ tends to infinity.

Remark

Because the radio-frequency field is described by the state $|\alpha(t)\rangle$, the density matrix describing a system prepared at time t is

$$\sigma_{ex}(t) = |+\rangle\langle+| \otimes |\alpha(t)\rangle\langle\alpha(t)|$$

$$= \sum_{N,N'} |+,N\rangle\langle N|\alpha(t)\rangle\langle\alpha(t)|N'\rangle\langle+,N'|,$$ (57)

or else by approximating the evolution of the field between 0 and t by the evolution of a free field (the modification of the field due to the spins is assumed to be negligible):

$$\sigma_{ex}(t) = \sum_{N,N'} \left[\langle N|\alpha(0)\rangle\langle\alpha(0)|N'\rangle \times \right.$$

$$\left. \times e^{-i(N-N')\omega t}\right] |+,N\rangle\langle+,N'|.$$ (58)

Thus, the source term of the master equation describing the evolution of the spins interacting with the field contains modulated terms at frequencies $(N - N')\omega$. These terms give rise to resonant signals, detectable for $N - N' = \pm 1$. Here we are interested only in static signals, for which it is sufficient to consider the time-independent term in (58):

$$\sigma_{ex}^{(stat)} = \sum_N |\langle N|\alpha(0)\rangle|^2 |+,N\rangle\langle+,N|.$$ (59)

This distribution of states $|+, N\rangle$ with weights $|\langle N|\alpha(0)\rangle|^2$ is precisely the one we considered for the derivation of (56).

c) DRESSED STATE LEVEL-CROSSING RESONANCES

In the absence of the radio-frequency field, the two energy levels of the spin in the magnetic field cross each other at $\omega_0 = 0$. A level-crossing resonance can be observed on spins prepared and detected, respectively, in the states $|\psi_i\rangle$ and $|\psi_f\rangle$ given by (43) and (45). The correspondence $|a\rangle = |-\rangle$, $|b\rangle = |+\rangle$ allows one to apply formula (25) with $\lambda_a = -1/\sqrt{2}$, $\lambda_b = \mu_a = \mu_b = 1/\sqrt{2}$, $\omega_{ab} = -\omega_0$. This gives the shape of the level-crossing resonance (*)

$$\bar{P}_{fi} = \frac{1}{4} + \frac{1}{4} - \frac{1}{4}\left[\frac{\Gamma}{\Gamma - i\omega_0} + \text{c.c.}\right]$$

$$= \frac{1}{2}\frac{\omega_0^2}{\Gamma^2 + \omega_0^2}. \tag{60}$$

In the presence of the radio-frequency field, but in the absence of interaction, the energies of the levels $|+, N\rangle$ and $|-, N\rangle$ cross each other at $\omega_0 = 0$. When the interaction is taken into account, the straight lines of Figure 3 become the hyperbolas of Figure 4, which no longer cross at $\omega_0 = 0$. More precisely, the energy splitting between the states $E_1(N-1)$ and $E_2(N)$ equals, according to (51),

$$\hbar\tilde{\omega}_{ab} = E_1(N-1) - E_2(N) = \hbar\left[\sqrt{(\omega_0 - \omega)^2 + \Omega_1^2} - \omega\right]. \tag{61}$$

It cancels out at

$$\omega_0 = \omega - \sqrt{\omega^2 - \Omega_1^2}. \tag{62}$$

The interaction thus shifts the crossing point. For low values of Ω_1, it is found at $\omega_0 \simeq \Omega_1^2/2\omega$ and approaches $\omega_0 = \omega$ when Ω_1 increases. It disappears for $\Omega_1 \geq \omega$.

To analytically determine the expression for the level-crossing resonance, we can express the unperturbed states as a function of the dressed

---

(*) This crossing resonance in zero magnetic field corresponds to the Hanle effect [see W. Hanle, Z. Phys., **30**, 93 (1924) and also Brossel].

states by inverting formulas (52):

$$|-, N\rangle = \sin\theta|\chi_1(N-1)\rangle + \cos\theta|\chi_2(N-1)\rangle \qquad (63.a)$$

$$|+, N\rangle = \cos\theta|\chi_1(N)\rangle - \sin\theta|\chi_2(N)\rangle. \qquad (63.b)$$

The states that are prepared and detected are, respectively,

$$|\psi_i, N\rangle = \frac{1}{\sqrt{2}}(|+\rangle - |-\rangle) \otimes |N\rangle$$

$$= \frac{1}{\sqrt{2}}\{\cos\theta|\chi_1(N)\rangle - \sin\theta|\chi_2(N)\rangle -$$

$$- \sin\theta|\chi_1(N-1)\rangle - \cos\theta|\chi_2(N-1)\rangle\} \quad (64.a)$$

$$|\psi_f, N\rangle = \frac{1}{\sqrt{2}}(|+\rangle + |-\rangle) \otimes |N\rangle$$

$$= \frac{1}{\sqrt{2}}\{\cos\theta|\chi_1(N)\rangle - \sin\theta|\chi_2(N)\rangle +$$

$$+ \sin\theta|\chi_1(N-1)\rangle + \cos\theta|\chi_2(N-1)\rangle\}. \quad (64.b)$$

To apply formula (25), we must consider the projections of $|\psi_i, N\rangle$ and $|\psi_f, N\rangle$ on the levels that cross each other, in this case $|\chi_1(N-1)\rangle$ and $|\chi_2(N)\rangle$. They equal $-\sin\theta/\sqrt{2}$, $-\sin\theta/\sqrt{2}$, $+\sin\theta/\sqrt{2}$, $-\sin\theta/\sqrt{2}$, which, after the summation over N, results in

$$\bar{P}_{fi} = \frac{\sin^4\theta}{2}\frac{\tilde{\omega}_{ab}^2}{\Gamma^2 + \tilde{\omega}_{ab}^2} \qquad (65)$$

where θ is given by (53) and $\tilde{\omega}_{ab}$ is given by (61). In the limit of low values of Ω_1, the crossing remains near $\omega_0 \simeq 0$, and $\sin\theta \simeq 1$. The circular radio-frequency field then acts like an effective static field parallel to $0z$ which shifts the position of the level crossing by a quantity equal to $\Omega_1^2/2\omega$.

The energy diagram in Figure 4 has many other crossings. Nevertheless, a small number of them can be observed in the detection signals. Indeed, the states prepared and detected are linear combinations of the four dressed states appearing in the expansions (64.a) and (64.b). Only the crossing points between the corresponding energy levels are observable in the detection signals. There are only two of them and they correspond to the points of intersection of $E_2(N)$ and $E_1(N-1)$ marked C_N and C'_N in Figure 4. The first is the one we just discussed. The second one is located at $\omega_0 = 2\omega$ at the weak coupling limit, and approaches $\omega_0 = \omega$ at high intensity. The expression for the corresponding resonance is still given by formula (65), but for values of ω_0 close to the crossing point given by the second solution of $\tilde{\omega}_{ab} = 0$, or, according to (61),

$$\omega_0 = \omega + \sqrt{\omega^2 - \Omega_1^2}. \tag{66}$$

In the perturbative limit and near $\omega_0 = 2\omega$, formulas (61) and (53) give $\tilde{\omega}_{ab} = [\omega_0 - (2\omega - \Omega_1^2/2\omega)]$, and $\tan 2\theta = \Omega_1/\omega$, or $2\theta \simeq \Omega_1/\omega$. The resonance appears in the form

$$\bar{P}_{fi}(2\omega) = \frac{1}{2}\left(\frac{\Omega_1}{2\omega}\right)^4 \left\{1 - \frac{\Gamma^2}{\Gamma^2 + \left[\omega_0 - (2\omega - \Omega_1^2/2\omega)\right]^2}\right\}. \tag{67}$$

It is observable only in the presence of the coupling with the radio frequency, which, by contaminating the states $|-, N+1\rangle$ by $|+, N\rangle$ and $|+, N-1\rangle$ by $|-, N\rangle$, allows the excitation and the coherent detection of the two levels that cross each other.

4. Linearly Polarized Radio-Frequency Photons

The energy diagram for a spin $\frac{1}{2}$ dressed by σ_+ (or σ_-) radio-frequency photons causes only one anticrossing resonance and two crossing resonances to appear. We now consider the case where the radio-frequency photons have a linear polarization, perpendicular to the magnetic field (polarization σ), and show that such a situation is richer in physical phenomena.

a) SURVEY OF THE NEW EFFECTS

Because $S_x = (S_+ + S_-)/2$, the interaction Hamiltonian (40) can be written in the form

$$V_\sigma = V_+ + V_- \tag{68}$$

where

$$V_+ = \frac{\lambda}{2}(S_+ a + S_- a^+) \tag{69.a}$$

$$V_- = \frac{\lambda}{2}(S_- a + S_+ a^+). \tag{69.b}$$

V_+ couples the level $|+, N\rangle$ to the level $|-, N + 1\rangle$, which gives rise to a level anticrossing at $\omega_0 = \omega$ as in the preceding subsection. V_- couples $|+, N\rangle$ to $|-, N - 1\rangle$ and gives rise to one anticrossing at $\omega_0 = -\omega$. The corresponding matrix elements are

$$\langle +, N|V_\sigma| -, N + 1\rangle = \frac{\lambda}{2}\sqrt{N + 1}$$

$$\langle +, N|V_\sigma| -, N - 1\rangle = \frac{\lambda}{2}\sqrt{N} \qquad (70)$$

and can be approximated by

$$\langle +, N|V_\sigma| -, N \pm 1\rangle = \frac{\hbar\Omega_1}{2} \qquad (71)$$

with (*)

$$\hbar\Omega_1 = \lambda\sqrt{\langle N\rangle}. \qquad (72)$$

To lowest order in Ω_1, the energy diagram presents anticrossings at $\omega_0 = \pm\omega$, with which are associated two resonances symmetrical with respect to $\omega_0 = 0$.

In higher orders, it is necessary to consider, in addition to the saturation effects produced separately by V_+ and V_-, the crossed effects between these two terms. For example, the anticrossing created by V_- near $\omega_0 = -\omega$ shifts the levels that are resonantly coupled by V_+, at $\omega_0 = \omega$. This latter anticrossing is then shifted, as well as the associated magnetic resonance: this is the Bloch–Siegert shift that we will calculate in subsection 4-b.

Moreover, levels which, for a pure coupling V_+ (or V_-) would belong to subspaces unconnected by V_+ (or V_-), are now coupled in higher orders. For example, the state $|+, N\rangle$ is coupled in first order to $|-, N \pm 1\rangle$, in second order to $|+, N \pm 2\rangle$ and $|+, N\rangle$, etc. In general, $|+, N\rangle$ is coupled to all the states $|-, N \pm (2p + 1)\rangle$ and $|+, N \pm 2p\rangle$, where p is an integer. The unperturbed states can then be regrouped into two subspaces that are not connected by V_σ. These two subspaces actually

(*) With this definition, Ω_1 represents, as it does in the chapter, the resonance Rabi frequency for the one-photon resonance ($\omega_0 = \pm\omega$). It should be noted that most of the original articles use another parameter, $\omega_1 = -\gamma B_1$, for a linearly polarized radio-frequency field $B_1 \cos \omega t \, \mathbf{e}_x$. The relationship between these two parameters can be obtained by decomposing a linear field into left and right circular fields and is written $\omega_1 = 2\Omega_1$.

correspond to the values $+1$ and -1 of the quantum number $\eta(m_s, N)$ conserved by V_σ, which is given in (41). The crossings between levels of the same subspace are transformed into anticrossings by V_σ. This is the case for the states $|+, N\rangle$ and $|-, N + (2p + 1)\rangle$ which cross each other at $\omega_0 = (2p + 1)\omega$. By contrast, the intersections between levels of different subspaces remain crossings. This is the case for the states $|+, N\rangle$ and $|-, N + 2p\rangle$ which cross each other at $\omega_0 = 2p\omega$. The dressed atom energy diagram thus has an odd spectrum of anticrossings and an even spectrum of crossings, with which are associated resonances that will be discussed in subsections 4-c and 4-d. As the resonances $\omega_0 = \pm\omega$, all these resonances undergo radiative shifts.

As a result of the invariance of V_σ and H_R by time reversal and the odd character of H_A in this operation, the energy levels are symmetric relative to $\omega_0 = 0$. Consequently, the zero-field level crossings are not shifted (*). When the intensity of the field varies, the slope of the energy levels in zero field, which represents the Landé factor of the dressed atom, also varies. This variation can be precisely determined for any value of Ω_1/ω, as we will show in subsection 4-e.

b) BLOCH–SIEGERT SHIFT

We consider the levels $|\varphi_a\rangle = |-, N + 1\rangle$ and $|\varphi_b\rangle = |+, N\rangle$ between which magnetic resonance transitions occur near $\omega_0 = \omega$, and we use the results from subsection 1-b. The matrix elements of the operator $R(E_0)$ taken for

$$E_0 = \frac{E_a + E_b}{2} = \left(N + \frac{1}{2}\right)\hbar\omega \tag{73}$$

will be determined to lowest nonvanishing order. The nondiagonal element has actually only one first-order term, which, according to (71), equals

$$R_{ab}(E_0) = \langle -, N + 1|V_\sigma|+, N\rangle = \frac{\hbar\Omega_1}{2}. \tag{74}$$

For the diagonal matrix elements, the first-order term is zero, so we must

---

(*) The fact that the energy levels are doubly degenerate in zero field is a consequence of the Kramers theorem: see Messiah, Chapter XV.

consider the second-order term. For example,

$$R_{aa}(E_0) = \sum_{c \neq a, b} \frac{|\langle \varphi_a |V_\sigma| \varphi_c \rangle|^2}{E_0 - E_c}. \tag{75}$$

The only nonresonant intermediate state that contributes is the state $| +, N + 2 \rangle$. By using (73) and by approximating ω_0 by ω, we obtain

$$R_{aa}(E_0) = \frac{|\langle -, N + 1|V_\sigma| +, N + 2 \rangle|^2}{(N + \frac{1}{2})\hbar\omega - [\hbar\omega_0/2 + (N + 2)\hbar\omega]}$$

$$\simeq -\frac{\hbar\Omega_1^2}{8\omega}. \tag{76}$$

Similarly,

$$R_{bb}(E_0) = \frac{\hbar\Omega_1^2}{8\omega}. \tag{77}$$

Thus the anticrossing is shifted, according to (16), by the quantity

$$\delta\omega_0 = -\delta\omega_{ab} = -\frac{\Omega_1^2}{4\omega} \tag{78}$$

which is the Bloch–Siegert shift (*). The magnetic resonance now occurs for a weaker static field. It is clear from the foregoing calculation that such an effect is due to nonresonant couplings induced by V_σ. These couplings disappear when the rotating-wave approximation is made, as in the chapter (§B-3-b). An analogous calculation shows that the shift of the resonance $\omega_0 = -\omega$ is the opposite of (78), so that the two resonances approach the zero field.

Remark

We calculated the Bloch–Siegert shift to lowest order in Ω_1/ω. To derive the higher-order terms, we must consider the next terms of the expansion of $R(z)$ and not make the approximation consisting of replacing z by E_0. The energies

(*) See, for example Abragam (Chapter II).

$E_1(N)$ and $E_2(N)$ are derived as solutions of an implicit equation giving the poles of $G(z)$. The center of the anticrossing, defined as the point where the energy levels have a horizontal tangent, can then be expressed in the form of a power expansion of Ω_1/ω (*).

c) The Odd Spectrum of Level-Anticrossing Resonances

Consider two unperturbed levels that cross at $\omega_0 = (2p + 1)\omega$, for example, $|\varphi_a\rangle = |-, N + 1\rangle$ and $|\varphi_b\rangle = |+, N - 2p\rangle$, and once more apply the results of subsection 1-b. The shift $R_{aa}(E_0)$ of the level $|\varphi_a\rangle$ is produced by its coupling with the levels $|+, N\rangle$ and $|+, N + 2\rangle$ with the corresponding matrix elements being equal to $\hbar\Omega_1/2$. The average energy E_0 equals

$$E_0 = \frac{E_a + E_b}{2} = \left(N - p + \frac{1}{2}\right)\hbar\omega, \qquad (79)$$

so that by approximating ω_0 by $(2p + 1)\omega$, the energy denominators corresponding to the states $|+, N\rangle$ and $|+, N + 2\rangle$ are, respectively,

$$(N - p + 1/2)\hbar\omega - \left[(\hbar\omega_0/2) + N\hbar\omega\right] \approx -2p\hbar\omega$$

and

$$(N - p + 1/2)\hbar\omega - \left[(\hbar\omega_0/2) + (N + 2)\hbar\omega\right] \approx -(2p + 2)\hbar\omega.$$

Finally

$$R_{aa}(E_0) = -\left(\frac{\hbar\Omega_1}{2}\right)^2\left[\frac{1}{2p\hbar\omega} + \frac{1}{(2p + 2)\hbar\omega}\right]$$

$$= -\frac{\hbar\Omega_1^2}{8\omega}\frac{2p + 1}{p(p + 1)}. \qquad (80)$$

The shift of the level $|\varphi_b\rangle$ is the opposite of (80), which, according to (16), leads to a shift of the resonance center:

$$\delta\omega_0 = -\frac{\Omega_1^2}{4\omega}\frac{2p + 1}{p(p + 1)}. \qquad (81)$$

(*) See, for example: C. Cohen-Tannoudji, J. Dupont-Roc, and C. Fabre, *J. Phys. B*, **6**, L214 (1973).

We now derive the nondiagonal element $R_{ab}(E_0)$ that produces the anticrossing. Because the states $|\varphi_a\rangle$ and $|\varphi_b\rangle$ differ by $2p + 1$ photons, the first nonzero term of the expansion of $R_{ab}(E_0)$ is of order $2p + 1$:

$$R_{ab}(E_0) = \frac{\langle -, N + 1|V_\sigma| +, N\rangle\langle +, N|V_\sigma| -, N - 1\rangle \cdots}{(E_0 - E_{+,N})(E_0 - E_{-,N-1}) \cdots}$$

$$\times \frac{\cdots \langle -, N - 2p + 1|V_\sigma| +, N - 2p\rangle}{\cdots (E_0 - E_{-,N-2p+1})} + \cdots \quad (82)$$

and equals, after all the calculations are done

$$R_{ab}(E_0) = \frac{(-)^P \hbar \Omega_1^{2p+1}}{\omega^{2p} 2^{4p+1} (p!)^2} + \cdots. \quad (83)$$

A resonant transfer between the two states $|\varphi_a\rangle$ and $|\varphi_b\rangle$ is associated with this anticrossing. When the resonance is not saturated, i.e., for $|R_{ab}| \ll [\omega_{ab}^2 + \Gamma^2]^{1/2}$, its intensity is proportional to R_{ab}^2, and thus to Ω_1^{4p+2}. Its shift relative to $\omega_0 = (2p + 1)\omega$ is given by (81). Note that the shift varying as Ω_1^2 increases more rapidly than the intensity varying as Ω_1^{4p+2}, so that when the resonances are significant, they are already considerably shifted.

We return now to the physical interpretation of the odd resonances. These are multiphoton transitions, as shown clearly by the comparison of the initial state $|\varphi_a\rangle$ and the final state $|\varphi_b\rangle$ of the system. More precisely, the numerator of (82) is a product of matrix elements of V_σ, which, from left to right, causes the number of photons in each intermediate state to decrease by one unit, the spin going alternatively to the state $|+\rangle$ or $|-\rangle$. The photons σ are linear superpositions of σ_+ and σ_- photons. As a result of the conservation of angular momentum along $0z$, the transition from the state $|-, N + 1\rangle$ to the state $|+, N - 2p\rangle$ is realized by the successive absorption of a σ_+ photon, then a σ_- photon, etc., for a total of $(p + 1)$ σ_+ photons and $p\sigma_-$ photons (see Fig. 25 in Chap. II). For the same reason, a $2p$-photon transition between the states $|-\rangle$ and $|+\rangle$ is forbidden for σ polarization, because the atomic angular momentum cannot have changed by one unit of $\hbar$ at the end of the process. By contrast, such a transition would be allowed if the polarization of the field were no longer perpendicular to $0z$, with the three types of photons, σ_+, σ_-, and π then being present.

d) THE EVEN SPECTRUM OF LEVEL-CROSSING RESONANCES

The levels $|+, N\rangle$ and $|-, N + 2p\rangle$ cross at $\omega_0 = 2p\omega$. In the presence of coupling, the perturbed levels continue to cross each other, but are shifted. The calculation of the diagonal matrix elements of $R(E_0)$ is analogous to those in the foregoing paragraph. The result is identical, except for the substitution of $2p + 1$ for $2p$, which gives for the shift of the crossing

$$\delta\omega_0 = -\frac{\Omega_1^2}{4\omega}\frac{2p}{\left(p - \frac{1}{2}\right)\left(p + \frac{1}{2}\right)}. \tag{84}$$

As in the case discussed in subsection 3-c, the level-crossing resonance is observable only if the perturbed states $|+, N\rangle$ and $|-, N + 2p\rangle$ contain states $|+, N'\rangle$ and $|-, N'\rangle$ with the same number of photons. The initial and final states $|\psi_i\rangle$ and $|\psi_f\rangle$ are indeed linear combinations of $|+, N'\rangle$ and $|-, N'\rangle$. The amplitudes $\langle\psi_i, N'|+, N\rangle$, $\langle\psi_f, N'|+, N\rangle$, and $\langle\psi_i, N'|-, N + 2p\rangle$, $\langle\psi_f, N'|-, N + 2p\rangle$, being, respectively, of order $(\Omega_1/\omega)^{|N-N'|}$ and $(\Omega_1/\omega)^{|N+2p-N'|}$, the crossing resonance is excited in lowest order for $N \leq N' \leq N + 2p$, and its intensity is on the order of

$$\left[\left(\frac{\Omega_1}{\omega}\right)^{N'-N} \times \left(\frac{\Omega_1}{\omega}\right)^{N+2p-N'}\right]^2 = \left(\frac{\Omega_1}{\omega}\right)^{4p}. \tag{85}$$

These level-crossing resonances are interesting in that they do not undergo any radiative broadening and that their radiative shift is measurable up to values of Ω_1 on the order of ω. The quadratic approximation (84) is then no longer sufficient, and the higher-order terms must then be taken into account (*).

e) A NONPERTURBATIVE CALCULATION: THE LANDÉ FACTOR
   OF THE DRESSED ATOM

Until now, we have adopted a perturbative approach. The Hamiltonian for the atom dressed by σ photons cannot actually be explicitly diagonalized, except at one point, $\omega_0 = 0$. It then equals

$$H(\omega_0 = 0) = \hbar\omega a^+ a + \lambda S_x(a + a^+). \tag{86}$$

(*) See, for example, C. Cohen-Tannoudji, J. Dupont-Roc, and C. Fabre, *J. Phys. B*, **6**, L218 (1973).

In this expression, S_x is the only atomic operator. H thus commutes with S_x and its diagonalization is reduced to that of its projection H_ε onto each of the two eigensubspaces of S_x,

$$H_\varepsilon = \hbar\omega a^+ a + \frac{\varepsilon}{2}\lambda(a + a^+)$$

$$= \hbar\omega\left(a^+ + \frac{\varepsilon\lambda}{2\hbar\omega}\right)\left(a + \frac{\varepsilon\lambda}{2\hbar\omega}\right) - \frac{\lambda^2}{4\hbar\omega} \qquad (87)$$

where $\varepsilon = \pm 1$. The application to H_ε of the translation operator $\exp[\varepsilon\lambda(a^+ - a)/2\hbar\omega]$ [see Appendix, formula (66)] gives $\hbar\omega a^+ a - (\lambda^2/4\hbar\omega)$, for which the diagonalization is immediate, its eigenvalues being

$$E_N = N\hbar\omega - \frac{\lambda^2}{4\hbar\omega}. \qquad (88)$$

Finally, the eigenstates of H corresponding to the eigenvalue E_N are

$$|\widetilde{\varepsilon_x; N}\rangle = \exp[-\varepsilon\lambda(a^+ - a)/2\hbar\omega]|\varepsilon_x\rangle|N\rangle \qquad (89)$$

(the states $|-_x\rangle$ and $|+_x\rangle$ are given, respectively, by (43) and (45) as a function of the eigenstates $|\pm\rangle$ of S_z).

The Zeeman effect for the dressed atom corresponds to the lifting of the degeneracy of the states $|\widetilde{\pm_x; N}\rangle$ by the Zeeman Hamiltonian

$$H_z = \hbar\omega_0 S_z. \qquad (90)$$

For weak magnetic fields ($\omega_0 \ll \omega$), first-order perturbation theory in H_z yields the term linear in ω_0, and thus the Landé factor of the dressed atom. By using the matrix elements of S_z in the basis $|\varepsilon_x\rangle$

$$\langle\varepsilon_x|S_z|\varepsilon_x\rangle = 0 \qquad \langle +_x|S_z|-_x\rangle = \tfrac{1}{2} \qquad (91)$$

we find that, in the subspace $|\widetilde{\pm_x; N}\rangle$, H_z is represented by a 2×2 matrix whose matrix elements are

$$\langle\widetilde{\varepsilon_x; N}|H_z|\widetilde{\varepsilon_x; N}\rangle = 0$$

$$\langle\widetilde{+_x; N}|H_z|\widetilde{-_x; N}\rangle = \frac{\hbar\omega_0}{2}\langle N|\exp\frac{\lambda}{\hbar\omega}(a^+ - a)|N\rangle. \qquad (92)$$

The factor $\langle N|\exp[\lambda(a^+ - a)/\hbar\omega]|N\rangle$ can be calculated explicitly (see the remark below). In the limit where $N \approx \bar{N} \gg 1$, it is expressed as a function of the zero-order Bessel function, J_0:

$$\langle N | \exp[\lambda(a^+ - a)/\hbar\omega] | N \rangle \approx J_0(2\Omega_1/\omega). \tag{93}$$

Thus, the Larmor frequency ω_0 of the free atom is, for the dressed atom, multiplied by $J_0(2\Omega_1/\omega)$. This function decreases from the value 1, cancels out and changes sign for $2\Omega_1/\omega = 2.4$, then continues to oscillate while slowly decreasing. The Landé factor of the dressed atom can thus be markedly different from that of the free atom.

Remark (*)

We now evaluate the matrix element (93) for $N \approx \bar{N} \gg 1$. By using the Glauber formula, we can put it in the form

$$\langle N | \exp\left[\frac{\lambda}{\hbar\omega}(a^+ - a)\right] | N \rangle =$$

$$\exp \frac{1}{2}\left(\frac{\lambda}{\hbar\omega}\right)^2 \langle N | \exp\left(\frac{\lambda a^+}{\hbar\omega}\right)\exp\left(\frac{-\lambda a}{\hbar\omega}\right) | N \rangle. \tag{94}$$

Because $\lambda/\hbar\omega \ll 1$, the first factor equals 1. The second factor involves the vector

$$\exp\left(\frac{-\lambda a}{\hbar\omega}\right) | N \rangle = \sum_{p=0}^{N} \frac{1}{p!}\left(\frac{-\lambda}{\hbar\omega}\right)^p a^p | N \rangle. \tag{95}$$

The values of p that contribute to (95) are on the order of

$$p \approx \frac{\lambda\sqrt{N}}{\hbar\omega} \approx \frac{\Omega_1}{\omega} \ll N. \tag{96}$$

In the calculation of $a^p | N \rangle$, we can thus take $N - p \approx N \approx \langle N \rangle$. This leads to the approximation

$$a^p | N \rangle \approx \sqrt{\langle N \rangle^p} | N - p \rangle. \tag{97}$$

We then have

$$\exp\left(\frac{-\lambda a}{\hbar\omega}\right) | N \rangle \approx \sum_{p=0}^{N} \frac{1}{p!}\left(\frac{-\lambda\sqrt{\langle N \rangle}}{\hbar\omega}\right)^p | N - p \rangle. \tag{98}$$

(*) See N. Polonsky and C. Cohen-Tannoudji, J. Phys. (Paris), **26**, 409 (1965).

Similarly,

$$\langle N|\exp\left(\frac{\lambda a^+}{\hbar\omega}\right) \simeq \sum_{p=0}^{N} \frac{1}{p!}\left(\frac{\lambda\sqrt{\langle N\rangle}}{\hbar\omega}\right)^{p}\langle N-p| \tag{99}$$

and

$$\langle N|\exp\left[\frac{\lambda}{\hbar\omega}(a^+ - a)\right]|N\rangle = \sum_{p=0}^{N} \frac{(-)^{p}}{(p!)^{2}}\left(\frac{\lambda\sqrt{\langle N\rangle}}{\hbar\omega}\right)^{2p}. \tag{100}$$

The sum appearing in (100) is simply the series expansion of the Bessel function J_0 so that

$$\langle N|\exp\left[\frac{\lambda}{\hbar\omega}(a^+ - a)\right]|N\rangle = J_0\left(\frac{2\lambda\sqrt{\langle N\rangle}}{\hbar\omega}\right) = J_0\left(\frac{2\Omega_1}{\omega}\right). \tag{101}$$

f) QUALITATIVE EVOLUTION OF THE ENERGY DIAGRAM AT HIGH INTENSITY

All the results that we discussed in subsection 4 clearly appear in the energy diagram for the dressed spin in Figure 5. In this figure, drawn for $\Omega_1 < \omega$, we can distinguish anticrossings near $\omega_0 = \omega$, $3\omega\ldots$, and cross-

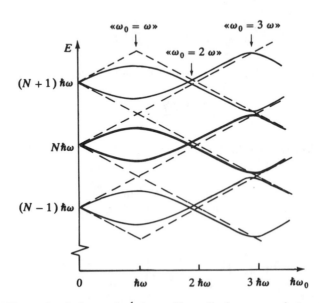

Figure 5. Energy levels for a spin $\frac{1}{2}$ dressed by radio-frequency photons having a σ polarization.

ings at $\omega_0 = 0, 2\omega, \ldots$. The radiative shifts are also apparent, as well as the decrease in the slope of the levels in zero field.

When Ω_1 increases, the anticrossings $\omega_0 = \pm\omega$ and the crossings $\omega_0 = \pm 2\omega$ arrive at $\omega_0 = 0$, at the same time that the slopes of the levels cancel out at the origin, i.e., for $2\Omega_1/\omega = 2.40$. If Ω_1 increases even more, the Landé factor changes sign, and then the three-quantum resonance and the crossing $\omega_0 = 4\omega$ each progressively come to and disappear at the origin for the second zero of the Bessel function ($2\Omega_1/\omega = 5.52$), and so on ... . The study of the energy diagram for the dressed spin thus provides a global view of the phenomena that allows us to understand their evolution when the intensity of the radio frequency increases and when a perturbative treatment is no longer valid.

REFERENCES

C. Cohen-Tannoudji, in *Optical Pumping and Interactions of Atoms with the Electromagnetic Field* Vol. 2 of *Cargese Lectures in Physics*, edited by M. Levy (Gordon and Breach, New York, 1968), p. 347.

S. Haroche, Doctoral thesis, Paris, 1971, published in *Ann. Phys. (Paris)*, **6**, 189 and 327 (1971). In this reference the reader can find a description of the experimental results and a generalization of the treatment presented in this complement to other situations (other polarizations of the radio-frequency field, spins $J > \frac{1}{2}$).

An interpretation of level-crossing resonances ($\omega_0 = 2\omega, 4\omega, \ldots$) in terms of interferences between multiphoton transition amplitudes involving both optical photons and radio-frequency photons can be found in C. Cohen-Tannoudji and S. Haroche, *J. Phys. (Paris)*, **30**, 125 (1969).

COMPLEMENT B$_{VI}$

COLLISIONAL PROCESSES IN THE PRESENCE OF LASER IRRADIATION

Several differences exist between the emission spectra of an isolated atom and those of an atom undergoing collisions with other atoms. Consider, for example, the fluorescence spectrum of a two-level atom. In the absence of collisions, such a spectrum is composed of a symmetrical triplet in which the two sidebands have the same intensity (see §E-1). In addition, in the limit of large detunings from resonance, the intensity of the central line of the triplet is proportional to Ω_1^2 (where Ω_1 is the Rabi frequency), whereas the intensity of the sidebands varies as Ω_1^4. In this complement we will show that, in the presence of collisions, the fluorescence triplet becomes asymmetrical. For example, in the large detuning limit, the sideband whose frequency is close to the atomic frequency ω_0 changes its behavior, with its intensity also becoming proportional to Ω_1^2. The emission spectrum then contains two dominant lines, one at the laser frequency ω_L and the other at the atomic frequency ω_0 (Figure 1). This line at the frequency ω_0 results from the excitation of the level b in a collisionally aided process, with the collision providing the energy defect $\hbar(\omega_0 - \omega_L)$ between the energy of level b (solid line in Figure 1) and the energy reached as a result of the absorption of a photon from level a (dashed line in Figure 1). We know that this intermediate step of the

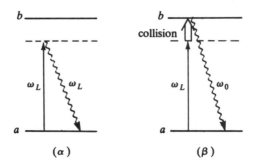

Figure 1. Dominant lines observed in the emission spectrum of an atom subjected to collisions. In addition to the Rayleigh scattering at the frequency ω_L (α), an emission at the frequency ω_0 due to the excitation of the level b in a collisionally aided process (β) is also observed.

excitation process can actually be interpreted as an energy level of the atom + photons global system (see the discussion concerning Figure 13 in subsection C-1 of Chapter II). The dressed atom approach, which will be used in this complement, is thus particularly appropriate for describing collisional processes in the presence of radiation (*).

After having introduced the parameters describing the collisional relaxation of an atom in the absence of radiation (§1), we describe the effect of collisions in the dressed state basis (§2). We then study the modifications of absorption and emission spectra of the atom resulting from collisions (§3). Finally, we discuss how the transfer rates between dressed states can be calculated in two different limits corresponding to detunings which are either small or large compared with the inverse of the duration τ_{coll} of a collision (§4).

1. Collisional Relaxation in the Absence of Laser Irradiation

a) SIMPLIFYING ASSUMPTIONS

Consider an atom A with two levels a and b in a gas composed of atoms X. Atom A undergoes random collisions with these atoms. The duration τ_{coll} of each collision is assumed to be small compared with the time interval T_{coll} separating two successive collisions: The collisions are thus binary and well separated in time. The average kinetic energy of the atoms $(3k_B T/2)$ is assumed to be much smaller than $\hbar\omega_0 = E_b - E_a$, so that the excitation of the level b from the level a during a collision is energetically impossible. The energy of the first excited level of the atom X is also assumed to be very large compared with $\hbar\omega_0$. Consequently, during a collision, it is not possible to transfer the energy from the excited atom A to the atom X. More generally, we assume that there is no nonradiative deexcitation ("quenching") of b to a during a collision.

To summarize, the only effect of the collisions considered here is to modify the energies of the levels of A during the collision. We call $E_a(r)$ and $E_b(r)$ the energies of the levels a and b of the atom A in the presence of the X atom at a distance r (see Figure 2) (**). During the collision, the Bohr frequency associated with the transition $a \leftrightarrow b$ is modified so that the essential effect of the collision is to induce a phase

---

(*) Such a point of view is used by S. Reynaud and C. Cohen-Tannoudji, *J. Phys.* (*Paris*), **43**, 1021 (1982).

(**) By contrast, we assume that the wave functions of the levels a and b are only slightly modified by the presence of the X atom. In particular, the electric dipole matrix element of the atom A between the states a and b is assumed to be the same as for an isolated atom.

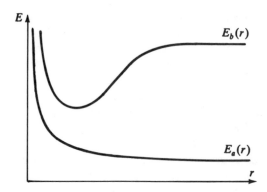

Figure 2. Energy levels of the atom A in the presence of an atom X at a distance r.

shift of the oscillation of the atomic dipole. For this reason, this type of collision is known as a "dephasing collision".

For the sake of simplicity, we assume that the atom A is at rest and that the X atoms follow classical linear trajectories (*).

b) MASTER EQUATION DESCRIBING THE EFFECT OF COLLISIONS ON THE EMITTING ATOM

First we discuss the effect of a single collision. According to the assumptions made in the preceding subsection, the diagonal elements σ_{aa} and σ_{bb} of the density matrix σ_A of the atom A do not change. On the other hand, the nondiagonal element is modified by a quantity which, in the interaction representation, is written

$$\Delta\tilde{\sigma}_{ba} = -(1 - e^{-i\phi})\tilde{\sigma}_{ba} \tag{1}$$

where ϕ is the phase shift accumulated during the collision. It is equal to

$$\phi = \int_{-\infty}^{+\infty} dt[\omega_{ba}(t) - \omega_0] \tag{2}$$

(*) The effects associated with the changes in velocity of the atom resulting from collision are discussed, for example, by P. R. Berman, in *Les Houches XXXVIII "New Trends in Atomic Physics"*, edited by G. Grynberg and R. Stora, North-Holland, Amsterdam, 1984, p. 451.

where $\omega_{ba}(t)$ represents the instantaneous oscillation frequency of the dipole:

$$\omega_{ba}(t) = \frac{E_b(r(t)) - E_a(r(t))}{\hbar}. \tag{3}$$

Consider a time interval Δt which is long compared with the duration τ_{coll} of a collision and short compared with the time interval T_{coll} separating two collisions ($\tau_{coll} \ll \Delta t \ll T_{coll}$). Because $\Delta t \ll T_{coll}$, the probability that an atom A has undergone a collision during Δt is very low, so that the modification $\Delta\tilde{\sigma}_{ba}$ of $\tilde{\sigma}_{ba}$ during Δt is small. By contrast, because $\Delta t \gg \tau_{coll}$, any collision that occurred during the interval Δt has had enough time to take place from start to finish. To find $\Delta\tilde{\sigma}_{ba}$, we must sum the relation (1) over all the possible collisions. During Δt, the number of collisions the atom A undergoes with an atom X whose velocity is included in the volume element d^3v about $\mathbf{v}$ and whose trajectory has an impact parameter between b and $b + db$, is

$$\Delta N(b, \mathbf{v}) = N2\pi b\, db|\mathbf{v}|f(\mathbf{v})\, d^3v\, \Delta t \tag{4}$$

where N is the number of atoms X per unit volume and $f(\mathbf{v})$ is their velocity distribution. The variation $\Delta\tilde{\sigma}_{ba}/\Delta t$ can then be written

$$\frac{\Delta\tilde{\sigma}_{ba}}{\Delta t} = -\langle 1 - e^{-i\phi}\rangle_{coll}\tilde{\sigma}_{ba} \tag{5}$$

with

$$\langle 1 - e^{-i\phi}\rangle_{coll} = N\int_0^\infty 2\pi b\, db \int |\mathbf{v}|f(\mathbf{v})[1 - e^{-i\phi(b, \mathbf{v})}]\, d^3v \tag{6}$$

where $\phi(b, \mathbf{v})$ is the phase shift for a collision associated with a trajectory having an impact parameter b and relative velocity $\mathbf{v}$. We call γ and η the real and imaginary parts of $\langle 1 - e^{-i\phi}\rangle_{coll}$:

$$\gamma = \langle 1 - \cos\phi\rangle_{coll} \tag{7.a}$$

$$\eta = \langle \sin\phi\rangle_{coll}. \tag{7.b}$$

For times long compared with τ_{coll}, the evolution of $\tilde{\sigma}_{ba}$ can be described by the coarse-grained average (5), so that

$$\frac{d\tilde{\sigma}_{ba}}{dt} \simeq \frac{\Delta\tilde{\sigma}_{ba}}{\Delta t} = -(\gamma + i\eta)\tilde{\sigma}_{ba}. \tag{8}$$

By contrast, as we have mentioned above, the dephasing collisions do not introduce any transfer from one atomic level to another, which leads to the following equations for the populations

$$\frac{d\tilde{\sigma}_{aa}}{dt} = \frac{d\tilde{\sigma}_{bb}}{dt} = 0. \tag{9}$$

Finally, Equations (8) and (9) can be combined into a single operator equation

$$\frac{d}{dt}\tilde{\sigma}_A = -\frac{\gamma}{2}\tilde{\sigma}_A + 2\gamma S_z\tilde{\sigma}_A S_z - i\eta(S_z\tilde{\sigma}_A - \tilde{\sigma}_A S_z) \tag{10}$$

where S_z is the operator defined by $S_z|a\rangle = -(\frac{1}{2})|a\rangle$, and $S_z|b\rangle = (\frac{1}{2})|b\rangle$ [see Equation (A.15.c) in Chapter V].

Remark

Every two-level system can be considered as a fictitious $\frac{1}{2}$ spin (see §A-4 in Chapter V). Because a dephasing collision modifies the energy difference between the two levels, its effect on the fictitious spin is equivalent to the effect of a magnetic field $\mathbf{b}(t)$ parallel to $0z$ applied to the spin during the collision. The angle ϕ then represents the angle of rotation of the spin about $0z$ under the influence of $\mathbf{b}(t)$. Note that we have made no assumptions about ϕ and that this angle can correspond to a rotation greater than 2π.

2. Collisional Relaxation in the Presence of Laser Irradiation

a) THE DRESSED ATOM APPROACH

Collisions also appear as relaxation processes for the dressed atom, and it is possible to describe their effect by using a master equation. Despite the large number of levels of the dressed atom, the number of parameters necessary to describe the effect of dephasing collisions is relatively small. This is a result of, on the one hand, our choice of a simple collisional model (the model described in subsection 1-a), and, on the other hand, the fact that it is usually justified to use the secular approximation that allows the evolution of the populations to be decoupled from the evolution of the coherences (and also the evolution of the coherences having different eigenfrequencies to be decoupled from each other).

b) EVOLUTION OF POPULATIONS: COLLISIONAL TRANSFERS
   BETWEEN DRESSED STATES

First note that the dephasing collisions cannot induce population transfer from one manifold to another: $\Gamma^{coll}_{i(N) \to j(N')}$ is zero if $N \neq N'$. Indeed, we assumed in subsection 1-a that the energy likely to be transferred during a collision is on the order of $k_B T$, a quantity that is small compared with the optical energy. By contrast, $k_B T$ is, in general, large compared with the splitting $\hbar\Omega$ between the levels $|1(N)\rangle$ and $|2(N)\rangle$ of the same manifold, and the collisions can induce transfers from $|1(N)\rangle$ to $|2(N)\rangle$ with a rate $\Gamma^{coll}_{1(N)\rightarrow 2(N)}$ that we write as w (for N near $\langle N\rangle$, w can be considered as being independent of N). Note that the relation $\hbar\Omega \ll k_B T$ leads to [see Equation (C.10) in Chapter IV]

$$\Gamma^{coll}_{2(N)\rightarrow 1(N)} = \Gamma^{coll}_{1(N)\rightarrow 2(N)} = w. \tag{11}$$

Furthermore, we assume that Ω is large compared with γ, which allows the secular approximation to be applied and the couplings between populations and coherences to be neglected. The evolution equation for populations $\pi_{i(N)}$ of the dressed atom under the influence of collisions is thus written

$$\frac{d}{dt}\pi_{1(N)} = -w\big(\pi_{1(N)} - \pi_{2(N)}\big) \tag{12.a}$$

$$\frac{d}{dt}\pi_{2(N)} = -\frac{d}{dt}\pi_{1(N)}. \tag{12.b}$$

The dephasing collisions thus tend to equalize the populations between sublevels of the same manifold of the dressed atom (Figure 3).

Remark

Equations (12) indicate that, in the dressed state basis, collisional relaxation affects not only the coherences, as is the case for the bare atom [see Equations (8) and (9)], but also the populations. Using the language of nuclear magnetic resonance, we can say that the relaxation, which is only of type T_2 for the bare atom, becomes partially of type T_1 for the dressed atom. It is easy to understand such a result by using the notion of fictitious spin. For the bare atom, the

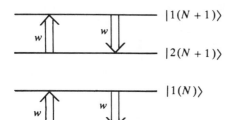

Figure 3. Transfers between levels of the dressed atom induced by dephasing collisions.

random field $\mathbf{b}(t)$ representing the effect of the dephasing collision is aligned along $0z$, as is the field $\mathbf{B}_0$ associated with the energy separation $\hbar\omega_0$ between the levels b and a. By contrast, in the presence of a laser wave, $\mathbf{b}(t)$ is no longer aligned with the total field $\mathbf{B}_e$ in the rotating reference frame. Recall that the components of $\mathbf{B}_e$ along $0Z$ and $0X$ are, respectively, proportional to $(\omega_0 - \omega_L)$ and Ω_1 (see Figure 1 in Chapter V). Thus, in the rotating reference frame, the field $\mathbf{b}(t)$ has a component orthogonal to $\mathbf{B}_e$ which can induce transfers between the eigenstates $|+\rangle$ and $|-\rangle$ of the component of the spin on $\mathbf{B}_e$. Equations (12) do express quantitatively the existence of such transfers between dressed states.

c) Evolution of Coherences. Collisional Damping and Collisional Shift

We now consider the coherences $\langle 1(N)|\bar{\sigma}|2(N)\rangle$ and $\langle 2(N)|\bar{\sigma}|1(N)\rangle$ between two sublevels of the same manifold $\mathscr{E}(N)$. The relaxation of these coherences can be described by using two parameters κ and ξ which describe, respectively, the damping of the coherence and the shift of its frequency of evolution:

$$\frac{d}{dt}\langle 1(N)|\bar{\sigma}|2(N)\rangle = -(\kappa + i\xi)\langle 1(N)|\bar{\sigma}|2(N)\rangle \qquad (13.a)$$

$$\frac{d}{dt}\langle 2(N)|\bar{\sigma}|1(N)\rangle = -(\kappa - i\xi)\langle 2(N)|\bar{\sigma}|1(N)\rangle. \qquad (13.b)$$

By contrast with the situation encountered in §D-2-c for spontaneous emission, here there is no transfer of coherence between two manifolds. This is a consequence of the collision model described here, which ignores any collisional transfer phenomenon between two levels separated by optical energies.

Remark

The effect of collisions on the evolution of the coherence $\langle 1(N)|\bar{\sigma}|2(N')\rangle$ with $N \neq N'$ is given by an equation similar to (13.a)

$$\frac{d}{dt}\langle 1(N)|\bar{\sigma}|2(N')\rangle = -(\kappa + i\xi)\langle 1(N)|\bar{\sigma}|2(N')\rangle \qquad (14)$$

This point will be justified in subsection 4.

d) Explicit Form of the Master Equation in the Impact Limit

A particularly important experimental situation occurs when $\Omega \tau_{coll} \ll 1$, with Ω being the Rabi frequency (B.19). This condition, which implies that the relations $|\omega_L - \omega_0|\tau_{coll} \ll 1$ and $\Omega_1 \tau_{coll} \ll 1$ are simultaneously verified, corresponds to the "impact limit". In this case, the Rabi oscillation at the frequency Ω does not have time to occur over the duration τ_{coll} of the collision. We then have a situation similar to the one encountered for the case of spontaneous emission in subsection D-1-a of the chapter. Here again, it is therefore possible to use the approximation of independent rates of variation to find the explicit form of the master equation for the dressed atom. This approximation consists of writing the rate of variation of σ_{AL} as the sum of the rate of variation (10) associated with the collisions (the quantum numbers associated with the laser photons remain "spectators") and of the rate of variation associated with the Hamiltonian H_{AL} for the dressed atom

$$\frac{d}{dt}\sigma_{AL} = -\frac{i}{\hbar}[H_{AL}, \sigma_{AL}] - \frac{\gamma}{2}\sigma_{AL}$$

$$+ 2\gamma S_z \sigma_{AL} S_z - i\eta(S_z \sigma_{AL} - \sigma_{AL} S_z). \qquad (15)$$

We will now project Equation (15) over the basis $|i(N)\rangle$ of the dressed states and show that it is possible in the impact limit to express the three parameters w, κ, and ξ describing the collisional relaxation of the dressed atom as a function of the two parameters γ and η introduced for the bare atom. To do this, we use the following matrix elements deduced from (B.20):

$$\langle 1(N)|S_z|1(N')\rangle = \frac{1}{2}(\cos^2\theta - \sin^2\theta)\delta_{N,N'} = \delta_{N,N'}\frac{\cos 2\theta}{2} \qquad (16.a)$$

$$\langle 2(N)|S_z|2(N')\rangle = \frac{1}{2}(\sin^2\theta - \cos^2\theta)\delta_{N,N'} = -\delta_{N,N'}\frac{\cos 2\theta}{2} \qquad (16.b)$$

$$\langle 1(N)|S_z|2(N')\rangle = -\sin\theta\cos\theta\,\delta_{N,N'} = -\delta_{N,N'}\frac{\sin 2\theta}{2}. \qquad (16.c)$$

We first consider (*) $\langle 1(N)|\sigma|1(N')\rangle$. By using the secular approximation,

(*) We omit the index AL of σ_{AL} when there is no ambiguity.

we deduce from (15)

$$\frac{d}{dt}\langle 1(N)|\sigma|1(N')\rangle =$$

$$= -i(N - N')\omega_L\langle 1(N)|\sigma|1(N')\rangle - \frac{\gamma}{2}\langle 1(N)|\sigma|1(N')\rangle +$$

$$+ \gamma\frac{\cos^2 2\theta}{2}\langle 1(N)|\sigma|1(N')\rangle + \gamma\frac{\sin^2 2\theta}{2}\langle 2(N)|\sigma|2(N')\rangle$$

$$= -i(N - N')\omega_L\langle 1(N)|\sigma|1(N')\rangle -$$

$$- \frac{\gamma}{2}\sin^2 2\theta[\langle 1(N)|\sigma|1(N')\rangle - \langle 2(N)|\sigma|2(N')\rangle]. \tag{17}$$

For $N = N'$, we then find, by comparing (17) and (12.a), that the value of w for the impact limit is simply

$$w = \frac{\gamma}{2}\sin^2 2\theta. \tag{18}$$

By using (15) again, as well as the secular approximation, we obtain for the evolution equation of $\langle 1(N)|\sigma|2(N')\rangle$

$$\frac{d}{dt}\langle 1(N)|\sigma|2(N')\rangle = -i[\Omega + (N - N')\omega_L]\langle 1(N)|\sigma|2(N')\rangle -$$

$$- [\gamma(\cos^4 \theta + \sin^4 \theta) + i\eta \cos 2\theta]\langle 1(N)|\sigma|2(N')\rangle. \tag{19}$$

Comparing with (13.a) then gives

$$\kappa = \gamma(\cos^4 \theta + \sin^4 \theta) \tag{20}$$

$$\xi = \eta \cos 2\theta. \tag{21}$$

Remarks

(i) It is possible to find the coefficient w in the impact limit by using a more physical approach. To do this, we assume that the system is at the initial time $t_i = -T$ in the state $|1(N)\rangle$ and study the effect of a collision with an atom X. Because $\Omega\tau_{coll} \ll 1$, we can ignore the atom-laser coupling during the collision. The effect of the collision is simply to phase shift the coefficients of the expansion of the state $|1(N)\rangle$ over the states $|a, N + 1\rangle$ and $|b, N\rangle$. Starting at the initial time from the state

$$|\psi(-T)\rangle = |1(N)\rangle = \sin\theta|a, N+1\rangle + \cos\theta|b, N\rangle \tag{22}$$

the system is found at time T in the state

$$|\psi(T)\rangle =$$

$$= \sin\theta|a, N+1\rangle\exp\left\{-\frac{i}{\hbar}\int_{-T}^{+T}[E_a(r(t)) + (N+1)\hbar\omega_L]\,dt\right\} +$$

$$+ \cos\theta|b, N\rangle\exp\left\{-\frac{i}{\hbar}\int_{-T}^{+T}[E_b(r(t)) + N\hbar\omega_L]\,dt\right\}. \tag{23}$$

By assuming that the collision occurs within the interval $(-T, T)$, we can, by using (2), rewrite $|\psi(T)\rangle$ (except for a global phase factor) in the form:

$$|\psi(T)\rangle = \sin\theta|a, N+1\rangle + \cos\theta\, e^{-i\phi}\, e^{-2i(\omega_0 - \omega_L)T}|b, N\rangle. \tag{24}$$

In the impact limit, we can take an interval T which is large compared with τ_{coll} while still having $|\delta_L|T \ll 1$ so that

$$|\psi(T)\rangle \simeq \sin\theta|a, N+1\rangle + \cos\theta\, e^{-i\phi}|b, N\rangle. \tag{25}$$

The state $|\psi(T)\rangle$ at the end of the collision then differs from $|1(N)\rangle$ because of the phase shift $e^{-i\phi}$ of the component along $|b, N\rangle$. Using (B.20) then gives for the transition probability

$$|\langle 2(N)|\psi(T)\rangle|^2 = 2\sin^2\theta\cos^2\theta(1 - \cos\phi). \tag{26}$$

The transition rate from $|1(N)\rangle$ to $|2(N)\rangle$ is obtained by summing (26) over all the possible collisions. It is then sufficient to use (7.a) to find again expression (18) giving w.

(ii) An argument analogous to that used in the preceding remark also allows us to understand certain collisional effects in optics and nonlinear spectroscopy. Consider, for example, a three-level atom $\{a, b, b'\}$ identical to the one in Figure 14 of this chapter. We assume that the two incident beams having frequencies ω_L and ω'_L and exciting the transitions $a \leftrightarrow b$ and $a \leftrightarrow b'$ are nonresonant and that $\Omega_1/|\delta_L|$ and $\Omega'_1/|\delta'_L|$ are small compared with 1 (with $\delta_L = \omega_L - \omega_0$ and $\delta'_L = \omega'_L - \omega'_0$). The atom being studied also undergoes collisions with the atoms X and the conditions defining the impact limit $(|\delta_L|, |\delta'_L| \ll \tau_{\text{coll}}^{-1})$ are assumed to be fulfilled. The dressed states in the manifold $\mathcal{E}(N, N')$ introduced in (E.40) are:

$$|1(N, N')\rangle \simeq -\frac{\Omega_1}{2\delta_L} |a, N + 1, N' + 1\rangle + |b, N, N' + 1\rangle \qquad (27.a)$$

$$|2(N, N')\rangle \simeq -\frac{\Omega'_1}{2\delta'_L} |a, N + 1, N' + 1\rangle + |b', N + 1, N'\rangle \qquad (27.b)$$

$$|3(N, N')\rangle \simeq |a, N + 1, N' + 1\rangle +$$
$$+ \frac{\Omega_1}{2\delta_L} |b, N, N' + 1\rangle + \frac{\Omega'_1}{2\delta'_L} |b', N + 1, N'\rangle. \qquad (27.c)$$

If the system is initially in the state $|3(N, N')\rangle$ and then undergoes a collision with an atom X, a procedure analogous to the one made in the previous remark allows us to show that its state after the collision is

$$|\psi(T)\rangle = |a, N + 1, N' + 1\rangle + \frac{\Omega_1}{2\delta_L} e^{-i\phi}|b, N, N' + 1\rangle +$$

$$+ \frac{\Omega'_1}{2\delta'_L} e^{-i\phi'}|b', N + 1, N'\rangle \qquad (28)$$

where $\phi' = \int_{-\infty}^{+\infty} dt(\omega_{b'a}(t) - \omega'_0)$ is the phase shift of the coherence between the states a and b' induced by the collision. Because the state $|\psi(T)\rangle$ differs from the state $|3(N, N')\rangle$, we find transfers to the levels $|2(N, N')\rangle$ and $|1(N, N')\rangle$ analogous to those discussed in the preceding remark. Moreover, a coherence between the states $|1(N, N')\rangle$ and $|2(N, N')\rangle$ is created by the collision

$$\langle 1(N, N')|\psi(T)\rangle\langle\psi(T)|2(N, N')\rangle =$$

$$= \frac{\Omega_1\Omega'_1}{4\delta_L\delta'_L}(1 - e^{-i\phi})(1 - e^{i\phi'})$$

$$= \frac{\Omega_1\Omega'_1}{4\delta_L\delta'_L}\left[(1 - e^{-i\phi}) + (1 - e^{i\phi'}) - (1 - e^{-i(\phi-\phi')})\right]. \qquad (29)$$

The average over the collisions leads to an average excitation rate of the coherence equal to

$$w^{coh} = \frac{\Omega_1\Omega'_1}{4\delta_L\delta'_L}(\gamma_{ba} + \gamma^*_{b'a} - \gamma_{bb'}) \qquad (30)$$

where $\gamma_{ba} = \gamma + i\eta$, $\gamma_{b'a}$ and $\gamma_{bb'}$ being similarly associated with the relaxation of the coherences $\sigma_{b'a}$ and $\sigma_{bb'}$ of the bare atom. The right-hand side of (30) is generally nonzero, which shows that the collisions are able to create coherences between dressed states. We also know (see Complement A_{VI}, §1-c) that a coherence excited in this way goes through a resonant value when its characteristic evolution frequency is zero. The evolution frequency of the coherence between $|2(N, N')\rangle$ and $|1(N, N')\rangle$ being $(\delta_L - \delta'_L)$, the process induced by collision that we consider here must be resonant when $\delta_L = \delta'_L$, i.e., when $E_b - E_{b'} = \hbar(\omega_L - \omega'_L)$. The dressed atom approach thus allows us to understand how collisions can cause the appearance of resonances associated with Bohr frequencies between excited levels (*). Later on [see Remark (ii) in subsection 3-d] we will present an example of a physical situation where such resonances can appear.

3. Collision-Induced Modifications of the Emission and Absorption of Light by the Atom. Collisional Redistribution

We now discuss how the resonance fluorescence of an atom excited by a laser wave is modified when this atom undergoes collisions.

a) TAKING INTO ACCOUNT SPONTANEOUS EMISSION

In the dressed atom approach, it is necessary to take into account two relaxation processes, the "radiative" relaxation resulting from spontaneous emission processes and the "collisional" relaxation associated with collisions. The first process, discussed in this chapter (§C-1 and Section D), is characterized by a correlation time τ_{c1}, at most of the order of an optical period $1/\omega_0$, and a relaxation time T_{R1} equal to Γ^{-1}. For the second process, discussed in the preceding subsection, the correlation time τ_{c2} is on the order of the collision time τ_{coll} and the relaxation time T_{R2} is on the order of the time between collisions T_{coll}.

We are able to separately describe each of these relaxation processes by a master equation as a result of the existence of two distinct time scales for each process, which is manifested by the two conditions

$$\tau_{c1} \ll T_{R1} \quad \text{or} \quad \frac{1}{\omega_0} \ll \Gamma^{-1} \tag{31.a}$$

$$\tau_{c2} \ll T_{R2} \quad \text{or} \quad \tau_{coll} \ll T_{coll}. \tag{31.b}$$

When the two relaxation processes act together, as we assume here, it is necessary, in order to be able to calculate a coarse-grained relaxation rate

---

(*) For more details, see G. Grynberg, *J. Phys. B*, **14**, 2089 (1981). The first observation of such a resonance was obtained in a four-wave mixing experiment by Y. Prior, A. R. Bogdan, M. Dagenais, and N. Bloembergen, *Phys. Rev. Lett.* **46**, 111 (1981).

for σ_{AL}, to introduce a time interval Δt that is both long compared to τ_{c1} and τ_{c2} and short compared to T_{R1} and T_{R2}, which implies, in addition to the conditions (31), that

$$\tau_{c1} \ll T_{R2} \quad \text{or} \quad \frac{1}{\omega_0} \ll T_{\text{coll}} \tag{32.a}$$

$$\tau_{c2} \ll T_{R1} \quad \text{or} \quad \tau_{\text{coll}} \ll \frac{1}{\Gamma}. \tag{32.b}$$

We assume in what follows that conditions (32) are also satisfied.

Finally, note that, because τ_{c1} is in general very small compared to τ_{c2}, a spontaneous emission process can take place from start to finish during a collision. The probability that such an event will occur is on the order of $\tau_{\text{coll}}/T_{\text{coll}}$, which is very small compared to 1, according to (31.b) (*). This condition then allows us to neglect any modification of one relaxation process by the other. Finally, to obtain the master equation for the dressed atom, it is sufficient to independently add the radiative and collisional relaxation terms, calculated as if each relaxation process acted alone. We then obtain

$$\frac{d}{dt}\sigma_{AL} = -\frac{i}{\hbar}[H_{AL}, \sigma_{AL}] + \left\{\frac{d}{dt}\sigma_{AL}\right\}_{\text{rad}} + \left\{\frac{d}{dt}\sigma_{AL}\right\}_{\text{coll}} \tag{33}$$

where $\{d\sigma_{AL}/dt\}_{\text{rad}}$ is given by Equation (D.3) in this chapter and where $\{d\sigma_{AL}/dt\}_{\text{coll}}$ is given by Equations (12) and (13) of this complement.

Remarks

(i) Conditions (31) and (32), which allow us to justify the structure of Equation (33), do not involve the generalized Rabi frequency $\Omega = [\Omega_1^2 + \delta_L^2]^{1/2}$. The comparison between Ω and τ_{ci} or between Ω and T_{Ri} ($i = 1, 2$) appears at a later stage. For example, when $\Omega \tau_{c2} \approx \Omega \tau_{\text{coll}} \ll 1$ (impact limit), the terms $\{d\sigma_{AL}/dt\}_{\text{coll}}$ have the same form as they do for the bare atom, the laser photons remaining spectators, because we can neglect the atom-laser interaction over the duration of the collision. Moreover, the conditions $\Omega T_{R1} \gg 1$ and $\Omega T_{R2} \gg 1$ are the basis of the secular approximation.

(*) This assumes, of course, that the collision does not allow a spontaneous transition that would otherwise be quasi-forbidden (for example, if a and b have the same parity). In fact, such collisions cannot be described within the framework of the model chosen here because we have assumed that the matrix element of the dipole between a and b does not vary significantly during the collision (see §1-a).

(ii) Equation (33) and the quantum regression theorem do not allow one to correctly calculate the frequency of the photons emitted spontaneously during a collision, because the atomic frequency is then strongly perturbed (see Figure 2). The error made is, however, negligible, because it concerns only a small fraction τ_{coll}/T_{coll} of the emitted photons.

b) REDUCED STEADY-STATE POPULATIONS

Starting with Equation (33), it is possible to determine the values of the steady-state populations and thus the intensity of the three lines of the fluorescence spectrum having frequencies ω_L, $\omega_L - \Omega$, and $\omega_L + \Omega$.

Thus, a procedure similar to the one used in the chapter allows the evolution equations for reduced populations to be deduced from (33). More precisely, taking into account the collisional term (12) leads us to generalize equations (D.20) in the following form:

$$\dot{\pi}_1 = -\pi_1\Gamma_{1\to 2} + \pi_2\Gamma_{2\to 1} - w(\pi_1 - \pi_2) \qquad (34.a)$$

$$\dot{\pi}_2 = -\pi_2\Gamma_{2\to 1} + \pi_1\Gamma_{1\to 2} - w(\pi_2 - \pi_1). \qquad (34.b)$$

The steady-state solution of these equations is

$$\pi_1^{st} = \frac{\Gamma_{2\to 1} + w}{\Gamma_{1\to 2} + \Gamma_{2\to 1} + 2w} \qquad (35.a)$$

$$\pi_2^{st} = \frac{\Gamma_{1\to 2} + w}{\Gamma_{1\to 2} + \Gamma_{2\to 1} + 2w} \qquad (35.b)$$

or also, using (D.11)

$$\pi_1^{st} = \frac{\Gamma \sin^4\theta + w}{\Gamma(\cos^4\theta + \sin^4\theta) + 2w} \qquad (36.a)$$

$$\pi_2^{st} = \frac{\Gamma \cos^4\theta + w}{\Gamma(\cos^4\theta + \sin^4\theta) + 2w}. \qquad (36.b)$$

It is easy to verify in expressions (35) or (36) that the collisions tend to decrease the difference $|\pi_1^{st} - \pi_2^{st}|$ between the steady-state populations in agreement with the picture in Figure 3.

c) Intensity of the Three Components of the Fluorescence Triplet

The two sidebands centered at $(\omega_L - \Omega)$ and $(\omega_L + \Omega)$ have equal weights in the absence of collisions. Such a result is no longer true in the presence of collisions. The weight of the line centered at $(\omega_L + \Omega)$, equal to $\pi_1^{st}\Gamma_{1\rightarrow 2}$ according to (E.17), is different from the weight of the line at $(\omega_L - \Omega)$ which, according to (E.18), is equal to $\pi_2^{st}\Gamma_{2\rightarrow 1}$. By using (D.11) and (36), we indeed find

$$I(\omega_L + \Omega) = \Gamma \cos^4 \theta \frac{\Gamma \sin^4 \theta + w}{\Gamma(\cos^4 \theta + \sin^4 \theta) + 2w} \qquad (37.a)$$

$$I(\omega_L - \Omega) = \Gamma \sin^4 \theta \frac{\Gamma \cos^4 \theta + w}{\Gamma(\cos^4 \theta + \sin^4 \theta) + 2w} \qquad (37.b)$$

which do not coincide when $w \neq 0$ (except for resonant excitation where $\sin^2 \theta = \cos^2 \theta = \frac{1}{2}$).

The total weight of the central line, which is deduced from (E.23) and (D.11), is unchanged:

$$I(\omega_L) = \Gamma \cos^2 \theta \sin^2 \theta \qquad (37.c)$$

but the ratio between coherent and incoherent contributions is modified. Indeed, the weight of the coherent line which, according to (E.25), is proportional to $\Gamma_{1\rightarrow 1}(\pi_1^{st} - \pi_2^{st})^2$ decreases in the presence of collisions.

In the steady-state regime, the number of absorbed photons is equal to the number of photons emitted spontaneously. As a result, the total absorption A varies as the sum of the three weights (37.a), (37.b), and (37.c) which can be put in the form

$$A = \frac{\Gamma}{2}\left[1 - \frac{\Gamma \cos^2 2\theta}{\Gamma(\cos^4 \theta + \sin^4 \theta) + 2w}\right]. \qquad (38)$$

It is then clear that A is an increasing function of w, i.e., the absorption increases in the presence of collisions.

d) Physical Discussion in the Limit $\Omega_1 \ll |\delta_L| \ll \tau_{coll}^{-1}$

In this subsection, we consider the case of nonresonant excitation in the perturbative limit $(\Omega_1/|\delta_L| \ll 1)$ and in the impact limit $(|\delta_L| \ll \tau_{coll}^{-1})$. Under these conditions, and by also assuming $\delta_L < 0$ (*), we have,

---

(*) In the case where $\delta_L > 0$, we have $\theta \simeq (\pi/2) - (\Omega_1/2\delta_L)$ instead of (39). Starting with this result, it is easy to show that formulas (40)–(43) demonstrated in this subsection remain valid when $\delta_L > 0$.

according to (B.21):

$$\theta \simeq -\frac{\Omega_1}{2\delta_L}. \tag{39}$$

As a result of $\cos^2\theta \sim 1$, $\sin^2\theta \sim \Omega_1^2/4\delta_L^2$, and according to (18)

$$w \simeq \gamma\frac{\Omega_1^2}{2\delta_L^2}. \tag{40}$$

The intensities of the different components of the triplet, to lowest order in $\Omega_1/|\delta_L|$ where they appear, then equal, according to (37),

$$I(\omega_0) \simeq \frac{\gamma}{2}\frac{\Omega_1^2}{\delta_L^2} \tag{41.a}$$

$$I(\omega_L) \simeq \frac{\Gamma}{4}\frac{\Omega_1^2}{\delta_L^2} \tag{41.b}$$

$$I(2\omega_0 - \omega_L) \simeq \frac{\Gamma}{16}\frac{\Omega_1^4}{\delta_L^4}. \tag{41.c}$$

The asymmetry among the components of the fluorescence triplet clearly appears in formulas (41.a) and (41.c). Whereas the line of frequency $(2\omega_0 - \omega_L)$ varies as $(\Omega_1/\delta_L)^4$, the line of frequency ω_0 is more intense because it is proportional to $(\Omega_1/\delta_L)^2$. Moreover, at this order of perturbation, this line depends exclusively on the collisional processes as shown by the coefficient γ of formula (41.a). Also note that the ratio $I(\omega_0)/I(\omega_L)$ neither depends on the intensity of the incident field nor on the detuning from resonance (if the impact condition remains valid) and equals

$$\frac{I(\omega_0)}{I(\omega_L)} = \frac{2\gamma}{\Gamma}. \tag{42}$$

This formula shows that we can determine the coefficient of collisional relaxation γ by measuring the ratio between the intensities of the fluorescence line of frequency ω_0 and that of the Rayleigh scattering line at ω_L. The absence of dependence of (42) with the incident field is due to the fact that in the perturbative limit, the two processes under consideration are associated with the absorption of a single incident photon (see Figures 1b and 1a).

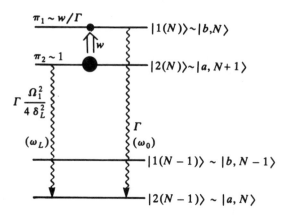

Figure 4. Diagram of the radiative and collisional processes in the perturbative limit.

The foregoing results can be represented on the energy-level diagram of the dressed atom (Figure 4). The system has a relative probability close to 1 of being in the state $|2(N)\rangle$, which, in the limit considered, is only slightly different from $|a, N + 1\rangle$ (this large probability is represented by the large filled circle in Figure 4). Because the level $|2(N)\rangle$ is only slightly contaminated by $|b, N\rangle$, the probability of spontaneous emission from this level is very low. By using (D.11) and (39), we find $\Gamma\Omega_1^2/4\delta_L^2$ for the probability of radiative transition from $|2(N)\rangle$ to $|2(N - 1)\rangle$ and an even smaller quantity, $\Gamma\Omega_1^4/16\delta_L^4$ for the probability of transition from $|2(N)\rangle$ to $|1(N - 1)\rangle$. By contrast, the level $|1(N)\rangle$, which is only slightly different from $|b, N\rangle$, decays easily to $|2(N - 1)\rangle \sim |a, N\rangle$ with a probability equal to Γ. The relative population in the level $|1(N)\rangle$ results from competition between the supply resulting from collisions whose probability is w [given by formula (40)] and the disappearance resulting from spontaneous emission. The probability of finding the system in the state $|1(N)\rangle$ is thus simply w/Γ.

Finally, note that in the limit considered in this subsection, the absorption signal (38) is equal (to second order in Ω_1/δ_L) to

$$A = \frac{\Omega_1^2}{4\delta_L^2}(\Gamma + 2\gamma) \tag{43}$$

and corresponds to the sum of the intensity of the Rayleigh line and the

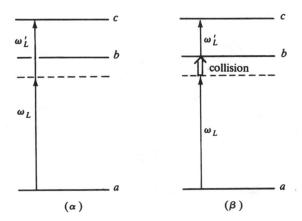

Figure 5. Absorption of a probe beam of frequency ω'_L by an atom interacting with a nonresonant beam of frequency ω_L and undergoing collisions. The excitation spectrum has two resonances associated with the processes represented in (α) and (β).

line at the atomic frequency ω_0. The absorption increases in the presence of collisions. More precisely, A increases linearly as a function of the number of perturbing atoms X because γ is proportional to N, according to formulas (6) and (7).

Remarks

(i) Instead of observing the fluorescence on the transition $b \leftrightarrow a$, one can also probe the atom with a second laser beam of frequency ω'_L on a transition starting from the level b to a level c having higher energy. The excitation spectrum of the level c then has two resonances: The first resonance is obtained when $\omega'_L + \omega_L = \omega_{ca}$ and is associated with the two-photon excitation of the level c starting from the level a (Figure 5α). The second resonance $\omega'_L = \omega_{cb}$ corresponds to a process where the atom is first brought to the level b in a collisionally aided process, then to the level c by the absorption of a second resonant photon on the transition $c \leftrightarrow b$ (Figure 5β). Note that the two processes represented in Figures 5α and 5β appear at the same order only in the presence of collisions (*). The ratio of the intensity of the line at $\omega'_L = \omega_{cb}$ to the intensity of the line at $\omega'_L = \omega_{ca} - \omega_L$ increases with the collision rate. If

---

(*) In the absence of collisions, the atoms can be brought to the level b only by nonlinear processes involving two laser photons ω_L (see Figure 28 in Chapter II).

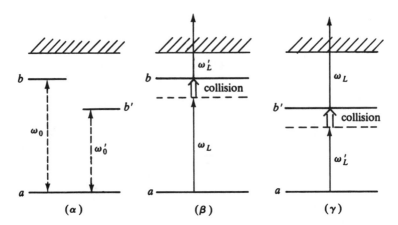

Figure 6. Diagram of the levels considered in Remark (ii). Ionization by absorption of two photons in a collision-assisted process. The excited level in the intermediate step can be either b or b', depending on whether the first absorbed photon is ω_L or ω_L'.

it is possible to neglect the influence of intermediate levels other than b in the two-photon excitation, then the ratio between the intensities of the two lines is, in the impact limit, equal to $2\gamma/\Gamma$ (*).

(ii) Consider now an atom having three discrete levels a, b, and b' and an ionization continuum (Figure 6α). This atom interacts with two nonresonant incident laser beams, with frequencies ω_L and $\omega_{L'}$ close to ω_0 and ω_0'. We study the ionization resulting from the absorption of two photons, one of frequency ω_L, and the other of frequency ω_L'. In addition to the ionization resulting from the simultaneous absorption of these two photons, there exists, in a collisional environment, a stepwise ionization where the atom is brought in an intermediate step into one of the levels b or b' by a collisionally aided excitation process (Figures 6β and 6γ).

Consider then the probability amplitudes associated with the ionization processes represented in Figures 6β and 6γ. If the transfer of energy and momentum resulting from the collision is the same for these two processes, it is not possible to determine the path actually taken by the system and an interference between the two probability amplitudes is then possible. The energy transferred during the collision is equal to $\hbar\delta_L$ for the process (6β) and to $\hbar\delta_L'$ for the process (6γ). A necessary condition for the two diagrams to

(*) For more details, see, for example, P. F. Liao, J. E. Bjorkholm, and P. R. Berman, *Phys. Rev. A*, **21**, 1927 (1980).

interfere is thus $\hbar\delta_L = \hbar\delta'_L$, i.e., again

$$E_b - E_{b'} = \hbar(\omega_L - \omega'_L). \tag{44}$$

We then see the appearance of the resonance condition mentioned in Remark (ii) in subsection 2-d. The collisions indeed lead to a resonant variation of the photoionization probability when $\omega_L - \omega'_L$ varies around the value corresponding to the energy separation between the excited levels b and b'. One can actually show that the intensity of this resonance depends on the collisional factor (30) calculated above (*).

(iii) Up to this point, we have considered only the case where the incident field is a plane wave. If the incident field is the superposition of two plane waves propagating in different directions, the resulting field is spatially modulated. This has several consequences. First, because the wave functions of the dressed atom change from point to point, the dipole moment of the transition $|i(N)\rangle \rightarrow |i(N + 1)\rangle$ (with $i = 1, 2$) varies spatially. Consequently, a grating of transition moments is created, and thus an index grating, which will instantaneously follow the variations of the applied field. Second, in the presence of collisions, the transition probability w between the levels $|2(N)\rangle$ and $|1(N)\rangle$ is larger for atoms located at the antinodes of the wave [see formula (40)]. A grating of excited atoms is thereby created. The characteristic evolution time for this second grating is the radiative lifetime of the level b i.e., $1/\Gamma$ (see Figure 4).

These atomic gratings give rise to many effects in nonlinear optics. For example, consider the situation where the dressing field results from the superposition of two plane waves with frequencies ω_L and ω'_L and wave vectors **k** and **k'**. The dressing field is then modulated, not only spatially, but also temporally (at the frequency $\omega_L - \omega'_L$ that we assume to be small compared to $|\delta_L|$). The mechanisms giving rise to the gratings of transition moments and of excited atoms considered above are then modulated in time. Contrary to the grating of transition moments that instantaneously follows the temporal variations of the field, the grating of excited atoms follows with a time constant Γ^{-1}, which results in a phase shift between the light modulation and the grating of excited atoms (**) and to a complete smearing of the latter when $|\omega_L - \omega'_L| \gg \Gamma$.

We now introduce a probe wave. Several wave-mixing phenomena occur that can be interpreted as resulting from the diffraction of the probe wave on the atomic gratings created by the waves **k**, ω_L and **k'**, ω'_L. The fact that the amplitude of the grating of excited atoms induced by collision varies in a

(*) For more details, see G. Grynberg and P. R. Berman, *Phys. Rev. A*, **41**, 2677 (1990); **43**, 3994 (1991).

(**) Such a phase shift gives rise to an energy exchange between the two waves; see D. Grand-Clement, G. Grynberg, and M. Pinard, *Phys. Rev. Lett.*, **59**, 40 (1987).

resonant fashion about $\omega_L = \omega'_L$ allows us to understand why, in the presence of collisions, the generation of a new wave can vary in a resonant fashion over an interval of width Γ about $\omega_L - \omega'_L = 0$ (*).

4. Sketch of the Calculation of the Collisional Transfer Rate

a) EXPRESSION OF THE TRANSFER RATE AS A FUNCTION OF THE COLLISION *S*-MATRIX

We now return to the general case and introduce the *S*-matrix describing the internal evolution of the atom A + laser mode system in a collision with the perturber X characterized by the initial parameters b and $\mathbf{v}$. Such a collision replaces the initial state, described by $\tilde{\sigma}_{AL}$, by the final state $S\tilde{\sigma}_{AL}S^+$. If we consider time intervals Δt such that $\tau_{\text{coll}} \ll \Delta t \ll T_{\text{coll}}$, the variation $\Delta\tilde{\sigma}_{AL}$ of the density matrix of the global system during Δt is equal to

$$\frac{\Delta\tilde{\sigma}_{AL}}{\Delta t} = \langle S\tilde{\sigma}_{AL}S^+ - \tilde{\sigma}_{AL}\rangle_{\text{coll}} \tag{45}$$

where the average over the collisions $\langle \ \rangle_{\text{coll}}$ was defined by formula (6). The quantity $\Delta\tilde{\sigma}_{AL}/\Delta t$ is the average of $d\tilde{\sigma}_{AL}/dt$ over times that are long compared with τ_{coll}. The identification of these two quantities is equivalent to making a coarse-grained average, i.e., to ignoring the behavior of $\tilde{\sigma}_{AL}$ during very short times. Within the framework of this approximation, the master equation is written

$$\frac{d\tilde{\sigma}_{AL}}{dt} = \langle S\tilde{\sigma}_{AL}S^+ - \tilde{\sigma}_{AL}\rangle_{\text{coll}}. \tag{46}$$

We now project this equation onto the basis $|i(N)\rangle$ of the dressed states. Because we assumed that the collisions cannot couple two levels of different manifolds, the matrix elements of S satisfy the relation

$$\langle i(N)|S|j(N')\rangle = \delta_{NN'}S_{ij}. \tag{47}$$

For example, we write the evolution equation for $\langle 1(N)|\tilde{\sigma}|1(N)\rangle$. By

(*) See A. R. Bogdan, M. W. Downer, and N. Bloembergen, *Opt. Lett.*, **6**, 348 (1981).

using (46), (47), and the secular approximation, we find

$$\frac{d}{dt}\langle 1(N)|\bar{\sigma}|1(N)\rangle = \langle (S_{11}S_{11}^* - 1)\rangle_{\text{coll}}\langle 1(N)|\bar{\sigma}|1(N)\rangle +$$

$$+ \langle S_{12}S_{12}^*\rangle_{\text{coll}}\langle 2(N)|\bar{\sigma}|2(N)\rangle. \tag{48}$$

The unitarity of the S matrix results in

$$S_{11}S_{11}^* + S_{12}S_{12}^* = 1 \tag{49}$$

which allows one to show that Equation (48) is identical to Equation (12) for the evolution of populations of the dressed atom and gives the expression for w:

$$w = \langle S_{12}S_{12}^*\rangle_{\text{coll}}. \tag{50}$$

For the physical discussion that follows, it is convenient to reexpress w as a function of the matrix elements of S in the uncoupled basis $\{|a, N + 1\rangle, |b, N\rangle\}$. We then transform $S_{12} = \langle 1(N)|S|2(N)\rangle$ by using (B.20). We then have

$$\langle 1(N)|S|2(N)\rangle =$$

$$= \langle 1(N)|a, N + 1\rangle\langle a, N + 1|S|a, N + 1\rangle\langle a, N + 1|2(N)\rangle +$$

$$+ \langle 1(N)|b, N\rangle\langle b, N|S|b, N\rangle\langle b, N|2(N)\rangle +$$

$$+ \langle 1(N)|a, N + 1\rangle\langle a, N + 1|S|b, N\rangle\langle b, N|2(N)\rangle +$$

$$+ \langle 1(N)|b, N\rangle\langle b, N|S|a, N + 1\rangle\langle a, N + 1|2(N)\rangle \tag{51}$$

or also

$$\langle 1(N)|S|2(N)\rangle =$$

$$= \sin\theta\cos\theta[\langle a, N + 1|S|a, N + 1\rangle - \langle b, N|S|b, N\rangle] -$$

$$- \sin^2\theta\langle a, N + 1|S|b, N\rangle + \cos^2\theta\langle b, N|S|a, N + 1\rangle. \tag{52}$$

It is sometimes possible to neglect certain terms in Equation (52). Consider, for example, the impact limit. The amplitude $\langle b, N|S|a, N + 1\rangle$, which is associated with the absorption of a laser photon during the collision, is on the order of $\Omega_1\tau_{\text{coll}}$ that we have assumed to be very small compared to 1. The last two terms of (52) are thus at most on the order of $\Omega_1\tau_{\text{coll}}$. The first term of (52) is, in the perturbative limit, on the order of

$\Omega_1/|\delta_L|$. It is larger than the last two terms because $1/|\delta_L| \gg \tau_{coll}$ in the impact limit. It is thus possible in this case to calculate the effect of the collisions by taking into account only the phase shifts $\langle a, N+1|S|a, N+1\rangle$ and $\langle b, N|S|b, N\rangle$. This is, in fact, the procedure we followed in subsection 2-d [see, in particular, Remark (i)]. In the next subsection, we analyze another situation where, on the contrary, it is the last term of (52) that is dominant.

b) Case Where the Laser Frequency Becomes Resonant during the Collision. Limit of Large Detunings

In this subsection, we consider the limit where $\Omega_1/|\delta_L| \ll 1$. We assume that $\delta_L < 0$, and we rewrite formula (52) limiting ourselves to terms up to order one in Ω_1/δ_L:

$$\langle 1(N)|S|2(N)\rangle = -\frac{\Omega_1}{2\delta_L}[\langle a, N+1|S|a, N+1\rangle -$$

$$-\langle b, N|S|b, N\rangle] + \langle b, N|S|a, N+1\rangle. \quad (53)$$

We saw in the preceding subsection that, in the impact limit, the last term of the right-hand side of (53) is smaller than the first term. We now show how this result is no longer correct for large detunings $|\delta_L| > \tau_{coll}^{-1}$.

Consider the energy levels in Figure 2 and draw the energy levels of the atom A + laser mode system as a function of the distance r between A and X. We see in Figure 7 that there are two possible situations, depending on the sign of δ_L. When $\omega_L < \omega_0$, the curves representing $E_b(r) + N\hbar\omega_L$ and $E_a(r) + (N+1)\hbar\omega_L$ cross each other at $r = r_0$ (Figure 7α), whereas there is no crossing between these potential curves for $\omega_L > \omega_0$ (Figure 7β). We first consider the case where the laser frequency becomes resonant during the collision, i.e., the case where there are level crossings. Because the energies vary by a quantity greater than or equal to $\hbar\delta_L$ during the time τ_{coll} of the collision, the distance ΔE between the energy levels near the crossing point has an order of magnitude given by $\Delta E = \hbar\delta_L(t - t_0)/\tau_{coll}$, where t_0 is the time at which the crossing occurs (see Figure 8). The transition essentially occurs during a time interval such that $|t - t_0| \lesssim \hbar/\Delta E$, which gives $|t - t_0| \sim (\tau_{coll}/|\delta_L|)^{1/2}$. The transition amplitude $\langle b, N|S|a, N+1\rangle$ is, in order of magnitude, equal to $\Omega_1|t - t_0|$, i.e., $(|\delta_L|\tau_{coll})^{1/2}\Omega_1/|\delta_L|$. In the large detuning limit $(|\delta_L|\tau_{coll} \gg 1)$, the last term of the right-hand side of (53) is thus larger than the first term, whose order of magnitude is $\Omega_1/|\delta_L|$. In this limit, the

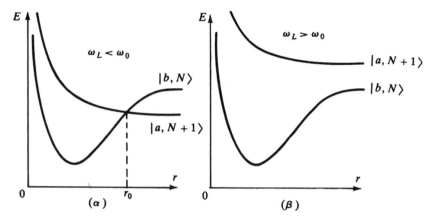

Figure 7. Variation of the energy levels $|a, N + 1\rangle$ and $|b, N\rangle$ as a function of the distance r between the atoms A and X. The two schemes correspond to the two possible signs of the detuning from resonance: $\delta_L < 0$ (α) and $\delta_L > 0$ (β).

transition probability between the dressed levels is thus simply equal to $|\langle b, N|S|a, N + 1\rangle|^2$. Because this term is larger than the term retained in the impact limit, we deduce that the transition probability decreases more slowly as a function of $|\delta_L|$ than the Lorentzian (41a) found in the impact limit. Such a result is related to the existence of a crossing between the potential curves in Figure 7α. Consequently, the transition probability takes different values for opposite values of δ_L. In the situation depicted

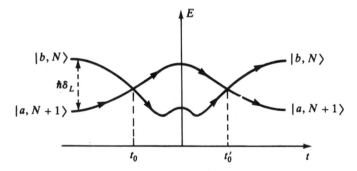

Figure 8. Variation of the uncoupled energy levels as a function of time during a collision in the case where $\omega_L < \omega_0$. Under the influence of the coupling, it is possible for a system prepared in the state $|a, N + 1\rangle$ before the collision to move to the level $|b, N\rangle$ near the crossing points located at times t_0 and t_0'.

in Figure 7, the transition probability will be larger for $\omega_L < \omega_0$ than for $\omega_L > \omega_0$ (*).

Remark

We return to the situation encountered in Figure 7α where there is a crossing between the potential curves, and we briefly consider how it is possible to calculate the transition probability between $|a, N + 1\rangle$ and $|b, N\rangle$ in this situation. Because the transition results from the existence of the level crossing located at $r = r_0$, we can calculate the transition probability for each value of the impact parameter b and of the relative velocity **v** by using the Landau formula

$$|\langle b, N|S|a, N + 1\rangle|^2 = \frac{\pi \hbar \Omega_1^2}{|\Delta \dot{E}|} \tag{54}$$

with $\Delta \dot{E}$ being the time derivative of $E_b(r(t)) - E_a(r(t))$ taken at $t = t_0$. Formula (54) takes into account the two crossings located at t_0 and t_0'. To obtain w, it is necessary, according to (50), to average (54) over b and **v**. Note, however, that for large values of $|\delta_L|$, the level crossing at r_0 is associated with a very small impact parameter, so that the approximation of linear classical trajectories is no longer justified (**).

(*) Such an asymmetry between the wings of the absorption line was observed, for example, by J. L. Carlsten, A. Szöke, and M. G. Raymer, *Phys. Rev. A*, **15**, 1029 (1977).

(**) For more details, see V. S. Lisitsa and S. I. Yakovlenko, *JETP Lett.*, **39**, 759 (1974); **41**, 233 (1975).

Exercises

1. CALCULATION OF THE RADIATIVE LIFETIME OF AN EXCITED ATOMIC LEVEL. COMPARISON WITH THE DAMPING TIME OF A CLASSICAL DIPOLE MOMENT

A particle of mass m and charge q is bound near the origin 0 by a static potential $V(\mathbf{r})$. Let H_P be the Hamiltonian for this particle:

$$H_P = \frac{\mathbf{p}^2}{2m} + V(\mathbf{r}) \tag{1}$$

$|a\rangle$ be the ground state of energy E_a, and $|b\rangle$ be the first excited state, having energy E_b. These two states are assumed to be discrete and we set

$$E_b - E_a = \hbar\omega_0. \tag{2}$$

a) Let H_I be the interaction Hamiltonian in the Coulomb gauge between this particle and the radiation field. Calculate the matrix element of H_I between the state $|b; 0\rangle$ (particle in the state b; 0 photon) and the state $|a; \mathbf{k}\boldsymbol{\varepsilon}\rangle$ (particle in the state a in the presence of one photon $\mathbf{k}\boldsymbol{\varepsilon}$). The spins of the particles will be ignored and the long-wavelength approximation will be used, which consists of neglecting the spatial variations of the vector potential $\mathbf{A}$ over the range of the wave functions of states a and b.

515

b) Calculate the commutator $[\mathbf{r}, H_P]$ and derive from it a relation between the matrix elements $\langle a|\mathbf{p}|b\rangle$ and $\langle a|\mathbf{r}|b\rangle$. Express the matrix element $\langle a; \mathbf{k}\varepsilon|H_I|b; 0\rangle$ as a function of $\mathbf{d} = \langle a|q\mathbf{r}|b\rangle$, the matrix element of the electric dipole of the particle. It will be assumed that a and b are levels having a magnetic quantum number $m_l = 0$, so that only the component along the $0z$ axis of $\mathbf{d}$ (written d and assumed to be real) is different from zero.

c) Calculate the radiative lifetime $\tau = \Gamma^{-1}$ of the excited level, where Γ is the spontaneous emission rate of a photon from b. The summation is to be carried out over all the polarizations and directions of the emitted photon. Express Γ as a function of d, ω_0, and various fundamental constants such as $\hbar$, c, and ε_0.

d) Assume that the value of ω_0 and the spatial ranges of the wave functions of states a and b are approximately those of a hydrogen atom in states $1s$ and $2p$. Calculate an order of magnitude for $1/\omega_0\tau$. Express the result as a function of the fine-structure constant $\alpha = q^2/4\pi\varepsilon_0\hbar c$. Comment on this result.

e) Let Γ_{cl} be the inverse of the damping time for the motion of a classical particle with the same charge, oscillating parallel to $0z$ at the frequency ω_0. We recall that (*)

$$\Gamma_{cl} = \frac{q^2}{6\pi\varepsilon_0} \frac{\omega_0^2}{mc^3}. \tag{3}$$

Show that the spontaneous emission rate Γ calculated above can be written

$$\Gamma = \Gamma_{cl} f_{ab} \tag{4}$$

where f_{ab} is a dimensionless parameter characteristic of the atomic transition $a \leftrightarrow b$, known as the "oscillator strength" of the transition. Give the expression of f_{ab}.

Solution

a) Within the long-wavelength approximation, H_I is equal to $-(q/m)\mathbf{p} \cdot \mathbf{A}(0)$. The expansion of $\mathbf{A}(\mathbf{r})$ in modes [see expression (28) in the Appendix] allows us to rewrite H_I in the form

$$H_I = -\frac{q}{m} \sum_i \mathscr{A}_{\omega_i}[a_i\mathbf{p} \cdot \boldsymbol{\varepsilon}_i + a_i^+ \mathbf{p} \cdot \boldsymbol{\varepsilon}_i] \tag{5}$$

(*) See, for example, *Photons and Atoms—Introduction to Quantum Electrodynamics*, Exercise 7 in Complement C_I.

where $\mathscr{A}_{\omega_i} = (\hbar/2\varepsilon_0 \omega_i L^3)^{1/2}$, which gives

$$\langle a; \mathbf{k}\varepsilon | H_I | b; 0 \rangle = -\frac{q}{m} \mathscr{A}_\omega \langle a | \mathbf{p} \cdot \boldsymbol{\varepsilon} | b \rangle. \tag{6}$$

b) To calculate the commutator of $\mathbf{r}$ and H_P, note that $\mathbf{r}$ and $V(\mathbf{r})$ commute with each other, which leads to

$$[\mathbf{r}, H_P] = i\hbar \frac{\mathbf{p}}{m}. \tag{7}$$

We then find for the matrix element of $\mathbf{p}$

$$\langle a | \mathbf{p} | b \rangle = \frac{m}{i\hbar} \langle a | [\mathbf{r}, H_P] | b \rangle = \frac{E_b - E_a}{i\hbar} m \langle a | \mathbf{r} | b \rangle \tag{8}$$

which gives, using (2) and (6)

$$\langle a; \mathbf{k}\varepsilon | H_I | b; 0 \rangle = i\omega_0 \langle a | q \mathbf{r} \cdot \boldsymbol{\varepsilon} | b \rangle \mathscr{A}_\omega. \tag{9}$$

Because *a* and *b* have quantum numbers $m_l = 0$, the only component of $\mathbf{r}$ coupling these levels is z. By using $d = q \langle a | z | b \rangle$, we obtain

$$\langle a; \mathbf{k}\varepsilon | H_I | b; 0 \rangle = i\omega_0 (\boldsymbol{\varepsilon} \cdot \mathbf{e}_z) d \mathscr{A}_\omega. \tag{10}$$

c) The probability of emission per unit time and per unit solid angle of a photon with polarization $\boldsymbol{\varepsilon}$ is given by the Fermi golden rule

$$\frac{dw}{d\Omega} = \frac{2\pi}{\hbar} \omega_0^2 (\boldsymbol{\varepsilon} \cdot \mathbf{e}_z)^2 d^2 \mathscr{A}_{\omega_0}^2 \rho(E = \hbar\omega_0) \tag{11}$$

where $\rho(E)$ is given by formula (46) in Complement A_I. By replacing $\rho(E)$ and $\mathscr{A}_{\omega_0}$ by their values, we find

$$\frac{dw}{d\Omega} = \frac{1}{8\pi^2 \varepsilon_0} \frac{\omega_0^3}{\hbar c^3} d^2 (\boldsymbol{\varepsilon} \cdot \mathbf{e}_z)^2. \tag{12}$$

To find the probability for the emission of a photon in any direction and with any polarization, it is necessary to sum over $\boldsymbol{\varepsilon}$ and integrate over $d\Omega$, which gives, using relation (55) in Complement A_I,

$$\Gamma = \frac{1}{3\pi\varepsilon_0} \frac{\omega_0^3}{\hbar c^3} d^2. \tag{13}$$

d) We replace *d* by qa_0, where a_0 is the Bohr radius. We then have

$$\frac{1}{\omega_0 \tau} \sim \frac{q^2}{3\pi\varepsilon_0 \hbar c} \frac{\omega_0^2 a_0^2}{c^2} \sim \alpha \frac{\omega_0^2 a_0^2}{c^2}. \tag{14}$$

In the hydrogen atom, $\omega_0 a_0/c \sim \alpha$, which gives

$$\frac{1}{\omega_0 \tau} \sim \alpha^3. \tag{15}$$

From this we conclude that the lifetime of the excited state is much longer, by a factor $(1/\alpha)^3$, than the characteristic evolution time of the electron around the nucleus. Relation (15) also expresses that the width of the transition, on the order of $1/\tau$, is very small compared with the emission frequency ω_0.

  e) By comparing (13) to (4) and (3), we see that

$$f_{ab} = \frac{2m\omega_0}{\hbar} \frac{d^2}{q^2}. \tag{16}$$

That is, again, using the definition of $d = q\langle a|z|b \rangle$:

$$f_{ab} = \frac{2m\omega_0}{\hbar} |\langle a|z|b \rangle|^2. \tag{17}$$

2. Spontaneous Emission of Photons by a Trapped Ion. Lamb–Dicke Effect

  The purpose of this exercise is to study the structure of the lines emitted by an excited atomic system for which the motion of the center of mass is not free, but is confined to a finite region of space by a potential $V(\mathbf{R})$. This type of situation appears in different contexts: photon emission from a nucleus bound in a crystalline matrix (Mössbauer effect), or from an atom, or a molecule for which the mean free path is small compared with the wavelength of the emitted radiation (Dicke effect) (*). We will consider here the spontaneous emission of photons by a trapped ion.

  The state of the ion can be expanded over a basis $|i\rangle \otimes |\chi\rangle$, where $|i\rangle$ corresponds to the internal degrees of freedom (excitation of electrons in the center-of-mass reference frame) and $|\chi\rangle$ corresponds to the external degrees of freedom (relative to the center of mass). We assume, for the sake of simplicity, that we have a hydrogen-like ion composed of an electron with charge $q_1 = q$ and mass m_1 and a nucleus with charge $q_2 = -Zq$ and mass m_2. The total charge and mass are $Q = (1 - Z)q$ and $M = m_1 + m_2$. The variables of the center of mass $\mathbf{R}$ and $\mathbf{P}$ and the internal variables $\mathbf{r}$ and $\mathbf{p}$ are related to the variables of the two particles

---

  (*) See, for example, A. Abragam, *The Mössbauer Effect*, Gordon and Breach, New York, 1964; R. H. Dicke, *Phys. Rev.*, **89**, 472 (1953).

by the equations:

$$R = \frac{m_1 \mathbf{r}_1 + m_2 \mathbf{r}_2}{m_1 + m_2} \tag{1.a}$$

$$\mathbf{P} = \mathbf{p}_1 + \mathbf{p}_2 \tag{1.b}$$

$$\mathbf{r} = \mathbf{r}_1 - \mathbf{r}_2 \tag{1.c}$$

$$\frac{\mathbf{p}}{m} = \frac{\mathbf{p}_1}{m_1} - \frac{\mathbf{p}_2}{m_2} \tag{1.d}$$

where m is the reduced mass:

$$\frac{1}{m} = \frac{1}{m_1} + \frac{1}{m_2}. \tag{2}$$

a) Show that, within the long-wavelength approximation, the Hamiltonian in the Coulomb gauge for a free ion is

$$H = H_0 + \frac{\mathbf{P}^2}{2M} + V_{\text{int}} + V_{\text{ext}} + H_R \tag{3}$$

where H_R is the radiation field Hamiltonian and where

$$H_0 = \frac{\mathbf{p}^2}{2m} - \frac{Zq^2}{4\pi\varepsilon_0 r} + \varepsilon_{\text{Coul}} \tag{4.a}$$

$$V_{\text{int}} = -\frac{q}{\mu}\mathbf{p} \cdot \mathbf{A}(\mathbf{R}) \tag{4.b}$$

$$V_{\text{ext}} = -\frac{Q}{M}\mathbf{P} \cdot \mathbf{A}(\mathbf{R}) + \left(\frac{q^2}{2m_1} + \frac{Z^2q^2}{2m_2}\right)\mathbf{A}^2(\mathbf{R}). \tag{4.c}$$

We have called $\varepsilon_{\text{Coul}}$ the sum of the Coulomb self-energies and set

$$\frac{1}{\mu} = \frac{1}{m_1} + \frac{Z}{m_2}. \tag{5}$$

b) The center of mass of the ion is assumed to evolve in an external potential $V(\mathbf{R})$ so that the new Hamiltonian is $H + V(\mathbf{R})$. $V(\mathbf{R})$ is assumed to be attractive and to have a minimum at $\mathbf{R} = \mathbf{0}$ so that the eigenstates of the Hamiltonian

$$H_{\text{ext}} = \frac{\mathbf{P}^2}{2M} + V(\mathbf{R}) \tag{6}$$

are discrete bound states $|\chi_n\rangle$ with energy $\mathscr{E}_n$.

At the initial time, the ion is in an excited electronic state $|b\rangle$ in the absence of photons. Its state vector is $|b, \chi_n; 0\rangle$ where $|\chi_n\rangle$ describes the initial vibrational state of the center of mass. After a photon $\mathbf{k}\boldsymbol{\varepsilon}$ is spontaneously emitted, the ion is found in its electronic ground state $|a\rangle$. Show that the transition amplitude, associated with a process where the final vibrational state is $|\chi_l\rangle$, is proportional to $\langle a|\mathbf{p} \cdot \boldsymbol{\varepsilon}|b\rangle\langle\chi_l|e^{-i\mathbf{k}\cdot\mathbf{R}}|\chi_n\rangle$.

c) Calculate the frequencies of the photons spontaneously emitted in a given direction $\boldsymbol{\kappa}_0$ by an ion assumed to be initially in the state $|b, \chi_n\rangle$, as well as the relative intensities of the corresponding lines. We assume that $|\mathscr{E}_n - \mathscr{E}_l| \ll \hbar\omega_0$, where $\hbar\omega_0 = E_b - E_a$, and we set

$$I_{nl}(\boldsymbol{\kappa}_0) = |\langle\chi_l|e^{-i\mathbf{k}_0\cdot\mathbf{R}}|\chi_n\rangle|^2 \tag{7}$$

where $\mathbf{k}_0 = (\omega_0/c)\boldsymbol{\kappa}_0$.

Show that the total intensity emitted over all these lines is independent of $V(\mathbf{R})$ and is thus equal to the total intensity emitted by a free ion.

Show, in the case where the spatial extent D of the center-of-mass motion is very small compared with the wavelength of the emitted radiation, that the line of frequency ω_0 is much more intense than the other lines (Dicke effect).

d) The emission of a photon on the transition $b, \chi_n \rightarrow a, \chi_l$ causes the vibrational energy of the ion to vary. Show that, for an ion leaving the state b, χ_n and emitting a photon in the direction $\boldsymbol{\kappa}_0$, the average value $\delta\mathscr{E}$ of this energy variation is equal to

$$\delta\mathscr{E} = \langle\chi_n|e^{i\mathbf{k}_0\cdot\mathbf{R}} H_{\text{ext}} e^{-i\mathbf{k}_0\cdot\mathbf{R}} - H_{\text{ext}}|\chi_n\rangle. \tag{8}$$

By using the fact that the average value of $\mathbf{P}$ in the bound state $|\chi_n\rangle$ is zero, show that $\delta\mathscr{E}$ coincides with the recoil energy of a free ion.

The energy difference $|\mathscr{E}_l - \mathscr{E}_n|$ between vibrational levels with $l \neq n$ is assumed to be very large compared to $\delta\mathscr{E}$. What can be deduced from this with regard to the intensity of the unshifted line of frequency ω_0?

Solution

a) Start with the Hamiltonian in the Coulomb gauge

$$H = \sum_{\alpha} \frac{\mathbf{p}_\alpha^2}{2m_\alpha} + \sum_{\alpha > \beta} \frac{q_\alpha q_\beta}{4\pi\varepsilon_0 |\mathbf{r}_\alpha - \mathbf{r}_\beta|} + \sum_{\alpha} \varepsilon_{\text{Coul}}^\alpha -$$

$$- \sum_{\alpha} \frac{q_\alpha}{m_\alpha} \mathbf{p}_\alpha \cdot \mathbf{A}(\mathbf{r}_\alpha) + \sum_{\alpha} \frac{q_\alpha^2}{2m_\alpha} \mathbf{A}^2(\mathbf{r}_\alpha) + H_R. \tag{9}$$

Within the long-wavelength approximation, $\mathbf{A}(\mathbf{r}_\alpha)$ can be replaced by $\mathbf{A}(\mathbf{R})$. The term for the interaction with the field, which is linear in q_α, then becomes

$$-\left(\frac{q}{m_1} \mathbf{p}_1 - \frac{Zq}{m_2} \mathbf{p}_2 \right) \cdot \mathbf{A}(\mathbf{R}) = -q \left(\frac{\mathbf{p}_1}{m_1} - \frac{Z\mathbf{p}_2}{m_2} \right) \cdot \mathbf{A}(\mathbf{R}). \tag{10}$$

Rewrite $\mathbf{p}_1$ and $\mathbf{p}_2$ as a function of $\mathbf{p}$ and $\mathbf{P}$ by using (1.b) and (1.d):

$$\mathbf{p}_1 = \mathbf{p} + \frac{m_1}{M} \mathbf{P} \tag{11.a}$$

$$\mathbf{p}_2 = -\mathbf{p} + \frac{m_2}{M} \mathbf{P}. \tag{11.b}$$

The interaction terms linear in q are

$$-q\left(\frac{1}{m_1} + \frac{Z}{m_2} \right) \mathbf{p} \cdot \mathbf{A}(\mathbf{R}) - \frac{Q}{M} \mathbf{P} \cdot \mathbf{A}(\mathbf{R}) \tag{12}$$

which correspond, respectively, to V_{int} and to the first term of V_{ext}. The second term of V_{ext} is obtained directly from (9), when $\mathbf{A}^2(\mathbf{r}_\alpha)$ is replaced by $\mathbf{A}^2(\mathbf{R})$.

Finally, by using the relation

$$\sum_{\alpha} \frac{\mathbf{p}_\alpha^2}{2m_\alpha} = \frac{\mathbf{p}^2}{2m} + \frac{\mathbf{P}^2}{2M} \tag{13}$$

we recover, after having regrouped terms, the form (3) of the Hamiltonian.

b) The states $|b, \chi_n; 0\rangle$ and $|a, \chi_l; \mathbf{k}\boldsymbol{\varepsilon}\rangle$ are the eigenstates of $H_0 + H_{\text{ext}} + H_R$ having energy $E_b + \mathscr{E}_n$ and $E_a + \mathscr{E}_l + \hbar ck$, respectively. The only term for the Hamiltonian coupling these states is V_{int} (because V_{ext} does not depend on internal variables, the matrix elements of V_{ext} between $|a\rangle$ and $|b\rangle$ are zero). By using the expansion of $\mathbf{A}$ into a and a^+ [see Appendix, formula (21)], we find

$$\langle a, \chi_l; \mathbf{k}\boldsymbol{\varepsilon} | V_{\text{int}} | b, \chi_n; 0 \rangle =$$

$$= \sqrt{\frac{\hbar}{2\varepsilon_0 \omega (2\pi)^3}} \left(-\frac{q}{\mu} \right) \langle a | \mathbf{p} \cdot \boldsymbol{\varepsilon} | b \rangle \langle \chi_l | e^{-i\mathbf{k} \cdot \mathbf{R}} | \chi_n \rangle. \tag{14}$$

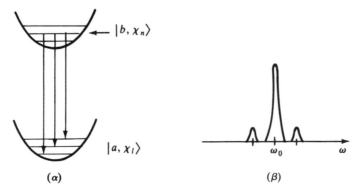

(α) (β)

Figure 1. (α) Representation of possible transitions from the excited level $|b, \chi_n\rangle$. (β) Shape of the emission spectrum in the case where $|\mathscr{E}_n - \mathscr{E}_l| \gg \hbar\Gamma$ (radiative width of the transition).

c) Because of energy conservation, the emitted photon has a frequency

$$\omega = \omega_0 + \frac{\mathscr{E}_n - \mathscr{E}_l}{\hbar}. \tag{15}$$

As a consequence of the discretization of the levels associated with the center-of-mass motion, the spontaneous emission spectrum has a vibrational structure (see Figure 1). Because $|\mathscr{E}_n - \mathscr{E}_l| \ll \hbar\omega_0$, the density of states of the electromagnetic field varies slightly from one emission line to another and the probability that a photon will be emitted in the line $|b, \chi_n\rangle \to |a, \chi_l\rangle$ is proportional to the square of the modulus of (14), the only term depending on the center-of-mass variables being $I_{nl}(\kappa_0)$.

Because $|\mathscr{E}_n - \mathscr{E}_l| \ll \hbar\omega_0$, we will also neglect the variation of the modulus of the wave vector $\mathbf{k}$ as a function of the final vibrational level and replace $\mathbf{k}$ with $\mathbf{k}_0$, because we are interested in the emission in the direction κ_0. We then calculate $\sum_l I_{nl}(\kappa_0)$

$$\sum_l I_{nl}(\kappa_0) = \sum_l \langle \chi_n | e^{i\mathbf{k}_0 \cdot \mathbf{R}} | \chi_l \rangle \langle \chi_l | e^{-i\mathbf{k}_0 \cdot \mathbf{R}} | \chi_n \rangle$$

$$= \langle \chi_n | \chi_n \rangle = 1 \tag{16}$$

(we used the closure relation $\sum_l |\chi_l\rangle \langle \chi_l| = 1$). The normalization condition (16) shows that $I_{nl}(\kappa_0)$ can be considered to be the probability that the atom will go from $|b, \chi_n\rangle$ to $|a, \chi_l\rangle$ by emitting a photon of frequency $\omega_0 + (\mathscr{E}_n - \mathscr{E}_l)/\hbar$ in the direction κ_0, with any polarization (I_{nl} is indeed independent of $\boldsymbol{\varepsilon}$). Condition (16) also shows that the total intensity emitted in the direction κ_0 is independent of the external quantum numbers and therefore independent of $V(\mathbf{R})$.

In the case where $|k_0 D| \ll 1$, $e^{i\mathbf{k}_0 \cdot \mathbf{R}}$ can be replaced, to a first approximation, by 1. $I_{nl}(\boldsymbol{\kappa}_0)$ is then close to

$$I_{nl}(\boldsymbol{\kappa}_0) \simeq |\langle \chi_l | \chi_n \rangle|^2 = \delta_{nl}. \tag{17}$$

In this case, the transition where the vibrational excitation does not change ($n = l$) is much more intense than the other transitions. The central frequency for this line coincides with ω_0 and is not shifted.

d) To obtain $\delta\mathscr{E}$, it is necessary to first weight $(\mathscr{E}_l - \mathscr{E}_n)$ by the probability $I_{nl}(\boldsymbol{\kappa}_0)$ of the corresponding transition, then to sum over all the possible final states χ_l

$$\delta\mathscr{E} = \sum_l (\mathscr{E}_l - \mathscr{E}_n) I_{nl}(\boldsymbol{\kappa}_0). \tag{18}$$

By writing $I_{nl}(\boldsymbol{\kappa}_0) = \langle \chi_n | e^{i\mathbf{k}_0 \cdot \mathbf{R}} | \chi_l \rangle \langle \chi_l | e^{-i\mathbf{k}_0 \cdot \mathbf{R}} | \chi_n \rangle$, we find

$$\delta\mathscr{E} = \sum_l \langle \chi_n | e^{i\mathbf{k}_0 \cdot \mathbf{R}} | \chi_l \rangle \langle \chi_l | [H_{\text{ext}}, e^{-i\mathbf{k}_0 \cdot \mathbf{R}}] | \chi_n \rangle \tag{19}$$

an expression that can be simplified by using the closure relation over $|\chi_l\rangle$:

$$\delta\mathscr{E} = \langle \chi_n | e^{i\mathbf{k}_0 \cdot \mathbf{R}} [H_{\text{ext}}, e^{-i\mathbf{k}_0 \cdot \mathbf{R}}] | \chi_n \rangle$$
$$= \langle \chi_n | e^{i\mathbf{k}_0 \cdot \mathbf{R}} H_{\text{ext}} e^{-i\mathbf{k}_0 \cdot \mathbf{R}} - H_{\text{ext}} | \chi_n \rangle. \tag{20}$$

The operator $e^{i\mathbf{k}_0 \cdot \mathbf{R}}$ is a translation operator, by the amount $-\hbar\mathbf{k}_0$, in momentum space. We can then rewrite (20) in the form

$$\delta\mathscr{E} = \langle \chi_n | \frac{1}{2M} (\mathbf{P} - \hbar\mathbf{k}_0)^2 + V(\mathbf{R}) - \frac{\mathbf{P}^2}{2M} - V(\mathbf{R}) | \chi_n \rangle$$
$$= \langle \chi_n | \frac{\hbar^2 \mathbf{k}_0^2}{2M} - \frac{\hbar\mathbf{k}_0 \cdot \mathbf{P}}{M} | \chi_n \rangle. \tag{21}$$

Because the average value of $\mathbf{P}$ in the bound state $|\chi_n\rangle$ is zero, we obtain

$$\delta\mathscr{E} = \frac{\hbar^2 \mathbf{k}_0^2}{2M} \tag{22}$$

which is indeed the recoil energy of a free ion. We also observe that, on the average, the vibrational energy increases during the spontaneous emission process (regardless of the initial vibrational state). Note that $\delta\mathscr{E}$ is independent of $\boldsymbol{\kappa}_0$ and therefore remains equal to the recoil energy when we average over the photon's direction of emission.

If $\delta\mathscr{E} \ll |\mathscr{E}_l - \mathscr{E}_n|$ for all $l \neq n$, to satisfy Equation (18), it is necessary for the line associated with the emission of an unshifted photon ω_0 to be more intense than the other lines. The condition $|\delta\mathscr{E}| \ll |\mathscr{E}_l - \mathscr{E}_n|$ for $l \neq n$ means that the emitting ion is very rigidly bound to the trap. It cannot by itself absorb the recoil associated with the emission of the photon. It is the ion + trap global system that recoils. This demonstrates the very close similarity between this phenomenon and the emission, without recoil effect, of a γ ray by a nucleus rigidly bound in a crystalline potential (Mössbauer effect).

3. RAYLEIGH SCATTERING

The purpose of this exercise is to calculate the cross section for the scattering of a photon by an atomic system in the limit where the energy $\hbar\omega$ of the photon is very small compared with the excitation energies $(E_b - E_a)$ of the system with a designating the ground state and b the excited levels. The system is assumed to be located at the coordinate system origin and the long-wavelength approximation is made, so that the electric dipole Hamiltonian $H_I' = -\mathbf{d} \cdot \mathbf{E}_\perp(0)$ can be used (see Appendix, §5).

a) The initial state of the system being $|\varphi_i\rangle = |a; \mathbf{k}\varepsilon\rangle$ (atom in the ground state in the presence of a photon $\mathbf{k}\varepsilon$), calculate, to second order in H_I', the element of the transition matrix $\mathcal{T}_{fi}$ between the states $|\varphi_i\rangle$ and $|\varphi_f\rangle = |a; \mathbf{k}'\varepsilon'\rangle$. The calculation of $\mathcal{T}_{fi}$ will be done to zero order in $\hbar\omega/(E_b - E_a)$ and the result will be expressed as a function of the static polarizability tensor α_{mn} of the atomic system:

$$\alpha_{mn} = -\frac{1}{\varepsilon_0} \sum_b \frac{\langle a|\mathbf{d} \cdot \mathbf{e}_n|b\rangle\langle b|\mathbf{d} \cdot \mathbf{e}_m|a\rangle + \langle a|\mathbf{d} \cdot \mathbf{e}_m|b\rangle\langle b|\mathbf{d} \cdot \mathbf{e}_n|a\rangle}{E_a - E_b}$$

$$(1)$$

where $\mathbf{d} \cdot \mathbf{e}_m$ is the component of the atomic dipole $\mathbf{d}$ in the direction $0m$ $(m = x, y, z)$.

b) Calculate the differential cross section for the scattering of a photon $\mathbf{k}\varepsilon$ in a solid angle $d\Omega'$ around $\mathbf{k}'$ with polarization ε'. To do this, first calculate the transition rate to all the relevant final states.

c) Consider the usual situation in atomic physics where the static polarizability tensor is isotropic ($\alpha_{mn} = \alpha_0 \delta_{mn}$). In this case evaluate the differential and total cross sections. Compare this last result to the classical cross section

$$\sigma_{cl} = \frac{8\pi}{3} r_0^2 \left(\frac{\omega}{\omega_0}\right)^4$$

$$(2)$$

obtained (*) for a classical elastically bound electron whose resonance

(*) See, for example, *Photons and Atoms—Introduction to Quantum Electrodynamics*, Exercise 7 in Complement C_I.

frequency is ω_0 ($r_0 = q^2/4\pi\varepsilon_0 mc^2$ is the classical radius of the electron). It will be useful to reexpress α_0 as a function of the oscillator strength defined by

$$f_{ab} = \frac{2m(E_b - E_a)}{\hbar^2}|\langle a|z|b\rangle|^2. \tag{3}$$

Solution

a) The intermediate state of the scattering process is either a zero-photon state $|b;0\rangle$ or a two-photon state $|b;\mathbf{k}\boldsymbol{\varepsilon},\mathbf{k}'\boldsymbol{\varepsilon}'\rangle$. The matrix elements of H_I' between these states and $|\varphi_i\rangle$ can be calculated from the expansion in modes of $\mathbf{E}_\perp(0)$ [see Appendix, formula (89)]:

$$\langle b;0|H_I'|a;\mathbf{k}\boldsymbol{\varepsilon}\rangle = -i\sqrt{\frac{\hbar\omega}{2\varepsilon_0 L^3}}\,\langle b|\mathbf{d}\cdot\boldsymbol{\varepsilon}|a\rangle \tag{4.a}$$

$$\langle b;\mathbf{k}\boldsymbol{\varepsilon},\mathbf{k}'\boldsymbol{\varepsilon}'|H_I'|a;\mathbf{k}\boldsymbol{\varepsilon}\rangle = i\sqrt{\frac{\hbar\omega'}{2\varepsilon_0 L^3}}\,\langle b|\mathbf{d}\cdot\boldsymbol{\varepsilon}'|a\rangle. \tag{4.b}$$

In the Born expansion of $\mathcal{T}_{fi}$ [see Chapter I, formula (B.14)], we retain only the second-order term (the first-order term is, of course, zero, because H_I' does not directly couple $|a;\mathbf{k}\boldsymbol{\varepsilon}\rangle$ and $|a;\mathbf{k}'\boldsymbol{\varepsilon}'\rangle$):

$$\mathcal{T}_{fi} = \frac{\hbar}{2\varepsilon_0 L^3}\sqrt{\omega\omega'}\sum_b\left[\frac{\langle a|\mathbf{d}\cdot\boldsymbol{\varepsilon}'|b\rangle\langle b|\mathbf{d}\cdot\boldsymbol{\varepsilon}|a\rangle}{E_a + \hbar\omega - E_b} + \frac{\langle a|\mathbf{d}\cdot\boldsymbol{\varepsilon}|b\rangle\langle b|\mathbf{d}\cdot\boldsymbol{\varepsilon}'|a\rangle}{E_a - E_b - \hbar\omega'}\right]. \tag{5}$$

In the low-frequency limit, i.e., for $\hbar\omega$, $\hbar\omega' \ll (E_b - E_a)$, the foregoing expression can be approximated by

$$\mathcal{T}_{fi} = \frac{\hbar}{2\varepsilon_0 L^3}\sqrt{\omega\omega'}\sum_b\left[\frac{\langle a|\mathbf{d}\cdot\boldsymbol{\varepsilon}'|b\rangle\langle b|\mathbf{d}\cdot\boldsymbol{\varepsilon}|a\rangle + \langle a|\mathbf{d}\cdot\boldsymbol{\varepsilon}|b\rangle\langle b|\mathbf{d}\cdot\boldsymbol{\varepsilon}'|a\rangle}{E_a - E_b}\right] \tag{6}$$

which can also be written

$$\mathcal{T}_{fi} = -\frac{\hbar}{2L^3}\sqrt{\omega\omega'}\sum_{m,n}\alpha_{mn}(\boldsymbol{\varepsilon}'\cdot\mathbf{e}_m)(\boldsymbol{\varepsilon}\cdot\mathbf{e}_n). \tag{7}$$

b) By using formulas (47) and (51) from Complement A_I (the second formula being generalized to the case of an indirect coupling), we obtain for the transition probability per unit time and unit solid angle:

$$\frac{\delta w_{fi}}{\delta\Omega'} = \frac{2\pi}{\hbar}\left(\frac{\hbar}{2L^3}\sqrt{\omega\omega'}\right)^2\left(\sum_{m,n}\alpha_{mn}(\boldsymbol{\varepsilon}'\cdot\mathbf{e}_m)(\boldsymbol{\varepsilon}\cdot\mathbf{e}_n)\right)^2\frac{L^3}{8\pi^3}\frac{(\hbar ck')^2}{\hbar^3 c^3} \tag{8}$$

or also, because the conservation of energy requires $\omega = \omega'$,

$$\frac{\delta w_{fi}}{d\Omega'} = \frac{\omega^4}{16\pi^2 c^3 L^3} \left(\sum_{m,n} \alpha_{mn}(\boldsymbol{\varepsilon}' \cdot \mathbf{e}_m)(\boldsymbol{\varepsilon} \cdot \mathbf{e}_n) \right)^2. \tag{9}$$

The differential cross section is obtained by dividing (9) by the photon flux, which equals c/L^3 [see Complement A_I, formula (57)]:

$$\frac{d\sigma}{d\Omega'} = \frac{\omega^4}{16\pi^2 c^4} \left[\sum_{m,n} \alpha_{mn}(\boldsymbol{\varepsilon}' \cdot \mathbf{e}_m)(\boldsymbol{\varepsilon} \cdot \mathbf{e}_n) \right]^2. \tag{10}$$

The differential cross section varies as ω^4, i.e., increases when the frequency increases. For example, for the visible part of the spectrum, the blue wavelengths are more scattered than the red wavelengths.

c) Replace α_{mn} by $\alpha_0 \delta_{mn}$ in (10). This gives

$$\frac{d\sigma}{d\Omega'} = \frac{\omega^4}{16\pi^2 c^4} \alpha_0^2 \left(\sum_n (\boldsymbol{\varepsilon}' \cdot \mathbf{e}_n)(\boldsymbol{\varepsilon} \cdot \mathbf{e}_n) \right)^2 = \frac{\omega^4}{16\pi^2 c^4} \alpha_0^2 (\boldsymbol{\varepsilon} \cdot \boldsymbol{\varepsilon}')^2. \tag{11}$$

To calculate the total cross section, we select the direction $0z$ along $\boldsymbol{\varepsilon}$. It is then necessary to carry out the summation over $\boldsymbol{\varepsilon}'$ perpendicular to $\boldsymbol{\kappa}'$ and the angular average. By using relation (54) from Complement A_I, we find

$$\sigma = \frac{\omega^4}{16\pi^2 c^4} \alpha_0^2 \int d\Omega (1 - \cos^2 \theta) = \frac{\omega^4}{6\pi c^4} \alpha_0^2. \tag{12}$$

To calculate α_0, we express the square of the matrix element $|\langle a|z|b\rangle|^2$ as a function of the oscillator strength

$$\alpha_0 = -\frac{2q^2}{\varepsilon_0} \sum_b \frac{|\langle a|z|b\rangle|^2}{E_a - E_b} = \frac{q^2}{\varepsilon_0 m} \sum_b \frac{f_{ab}}{\omega_{ab}^2}. \tag{13}$$

By substituting this value for α_0 into (12), we find the expression

$$\sigma = \frac{8\pi}{3} r_0^2 \left(\sum_b f_{ab} \frac{\omega^2}{\omega_{ab}^2} \right)^2 \tag{14}$$

which can be rewritten in the form (2) if we set

$$\frac{1}{\omega_0^2} = \sum_b \frac{f_{ab}}{\omega_{ab}^2}.$$

Because $\Sigma_b f_{ab} = 1$, according to the Reiche–Thomas–Kuhn sum rule, $1/\omega_0^2$ can be considered to be the average of the quantities $1/\omega_{ba}^2$ relative to the various transitions $a \to b$ starting from a, weighted by the corresponding oscillator strengths.

Remark on the Choice of the Interaction Hamiltonian

The calculations made in this exercise use the electric dipole Hamiltonian. It is also possible to use the Coulomb-gauge Hamiltonian

$$-\frac{q}{m} \mathbf{p} \cdot \mathbf{A}(0) + \frac{q^2}{2m} \mathbf{A}^2(0). \tag{15}$$

We know from general arguments (see *Photons and Atoms—Introduction to Quantum Electrodynamics*, Exercise 7 in Complement E_{IV}) that the results obtained are the same in the two representations. This equivalence can be verified directly in the case of Rayleigh scattering. With the Coulomb-gauge Hamiltonian, $\mathcal{S}_{fi}$ appears as the sum of two terms. The first term is similar to (5), except that the matrix elements of $\mathbf{d}$ are replaced by those of $\mathbf{p}$. The second term is derived from the $\mathbf{A}^2$ term that directly couples $|\varphi_i\rangle$ to $|\varphi_f\rangle$. If, in the sum over the intermediate levels, we ignore $\hbar\omega$ compared with $(E_b - E_a)$ in the denominator, the result is zero, because the term in $\mathbf{A}^2$ exactly compensates for the term involving a summation over b (to verify this point, one has just to replace $\langle b|\mathbf{p} \cdot \boldsymbol{\varepsilon}|a\rangle$ by $(m/i\hbar)(E_a - E_b)\langle b|\mathbf{r} \cdot \boldsymbol{\varepsilon}|a\rangle$ so that the closure relation over b can be introduced). This proves that the approximation made with the electric dipole Hamiltonian [zeroth-order calculation in $\hbar\omega/(E_b - E_a)$] cannot be so hastily used with the Coulomb-gauge Hamiltonian. It is actually necessary in this representation to carry out the expansion of the energy denominators $(E_a + \hbar\omega - E_b)^{-1}$ and $(E_a - \hbar\omega - E_b)^{-1}$ to higher orders in $\hbar\omega/(E_b - E_a)$. If, in the first-order term, we replace all the matrix elements of $\mathbf{p} \cdot \boldsymbol{\varepsilon}$ by those of $\mathbf{r} \cdot \boldsymbol{\varepsilon}$, and if we introduce a closure relation, we can show that this term is also equal to zero. It is in fact the second-order term in $\hbar\omega/(E_b - E_a)$ of the expansion that coincides with (6). Using the Coulomb-gauge Hamiltonian thus leads, for this problem, to longer calculations, although, of course, the final result is the same.

4. THOMSON SCATTERING

The purpose of this exercise is to calculate the elastic cross section for the scattering of a photon by an atomic electron in the limit where the photon energy $\hbar\omega$ is large compared with the ionization energy E_I of the atom ($\hbar\omega \gg E_I$). By contrast, $\hbar\omega$ is assumed to be small compared with αmc^2 (where α is the fine-structure constant), so that the long-wavelength approximation can be applied.

a) The system, placed at the coordinate system origin, is assumed to be initially in the state $|\varphi_i\rangle = |a; \mathbf{k}\boldsymbol{\varepsilon}\rangle$ (atom in the ground state $|a\rangle$ in the presence of a photon $\mathbf{k}\boldsymbol{\varepsilon}$). Use the Coulomb-gauge interaction Hamiltonian

$$H_I = H_{I1} + H_{I2} \tag{1}$$

with $H_{I1} = -(q/m)\mathbf{p} \cdot \mathbf{A(0)}$ and $H_{I2} = q^2\mathbf{A}^2(0)/2m$. Calculate to second order in q the contributions of H_{I1} and H_{I2} to the transition matrix element $\mathscr{T}_{fi}$ between the initial state $|\varphi_i\rangle$ and the final state $|\varphi_f\rangle = |a; \mathbf{k}'\boldsymbol{\varepsilon}'\rangle$ (the excited states of the atomic Hamiltonian are called b).

b) Show that the condition $\hbar\omega \gg E_I$ results in the contribution of H_{I1} to $\mathscr{T}_{fi}$ being small compared with the contribution of H_{I2}.

c) By keeping only the contribution of H_{I2} to $\mathscr{T}_{fi}$, calculate the differential cross section for the scattering of a photon $\mathbf{k}\boldsymbol{\varepsilon}$ in a solid angle $d\Omega'$ about $\mathbf{k}'$ and with a polarization $\boldsymbol{\varepsilon}'$. Calculate the total cross section and show that this result coincides with the result obtained for a classical elastically bound electron (*):

$$\sigma_{\text{tot}} = \frac{8\pi}{3}r_0^2 \tag{2}$$

with r_0 being the classical electron radius ($r_0 = q^2/4\pi\varepsilon_0 mc^2$).

Solution

a) By using the expansion in modes of $\mathbf{A(0)}$ [see Appendix, formula (28)], we find

$$H_{I2} = \frac{q^2}{2m}\frac{\hbar}{2\varepsilon_0 L^3}\sum_{\mathbf{k},\boldsymbol{\varepsilon}}\sum_{\mathbf{k}',\boldsymbol{\varepsilon}'}\frac{1}{\sqrt{\omega\omega'}}(\boldsymbol{\varepsilon}\cdot\boldsymbol{\varepsilon}')(a_{\mathbf{k}\boldsymbol{\varepsilon}} + a_{\mathbf{k}\boldsymbol{\varepsilon}}^+)(a_{\mathbf{k}'\boldsymbol{\varepsilon}'} + a_{\mathbf{k}'\boldsymbol{\varepsilon}'}^+) \tag{3}$$

which contains terms in $a_{\mathbf{k}\boldsymbol{\varepsilon}}a_{\mathbf{k}'\boldsymbol{\varepsilon}'}^+$ and $a_{\mathbf{k}'\boldsymbol{\varepsilon}'}^+a_{\mathbf{k}\boldsymbol{\varepsilon}}$ allowing $|\varphi_i\rangle$ to be directly coupled to $|\varphi_f\rangle$. It follows that the contribution $\mathscr{T}_{fi}^{(2)}$ of H_{I2} to $\mathscr{T}_{fi}$ is equal to

$$\mathscr{T}_{fi}^{(2)} = \frac{q^2}{2m}\frac{\hbar}{\varepsilon_0 L^3\sqrt{\omega\omega'}}(\boldsymbol{\varepsilon}\cdot\boldsymbol{\varepsilon}'). \tag{4}$$

Now calculate the second-order contribution of H_{I1}. The intermediate state of the process can be a zero-photon state $|b; 0\rangle$ or a two-photon state $|b; \mathbf{k}\boldsymbol{\varepsilon}, \mathbf{k}'\boldsymbol{\varepsilon}'\rangle$, so that the contribution $\mathscr{T}_{fi}^{(1)}$ of H_{I1} to the transition matrix is equal to:

$$\mathscr{T}_{fi}^{(1)} = \frac{q^2}{m^2}\frac{\hbar}{2\varepsilon_0 L^3\sqrt{\omega\omega'}}\sum_b\left[\frac{\langle a|\mathbf{p}\cdot\boldsymbol{\varepsilon}'|b\rangle\langle b|\mathbf{p}\cdot\boldsymbol{\varepsilon}|a\rangle}{E_a + \hbar\omega - E_b + i\varepsilon} + \frac{\langle a|\mathbf{p}\cdot\boldsymbol{\varepsilon}|b\rangle\langle b|\mathbf{p}\cdot\boldsymbol{\varepsilon}'|a\rangle}{E_a - E_b - \hbar\omega'}\right] \tag{5}$$

where $\omega' = \omega$ by conservation of energy.

(*) See *Photons and Atoms—Introduction to Quantum Electrodynamics*, Exercise 7 in Complement C_1.

b) The levels b for which the matrix element is sufficiently large to give a significant contribution to (5) are those satisfying $|E_b - E_a| \sim E_I \ll \hbar\omega$. If we then neglect $(E_b - E_a)$ compared with $\hbar\omega$ and $-\hbar\omega'$ $(= -\hbar\omega)$, the two terms of (5) exactly cancel each other. To find the order of magnitude of $\mathscr{T}_{fi}^{(1)}$, it is thus necessary to carry out the expansion at a higher order in $(E_b - E_a)/\hbar\omega$. We then find

$$\mathscr{T}_{fi}^{(1)} \sim \frac{q^2}{m^2} \frac{\hbar}{2\varepsilon_0 L^3 \sqrt{\omega\omega'}} \sum_b \frac{(E_b - E_a)|\langle a|\mathbf{p}\cdot\boldsymbol{\varepsilon}|b\rangle|^2}{\hbar^2\omega^2}. \tag{6}$$

By replacing $(E_b - E_a)$ by E_I in the numerator and by then using the closure relation over b, we obtain

$$\mathscr{T}_{fi}^{(1)} \sim \frac{q^2}{m} \frac{\hbar}{2\varepsilon_0 L^3 \sqrt{\omega\omega'}} \frac{E_I}{\hbar\omega} \frac{\langle a|(p^2/m)|a\rangle}{\hbar\omega}. \tag{7}$$

Now $\langle a|(p^2/m)|a\rangle$ is twice the average kinetic energy of the ground state, and its order of magnitude is E_I. Finally, the comparison of (4) with (7) shows that

$$\mathscr{T}_{fi}^{(1)} \sim \mathscr{T}_{fi}^{(2)}\left(\frac{E_I}{\hbar\omega}\right)^2. \tag{8}$$

$\mathscr{T}_{fi}^{(2)}$ is thus larger than $\mathscr{T}_{fi}^{(1)}$ by a factor on the order of $(\hbar\omega/E_I)^2$.

c) To calculate the differential cross section, we proceed as we did in Exercise 3 (Question b). The transition probability per unit time and per unit solid angle is equal to

$$\frac{\delta w_{fi}}{\delta\Omega'} = \frac{2\pi}{\hbar}\left(\frac{q^2}{2m}\frac{\hbar}{\varepsilon_0 L^3\omega}\right)^2 (\boldsymbol{\varepsilon}\cdot\boldsymbol{\varepsilon}')^2 \frac{L^3}{8\pi^3}\frac{(\hbar c k)^2}{\hbar^3 c^3}$$

$$= \frac{e^4}{m^2 c^3 L^3}(\boldsymbol{\varepsilon}\cdot\boldsymbol{\varepsilon}')^2 \tag{9}$$

where $e^2 = q^2/4\pi\varepsilon_0$. If we divide by the photon flux, which is equal to c/L^3, we find that the differential cross section is

$$\frac{d\sigma}{d\Omega'} = r_0^2(\boldsymbol{\varepsilon}\cdot\boldsymbol{\varepsilon}')^2. \tag{10}$$

To find the total cross section, it is necessary to sum $(\boldsymbol{\varepsilon}\cdot\boldsymbol{\varepsilon}')^2$ over the two polarizations orthogonal to $\mathbf{k}'$ and to make the angular average. Such a calculation was made in Exercise 3 (Question c) and gives $8\pi/3$. The total cross section is thus the same as the one obtained in classical electrodynamics for an elastically bound electron.

Remark on the Choice of the Interaction Hamiltonian

It is, of course, possible to find (2) by using the electric dipole Hamiltonian. The calculations are, however, slightly more difficult. There is no direct coupling between $|\varphi_i\rangle$ and $|\varphi_f\rangle$, and $\mathcal{T}_{fi}$ is equal, on the energy shell, to

$$\mathcal{T}_{fi} = \frac{\hbar\omega}{2\varepsilon_0 L^3} \sum_b \left[\frac{\langle a|\mathbf{d} \cdot \mathbf{\varepsilon}'|b\rangle\langle b|\mathbf{d} \cdot \mathbf{\varepsilon}|a\rangle}{E_a + \hbar\omega - E_b + i\varepsilon} + \frac{\langle a|\mathbf{d} \cdot \mathbf{\varepsilon}|b\rangle\langle b|\mathbf{d} \cdot \mathbf{\varepsilon}'|a\rangle}{E_a - E_b - \hbar\omega'} \right]. \quad (11)$$

We see that, if, in the energy denominators, $E_b - E_a$ is neglected as compared with $\hbar\omega$, the two terms in brackets cancel out each other because $\omega' = \omega$. It is then necessary to carry out the expansion of the energy denominators at the next order in $(E_b - E_a)/\hbar\omega$. By replacing $(E_a - E_b)\langle b|\mathbf{d} \cdot \mathbf{\varepsilon}|a\rangle$ by $(i\hbar q/m)\langle b|\mathbf{p} \cdot \mathbf{\varepsilon}|a\rangle$, we can then introduce the closure relation over b and thus arrive at a value for $\mathcal{T}_{fi}$ which is identical to the value found directly with H_{I2} [formula (4) with $\omega = \omega'$].

5. RESONANT SCATTERING

Consider an atom A whose center of mass is assumed to be fixed at a point 0, taken as the origin. Let $|a\rangle$ be the ground state of this atom and $|b\rangle$ be the first excited state, both assumed to be nondegenerate. The Bohr frequency of the transition $b \leftrightarrow a$ is written as ω_0. For the sake of simplicity, only the component d_z of the electric dipole operator of the atom is assumed to have a nonzero matrix element between a and b. It is assumed to be real and is written d_{ba}.

$$\langle b|d_z|a\rangle = d_{ba} \quad (1.a)$$

$$\langle b|d_x|a\rangle = \langle b|d_y|a\rangle = 0. \quad (1.b)$$

The spontaneous emission rate from the state b equals (see Exercise 1):

$$\Gamma = \frac{1}{3\pi} \left(\frac{\omega_0}{c} \right)^3 \frac{d_{ba}^2}{\varepsilon_0 \hbar}. \quad (2)$$

We consider a photon with wave vector $\mathbf{k}$ and polarization $\mathbf{\varepsilon}$, whose frequency ω is close to the resonance frequency ω_0. The transition rate for the system to go from the initial state

$$|\varphi_i\rangle = |a; \mathbf{k}\mathbf{\varepsilon}\rangle \quad (3.a)$$

to the final state

$$|\varphi_f\rangle = |a; \mathbf{k}'\mathbf{\varepsilon}'\rangle \quad (3.b)$$

is equal to [formula (B.18) in Chapter I]:

$$w_{fi} = \frac{2\pi}{\hbar} |\mathcal{F}_{fi}|^2 \delta^{(T)}(E_f - E_i) \tag{4}$$

where $\mathcal{F}_{fi}$ is given by expression (C.5) in Chapter II:

$$\mathcal{F}_{fi} = \frac{\langle a; \mathbf{k'}\boldsymbol{\varepsilon'}|H_I'|b; 0\rangle\langle b; 0|H_I'|a; \mathbf{k}\boldsymbol{\varepsilon}\rangle}{\hbar\omega - \hbar\omega_0 + i\hbar(\Gamma/2)}. \tag{5}$$

H_I' is the interaction Hamiltonian in the long-wavelength approximation and in the electric dipole representation [see Appendix, formula (91)]

$$H_I' = -\mathbf{d} \cdot \mathbf{E}_\perp(0). \tag{6}$$

a) Find an explicit expression for $\mathcal{F}_{fi}$. Discuss its amplitude and its phase variation near resonance.

b) Show that the differential cross section for the scattering of a photon $\mathbf{k}\boldsymbol{\varepsilon}$ in a solid angle $d\Omega'$ about the direction $\boldsymbol{\kappa'}$, with a polarization $\boldsymbol{\varepsilon'}$ (perpendicular to $\boldsymbol{\kappa'}$) is equal to

$$\sigma(\mathbf{k}\boldsymbol{\varepsilon} \to \boldsymbol{\kappa'}\boldsymbol{\varepsilon'}) = \frac{9}{16\pi^2} \lambda_0^2 \varepsilon_z'^2 \varepsilon_z^2 \frac{(\Gamma/2)^2}{(\omega - \omega_0)^2 + (\Gamma/2)^2} \tag{7}$$

with $\lambda_0 = 2\pi c/\omega_0$.

c) Show that, at resonance, the total cross section for the scattering of a photon with polarization $\mathbf{e}_z$ is on the order of λ_0^2, with a numerical coefficient to be calculated.

Solution

a) Using expression (6) for H_I', the expression for $\mathbf{E}_\perp(0)$ [Appendix, formula (89)], and relations (1.a) and (1.b), we get

$$\langle b; 0|H_I'|a; \mathbf{k}\boldsymbol{\varepsilon}\rangle = -i\sqrt{\frac{\hbar\omega}{2\varepsilon_0 L^3}} \langle b|\boldsymbol{\varepsilon} \cdot \mathbf{d}|a\rangle = -i\varepsilon_z\sqrt{\frac{\hbar\omega}{2\varepsilon_0 L^3}} d_{ba}. \tag{8}$$

Thus

$$\mathcal{F}_{fi} = \varepsilon_z\varepsilon_z' \frac{d_{ba}^2}{2\varepsilon_0 L^3}\sqrt{\omega\omega'} \frac{1}{\omega - \omega_0 + i(\Gamma/2)}. \tag{9}$$

The transition matrix element $\mathcal{S}_{fi}$ is a complex number which exhibits a resonance for $\omega = \omega_0$. Its amplitude is maximal at resonance, and its phase varies from π to 0 when ω goes through ω_0. $\mathcal{S}_{fi}$ is purely imaginary at resonance. Because we are interested in frequencies near ω_0, in the coefficient of the resonant term, we can replace ω and ω' by ω_0, and d_{ba}^2 by its expression as a function of Γ taken from (2). This yields

$$\mathcal{S}_{fi} \simeq \frac{3\pi\hbar c^3}{\omega_0^2 L^3} \varepsilon_z \varepsilon'_z \frac{\Gamma/2}{\omega - \omega_0 + i(\Gamma/2)}. \tag{10}$$

b) The number of final states with energy between $\hbar c k'$ and $\hbar c(k' + dk')$ whose wave vector points inside the solid angle $d\Omega'$ about $\mathbf{\kappa}'$ equals, using formula (46) of Complement A_I,

$$\rho(\hbar c k')\hbar c \, dk' \, d\Omega' = \frac{L^3}{8\pi^3} k'^2 \, dk' \, d\Omega'. \tag{11}$$

The total transition rate to all the relevant final states is then equal to

$$\sum_f w_{fi} = \frac{2\pi}{\hbar} d\Omega' \int_0^\infty \frac{k'^2 \, dk'}{(2\pi/L)^3} |\mathcal{S}_{fi}|^2 \delta(\hbar c k' - \hbar c k)$$

$$= d\Omega' \frac{9}{4} \frac{k^2 c^5}{\omega_0^4 L^3} \varepsilon_z^2 \varepsilon_z'^2 \frac{(\Gamma/2)^2}{(\omega - \omega_0)^2 + (\Gamma/2)^2}. \tag{12}$$

As we did previously, we now replace ck by ω_0 in the factors preceding the resonant term.

The initial state corresponds to a photon in the volume L^3, moving at velocity c, and corresponds to a flux

$$\phi = \frac{c}{L^3}. \tag{13}$$

The differential cross section is then

$$\sigma(\mathbf{\kappa}\varepsilon \to \mathbf{\kappa}'\varepsilon') = \frac{\sum_f w_{fi}}{d\Omega' \, \phi} = \frac{9}{16\pi^2} \lambda_0^2 \varepsilon_z^2 \varepsilon_z'^2 \frac{(\Gamma/2)^2}{(\omega - \omega_0)^2 + (\Gamma/2)^2}. \tag{14}$$

It depends implicitly on $\mathbf{\kappa}'$ through ε', which is perpendicular to $\mathbf{\kappa}'$.

c) If we do not select the polarization in the detection, it is necessary to sum the probabilities corresponding to two polarizations ε'_1 and ε'_2 that are orthogonal to each other (and to $\mathbf{\kappa}'$). We obtain [formula (54) from Complement A_I]

$$\sum_{i=1,2} \varepsilon_{iz}'^2 = 1 - \kappa_z'^2. \tag{15}$$

The total scattering cross section is obtained by then summing the differential cross section over all the directions $\mathbf{\kappa}'$.

$$\sigma_{\text{tot}} = \int d\Omega' \sum_{i=1,2} \sigma(\mathbf{k}\varepsilon_z \to \mathbf{\kappa}'\varepsilon'_i). \tag{16}$$

Using $(e_z)_z = 1$, and

$$\int d\Omega' (1 - \kappa_z'^2) = \frac{8\pi}{3}. \tag{17}$$

We get, for $\omega = \omega_0$,

$$\sigma_{tot}^{res} = \frac{3}{2\pi}\lambda_0^2. \tag{18}$$

This result is identical to the one obtained for the resonant scattering of a classical wave by an elastically bound classical electron (see *Photons and Atoms—Introduction to Quantum Electrodynamics*, Exercise 7 in Complement C_1).

6. OPTICAL DETECTION OF A LEVEL CROSSING BETWEEN TWO EXCITED ATOMIC STATES (*)

An atom A, placed at the coordinate system origin 0, has two excited states $|b\rangle$ and $|c\rangle$, with energies E_b and E_c that are close to each other. These states are connected to the ground state $|a\rangle$ by electric dipole transitions characterized by the matrix elements

$$\mathbf{d}_{ab} = \langle a|\mathbf{d}|b\rangle \qquad \mathbf{d}_{ac} = \langle a|\mathbf{d}|c\rangle. \tag{1}$$

The states $|b\rangle$ and $|c\rangle$ have radiative lifetimes $1/\Gamma_b$ and $1/\Gamma_c$. We assume that the energies E_b and E_c can be modified by means of an external parameter x (a magnetic field, for example) and that the levels cross each other for a particular value x_0 of the parameter (Figure 1).

Starting with the atom in state $|a\rangle$ in the presence of a photon $\mathbf{k}_1\boldsymbol{\varepsilon}_1$, we want to study the resonant scattering process that consists of going from the initial state $|a; \mathbf{k}_1\boldsymbol{\varepsilon}_1\rangle$ of the global system to the final state $|a; \mathbf{k}_2\boldsymbol{\varepsilon}_2\rangle$. Due to the quasi-resonance of the incident photon with the Bohr frequencies $\omega_{ba} = (E_b - E_a)/\hbar$ and $\omega_{ca} = (E_c - E_a)/\hbar$, the states $|b\rangle$ and $|c\rangle$ play a preponderant role in the scattering. An obvious generalization of formula (C.5) from Chapter II gives the following expression for the transition matrix element:

$$\mathscr{T}_{fi} = \frac{\langle a; \mathbf{k}_2\boldsymbol{\varepsilon}_2|H_I'|b; 0\rangle\langle b; 0|H_I'|a; \mathbf{k}_1\boldsymbol{\varepsilon}_1\rangle}{\hbar\omega_1 - \hbar\omega_{ba} + i\hbar(\Gamma_b/2)} +$$
$$+ \frac{\langle a; \mathbf{k}_2\boldsymbol{\varepsilon}_2|H_I'|c; 0\rangle\langle c; 0|H_I'|a; \mathbf{k}_1\boldsymbol{\varepsilon}_1\rangle}{\hbar\omega_1 - \hbar\omega_{ca} + i\hbar(\Gamma_c/2)} \tag{2}$$

(*) For the experimental observation of such effects, see W. Hanle, *Z. Phys.*, **30**, 93 (1924); P. A. Franken, *Phys. Rev.*, **121**, 508 (1961).

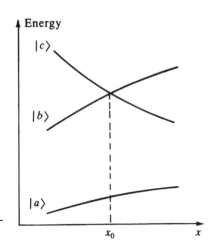

Figure 1. Level crossing between two ex-
cited states $|b\rangle$ and $|c\rangle$.

where

$$H'_I = -\mathbf{d} \cdot \mathbf{E}_\perp(0) \tag{3}$$

is the electric dipole Hamiltonian.

a) Find the expression for $\mathcal{T}_{fi}$ as a function of $\mathbf{d}_{ab}$ and $\mathbf{d}_{ac}$.

b) Following a procedure similar to that used in Exercise 5, show that the scattering cross section is the sum of resonant cross sections $\sigma(b)$ and $\sigma(c)$ relative to the states b and c, and of an additional term $\delta\sigma$ that is to be calculated.

c) The atom A interacts with a light beam having a spectral width Δ that is large compared with Γ_b and Γ_c. This beam is assumed to be described by a flux ϕ of photons whose frequencies are randomly distributed according to a probability distribution $f(\omega)$, centered at $\omega = \omega_{ba}$ and of width Δ with $\omega_{ba} \gg \Delta \gg \Gamma_b, \Gamma_c$.

α) The light intensity I, scattered in a solid angle $\delta\Omega_2$ about $\kappa_2 = k_2/k_2$, with the polarization ε_2, consists of three parts $I(b)$, $I(c)$, and δI corresponding, respectively, to $\sigma(b)$, $\sigma(c)$, and $\delta\sigma$. Calculate these three quantities. Show that the first two quantities vary only slightly when E_b and E_c vary over an interval which is large compared to Γ_b and Γ_c, but small compared to Δ.

β) Show that δI varies in a resonant fashion when x is varied about x_0. What is the width of this resonance? What conditions must the polarizations ε_1 and ε_2 satisfy for this resonance to be observed?

Solution

a) The matrix elements of H_I' are calculated in Exercise 5. From them we deduce

$$\mathscr{T}_{fi} = \frac{\sqrt{\omega_1\omega_2}}{2\varepsilon_0 L^3}\left\{\frac{(\boldsymbol{\varepsilon}_2\cdot\mathbf{d}_{ab})(\boldsymbol{\varepsilon}_1\cdot\mathbf{d}_{ab}^*)}{\omega_1 - \omega_{ba} + i(\Gamma_b/2)} + \frac{(\boldsymbol{\varepsilon}_2\cdot\mathbf{d}_{ac})(\boldsymbol{\varepsilon}_1\cdot\mathbf{d}_{ac}^*)}{\omega_1 - \omega_{ca} + i(\Gamma_c/2)}\right\}. \tag{4}$$

b) Using Exercise 5, the scattering cross section is written

$$\sigma(\mathbf{k}_1\boldsymbol{\varepsilon}_1 \to \boldsymbol{\kappa}_2\boldsymbol{\varepsilon}_2) = \frac{\Sigma_f w_{fi}}{\phi\,\delta\Omega_2} \tag{5}$$

where $\phi = c/L^3$ and

$$\sum_f w_{fi} = \frac{2\pi}{\hbar}\delta\Omega_2\int_0^\infty \frac{k_2^2\,dk_2}{(2\pi/L)^3}|\mathscr{T}_{fi}|^2\delta(\hbar ck_2 - \hbar ck_1). \tag{6}$$

This gives

$$\sigma = \frac{k_1^4}{16\pi^2\hbar^2\varepsilon_0^2}\left|\frac{S_b}{\omega_1 - \omega_{ba} + i(\Gamma_b/2)} + \frac{S_c}{\omega_1 - \omega_{ca} + i(\Gamma_c/2)}\right|^2 \tag{7}$$

where

$$S_b = (\boldsymbol{\varepsilon}_2\cdot\mathbf{d}_{ab})(\boldsymbol{\varepsilon}_1\cdot\mathbf{d}_{ab}^*) \tag{8.a}$$

$$S_c = (\boldsymbol{\varepsilon}_2\cdot\mathbf{d}_{ac})(\boldsymbol{\varepsilon}_1\cdot\mathbf{d}_{ac}^*). \tag{8.b}$$

Expanding the squared modulus of the sum yields

$$\sigma = \sigma(b) + \sigma(c) + \delta\sigma \tag{9}$$

where

$$\sigma(b) = \frac{k_1^4}{16\pi^2\hbar^2\varepsilon_0^2}\frac{|S_b|^2}{(\omega_1 - \omega_{ba})^2 + (\Gamma_b/2)^2} \tag{10}$$

and an analogous expression for $\sigma(c)$, and

$$\delta\sigma = \frac{k_1^4}{16\pi^2\hbar^2\varepsilon_0^2}\left\{\frac{S_b S_c^*}{[\omega_1 - \omega_{ba} + i(\Gamma_b/2)][\omega_1 - \omega_{ca} - i(\Gamma_c/2)]} + \text{c.c.}\right\}. \tag{11}$$

c) According to the definition of the differential cross section, the light intensity scattered in the solid angle $\delta\Omega_2$ about $\boldsymbol{\kappa}_2$ is

$$I = \phi\,\delta\Omega_2\int d\omega_1\,\sigma(\mathbf{k}_1\boldsymbol{\varepsilon}_1 \to \boldsymbol{\kappa}_2\boldsymbol{\varepsilon}_2)f(\omega_1). \tag{12}$$

α) By replacing σ by its expression (9), we obtain

$$I = I(b) + I(c) + \delta I \tag{13}$$

where $I(b)$, $I(c)$, and δI are expressions analogous to (12), except for the replacement of σ by $\sigma(b)$, $\sigma(c)$, or $\delta\sigma$.

Expression (10) shows that $\sigma(b)$ is nonzero only near $\omega_1 = \omega_{ba}$ over a width Γ_b. When the integral over ω_1 is evaluated, the fact that the width Δ of $f(\omega_1)$ is large compared to Γ_b allows one to approximate the Lorentzian by a function $\delta(\omega_1 - \omega_{ba})$:

$$I(b) \simeq \frac{k_{ba}^4}{8\pi\hbar^2\varepsilon_0^2} \frac{|S_b|^2}{\Gamma_b} f(\omega_{ba}). \tag{14}$$

The expression for $I(c)$ is similar. When E_b is swept by some $\hbar\Gamma_b$, $I(b)$ varies in relative value by a quantity on the order of Γ_b/Δ, which is very small.

β) Similarly, $\delta\sigma$ is important only in the neighborhood of $\omega_1 \simeq \omega_{ba} \simeq \omega_{ca}$. In this domain, we can take $f(\omega_1)$ as constant and equal to $f(\omega_{ba})$. The integral of $\delta\sigma$ over ω_1 is done by the residue method, and gives

$$\delta I = \frac{k_{ba}^4}{8\pi\hbar^2\varepsilon_0^2} f(\omega_{ba}) \left\{ \frac{iS_b S_c^*}{\omega_{cb} + i[(\Gamma_b + \Gamma_c)/2]} + \text{c.c.} \right\} \tag{15}$$

where $\omega_{cb} = (E_c - E_b)/\hbar$. δI varies resonantly around $\omega_{cb} = 0$, with a width on the order of $(\Gamma_b + \Gamma_c)$. When x is swept about x_0, the shape of the resonance is Lorentzian, a mixture of absorption and dispersion depending on the value of the complex number $S_b S_c^*$. For δI to be nonzero, it is necessary that

$$S_b S_c^* = (\boldsymbol{\varepsilon}_2 \cdot \mathbf{d}_{ab})(\boldsymbol{\varepsilon}_1 \cdot \mathbf{d}_{ab}^*)(\boldsymbol{\varepsilon}_2 \cdot \mathbf{d}_{ac}^*)(\boldsymbol{\varepsilon}_1 \cdot \mathbf{d}_{ac}) \tag{16}$$

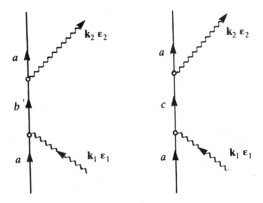

Figure 2. Scattering amplitudes whose interference gives rise to the level-crossing resonance.

be nonzero, which implies that each of the factors must be different from zero. Physically, it is then necessary that each polarization $\boldsymbol{\varepsilon}_1$ and $\boldsymbol{\varepsilon}_2$ be able to excite both states b and c. They respectively excite and detect *linear superpositions* of the states $|b\rangle$ and $|c\rangle$. At the crossing point, the superposition created by the incident photons is stationary. It persists until the photon $\mathbf{k}_2\boldsymbol{\varepsilon}_2$ is reemitted. Far from this point, the superposition of states evolves before the reemission of the photon, which thus occurs from a different state, so that the detected light intensity is, in general, different.

This phenomenon can also be understood as an interference between *scattering amplitudes* allowing the passage from the same initial state $|a;\mathbf{k}_1\boldsymbol{\varepsilon}_1\rangle$ to the same final state $|a;\mathbf{k}_2\boldsymbol{\varepsilon}_2\rangle$ via two different "paths," going through either state $|b;0\rangle$ or state $|c;0\rangle$ (Figure 2). The scattered intensity is maximal when the two amplitudes are in phase. The phase difference between these two amplitudes depends on the energy splitting between $|b\rangle$ and $|c\rangle$ and varies rapidly versus $E_b - E_c$ for resonant excitation. This is manifested by a rapid variation of the scattered intensity over an interval on the order of $\hbar(\Gamma_b + \Gamma_c)$ around the crossing point.

7. RADIATIVE SHIFT OF AN ATOMIC LEVEL. BETHE FORMULA FOR THE LAMB SHIFT

The purpose of this exercise is to calculate the radiative shifts of the levels of the hydrogen atom, and to obtain the expression that was first derived by Bethe (*) for the $2s$ level (Lamb shift). It is interesting to simultaneously carry out the calculation in both the $\mathbf{A} \cdot \mathbf{p}$ and $\mathbf{E} \cdot \mathbf{r}$ representations, because, although they both lead to the same result, the two calculations look rather different. A third representation, the Pauli–Fierz representation, is discussed in the last question. It is particularly convenient because the correction to the electron mass already appears explicitly in the Hamiltonian.

We introduce the following simplifications: (a) Only the modes having a wave vector with a modulus less than a bound k_M such as $k_M a_0 \ll 1$ (a_0 is the Bohr radius) are taken into account. The long-wavelength approximation is made for these modes. (b) The electron is considered as a spinless particle of mass m and charge q. The proton is assumed to be infinitely heavy, fixed at the origin, so that the potential energy at point $\mathbf{r}$ is

$$V_0(\mathbf{r}) = -\frac{q^2}{4\pi\varepsilon_0 r}. \tag{1}$$

The operators $\mathbf{r}$ and $\mathbf{p}$, being the usual quantum operators of the electron, the Hamiltonians of the system in the $\mathbf{A} \cdot \mathbf{p}$ and $\mathbf{E} \cdot \mathbf{r}$ representations are

(*) See H. A. Bethe, *Phys. Rev.*, **72**, 339 (1947); Bethe and Salpeter (§19).

written, respectively,

$$H = H_0 + H_I \tag{2}$$

and

$$H' = H_0 + \varepsilon'_{dip} + H'_I \tag{3}$$

where

$$H_0 = H_A + H_R \tag{4}$$

with

$$H_A = \frac{\mathbf{p}^2}{2m} + V_0(\mathbf{r}) + \varepsilon_{Coul} \qquad \varepsilon_{Coul} = \frac{q^2}{4\pi^2\varepsilon_0} \int_0^{k_M} dk = \frac{q^2 k_M}{4\pi^2\varepsilon_0} \tag{5}$$

$$H_R = \sum_j \hbar\omega_j \left(a_j^+ a_j + \tfrac{1}{2}\right). \tag{6}$$

Within the long-wavelength approximation, the interaction Hamiltonians are written

$$H_I = H_{I1} + H_{I2} = -\frac{q}{m}\mathbf{p} \cdot \mathbf{A}(\mathbf{0}) + \frac{q^2}{2m}\mathbf{A}^2(\mathbf{0}) \tag{7}$$

$$H'_I = -q\mathbf{r} \cdot \mathbf{E}_\perp(\mathbf{0}). \tag{8}$$

The dipole self-energy equals

$$\varepsilon'_{dip} = \frac{q^2}{2\varepsilon_0 L^3} \sum_j (\boldsymbol{\varepsilon}_j \cdot \mathbf{r})^2. \tag{9}$$

Finally the mode expansions of $\mathbf{A}(\mathbf{0})$ and $\mathbf{E}_\perp(\mathbf{0})$ are

$$\mathbf{A}(\mathbf{0}) = \sum_j \sqrt{\frac{\hbar}{2\varepsilon_0\omega_j L^3}}\, \boldsymbol{\varepsilon}_j \left(a_j + a_j^+\right) \tag{10}$$

$$\mathbf{E}_\perp(\mathbf{0}) = i\sum_j \sqrt{\frac{\hbar\omega_j}{2\varepsilon_0 L^3}}\, \boldsymbol{\varepsilon}_j \left(a_j - a_j^+\right). \tag{11}$$

Let $|a; 0\rangle$ be the ground state of H_A in the presence of the vacuum. The corrections to the energy E_a of this state, produced by the interaction of the electron with the transverse field, are written, respectively, in the two representations, and to second order in q:

$$\hbar\Delta_a = \langle a;0|H_{I2}|a;0\rangle +$$

$$+ \sum_j \sum_b \frac{\langle a;0|H_{I1}|b;k_j\varepsilon_j\rangle\langle b;k_j\varepsilon_j|H_{I1}|a;0\rangle}{E_a - (E_b + \hbar\omega_j)} \qquad (12)$$

$$\hbar\Delta'_a = \langle a;0|\varepsilon'_{dip}|a;0\rangle +$$

$$+ \sum_j \sum_b \frac{\langle a;0|H'_I|b;k_j\varepsilon_j\rangle\langle b;k_j\varepsilon_j|H'_I|a;0\rangle}{E_a - (E_b + \hbar\omega_j)}. \qquad (13)$$

a) By using the following notation:

$$k_{ba} = (E_b - E_a)/\hbar c \qquad (14.a)$$

$$\mathbf{p}_{ab} = \langle a|\mathbf{p}|b\rangle \qquad (14.b)$$

$$\mathbf{r}_{ab} = \langle a|\mathbf{r}|b\rangle \qquad (14.c)$$

show that Δ_a and Δ'_a can be put in the form

$$\hbar\Delta_a = \frac{r_0}{\pi}\int_0^{k_M}\hbar ck\,dk - \frac{2r_0}{3\pi m}\int_0^{k_M}k\,dk \sum_b \frac{|\mathbf{p}_{ab}|^2}{k_{ba}+k} \qquad (15)$$

$$\hbar\Delta'_a = \frac{2r_0}{3\pi}mc^2\sum_b |\mathbf{r}_{ab}|^2\int_0^{k_M}k^2\,dk - \frac{2r_0}{3\pi}mc^2\int_0^{k_M}k^3\,dk \sum_b \frac{|\mathbf{r}_{ab}|^2}{k_{ba}+k} \qquad (16)$$

where r_0 is a constant to be calculated.

b) We assume that k_M is large compared with all the quantities k_{ba} relative to the states that contribute appreciably to the sums over $|b\rangle$ in expressions (12) and (13). Expand $\hbar\Delta_a$ and $\hbar\Delta'_a$ in powers of k_M in the form

$$\hbar\Delta_a = D_a + P_a + L_a \qquad (17)$$

$$\hbar\Delta'_a = T'_a + D'_a + P'_a + L'_a \qquad (18)$$

where T, D, P, and L are expressions, respectively, of order k_M^3, k_M^2, k_M, and $\ln k_M$. The terms of order $1/k_M, (1/k_M)^2, \ldots$, are neglected. Calculate these different terms and show that T'_a is zero.

c) Derive the relation

$$imck_{ab}\mathbf{r}_{ab} = \mathbf{p}_{ab} \tag{19}$$

and demonstrate that $D'_a = D_a$, $P'_a = P_a$, and $L'_a = L_a$. By introducing the quantity

$$\delta m = \frac{4}{3}\frac{\varepsilon_{Coul}}{c^2}, \tag{20}$$

physically interpret the term P_a.

d) The Bethe formula is the expression for the last term L_a of (17). By introducing the average excitation energy $\hbar c K_a$ with the formula

$$\left(\sum_b |\mathbf{p}_{ab}|^2 k_{ba}\right)\ln\frac{k_M}{K_a} = \sum_b |\mathbf{p}_{ab}|^2 k_{ba}\ln\frac{k_M}{k_{ba}} \tag{21}$$

and by using the double commutator of $\mathbf{p}$ with H_0, show that L_a can be written in the form

$$L_a = \frac{\alpha}{3\pi}\left(\frac{\hbar}{mc}\right)^2\left(\ln\frac{k_M}{K_a}\right)\langle a|\frac{q^2}{\varepsilon_0}\delta(\mathbf{r})|a\rangle \tag{22}$$

($\alpha = q^2/4\pi\varepsilon_0\hbar c$ is the fine-structure constant).

e) Generalize expression (17) to the energy levels $|nlm\rangle$ other than the ground state of the hydrogen atom. Show that, within a manifold n, all the levels have the same shift $\hbar\Delta_{nl}$, except for the s states ($l = 0$).

f) In the Pauli–Fierz representation, the Hamiltonian is written [see formula (34′) in Complement B_{II}]

$$H'' = \frac{\mathbf{p}^2}{2m^*} + V_0(\mathbf{r} + \boldsymbol{\xi}) + H_{I2} + \varepsilon_{Coul} + H_R \tag{23}$$

where $m^* = m + \delta m$ and where $\boldsymbol{\xi}$ is given by

$$\boldsymbol{\xi} = \frac{q}{m}\mathbf{Z}(0) = \frac{q}{m}\sum_j\sqrt{\frac{\hbar}{2\varepsilon_0\omega_j L^3}}\left[\frac{\boldsymbol{\varepsilon}_j}{i\omega_j}(a_j - a_j^+)\right]. \tag{24}$$

α) Show that, to order 2 in the coupling with the transverse field, the shift $\hbar \Delta''_a$ of the state $|a; 0\rangle$ is written

$$\hbar \Delta''_a = D_a + L''_a \tag{25}$$

where

$$L''_a = \frac{1}{2} \sum_{\mu\nu} \langle 0|\xi_\mu \xi_\nu|0\rangle \langle a|\nabla_\mu \nabla_\nu V_0(\mathbf{r})|a\rangle +$$

$$+ \sum_j \sum_{\mu\nu} \sum_b \langle 0|\xi_\mu|\mathbf{k}_j \boldsymbol{\epsilon}_j\rangle\langle \mathbf{k}_j \boldsymbol{\epsilon}_j|\xi_\nu|0\rangle \frac{\langle a|\nabla_\mu V_0|b\rangle\langle b|\nabla_\nu V_0|a\rangle}{-\hbar c(k_{ba} + k_j)} \tag{26}$$

with μ and ν indicating the x, y, z components of the vectors $\boldsymbol{\xi}$ and ∇.
β) Derive the following formulas

$$\langle 0|\xi_\mu \xi_\nu|0\rangle = \frac{2\alpha}{3\pi}\left(\frac{\hbar}{mc}\right)^2 \delta_{\mu\nu} \int_{k_m}^{k_M} \frac{dk}{k} = \frac{2\alpha}{3\pi}\left(\frac{\hbar}{mc}\right)^2 \delta_{\mu\nu}\left(\ln \frac{k_M}{k_m}\right) \tag{27}$$

and

$$\sum_j \frac{\langle 0|\xi_\mu|\mathbf{k}_j \boldsymbol{\epsilon}_j\rangle\langle \mathbf{k}_j \boldsymbol{\epsilon}_j|\xi_\nu|0\rangle}{\hbar c k_{ba} + \hbar c k_j} = \frac{2\alpha}{3\pi}\left(\frac{\hbar}{mc}\right)^2 \delta_{\mu\nu} \int_{k_m}^{k_M} \frac{dk}{k} \frac{1}{\hbar c(k_{ba} + k)}$$

$$\simeq \frac{2\alpha}{3\pi}\left(\frac{\hbar}{mc}\right)^2 \frac{\delta_{\mu\nu}}{\hbar c k_{ba}} \ln \frac{k_{ba}}{k_m} \tag{28}$$

where k_{ba} is positive, small compared with k_M, and where k_m is a lower bound assumed to be very small, but nonzero.
γ) Using the foregoing results, show that L''_a can be written in the form

$$L''_a = \frac{\alpha}{3\pi}\left(\frac{\hbar}{mc}\right)^2 \left\{ \langle a|\Delta V_0(\mathbf{r})|a\rangle\left(\ln \frac{k_M}{k_m}\right) - \right.$$

$$\left. -2 \sum_{b \neq a} \frac{\langle a|\nabla V_0|b\rangle \cdot \langle b|\nabla V_0|a\rangle}{\hbar c k_{ba}} \ln \frac{k_{ba}}{k_m} \right\}. \tag{29}$$

Physically interpret the first term and the correction associated with the second term.

Using formula (21) and the commutator of $\mathbf{p}$ with H_0, show that L_a'' is equal to L_a.

Solution

a) Using (7) and (10)

$$\langle a;0|H_{I2}|a;0\rangle = \frac{q^2}{2m} \sum_{ij} \frac{\hbar}{2\varepsilon_0 L^3 \sqrt{\omega_i \omega_j}} \langle 0|(a_i + a_i^+)(a_j + a_j^+)|0\rangle. \qquad (30)$$

The last factor equals δ_{ij} and when the discrete sum is replaced by an integral, we have

$$\langle a;0|H_{I2}|a;0\rangle = \frac{q^2 \hbar}{4\varepsilon_0 m} \int \frac{d^3 k}{(2\pi)^3} \sum_{\varepsilon} \frac{1}{\omega} = \frac{r_0}{\pi} \int_0^{k_M} \hbar c k \, dk \qquad (31)$$

where

$$r_0 = \frac{q^2}{4\pi\varepsilon_0 m c^2} = \alpha\left(\frac{\hbar}{mc}\right) \qquad (32)$$

is the classical electron radius (α is the fine-structure constant). The matrix element of H_{I1} equals

$$\langle a;0|H_{I1}|b;\mathbf{k}_j\varepsilon_j\rangle = \sqrt{\frac{\hbar}{2\varepsilon_0 \omega_j L^3}}\left(\frac{-q}{m}\right)\langle a|\varepsilon_j \cdot \mathbf{p}|b\rangle \qquad (33)$$

and the second term of (12) is written

$$\sum_b \sum_j \frac{\hbar}{2\varepsilon_0 \omega_j L^3} \frac{q^2}{m^2} \frac{|\varepsilon_j \cdot \mathbf{p}_{ab}|^2}{\left[-\hbar c(k_{ba} + k_j)\right]}. \qquad (34)$$

Note that $\mathbf{p}_{aa} = \mathbf{0}$, so that the term $b = a$ does not contribute. The sum over the transverse polarizations is carried out by using formula (54) from Complement A_I. Taking the continuous limit then gives

$$\sum_b \frac{-q^2}{2\varepsilon_0 m^2 c^2} \int \frac{d^3 k}{(2\pi)^3} \frac{|\mathbf{p}_{ab}|^2 - |\mathbf{\kappa} \cdot \mathbf{p}_{ab}|^2}{k(k_{ba} + k)} = -\frac{2r_0}{3\pi m} \int_0^{k_M} k \, dk \sum_b \frac{|\mathbf{p}_{ab}|^2}{k_{ba} + k}. \qquad (35)$$

By inserting (31) and (35) into (12), we obtain expression (15) for $\hbar\Delta_a$.

The first term of expression (13) for $\hbar \Delta_a'$ is written, using (9),

$$\langle a;0|\varepsilon_{dip}'|a;0\rangle = \frac{q^2}{2\varepsilon_0 L^3} \sum_j \langle a|(\varepsilon_j \cdot \mathbf{r})^2|a\rangle$$

$$= \frac{q^2}{2\varepsilon_0} \int \frac{d^3k}{(2\pi)^3} \langle a|\mathbf{r}^2 - (\boldsymbol{\kappa} \cdot \mathbf{r})^2|a\rangle$$

$$= \frac{2}{3\pi} mc^2 r_0 \langle a|\mathbf{r}^2|a\rangle \int_0^{k_M} k^2 \, dk$$

$$= \frac{2}{3\pi} mc^2 r_0 \left(\sum_b |\mathbf{r}_{ab}|^2 \right) \int_0^{k_M} k^2 \, dk. \tag{36}$$

The calculation of the second term of (13) is quite similar to the calculation of the second term of (12), except for the replacement of $\varepsilon_j \cdot \mathbf{p}$ by $im\omega_j \varepsilon_j \cdot \mathbf{r}$. By making this substitution in (35), we directly find the second term of (16).

b) Consider the fractions in k appearing in the second terms of (15) and (16) and develop them into simple elements:

$$k\frac{1}{k_{ba} + k} = 1 - k_{ba}\frac{1}{k_{ba} + k} \tag{37}$$

$$k^3\frac{1}{k_{ba} + k} = \left[k^2(k_{ba} + k) - k_{ba}(k_{ba} + k)k + k_{ba}^2(k_{ba} + k) - k_{ba}^3\right]\frac{1}{k_{ba} + k}$$

$$= k^2 - k_{ba}k + k_{ba}^2 - k_{ba}^3\frac{1}{k_{ba} + k}. \tag{38}$$

When the first term of (38) is integrated, it gives a contribution in k_M^3 to $\hbar \Delta_a'$, which must be regrouped with the first term of (16). Note then that they cancel out so that $T_a' = 0$. The integration of the other terms of (38) or of (37), as well as the integration of the first term of (15), gives

$$D_a = \frac{r_0}{2\pi} \hbar c k_M^2 \qquad D_a' = \frac{r_0}{3\pi} mc^2 k_M^2 \left(\sum_b k_{ba}|\mathbf{r}_{ab}|^2 \right) \tag{39}$$

$$P_a = -\frac{2r_0}{3\pi m} k_M \left(\sum_b |\mathbf{p}_{ab}|^2 \right) \qquad P_a' = -\frac{2r_0}{3\pi} mc^2 k_M \left(\sum_b k_{ba}^2|\mathbf{r}_{ba}|^2 \right) \tag{40}$$

$$L_a = \frac{2r_0}{3\pi m} \sum_b k_{ba}|\mathbf{p}_{ab}|^2 \ln \frac{k_{ba} + k_M}{k_{ba}} \qquad L_a' = \frac{2r_0}{3\pi} mc^2 \sum_b k_{ab}^3|\mathbf{r}_{ab}|^2 \ln \frac{k_{ba} + k_M}{k_{ba}}. \tag{41}$$

Because $k_M \gg k_{ba}$, in expressions (41) we can make the approximation

$$\ln \frac{k_{ba} + k_M}{k_{ba}} \simeq \ln \frac{k_M}{k_{ba}}. \tag{42}$$

c) Start with the commutator

$$[H_0, \mathbf{r}] = [\mathbf{p}^2/2m, \mathbf{r}] = -i\hbar\mathbf{p}/m. \tag{43}$$

By taking the matrix element of this equation between the states $|a\rangle$ and $|b\rangle$, we obtain (19). When $k_{ba}^2 |\mathbf{r}_{ba}|^2$ is replaced by $|\mathbf{p}_{ab}|^2/m^2c^2$ in expressions (41) and (40) for L_a' and P_a', we immediately obtain the corresponding expressions for L_a and P_a.

For D_a', it should be noted that

$$\sum_b k_{ab} \mathbf{r}_{ab}^* \cdot \mathbf{r}_{ab} = \sum_b \frac{1}{2imc} [-\mathbf{p}_{ab}^* \cdot \mathbf{r}_{ab} + \mathbf{r}_{ab}^* \cdot \mathbf{p}_{ab}] =$$
$$= \frac{1}{2imc} \langle a| - \mathbf{r} \cdot \mathbf{p} + \mathbf{p} \cdot \mathbf{r}|a\rangle = \frac{-3\hbar}{2mc}. \tag{44}$$

By inserting (44) into expression (39) for D_a', we find the expression of D_a.

We return to expression (40) for P_a. Using the closure relation over $|b\rangle$, we can put it in the form

$$P_a = -\frac{\langle a|\mathbf{p}^2|a\rangle}{2m} \frac{\delta m}{m} \tag{45}$$

where

$$\frac{\delta m}{m} = \frac{4r_0}{3\pi} k_M = \frac{4}{3} \frac{q^2}{4\pi^2 \varepsilon_0} k_M \times \frac{1}{mc^2} = \frac{4}{3} \frac{\varepsilon_{\text{Coul}}}{mc^2}. \tag{46}$$

The term P_a is interpreted as the variation of the average in state $|a\rangle$ of the kinetic-energy term $\mathbf{p}^2/2m$, due to a mass variation δm. From (46), we see that δm is (except for a factor $\frac{4}{3}$ resulting from the noncovariant cutoff procedure) the contribution of the Coulomb field energy of the electron to its mass, calculated by limiting the integral over k to k_M.

d) By making approximation (42) in the expression (41) for L_a, and by using formula (21), we obtain

$$L_a = \frac{2r_0}{3\pi m} \left(\ln \frac{k_M}{K_a}\right) \sum_b |\mathbf{p}_{ab}|^2 k_{ba}. \tag{47}$$

Note that

$$\langle a|[p_x, [p_x, H_0]]|a\rangle = \sum_b \{(p_x)_{ab}\langle b|[p_x, H_0]|a\rangle - \langle a|[p_x, H_0]|b\rangle(p_x)_{ba}\}$$
$$= \sum_b \{|(p_x)_{ab}|^2(E_a - E_b) - (E_b - E_a)|(p_x)_{ab}|^2\}$$
$$= 2\hbar c \sum_b k_{ab}|(p_x)_{ab}|^2. \tag{48}$$

Moreover,

$$[p_x, [p_x, H_0]] = -\hbar^2 \frac{\partial^2}{\partial x^2} V_0(\mathbf{r}) \tag{49}$$

so that

$$\sum_b |\mathbf{p}_{ab}|^2 k_{ba} = \frac{\hbar}{2c} \langle a | \Delta V_0(\mathbf{r}) | a \rangle = \frac{\hbar}{2c} \langle a | \frac{q^2}{\varepsilon_0} \delta(\mathbf{r}) | a \rangle. \tag{50}$$

Finally, by using (50) and (32) in (47), L_a is put in the form (22).

e) Levels other than the ground state are degenerate and the application of perturbation theory requires some caution. Indeed, the radiation field is isotropic, and when we sum over all the modes, no second-order coupling is introduced between the levels of different quantum numbers (l, m). In each of the subspaces (l, m), H_A has a nondegenerate spectrum characterized by the quantum number n. We can then apply the same calculations as we did previously.

A second difference appears because the denominators $k_{ba} + k$ can now cancel out for $k_{ba} < 0$, showing that when the atom is in excited states, it can emit a real photon. If we are concerned only with the positions of the energy levels, expressions (12) and (13) remain valid, on the condition that the fraction be replaced by its principal part. When we integrate over k, we obtain

$$\int_0^{k_M} dk \, \mathscr{P} \frac{1}{k_{ba} + k} = \int_0^{|k_{ba}| - \eta} \frac{dk}{k - |k_{ba}|} + \int_{|k_{ba}| + \eta}^{k_M} \frac{dk}{k - |k_{ba}|} =$$

$$= \ln \frac{\eta}{|k_{ba}|} + \ln \frac{k_M - |k_{ba}|}{\eta} \simeq \ln \frac{k_M}{|k_{ba}|}. \tag{51}$$

The result is the same as for positive k_{ba}, on the condition that k_{ba} be replaced by $|k_{ba}|$ in the logarithm. We then obtain

$$\hbar \Delta_{nlm} = \frac{r_0}{2\pi} \hbar c k_M^2 - \frac{\delta m}{m} \langle nlm | \frac{\mathbf{p}^2}{2m} | nlm \rangle + L_{nl} \tag{52}$$

with

$$L_{nl} = \frac{\alpha}{3\pi} \left(\frac{\hbar}{mc} \right)^2 \left(\ln \frac{k_M}{K_{nl}} \right) \langle nlm | \frac{q^2}{\varepsilon_0} \delta(\mathbf{r}) | nlm \rangle \tag{53}$$

where K_{nl} is defined similarly to K_a [see formula (21)]. The first term of the right-hand side of (52) is independent of the level and the second term, according to the virial theorem, depends only on n. L_{nl} is proportional to the average of $\delta(\mathbf{r})$, and therefore to $|\varphi_{nlm}(0)|^2$, which is nonzero only for the s states $(l = 0)$. The shift of the s states relative to the other levels of the same manifold is positive. Its physical interpretation was given in Chapter II (§E-1-b).

f) We use the Pauli–Fierz representation and we first consider the zero-order terms with regard to the interaction with the transverse field ($\xi = 0$, and therefore $H_{I2} = 0$). The eigenvalues of the Hamiltonian

$$H_0'' = \frac{\mathbf{p}^2}{2m^*} + V_0(\mathbf{r}) + H_R \tag{54}$$

are not identical to the eigenvalues of H_0. Indeed, the contribution of the electromagnetic mass δm of the particle to the kinetic energy is already taken into account through m^* (the physical significance of the eigenstates is also different as a consequence of the transformation of the observables; see Complement B_{II}). We use the notation E_a^* for the energy of the ground state $|a;0\rangle$ of H_0''. In the expression for the shift $\hbar\Delta_a''$ of this level, we will not find the term P_a because it has been taken into account in E_a^*.

α) To second order in the interaction with the transverse field, the Hamiltonian H'' is written

$$H'' = H_0'' + H_I'' \tag{55}$$

where

$$H_I'' = H_{I1}'' + H_{I2}'' + \cdots \tag{56}$$

with

$$H_{I1}'' = (\boldsymbol{\xi} \cdot \boldsymbol{\nabla})V_0(\mathbf{r}) \tag{57.a}$$

$$H_{I2}'' = \tfrac{1}{2}(\boldsymbol{\xi} \cdot \boldsymbol{\nabla})^2 V_0(\mathbf{r}) + H_{I2}. \tag{57.b}$$

The shift $\hbar\Delta_a''$ then equals, in second order,

$$\hbar\Delta_a'' = \langle a;0|H_{I2}''|a;0\rangle +$$

$$+ \sum_j \sum_b \frac{\langle a;0|H_{I1}''|b;\mathbf{k}_j\boldsymbol{\varepsilon}_j\rangle\langle b;\mathbf{k}_j\boldsymbol{\varepsilon}_j|H_{I1}''|a;0\rangle}{E_a^* - (E_b^* + \hbar ck_j)}, \tag{58}$$

or also, after using (31), (57), and with the approximation $E_a^* - E_b^* \simeq -\hbar ck_{ba}$:

$$\hbar\Delta_a'' = D_a + L_a'', \tag{59}$$

with L_a'' being given by (26).

β) Using (24), we get

$$\langle 0|\xi_\mu|\mathbf{k}_j\boldsymbol{\varepsilon}_j\rangle = \frac{q}{m}\sqrt{\frac{\hbar}{2\varepsilon_0\omega_j L^3}}\frac{(\varepsilon_j)_\mu}{i\omega_j}. \tag{60}$$

By using this matrix element and formula (54) from Complement A_I, we find

$$\langle 0|\xi_\mu\xi_\nu|0\rangle = \sum_j \langle 0|\xi_\mu|\mathbf{k}_j\boldsymbol{\varepsilon}_j\rangle\langle\mathbf{k}_j\boldsymbol{\varepsilon}_j|\xi_\nu|0\rangle = \frac{q^2\hbar}{2\varepsilon_0 m^2}\sum_j \frac{(\varepsilon_j)_\mu(\varepsilon_j)_\nu}{\omega_j^3 L^3}$$

$$= \frac{q^2}{4\pi\varepsilon_0\hbar c}\left(\frac{\hbar^2}{m^2 c^2}\right)(2\pi)\int \frac{d^3k}{(2\pi)^3}\frac{1}{k^3}\sum_\varepsilon \varepsilon_\mu\varepsilon_\nu$$

$$= \alpha\left(\frac{\hbar}{mc}\right)^2\int\frac{4\pi k^2\,dk}{(2\pi)^2}\frac{1}{k^3}\frac{2}{3}\delta_{\mu\nu} = \frac{2\alpha}{3\pi}\left(\frac{\hbar}{mc}\right)^2\delta_{\mu\nu}\int_{k_m}^{k_M}\frac{dk}{k} \qquad (61)$$

which leads to formula (27).

For (28), the calculation is quite similar, and we obtain

$$\sum_j \frac{\langle 0|\xi_\mu|\mathbf{k}_j\boldsymbol{\varepsilon}_j\rangle\langle\mathbf{k}_j\boldsymbol{\varepsilon}_j|\xi_\nu|0\rangle}{\hbar c k_{ba} + \hbar c k_j} = \alpha\left(\frac{\hbar}{mc}\right)^2\int\frac{4\pi k^2\,dk}{(2\pi)^2}\frac{1}{k^3}\frac{2}{3}\delta_{\mu\nu}\frac{1}{\hbar c(k_{ba}+k)}$$

$$= \frac{2\alpha}{3\pi}\left(\frac{\hbar}{mc}\right)^2\delta_{\mu\nu}\int_{k_m}^{k_M}\frac{dk}{k}\frac{1}{\hbar c(k_{ba}+k)}$$

$$= \frac{2\alpha}{3\pi}\left(\frac{\hbar}{mc}\right)^2\delta_{\mu\nu}\frac{1}{\hbar c k_{ba}}\int_{k_m}^{k_M}dk\left[\frac{1}{k}-\frac{1}{k_{ba}+k}\right]. \qquad (62)$$

The integral equals

$$\ln\left(\frac{k_M(k_{ba}+k_m)}{k_m(k_{ba}+k_M)}\right) \simeq \ln\left(\frac{k_M k_{ba}}{k_m k_M}\right) \simeq \ln\frac{k_{ba}}{k_m}.$$

This leads to formula (28).

γ) By inserting (27) and (28) into (26), we immediately obtain (29) [the term $b = a$ is zero in (26) because $\langle a|\nabla V_0|a\rangle = 0$ and it can then be eliminated from the sum]. As explained in Complement B_{II} (§3-b), the first term of the right-hand side represents the averaging of the potential $V_0(r)$ by the particle vibrating with an amplitude ξ under the influence of vacuum fluctuations. However, ξ actually represents the motion of the particle only for a free particle. This is still a good approximation for the modes of frequencies ck_j that are large compared with the eigenfrequencies ck_{ba} of the particle in the potential $V_0(\mathbf{r})$. For the low-frequency modes, the response of the particle is weakened by the presence of the potential. This is expressed by the second term of the right-hand side of (29).

By taking the matrix element of $[\mathbf{p}, H_0] = (\hbar/i)\nabla V_0(\mathbf{r})$ between $|a\rangle$ and $|b\rangle$, we obtain

$$\mathbf{p}_{ab}\hbar c k_{ba} = (\hbar/i)\langle a|\nabla V_0|b\rangle. \qquad (63)$$

This formula allows us to write the second term inside the brackets in (29) in the form

$$-2 \sum_{b \neq a} |\mathbf{p}_{ab}ck_{ba}|^2 \frac{1}{\hbar ck_{ba}} \ln \frac{k_{ba}}{k_m} = -\frac{2c}{\hbar} \sum_b |\mathbf{p}_{ab}|^2 k_{ba} \ln \frac{k_{ba}}{k_m}. \tag{64}$$

Using (21) (k_M being changed to k_m) and (50), it can then be written in the form

$$-\langle a|\Delta V_0(\mathbf{r})|a\rangle \ln \frac{K_a}{k_m} \tag{65}$$

and is regrouped with the first term inside the brackets of (29) to give $\ln(k_M/K_a)$. Replacing $\Delta V_0(\mathbf{r})$ by $(q^2/\varepsilon_0)\delta(\mathbf{r})$ results in an expression for L''_a that is identical to expression (22) for L_a.

8. BREMSSTRAHLUNG. RADIATIVE CORRECTIONS TO ELASTIC SCATTERING BY A POTENTIAL

The goal of this exercise and the following one is to study a few radiative processes that appear during the scattering of a charged particle by a potential. Part A is devoted to a discussion of the cross section for the emission of a photon by a particle decelerated by a potential ("Bremsstrahlung"). Part B examines the first radiative corrections to the elastic scattering cross section (scattering without photon emission). Finally, the following exercise shows how low-frequency divergences of the cross sections calculated in A and B can be eliminated by an appropriate redefinition of the measured quantities, which takes into account the finite energy resolution in energy of the detector observing the scattered particle. The Pauli–Fierz representation, which is particularly convenient for approaching these various problems, will be used in these two exercises. The reader may find it useful to refer to Complement B_{II}, which presents a detailed description of this point of view, as well as Complement A_I, which contains the essential results concerning the perturbative calculation of scattering amplitudes and cross sections.

Notation

In the Pauli–Fierz representation, the Hamiltonian H of a particle of charge q, coupled to a static external potential and to the transverse radiation field, is written

$$H = \frac{\mathbf{p}^2}{2m} + V_e(\mathbf{r} + \boldsymbol{\xi}) + H_R. \tag{1}$$

In (1), m is the mass of the particle (corrected for electromagnetic inertia), $\mathbf{r}$ is its position (assumed to be close to the origin $\mathbf{0}$), $\mathbf{p}$ is its momentum; $V_e(\mathbf{r} + \boldsymbol{\xi})$ is the potential energy of the particle in the external potential, evaluated at the point $\mathbf{r} + \boldsymbol{\xi}$ where

$$\boldsymbol{\xi} = \frac{q}{m} \mathbf{Z}(\mathbf{0}), \qquad (2)$$

$\mathbf{Z}(\mathbf{r})$ is a field operator ("Hertz vector") defined by

$$\mathbf{Z}(\mathbf{r}) = \sum_{\mathbf{k}\boldsymbol{\varepsilon}} \sqrt{\frac{\hbar}{2\varepsilon_0 \omega L^3}} \left[\boldsymbol{\varepsilon} \frac{a_{\mathbf{k}\boldsymbol{\varepsilon}}}{i\omega} e^{i\mathbf{k}\cdot\mathbf{r}} - \boldsymbol{\varepsilon} \frac{a_{\mathbf{k}\boldsymbol{\varepsilon}}^+}{i\omega} e^{-i\mathbf{k}\cdot\mathbf{r}} \right]. \qquad (3)$$

Finally, H_R is the Hamiltonian of the radiation field whose expansion in a and a^+ is given by formula (45) in the Appendix.

In comparison with expression (34') in Complement B_{II}, in (1) we ignored the characteristic Coulomb self-energy ε_{Coul} of the particle (which is a constant) and a radiation field term equal to $q^2 A_\perp^2(\mathbf{0})/2m$ which can be shown to play a negligible role in problems dealing with scattering by a potential V_e considered in this exercise and in the following one.

A. Brehmsstrahlung

In this first part, the initial state of the particle + field system is the state

$$|\psi_{in}\rangle = |\mathbf{p}_1; 0\rangle \qquad (4.a)$$

representing a particle with momentum $\mathbf{p}_1$ in the presence of the photon vacuum. The final state is the state

$$|\psi_{fin}\rangle = |\mathbf{p}_2; \mathbf{k}\boldsymbol{\varepsilon}\rangle \qquad (4.b)$$

representing the particle of momentum $\mathbf{p}_2$ in the presence of a photon $\mathbf{k}\boldsymbol{\varepsilon}$ of energy $\hbar\omega$. We write $E_1 = \mathbf{p}_1^2/2m$ and $E_2 = \mathbf{p}_2^2/2m$. The states $|\mathbf{p}_i\rangle$ are taken as discrete, as are the modes of the field, in a cube having sides of length L.

a) Show, in the nonrelativistic limit ($p_1/m \ll c$), that the momentum $\hbar\mathbf{k}$ of the emitted photon is very small compared with $\mathbf{p}_1$. In all of what follows, the momentum of the emitted photons is ignored and the variation in momentum of the particle during collision is written as

$$\hbar Q = p_1 - p_2. \tag{5}$$

b) Show that the interaction Hamiltonian for going from $|\psi_{in}\rangle$ to $|\psi_{fin}\rangle$ originates only from the term $V_e(r + \xi)$ of H. Calculate, to first order in V_e and in q, the amplitude $\mathscr{A}(p_1 \to p_2 + k\varepsilon)$ for the particle to be scattered from p_1 to p_2 with the emission of a photon $k\varepsilon$ (*S* matrix element between $|\psi_{in}\rangle$ and $|\psi_{fin}\rangle$). Express the result obtained as a function of

$$\mathscr{V}_e(Q) = \int d^3r \, e^{iQ \cdot r} V_e(r). \tag{6}$$

c) Without making detailed calculations, indicate how the calculation of $\mathscr{A}(p_1 \to p_2 + k\varepsilon)$, to first order in V_e and in q, would appear in the Coulomb representation. More precisely, discuss the form taken in the Coulomb representation by the asymptotic states of the scattering processes and give the diagrams which contribute to $\mathscr{A}(p_1 \to p_2 + k\varepsilon)$ in this representation. Why is the Pauli–Fierz representation simpler?

d) The momentum p_1 and the direction $n_2 = p_2/p_2$ of the scattered particle are kept constant. Calculate, from the results in *b*, the probability per unit time $dw(p_1 \to p_2 + k\varepsilon)$ for the particle to be scattered in a solid angle $d\Omega_2$ about n_2, with the emission of a photon with a frequency between ω and $\omega + d\omega$, in the solid angle $d\Omega$ about $\kappa = k/k$. Deduce from this the value of the differential scattering cross section $d\sigma(p_1 \to p_2 + k\varepsilon)/d\Omega_2 \, d\Omega \, d\omega$.

The particle is assumed to be a nucleus with a charge $q = Zq_p$ (where q_p is the charge of the proton), the scattering potential being the one created by a nucleus of charge $Z'q_p$ fixed at the origin. Express the preceding differential cross section as a function of Z, Z', p_1, p_2, ω and the fine-structure constant α.

e) This question examines the asymptotic form of the amplitude $\mathscr{A}(p_1 \to p_2 + k\varepsilon)$ calculated in *b* when the energy $\hbar\omega$ of the emitted photon is very small compared with E_1 and E_2 and can thus be ignored. Show that $\mathscr{A}(p_1 \to p_2 + k\varepsilon)$ can then be written

$$\mathscr{A}(p_1 \to p_2 + k\varepsilon) = \mathscr{A}_{el}^{(0)}(p_1 \to p_2) \times q\eta_1(Q, k\varepsilon) \tag{7}$$

where $\mathscr{A}_{el}^{(0)}(p_1 \to p_2)$ is the elastic scattering amplitude (without photon emission), calculated to first order in V_e and zero order in q, and where $\eta_1(Q, k\varepsilon)$ is a correction coefficient which depends only on Q and the emitted photon $k\varepsilon$. Give the expression for η_1.

B. Radiative Corrections to Elastic Scattering

The initial state is still (4.a), but the final state is now

$$|\psi_{\text{fin}}\rangle = |\mathbf{p}_2; 0\rangle. \tag{8}$$

In the absence of interaction with the transverse field, i.e., to zero order in q, the scattering amplitude to first order in V_e is written $\mathscr{A}_{\text{el}}^{(0)}(\mathbf{p}_1 \to \mathbf{p}_2)$ and was previously introduced in e. The purpose of Part B is to calculate the first corrections (in q^2) to this amplitude.

f) Show that it is necessary to expand $V_e(\mathbf{r} + \boldsymbol{\xi})$ up to the second order in $\boldsymbol{\xi}$ to obtain the first radiative correction to $\mathscr{A}_{\text{el}}^{(0)}(\mathbf{p}_1 \to \mathbf{p}_2)$. Calculate this correction and show, by using the properties of the Fourier transform, that this correction is proportional to $\mathscr{A}_{\text{el}}^{(0)}(\mathbf{p}_1 \to \mathbf{p}_2)$. Deduce from this result that the elastic scattering amplitude $\mathscr{A}_{\text{el}}(\mathbf{p}_1 \to \mathbf{p}_2)$, to first order in V_e and to second order in q, can be written

$$\mathscr{A}_{\text{el}}(\mathbf{p}_1 \to \mathbf{p}_2) = \mathscr{A}_{\text{el}}^{(0)}(\mathbf{p}_1 \to \mathbf{p}_2)\mathscr{F}(\mathbf{Q}) \tag{9}$$

where the form factor $\mathscr{F}(\mathbf{Q})$ is equal to

$$\mathscr{F}(\mathbf{Q}) = 1 - \frac{q^2}{2} \sum_{\mathbf{k}\varepsilon} \eta_1^2(\mathbf{Q}, \mathbf{k}\varepsilon), \tag{10}$$

with η_1 being the coefficient introduced in (7). What is the physical effect of $\mathscr{F}(\mathbf{Q})$?

g) After having replaced the discrete sum by an integral, show that the sum over $\mathbf{k}\varepsilon$ that appears in the correction term in (9) diverges logarithmically as $\ln k_m$ where k_m is the lower bound of the integral over $k = |\mathbf{k}|$. We will then keep k_m finite and nonzero, and postpone to the next exercise the solution of this "infrared divergence" problem.

h) Without making detailed calculations, indicate the diagrams that must be considered to study the same physical phenomenon in the Coulomb representation.

i) Let $d\sigma_{\text{el}}(\mathbf{p}_1 \to \mathbf{p}_2)/d\Omega_2$ be the differential cross section for elastic scattering to the second order in q. Show that

$$d\sigma_{\text{el}}(\mathbf{p}_1 \to \mathbf{p}_2)/d\Omega_2 = \left[1 - q^2 \sum_{\mathbf{k}\varepsilon} \eta_1^2(\mathbf{Q}, \mathbf{k}\varepsilon)\right] d\sigma_{\text{el}}^{(0)}(\mathbf{p}_1 \to \mathbf{p}_2)/d\Omega_2 \tag{11}$$

where $d\sigma_{\text{el}}^{(0)}(\mathbf{p}_1 \to \mathbf{p}_2)/d\Omega_2$ is the differential elastic scattering cross section at zero order in q.

j) Consider a set $\mathscr{E}$ of very low-frequency modes of the field. Show, by using (7), (9), and (10) that the decrease in the elastic scattering cross section resulting from the emission and the reabsorption of a photon in one of the modes of the set $\mathscr{E}$ is equal in absolute value to the cross section for the emission of a photon in any one of the modes of the set $\mathscr{E}$. What can be deduced about the cross sections for the scattering of the particle with the emission of 0 *or* 1 photon?

Solution

a) The conservation of global energy results in

$$E_1 = E_2 + \hbar\omega. \tag{12}$$

Because E_2 is positive, it follows from (12) that

$$\hbar|\mathbf{k}| \leq \frac{E_1}{c} = |\mathbf{p}_1| \frac{|\mathbf{p}_1|}{2mc} = |\mathbf{p}_1| \frac{v_1}{2c} \ll |\mathbf{p}_1|. \tag{13}$$

The momentum of the emitted photon therefore has a modulus that is very small compared with the momentum of the particle.

b) Expand $V_e(\mathbf{r} + \boldsymbol{\xi})$ in powers of $\boldsymbol{\xi} = q\mathbf{Z}(0)/m$

$$V_e(\mathbf{r} + \boldsymbol{\xi}) = V_e(\mathbf{r}) + \boldsymbol{\xi} \cdot \nabla V_e(\mathbf{r}) + \frac{1}{2}\sum_{i,j} \xi_i \xi_j \, \partial^2 V_e/\partial r_i \, \partial r_j + \cdots . \tag{14}$$

The first-order term in $\boldsymbol{\xi}$ relates $|\psi_{in}\rangle$ to $|\psi_{fin}\rangle$:

$$\langle \psi_{fin}|\boldsymbol{\xi} \cdot \nabla V_e(\mathbf{r})|\psi_{in}\rangle = \frac{q}{m}\langle k\boldsymbol{\varepsilon}|\mathbf{Z}(0)|0\rangle \cdot \langle \mathbf{p}_2|\nabla V_e(\mathbf{r})|\mathbf{p}_1\rangle =$$

$$= \frac{\mathscr{V}_e(\mathbf{Q})}{L^3} \frac{q}{m} \sqrt{\frac{\hbar}{2\varepsilon_0 \omega^3 L^3}} \, \boldsymbol{\varepsilon} \cdot \mathbf{Q}. \tag{15}$$

We used (3) and (5) and the fact that if $\mathscr{V}_e(\mathbf{Q})$ is the Fourier transform of $V_e(\mathbf{r})$ defined in (6), the Fourier transform of $\nabla V_e(\mathbf{r})$ appearing in $L^3\langle \mathbf{p}_2|\nabla V_e(\mathbf{r})|\mathbf{p}_1\rangle$ is equal to $-i\mathbf{Q}\mathscr{V}_e(\mathbf{Q})$.

By using formula (19) from Complement A_I, we then obtain, to first order in q and in V_e:

$$\mathscr{A}(\mathbf{p}_1 \rightarrow \mathbf{p}_2 + k\boldsymbol{\varepsilon}) = -2\pi i \, \delta^{(T)}(E_1 - E_2 - \hbar\omega)\langle \mathbf{p}_2; k\boldsymbol{\varepsilon}|\boldsymbol{\xi} \cdot \nabla V_e(\mathbf{r})|\mathbf{p}_1; 0\rangle$$

$$= -2\pi i \, \delta^{(T)}(E_1 - E_2 - \hbar\omega)\frac{\mathscr{V}_e(\mathbf{Q})}{L^3} \frac{q}{m} \sqrt{\frac{\hbar}{2\varepsilon_0 \omega^3 L^3}} \, \boldsymbol{\varepsilon} \cdot \mathbf{Q} \tag{16}$$

where $\delta^{(T)}$ is a delta function having a width on the order of $\hbar/T$, with T being the interaction time [formula (20) in Complement A_I].

c) First note that the states (4.a) and (4.b) are correct asymptotic states in the Pauli–Fierz representation. Indeed outside the range of the potential, the term $V_e(\mathbf{r} + \boldsymbol{\xi})$ in (1) can be ignored and the Hamiltonian H is then reduced to the Hamiltonian $(\mathbf{p}^2/2m) + H_R$ so that states (4.a) and (4.b) can be considered as eigenstates of H. The asymptotic states corresponding to (4.a) and (4.b) in the Coulomb representation are obtained from (4.a) and (4.b) by the unitary transformation

$$T^+ = \exp\left[-\frac{i}{\hbar}\frac{q}{m}\mathbf{p} \cdot \mathbf{Z}(\mathbf{0})\right] \tag{17}$$

and they have a form that is more complex than (4.a) and (4.b) because they represent the incident or scattered particle "dressed" by a cloud of virtual photons that this particle emits and reabsorbs.

In the calculation of the amplitude for the emission of a real photon by the particle, we can, at the lowest order in q, replace T^+ by 1 and use (4.a) and (4.b) to describe the initial and final states of the scattering process in the Coulomb representation. However, in this representation, it is necessary to have two interactions, one with $V_e(\mathbf{r})$ and the other with $H_{I1} = -q\mathbf{p} \cdot \mathbf{A}_\perp(\mathbf{0})/m$ to go from $|\psi_{\text{in}}\rangle$ to $|\psi_{\text{fin}}\rangle$. Because these two interactions can occur with two different time orderings, there are two diagrams (Figure 1). The calculation for $\mathscr{A}$ is then more complicated because it is a second-order calculation with an intermediate state [see formula (28) in Complement A_I].

d) To obtain the transition rate from the amplitude (16), one must follow the procedure presented in subsection 3 in Complement A_I: square the modulus of (16), use the fact that the square of $\delta^{(T)}$ is proportional to T [see formula (49) in A_I], which, after dividing by T, yields the transition rate

$$w(\mathbf{p}_1 \to \mathbf{p}_2 + \mathbf{k}\boldsymbol{\varepsilon}) = |\mathscr{A}(\mathbf{p}_1 \to \mathbf{p}_2 + \mathbf{k}\boldsymbol{\varepsilon})|^2 / T =$$

$$= \frac{2\pi}{\hbar}\delta^{(T)}(E_1 - E_2 - \hbar\omega)\frac{|\mathscr{V}_e(\mathbf{Q})|^2}{L^6}\frac{q^2}{m^2}\frac{\hbar}{2\varepsilon_0\omega^3 L^3}(\boldsymbol{\varepsilon} \cdot \mathbf{Q})^2. \tag{18}$$

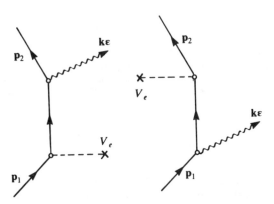

Figure 1. Diagrams representing Bremsstrahlung emission in the Coulomb representation.

It is then necessary to sum (18) over the final states of the particle and of the photon. Formulas (42) and (46) in A_I give, for the number of states of the particle in $dE_2 \, d\Omega_2$, the result $L^3 m p_2 \, dE_2 \, d\Omega_2/(2\pi\hbar)^3$ and for the number of states of the photon in $d\omega \, d\Omega$ the result $L^3\omega^2 \, d\omega \, d\Omega/(2\pi c)^3$. The integral over E_2 eliminates the function $\delta^{(T)}$ in (18), provided that E_2 is replaced by $E_1 - \hbar\omega$, so that finally

$$dw(\mathbf{p}_1 \to \mathbf{p}_2 + \mathbf{k}\boldsymbol{\varepsilon}) = \frac{1}{(2\pi)^5 2\varepsilon_0 \hbar^3 c^3 L^3} \frac{q^2}{m} (\boldsymbol{\varepsilon} \cdot \mathbf{Q})^2 |\mathscr{V}_e(\mathbf{Q})|^2 p_2 \frac{d\omega}{\omega} d\Omega \, d\Omega_2. \quad (19)$$

The flux associated with the incident particle is $\phi = L^{-3} p_1/m$, which gives the differential cross section

$$\frac{d\sigma}{d\omega \, d\Omega \, d\Omega_2} = \frac{1}{\phi} \frac{dw(\mathbf{p}_1 \to \mathbf{p}_2 + \mathbf{k}\boldsymbol{\varepsilon})}{d\omega \, d\Omega \, d\Omega_2} = \frac{1}{(2\pi)^5 2\varepsilon_0 \hbar^3 c^3} q^2 (\boldsymbol{\varepsilon} \cdot \mathbf{Q})^2 |\mathscr{V}_e(\mathbf{Q})|^2 \frac{p_2}{p_1} \frac{1}{\omega}. \quad (20)$$

If $q = Zq_p$ and if $V_e(\mathbf{r}) = ZZ'q_p^2/4\pi\varepsilon_0 r$, we have, according to (6), $\mathscr{V}_e(\mathbf{Q}) = ZZ'q_p^2/\varepsilon_0 Q^2$, which gives, for (20),

$$\frac{d\sigma}{d\omega \, d\Omega \, d\Omega_2} = Z^4 Z'^2 \frac{\alpha^3}{\pi^2} \frac{(\boldsymbol{\varepsilon} \cdot \mathbf{Q})^2}{Q^4} \frac{p_2}{p_1} \frac{1}{\omega} \quad (21)$$

where $\alpha = q_p^2/4\pi\varepsilon_0 \hbar c = \frac{1}{137}$.

The presence of ω in the denominators of (20) and (21) shows that $d\sigma/d\omega \, d\Omega \, d\Omega_2$ diverges when $\omega \to 0$.

e) To first order in V_e and zero order in q, only the first term of (14) can be retained, which gives

$$\mathscr{A}_{el}^{(0)}(\mathbf{p}_1 \to \mathbf{p}_2) = -2\pi i \delta^{(T)}(E_1 - E_2) \langle \mathbf{p}_2; 0 | V_e(\mathbf{r}) | \mathbf{p}_1; 0 \rangle$$

$$= -2\pi i \delta^{(T)}(E_1 - E_2)\mathscr{V}_e(\mathbf{Q})/L^3. \quad (22)$$

If $\hbar\omega \ll E_1, E_2$, one can ignore $\hbar\omega$ compared with $E_1 - E_2$ in the function $\delta^{(T)}$ of (16) (in the following calculations, this function $\delta^{(T)}$ indeed acts on slowly varying functions). Moreover, if $\hbar\omega \ll E_1, E_2$, one can, by using (12), ignore the difference between $|\mathbf{p}_1|$ and $|\mathbf{p}_2|$ in (16), so that for constant $\mathbf{p}_1$ and $\mathbf{n}_2 = \mathbf{p}_2/p_2$, the wave vectors $\mathbf{Q}$ appearing in (16) and (22) can be considered to be identical (recall that in (22), $|\mathbf{p}_1|$ and $|\mathbf{p}_2|$ are rigorously equal because $E_1 = E_2$). Using these approximations, Equation (7) is a consequence of (16) and (22), with

$$\eta_1(\mathbf{Q}, \mathbf{k}\boldsymbol{\varepsilon}) = \frac{1}{m} \sqrt{\frac{\hbar}{2\varepsilon_0 \omega^3 L^3}} \, \boldsymbol{\varepsilon} \cdot \mathbf{Q}. \quad (23)$$

f) The term in $\boldsymbol{\xi} \cdot \nabla V_e$ of (14) relates $|\mathbf{p}_1; 0\rangle$ to $|\mathbf{p}_2; 0\rangle$ only at second order and thus contributes to the scattering amplitude only at second order in V_e. By contrast, the next term, $\xi_i \xi_j$, directly relates $|\mathbf{p}_1; 0\rangle$ to $|\mathbf{p}_2; 0\rangle$ (one of the two operators ξ_i creates a photon that the other operator destroys), so that this term gives the first correction to (22), which is first order in V_e and second order in q.

Using (2) and (3), the matrix element of the last term of (14) between $|\mathbf{p}_1;0\rangle$ and $|\mathbf{p}_2;0\rangle$ equals

$$\frac{1}{2}\frac{q^2}{m^2}\sum_{i,j=x,y,z}\langle 0|Z_i(0)Z_j(0)|0\rangle\langle\mathbf{p}_2|\nabla_i\nabla_j V_e(\mathbf{r})|\mathbf{p}_1\rangle. \tag{24}$$

Using (3), the first matrix element of (24) equals

$$\langle 0|Z_i(0)Z_j(0)|0\rangle = \sum_{\mathbf{k}\varepsilon}\frac{\hbar}{2\varepsilon_0\omega^3 L^3}\varepsilon_i\varepsilon_j. \tag{25}$$

For the second matrix element, the properties of the Fourier transform result in

$$\langle\mathbf{p}_2|\nabla_i\nabla_j V_e(\mathbf{r})|\mathbf{p}_1\rangle = -Q_iQ_j\frac{\mathscr{V}_e(\mathbf{Q})}{L^3}. \tag{26}$$

Finally, the first radiative correction to (22) is written

$$-2\pi i\delta^{(T)}(E_i-E_f)\frac{\mathscr{V}_e(\mathbf{Q})}{L^3}\times\left[-\frac{q^2}{2m^2}\sum_{\mathbf{k}\varepsilon}\frac{\hbar}{2\varepsilon_0\omega^3 L^3}(\boldsymbol{\varepsilon}\cdot\mathbf{Q})^2\right]. \tag{27}$$

Regrouping (22) and (27) gives (9) if we set

$$\mathscr{F}(\mathbf{Q}) = 1 - \frac{q^2}{2m^2}\sum_{\mathbf{k}\varepsilon}\frac{\hbar}{2\varepsilon_0\omega^3 L^3}(\boldsymbol{\varepsilon}\cdot\mathbf{Q})^2 \tag{28}$$

which, using (23), gives Equation (10).

Finally, the radiative corrections are equivalent to replacing $\mathscr{V}_e(\mathbf{Q})$ by $\mathscr{V}_e(\mathbf{Q})\mathscr{F}(\mathbf{Q})$. This is equivalent to replacing the scattering potential $V_e(\mathbf{r})$ by the convolution product of $V_e(\mathbf{r})$ by the Fourier transform $F(\mathbf{r})$ of $\mathscr{F}(\mathbf{Q})$. This result shows that the particle, vibrating under the influence of vacuum fluctuations, averages the potential $V_e(\mathbf{r})$ over the range of its vibrational motion.

g) When $\Sigma_{\mathbf{k}\varepsilon}$ is replaced by $(L/2\pi)^3\int k^2\,dk\,d\Omega\,\Sigma_\varepsilon$ in the correction term of (9), the integral $\int d\omega/\omega$ appears, which diverges logarithmically as $\ln k_m$ when the lower bound tends to zero.

h) In the Coulomb representation, the foregoing radiative corrections appear only in the third order, V_e being used once and H_{I1} twice. They are represented by the three diagrams in Figure 2. Such a third-order calculation is much more complicated than the first-order calculation made here using the last term of (14). Moreover, it is no longer possible, in the second-order calculation in q made here, to replace T^+ with 1 in (17) and to use (4.a) and (8) as the initial and final states of the scattering process in the Coulomb representation (which gives rise to normalization corrections for the final and initial states).

i) The calculations for $d\sigma_{\mathrm{el}}$ and $d\sigma_{\mathrm{el}}^{(0)}$ made starting from $\mathscr{A}_{\mathrm{el}}$ and $\mathscr{A}_{\mathrm{el}}^{(0)}$ involve the same functions $[\delta^{(T)}]^2$, the same summations over the final states and the same incident fluxes. Because, according to (9), $\mathscr{A}_{\mathrm{el}}$ and $\mathscr{A}_{\mathrm{el}}^{(0)}$ are proportional, we can deduce that the ratio $d\sigma_{\mathrm{el}}/d\sigma_{\mathrm{el}}^{(0)}$ is just the square of the term in brackets of (9), which, at second order in q, indeed coincides with the term in brackets of (11).

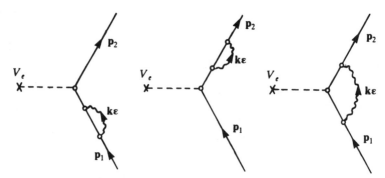

Figure 2. Diagrams representing the first radiative corrections to elastic scattering by a potential in the Coulomb representation.

j) The decrease in the elastic scattering cross section as a result of the emission and the virtual reabsorption of a photon in any one of the modes of the set $\mathscr{E}$ is, using (11), equal to

$$\left[-q^2 \sum_{\mathbf{k}\varepsilon \in \mathscr{E}} \eta_1^2(\mathbf{Q}, \mathbf{k}\varepsilon) \right] d\sigma_{el}^{(0)}(\mathbf{p}_1 \rightarrow \mathbf{p}_2)/d\Omega_2. \tag{29}$$

Besides, using (7), the cross section for the emission of a photon in any one of the modes of $\mathscr{E}$ can be written as

$$\left[q^2 \sum_{\mathbf{k}\varepsilon \in \mathscr{E}} \eta_1^2(\mathbf{Q}, \mathbf{k}\varepsilon) \right] d\sigma_{el}^{(0)}(\mathbf{p}_1 \rightarrow \mathbf{p}_2)/d\Omega_2. \tag{30}$$

The sum of (29) and (30) indeed cancels out.

We now reexamine, in light of the preceding result, the problem of infrared divergences encountered above, in questions *d* and *g*. To do this, we concentrate on the low-frequency modes and assume that the particle interacts only with the modes of the set $\mathscr{E}$. The cancellation of (29) by (30) shows that the sum of the scattering cross sections with the emission of 0 or 1 photon does not depend on the modes of $\mathscr{E}$ and thus has no infrared divergence when the set $\mathscr{E}$ is enlarged so as to contain modes with lower and lower frequencies.

The foregoing result then allows us to show that the experimentally relevant quantities do not exhibit any infrared divergence. Indeed, the detector observing the scattered particle always has a finite energy resolution. If the energy $\hbar\omega$ of the emitted photon is smaller than the energy sensitivity $\hbar\delta$ of the detector, the observation carried out on the scattered particle will not allow us to know whether or not a very soft photon (frequency less than δ) has been emitted during the scattering process. The measured quantity is thus indeed the sum of the elastic scattering cross section and the inelastic scattering cross section with the emission of a photon having a frequency less than δ. This sum remains finite, whereas each cross section separately presents an infrared divergence.

It should, however, be noted that expressions (29) and (30) were obtained here in a perturbative fashion and thus no longer have any meaning in certain limits. For example, when the frequency of the modes of $\mathscr{E}$ tends to zero, the absolute value of the term in brackets of (29) can become greater than 1, as a result of infrared divergence. Such a result is manifestly absurd (it leads to a negative elastic scattering cross section!). This shows the necessity of taking into account higher-order terms in q. Consequently, a satisfactory discussion of the "infrared catastrophe" can be given only within the framework of a nonperturbative treatment in q (see the following exercise).

9. Low-Frequency Bremsstrahlung. Nonperturbative Treatment of the Infrared Catastrophe

The purpose of this exercise is to study the low-frequency radiation emitted by a charged particle during the scattering of this particle by a static potential V_e.

The problem is first analyzed within the framework of classical radiation theory (Part A). One then considers in Part B the amplitudes describing the scattering of the particle with the emission of an arbitrary number of low-frequency photons. Using the Pauli–Fierz representation, the calculation is carried out in a nonperturbative fashion relative to the coupling between the charged particle and the transverse field. It is possible to establish in this way a connection between the classical and quantum descriptions for the radiated field and to show how the infrared divergences appearing in the perturbative calculation of scattering cross sections can be eliminated.

This exercise is an extension of the preceding exercise, which the reader is advised to study beforehand. For this reason, any previously introduced notation will not be redefined.

A. Classical Treatment of the Radiated Field

Let $\alpha_{k\varepsilon}(t)$ be the normal classical variable describing the state of the classical transverse field in the mode $k\varepsilon$. Recall that the equation of motion for $\alpha_{k\varepsilon}(t)$ is written [see Appendix, formulas (18.b) and (26)]

$$\dot{\alpha}_{k\varepsilon}(t) + i\omega\alpha_{k\varepsilon}(t) = \frac{i}{\sqrt{2\varepsilon_0\hbar\omega}}\boldsymbol{\varepsilon}\cdot\boldsymbol{\jmath}(\mathbf{k}, t) \tag{1}$$

with $\boldsymbol{\jmath}(\mathbf{k}, t)$ being defined by

$$\boldsymbol{\jmath}(\mathbf{k}, t) = \frac{1}{\sqrt{L^3}}\int d^3\rho \, e^{-i\mathbf{k}\cdot\boldsymbol{\rho}}\,\mathbf{j}(\boldsymbol{\rho}, t) \tag{2}$$

where

$$\mathbf{j}(\boldsymbol{\rho}, t) = q\dot{\mathbf{r}}(t)\delta(\boldsymbol{\rho} - \mathbf{r}(t)) \tag{3}$$

is the current associated with the charge q located at $\mathbf{r}(t)$ with a velocity $\dot{\mathbf{r}}(t)$.

The range of the scattering potential V_e is sufficiently small so that it is possible to use the long-wavelength approximation $|\mathbf{k} \cdot \mathbf{r}(t)| \ll 1$, not only during the collision ($t_1 \le t \le t_2$), but also before the collision ($t \le t_1$) and after the collision ($t \ge t_2$). Moreover, the reaction of the emitted radiation upon the motion of the particle is neglected, which is equivalent to considering $\mathbf{r}(t)$ as a given function of time.

a) Let $\mathbf{v}_1$ be the velocity of the particle before the collision [$\dot{\mathbf{r}}(t) = \mathbf{v}_1$ if $t < t_1$]. Show that, within the framework of the long-wavelength approximation, Equation (1) has for $t \le t_1$ a particular time-independent solution $\beta_{\mathbf{k}\varepsilon}(\mathbf{v}_1)$. Calculate $\beta_{\mathbf{k}\varepsilon}(\mathbf{v}_1)$. What is the physical interpretation of the field described by the normal variables $\{\beta_{\mathbf{k}\varepsilon}(\mathbf{v}_1)\}$?

b) Show that the solution to Equation (1), which is equal to $\beta_{\mathbf{k}\varepsilon}(\mathbf{v}_1)$ for $t \le t_1$, can be written, for $t \ge t_2$,

$$\alpha_{\mathbf{k}\varepsilon}(t) = \beta_{\mathbf{k}\varepsilon}(\mathbf{v}_2) + \gamma_{\mathbf{k}\varepsilon} e^{-i\omega t} \tag{4}$$

where $\mathbf{v}_2$ is the final velocity of the particle [$\dot{\mathbf{r}}(t) = \mathbf{v}_2$ for $t \ge t_2$]. Calculate $\gamma_{\mathbf{k}\varepsilon}$ and show that $\gamma_{\mathbf{k}\varepsilon}$ is proportional to the temporal Fourier transform of the acceleration of the particle. What is the field radiated by the particle at the end of the scattering process?

c) Consider the low-frequency limit $\omega\tau_c \ll 1$, where τ_c is the duration of the scattering process, on the order of $t_2 - t_1$. Show that $\gamma_{\mathbf{k}\varepsilon}$ then depends only on the variation in velocity $\delta\mathbf{v} = \mathbf{v}_2 - \mathbf{v}_1$ of the particle during the scattering process.

d) In the low-frequency limit, calculate the total energy radiated by the particle in the modes of frequency less than ω_m. This total energy will be reexpressed as a function of $\hbar\omega_m$, $\delta\mathbf{v}/c$ and the fine-structure constant (the charged particle is assumed to be an electron).

B. Quantum Nonperturbative Treatment of the Radiated Field

In this entire section, two categories of modes are considered: the low-frequency modes $\mathbf{k}_\lambda\varepsilon_\lambda$, written with a lower-case Greek subscript and such that $\omega_\lambda < \omega_m$, and the high-frequency modes $\mathbf{k}_\Lambda\varepsilon_\Lambda$, written with an upper-case Greek subscript and such that $\omega_m < \omega_\Lambda$ (*). The contributions of these two types of modes to a field operator, for example the Hertz vector $\mathbf{Z}(\mathbf{r})$, will be written $\mathbf{Z}^{l \cdot f}(\mathbf{r})$ and $\mathbf{Z}^{h \cdot f}(\mathbf{r})$ [to obtain $\mathbf{Z}^{l \cdot f}$ or $\mathbf{Z}^{h \cdot f}$, it is sufficient to replace $\mathbf{k}\varepsilon$ by $\mathbf{k}_\lambda \varepsilon_\lambda$ or $\mathbf{k}_\Lambda \varepsilon_\Lambda$ in formula (3) in Exercise 8]. Finally, the state $|\{n_\lambda\}, \{n_\Lambda\}\rangle$ represents a state of the field where the n_λ are the numbers of photons in the modes $\mathbf{k}_\lambda \varepsilon_\lambda$, and the n_Λ are the numbers of photons in the modes $\mathbf{k}_\Lambda \varepsilon_\Lambda$, with the energy of such a state being $\sum_\lambda n_\lambda \hbar \omega_\lambda + \sum_\Lambda n_\Lambda \hbar \omega_\Lambda$.

e) In the Pauli–Fierz Hamiltonian [see formula (1) in Exercise 8], the particle-transverse field interaction appears only in the term $V_e(\mathbf{r} + \boldsymbol{\xi})$. For the nonperturbative treatment presented here, it is preferable, instead of expanding $V_e(\mathbf{r} + \boldsymbol{\xi})$ in powers of $\boldsymbol{\xi}$, to express it as a function of the Fourier transform $\mathcal{V}_e(\mathbf{k})$ of $V_e(\mathbf{r})$ [see Equation (6) in Exercise 8]. Express $V_e(\mathbf{r} + \boldsymbol{\xi})$ as a function of $\mathcal{V}_e(\mathbf{k})$ and the operators $\exp(-i\mathbf{k} \cdot \mathbf{r})$ and $\exp(-i\mathbf{k} \cdot \boldsymbol{\xi})$.

f) Calculate, to first order in V_e and to all orders in q, the amplitude $\mathscr{A}(\mathbf{p}_1, \{0_\Lambda\}, \{0_\lambda\} \to \mathbf{p}_2, \{0_\Lambda\}, \{n_\lambda\})$ for the particle to be scattered from $\mathbf{p}_1$ to $\mathbf{p}_2$, with the emission of 0 high-frequency photons and a given number of low-frequency photons characterized by the set $\{n_\lambda\}$. Use the notation $\mathbf{p}_1 - \mathbf{p}_2 = \hbar \mathbf{Q}$, and express this amplitude as a function of $\mathcal{V}_e(\mathbf{Q})$ and the matrix elements of radiation operators involving the high- and low-frequency parts, $\boldsymbol{\xi}^{h \cdot f}$ and $\boldsymbol{\xi}^{l \cdot f}$, of $\boldsymbol{\xi}$.

g) The ω_λ are assumed to be so small that one can completely neglect $\sum_\lambda n_\lambda \hbar \omega_\lambda$ compared with E_1 and E_2, and also neglect the difference between $|\mathbf{p}_1|$ and $|\mathbf{p}_2|$ in the amplitude calculated in Question *f*). Show that

$$\mathscr{A}(\mathbf{p}_1, \{0_\Lambda\}, \{0_\lambda\} \to \mathbf{p}_2, \{0_\Lambda\}, \{n_\lambda\}) =$$

$$= \mathscr{A}_{el}(\mathbf{p}_1, \{0_\Lambda\} \to \mathbf{p}_2, \{0_\Lambda\}) \langle \{n_\lambda\} | \exp[-i\mathbf{Q} \cdot \boldsymbol{\xi}^{l \cdot f}] | \{0_\lambda\} \rangle \quad (5)$$

where the first amplitude appearing in the right-hand side of (5) is the elastic scattering amplitude, to first order in V_e and to all orders in q, which is calculated by completely ignoring the low-frequency modes. Show that

$$\sum_{\{n_\lambda\}} |\langle \{n_\lambda\} | \exp[-i\mathbf{Q} \cdot \boldsymbol{\xi}^{l \cdot f}] | \{0_\lambda\} \rangle|^2 = 1. \quad (6)$$

(*) One nevertheless assumes that ω_Λ is not too large in order to allow the long-wavelength approximation even for high-frequency modes ($\omega_\Lambda < \omega_M$).

h) Show that the final state of the low-frequency field is

$$\exp\left[-i\mathbf{Q}\cdot\boldsymbol{\xi}^{l\cdot f}\right]|\{0_\lambda\}\rangle$$

at the end of a scattering process in which the particle goes from $\mathbf{p}_1$ to $\mathbf{p}_2$, without any emission of high-frequency photon.

i) Show that the state of the quantum field radiated at the end of the scattering process $\mathbf{p}_1, \{0_\lambda\} \to \mathbf{p}_2, \{0_\lambda\}$ is the coherent state associated with the classical radiated field during the same scattering process. What can be concluded from this result about the average energy of the quantum field radiated at low frequency and for the distribution of the number of photons emitted in each low-frequency mode?

j) The detector observing the particle scattered in the direction $\mathbf{n}_2 = \mathbf{p}_2/p_2$ has an energy resolution that is too low to allow the observation of the energy loss $\sum_\lambda n_\lambda \hbar\omega_\lambda$ of the particle due to low-frequency radiation (actually, it is this energy resolution that fixes the bound ω_m selected to separate the two types of modes). The transition rate, measured by the detector, is then written

$$dw_{\text{measured}}(\mathbf{p}_1 \to \mathbf{p}_2) = \sum_{\{n_\lambda\}} dw(\mathbf{p}_1, \{0_\lambda\}, \{0_\lambda\} \to \mathbf{p}_2, \{0_\lambda\}, \{n_\lambda\}). \quad (7)$$

By using (5) and (6), show that dw_{measured} is simply equal to the elastic transition rate $dw_{\text{el}}(\mathbf{p}_1, \{0_\lambda\} \to \mathbf{p}_2, \{0_\lambda\})$ calculated by completely ignoring the low-frequency modes.

Generalize the preceding result to other processes in which high-frequency photons are emitted, for example, to the scattering process $\mathbf{p}_1, \{0_\lambda\}, \{0_\lambda\} \to \mathbf{p}_2, \{n_\lambda\}, \{n_\lambda\}$.

Solution

a) Substitute (3) into (2), then substitute (2) into (1). By approximating $\exp[-i\mathbf{k}\cdot\mathbf{r}(t)]$ by 1, we then obtain

$$\dot{\alpha}_{\mathbf{k}\varepsilon}(t) + i\omega\alpha_{\mathbf{k}\varepsilon}(t) = \frac{iq}{\sqrt{2\varepsilon_0\hbar\omega L^3}}\boldsymbol{\varepsilon}\cdot\dot{\mathbf{r}}(t). \quad (8)$$

Prior to the collision ($t \leq t_1$), the particle is not in the range of the scattering potential V_e. Its velocity $\dot{\mathbf{r}}(t)$ is therefore constant and equal to $\mathbf{v}_1$. The right-hand side of (8) is thus time independent, which demonstrates that Equation (8) has, for $t \leq t_1$, the particular solution

$$\beta_{\mathbf{k}\varepsilon}(\mathbf{v}_1) = \frac{q}{\sqrt{2\varepsilon_0\hbar\omega^3 L^3}}\boldsymbol{\varepsilon}\cdot\mathbf{v}_1 \quad (9)$$

which is time independent. This solution represents the value of the normal variables associated with the transverse field bound to the incident particle, which is in uniform translational motion (see Complement B_{II}, §1-b).

b) After the collision ($t \geq t_2$), the particle is free again, and it has a velocity v_2. The general solution to Equation (8) is then the sum of the particular solution of the equation with the right-hand side, obtained from (9) by replacing v_1 by v_2, and the general solution to the equation without the right-hand side, which varies as $\exp(-i\omega t)$, and can therefore be written $\gamma_{k\varepsilon} \exp(-i\omega t)$:

$$\alpha_{k\varepsilon}(t) = \beta_{k\varepsilon}(v_2) + \gamma_{k\varepsilon} e^{-i\omega t}. \tag{10}$$

The two terms in the right-hand side of (10) describe, respectively, the transverse field "bound" to the scattered particle moving away from the range of V_e and the free transverse field that appeared after the collision and is thus the transverse field radiated by the charged particle.

To calculate $\gamma_{k\varepsilon}$, integrate Equation (8) between t_1 and $t > t_2$ to get

$$\alpha_{k\varepsilon}(t) = \alpha_{k\varepsilon}(t_1) e^{-i\omega(t-t_1)} + \frac{iq}{\sqrt{2\varepsilon_0 \hbar \omega L^3}} \int_{t_1}^{t} \mathbf{\varepsilon} \cdot \dot{\mathbf{r}}(t') e^{i\omega(t'-t)} dt'. \tag{11}$$

Expressing $\alpha_{k\varepsilon}(t_1)$ in the form of Equation (9) and $\alpha_{k\varepsilon}(t)$ in the form of Equation (10) then gives

$$\gamma_{k\varepsilon} = \frac{-q}{\sqrt{2\varepsilon_0 \hbar \omega^3 L^3}} \left[\mathbf{\varepsilon} \cdot v_2 e^{i\omega t} - \mathbf{\varepsilon} \cdot v_1 e^{i\omega t_1} \right] + \frac{iq}{\sqrt{2\varepsilon_0 \hbar \omega L^3}} \int_{t_1}^{t} \mathbf{\varepsilon} \cdot \dot{\mathbf{r}}(t') e^{i\omega t'} dt'$$

$$= -\frac{q}{\sqrt{2\varepsilon_0 \hbar \omega^3 L^3}} \int_{t_1}^{t} \mathbf{\varepsilon} \cdot \ddot{\mathbf{r}}(t') e^{i\omega t'} dt'. \tag{12}$$

Because $\ddot{\mathbf{r}}(t')$ is zero for $t' < t_1$ and $t' > t_2$ (the particle is in uniform motion), it is possible to replace $\int_{t_1}^{t}$ by $\int_{-\infty}^{+\infty}$ in (12), which results in the appearance of the Fourier component at the frequency ω of the acceleration of the particle. Such a result confirms the interpretation given above for the last term in (10), which describes the field radiated by the particle.

c) If $\omega(t_2 - t_1) \ll 1$, we can approximate the exponential of the integral of (12) by 1, which allows us to replace this integral by $\mathbf{\varepsilon} \cdot (v_2 - v_1) = \mathbf{\varepsilon} \cdot \delta v$. We then have

$$\gamma_{k\varepsilon} = -\frac{q}{\sqrt{2\varepsilon_0 \hbar \omega^3 L^3}} \mathbf{\varepsilon} \cdot \delta v. \tag{13}$$

d) The total energy of the free field described by the normal variables $\{\gamma_{k\varepsilon} \exp(-i\omega t)\}$ and having a frequency $\omega \leq \omega_m$ is written, using (13),

$$\mathcal{H}_{1 \cdot f} = \sum_{k\varepsilon} \hbar \omega \gamma_{k\varepsilon}^* \gamma_{k\varepsilon} = \int_0^{k_m} \frac{k^2 \, dk}{(2\pi/L)^3} \int d\Omega \sum_{\varepsilon} \hbar \omega \gamma_{k\varepsilon}^* \gamma_{k\varepsilon} =$$

$$= \frac{q^2}{2\varepsilon_0 (2\pi)^3 c^3} \int_0^{\omega_m} d\omega \int d\Omega \sum_{\varepsilon} \sum_{\substack{i,j= \\ x,y,z}} \varepsilon_i \varepsilon_j \, \delta v_i \, \delta v_j. \tag{14}$$

Expression (14) represents the energy radiated at low frequency ($\omega < \omega_m$) by the particle. Using relation (54) from Complement A_I gives

$$\int d\Omega \sum_\varepsilon \varepsilon_i \varepsilon_j = \int d\Omega \left(\delta_{ij} - \frac{k_i k_j}{k^2} \right) = \frac{8\pi}{3} \delta_{ij} \tag{15}$$

and consequently

$$\mathscr{H}_{l \cdot f} = \frac{q^2 \omega_m}{6 \varepsilon_0 \pi^2 c^3} (\delta \mathbf{v})^2 = \frac{2\alpha}{3\pi} \left(\frac{\delta \mathbf{v}}{c} \right)^2 \hbar \omega_m \tag{16}$$

where $\alpha = q^2 / 4\pi \varepsilon_0 \hbar c$ is the fine-structure constant. It is thus clear from (16) that $\mathscr{H}_{l \cdot f} \ll \hbar \omega_m$.

e) Inverting formula (6) of Exercise 8 gives

$$V_e(\mathbf{r}) = \frac{1}{L^3} \sum_k e^{-i k \cdot \mathbf{r}} \mathscr{V}_e(\mathbf{k}). \tag{17}$$

It is then sufficient to replace $\mathbf{r}$ by $\mathbf{r} + \boldsymbol{\xi}$ and to use the fact that $\mathbf{r}$ and $\boldsymbol{\xi} = q\mathbf{Z}(0)/m$ commute to obtain

$$V_e(\mathbf{r} + \boldsymbol{\xi}) = \frac{1}{L^3} \sum_k \mathscr{V}_e(\mathbf{k}) \exp(-i \mathbf{k} \cdot \mathbf{r}) \exp(-i \mathbf{k} \cdot \boldsymbol{\xi}). \tag{18}$$

f) We first calculate the matrix element of $V_e(\mathbf{r} + \boldsymbol{\xi})$ between $|\mathbf{p}_1, \{0_\lambda\}, \{0_\lambda\}\rangle$ and $|\mathbf{p}_2, \{0_\lambda\}, \{n_\lambda\}\rangle$. Because $\boldsymbol{\xi} = \boldsymbol{\xi}^{h \cdot f} + \boldsymbol{\xi}^{l \cdot f}$ and $\boldsymbol{\xi}^{h \cdot f}$ and $\boldsymbol{\xi}^{l \cdot f}$ commute, this matrix element can be factored into three parts. The part relative to the particle is written

$$\frac{1}{L^3} \sum_k \mathscr{V}_e(\mathbf{k}) \langle \mathbf{p}_2 | \exp(-i \mathbf{k} \cdot \mathbf{r}) | \mathbf{p}_1 \rangle = \frac{1}{L^3} \sum_k \mathscr{V}_e(\mathbf{k}) \delta_{\mathbf{p}_1, \mathbf{p}_2 + \hbar \mathbf{k}} = \frac{\mathscr{V}_e(\mathbf{Q})}{L^3}. \tag{19}$$

The other two parts relative to $\boldsymbol{\xi}^{h \cdot f}$ and $\boldsymbol{\xi}^{l \cdot f}$ equal, respectively, using the $\delta_{\mathbf{k}, \mathbf{Q}}$ appearing in (19), $\langle \{0_\lambda\} | \exp(-i \mathbf{Q} \cdot \boldsymbol{\xi}^{h \cdot f}) | \{0_\lambda\} \rangle$ and $\langle \{n_\lambda\} | \exp(-i \mathbf{Q} \cdot \boldsymbol{\xi}^{l \cdot f}) | \{0_\lambda\} \rangle$.

Regrouping the preceding results gives

$$\mathscr{A}(\mathbf{p}_1, \{0_\lambda\}, \{0_\lambda\} \to \mathbf{p}_2, \{0_\lambda\}, \{n_\lambda\}) = -2\pi i \, \delta^{(T)} \left(E_1 - E_2 - \sum_\lambda n_\lambda \hbar \omega_\lambda \right) \times$$

$$\times \frac{\mathscr{V}_e(\mathbf{Q})}{L^3} \langle \{0_\lambda\} | \exp(-i \mathbf{Q} \cdot \boldsymbol{\xi}^{h \cdot f}) | \{0_\lambda\} \rangle \langle \{n_\lambda\} | \exp(-i \mathbf{Q} \cdot \boldsymbol{\xi}^{l \cdot f}) | \{0_\lambda\} \rangle. \tag{20}$$

g) If $\sum_\lambda n_\lambda \hbar \omega_\lambda$ is negligible compared with E_1 and E_2, we can replace the function $\delta^{(T)}$ in (20) by $\delta^{(T)}(E_1 - E_2)$. We can also neglect the difference between $|\mathbf{p}_1|$ and $|\mathbf{p}_2|$ and replace $\mathbf{Q}$ by the value corresponding to the elastic scattering. The low-frequency modes are no longer involved, except in the last matrix element in (20). We then obtain the equality (5),

where

$$\mathscr{A}_{el}(\mathbf{p}_1, \{0_\Lambda\} \to \mathbf{p}_2, \{0_\Lambda\}) = -2\pi i\, \delta^{(T)}(E_1 - E_2) \times$$
$$\times \langle\{0_\Lambda\}|\exp(-i\mathbf{Q} \cdot \boldsymbol{\xi}^{h \cdot f})|\{0_\Lambda\}\rangle \mathscr{V}_e(\mathbf{Q})/L^3 \qquad (21)$$

does not depend on low-frequency modes.

To prove (6), it is sufficient to write the left-hand side of (6) in the form

$$\langle\{0_\Lambda\}|\exp(i\mathbf{Q} \cdot \boldsymbol{\xi}^{l \cdot f})\left[\sum_{\{n_\lambda\}}|\{n_\lambda\}\rangle\langle\{n_\lambda\}|\right]\exp(-i\mathbf{Q} \cdot \boldsymbol{\xi}^{l \cdot f})|\{0_\Lambda\}\rangle. \qquad (22)$$

Because the operators $\exp(\pm i\mathbf{Q} \cdot \boldsymbol{\xi}^{l \cdot f})$ act only on the low-frequency photon subspace, and because the operator appearing in brackets in (22) is just the unit operator in this subspace (closure relation), expression (22) is reduced to $\langle\{0_\Lambda\}|\{0_\Lambda\}\rangle = 1$.

h) Equations (5) shows that if the particle is scattered from $\mathbf{p}_1$ to $\mathbf{p}_2$ without emitting any high-frequency photon, the coefficients for the expansion of the final state of the low-frequency field over the states $|\{n_\lambda\}\rangle$ are proportional to $\langle\{n_\lambda\}|\exp(-i\mathbf{Q} \cdot \boldsymbol{\xi}^{l \cdot f})|\{0_\Lambda\}\rangle$. The normalization condition (6) then shows that $\langle\{n_\lambda\}|\exp(-i\mathbf{Q} \cdot \boldsymbol{\xi}^{l \cdot f})|\{0_\Lambda\}\rangle$ is indeed the probability amplitude of finding the low-frequency field in state $|\{n_\lambda\}\rangle$. We deduce that the final state of the low-frequency field after the scattering process $(\mathbf{p}_1, \{0_\Lambda\} \to \mathbf{p}_2, \{0_\Lambda\})$ can be written

$$|\psi_{fin}^{l \cdot f}\rangle = \sum_{\{n_\lambda\}}|\{n_\lambda\}\rangle\langle\{n_\lambda\}|\exp(-i\mathbf{Q} \cdot \boldsymbol{\xi}^{l \cdot f})|\{0_\Lambda\}\rangle$$
$$= \exp(-i\mathbf{Q} \cdot \boldsymbol{\xi}^{l \cdot f})|\{0_\Lambda\}\rangle \qquad (23)$$

because the states $|\{n_\lambda\}\rangle$ form a basis in the low-frequency photon subspace.

i) In (23) we replace $\mathbf{Q}$ by $(\mathbf{p}_1 - \mathbf{p}_2)/\hbar = -m\delta\mathbf{v}/\hbar$ where $\delta\mathbf{v} = \mathbf{v}_2 - \mathbf{v}_1$, then replace $\boldsymbol{\xi}^{l \cdot f}$ by $q\mathbf{Z}^{l \cdot f}(0)/m$. By using the expansion of $\mathbf{Z}^{l \cdot f}(0)$ in $a_{\mathbf{k}\varepsilon}$ and $a_{\mathbf{k}\varepsilon}^+$ with $\omega < \omega_m$ [see formula (3) in Exercise 8] we obtain

$$\exp[-i\mathbf{Q} \cdot \boldsymbol{\xi}^{l \cdot f}] = \exp\left\{-\sum_{\mathbf{k}\varepsilon}^{l \cdot f}[\gamma_{\mathbf{k}\varepsilon}^* a_{\mathbf{k}\varepsilon} - \gamma_{\mathbf{k}\varepsilon}a_{\mathbf{k}\varepsilon}^+]\right\} \qquad (24)$$

where $\gamma_{\mathbf{k}\varepsilon}$ is given in (13). From this we deduce [see formulas (65)–(67) in the Appendix] that $\exp[-i\mathbf{Q} \cdot \boldsymbol{\xi}^{l \cdot f}]$ is a translation operator for $a_{\mathbf{k}\varepsilon}$:

$$\exp[+i\mathbf{Q} \cdot \boldsymbol{\xi}^{l \cdot f}]a_{\mathbf{k}\varepsilon}\exp[-i\mathbf{Q} \cdot \boldsymbol{\xi}^{l \cdot f}] = a_{\mathbf{k}\varepsilon} + \gamma_{\mathbf{k}\varepsilon} \qquad (25)$$

and that $|\psi_{fin}^{l \cdot f}\rangle$ is the coherent state corresponding to the classical field radiated at low frequency

$$|\psi_{fin}^{l \cdot f}\rangle = |\{\gamma_{\mathbf{k}\varepsilon}\}\rangle. \qquad (26)$$

As a consequence of (26) and using the properties of coherent states, the average value of the energy of the quantum field radiated at low frequency is equal to the corresponding classical energy given by Equation (16). Similarly, the distribution of the number of photons $\mathbf{k}\varepsilon$ emitted is a Poisson distribution with average $|\gamma_{\mathbf{k}\varepsilon}|^2$.

Finally, in the nonperturbative calculation presented in this exercise, no infrared divergence appears in the physical predictions concerning the low-frequency photons that are emitted. Moreover, Equation (26) provides a clear connection with classical theory.

j) Take the square of the modulus of (5) and divide by the duration of the interaction. One obtains for the transition rate

$$dw(\mathbf{p}_1, \{0_\Lambda\}, \{0_\lambda\} \to \mathbf{p}_2, \{0_\Lambda\}, \{n_\lambda\}) =$$

$$dw_{\mathrm{el}}(\mathbf{p}_1, \{0_\Lambda\} \to \mathbf{p}_2, \{0_\Lambda\}) \big| \langle\{n_\lambda\}| \exp(-i\mathbf{Q} \cdot \boldsymbol{\xi}^{l \cdot f})|\{0_\lambda\}\rangle\big|^2 \tag{27}$$

where the first term of the right-hand side, dw_{el}, is independent of the low-frequency modes. The sum over $\{n_\lambda\}$ in (27) then gives, using (6)

$$dw_{\mathrm{measured}}(\mathbf{p}_1 \to \mathbf{p}_2) = dw_{\mathrm{el}}(\mathbf{p}_1, \{0_\Lambda\} \to \mathbf{p}_2, \{0_\Lambda\}). \tag{28}$$

The measured rate therefore no longer depends on the low-frequency modes.

The foregoing arguments are easily generalized to processes in which high-frequency photons are also emitted. It is sufficient in (5) to replace $\mathscr{A}_{\mathrm{el}}(\mathbf{p}_1, \{0_\Lambda\} \to \mathbf{p}_2, \{0_\Lambda\})$ by the amplitude $\mathscr{A}_{\mathrm{el}}(\mathbf{p}_1, \{0_\Lambda\} \to \mathbf{p}_2, \{n_\Lambda\})$ given by the following equation generalizing (21):

$$\mathscr{A}(\mathbf{p}_1, \{0_\Lambda\} \to \mathbf{p}_2, \{n_\Lambda\}) = -2\pi i\, \delta^{(T)}\left(E_1 - E_2 - \sum_\Lambda n_\Lambda \hbar\omega_\Lambda\right) \times$$

$$\times \langle\{n_\Lambda\}|\exp(-i\mathbf{Q} \cdot \boldsymbol{\xi}^{h \cdot f})|\{0_\Lambda\}\rangle \mathscr{V}_e(\mathbf{Q})/L^3. \tag{29}$$

In (29), $|\mathbf{p}_1|$ is of course not equal to $|\mathbf{p}_2|$. Because the amplitude (29) depends only on the high-frequency modes, we can, as we did above, use (6) to show that the measured rate

$$\sum_{\{n_\lambda\}} dw(\mathbf{p}_1, \{0_\Lambda\}, \{0_\lambda\} \to \mathbf{p}_2, \{n_\Lambda\}, \{n_\lambda\}) \tag{30}$$

does not depend on low-frequency modes.

Thus, the foregoing nonperturbative treatment rigorously proves that the experimentally relevant quantities do not depend on low-frequency modes.

10. MODIFICATION OF THE CYCLOTRON FREQUENCY OF A PARTICLE DUE TO ITS INTERACTION WITH THE RADIATION FIELD

The purpose of this exercise is to study the way the interaction with the radiation field modifies the cyclotron motion of a charged particle in a static magnetic field B_0. An effective Hamiltonian will be used to describe the modification of the slow motion of the particle due to its interaction with the high-frequency modes of the radiation field.

Let m be the mass of the particle and q be its charge. The uniform magnetic field $\mathbf{B}_0$ is described by the vector potential

Let m be the mass of the particle and q be its charge. The uniform magnetic field $\mathbf{B}_0$ is described by the vector potential

$$\mathbf{A}_0(\mathbf{r}) = \tfrac{1}{2}(\mathbf{B}_0 \times \mathbf{r}) \tag{1}$$

and is parallel to the $0z$ axis. The Hamiltonian for the particle (assumed to have no spin) is

$$H_P = \frac{1}{2m}[\mathbf{p} - q\mathbf{A}_0(\mathbf{r})]^2. \tag{2}$$

a) Derive the expression for the three components v_x, v_y, and v_z for the velocity operator as a function of $\mathbf{p}$ and $\mathbf{r}$. Calculate their commutation relations. Write the Heisenberg equations for the velocity components. Deduce from these equations the motion of v_x, v_y, v_z and the expression for the cyclotron frequency ω_c.

b) First consider the effect of the interaction of the particle with a particular mode of the transverse field, with wave vector $\mathbf{k}$ and polarization $\boldsymbol{\varepsilon}$, and whose frequency ω is assumed to be large compared with the cyclotron frequency

$$\omega_c \ll \omega. \tag{3}$$

The Hamiltonian of the system is then

$$H = H_0 + H_I \tag{4}$$

where

$$H_0 = H_P + \hbar\omega\left(a^+a + \tfrac{1}{2}\right) \tag{5}$$

$$H_I = -q\mathbf{v} \cdot \mathbf{A}(\mathbf{r}) + \frac{q^2}{2m}\mathbf{A}^2(\mathbf{r}) \tag{6}$$

with

$$\mathbf{A}(\mathbf{r}) = \mathscr{A}_\omega(\boldsymbol{\varepsilon}\, e^{i\mathbf{k}\cdot\mathbf{r}}\, a + \boldsymbol{\varepsilon}\, e^{-i\mathbf{k}\cdot\mathbf{r}}\, a^+) \tag{7}$$

$\mathscr{A}_\omega$ being a constant equal to $(\hbar/2\varepsilon_0\omega L^3)^{1/2}$.

Show that, for the low-energy states of the particle, a large frequency range ω exists for which the long-wavelength approximation is justified. Such an approximation will be used in what follows.

c) By using the results in Complement B_I, derive the effective Hamiltonian that describes the dynamics of the particle in the presence of the radiation vacuum, to second order in q and in the approximation $\omega_c/\omega \ll 1$. Calculate the projection of the velocity of the particle over the polarization vector $\boldsymbol{\varepsilon}$ as well as the kinetic energy of the corresponding motion. Give a physical interpretation for the result obtained.

d) The particle now interacts with all the modes of frequency between ck_m and ck_M:

$$ck_m < \omega < ck_M. \tag{8}$$

The upper bound is fixed by the long-wavelength approximation; the lower bound is taken to be greater than ω_c, so that the results of c can be used for each of these modes.

Show that one can independently sum the effective Hamiltonians associated with each of the modes to describe the motion of the particle in the photon vacuum. How is the cyclotron frequency modified?

Solution

a) The velocity operator for the particle in the presence of the vector potential $\mathbf{A}_0(\mathbf{r})$ is

$$\mathbf{v} = \frac{1}{m}(\mathbf{p} - q\mathbf{A}_0(\mathbf{r})) \tag{9}$$

or

$$v_x = \frac{1}{m}\left(p_x + \frac{q}{2}B_0 y\right) \qquad v_y = \frac{1}{m}\left(p_y - \frac{q}{2}B_0 x\right) \qquad v_z = \frac{p_z}{m}. \tag{10}$$

v_z commutes with v_x and v_y, and

$$[v_x, v_y] = i\hbar\frac{q}{m^2}B_0. \tag{11}$$

The Hamiltonian for the particle, which describes its kinetic energy, can be written

$$H_P = \tfrac{1}{2}mv^2. \tag{12}$$

We then consider the Heisenberg equations for the components of $\mathbf{v}$

$$\dot{v}_x = \frac{i}{\hbar}[H_P, v_x] = \frac{im}{2\hbar}[\mathbf{v}^2, v_x] = \frac{im}{2\hbar}[v_y^2, v_x] = \frac{q}{m}B_0 v_y \tag{13.a}$$

$$\dot{v}_y = \frac{i}{\hbar}[H_P, v_y] = \frac{im}{2\hbar}[v_x^2, v_y] = -\frac{q}{m}B_0 v_x. \tag{13.b}$$

v_x and v_y oscillate with a $\pi/2$ phase difference at the cyclotron frequency

$$\omega_c = -\frac{q}{m} B_0.$$ (14)

The equation for v_z is written

$$\dot{v}_z = \frac{i}{\hbar}[H_p, v_z] = 0$$ (15)

and shows that the motion along $0z$ is uniform.

b) The commutation relation (11) leads to the uncertainty relation

$$\langle v_x^2 \rangle \langle v_y^2 \rangle \geq \left(\frac{\hbar q B_0}{2m^2}\right)^2.$$ (16)

This relation fixes the order of magnitude of the velocity in the lowest energy levels (called Landau levels) (*)

$$v^4 \sim \left(\frac{\hbar q B_0}{m^2}\right)^2$$ (17)

or

$$v \sim \sqrt{\frac{\hbar \omega_c}{m}}.$$ (18)

Because the motion is periodic, with a period $2\pi/\omega_c$, the spatial extension of the wave functions is on the order of

$$a \sim v\frac{1}{\omega_c} = \sqrt{\frac{\hbar}{m\omega_c}}.$$ (19)

For the long-wavelength approximation to be valid, it is necessary that

$$ka \ll 1$$ (20)

or

$$\frac{\omega}{c}\sqrt{\frac{\hbar}{m\omega_c}} \ll 1.$$ (21)

(*) See, for example, C. Cohen-Tannoudji, Diu, and Laloë, Complement E_{VI}.

Conditions (3) and (21) then imply that ω must satisfy

$$\omega_c \ll \omega \ll \sqrt{\omega_c \frac{(mc^2)}{\hbar}} . \tag{22}$$

For a magnetic field of 1 Tesla, this range extends from microwaves to visible light.

In the direction $0z$, the particle is free, and therefore nonlocalized. One can assume that an appropriate potential keeps it in the neighborhood of $z = 0$ with an extension also satisfying condition (20).

c) Assume that the particle is localized in the neighborhood of the origin: $r \simeq 0$. In the long-wavelength approximation, the term $q^2 A^2(0)/2m$ depends only on the field variables, and the coupling Hamiltonian between the field and the particle reduces to

$$H_{I1} = -q\mathbf{v} \cdot \mathbf{A}(0) = -q(\boldsymbol{\varepsilon} \cdot \mathbf{v})\mathscr{A}_\omega(a + a^+). \tag{23}$$

In the manifold corresponding to the vacuum $|0\rangle$, $\langle a + a^+ \rangle$ is zero. The motion of the particle is characterized by the frequency ω_c, which is very small compared with the frequency of the mode $(\mathbf{k}\boldsymbol{\varepsilon})$. We can then directly apply the results of subsection 4 in Complement B_I.

The expression for the effective Hamiltonian is

$$(H_P')_{\text{vac}}^{\mathbf{k},\boldsymbol{\varepsilon}} = H_P + \frac{q^2}{2m}\langle 0|A^2(0)|0\rangle +$$

$$+ q^2\mathscr{A}_\omega^2\langle 0|(a + a^+)\frac{1}{(\hbar\omega/2) - \hbar\omega(a^+a + \frac{1}{2})}(a + a^+)|0\rangle(\boldsymbol{\varepsilon} \cdot \mathbf{v})^2 -$$

$$- \frac{q^2\mathscr{A}_\omega^2}{2}\langle 0|(a + a^+)\frac{1}{[(\hbar\omega/2) - \hbar\omega(a^+a + \frac{1}{2})]^2}(a + a^+)|0\rangle[\boldsymbol{\varepsilon} \cdot \mathbf{v}, [\boldsymbol{\varepsilon} \cdot \mathbf{v}, H_P]] + \cdots$$

$$= \tfrac{1}{2}m v^2 + \frac{q^2}{2m}\mathscr{A}_\omega^2 - \frac{q^2\mathscr{A}_\omega^2}{\hbar\omega}(\boldsymbol{\varepsilon} \cdot \mathbf{v})^2 - \frac{mq^2\mathscr{A}_\omega^2}{4\hbar^2\omega^2}[\boldsymbol{\varepsilon} \cdot \mathbf{v}, [\boldsymbol{\varepsilon} \cdot \mathbf{v}, \mathbf{v}^2]] + \cdots . \tag{24}$$

The double commutator reduces to

$$\left[\varepsilon_x v_x + \varepsilon_y v_y, \left[\varepsilon_x v_x + \varepsilon_y v_y, v_x^2 + v_y^2\right]\right] = \varepsilon_x^2\left[v_x, \left[v_x, v_y^2\right]\right] + \varepsilon_y^2\left[v_y, \left[v_y, v_x^2\right]\right]$$

$$= -2(\varepsilon_x^2 + \varepsilon_y^2)(\hbar q B_0/m^2)^2 \tag{25}$$

and we finally obtain

$$(H_P')_{\text{vac}}^{\mathbf{k},\boldsymbol{\varepsilon}} = \tfrac{1}{2}m v^2 - \tfrac{1}{2}\delta m_{\mathbf{k}\boldsymbol{\varepsilon}}(\boldsymbol{\varepsilon} \cdot \mathbf{v})^2 + \frac{q^2}{2m}\mathscr{A}_\omega^2\left[1 + (\varepsilon_x^2 + \varepsilon_y^2)(\omega_c/\omega)^2\right] + \cdots \tag{26}$$

with

$$\delta m_{\mathbf{k}\varepsilon} = \frac{2q^2 \mathscr{A}_\omega^2}{\hbar\omega}. \tag{27}$$

The coupling with the mode $\mathbf{k}, \varepsilon$ introduces a correction to the kinetic energy in the effective Hamiltonian that describes the slow motion of the particle. This correction is anisotropic, zero for the components of $\mathbf{v}$ perpendicular to ε, and nonzero for the motion along the direction ε. The velocity of the slow motion along ε, which is written

$$\varepsilon \cdot \dot{\mathbf{r}} = \frac{i}{\hbar}\left[\frac{1}{2}(m - \delta m_{\mathbf{k}\varepsilon})(\varepsilon \cdot \mathbf{v})^2, (\varepsilon \cdot \mathbf{r})\right] = \left(1 - \frac{\delta m_{\mathbf{k}\varepsilon}}{m}\right)(\varepsilon \cdot \mathbf{v}) \tag{28}$$

is represented by the operator

$$v_\varepsilon' = \left(1 - \frac{\delta m_{\mathbf{k}\varepsilon}}{m}\right)v_\varepsilon. \tag{29}$$

In the effective Hamiltonian, the corresponding kinetic energy is written (to second order in q)

$$\tfrac{1}{2}(m - \delta m_{\mathbf{k}\varepsilon})v_\varepsilon^2 = \tfrac{1}{2}(m + \delta m_{\mathbf{k}\varepsilon})(v_\varepsilon')^2. \tag{30}$$

Everything occurs as if the inertia of the particle had increased in the direction ε.

For the last term of (26), one can suggest the following interpretation. In the initial Hamiltonian, the term $(q^2/2m)\mathscr{A}_\omega^2$ represents the vibrational kinetic energy of the particle under the influence of the vacuum fluctuations of the mode $\mathbf{k}\varepsilon$. In the presence of the magnetic field, the vibrational motion of the particle in the xy plane is modified by the magnetic field which curves the trajectory and results in the appearance of an orbital magnetic moment proportional to B_0. The last term in (26) represents the diamagnetic energy associated with this induced motion (*).

d) The coupling H_{I1} is now written

$$H_{I1} = -q\sum_i (\varepsilon_i \cdot \mathbf{v})\mathscr{A}_{\omega_i}(a_i + a_i^+). \tag{31}$$

It is linear in a and a^+ so that only the one-photon intermediate states $|\mathbf{k}_i\varepsilon_i\rangle$ appear in the second-order term of the effective Hamiltonian. The calculation of the contribution corresponding to this state can be made by taking into account only the coupling with the mode

(*) For a more detailed discussion, see, for example, P. Avan, C. Cohen-Tannoudji, J. Dupont-Roc, and C. Fabre, J. Phys. (Paris), 37, 993 (1976).

$\mathbf{k}_i \boldsymbol{\varepsilon}_i$. This is the same as summing the effective Hamiltonian $(H_P')_{\text{vac}}^i$ over i:

$$(H_P')_{\text{vac}} = \sum_{\substack{i \\ k_m < k_i < k_M}} (H_P')_{\text{vac}}^i \tag{32}$$

The sum over $\boldsymbol{\varepsilon}$ and $\boldsymbol{\varepsilon}'$ and the angular average over the polar angles of $\mathbf{k}$ gives, in the continuous limit [see formula (55) in Complement A_I]

$$\int \frac{d\Omega}{4\pi} \sum_{\varepsilon} (\boldsymbol{\varepsilon} \cdot \mathbf{v})^2 = \frac{2}{3} \mathbf{v}^2 \tag{33}$$

$$\int \frac{d\Omega}{4\pi} \sum_{\varepsilon} \varepsilon_x^2 = \frac{2}{3} \tag{34}$$

and an analogous formula for the average of ε_y^2. We then have

$$(H_P')_{\text{vac}} = \frac{1}{2}(m - \delta m)\mathbf{v}^2 + \frac{q^2}{2m} \int_{k_m}^{k_M} 4\pi k^2 \, dk \, \frac{\mathcal{A}_\omega^2}{(2\pi/L)^3} \left[1 + \frac{4}{3} \frac{\omega_c^2}{\omega^2}\right] \tag{35}$$

where

$$\delta m = \frac{2}{3} \int \frac{4\pi k^2 \, dk}{(2\pi/L)^3} \delta m_{\mathbf{k}\boldsymbol{\varepsilon}} = \frac{16\pi}{3} \int_{k_m}^{k_M} k^2 \, dk \, \frac{q^2}{\hbar\omega} \left[\frac{\hbar}{2\varepsilon_0 \omega (2\pi)^3}\right]$$

$$= \frac{1}{3\pi^2} \int_{k_m}^{k_M} dk \, \frac{q^2}{\varepsilon_0 c^2} = \frac{q^2(k_M - k_m)}{3\varepsilon_0 \pi^2 c^2} \simeq \frac{q^2 k_M}{3\varepsilon_0 \pi^2 c^2} \tag{36}$$

(k_m is very small compared with k_M and can be neglected). We obtain the mass correction δm previously calculated in Complement B_{II} [see formula (32)].

The second term of (35) is constant and does not change the dynamics of the particle. Returning to the operators $\mathbf{r}$ and $\mathbf{p}$, the first term is written

$$\frac{1}{2} \frac{m - \delta m}{m^2} [\mathbf{p} - q\mathbf{A}_0(\mathbf{r})]^2 \tag{37}$$

that is, again, to first order in δm

$$\frac{1}{2m}\left(1 - \frac{\delta m}{m}\right) [\mathbf{p} - q\mathbf{A}_0(\mathbf{r})]^2 \simeq \frac{1}{2(m + \delta m)} [\mathbf{p} - q\mathbf{A}_0(\mathbf{r})]^2. \tag{38}$$

Everything occurs as if the mass of the particle had increased by δm so that the cyclotron frequency becomes

$$\omega_c' = -\frac{q}{m + \delta m} B_0 \simeq \omega_c \left(1 - \frac{\delta m}{m}\right). \tag{39}$$

The cyclotron motion is thus slowed down by the increase in mass.

11. MAGNETIC INTERACTIONS BETWEEN SPINS

The purpose of this exercise is to study a unitary transformation that causes the Fermi contact interaction and the dipole-dipole interaction between two spins to appear explicitly in the particle Hamiltonian. In the usual picture, these interactions are due to an exchange of transverse photons between the two spins. To simplify the calculations as much as possible, the particles carrying the spins will be assumed here to be fixed (we thereby ignore the magnetic interactions associated with the translational motion of the particles, such as the current-current interaction, previously discussed in subsection F-1 in Chapter II, or the spin-other orbit interaction).

Consider the spins $S_\alpha, S_\beta, \ldots$ located at fixed points $r_\alpha, r_\beta, \ldots$. These spins interact, on the one hand with the quantized radiation field, and, on the other hand, with a static magnetic field B_0. The Hamiltonian of this system is, in the usual representation,

$$H = H_R - \sum_\alpha \gamma_\alpha S_\alpha \cdot [B_0 + B(r_\alpha)] \tag{1}$$

where H_R is the free radiation Hamiltonian:

$$H_R = \int d^3k \sum_\varepsilon \hbar\omega \left[a_\varepsilon^+(k) a_\varepsilon(k) + \tfrac{1}{2} \right]. \tag{2}$$

The gyromagnetic ratio γ_α is equal to $g_\alpha q_\alpha / 2m_\alpha$ where g_α is the g factor of the particle α, having a mass m_α and a charge q_α. $B(r_\alpha)$ is the magnetic field of the radiation at point r_α:

$$B(r_\alpha) = i \int d^3k \sum_\varepsilon \left(\frac{\hbar\omega}{2\varepsilon_0 c^2 (2\pi)^3} \right)^{1/2} \times$$
$$\times \left[a_\varepsilon(k) e^{ik \cdot r_\alpha} - a_\varepsilon^+(k) e^{-ik \cdot r_\alpha} \right] (\kappa \times \varepsilon). \tag{3}$$

H is split into three terms: H_R, which is of zero order in electric charge q_α, $H_I = -\sum_\alpha \gamma_\alpha S_\alpha \cdot B(r_\alpha)$, corresponding to the interaction between the spins and the radiation field $B(r)$, which is first order in q_α, and $H_S = -\sum_\alpha \gamma_\alpha S_\alpha \cdot B_0$, corresponding to the interaction between the spins and the external magnetic field.

a) First consider the case for $B_0 = 0$. One looks for a unitary transformation $T = \exp(iF/\hbar)$, F being first order in q_α, on the Hamiltonian H such that the linear term in q_α of the transformed Hamiltonian $H' =$

THT^+ is zero. What is the relation verified by F? One can use the identity

$$e^{iA} B e^{-iA} = B + i[A, B] + \frac{i^2}{2!}[A, [A, B]] + \cdots. \qquad (4)$$

b) The operator F is taken in the form

$$F = i\hbar \sum_\alpha \int d^3k \sum_\varepsilon [\beta_\alpha(\mathbf{k}, \varepsilon) a_\varepsilon^+(\mathbf{k}) - \beta_\alpha^+(\mathbf{k}, \varepsilon) a_\varepsilon(\mathbf{k})]. \qquad (5)$$

Calculate $\beta_\alpha(\mathbf{k}, \varepsilon)$ and check that

$$F = \sum_\alpha \gamma_\alpha [\nabla \times \mathbf{Z}(\mathbf{r}_\alpha)] \cdot \mathbf{S}_\alpha \qquad (6)$$

where $\mathbf{Z}(\mathbf{r})$ is the vector quantum field

$$\mathbf{Z}(\mathbf{r}) = \int d^3k \sum_\varepsilon \left[\frac{\hbar}{2\varepsilon_0 \omega (2\pi)^3}\right]^{1/2} \varepsilon \left[\frac{a_\varepsilon(\mathbf{k}) e^{i\mathbf{k}\cdot\mathbf{r}}}{i\omega} - \frac{a_\varepsilon^+(\mathbf{k}) e^{-i\mathbf{k}\cdot\mathbf{r}}}{i\omega}\right]. \qquad (7)$$

c) Give the terms of order less than or equal to 2 in q_α of the transformed Hamiltonian H'. Show that the magnetic interactions between different spins (Fermi contact interaction and dipole-dipole interaction) appear explicitly in H'. Show that self-energy terms also appear for each spin. To calculate these terms, a cutoff ($|\mathbf{k}| \leq k_c$) will be introduced in the mode expansion of the transverse fields.

What is the new interaction Hamiltonian between the spins and the transverse field?

d) The field B_0 is now assumed to be nonzero. By using the same unitary transformation as before and limiting the calculation to the first two terms of (4), calculate the operator H'_S transformed of H_S. Show that a new interaction term between the spins and the radiation field now appears, which is linear in a and a^+. Give a physical interpretation for the dependence of this term on B_0.

Solution

a) By using (4), we find that H' can be expressed as a function of F in the form of an expansion:

$$H' = e^{iF/\hbar} H e^{-iF/\hbar}$$

$$= H + \frac{i}{\hbar}[F, H] + \frac{1}{2}\left(\frac{i}{\hbar}\right)^2 [F, [F, H]] + \cdots. \qquad (8)$$

We regroup the terms of the same order in powers of q_α. We then have, because F is first order in q_α,

zero order $\quad H_R$

first order $\quad H_I + \dfrac{i}{\hbar}[F, H_R]$

second order $\quad \dfrac{i}{\hbar}[F, H_I] + \dfrac{1}{2}\left(\dfrac{i}{\hbar}\right)^2 [F, [F, H_R]]$.

For the first-order term to be zero, it is necessary that

$$H_I + \frac{i}{\hbar}[F, H_R] = 0. \tag{9}$$

b) Using the form of F given in (5) along with the commutators

$$[a_\varepsilon(\mathbf{k}), a_{\varepsilon'}^+(\mathbf{k}')] = \delta_{\varepsilon\varepsilon'}\delta(\mathbf{k} - \mathbf{k}') \tag{10}$$

we obtain

$$i\left[\frac{F}{\hbar}, H_R\right] = \sum_\alpha \int d^3k \sum_\varepsilon [\beta_\alpha^+(\mathbf{k}, \varepsilon)a_\varepsilon(\mathbf{k}) + \beta_\alpha(\mathbf{k}, \varepsilon)a_\varepsilon^+(\mathbf{k})]\hbar\omega. \tag{11}$$

But, from (1) and (3), H_I is equal to

$$H_I = -i\sum_\alpha \int d^3k \sum_\varepsilon \gamma_\alpha \left[\frac{\hbar\omega}{2\varepsilon_0 c^2 (2\pi)^3}\right]^{1/2} \times$$
$$\times [\mathbf{S}_\alpha \cdot (\boldsymbol{\kappa} \times \boldsymbol{\varepsilon})][a_\varepsilon(\mathbf{k}) e^{i\mathbf{k}\cdot\mathbf{r}_\alpha} - a_\varepsilon^+(\mathbf{k}) e^{-i\mathbf{k}\cdot\mathbf{r}_\alpha}]. \tag{12}$$

As a consequence of condition (9), the factors of the annihilation and creation operators are opposite in (11) and (12), which leads to

$$\beta_\alpha(\mathbf{k}, \varepsilon) = \frac{-i\gamma_\alpha \mathbf{S}_\alpha \cdot (\boldsymbol{\kappa} \times \boldsymbol{\varepsilon}) e^{-i\mathbf{k}\cdot\mathbf{r}_\alpha}}{\sqrt{2\varepsilon_0 c^2 \hbar\omega(2\pi)^3}}. \tag{13}$$

We substitute this expression into formula (5) giving F. We see that the expansion of the vector field $\nabla \times \mathbf{Z}(\mathbf{r}_\alpha)$ appears, $\mathbf{Z}(\mathbf{r}_\alpha)$ being defined by (7), and we finally obtain relation (6).

c) The zero-order term of H' is the radiation Hamiltonian H_R; the first-order term is zero as a consequence of condition (9) imposed on the unitary transformation. The second-order term is equal to

$$H_2' = \left[\frac{iF}{\hbar}, H_I\right] + \frac{1}{2}\left[\frac{iF}{\hbar}, \left[\frac{iF}{\hbar}, H_R\right]\right] \tag{14.a}$$

that is also, by using (9) to transform the second commutator

$$H'_2 = \frac{1}{2} \left[\frac{iF}{\hbar}, H_I \right].$$ (14.b)

Equations (1) and (6) allow H'_2 to be rewritten in the form

$$H'_2 = -\frac{i}{2} \sum_{\alpha, \beta} \sum_{i, j} \frac{\gamma_\alpha \gamma_\beta}{\hbar} \left[(\nabla \times \mathbf{Z}(\mathbf{r}_\alpha))_i (\mathbf{S}_\alpha)_i, B_j(\mathbf{r}_\beta)(\mathbf{S}_\beta)_j \right]$$

$$= -\frac{i}{2} \sum_{\alpha, \beta} \sum_{i, j} \frac{\gamma_\alpha \gamma_\beta}{\hbar} \left\{ \left[(\nabla \times \mathbf{Z}(\mathbf{r}_\alpha))_i, B_j(\mathbf{r}_\beta) \right] (\mathbf{S}_\alpha)_i (\mathbf{S}_\beta)_j + \right.$$

$$\left. + (\nabla \times \mathbf{Z}(\mathbf{r}_\alpha))_i B_j(\mathbf{r}_\beta) \left[(\mathbf{S}_\alpha)_i, (\mathbf{S}_\beta)_j \right] \right\}.$$ (15)

The commutator associated with the particle operators is

$$\left[(\mathbf{S}_\alpha)_i, (\mathbf{S}_\beta)_j \right] = i\hbar \sum_k \varepsilon_{ijk} \delta_{\alpha\beta} (\mathbf{S}_\alpha)_k$$ (16)

with ε_{ijk} being the antisymmetric tensor. Using (3), (7), and (10), as well as formula (33) from the Appendix, we calculate the commutator between the radiation operators

$$\left[(\nabla \times \mathbf{Z}(\mathbf{r}_\alpha))_i, B_j(\mathbf{r}_\beta) \right] = i \int d^3k \int d^3k' \sum_{\varepsilon, \varepsilon'} \frac{\hbar}{2\varepsilon_0 c^2 (2\pi)^3} \left(\frac{\omega'}{\omega} \right)^{1/2} \times$$

$$\times \left\{ - \left[a_\varepsilon(\mathbf{k}), a_{\varepsilon'}^+(\mathbf{k}') \right] (\mathbf{\kappa} \times \mathbf{\varepsilon})_i (\mathbf{\kappa}' \times \mathbf{\varepsilon}')_j e^{i\mathbf{k} \cdot \mathbf{r}_\alpha} e^{-i\mathbf{k}' \cdot \mathbf{r}_\beta} + \right.$$

$$\left. + \left[a_\varepsilon^+(\mathbf{k}), a_{\varepsilon'}(\mathbf{k}') \right] (\mathbf{\kappa} \times \mathbf{\varepsilon})_i (\mathbf{\kappa}' \times \mathbf{\varepsilon}')_j e^{-i\mathbf{k} \cdot \mathbf{r}_\alpha} e^{i\mathbf{k}' \cdot \mathbf{r}_\beta} \right\}$$

$$= \frac{-i\hbar}{\varepsilon_0 c^2 (2\pi)^3} \int d^3k \sum_\varepsilon (\mathbf{\kappa} \times \mathbf{\varepsilon})_i (\mathbf{\kappa} \times \mathbf{\varepsilon})_j e^{i\mathbf{k} \cdot (\mathbf{r}_\alpha - \mathbf{r}_\beta)}$$

$$= \frac{-i\hbar}{\varepsilon_0 c^2 (2\pi)^3} \int d^3k \left(\delta_{ij} - \frac{k_i k_j}{k^2} \right) e^{i\mathbf{k} \cdot (\mathbf{r}_\alpha - \mathbf{r}_\beta)}.$$ (17)

By using the transverse delta function (see the Appendix and Complement A_I of *Photons and Atoms—Introduction to Quantum Electrodynamics*), we can rewrite (17) in the form

$$\left[(\nabla \times \mathbf{Z}(\mathbf{r}_\alpha))_i, B_j(\mathbf{r}_\beta) \right] = -\frac{i\hbar}{\varepsilon_0 c^2} \delta_{ij}^\perp (\mathbf{r}_\alpha - \mathbf{r}_\beta).$$ (18)

Finally, formula (15) giving the Hamiltonian H'_2 is simplified by using (16) and (18):

$$H'_2 = -\sum_{\alpha, \beta} \sum_{i, j} \frac{\gamma_\alpha \gamma_\beta}{2\varepsilon_0 c^2} \delta_{ij}^\perp (\mathbf{r}_\alpha - \mathbf{r}_\beta)(\mathbf{S}_\alpha)_i (\mathbf{S}_\beta)_j +$$

$$+ \sum_\alpha \frac{\gamma_\alpha^2}{2} [\nabla \times \mathbf{Z}(\mathbf{r}_\alpha)] \cdot [\mathbf{B}(\mathbf{r}_\alpha) \times \mathbf{S}_\alpha].$$ (19)

We first study, in the first term of H_2', the contribution of terms corresponding to different particles ($\alpha \neq \beta$). By using (34) from the Appendix, we can rewrite the term relating to the pair α, β:

$$H_2'(\alpha, \beta) = -\frac{\gamma_\alpha \gamma_\beta (\mathbf{S}_\alpha)_i (\mathbf{S}_\beta)_j}{\varepsilon_0 c^2} \delta_{ij}^\perp (\mathbf{r}_\alpha - \mathbf{r}_\beta)$$

$$= -\frac{2}{3} \frac{\gamma_\alpha \gamma_\beta}{\varepsilon_0 c^2} \mathbf{S}_\alpha \cdot \mathbf{S}_\beta \delta(\mathbf{r}_\alpha - \mathbf{r}_\beta) +$$

$$+ \frac{\gamma_\alpha \gamma_\beta}{4\pi\varepsilon_0 c^2} \left[\frac{\mathbf{S}_\alpha \cdot \mathbf{S}_\beta}{|\mathbf{r}_\alpha - \mathbf{r}_\beta|^3} - 3 \frac{[\mathbf{S}_\alpha \cdot (\mathbf{r}_\alpha - \mathbf{r}_\beta)][\mathbf{S}_\beta \cdot (\mathbf{r}_\alpha - \mathbf{r}_\beta)]}{|\mathbf{r}_\alpha - \mathbf{r}_\beta|^5} \right]. \quad (20)$$

We recognize two contributions in $H_2'(\alpha, \beta)$: the first is the Fermi contact interaction, and the second is the dipole-dipole interaction.

The first term in (19), for the case where $\alpha = \beta$, gives a contribution ε_{S_α} to the magnetic moment self-energy

$$\varepsilon_{S_\alpha} = -\frac{\gamma_\alpha^2}{2\varepsilon_0 c^2} \sum_{i,j} \delta_{ij}^\perp(\mathbf{0})(\mathbf{S}_\alpha)_i (\mathbf{S}_\alpha)_j. \quad (21)$$

By introducing a cutoff $|\mathbf{k}| \leq k_c$, we obtain, for $\delta_{ij}^\perp(\mathbf{0})$,

$$\delta_{ij}^\perp(\mathbf{0}) = \int_{|\mathbf{k}|<k_c} \frac{d^3k}{(2\pi)^3} \left(\delta_{ij} - \frac{k_i k_j}{k^2} \right) = \frac{k_c^3}{9\pi^2} \delta_{ij} \quad (22)$$

which gives for ε_{S_α}

$$\varepsilon_{S_\alpha} = -\frac{\gamma_\alpha^2 k_c^3}{18\pi^2 \varepsilon_0 c^2} \mathbf{S}_\alpha^2. \quad (23)$$

The interaction of the spins with the transverse field is given by the last term in (19) which is quadratic as a function of the annihilation and creation operators $a_\varepsilon(\mathbf{k})$ and $a_\varepsilon^+(\mathbf{k})$. Note that the operator $\mathbf{B}(\mathbf{r}_\alpha)$ which appears in this formula no longer represents the total magnetic field in the new representation. Instead, it now represents the difference between the total magnetic field and the magnetic field produced by the magnetic moments (discussion analogous to the one in subsection 2-a in Complement B_{II}).

d) In the unitary transformation, H_S becomes

$$H_S' = e^{iF/\hbar} H_S e^{-iF/\hbar} = \sum_\alpha -\gamma_\alpha \mathbf{B}_0 \cdot \left(e^{iF/\hbar} \mathbf{S}_\alpha e^{-iF/\hbar} \right). \quad (24)$$

If we keep only the first two terms of (4), we obtain, for the component $(\mathbf{S}_\alpha)_i$,

$$e^{iF/\hbar}(\mathbf{S}_\alpha)_i e^{-iF/\hbar} = (\mathbf{S}_\alpha)_i + \frac{i}{\hbar}[F, (\mathbf{S}_\alpha)_i] \quad (25)$$

an expression which can be transformed by using (6) and (16)

$$[F, (S_\alpha)_i] = \sum_\beta \sum_j \gamma_\beta (\boldsymbol{\nabla} \times \mathbf{Z}(\mathbf{r}_\beta))_j [(S_\beta)_j, (S_\alpha)_i]$$

$$= \sum_{j,k} \gamma_\alpha (\boldsymbol{\nabla} \times \mathbf{Z}(\mathbf{r}_\alpha))_j i\hbar \varepsilon_{jik} (S_\alpha)_k. \tag{26}$$

Using (24) and (26), we then obtain for H_S'

$$H_S' = -\sum_\alpha \gamma_\alpha \mathbf{S}_\alpha \cdot \mathbf{B}_0 + \sum_\alpha \gamma_\alpha^2 [\boldsymbol{\nabla} \times \mathbf{Z}(\mathbf{r}_\alpha)] \cdot (\mathbf{B}_0 \times \mathbf{S}_\alpha). \tag{27}$$

The last term of (27) is a new spin-transverse field interaction Hamiltonian, which is linear in a and a^+ (one-photon term):

$$H_I' = \sum_\alpha \gamma_\alpha^2 [\boldsymbol{\nabla} \times \mathbf{Z}(\mathbf{r}_\alpha)] \cdot (\mathbf{B}_0 \times \mathbf{S}_\alpha). \tag{28}$$

It is obviously associated with the precession of the magnetic moment about the static field $\mathbf{B}_0$. The spin can radiate and then emit photons only because it precesses about $\mathbf{B}_0$.

Remark

The interaction Hamiltonian in the usual representation H_I differs from the Hamiltonian H_I' found after applying the unitary transformation. Of course, all the physical predictions must be identical in the two representations. If we consider, for example, a spin $\frac{1}{2}$ in a magnetic field $\mathbf{B}_0$ aligned along the axis $0z$, this spin can emit a photon into a mode $\mathbf{k}\boldsymbol{\varepsilon}$ of the field for which $ck = -\gamma_\alpha B_0$. When this resonance condition is satisfied, it is easy to prove the following equality:

$$\langle +\tfrac{1}{2}; 0 | H_I' | -\tfrac{1}{2}; \mathbf{k}\boldsymbol{\varepsilon} \rangle = \langle +\tfrac{1}{2}; 0 | H_I | -\tfrac{1}{2}; \mathbf{k}\boldsymbol{\varepsilon} \rangle \tag{29}$$

(with $|\mathbf{k}\boldsymbol{\varepsilon}\rangle = a_\varepsilon^+(\mathbf{k})|0\rangle$). Equality (29) ensures that the transition probabilities (absorption or emission of photons) are the same in the two representations on the energy shell.

12. MODIFICATION OF AN ATOMIC MAGNETIC MOMENT DUE TO ITS COUPLING WITH MAGNETIC FIELD VACUUM FLUCTUATIONS

Consider a paramagnetic atom in its ground state $|a\rangle$. This atom is supposed to be a neutral atom having a magnetic moment $\mathbf{M} = \gamma \mathbf{S}$ where $\mathbf{S}$ is the spin of the particle and γ is the gyromagnetic ratio that can be written in the form $\gamma = g_a q / 2m$, where q and m are the charge and the mass of the electron and g_a is the Landé factor for the state a. This atom is placed in a static magnetic field $\mathbf{B}_0$ parallel to $0z$. The free evolution of the magnetic moment coupled to $\mathbf{B}_0$ is a Larmor precession motion, with frequency $\omega_L = -\gamma B_0$. The purpose of this exercise is to show that this frequency is slightly modified as a result of the interaction of the atomic magnetic moment with the magnetic field vacuum fluctuations. We will use for this the same unitary transformation as the one introduced in Exercise 11.

To study the influence of the coupling with the magnetic field $\mathbf{B}(\mathbf{r})$ of the radiation, we take the following model Hamiltonian (*):

$$H = H_R + H_S + H_I \tag{1}$$

where H_R is the radiation Hamiltonian, $H_S = -\gamma \mathbf{S} \cdot \mathbf{B}_0$ describes the interaction of the magnetic moment with the static field and $H_I = -\gamma \mathbf{S} \cdot \mathbf{B}(\mathbf{r})$ describes the interaction with the radiation field.

a) Calculate at the third order in q the operator H_S' transformed from H_S by the unitary transformation $T = \exp(iF/\hbar)$ where $F = \gamma(\mathbf{\nabla} \times \mathbf{Z}(\mathbf{r})) \cdot \mathbf{S}$. The fields $\mathbf{B}(\mathbf{r})$ and $\mathbf{Z}(\mathbf{r})$ are given by formulas (3) and (7) in Exercise 11. One can use formula (4) from this exercise with $B = H_S$, $A = F/\hbar$ to calculate the first three terms of the expansion.

b) Calculate the average value of H_S' in the state $|m_s; 0\rangle$ (with $m_s \hbar$ being the eigenvalue of S_z and $|0\rangle$ being the photon vacuum). Show that the result can be interpreted as a correction $\delta\omega_L$ to the Larmor frequency ω_L. Express $\delta\omega_L/\omega_L$ as a function of $x_M = \hbar\omega_M/mc^2$ (where $\omega_M = ck_M$ is the cutoff frequency for the mode expansion) and of the fine-structure constant $\alpha = q^2/4\pi\varepsilon_0\hbar c$.

c) What is the physical origin of $\delta\omega_L$? Can its sign be interpreted?

d) In the same magnetic field $\mathbf{B}_0$, a particle with charge q and mass m has a cyclotron motion of frequency $\omega_c = -qB_0/m$. The interaction of this particle with the transverse field is responsible for a variation in mass

---

(*) This Hamiltonian is not sufficient to quantitatively predict the modification of the magnetic moment of a free electron. There are terms of relativistic origin in the electron-radiation interaction, which, for field frequencies large compared with the atomic frequencies, have the same order of magnitude as the terms retained in this model Hamiltonian. They do not, however, qualitatively change the conclusions drawn from (1). A more complete and more rigorous approach to this problem may be found in the following references: P. Avan, C. Cohen-Tannoudji, J. Dupont-Roc, and C. Fabre, *J. Phys. (Paris)*, **37**, 993 (1976); J. Dupont-Roc, C. Fabre, and C. Cohen-Tannoudji, *J. Phys. B*, **11**, 563 (1978); J. Dupont-Roc and C. Cohen-Tannoudji, in *New Trends in Atomic Physics*, Les Houches XXXVIII, edited by G. Grynberg and R. Stora, Elsevier, New York, 1984, p. 157.

δm and thus a variation $\delta \omega_c$ in the cyclotron frequency given by

$$\frac{\delta \omega_c}{\omega_c} = -\frac{\delta m}{m} \tag{2}$$

with δm being equal to $q^2 k_M / 3 \varepsilon_0 \pi^2 c^2$ (see Exercise 10). Compare $\delta \omega_L / \omega_L$ to $\delta \omega_c / \omega_c$.

If the corrected Landé factor g'_a is defined by the relation

$$\frac{g'_a}{2} = \frac{\omega'_L}{\omega'_c} \tag{3}$$

with $\omega'_L = \omega_L + \delta \omega_L$ and $\omega'_c = \omega_c + \delta \omega_c$, what is the sign of the correction $\delta g_a = g'_a - g_a$? What is its origin?

Solution

a) To determine H'_S

$$H'_S = e^{iF/\hbar} H_S e^{-iF/\hbar} = -\gamma B_0 e^{iF/\hbar} S_z e^{-iF/\hbar} \tag{4}$$

to third order in γ, it is necessary to calculate the transform of S_z to second order. The zero- and first-order terms were calculated in Exercise 11 [formulas (25) and (26)]. We now calculate the second-order term:

$$S_z'^{(2)} = \frac{1}{2}\left[\frac{iF}{\hbar}, \left[\frac{iF}{\hbar}, S_z\right]\right]. \tag{5}$$

Starting with relation (26) from Exercise 11, we obtain

$$S_z'^{(2)} = \frac{i\gamma^2}{2\hbar} \sum_{i,j,k} \varepsilon_{zjk}\left[(\nabla \times \mathbf{Z(r)})_i S_i, (\nabla \times \mathbf{Z(r)})_j S_k\right]. \tag{6}$$

The components of the field $\nabla \times \mathbf{Z(r)}$ commute with each other. We transform (6) by using the commutation relations between the components of $\mathbf{S}$ [see formula (16) in Exercise 11]:

$$S_z'^{(2)} = -\frac{\gamma^2}{2} \sum_{i,j,k,l} \varepsilon_{zjk} \varepsilon_{ikl} S_l (\nabla \times \mathbf{Z(r)})_i (\nabla \times \mathbf{Z(r)})_j$$

$$= -\frac{\gamma^2}{2} \sum_{j \neq z} \sum_{i,l} (\delta_{zl} \delta_{ji} - \delta_{zi} \delta_{jl}) S_l (\nabla \times \mathbf{Z(r)})_i (\nabla \times \mathbf{Z(r)})_j$$

$$= -\frac{\gamma^2}{2} \sum_{j \neq z} \left[S_z (\nabla \times \mathbf{Z(r)})_j (\nabla \times \mathbf{Z(r)})_j - S_j (\nabla \times \mathbf{Z(r)})_j (\nabla \times \mathbf{Z(r)})_z \right]. \tag{7}$$

Finally we obtain H'_S by regrouping $-\gamma B_0 S_z'^{(2)}$ with the previously calculated zero- and first-order terms.

$$H'_s = -\gamma B_0 S_z + \gamma^2 (\mathbf{B}_0 \times \mathbf{S}) \cdot (\nabla \times \mathbf{Z(r)}) +$$

$$+ \frac{\gamma^3}{2} B_0 \sum_{j \neq z} \left[S_z (\nabla \times \mathbf{Z(r)})_j (\nabla \times \mathbf{Z(r)})_j - S_j (\nabla \times \mathbf{Z(r)})_j (\nabla \times \mathbf{Z(r)})_z \right]. \quad (8)$$

b) Calculate the average value of the zero- and first-order terms of H'_s in the state $|m_s; 0\rangle$. We find $-m_s \hbar \gamma B_0$ for the zero-order term and 0 for the first-order term, because this term is linear in $a_\varepsilon(\mathbf{k})$ and $a_\varepsilon^+(\mathbf{k})$, whose average value in the vacuum is zero. To find the average value of the next term, first calculate

$$\langle 0|(\nabla \times \mathbf{Z(r)})_i (\nabla \times \mathbf{Z(r)})_j|0\rangle = \int d^3k' \, d^3k \sum_{\varepsilon, \varepsilon'} \frac{\hbar}{2\varepsilon_0 c^2 (2\pi)^3 \sqrt{\omega \omega'}} \times$$

$$\times (\mathbf{\kappa} \times \mathbf{\varepsilon})_i (\mathbf{\kappa}' \times \mathbf{\varepsilon}')_j \langle 0|a_\varepsilon(\mathbf{k}) a_{\varepsilon'}^+(\mathbf{k}')|0\rangle e^{i(\mathbf{k} - \mathbf{k}') \cdot \mathbf{r}}. \quad (9)$$

The other combinations of annihilation and creation operators have a zero average value in the vacuum. By using the commutation relation

$$\left[a_\varepsilon(\mathbf{k}), a_{\varepsilon'}^+(\mathbf{k}') \right] = \delta_{\varepsilon \varepsilon'} \delta(\mathbf{k} - \mathbf{k}') \quad (10)$$

and formula (33) from the Appendix, we obtain

$$\langle 0|(\nabla \times \mathbf{Z(r)})_i (\nabla \times \mathbf{Z(r)})_j|0\rangle = \int d^3k \sum_\varepsilon \frac{\hbar}{2\varepsilon_0 c^2 \omega (2\pi)^3} (\mathbf{\kappa} \times \mathbf{\varepsilon})_i (\mathbf{\kappa} \times \mathbf{\varepsilon})_j =$$

$$= \int d^3k \frac{\hbar}{2\varepsilon_0 c^2 \omega (2\pi)^3} \left(\delta_{ij} - \frac{k_i k_j}{k^2} \right) = \frac{\hbar k_M^2}{12\pi^2 \varepsilon_0 c^3} \delta_{ij}. \quad (11)$$

Finally, the average value in the state $|m_s; 0\rangle$ of the operator H'_S written in (8) is given by

$$\langle m_s; 0|H'_s|m_s; 0\rangle = -m_s \hbar \gamma B_0 \left(1 - \gamma^2 \frac{\hbar k_M^2}{12\pi^2 \varepsilon_0 c^3} \right). \quad (12)$$

We thus find a correction $\delta \omega_L$ to the Larmor frequency $\omega_L = -\gamma B_0$ equal to

$$\delta \omega_L = -\omega_L \frac{g_a^2 q^2}{4m^2} \frac{\hbar k_M^2}{12\pi^2 \varepsilon_0 c^3} \quad (13)$$

We can finally express $\delta \omega_L / \omega_L$ as a function of g_a, x_M, and α. This gives

$$\frac{\delta \omega_L}{\omega_L} = -\frac{g_a^2}{4} \frac{\alpha}{3\pi} x_M^2. \quad (14)$$

c) The magnetic moment precesses (while keeping a constant modulus) in the total field, which is the static field $\mathbf{B}_0$ + the vacuum field $\mathbf{B}(\mathbf{r})$. The vacuum fluctuations of frequency ω (high compared with ω_L) cause the magnetic moment to have an angular oscillation at the frequency ω. For the coupling with the static field $\mathbf{B}_0$, the relevant quantity is the time average of the magnetic moment, which clearly has a smaller modulus than the instantaneous magnetic moment. We then expect the rapid angular oscillation of the magnetic moment in the vacuum fluctuations to reduce the effective magnetic moment of the atom and thus to produce a negative correction $\delta\omega_L/\omega_L$ according to formula (14).

d) Express the variation $\delta\omega_c/\omega_c$ of the cyclotron frequency as a function of x_M and α. Using relation (2), we find

$$\frac{\delta\omega_c}{\omega_c} = -4\frac{\alpha}{3\pi}x_M. \tag{15}$$

It thus appears $\delta\omega_c/\omega_c$ is linear in x_M, whereas $\delta\omega_L/\omega_L$ is quadratic [see (14)]. For the low-frequency modes ($x_M \ll 1$), described correctly in this theoretical approach, the modification of the cyclotron frequency is thus more important than the modification of the Larmor frequency. Such a result is not surprising. It indicates that, in the nonrelativistic domain, the magnetic interactions, which are responsible for $\delta\omega_L/\omega_L$, are less intense than the electric interactions, which are responsible for $\delta\omega_c/\omega_c$.

If we define g'_a by using relation (3), the preponderant correction term comes from the linear correction in x_M of $\delta\omega_c/\omega_c$. As a consequence of the sign of $\delta\omega_c/\omega_c$ (15), this term results in a value of g'_a that is greater than g_a. Because the cyclotron frequency ω_c decreases more than the Larmor frequency ω_L, the ratio $2\omega_L/\omega_c$ increases. Thus

$$\delta g_a > 0. \tag{16}$$

13. EXCITATION OF AN ATOM BY A WAVE PACKET: BROADBAND EXCITATION AND NARROW-BAND EXCITATION

Consider an atom in its ground state *a* interacting with a photon whose state is described by the wave packet $|\psi_R\rangle$. The purpose of this exercise is to calculate the probability for the atom to be excited to a level *b* by this photon and to show that the time evolution of this probability can differ a great deal depending on whether the spectral width of the incident field is large or small compared with the natural width Γ_b of the excited level.

a) Show that the matrix element $\langle b; 0|G(E + i\eta)|a; \mathbf{k}\varepsilon\rangle$ of the resolvent (where η is infinitely small and positive) can be taken, to a very good approximation, as equal to

$$\langle b; 0|G(E + i\eta)|a; \mathbf{k}\varepsilon\rangle = \frac{\langle b; 0|H_{I1}|a; \mathbf{k}\varepsilon\rangle}{\left(E - \tilde{E}_b + i\hbar(\Gamma_b/2)\right)\left(E + i\eta - \tilde{E}_a - \hbar\omega\right)} \tag{1}$$

where $\tilde{E}_a$ and $\tilde{E}_b$ are the energies of the levels a and b including the radiative shifts ($\tilde{E}_i = E_i + \hbar\Delta_i$ for $i = a, b$).

b) Deduce from this result the probability amplitude $\langle b; 0|U(t)|a; \mathbf{k\varepsilon}\rangle$ of finding the system in the state $|b; 0\rangle$ at the time t knowing that it is in the state $|a; \mathbf{k\varepsilon}\rangle$ at the initial time. One will set $\hbar\tilde{\omega}_{ba} = (\tilde{E}_b - \tilde{E}_a)$.

c) Assume that the initial state of the field $|\psi_R\rangle$ is a one-photon wave packet of the form

$$|\psi_R\rangle = \int dk\, g(k - k_1)|\mathbf{k\varepsilon}\rangle \tag{2}$$

where $\mathbf{k}_1$ is the average wave vector and where the summation concerns states having the same polarization ε and the same direction of propagation $\mathbf{\kappa} = \mathbf{k}/k$. Give the expression for the transition amplitude $\langle b; 0|U(t)|a; \psi_R\rangle = U_{ba}(t)$. Neglect the dependence on k of the matrix elements $\langle b; 0|H_{I1}|a; \mathbf{k\varepsilon}\rangle$ which can then be replaced by a unique matrix element $\langle b; 0|H_{I1}|a; \mathbf{k}_1\mathbf{\varepsilon}\rangle$.

In what follows, $\omega_1 = ck_1$ is assumed to coincide with the frequency $\tilde{\omega}_{ba}$ of the transition. In other words, the incident wave packet has a central frequency that is resonant with the atomic transition frequency.

d) If the curve giving $|g(k)|^2$ as a function of k has a width Δk that is small compared with the average value k_1 of k, one can show that the square of the modulus of the Fourier transform of $g(k)$ is proportional to the intensity of the field $\langle E^{(-)}E^{(+)}\rangle$ at the atom location (*). Let $f(t)$ be the Fourier transform of $g(k)$

$$f(t) = \int d\omega\, g(k)\, e^{-i\omega t}. \tag{3}$$

The curve $f(t)$ is shown in Figure 1. The function $f(t)$ is assumed to be maximum at $t = t_0$ and has a width at half-maximum equal to Δt. The time origin $t = 0$ is chosen well before the time when the incident wave packet arrives in the neighborhood of the atom so that $t_0 \gg \Delta t, \Gamma_b^{-1}$.

Show that, using these assumptions, the transition amplitude $U_{ba}(t)$ calculated in c) is proportional to the convolution product of $f(t)$ by the Fourier transform of $-[\omega + i(\Gamma_b/2)]^{-1}$.

e) First consider the situation where $|g(k)|$ is a very narrow function of k, with a width Δk much smaller than Γ_b (narrow-band excitation). Show that the probability of finding the atom in the level b then adiabatically

(*) See *Photons and Atoms—Introduction to Quantum Electrodynamics*, Chapter III, §C-2-c.

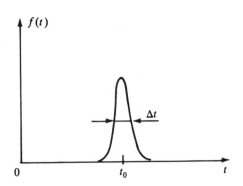

Figure 1. The curve $f(t)$.

follows the time evolution of $|f(t)|^2$, that is, the time evolution of the incident intensity.

f) Now consider the case for quasi-percussional excitation where $\Delta t \ll \Gamma_b^{-1}$. Show that the probability of finding the atom in the level b is then represented by a curve whose rise time is on the order of Δt and whose decay time is equal to Γ_b^{-1}.

Solution

a) Let $\mathscr{E}_0$ be the subspace subtended by $|b;0\rangle$ and P be the projector over this subspace. Relation (B.31) from Chapter III gives

$$\langle b;0|G(z)|a;\mathbf{k}\varepsilon\rangle = \langle b;0|G(z)|b;0\rangle\langle b;0|V\frac{1}{z - QH_0Q - QVQ}|a;\mathbf{k}\varepsilon\rangle. \quad (4)$$

The same type of reasoning as that leading from relation (C.26) to relation (C.29) in Chapter III then allows Equation (1) to be derived.

b) The transition amplitude $\langle b;0|U(t)|a;\mathbf{k}\varepsilon\rangle$ is deduced from (1) by the contour integral (A.22) of Chapter III. Integration by the residue method then gives

$$\langle b;0|U(t)|a;\mathbf{k}\varepsilon\rangle = \frac{\langle b;0|H_{I1}|a;\mathbf{k}\varepsilon\rangle}{\left(\tilde{E}_b - \tilde{E}_a - \hbar\omega - i\hbar(\Gamma_b/2)\right)} \times$$

$$\times\left[e^{-i\tilde{E}_b t/\hbar}\,e^{-\Gamma_b t/2} - e^{-i(\tilde{E}_a + \hbar\omega)t/\hbar}\right]. \quad (5)$$

c) Because $|\psi_R\rangle$ is a linear superposition of states $|\mathbf{k}\varepsilon\rangle$, we immediately find

$$\langle b;0|U(t)|a;\psi_R\rangle = \int d\mathbf{k}\,g(k - k_1)\langle b;0|U(t)|a;\mathbf{k}\varepsilon\rangle. \quad (6)$$

Using (5), as well as the assumption concerning $\langle b; 0|H_{I1}|a; \mathbf{k}\boldsymbol{\varepsilon}\rangle$, we then obtain

$$U_{ba}(t) = \langle b; 0|H_{I1}|a; \mathbf{k}_1\boldsymbol{\varepsilon}\rangle \times$$

$$\times \int dk\, g(k - k_1) \frac{\left[e^{-i\tilde{E}_b t/\hbar}\, e^{-\Gamma_b t/2} - e^{-i(\tilde{E}_a + \hbar\omega)t/\hbar} \right]}{\hbar(\tilde{\omega}_{ba} - \omega - i(\Gamma_b/2))} \tag{7.a}$$

that is, also, by carrying out the variable change $k' = k - k_1$ and by using the fact that $\omega_1 = ck_1 = \tilde{\omega}_{ba}$

$$U_{ba}(t) = \frac{1}{\hbar}\, e^{-i\tilde{E}_b t/\hbar} \langle b; 0|H_{I1}|a; \mathbf{k}_1\boldsymbol{\varepsilon}\rangle \int dk'\, g(k') \frac{\left[e^{-\Gamma_b t/2} - e^{-i\omega' t} \right]}{-(\omega' + i(\Gamma_b/2))}.$$

d) Let $h(t)$ be the Fourier transform of $-[\omega + i(\Gamma_b/2)]^{-1}$:

$$h(t) = \int d\omega \frac{e^{-i\omega t}}{-\omega - i(\Gamma_b/2)} = 2\pi i\theta(t)\exp(-\Gamma_b t/2) \tag{8}$$

where $\theta(t)$ is the Heaviside function.

The product of $g(k')$ and $-[\omega' + i(\Gamma_b/2)]^{-1}$ appears in the integral over k' of (7.b). The Fourier transform of this product gives the convolution product $f \circ h$ of $f(t)$ by $h(t)$, which will be written C:

$$C(t) = f \circ h(t) = \frac{1}{2\pi} \int_{-\infty}^{+\infty} f(t')h(t - t')\, dt'$$

$$= i \int_{-\infty}^{+\infty} f(t')\, e^{-(\Gamma_b/2)(t - t')}\, \theta(t - t')\, dt'. \tag{9}$$

The first term in brackets in (7.b), $\exp(-\Gamma_b t/2)$, does not depend on k'. It thus gives rise to a contribution proportional to $C(0)$:

$$\frac{1}{\hbar c} \langle b; 0|H_{I1}|a; \mathbf{k}_1\boldsymbol{\varepsilon}\rangle\, e^{-i\tilde{E}_b t/\hbar}\, e^{-\Gamma_b t/2}\, C(0). \tag{10}$$

According to the assumptions made concerning $f(t)$ (see Figure 1), $f(t')$ and $\theta(0 - t') = \theta(-t')$ do not overlap, so that Equation (9) gives a zero result for $C(0)$. Only the contribution of the second term in brackets of (7.b) remains, which leads to the following result for $U_{ba}(t)$

$$U_{ba}(t) = -\frac{1}{\hbar c} \langle b; 0|H_{I1}|a; \mathbf{k}_1\boldsymbol{\varepsilon}\rangle\, e^{-i\tilde{E}_b t/\hbar}\, C(t). \tag{11}$$

e) The convolution product $f \circ h$ is obtained by taking the integral over t' of the two functions $f(t')$ and $h(t - t')$ represented in Figure 2.

In the convolution product of the two functions having very different widths, it is the wider function that determines the shape of the curve for the time variation of the convolution product.

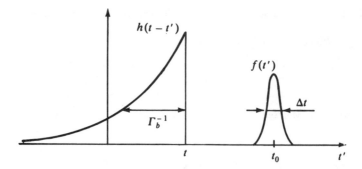

Figure 2. The functions $h(t - t')$ and $f(t')$.

If the width Δk of $g(k)$ is very small compared with Γ_b (narrow-band excitation), then the width $1/\Delta k$ of $f(t')$ is much larger than the width Γ_b^{-1} of $h(t - t')$. The convolution product $C(t)$ then closely resembles $f(t)$ and the excitation probability $|U_{ba}(t)|^2$ varies as the intensity $|f(t)|^2$ of the incident field at the location of the atom.

f) If, by contrast, $\Delta k \gg \Gamma_b$ (broadband excitation) it is then $f(t')$ that is much narrower than $h(t - t')$ and $C(t)$ then closely resembles $h(t - t_0)$, the rising edge at $t = t_0$ having a width on the order of Δt (Figure 3). $|U_{ba}(t)|^2$ thus increases over a time interval on the order of Δt, then decreases with a time constant equal to Γ_b^{-1}. In this case, the preparation phase of the system in the excited state, which lasts a time Δt, is clearly distinguishable from the decay phase, which lasts a time Γ_b^{-1}.

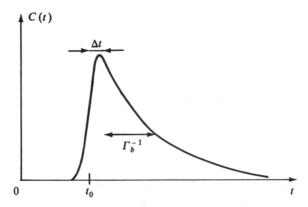

Figure 3. The function $C(t)$ in the case where $\Delta t \ll \Gamma_b^{-1}$.

14. SPONTANEOUS EMISSION BY A SYSTEM OF TWO NEIGHBORING ATOMS. SUPERRADIANT AND SUBRADIANT STATES (*)

Consider a system of two identical atoms. The first one, located at $\mathbf{R}_1$, is modeled as an electron 1, having position $\mathbf{r}_1$ and momentum $\mathbf{p}_1$, evolving in a central potential. The ground state of this atom is a_1 and b_1 is the first excited state. For the second atom, the corresponding variables are called $\mathbf{R}_2$, $\mathbf{r}_2$, and $\mathbf{p}_2$, and the states are written a_2 and b_2. The distance $R = |\mathbf{R}_1 - \mathbf{R}_2|$ between the two atoms is assumed to be large compared with the spatial extension a_0 of each atom, and small compared with the wavelength λ_0 of the transition $a_1 \leftrightarrow b_1$ (or $a_2 \leftrightarrow b_2$). The electrostatic interaction between the dipoles of the two atoms (van der Waals forces) will be neglected, but the coupling of each atom with the vector potential $\mathbf{A}$ of the radiation field is taken into consideration. The purpose of this exercise is to study the spontaneous emission of such a system. More precisely, it will be shown that there is an excitation state of the two-atom system in which it emits a photon more rapidly than an isolated atom (superradiant state), and another one in which it emits no photons at all (subradiant states).

Because $a_0 \ll R \ll \lambda_0$, it is possible to use the long-wavelength approximation and, in the interaction Hamiltonian H_I, to replace $\mathbf{A}(\mathbf{r}_1)$ and $\mathbf{A}(\mathbf{r}_2)$ by $\mathbf{A}(\mathbf{0})$, the origin of the coordinate system being selected in the neighborhood of $\mathbf{R}_1$ and $\mathbf{R}_2$. This gives for H_{I1} the following expression:

$$H_{I1} = \sum_{i=1,2} -\frac{q}{m}\mathbf{p}_i \cdot \mathbf{A}(\mathbf{0}). \tag{1}$$

One calls $|\varphi_1\rangle = |a_1, b_2; 0\rangle$ the state in which atom 1 is in the state a and atom 2 is in the state b in the presence of 0 photon and $|\varphi_2\rangle = |b_1, a_2; 0\rangle$ is the state in which the atomic excitations are exchanged. The restriction of the resolvent to the subspace $\mathcal{E}_0$ subtended by $|\varphi_1\rangle$ and $|\varphi_2\rangle$ will be calculated to determine the evolution of the system inside this subspace.

a) Let M be the operator $M = PH_0P + P\hat{R}(E_0 + i\eta)P$, where P is the projector onto the subspace $\mathcal{E}_0$, $\hat{R}$ is the shift operator calculated to second order in q, $E_0 = E_b + E_a$, and η is infinitely small and positive. Express the diagonal matrix elements $M_{11} = \langle \varphi_1|M|\varphi_1\rangle$ and $M_{22} = \langle \varphi_2|M|\varphi_2\rangle$ as a function of $\tilde{E}_a = E_a + \hbar\Delta_a$, and $\tilde{E}_b = E_b + \hbar\Delta_b$, where $\hbar\Delta_a$ and $\hbar\Delta_b$ are the radiative shifts of the states a and b.

(*) The superradiant states were introduced by R. H. Dicke, *Phys. Rev.*, **93**, 99 (1954).

b) Calculate the nondiagonal matrix elements $M_{12} = \langle \varphi_1 | M | \varphi_2 \rangle$ and $M_{21} = \langle \varphi_2 | M | \varphi_1 \rangle$. Show that the result obtained can be written

$$M_{12} = M_{21} = \hbar \Delta_a + \hbar \Delta_b - i\hbar \frac{\Gamma_b}{2} \tag{2}$$

if only levels *a* and *b* are taken into account in the calculation of the radiative shifts $\hbar \Delta_a$ and $\hbar \Delta_b$. In what follows, the atom is considered as a two-level system so that result (2) can be used.

c) Calculate the eigenvalues and eigenvectors of the operator M. Deduce from this result that if the system is in the state $| \varphi_s ; 0 \rangle$, where

$$| \varphi_s \rangle = \frac{1}{\sqrt{2}} [|a_1, b_2 \rangle + |b_1, a_2 \rangle] \tag{3}$$

then its radiative lifetime is shorter than the radiative lifetime of an isolated atom (this state is called superradiant).

Why is the state $| \varphi_a ; 0 \rangle$ where

$$| \varphi_a \rangle = \frac{1}{\sqrt{2}} [|a_1, b_2 \rangle - |b_1, a_2 \rangle] \tag{4}$$

called subradiant?

d) At the initial time, the system is assumed to be in the state $| \varphi_1 \rangle$. What is the probability of finding one of the two atoms in an excited state after a time interval that is long compared with $1/\Gamma_b$?

e) The Hamiltonians for atoms 1 and 2 are written $H_P^{(1)}$ and $H_P^{(2)}$. Calculate the average value of the dipole moment of atom 1, $\langle \varphi_s | q z_1(t) | \varphi_s \rangle$, in the superradiant state and the correlation function $q^2 \langle \varphi_s | z_1(t) z_2(0) | \varphi_s \rangle$ where $q z_1(t) = \exp(iH_P^{(1)}t/\hbar) q z_1 \exp(-iH_P^{(1)}t/\hbar)$ is the dipole moment operator in the Heisenberg representation. What are these same average values in the subradiant state $| \varphi_a \rangle$? What is the physical interpretation of superradiance and subradiance?

Solution

a) The matrix element M_{11} is equal to $E_a + E_b + \hat{R}_{11}(E_0 + i\eta)$. By proceeding as in subsection C-1 in Chapter III, we find at second order in q_α

$$\hat{R}_{11}(E_0 + i\eta) = \sum_c \sum_{k, \varepsilon} \frac{|\langle a_1, b_2; 0 | H_{I1} | c_1, b_2; k\varepsilon \rangle|^2}{E_a - E_c - \hbar\omega + i\eta}$$

$$+ \sum_c \sum_{k, \varepsilon} \frac{|\langle a_1, b_2; 0 | H_{I1} | a_1, c_2; k\varepsilon \rangle|^2}{E_b - E_c - \hbar\omega + i\eta}. \tag{5}$$

In the first sum, atom 2 remains a "spectator", as atom 1 does in the second sum. We can then write

$$\hat{R}_{11}(E_0 + i\eta) = \hat{R}_{aa}(E_a + i\eta) + \hat{R}_{bb}(E_b + i\eta) \tag{6}$$

where $\hat{R}_{aa}(E_a + i\eta)$ and $\hat{R}_{bb}(E_b + i\eta)$ are identical to the quantities calculated in subsection C-1 in Chapter III for an isolated atom. It follows that

$$\hat{R}_{11}(E_0 + i\eta) = \hbar \Delta_a + \hbar \Delta_b - i\hbar \frac{\Gamma_b}{2} \tag{7}$$

(because the ground state is stable, there is no imaginary term associated with $\hbar \Delta_a$). Consequently,

$$M_{11} = \tilde{E}_a + \tilde{E}_b - i\hbar \frac{\Gamma_b}{2}. \tag{8}$$

An identical calculation proves that $M_{22} = M_{11}$.

b) Now consider $M_{12} = R_{12}(E_0 + i\eta)$

$$\hat{R}_{12}(E_0 + i\eta) = \sum_{c,d} \sum_{\mathbf{k},\varepsilon} \frac{\langle a_1, b_2; 0|H_{I1}|c_1 d_2; \mathbf{k}\varepsilon\rangle\langle c_1, d_2; \mathbf{k}\varepsilon|H_{I1}|b_1, a_2; 0\rangle}{E_a + E_b - E_c - E_d - \hbar\omega + i\eta}. \tag{9}$$

Because H_{I1} can modify only the state of a single atom, it is clear that the only intermediate states leading to a nonzero result in (9) are $|b_1, b_2; \mathbf{k}\varepsilon\rangle$ and $|a_1, a_2; \mathbf{k}\varepsilon\rangle$ so that

$$\hat{R}_{12}(E_0 + i\eta) = \sum_{\mathbf{k},\varepsilon} \frac{|\langle a; 0|H_{I1}|b; \mathbf{k}\varepsilon\rangle|^2}{E_a - E_b - \hbar\omega + i\eta} + \sum_{\mathbf{k},\varepsilon} \frac{|\langle b; 0|H_{I1}|a; \mathbf{k}\varepsilon\rangle|^2}{E_b - E_a - \hbar\omega - i\eta}. \tag{10}$$

For two-level atoms, the two sums in the right-hand side of (10) are, respectively, equal to $R_{aa}(E_a + i\eta)$ and $R_{bb}(E_b + i\eta)$, which proves formula (2).

c) The effective Hamiltonian in the subspace $\mathscr{E}_0$ is represented by the matrix M

$$M = \begin{pmatrix} \tilde{E}_a + \tilde{E}_b - i\hbar\dfrac{\Gamma_b}{2} & \hbar\Delta_a + \hbar\Delta_b - i\hbar\dfrac{\Gamma_b}{2} \\ \hbar\Delta_a + \hbar\Delta_b - i\hbar\dfrac{\Gamma_b}{2} & \tilde{E}_a + \tilde{E}_b - i\hbar\dfrac{\Gamma_b}{2} \end{pmatrix}. \tag{11}$$

The eigenvectors of the matrix M are $|\varphi_s; 0\rangle$ and $|\varphi_a; 0\rangle$ and the corresponding eigenvalues are

$$\begin{aligned} \lambda_s &= \tilde{E}_a + \tilde{E}_b + \hbar\Delta_a + \hbar\Delta_b - i\hbar\Gamma_b \\ &= E_a + E_b + 2(\hbar\Delta_a + \hbar\Delta_b) - i\hbar\Gamma_b \end{aligned} \tag{12}$$

$$\begin{aligned} \lambda_a &= \tilde{E}_a + \tilde{E}_b - i\hbar\frac{\Gamma_b}{2} - (\hbar\Delta_a + \hbar\Delta_b) + i\hbar\frac{\Gamma_b}{2} \\ &= E_a + E_b. \end{aligned} \tag{13}$$

From this we conclude that the state $|\varphi_s\rangle$ has a radiative lifetime that is half the radiative lifetime of an isolated atom. For this reason, this state is called superradiant. By contrast, the state $|\varphi_a\rangle$ is stable and is thus called subradiant. It is also easy to check that $\langle a_1, a_2; \mathbf{k}\varepsilon|H_{I1}|\varphi_a; 0\rangle = 0$, which demonstrates that $|\varphi_a\rangle$ is not coupled to the one-photon continuum and thus cannot decay by spontaneous emission.

d) If at the initial time the system is in the state

$$|\varphi_1\rangle = \frac{1}{\sqrt{2}}\left[|\varphi_s; 0\rangle + |\varphi_a; 0\rangle\right] \tag{14}$$

the projection of the state vector on $\mathscr{E}_0$ is, at a later time t, equal to

$$P|\psi(t)\rangle = \frac{1}{\sqrt{2}} e^{-i(E_a+E_b)t/\hbar}\left[e^{-2i(\Delta_a+\Delta_b)t} e^{-\Gamma_b t}|\varphi_s; 0\rangle + |\varphi_a; 0\rangle\right]. \tag{15}$$

After a time that is long compared with $1/\Gamma_b$, $P|\psi(t)\rangle$ coincides, except for a phase factor, with $(1/\sqrt{2})|\varphi_a; 0\rangle$, so that the probability of finding one of the two atoms in the excited state is equal to $\frac{1}{2}$ (the probability for each atom is the same and is equal to $\frac{1}{4}$). The evolution of atom 1 is thus markedly different depending on whether it is isolated or close to another identical atom. This is a consequence of the fact that the two atoms 1 and 2 are indirectly coupled to each other by the radiation that they can emit. The two states $|\varphi_1\rangle$ and $|\varphi_2\rangle$ are coupled to the same continuum and it would be incorrect to apply the Fermi golden rule to only one of these two states.

e) When we evaluate $\langle \varphi_s|qz_1(t)|\varphi_s\rangle$, by using the expression (3) for the state $|\varphi_s\rangle$, we obtain two types of terms: the first terms, having the form $\langle a_1|qz_1(t)|b_1\rangle\langle b_2|a_2\rangle$, are zero because of the scalar product $\langle b_2|a_2\rangle$, and the second terms, having the form $\langle a_1|qz_1(t)|a_1\rangle\langle b_2|b_2\rangle$, are zero because of parity.

We now calculate the correlation function between the two dipoles. By introducing the closure relation between $z_1(t)$ and $z_2(0)$ and by keeping only the nonzero terms, we obtain

$$
\begin{aligned}
q^2\langle\varphi_s|z_1(t)z_2(0)|\varphi_s\rangle &= q^2\big[\langle\varphi_s|z_1(t)|b_1 b_2\rangle\langle b_1 b_2|z_2(0)|\varphi_s\rangle + \\
&\quad + \langle\varphi_s|z_1(t)|a_1 a_2\rangle\langle a_1 a_2|z_2(0)|\varphi_s\rangle\big] \\
&= q^2\big[\tfrac{1}{2}\langle a_1|z_1(t)|b_1\rangle\langle b_2|z_2(0)|a_2\rangle + \\
&\quad + \tfrac{1}{2}\langle b_1|z_1(t)|a_1\rangle\langle a_2|z_2(0)|b_2\rangle\big]
\end{aligned}
\tag{16}
$$

or, by using $z_1(t) = \exp[iH_P^{(1)}t/\hbar]\, z\, \exp[-iH_P^{(1)}t/\hbar]$,

$$q^2\langle\varphi_s|z_1(t)z_2(0)|\varphi_s\rangle = |d|^2\left(\frac{e^{-i\omega_0 t}}{2} + \frac{e^{i\omega_0 t}}{2}\right)$$

$$= |d|^2 \cos\omega_0 t \tag{17}$$

where $|d|^2 = q^2|z_{ba}|^2$. The same calculations made for the state $|\varphi_a\rangle$ give

$$\langle\varphi_a|qz_1(t)|\varphi_a\rangle = 0 \tag{18.a}$$

$$q^2\langle\varphi_a|z_1(t)z_2(0)|\varphi_a\rangle = -|d|^2 \cos\omega_0 t. \tag{18.b}$$

The average value of each electric dipole is zero for the superradiant state as well as for the subradiant state. By contrast, the dipoles of the two atoms are perfectly correlated. They oscillate in phase in the superradiant state and oscillate out of phase in the subradiant state [note that sign differs in (17) and (18.b) for $t = 0$]. Consequently, the waves emitted by the two atoms interfere in a perfectly constructive way in the superradiant state. As a result, the system radiates more efficiently than an isolated atom and it thus has a shorter lifetime. By contrast, in the subradiant state, the interference is constantly destructive, and the system does not radiate. Its lifetime is infinitely long.

15. RADIATIVE CASCADE OF A HARMONIC OSCILLATOR

The purpose of this exercise is to calculate the spectral distribution of photons emitted in a radiative cascade by an excited harmonic oscillator and, using this simple example, to show the importance of the interferences between transition amplitudes corresponding to different time orders for the emission of the various photons of the cascade.

A particle with charge q and mass m is assumed to move along the axis $0z$ and is bound to the origin 0 by an attractive force, $-m\omega_0^2 z$, proportional to the distance from the origin (one neglects any excitation of the motion perpendicular to $0z$ which is assumed to correspond to frequencies much greater than the frequency ω_0 of the oscillational motion along $0z$). Let b^+ and b be the creation and annihilation operators associated with such a one-dimensional harmonic oscillator. The particle Hamiltonian H_A is written

$$H_A = \hbar\omega_0\left(b^+b + \tfrac{1}{2}\right) \tag{1}$$

and has as its eigenstates the states $|u_n\rangle$ $(n = 0, 1, 2, \dots)$ with eigenvalues $E_n = (n + \tfrac{1}{2})\hbar\omega_0$. In the long-wavelength approximation, the interaction Hamiltonian between the oscillator and the radiation is written

$$H_I = H_{I1} + H_{I2} = -\frac{q}{m}p_z A_z(\mathbf{0}) + \frac{q^2}{2m}A_z^2(\mathbf{0}) \tag{2}$$

where

$$p_z = i\sqrt{\frac{m\hbar\omega_0}{2}}\,(b^+ - b) \tag{3}$$

is the momentum of the oscillator and $A_z(\mathbf{0})$ is the z component of the vector potential $\mathbf{A}(\mathbf{r})$ evaluated at $\mathbf{r} = \mathbf{0}$.

a) The particle + radiation system is initially in the state

$$|\varphi_n\rangle = |u_n;0\rangle \tag{4}$$

(oscillator in the state $|u_n\rangle$ in the presence of the photon vacuum). Calculate the probability per unit time Γ_n that the system will leave the state $|\varphi_n\rangle$ by spontaneous emission. Γ_n is called the natural width of the level u_n. Express Γ_n as a function of $\Gamma_1 = \Gamma$ and calculate Γ as a function of q, ε_0, m, c, and ω_0.

b) Give the expression, to second order in q, for the radiative shift $\hbar\Delta_n$ of the state $|\varphi_n\rangle$. Do not try to explicitly calculate Δ_n, as was done for Γ_n and assume that an adequate cutoff in the summation over the modes of the field allows the value of Δ_n to remain finite. Using the properties of the matrix elements of H_{I1} and H_{I2}, prove that $\Delta_n - \Delta_{n-1} = \Delta_1 - \Delta_0$. What can be concluded about the perturbed energies $\tilde{E}_n = E_n + \hbar\Delta_n$ of the oscillator? In what follows, set

$$\hbar\tilde{\omega}_0 = \hbar\omega_0 + \hbar(\Delta_1 - \Delta_0). \tag{5}$$

c) The system is initially in the state $|\varphi_2\rangle$. Study the amplitude

$$\langle u_0; \mathbf{k}_a\boldsymbol{\varepsilon}_a, \mathbf{k}_b\boldsymbol{\varepsilon}_b | U(\tau) | u_2;0\rangle \tag{6}$$

that, at the end of a period of time τ, the oscillator will go from level u_2 to level u_0 by spontaneously emitting two photons $\mathbf{k}_a\boldsymbol{\varepsilon}_a$ and $\mathbf{k}_b\boldsymbol{\varepsilon}_b$. In (6), $U(\tau)$ is the evolution operator. To calculate (6), first evaluate the matrix element corresponding to the resolvent $G(z)$

$$\langle u_0; \mathbf{k}_a\boldsymbol{\varepsilon}_a, \mathbf{k}_b\boldsymbol{\varepsilon}_b | G(z) | u_2;0\rangle \tag{7}$$

and use the result derived in subsection C-2-b in Chapter III according to which it is sufficient to replace, in the expression at the lowest order in q of (7), the unperturbed energies E_n of the states u_n by $\tilde{E}_n - i\hbar(\Gamma_n/2)$. Deduce from this result the value of the amplitude (6) for $\tau \gg \Gamma^{-1}$. Show that this amplitude is a sum of two terms that will be expressed as a function of $\tilde{\omega}_0$, ω_a, ω_b, and Γ and of the quantities μ_a and μ_b defined by

$$\langle u_0; \mathbf{k}_i\boldsymbol{\varepsilon}_i | H_{I1} | u_1;0\rangle = \mu_i \qquad (i = a, b). \tag{8}$$

d) Let $I(\omega_a, \omega_b)\,d\omega_a\,d\omega_b$ be the probability that the two emitted photons $\mathbf{k}_a\boldsymbol{\varepsilon}_a$ and $\mathbf{k}_b\boldsymbol{\varepsilon}_b$ have frequencies, respectively, between ω_a and

$\omega_a + d\omega_a$ and ω_b and $\omega_b + d\omega_b$. Prove that $I(\omega_a, \omega_b)$ can be factored in the form of a product of two functions of ω_a and ω_b, both having the same width. Deduce from this result that the frequencies of the two photons are independent and calculate the spectral distribution of the emitted radiation. Compare such a result with the one that would be obtained if the interferences between the two amplitudes calculated in c were neglected.

e) Can the results obtained in c and d be generalized to the case where the oscillator starts initially in the level $|u_3\rangle$ and then spontaneously emits three photons $\mathbf{k}_a\boldsymbol{\varepsilon}_a$, $\mathbf{k}_b\boldsymbol{\varepsilon}_b$, and $\mathbf{k}_c\boldsymbol{\varepsilon}_c$?

Solution

a) The Hamiltonian H_{I2} acts only on the radiation and cannot change the state of the oscillator. The probability per unit time Γ_n of leaving the state $|\varphi_n\rangle$ thus involves only H_{I1}, which couples $|\varphi_n\rangle = |u_n; 0\rangle$ to $|u_{n\pm1}; \mathbf{k}\boldsymbol{\varepsilon}\rangle$. In the expression of Γ_n, given by the Fermi golden rule, we can neglect the coupling of $|\varphi_n\rangle$ with $|u_{n+1}; \mathbf{k}\boldsymbol{\varepsilon}\rangle$ which does not conserve energy, and we then have

$$\Gamma_n = \frac{2\pi}{\hbar} \sum_{\mathbf{k}\boldsymbol{\varepsilon}} |\langle u_{n-1}; \mathbf{k}\boldsymbol{\varepsilon}|H_{I1}|u_n; 0\rangle|^2 \delta(\hbar\omega - \hbar\omega_0). \tag{9}$$

Equations (2) and (3) and the expansion of $A_z(0)$ in $a_{\mathbf{k}\boldsymbol{\varepsilon}}$ and $a^+_{\mathbf{k}\boldsymbol{\varepsilon}}$ lead to

$$\langle u_{n-1}; \mathbf{k}\boldsymbol{\varepsilon}|H_{I1}|u_n; 0\rangle = -\frac{q}{m}\langle u_{n-1}|p_z|u_n\rangle\langle \mathbf{k}\boldsymbol{\varepsilon}|A_z(0)|0\rangle$$

$$= i\frac{q}{m}\sqrt{\frac{m\hbar\omega_0}{2}}\sqrt{n}\sqrt{\frac{\hbar}{2\varepsilon_0\omega L^3}}\,\varepsilon_z \tag{10}$$

so that

$$|\langle u_{n-1}; \mathbf{k}\boldsymbol{\varepsilon}|H_{I1}|u_n; 0\rangle|^2 = n|\langle u_0; \mathbf{k}\boldsymbol{\varepsilon}|H_{I1}|u_1; 0\rangle|^2 \tag{11}$$

and, consequently,

$$\Gamma_n = n\Gamma_1 \tag{12}$$

To calculate $\Gamma_1 = \Gamma$, use (9) and (10) with $n = 1$, and replace the discrete sum $\sum_{\mathbf{k}\boldsymbol{\varepsilon}}$ by an integral. This leads to

$$\Gamma = \frac{2\pi}{\hbar}\frac{L^3}{(2\pi)^3}\int k^2\,dk\,d\Omega\,q^2\frac{\hbar\omega_0}{2m}\frac{\hbar}{2\varepsilon_0\omega L^3}\delta(\hbar\omega - \hbar\omega_0)\sum_{\boldsymbol{\varepsilon}\perp\mathbf{k}}\varepsilon_z^2. \tag{13}$$

The sum over the polarizations gives [see formula (54) in Complement A_I],

$$\sum_{\varepsilon \perp \mathbf{k}} \varepsilon_z^2 = 1 - \frac{k_z^2}{k^2} \tag{14}$$

and the integral over the solid angle leads to

$$\int d\Omega \left(1 - \frac{k_z^2}{k^2}\right) = \frac{8\pi}{3} \tag{15}$$

so that we finally have for Γ

$$\Gamma = \frac{q^2 \omega_0}{6\pi\varepsilon_0 mc^3} \int \omega \, d\omega \, \delta(\omega - \omega_0) = \frac{q^2 \omega_0^2}{6\pi\varepsilon_0 mc^3}. \tag{16}$$

Result (16) indeed coincides with the inverse of the energy damping time of a classical particle with charge q oscillating at the frequency ω_0 (*).

b) To second order in q, $\hbar\Delta_n$ is written

$$\hbar\Delta_n = \langle u_n; 0|H_{I2}|u_n; 0\rangle +$$

$$+ \mathscr{P} \sum_{\mathbf{k}\varepsilon} \frac{|\langle u_{n+1}; \mathbf{k}\varepsilon|H_{I1}|u_n; 0\rangle|^2}{-\hbar(\omega + \omega_0)} +$$

$$+ \mathscr{P} \sum_{\mathbf{k}\varepsilon} \frac{|\langle u_{n-1}; \mathbf{k}\varepsilon|H_{I1}|u_n; 0\rangle|^2}{\hbar(\omega_0 - \omega)}. \tag{17}$$

The first term in (17) represents the effect of H_{I2} at the first order. Because H_{I2} does not act on $|u_n\rangle$, this term is independent of n and represents a global shift. The two following terms represent the effect of the virtual emission and reabsorption of a photon $\mathbf{k}\varepsilon$ from the state $|u_n\rangle$, the oscillator passing through an intermediate state $|u_{n+1}\rangle$ or $|u_{n-1}\rangle$. By using (11) and an analogous identity, which is written

$$|\langle u_{n+1}; \mathbf{k}\varepsilon|H_{I1}|u_n; 0\rangle|^2 = (n+1)|\langle u_1; \mathbf{k}\varepsilon|H_{I1}|u_0; 0\rangle|^2 \tag{18}$$

we obtain for (17)

$$\hbar\Delta_n = \langle u_0; 0|H_{I2}|u_0; 0\rangle +$$

$$+ (n+1)\mathscr{P} \sum_{\mathbf{k}\varepsilon} \frac{|\langle u_1; \mathbf{k}\varepsilon|H_{I1}|u_0; 0\rangle|^2}{-\hbar(\omega + \omega_0)} +$$

$$+ n\mathscr{P} \sum_{\mathbf{k}\varepsilon} \frac{|\langle u_0; \mathbf{k}\varepsilon|H_{I1}|u_1; 0\rangle|^2}{\hbar(\omega_0 - \omega)} \tag{19}$$

(*) See, for example, *Photons and Atoms—Introduction to Quantum Electrodynamics,* Exercise 7 in Complement C_I.

an equation that leads to

$$\hbar(\Delta_n - \Delta_{n-1}) = \hbar(\Delta_1 - \Delta_0). \tag{20}$$

The shifted levels of the oscillator thus remain equidistant.

c) To lowest order in q, the matrix element (7) equals

$$\frac{1}{z - E_0 - \hbar\omega_a - \hbar\omega_b}\sqrt{1}\,\mu_b\frac{1}{z - E_1 - \hbar\omega_a}\sqrt{2}\,\mu_a\frac{1}{z - E_2} +$$

$$+ \frac{1}{z - E_0 - \hbar\omega_a - \hbar\omega_b}\sqrt{1}\,\mu_a\frac{1}{z - E_1 - \hbar\omega_b}\sqrt{2}\,\mu_b\frac{1}{z - E_2} \tag{21}$$

with the amplitude associated with (21) being represented by a sum of two diagrams

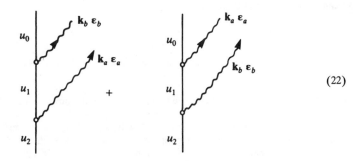

corresponding to different time orders for the emissions of the two photons. To derive (21), we used

$$\langle u_{n-1}|p_z|u_n\rangle = \sqrt{n}\,\langle u_0|p_z|u_1\rangle \tag{23}$$

and definition (8) for μ_a and μ_b.

As in subsection C-2-b in Chapter III, in (21) we replace E_n by $\tilde{E}_n - i\hbar(\Gamma_n/2)$ (with $\Gamma_0 = 0$, $\Gamma_1 = \Gamma$, $\Gamma_2 = 2\Gamma$) and integrate the expression obtained for (21) along the contour C_+ in Figure 1 of Chapter III (the only relevant contour for $\tau > 0$). For $\tau \gg \Gamma^{-1}$, only the contribution of the pole at $\tilde{E}_0 + \hbar\omega_a + \hbar\omega_b$ remains and we obtain for the amplitude (6) the result

$$\frac{\sqrt{2}}{\hbar^2}\mu_a\mu_b \exp\left\{-i\left(\tilde{E}_0 + \hbar\omega_a + \hbar\omega_b\right)t/\hbar\right\} \times$$

$$\times\left[\frac{1}{(\omega_b - \tilde{\omega}_0 + i(\Gamma/2))(\omega_a + \omega_b - 2\tilde{\omega}_0 + i\Gamma)} +\right.$$

$$\left. + \frac{1}{(\omega_a - \tilde{\omega}_0 + i(\Gamma/2))(\omega_a + \omega_b - 2\tilde{\omega}_0 + i\Gamma)}\right]. \tag{24}$$

Factor out $1/(\omega_a + \omega_b - 2\tilde{\omega}_0 + i\Gamma)$ from the expression inside the brackets of (24) and use

$$\frac{1}{\omega_b - \tilde{\omega}_0 + i(\Gamma/2)} + \frac{1}{\omega_a - \tilde{\omega}_0 + i(\Gamma/2)} = \frac{\omega_a + \omega_b - 2\tilde{\omega}_0 + i\Gamma}{\left(\omega_b - \tilde{\omega}_0 + i(\Gamma/2)\right)\left(\omega_a - \tilde{\omega}_0 + i(\Gamma/2)\right)}.$$

(25)

One finally gets for the amplitude (6)

$$\frac{\sqrt{2}}{\hbar^2} \frac{\mu_b}{\omega_b - \tilde{\omega}_0 + i(\Gamma/2)} \frac{\mu_a}{\omega_a - \tilde{\omega}_0 + i(\Gamma/2)} e^{-i(\bar{E}_0 + \hbar\omega_a + \hbar\omega_b)t/\hbar}.$$

(26)

d) The probability density $I(\omega_a, \omega_b)$ is proportional to the square of the modulus of (26). It can then be factored in the form

$$I(\omega_a, \omega_b) \sim \frac{2}{\hbar^4} \frac{|\mu_b|^2}{\left(\omega_b - \tilde{\omega}_0\right)^2 + \Gamma^2/4} \times \frac{|\mu_a|^2}{\left(\omega_a - \tilde{\omega}_0\right)^2 + \Gamma^2/4}$$

(27)

which proves that the frequencies ω_a and ω_b of the two photons emitted in the cascade are independent.

Because the variations of $|\mu_b|^2$ and $|\mu_a|^2$ with ω_b and ω_a can be neglected compared with the variations of the denominators, it is clear from (27) that each of the emitted photons has a Lorentzian spectral distribution centered at $\tilde{\omega}_0$ and having a full width at half-maximum Γ.

If we were to neglect the interferences between the two amplitudes (24), we would obtain for $I(\omega_a, \omega_b)$ the sum of the square modulus of the two terms inside the brackets in (24). A calculation identical to the one made at the end of subsection C-2-c in Chapter III would then give a spectral distribution that would be a superposition of two Lorentzians centered at $\tilde{\omega}_0$, but having different widths, respectively, equal to $\Gamma_1 + \Gamma_0 = \Gamma$ and $\Gamma_2 + \Gamma_1 = 3\Gamma$. The interference between the amplitudes corresponding to different time orders for the emissions of the two photons thus leads to the result that the spectral distribution of the emitted radiation is the same, whether the oscillator leaves from the state $|u_1\rangle$ or the state $|u_2\rangle$.

e) If the oscillator leaves from the state $|u_3\rangle$, there are now $3! = 6$ different amplitudes to take into account, corresponding to the six possible time orders for the emission of the three photons $\mathbf{k}_a \boldsymbol{\varepsilon}_a$, $\mathbf{k}_b \boldsymbol{\varepsilon}_b$, and $\mathbf{k}_c \boldsymbol{\varepsilon}_c$. The calculation of Question c above is easily generalized and gives, for example, for the amplitude associated with the process where $\mathbf{k}_c \boldsymbol{\varepsilon}_c$ is emitted first, then $\mathbf{k}_b \boldsymbol{\varepsilon}_b$ and finally $\mathbf{k}_a \boldsymbol{\varepsilon}_a$, the result

$$\frac{\sqrt{3!}\, \mu_a \mu_b \mu_c \exp\left\{-i\left(\bar{E}_0 + \hbar\omega_a + \hbar\omega_b + \hbar\omega_c\right)t/\hbar\right\}}{\hbar^3(\omega_a - \tilde{\omega}_0 + i(\Gamma/2))(\omega_a + \omega_b - 2\tilde{\omega}_0 + i\Gamma)(\omega_a + \omega_b + \omega_c - 3\tilde{\omega}_0 + 3i(\Gamma/2))}.$$

(28)

We must now sum (28) over all the possible permutations of the three frequencies ω_a, ω_b, and ω_c.

First we sum (28) over the 2! permutations, which keep unchanged the first emitted photon ω_c. From the results of the preceding question

$$\sum_{\substack{\text{permutations}\\ \omega_a, \omega_b}} \frac{1}{\left(\omega_a - \tilde{\omega}_0 + i\frac{\Gamma}{2}\right)(\omega_a + \omega_b - 2\tilde{\omega}_0 + i\Gamma)} =$$

$$= \frac{1}{\left(\omega_a - \tilde{\omega}_0 + i\frac{\Gamma}{2}\right)\left(\omega_b - \tilde{\omega}_0 + i\frac{\Gamma}{2}\right)}. \tag{29}$$

The probability amplitude associated with all processes where the first photon emitted is $\mathbf{k}_c \varepsilon_c$, i.e., the sum of (28) over all the permutations of $\mathbf{k}_a \varepsilon_a$ and $\mathbf{k}_b \varepsilon_b$, can then be written, using (29),

$$\frac{\sqrt{3!}\, \mu_a \mu_b \mu_c \exp\left\{-i\left(\bar{E}_0 + \hbar\omega_a + \hbar\omega_b + \hbar\omega_c\right)t/\hbar\right\}\left(\omega_c - \tilde{\omega}_0 + i\frac{\Gamma}{2}\right)}{\hbar^3\left(\omega_a - \tilde{\omega}_0 + i\frac{\Gamma}{2}\right)\left(\omega_b - \tilde{\omega}_0 + i\frac{\Gamma}{2}\right)\left(\omega_c - \tilde{\omega}_0 + i\frac{\Gamma}{2}\right)\left(\omega_a + \omega_b + \omega_c - 3\tilde{\omega}_0 + 3i\frac{\Gamma}{2}\right)}. \tag{30}$$

We multiplied the numerator and denominator of (28) by $[\omega_c - \tilde{\omega}_0 + i(\Gamma/2)]$.

We must now sum $n = 3$ expressions analogous to (30) that differ by the frequency of the first emitted photon, which can be ω_c or ω_b or ω_a. Because the three fractions have the same denominator, it is sufficient to add the numerators

$$\sum_{\omega_c}\left(\omega_c - \tilde{\omega}_0 + i\frac{\Gamma}{2}\right) = \left(\omega_a + \omega_b + \omega_c - 3\tilde{\omega}_0 + 3i\frac{\Gamma}{2}\right). \tag{31}$$

After simplification with the last term of the denominator of (30), the global amplitude becomes

$$\frac{\sqrt{3!}}{\hbar^3}\exp\left\{-i\left(\bar{E}_0 + \hbar\omega_a + \hbar\omega_b + \hbar\omega_c\right)t/\hbar\right\} \times$$

$$\times \frac{\mu_a}{\left(\omega_a - \tilde{\omega}_0 + i\frac{\Gamma}{2}\right)} \times \frac{\mu_b}{\left(\omega_b - \tilde{\omega}_0 + i\frac{\Gamma}{2}\right)} \times \frac{\mu_c}{\left(\omega_c - \tilde{\omega}_0 + i\frac{\Gamma}{2}\right)}. \tag{32}$$

The probability density $I(\omega_a, \omega_b, \omega_c)$ giving the frequencies of the three emitted photons, which is equal to the square of the modulus of (32), is still factored, which proves that the three emitted photons have independent frequencies, each distributed according to a Lorentzian centered at $\tilde{\omega}_0$ with a full width at half-maximum Γ.

The foregoing proof can be generalized by recurrence to the case where the oscillator starts from any excited state $|u_n\rangle$. It allows us to show that the interference between the $n!$

amplitudes corresponding to the $n!$ possible time orders of the n emitted photons starting from $|u_n\rangle$ leads to the result that the spectral distribution of the radiation emitted by the oscillator is the same, regardless of the initial excitation state of the oscillator.

16. PRINCIPLE OF THE DETAILED BALANCE

A system $\mathscr{A}$ interacts with a reservoir $\mathscr{R}$ so that the relaxation theory in Chapter IV is applicable. One calls $E_a, E_b, E_c, \ldots$ the energy levels of $\mathscr{A}$, and $E_\mu, E_\nu, \ldots$ the energy levels of the reservoir $\mathscr{R}$. If V is the coupling Hamiltonian, the transition probability per unit time that the system $\mathscr{A}$ will go from the state $|c\rangle$ to the state $|a\rangle$ is written [see formula (C.5) in Chapter IV]:

$$\Gamma_{c \to a} = \frac{2\pi}{\hbar} \sum_\mu p_\mu \sum_\nu |\langle \nu, a|V|\mu, c\rangle|^2 \delta(E_\mu + E_c - E_\nu - E_a) \quad (1)$$

where p_μ is the probability that the reservoir is in the state $|\mu\rangle$.

For the reservoir in thermodynamic equilibrium, verify the relation

$$\Gamma_{c \to a}\, e^{-E_c/k_B T} = \Gamma_{a \to c}\, e^{-E_a/k_B T}. \quad (2)$$

What consequence does this relation have on the flux of populations between the levels c and a of system $\mathscr{A}$ at thermodynamic equilibrium?

Solution

In thermodynamic equilibrium, the ratio of the populations of the microscopic states μ and ν of the reservoir is

$$\frac{p_\mu}{p_\nu} = \exp\left[-(E_\mu - E_\nu)/k_B T\right]. \quad (3)$$

In (1) we then replace p_μ as a function of p_ν and the exponential (3). This gives

$$\Gamma_{c \to a} = \frac{2\pi}{\hbar} \sum_\mu \sum_\nu p_\nu |\langle \nu, a|V|\mu, c\rangle|^2 \exp\left[-(E_\mu - E_\nu)/k_B T\right] \times$$

$$\times \delta\left[(E_\mu - E_\nu) - (E_a - E_c)\right]. \quad (4)$$

The δ function allows $E_\mu - E_\nu$ to be replaced by $E_a - E_c$ in the exponential. The exponential is then independent of μ and ν and can then be factored out of the sum. By inverting the order of the summations over ν and μ and by using the Hermiticity of V which implies $|\langle\mu,c|V|\nu,a\rangle| = |\langle\nu,a|V|\mu,c\rangle$, we can write $\Gamma_{c\to a}$ in the form

$$\Gamma_{c\to a} = \exp[-(E_a - E_c)/k_BT] \times \frac{2\pi}{\hbar}\sum_\nu P_\nu \sum_\mu |\langle\mu,c|V|\nu,a\rangle|^2 \,\delta(E_\nu + E_a - E_\mu - E_c)$$

$$= \exp[-(E_a - E_c)/k_BT]\Gamma_{a\to c} \tag{5}$$

from which relation (2) follows.

In thermodynamic equilibrium, the ratio of the populations π_a^{st} and π_c^{st} of the states $|a\rangle$ and $|c\rangle$ of the small system is

$$\frac{\pi_a^{\text{st}}}{\pi_c^{\text{st}}} = \exp[-(E_a - E_c)/k_BT]. \tag{6}$$

By regrouping (5) and (6), we obtain

$$\pi_a^{\text{st}}\Gamma_{a\to c} = \pi_c^{\text{st}}\Gamma_{c\to a}. \tag{7}$$

This relation expresses the fact that the flux of populations from a to c and from c to a equilibrate exactly. By generalizing this result to all transitions, the "detailed balance" principle is established. It expresses that the flux of populations between the levels is equilibrated separately over each transition.

17. Equivalence between a Quantum Field in a Coherent State and an External Field

Particles with charge q_α and mass m_α interact with the quantum electromagnetic field. The Coulomb gauge is chosen and at the initial time $t = 0$, each of the modes of the field (labeled by the subscript j) is assumed to be in a coherent state $|\alpha_j\rangle$ (this can be, in particular, the vacuum state of the mode, $|0_j\rangle$)

$$|\psi(0)\rangle = |\ldots,\alpha_j,\ldots\rangle = |\{\alpha_j\}\rangle. \tag{1}$$

The purpose of this exercise is to show that it is possible to apply a unitary transformation to the Hamiltonian of quantum electrodynamics such that, in the new representation, the particles are coupled on the one hand to a classical external field, and on the other hand, to the quantum field initially in the vacuum state (*).

(*) See, for example, B. R. Mollow, *Phys. Rev. A*, **12**, 1919 (1975).

a) Consider the unitary operator

$$T(t) = \prod_j T_j(t) \tag{2}$$

$$T_j(t) = \exp\left[\alpha_j^* \, e^{i\omega_j t} \, a_j - \alpha_j \, e^{-i\omega_j t} \, a_j^+\right] \tag{3}$$

where the α_j are the complex numbers appearing in (1).

Calculate $T(t)a_j T^+(t)$ and $T(t)a_j^+ T^+(t)$. Deduce from this that the transform $T(t)\mathbf{A}(\mathbf{r})T^+(t)$ of the transverse vector potential $\mathbf{A}(\mathbf{r})$ is the sum of the quantum field $\mathbf{A}(\mathbf{r})$ and a classical field $\mathbf{A}_{\mathrm{Cl}}(\{\alpha_j\}; \mathbf{r}, t)$ that will be written as a function of the α_j's.

In this exercise one can use the identity

$$\exp(A + B) = \exp(A)\exp(B)\exp\left(-\tfrac{1}{2}[A, B]\right) \tag{4}$$

which is valid if $[A, B]$ commutes with A and B.

b) What is the relation between the state vector $|\psi'(t)\rangle$ in the new representation and the state vector $|\psi(t)\rangle$ in the old representation? Show that the new initial state $|\psi'(0)\rangle$ is the vacuum $|0\rangle$.

c) The Hamiltonian that describes the dynamics of the field + particles system in the Coulomb gauge is written

$$H = \sum_\alpha \frac{1}{2m_\alpha}\left[\mathbf{p}_\alpha - q_\alpha \mathbf{A}(\mathbf{r}_\alpha)\right]^2 + V_{\mathrm{Coul}} + H_R \tag{5}$$

where V_{Coul} is the Coulomb interaction between the particles and where H_R is the radiation Hamiltonian

$$H_R = \sum_j \hbar\omega_j\left(a_j^+ a_j + \tfrac{1}{2}\right). \tag{6}$$

Prove that, in the new representation, the Hamiltonian $H'(t)$ that describes the dynamics of the global system is written

$$H'(t) = T(t)HT^+(t) + i\hbar\left[\frac{dT(t)}{dt}\right]T^+(t). \tag{7}$$

d) Show that

$$T(t)H_R T^+(t) + i\hbar\left[\frac{d}{dt}T(t)\right]T^+(t) = H_R. \tag{8}$$

Deduce from this the expression for $H'(t)$. Give a physical interpretation for this Hamiltonian.

e) Generalize the preceding proof to the case in which the charged particles form a neutral atom, located at **0**, and where the interaction Hamiltonian is the electric dipole Hamiltonian.

Solution

a) Using formula (67) from the Appendix, with $T_j = \exp(\lambda_j^* a_j - \lambda_j a_j^+)$, we get

$$T_j a_j T_j^+ = a_j + \lambda_j \tag{9.a}$$

$$T_j a_i T_j^+ = a_i \quad \text{for } j \neq i. \tag{9.b}$$

With $\lambda_j = \alpha_j e^{-i\omega_j t}$, Equations (9) become

$$T a_j T^+ = a_j + \alpha_j e^{-i\omega_j t} \tag{10.a}$$

$$T a_j^+ T^+ = a_j^+ + \alpha_j^* e^{i\omega_j t}. \tag{10.b}$$

When the operators $T a_j T^+$ and $T a_j^+ T^+$ in $T A(\mathbf{r}) T^+$ are replaced by (10.a) and (10.b), we obtain

$$A'(\mathbf{r}) = T A(\mathbf{r}) T^+ = A(\mathbf{r}) + A_{Cl}(\{\alpha_j\}; \mathbf{r}, t) \tag{11}$$

where

$$A_{Cl}(\{\alpha_j\}; \mathbf{r}, t) = \sum_j \sqrt{\frac{\hbar}{2\varepsilon_0 \omega_j L^3}} \left[\alpha_j \boldsymbol{\varepsilon}_j \, e^{i(\mathbf{k}_j \cdot \mathbf{r} - \omega_j t)} + \alpha_j^* \boldsymbol{\varepsilon}_j \, e^{-i(\mathbf{k}_j \cdot \mathbf{r} - \omega_j t)} \right]. \tag{12}$$

b) In the new representation, the state of the system is represented by

$$|\psi'(t)\rangle = T(t)|\psi(t)\rangle. \tag{13}$$

At the time $t = 0$

$$|\psi'(0)\rangle = T(0)|\psi(0)\rangle = T(0)|\{\alpha_j\}\rangle. \tag{14}$$

But, using formula (65) from the Appendix,

$$|\alpha_j\rangle = \exp(\alpha_j a_j^+ - \alpha_j^* a_j)|0_j\rangle = T_j^+(0)|0_j\rangle \tag{15}$$

or

$$|0_j\rangle = T_j(0)|\alpha_j\rangle \tag{16}$$

and by taking the product over all the modes

$$|0\rangle = T(0)\big|\{\alpha_j\}\big\rangle. \tag{17}$$

Comparing this with (14) proves that

$$|\psi'(0)\rangle\rangle = |0\rangle. \tag{18}$$

    c) The evolution equation for the new state vector is

$$i\hbar\frac{\mathrm{d}}{\mathrm{d}t}|\psi'(t)\rangle\rangle = i\hbar\left(\frac{\mathrm{d}}{\mathrm{d}t}T(t)\right)|\psi(t)\rangle\rangle + T(t)i\hbar\frac{\mathrm{d}}{\mathrm{d}t}|\psi(t)\rangle\rangle. \tag{19}$$

The Schrödinger equation in the old representation gives $i\hbar\,\mathrm{d}|\psi(t)\rangle\rangle/\mathrm{d}t = H|\psi(t)\rangle\rangle$. By inserting $T^+T = 1$ on the left of $|\psi(t)\rangle\rangle$, we then obtain

$$i\hbar\frac{\mathrm{d}}{\mathrm{d}t}|\psi'(t)\rangle\rangle = \left[i\hbar\left(\frac{\mathrm{d}}{\mathrm{d}t}T(t)\right)T^+(t) + T(t)HT^+(t)\right]T(t)|\psi(t)\rangle\rangle$$

$$= H'(t)|\psi'(t)\rangle\rangle. \tag{20}$$

    d) First calculate $i\hbar[\mathrm{d}T(t)/\mathrm{d}t]T^+(t)$. It should be noted that, if $X(t)$ does not commute with $\mathrm{d}X(t)/\mathrm{d}t$, then $\mathrm{d}\exp[X(t)]/\mathrm{d}t \neq [\mathrm{d}X(t)/\mathrm{d}t]\exp[X(t)]$. It is then necessary, before taking the derivative, to separate the factors containing a and a^+ by using formula (4). The calculation is then performed as follows:

$$T_j(t) = \exp\left(\alpha_j^* \, e^{i\omega_j t}\, a_j\right)\exp\left(-\alpha_j \, e^{-i\omega_j t}\, a_j^+\right)\exp\left(\tfrac{1}{2}\alpha_j\alpha_j^*\right) \tag{21}$$

$$\frac{\mathrm{d}}{\mathrm{d}t}T_j(t) = i\omega_j\alpha_j^* \, e^{i\omega_j t}\, a_j T_j(t) + i\omega_j\alpha_j \, e^{-i\omega_j t}\, T_j(t)a_j^+ \tag{22}$$

$$T_j(t)a_j^+ = T_j(t)a_j^+ T_j^+(t)T_j(t) = \left(a_j^+ + \alpha_j^* \, e^{i\omega_j t}\right)T_j(t) \tag{23}$$

$$\left[\frac{\mathrm{d}}{\mathrm{d}t}T_j(t)\right]T_j^+(t) = i\omega_j\left(\alpha_j^* \, e^{i\omega_j t}\, a_j + \alpha_j \, e^{-i\omega_j t}\, a_j^+ + \alpha_j^*\alpha_j\right) \tag{24}$$

$$i\hbar\left[\frac{\mathrm{d}}{\mathrm{d}t}T(t)\right]T^+(t) = -\sum_j \hbar\omega_j\left[\alpha_j^* \, e^{i\omega_j t}\, a_j + \alpha_j \, e^{-i\omega_j t}\, a_j^+ + \alpha_j^*\alpha_j\right]. \tag{25}$$

Using (10.a) and (10.b)

$$T(t)H_R T^+(t) = \sum_j \hbar\omega_j\left[\left(a_j^+ + \alpha_j^* \, e^{i\omega_j t}\right)\left(a_j + \alpha_j \, e^{-i\omega_j t}\right) + \tfrac{1}{2}\right]. \tag{26}$$

By adding (25) and (26), we obtain formula (8).

    The transformation $T(t)$ does not act on $\mathbf{r}_\alpha$ and $\mathbf{p}_\alpha$. It leaves V_{Coul} unchanged and, using (11) changes $\mathbf{p}_\alpha - q_\alpha\mathbf{A}(\mathbf{r}_\alpha)$ into $\mathbf{p}_\alpha - q_\alpha\mathbf{A}(\mathbf{r}_\alpha) - q_\alpha\mathbf{A}_{\mathrm{Cl}}(\{\alpha_j\};\mathbf{r}_\alpha,t)$, so that the new Hamiltonian $H'(t)$ is written

$$H'(t) = \sum_\alpha \frac{1}{2m_\alpha} \left[\mathbf{p}_\alpha - q_\alpha \mathbf{A}(\mathbf{r}_\alpha) - q_\alpha \mathbf{A}_{\text{Cl}}(\{\alpha_j\}; \mathbf{r}_\alpha, t) \right]^2 + V_{\text{Coul}} + H_R. \qquad (27)$$

This Hamiltonian describes the particles interacting with the quantum field $\mathbf{A}(\mathbf{r})$ and with the classical field $\mathbf{A}_{\text{Cl}}(\{\alpha_j\}; \mathbf{r}_\alpha, t)$. The initial state of the quantum field is now the vacuum, according to (18). Thus, we have proved a complete equivalence between a quantum field in a coherent state, and the corresponding classical field to which is added the vacuum field.

   e) In the electric dipole representation, the Hamiltonian for the interaction between the field and the particle is written (see Appendix, §5-d)

$$H'_I = -\mathbf{d} \cdot \mathbf{E}_\perp (0). \qquad (28)$$

The transformation $T(t)$ acts only on $\mathbf{E}_\perp (0)$ which it transforms into $\mathbf{E}_\perp (0) + \mathbf{E}_{\text{Cl}}(\{\alpha_j\}; 0, t)$ with

$$\mathbf{E}_{\text{Cl}}(\{\alpha_j\}; 0, t) = i \sum_j \sqrt{\frac{\hbar \omega_j}{2 \varepsilon_0 L^3}} \left[\alpha_j e^{-i\omega_j t} - \alpha_j^* e^{+i\omega_j t} \right] \boldsymbol{\varepsilon}_j. \qquad (29)$$

We then obtain

$$H''_I = -\mathbf{d} \cdot \left[\mathbf{E}_\perp (0) + \mathbf{E}_{\text{Cl}}(\{\alpha_j\}; 0, t) \right] \qquad (30)$$

whereas the particle Hamiltonian is unchanged and the radiation Hamiltonian is transformed as in the preceding question. We find the same result: In the new representation, the particles interact with the classical electrical field associated with the α_j's, and with the vacuum field.

18. ADIABATIC ELIMINATION OF COHERENCES AND TRANSFORMATION OF OPTICAL BLOCH EQUATIONS INTO RELAXATION EQUATIONS

   The purpose of this exercise is to study a physical situation where the nondiagonal elements of the density matrix of an atom (coherences) evolve much more rapidly than the populations of these levels. The coherences can then be reexpressed at each time as a function of the populations and the optical Bloch equations are then transformed into relaxation equations for the populations only.

   A two-level atom (*a*: ground state, *b*: excited state with a natural width Γ; $E_b - E_a = \hbar \omega_0$) interacts with an incident laser field with frequency ω_L and amplitude $\mathscr{E}_0$. The optical Bloch equations describing the evolution of the density matrix σ_{ij} ($i, j = a$ or b) of such an atom are written [see formula (A.13) in Chapter V]

$$\frac{d\sigma_{bb}}{dt} = i(\Omega_1/2)(\hat{\sigma}_{ba} - \hat{\sigma}_{ab}) - \Gamma\sigma_{bb} \tag{1.a}$$

$$\frac{d\sigma_{aa}}{dt} = -i(\Omega_1/2)(\hat{\sigma}_{ba} - \hat{\sigma}_{ab}) + \Gamma\sigma_{bb} \tag{1.b}$$

$$\frac{d\hat{\sigma}_{ab}}{dt} = -i\delta_L\hat{\sigma}_{ab} - i(\Omega_1/2)(\sigma_{bb} - \sigma_{aa}) - (\Gamma/2)\hat{\sigma}_{ab}. \tag{1.c}$$

We have set $\hat{\sigma}_{ab} = \sigma_{ab}\exp(-i\omega_L t) = (\hat{\sigma}_{ba})^*$, $\delta_L = \omega_L - \omega_0$. The Rabi frequency Ω_1 is defined by $\Omega_1 = -\mathbf{d}_{ab}\cdot\mathcal{E}_0/\hbar$, where $\mathbf{d}_{ab}$ is the matrix element of the atomic dipole $\mathbf{d}$ that is assumed to be real.

This atom is subjected to elastic collisions with a foreign gas which do not induce any transfer between a and b, and whose only effect is to dephase the atomic dipole. The effect of these collisions on σ_{ij} is thus manifested by an additional damping of the coherences $\hat{\sigma}_{ab}$, having the form

$$\left(\frac{d\hat{\sigma}_{ab}}{dt}\right)_{coll} = -\gamma\hat{\sigma}_{ab} \tag{2}$$

(the collisional shift of the line is assumed to be included in ω_0). The populations σ_{aa} and σ_{bb} are not affected by the collisions.

a) Assume that the rates of variation due, respectively, to the coupling with the laser wave and to the collisions can be added independently. Write the evolution equation for the coherence $\hat{\sigma}_{ab}$.

b) First consider the steady state. For this case, express $\hat{\sigma}_{ab}$ as a function of $(\sigma_{bb} - \sigma_{aa})$ and deduce from this the expression σ_{bb}^{st} for σ_{bb} in the steady state.

Knowing that the total intensity of the light emitted by the atom is proportional to $\Gamma\sigma_{bb}^{st}$, indicate how this emitted intensity varies with γ in the limit $|\delta_L| \gg \gamma, \Gamma, \Omega_1[1 + (2\gamma/\Gamma)]^{1/2}$. How can this result be interpreted?

c) Now consider a situation where $\gamma \gg \Gamma$. Also assume that the populations evolve very slightly over a time on the order of γ^{-1}. Show that the coherences adjust themselves at each time to the slowly varying populations. What is the corresponding expression for $\hat{\sigma}_{ab}$ as a function of σ_{aa} and σ_{bb}?

d) Substitute the preceding expression for $\hat{\sigma}_{ab}$ into the evolution equations of populations. Show that the equations obtained in this way can be interpreted in terms of absorption, stimulated emission, and spontaneous emission rates between the levels a and b. Calculate the rates Γ' of absorption and stimulated emission as a function of γ, δ_L, and Ω_1.

e) What condition must Ω_1 satisfy for the slow variation hypothesis for the populations, used in Question c, to be valid?

Solution

a) The evolution equation for $\hat{\sigma}_{ab}$ is

$$\frac{d\hat{\sigma}_{ab}}{dt} = -i\delta_L\hat{\sigma}_{ab} - i\frac{\Omega_1}{2}(\sigma_{bb} - \sigma_{aa}) - \frac{\Gamma}{2}\hat{\sigma}_{ab} - \gamma\hat{\sigma}_{ab}. \tag{3}$$

b) In the steady state, $d\hat{\sigma}_{ab}/dt = 0$ and Equation (3) gives

$$\hat{\sigma}_{ab} = \frac{-i(\Omega_1/2)(\sigma_{bb} - \sigma_{aa})}{(\gamma + (\Gamma/2)) + i\delta_L}. \tag{4}$$

By substituting this value and the value for $\hat{\sigma}_{ba} = \hat{\sigma}_{ab}^*$ into Equation (1.a) or (1.b), we find an equation whose steady-state solution is (using $\sigma_{aa}^{st} + \sigma_{bb}^{st} = 1$):

$$\sigma_{bb}^{st} = \frac{\Omega_1^2}{4} \frac{(1 + (2\gamma/\Gamma))}{\delta_L^2 + ((\Gamma/2) + \gamma)^2 + (\Omega_1^2/2)(1 + (2\gamma/\Gamma))}. \tag{5}$$

In the limit considered in this problem, we find

$$\Gamma\sigma_{bb}^{st} = \frac{\Omega_1^2}{4\delta_L^2}(\Gamma + 2\gamma). \tag{6}$$

The term proportional to Γ is associated with the Rayleigh scattering and the term proportional to γ corresponds to an emission at the frequency ω_0 which results from a collision-assisted excitation [see Complement B$_{VI}$ and, in particular, Figure 1 and formula (43)].

c) If $\sigma_{bb} - \sigma_{aa}$ were constant, then $\hat{\sigma}_{ab}$ would tend to the steady-state value (4) with a time constant which is equal, according to Equation (3), to the inverse of $\gamma + (\Gamma/2) \approx \gamma$. We can deduce from this result that if $\sigma_{bb} - \sigma_{aa}$ varies slowly on the time scale γ^{-1}, then $\hat{\sigma}_{ba}$ can quasi-instantaneously adapt to slow variations of $\sigma_{bb} - \sigma_{aa}$ and at each time take the value given by Equation (4).

d) By substituting Equation (4) into Equations (1.a) and (1.b), we find

$$d\sigma_{bb}/dt = \Gamma'(\sigma_{aa} - \sigma_{bb}) - \Gamma\sigma_{bb} \tag{7.a}$$

$$d\sigma_{aa}/dt = \Gamma'(\sigma_{bb} - \sigma_{aa}) + \Gamma\sigma_{bb} \tag{7.b}$$

with

$$\Gamma' = \frac{\Omega_1^2}{2} \frac{\gamma + (\Gamma/2)}{(\gamma + (\Gamma/2))^2 + \delta_L^2} \approx \frac{\Omega_1^2}{2} \frac{\gamma}{\gamma^2 + \delta_L^2}. \tag{8}$$

Because Equations (7) have the same structure as Equations (E.15) in Chapter IV, the interpretation of the absorption, induced emission, and spontaneous emission terms can be immediately deduced.

e) According to Equations (7), the characteristic evolution time of $\sigma_{bb} - \sigma_{aa}$ is $(\Gamma + 2\Gamma')^{-1}$. This time is long compared with γ^{-1} if

$$\Omega_1^2 \frac{\gamma}{\gamma^2 + \delta_L^2} \ll \gamma \tag{9}$$

that is, again, if

$$\Omega_1^2 \ll (\gamma^2 + \delta_L^2). \tag{10}$$

Equations (7) are therefore valid only if the saturation by the intense field is sufficiently weak.

19. Nonlinear Susceptibility for an Ensemble of Two-Level Atoms. A Few Applications

Consider a dilute medium made up of two-level atoms. These atoms, assumed to be at rest, interact with a plane wave $\mathscr{E}_0 \cos(\omega_L t - \mathbf{k}_L \cdot \mathbf{r})$ propagating in the plane $x0y$. The matrix element of the electric dipole between the levels a and b, $\mathbf{d}_{ab}$, is aligned along $0z$ as is the field $\mathscr{E}_0$ of the incident wave. For the sake of simplicity, these quantities will be treated as scalar variables.

The average value $\langle d \rangle$ of the atomic dipole can be written in the form

$$\langle d \rangle = \varepsilon_0 \operatorname{Re}(\alpha' + i\alpha'') \mathscr{E}_0 \, e^{-i(\omega_L t - \mathbf{k}_L \cdot \mathbf{r})} \tag{1}$$

where $\alpha = \alpha' + i\alpha''$ is the polarizability of the atom. The purpose of this exercise is to study the nonlinearities of the polarizability α and to apply the obtained results to the self-focusing problem and to the generation of a forward conjugated wave.

a) The susceptibility $\chi = \chi' + i\chi''$ of the medium is equal to the product $N\alpha$ of the number N of atoms per unit volume and the atomic polarizability α. By using the results from Chapter V, show that χ' and χ'' are, respectively, equal to

$$\chi' = -\frac{N|d_{ab}|^2}{\varepsilon_0 \hbar} \frac{\delta_L}{\delta_L^2 + (\Gamma^2/4) + (\Omega_1^2/2)} \tag{2.a}$$

$$\chi'' = \frac{N|d_{ab}|^2}{\varepsilon_0 \hbar} \frac{\Gamma/2}{\delta_L^2 + (\Gamma^2/4) + (\Omega_1^2/2)} \tag{2.b}$$

where Γ is the natural width of the level b, $\delta_L = \omega_L - \omega_0$ is the detuning from resonance, and $\Omega_1 = -d_{ab}\mathcal{E}_0/\hbar$ is the Rabi frequency.

b) Consider the situation where $|\delta_L| \gg \Gamma$ and recall that the refractive index of a dilute medium is equal to $1 + (\chi'/2)$. The distribution of the field in the plane perpendicular to the direction of propagation $\mathbf{k}_L$ is assumed not to be uniform and to have the form $\mathcal{E}_0 \exp(-r^2/r_0^2)$ when entering the medium ($\mathbf{k}_L$ coincides with $0x$ and r^2 is equal to $y^2 + z^2$). Without going into the details of the calculations, indicate why the wave has a tendency to become defocused when $\omega_L < \omega_0$ and focused when $\omega_L > \omega_0$.

c) Now consider the case where the incident field is made up of two plane waves, having the same frequency and the same polarization parallel to $0z$, but propagating in different directions in the $x0y$ plane. These two waves are written $\mathcal{E}_0 \cos(\omega_L t - \mathbf{k}_L \cdot \mathbf{r})$ and $\mathcal{E}_p \cos(\omega_L t - \mathbf{k}_p \cdot \mathbf{r})$ with $\mathcal{E}_p \ll \mathcal{E}_0$. Calculate $\langle d \rangle$ at the first order in $\mathcal{E}_p$ and deduce from this the polarizabilities α'_p and α''_p for the weak wave. Show that the susceptibilities for the wave $\mathcal{E}_p$, $\chi'_p = N\alpha'_p$ and $\chi''_p = N\alpha''_p$ do not depend on the direction of propagation of the weak wave.

Also show that $\langle d \rangle$ has an oscillating component varying as $\exp[-i(\omega_L t - \mathbf{k}_c \cdot \mathbf{r})]$ with $\mathbf{k}_c = 2\mathbf{k}_L - \mathbf{k}_p$, able to generate a field in the direction $\mathbf{k}_c$.

d) Assume $|\delta_L| \gg \Gamma$. By using the values of χ'_p and χ' found previously, calculate the angle θ that $\mathbf{k}_L$ and $\mathbf{k}_p$ must make in order to fulfill the phase-matching condition (the generation of the wave in the direction $\mathbf{k}_c$ is then the most intense). Is it possible to achieve this condition whatever of the sign of δ_L?

Solution

a) By using relations (A.21) and (C.3) from Chapter V, and by assuming $\mathbf{k}_L$ to be parallel to $0x$, we find

$$\langle \mathbf{d} \rangle = -\frac{\mathbf{d}_{ab}(\mathbf{d}_{ab} \cdot \mathcal{E}_0)}{\hbar\left[\delta_L^2 + \dfrac{\Gamma^2}{4} + \dfrac{\Omega_1^2}{2}\right]}\left[\delta_L \cos(\omega_L t - k_L x) - \frac{\Gamma}{2}\sin(\omega_L t - k_L x)\right]. \quad (3)$$

Because $\mathbf{d}_{ab}$ and $\mathcal{E}_0$ are both aligned along $0z$, (3) can be rewritten:

$$\langle d \rangle = \frac{|d_{ab}|^2}{\hbar\left[\delta_L^2 + \dfrac{\Gamma^2}{4} + \dfrac{\Omega_1^2}{2}\right]}\left[-\delta_L\mathcal{E}_0 \cos(\omega_L t - k_L x) + \frac{\Gamma}{2}\mathcal{E}_0 \sin(\omega_L t - k_L x)\right]. \quad (4)$$

Relation (1) leads to

$$\langle d \rangle = \varepsilon_0 \big[\alpha' \mathscr{E}_0 \cos(\omega_L t - k_L x) + \alpha'' \mathscr{E}_0 \sin(\omega_L t - k_L x) \big]. \tag{5}$$

When (4) is compared with (5), one obtains formulas (2). These formulas show that χ' and χ'' are decreasing functions of the intensity of the incident field (because of the factor $\Omega_1^2 = d_{ab}^2 \mathscr{E}_0^2 / \hbar^2$ of the denominator). At very high intensities, the dispersion and the absorption tend to 0. As a function of δ_L, χ' reaches its extrema when $\delta_L = \pm \sqrt{(\Gamma^2/4) + (\Omega_1^2/2)}$ and changes sign with δ_L. χ'' has a maximum for $\delta_L = 0$ and is divided by 2 for $\delta_L = \pm \sqrt{(\Gamma^2/4) + (\Omega_1^2/2)}$.

    b) The index of the medium, which is equal to $1 + (\chi'/2)$, is not the same at every point in space because Ω_1^2, which is proportional to $\mathscr{E}_0^2$, is not constant in the plane perpendicular to $\mathbf{k}_L$. In the case where $\delta_L = \omega_L - \omega_0 < 0$, χ' is positive, but as a result of the saturation effects, its value at the center of the beam is smaller than at the edges. The optical length of a slice δz of the medium is shorter for the central ray than for the lateral ones: The medium behaves as a divergent lens and there is self-defocusing of the beam (*).

    When $\omega_L > \omega_0$, the susceptibility χ' is negative. Because $|\chi'|$ is smaller in the center of the beam than at the edges as a result of saturation, the index is larger in the center of the beam than at the edges. Consequently, the medium behaves as a convergent lens and there is self-focusing. If the medium is sufficiently long, the transverse dimension of the beam decreases during its propagation but the saturation of the medium prevents this dimension from tending to 0. Indeed, as the focusing increases, the field at the center increases, as does the Rabi frequency. When the Rabi frequency is greater than $|\delta_L|$, the susceptibility χ' near the center of the beam becomes insensitive to transverse variations in intensity because of the saturation, and the lens effects are thus greatly attenuated. The competition with the diffraction effects then results in the propagation of a wave having a constant transverse dimension in the medium (**).

    c) The average value of the atomic dipole is now given by

$$\langle d \rangle = \varepsilon_0 \, \text{Re}(\tilde{\alpha}' + i\tilde{\alpha}'') \big(\mathscr{E}_0 \, e^{-i(\omega_L t - \mathbf{k}_L \cdot \mathbf{r})} + \mathscr{E}_p \, e^{-i(\omega_L t - \mathbf{k}_p \cdot \mathbf{r})} \big) \tag{6}$$

where $\tilde{\alpha}'$ and $\tilde{\alpha}''$ are given by formulas analogous to the formulas in Question *a*, except for the replacement of Ω_1^2 by $d_{ab}^2 |\mathscr{E}_0 \exp(i\mathbf{k}_L \cdot \mathbf{r}) + \mathscr{E}_p \exp(i\mathbf{k}_p \cdot \mathbf{r})|^2 / \hbar^2$. To first order in $\mathscr{E}_p$, this last term is written

$$\Omega_1^2 + \frac{d_{ab}^2}{\hbar^2} \big[\mathscr{E}_0 \mathscr{E}_p \, e^{i(\mathbf{k}_L \cdot \mathbf{r} - \mathbf{k}_p \cdot \mathbf{r})} + \mathscr{E}_0 \mathscr{E}_p \, e^{-i(\mathbf{k}_L \cdot \mathbf{r} - \mathbf{k}_p \cdot \mathbf{r})} \big]. \tag{7}$$

    (*) For more details on the effects of self-focusing and self-defocusing, see Shen (Chapter 17) or S. A. Akhmanov, R. V. Khokhlov, and A. P. Sukhorukov, in *Laser Handbook 2*, edited by F. T. Arecchi and E. O. Schulz-Dubois, North Holland, Amsterdam, 1972, p. 1151.

    (**) See A. Javan and P. L. Kelley, *IEEE J. Quantum. Electron.*, **QE-2**, 470 (1966).

By expanding $\tilde{\alpha}'$ and $\tilde{\alpha}''$ to first order in $\mathscr{E}_p$, we find

$$\langle d \rangle \simeq \varepsilon_0 \, \mathrm{Re}(\alpha' + i\alpha'')\mathscr{E}_0 \, e^{-i(\omega_L t - \mathbf{k}_L \cdot \mathbf{r})} +$$

$$+ \varepsilon_0 \, \mathrm{Re}\left[\alpha' + i\alpha'' + \left(\frac{d\alpha'}{d\Omega_1^2} + i\frac{d\alpha''}{d\Omega_1^2} \right)\frac{d_{ab}^2 \mathscr{E}_0^2}{\hbar^2} \right]\mathscr{E}_p \, e^{-i(\omega_L t - \mathbf{k}_p \cdot \mathbf{r})} +$$

$$+ \varepsilon_0 \, \mathrm{Re}\left(\frac{d\alpha'}{d\Omega_1^2} + i\frac{d\alpha''}{d\Omega_1^2} \right)\frac{d_{ab}^2 \mathscr{E}_0^2}{\hbar^2}\mathscr{E}_p \, e^{-i[\omega_L t - (2\mathbf{k}_L - \mathbf{k}_p)\cdot \mathbf{r}]}. \tag{8}$$

From the second term of (8), we deduce

$$\alpha'_p = \alpha' + \Omega_1^2 \frac{d\alpha'}{d\Omega_1^2} = -\frac{d_{ab}^2}{\varepsilon_0 \hbar} \frac{\delta_L\big(\delta_L^2 + (\Gamma^2/4)\big)}{\big(\delta_L^2 + (\Gamma^2/4) + (\Omega_1^2/2)\big)^2} \tag{9.a}$$

$$\alpha''_p = \alpha'' + \Omega_1^2 \frac{d\alpha''}{d\Omega_1^2} = \frac{d_{ab}^2}{\varepsilon_0 \hbar} \frac{(\Gamma/2)\big(\delta_L^2 + (\Gamma^2/4)\big)}{\big(\delta_L^2 + (\Gamma^2/4) + (\Omega_1^2/2)\big)^2}. \tag{9.b}$$

The susceptibility $\chi_p = N\alpha_p$ for the low-intensity wave $\mathscr{E}_p$ thus differs from the susceptibility (2) found for the intense wave. For instance, as Ω_1 increases, χ'_p and χ''_p tend to 0 more rapidly than χ' and χ''. Even for the situations where $\delta_L^2 + (\Gamma^2/4) > \Omega_1^2$, χ'_p and χ''_p coincide with χ' and χ'' only for the term independent of Ω_1^2 (usual linear susceptibility). The first nonlinear term of the susceptibility, which is proportional to Ω_1^2, is, in this limit, twice as large in modulus for the weak wave as for the intense wave. Finally, it is clear from Equations (9) that α_p, and therefore χ_p, do not depend on the direction of $\mathbf{k}_p$.

The third term in formula (8) corresponds to a source term in the Maxwell equations allowing a field to be generated that propagates in the direction $\mathbf{k}_c$.

d) For the generation of the wave in direction $\mathbf{k}_c$ to be important, it is necessary that the fields radiated by the successive planes of the atoms be in phase. This requires that the spatial period of the field of the dipoles, equal to $|\mathbf{k}_c|^{-1} = |2\mathbf{k}_L - \mathbf{k}_p|^{-1}$, coincides with the wavelength for a low-intensity wave of frequency ω_L, which is equal to $c/\omega_L[1 + (\chi'_p/2)]$. Because this last quantity is also equal to $|\mathbf{k}_p|^{-1}$, it should be possible to construct an isosceles triangle with sides $\mathbf{k}_c = 2\mathbf{k}_L - \mathbf{k}_p$ and $\mathbf{k}_p$ and height $\mathbf{k}_L$ (Figure 1b).

The geometry of Figure 1b is achieved when

$$k_p \cos \theta = k_L \tag{10}$$

i.e., for small angles θ, when

$$\frac{\omega_L}{c}\left(1 + \frac{\chi'_p}{2}\right)\left(1 - \frac{\theta^2}{2}\right) = \frac{\omega_L}{c}\left(1 + \frac{\chi'}{2}\right). \tag{11}$$

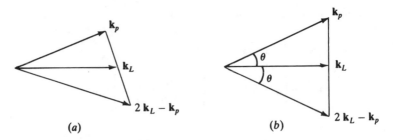

Figure 1. Arrangement of wave vectors: (a) in the general case; (b) when phase matching is fulfilled.

Angle θ is thus given by the equation

$$\theta^2 = (\chi'_p - \chi') \tag{12}$$

i.e., again, using (2.a) and (9.a) in the limit $|\delta_L| \gg \Gamma$

$$\theta^2 = \frac{N|d_{ab}|^2}{\varepsilon_0 \hbar} \frac{\delta_L(\Omega_1^2/2)}{\left[\delta_L^2 + (\Omega_1^2/2)\right]^2}. \tag{13}$$

We note that this equation has a solution only for $\delta_L > 0$. The angle θ is an increasing function of δ_L on the interval $0 < \delta_L < \Omega_1/\sqrt{6}$, and a decreasing function beyond that. As a function of the linear susceptibility $\chi_0 = -N|d_{ab}|^2/\varepsilon_0 \hbar \delta_L$

$$\theta^2 = -\chi_0 \frac{\delta_L^2(\Omega_1^2/2)}{\left[\delta_L^2 + (\Omega_1^2/2)\right]^2} \tag{14}$$

which proves that θ is always less than $\sqrt{|\chi_0|}$.

20. ABSORPTION OF A PROBE BEAM BY ATOMS INTERACTING WITH AN INTENSE BEAM. APPLICATION TO SATURATED ABSORPTION

Consider an ensemble of atoms with two levels a and b (resonance frequency ω_0, excited state b with natural width Γ) interacting with a monochromatic plane wave of frequency ω. The electric field of this wave is sufficiently intense for the condition $\Omega_1 \gg \Gamma$ to be always fulfilled (Ω_1 is the Rabi frequency). The dressed atom method introduced in Chapter VI will be used to describe the atom + laser mode system.

The purpose of this exercise is to describe the transmission of a weak wave which "probes" such a medium. The atoms are first assumed to be at rest (Part A) and one considers how the transmission of the probe beam varies versus its frequency ω'. One then considers the case of moving atoms (Part B) for which the Doppler broadening is large compared with Ω_1 and calculates the absorption of a probe beam having the same frequency as the intense wave and propagating in the opposite direction (saturated absorption spectrum). Throughout the exercise, the atomic dipoles as well as the polarizations of the fields are assumed to be parallel to a common axis $0x$, so that their vector character will be ignored.

A. Effect of an Intense Wave on the Absorption of a Weak Wave

a) Recall that the operator d associated with the dipole of the atom is equal to $d_+ + d_-$ with $d_\pm = d_{ab} \mathcal{S}_\pm$, where $\mathcal{S}_+ = |b\rangle\langle a|$, $\mathcal{S}_- = |a\rangle\langle b|$, and $d_{ab} = \langle a|d|b\rangle$ (d_{ab} is assumed to be real). Recall also that the reduced matrix elements of the density operator σ for the dressed atom are defined by $\sigma_{ij}^{(p)} = \sum_N \langle i(N-p)|\sigma|j(N)\rangle$, where the $|i(N)\rangle$ (with $i = 1, 2$) are the dressed states of the manifold $\mathcal{E}_N$ [see Chapter VI, formula (D.28)]. Express $\langle d_- \rangle$ as a function of d_{ab}, $\sigma_{ij}^{(p)}$ and of the angle θ defined by $\tan 2\theta = \Omega_1/(\omega_0 - \omega)$ ($0 \leq 2\theta < \pi$).

b) First consider the case where the atom interacts only with the intense wave. Calculate $\langle d_- \rangle$ in the quasi-steady-state regime and express the result as a function of d_{ab}, Ω_1, ω_0, and ω.

c) The atoms now interact with the intense wave and the probe beam. The latter is described classically by an electric field $\mathcal{E}' \cos(\omega't - \varphi')$ with amplitude $\mathcal{E}'$ and phase φ'. In the rotating-wave approximation, the Hamiltonian for the interaction between the atom and the probe beam is

$$V = \frac{\hbar\Omega_1'}{2}\left[e^{-i(\omega't-\varphi')}\mathcal{S}_+ + e^{i(\omega't-\varphi')}\mathcal{S}_-\right] \tag{1}$$

with $\hbar\Omega_1' = -d_{ab}\mathcal{E}'$.

Write the evolution equations of the reduced matrix elements $\sigma_{ij}^{(p)}$ (one can set $\Delta^{(p)} = \sigma_{22}^{(p)} - \sigma_{11}^{(p)}$). What is the quasi-steady-state solution of these equations to zero order in Ω_1'?

d) Now use perturbation theory to describe the solution at the first order in Ω_1' of the preceding equations. Calculate the reduced matrix elements $\sigma_{ij}^{(p)}$ proportional to $\Omega_1' e^{i\varphi'}$ whose time dependence is $e^{-i\omega't}$.

e) Calculate the component of $\langle d_- \rangle$ evolving as $\Omega'_1 e^{-i(\omega't-\varphi')}$. One sets

$$\langle d_- \rangle = \frac{\varepsilon_0}{2}(\alpha' + i\alpha'')\mathscr{E}' e^{-i(\omega't-\varphi')}. \tag{2}$$

α' is associated with the dispersion of the medium and α'' is associated with its absorption. Calculate α''. For what values of $(\omega' - \omega)$ is α'' resonant? Indicate whether these resonances correspond to an attenuation or an amplification and give the physical origin for the obtained results.

B. Saturated Absorption

Now assume that the intense wave and the probe beam have the same frequency ω_L in the laboratory reference frame, but that these waves propagate along $0z$ in opposite directions. Moreover, the atoms now have a nonzero velocity. As a consequence, in the rest frame of the atom the frequencies of the two waves undergo opposite Doppler shifts:

$$\omega = \omega_L + kv_z \tag{3.a}$$

$$\omega' = \omega_L - kv_z \tag{3.b}$$

where v_z is the projection of the atomic velocity on $0z$.

To obtain the average absorption (or dispersion) of the probe beam, it is necessary to average the result of formula (2) over the velocity distribution after having replaced ω and ω' by expressions (3.a) and (3.b). One thus obtains the average dipole

$$\overline{\langle d_- \rangle} = \frac{1}{u\sqrt{\pi}} \int_{-\infty}^{+\infty} \langle d_-(v_z) \rangle \exp\left[-\left(v_z^2/u^2\right)\right] dv_z \tag{4}$$

which is a function of $\delta_L = \omega_L - \omega_0$ (the parameter u equals $\sqrt{2k_BT/m}$). Recall that Ω_1 is assumed to be very small compared with the Doppler width ku. The purpose of this section is to study the imaginary part of the average polarizability (which will be written $\overline{\alpha''}$) for $|\delta_L| \ll ku$.

f) Assume that δ_L is fixed and fulfills the condition $|\delta_L| \ll ku$. What are the velocity groups that resonantly interact with the probe beam?

Show that, for these velocity groups, $|v_z| \ll u$ and from this deduce that $\exp[-(v_z^2/u^2)]$ can be replaced by 1 in (4).

g) Give the expression of $1/\tan 2\theta$ as a function of δ_L, kv_z, and Ω_1. One will set $s = \tan \theta$ (s varies from 0 to $+\infty$). Show that

$$\overline{\alpha''} = \frac{|d_{ab}|^2}{2\varepsilon_0\sqrt{\pi}\,ku} \int_0^{+\infty} \frac{ds}{s^2}\frac{1-s^2}{1+s^4}\left[-\frac{s^4\varepsilon(s)}{\varepsilon^2(s)+f_1^2(s)} + \frac{\varepsilon(s)}{\varepsilon^2(s)+f_2^2(s)}\right]$$

(5)

with

$$f_1(s) = 2\frac{\delta_L}{\Omega_1} + \frac{3-s^2}{2s}$$

(6.a)

and

$$f_2(s) = f_1\left(-\frac{1}{s}\right) = \frac{2\delta_L}{\Omega_1} + \frac{1-3s^2}{2s}$$

(6.b)

with $\varepsilon(s)$ being a small quantity equal to

$$\varepsilon(s) = \frac{\Gamma}{\Omega_1}\left[\frac{1}{2} + \frac{s^2}{(1+s^2)^2}\right].$$

(6.c)

h) Let s_1 be the positive root of $f_1(s) = 0$, and s_2 be the positive root of $f_2(s) = 0$. Evaluate s_1 and s_2 and show that $s_2 = s_1/3$. Express $(s_1/3) - (1/s_1)$, $(s_1/3)^2 + (1/s_1)^2$, and $(s_1/3)^4 + (1/s_1)^4$ as a function of δ_L/Ω_1.

i) Using the fact that the condition $\Omega_1 \gg \Gamma$ results in $\varepsilon(s) \ll 1$, one replaces the functions $\varepsilon(s)/[\varepsilon^2(s) + f_i^2(s)]$ with $i = 1, 2$ by $\pi\delta[(s-s_i)f_i'(s_i)] = [\pi/|f_i'(s_i)|]\delta(s-s_i)$. After integration over s, regroup the terms associated with s_1 and s_2 and show that

$$\overline{\alpha''} = \frac{\sqrt{\pi}\,d_{ab}^2}{\varepsilon_0 ku}\frac{1}{(s_1+(3/s_1))}\left[\frac{s_1-(1/s_1)}{1+(1/s_1^4)} - \frac{(s_1/3)-(3/s_1)}{1+(s_1/3)^4}\right]$$

(7.a)

Using the results from Question h, deduce from this result that

$$\overline{\alpha''} = \frac{\sqrt{\pi}\,d_{ab}^2}{\varepsilon_0\,ku}\left[1 - \frac{10+16(\delta_L/\Omega_1)^2}{25+48(\delta_L/\Omega_1)^2 + 64(\delta_L/\Omega_1)^4}\right].$$

(7.b)

j) Formula (7.b) gives the saturated absorption lineshape in the limit $\Gamma \ll \Omega_1 \ll ku$. Compare the absorption at the center ($\delta_L = 0$) and at the edges ($|\delta_L| \gg \Omega_1$). What is the order of magnitude of the width of the saturated absorption line?

In an elementary model of saturated absorption, one considers that the pump and probe beams interact, for $\delta_L = 0$, with the same velocity group $kv_z = 0$. The probe beam is then considered as propagating in a medium in which the populations of the levels a and b are equalized by the pump wave. Such a model thus predicts that $\overline{\alpha''} = 0$ for $\delta_L = 0$. Does this result coincide with the result of formula (7.b)? What is the physical origin for the difference?

k) Now assume that the atom undergoes a collisional relaxation due to dephasing collisions. Is it possible to simply predict how α'' varies for $\delta_L = 0$ as a function of the ratio γ/Γ (where γ is the damping rate of the coherence between levels a and b due to collisions)?

Solution

a) Using $\langle d_- \rangle = \text{Tr}\, \sigma d_-$ and inserting the closure relation between σ and d_-, gives

$$\langle d_- \rangle = d_{ab} \sum_{i,j,N,N'} \langle i(N)|\sigma|j(N')\rangle\langle j(N')|\mathscr{S}_-|i(N)\rangle. \tag{8}$$

The only nonzero matrix elements of $\mathscr{S}_-$ can be deduced from formulas (C.4) in Chapter VI and lead to

$$\langle d_- \rangle = d_{ab} \sum_N \left[\sin\theta\cos\theta(\langle 1(N+1)|\sigma|1(N)\rangle - \langle 2(N+1)|\sigma|2(N)\rangle) - \right.$$
$$\left. - \sin^2\theta\langle 2(N+1)|\sigma|1(N)\rangle + \cos^2\theta\langle 1(N+1)|\sigma|2(N)\rangle\right]$$
$$= d_{ab}\left[\sin\theta\cos\theta(\sigma_{11}^{(-1)} - \sigma_{22}^{(-1)}) - \sin^2\theta\sigma_{21}^{(-1)} + \cos^2\theta\sigma_{12}^{(-1)}\right]. \tag{9}$$

b) Using results (D.39), (D.40), and (D.21) from Chapter VI, we then have

$$\langle d_- \rangle = d_{ab}\sin\theta\cos\theta\frac{\sin^4\theta - \cos^4\theta}{\sin^4\theta + \cos^4\theta}e^{-i\omega t}. \tag{10}$$

It is then sufficient to express these trigonometric functions as a function of $\sin 2\theta$ and $\cos 2\theta$, and then of $\tan 2\theta = \Omega_1/(\omega_0 - \omega)$, to obtain

$$\langle d_- \rangle = \frac{1}{2}d_{ab}\frac{\Omega_1(\omega - \omega_0)}{(\omega_0 - \omega)^2 + (\Omega_1^2/2)}e^{-i\omega t} \tag{11}$$

[note that such a result coincides in the limit $\sqrt{\Omega_1^2 + (\omega_0 - \omega)^2} \gg \Gamma$ with the result of Exercise 19].

c) The rate of variation of the density operator σ for the dressed atom is given by the sum of two types of terms: terms that describe the effect of the atom-intense wave interaction and of the spontaneous emission, which were previously studied in Chapter VI; and terms that describe the interaction with the probe beam and which are written $[V, \sigma]/i\hbar$. From (1) and formulas (C.4), (D.11), and (D.31) from Chapter VI, we then find for the evolution equation of $\Delta^{(p)} = \sigma_{22}^{(p)} - \sigma_{11}^{(p)}$

$$\frac{d}{dt}\Delta^{(p)} - \left[ip\omega - \Gamma(\cos^4\theta + \sin^4\theta)\right]\Delta^{(p)} =$$

$$= -i\Omega_1'\cos^2\theta\left[e^{i(\omega't-\varphi')}\sigma_{12}^{(p-1)} - e^{-i(\omega't-\varphi')}\sigma_{21}^{(p+1)}\right] +$$

$$+ i\Omega_1'\sin^2\theta\left[e^{-i(\omega't-\varphi')}\sigma_{12}^{(p+1)} - e^{i(\omega't-\varphi')}\sigma_{21}^{(p-1)}\right] +$$

$$+ \Gamma(\cos^4\theta - \sin^4\theta)e^{ip\omega t}. \tag{12}$$

Similarly, by using formula (D.30) from Chapter VI, we find

$$\frac{d}{dt}\sigma_{12}^{(p)} - \left[i(p\omega - \Omega) - \Gamma_{\mathrm{coh}}\right]\sigma_{12}^{(p)} = -i\frac{\Omega_1'}{2}\cos^2\theta\, e^{-i(\omega't-\varphi')}\Delta^{(p+1)} +$$

$$+ i\frac{\Omega_1'}{2}\sin^2\theta\, e^{i(\omega't-\varphi')}\Delta^{(p-1)} \tag{13.a}$$

$$\frac{d}{dt}\sigma_{21}^{(p)} - \left[i(p\omega + \Omega) - \Gamma_{\mathrm{coh}}\right]\sigma_{21}^{(p)} = i\frac{\Omega_1'}{2}\cos^2\theta\, e^{i(\omega't-\varphi')}\Delta^{(p-1)} -$$

$$- i\frac{\Omega_1'}{2}\sin^2\theta\, e^{-i(\omega't-\varphi')}\Delta^{(p+1)}. \tag{13.b}$$

The solution to these equations to zero order in Ω_1', in the quasi-steady state, can be deduced from Equations (D.21), (D.39), and (D.40) of Chapter VI:

$$\Delta^{(p)} = \frac{\cos^4\theta - \sin^4\theta}{\cos^4\theta + \sin^4\theta}\, e^{ip\omega t} \tag{14.a}$$

$$\sigma_{12}^{(p)} = \sigma_{21}^{(p)} = 0. \tag{14.b}$$

d) To find the first-order solution in Ω_1', in the right-hand sides of Equations (12), (13.a), and (13.b), it is sufficient to replace $\Delta^{(p)}$, $\sigma_{12}^{(p)}$, and $\sigma_{21}^{(p)}$ by their zero-order solutions given in (14.a) and (14.b) [in Equation (12), we also omit the last term, which is proportional to $(\cos^4\theta - \sin^4\theta)$, and is a source term for the zero-order terms]. We then immediately find that the linear term in Ω_1' in the expansion of $\Delta^{(p)}$ in powers of Ω_1' is zero.

Consider now (13.a). To obtain an evolution in $e^{-i(\omega't-\varphi')}$, it is necessary to take $p = -1$ and keep only the first term in the right-hand side of the equation, which is approximated by $-i(\Omega_1'/2)\cos^2\theta\Delta^{(0)}\exp[-i(\omega't - \varphi')]$. By using (14.a) with $p = 0$, we then find

$$\sigma_{12}^{(-1)} = -i\frac{\Omega_1'}{2}\cos^2\theta\,\frac{\cos^4\theta - \sin^4\theta}{\cos^4\theta + \sin^4\theta}\,\frac{e^{-i(\omega't-\varphi')}}{\Gamma_{\mathrm{coh}} + i[\Omega + (\omega - \omega')]}. \tag{15}$$

Similarly, the solution of (13.b) evolving in $e^{-i(\omega't-\varphi')}$ is

$$\sigma_{21}^{(-1)} = -i\frac{\Omega'_1}{2}\sin^2\theta\,\frac{\cos^4\theta - \sin^4\theta}{\cos^4\theta + \sin^4\theta}\,\frac{e^{-i(\omega't-\varphi')}}{\Gamma_{coh} - i[\Omega + (\omega' - \omega)]}. \tag{16}$$

e) Substituting (15) and (16) into (9) gives

$$\langle d_-\rangle = -id_{ab}\frac{\Omega'_1}{2}\,e^{-i(\omega't-\varphi')}\,\frac{\cos^4\theta - \sin^4\theta}{\cos^4\theta + \sin^4\theta} \times$$

$$\times\left[\frac{\cos^4\theta}{\Gamma_{coh} + i(\Omega + \omega - \omega')} - \frac{\sin^4\theta}{\Gamma_{coh} - i(\Omega + \omega' - \omega)}\right]. \tag{17}$$

Using (2) and the definition $\Omega'_1 = -\mathbf{d}_{ab}\cdot\boldsymbol{\mathscr{E}}'/\hbar$ then leads to

$$\alpha'' = \frac{|d_{ab}|^2}{\varepsilon_0}\frac{\cos^4\theta - \sin^4\theta}{\cos^4\theta + \sin^4\theta}\Gamma_{coh}\left[\frac{\cos^4\theta}{\Gamma_{coh}^2 + (\Omega + \omega - \omega')^2} - \frac{\sin^4\theta}{\Gamma_{coh}^2 + (\Omega + \omega' - \omega)^2}\right]. \tag{18}$$

Assume that $\cos^4\theta > \sin^4\theta$ (which corresponds to $\delta_L < 0$). In this case, the resonance occurring at $\omega' = \omega + \Omega$ is associated with an attenuation process (α'' is positive) whereas the resonance occurring at $\omega' = \omega - \Omega$ is associated with an amplification (α'' is negative). The width at half-maximum of each of these resonances is $2\Gamma_{coh}$. The attenuation is greater than the amplification by a factor of $\tan^4\theta = (\Omega - \omega_0 + \omega)^2/(\Omega + \omega_0 - \omega)^2$. The physical interpretation of these resonances and their sign (attenuation or amplification of the probe beam) is given in subsection E-2-b of Chapter VI.

f) According to (3.a) and (3.b), the second term of (18) is resonant for the velocity group v_1 such that

$$2kv_1 = \sqrt{\Omega_1^2 + (\delta_L + kv_1)^2}. \tag{19.a}$$

Similarly, the first term of (18) is resonant when

$$2kv_2 = -\sqrt{\Omega_1^2 + (\delta_L + kv_2)^2}. \tag{19.b}$$

By squaring expressions (19.a) and (19.b), we note that kv_1 and kv_2 are solutions of the same second-degree equation and are equal to

$$kv_1 = \frac{\delta_L + \sqrt{3\Omega_1^2 + 4\delta_L^2}}{3} \tag{20.a}$$

$$kv_2 = \frac{\delta_L - \sqrt{3\Omega_1^2 + 4\delta_L^2}}{3}. \tag{20.b}$$

Because δ_L and Ω_1 are both small compared with ku, Equations (20) result in $|v_1| \ll u$ and $|v_2| \ll u$. Consequently, the velocity groups that make an important contribution to (4) are such that the Boltzmann exponential can be replaced by 1.

Remark

The velocity groups that interact resonantly with the probe beam can also be derived by a graphical method. First consider the frequency of the pump laser in the rest frame of an atom having velocity v_z. This frequency is, according to (3.a), equal to $\omega_L + kv_z$. Its variations with kv_z for constant δ_L are those of a straight line with slope 1, represented by the dashed line in Figure 1. Similarly, the Doppler-shifted frequency of the probe laser, which, according to (3.b), is equal to $\omega_L - kv_z$, is represented by the dashed straight line with a slope of -1, which intersects the first line at the point I having abscissa 0 and ordinate ω_L. From the discussion in Part A, for each value of kv_z, the absorption spectrum of the atom in its rest frame consists of two lines centered at the frequencies of the sidebands $\omega \pm [(\omega - \omega_0)^2 + \Omega_1^2]^{1/2}$ which, according to (3.a), equal

$$\omega_L + kv_z + \sqrt{\Omega_1^2 + (\omega_L - \omega_0 + kv_z)^2} \tag{21.a}$$

$$\omega_L + kv_z - \sqrt{\Omega_1^2 + (\omega_L - \omega_0 + kv_z)^2}. \tag{21.b}$$

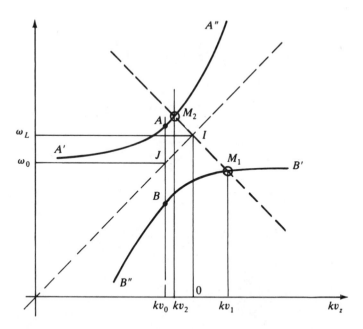

Figure 1. Graphical determination of the velocity groups which are resonant with the probe beam.

The variations of these two frequencies with kv_z for constant δ_L are represented by the two branches of the hyperbola drawn with solid lines in Figure 1, having as one asymptote the horizontal line at ω_0. For a given δ_L, the velocity groups kv_1 and kv_2 that are resonant with the probe beam, i.e., capable of absorbing (or amplifying) this wave, are given by the abscissas of the intersections M_1 and M_2 of this hyperbola with the straight line of slope -1.

To determine whether these velocity groups absorb or amplify the probe beam, we introduce the velocity group $kv_0 = \omega_0 - \omega_L$ for which the pump wave is resonant. The vertical line having the abscissa kv_0 intersects the hyperbola at the points A and B having ordinates $\omega_0 + \Omega_1$ and $\omega_0 - \Omega_1$ (the middle J of AB, with ordinate ω_0, is the center of the hyperbola). For $v < v_0$, the apparent laser frequency is smaller than ω_0. The branch $A'A$ of the hyperbola, which corresponds to the upper sideband, is then absorbing, whereas the branch $B''B$ is amplifying [see the discussion following formula (18) and the discussion in subsection E-2-b of Chapter VI]. For $v > v_0$, these conclusions are reversed, the branch BB' is absorbing whereas the branch AA'' is amplifying. From this we conclude that, for the value of δ_L chosen for Figure 1, the velocity group kv_1 is absorbing whereas the velocity class kv_2 is amplifying. For other situations, for example, $\omega_L - \omega_0 = 0$, for which $v_0 = 0$, the two resonant velocity groups are both on the absorbing branches.

g) From (3.a), we deduce that

$$\frac{1}{\tan 2\theta} = \frac{1 - s^2}{2s} = -\frac{\delta_L + kv_z}{\Omega_1} \tag{22}$$

and consequently

$$kv_z = -\Omega_1 \left[\frac{\delta_L}{\Omega_1} + \frac{1 - s^2}{2s} \right]. \tag{23}$$

When this relation is differentiated, we obtain

$$dv_z = \frac{\Omega_1}{2k} \frac{1 + s^2}{s^2} \, ds. \tag{24}$$

In expression (18) for α'', we replace $\cos^2 \theta$ by $1/(1 + s^2)$ and $\sin^2 \theta$ by $s^2/(1 + s^2)$. We proceed in the same way for Γ_{coh} by using Equation (D.27) from Chapter VI;

$$\Gamma_{\text{coh}} = \Gamma \left[\frac{1}{2} + \frac{s^2}{(1 + s^2)^2} \right] = \Omega_1 \varepsilon(s). \tag{25}$$

The denominators of the fractions appearing in (18) have, respectively, the form $\Gamma_{\text{coh}}^2 +$ $[\sqrt{\Omega_1^2 + (\delta_L + kv_z)^2} \pm 2kv_z]^2$. The value of kv_z is given by (23). Finally, $\sqrt{\Omega_1^2 + (\delta_L + kv_z)^2}$ is equal to

$$\sqrt{\Omega_1^2 + (\delta_L + kv_z)^2} = \frac{\Omega_1}{2s}(1 + s^2). \tag{26}$$

Using (4), (18), and (23)–(26) then leads to formula (5).

h) The function $f_1(s)$ cancels out for

$$s_1 = \frac{2\delta_L}{\Omega_1} + \sqrt{4\frac{\delta_L^2}{\Omega_1^2} + 3} \tag{27}$$

with the other negative root being outside the domain of integration of (5). Similarly, the only positive root of $f_2(s)$ is

$$s_2 = \frac{1}{3}\left[\frac{2\delta_L}{\Omega_1} + \sqrt{\frac{4\delta_L^2}{\Omega_1^2} + 3}\right] = \frac{s_1}{3}. \tag{28}$$

Start with

$$\frac{s_1}{3} - \frac{1}{s_1} = \frac{1}{3}\left(\frac{s_1^2 - 3}{s_1}\right) \tag{29}$$

and use the fact that $f_1(s_1) = 0$. Using (6.a), we obtain

$$\frac{s_1}{3} - \frac{1}{s_1} = -\frac{4}{3}\frac{\delta_L}{\Omega_1}. \tag{30}$$

By squaring both sides of this equation, we get

$$\left(\frac{s_1}{3}\right)^2 + \frac{1}{s_1^2} = \frac{16}{9}\frac{\delta_L^2}{\Omega_1^2} + \frac{2}{3} \tag{31}$$

and by again squaring both sides of (31)

$$\left(\frac{s_1}{3}\right)^4 + \frac{1}{s_1^4} = \frac{256}{81}\frac{\delta_L^4}{\Omega_1^4} + \frac{64}{27}\frac{\delta_L^2}{\Omega_1^2} + \frac{2}{9}. \tag{32}$$

i) According to (6.a) and (6.b)

$$\frac{1}{|f_1'(s_1)|} = \frac{2s_1^2}{3 + s_1^2} \tag{33.a}$$

$$\frac{1}{|f_2'(s_2)|} = \frac{2s_2^2}{1 + 3s_2^2}. \tag{33.b}$$

Integrating (5) then gives

$$\overline{\alpha}'' = \frac{\sqrt{\pi}\, d_{ab}^2}{\varepsilon_0 k u}\left[-\frac{(1 - s_1^2)s_1^4}{(3 + s_1^2)(1 + s_1^4)} + \frac{1 - s_2^2}{(1 + 3s_2^2)(1 + s_2^4)} \right]. \tag{34.a}$$

Using $s_2 = s_1/3$, we transform (34.a) into (7.a). When (7.a) is simplified by finding a common denominator, we get

$$\overline{\alpha}'' = \frac{\sqrt{\pi}\, d_{ab}^2}{\varepsilon_0 k u}\, \frac{1}{\left[s_1 + \dfrac{3}{s_1} \right]}$$

$$\times \frac{\left\{ \left(\dfrac{s_1}{3}\right)^4 + \left(\dfrac{1}{s_1}\right)^4 - \dfrac{4}{9}\left[\left(\dfrac{s_1}{3}\right)^2 + \dfrac{1}{s_1^2}\right] + \dfrac{22}{27}\right\}\left[s_1 + \dfrac{3}{s_1}\right]}{\dfrac{82}{81} + \left(\dfrac{s_1}{3}\right)^4 + \left(\dfrac{1}{s_1}\right)^4}. \tag{34.b}$$

Simplifying by $s_1 + (3/s_1)$ and using (31) and (32), we find (7.b).

j) The curve giving $\overline{\alpha}''$ as a function of δ_L has a minimum at $\delta_L = 0$, has a width on the order of Ω_1, and tends to a limit equal to $\sqrt{\pi}\, |d_{ab}|^2/\varepsilon_0 k u$ when $|\delta_L| \gg \Omega_1$. Note that this limit corresponds to the linear absorption for the probe beam alone. It should be noted that the width of this curve is smaller than ku, which shows that it is possible to perform Doppler free spectroscopy by using saturated absorption (*).

The value of $\overline{\alpha}''$ at $\delta_L = 0$ is equal to 0.6 times its asymptotic value. This shows that the medium is never totally transparent for the probe beam, contrary to the predictions of the

(*) For more details on saturated absorption spectroscopy, see, for example, T. W. Hänsch, in *Nonlinear Spectroscopy*, Proceedings of the International School of Physics "Enrico Fermi," Course LXIV, edited by N. Bloembergen, North-Holland, Amsterdam, 1977, p. 17; Lethokov and Chebotayev (Chapter 3).

elementary model described in Question j. Indeed, when $\delta_L = 0$, the velocity groups interacting resonantly with the probe beam are, according to (20.a) and (20.b), $kv_1 = \Omega_1/\sqrt{3}$ and $kv_2 = -\Omega_1/\sqrt{3}$. For these velocity groups, the population difference $\pi_2^{\mathrm{st}} - \pi_1^{\mathrm{st}}$ between the two levels of the same manifold of the dressed atom is very different from 0. More precisely from (27) and (28), s_1 and s_2 equal $\sqrt{3}$ and $1/\sqrt{3}$ when $\delta_L = 0$, and using formula (D.21) from Chapter VI, $|\pi_2^{\mathrm{st}} - \pi_1^{\mathrm{st}}|$ is equal to $\frac{4}{5}$. The probe beam thus interacts with velocity groups for which the populations of the dressed levels are not equalized.

k) In the presence of collisional relaxation, the population difference between the dressed levels can be deduced from formulas (36) and (18) in Complement B_{VI} applicable in the impact limit.

$$\pi_1^{\mathrm{st}} - \pi_2^{\mathrm{st}} = \frac{\sin^4\theta - \cos^4\theta}{\cos^4\theta + \sin^4\theta + 4(\gamma/\Gamma)\sin^2\theta\cos^2\theta}. \tag{35}$$

To obtain the saturated absorption line shape, it is thus necessary in formula (5) to replace $1/(1 + s^4)$ by

$$\frac{1}{1 + s^4 + (4\gamma/\Gamma)s^2}. \tag{36}$$

It is also necessary to replace $\varepsilon(s)$ by

$$\varepsilon'(s) = \frac{\Gamma}{\Omega_1}\left[\frac{1}{2} + \frac{s^2}{(1+s^2)^2}\right] + \frac{\gamma}{\Omega_1}\left[\frac{1+s^4}{(1+s^2)^2}\right] \tag{37}$$

[see Complement B_{VI}, formula (20.a)].

When the relaxation rate γ remains small compared with ku, we can replace as before the functions $\varepsilon'(s)/[\varepsilon'^2(s) + f_i^2(s)]$ by Dirac functions and thus obtain

$$\overline{\alpha''} = \frac{\sqrt{\pi}\,|d_{ab}|^2}{\varepsilon_0 ku}\left[-\frac{(1-s_1^2)s_1^4}{(3+s_1^2)(1+s_1^4+4(\gamma/\Gamma)s_1^2)}\right.$$

$$\left. + \frac{(1-s_2^2)}{(1+3s_2^2)(1+s_2^4+4(\gamma/\Gamma)s_2^2)}\right] \tag{38}$$

instead of (34.a).

To find the behavior at $\delta_L = 0$, we replace s_1 by $\sqrt{3}$ and s_2 by $1/\sqrt{3}$ in (38) to obtain

$$\overline{\alpha''} = \frac{\sqrt{\pi}}{\varepsilon_0} \frac{|d_{ab}|^2}{ku} \frac{3}{5 + 6(\gamma/\Gamma)}.$$ (39)

When the collision rate increases, $\overline{\alpha''}$ decreases and tends to 0. This is due to the fact that the collisions tend to equalize the populations of the dressed levels. Consequently, the medium becomes more and more transparent for the probe beam.

One might wonder whether the elementary model presented in Question j (which indeed predicts $\overline{\alpha''} = 0$ for $\delta_L = 0$) is better justified when $\gamma \gg \Gamma$. In fact, this model is associated with a description of the atom only in terms of the populations σ_{aa} and σ_{bb} of the atomic levels a and b. According to the optical Bloch equations, such a model is generally incorrect because it neglects effects connected with the optical coherence $\hat{\sigma}_{ab}$. Nevertheless, when γ increases, this optical coherence is increasingly damped and it might be justified to consider only the populations σ_{aa} and σ_{bb} (see Exercise 18). We must not forget, however, that the velocity groups that effectively interact with the probe beam remain determined by the energy diagram of the dressed atom [see Figure 1 and Equation (20)] and are independent of γ.

APPENDIX

QUANTUM ELECTRODYNAMICS IN THE COULOMB GAUGE—SUMMARY OF THE ESSENTIAL RESULTS

This appendix briefly summarizes the procedure for quantizing the electromagnetic field in the Coulomb gauge and gathers together the essential formulas that are used in this volume (*).

The system studied in electrodynamics is composed of two interacting subsystems: the electromagnetic field on the one hand, and an ensemble of charged particles on the other. Quantum electrodynamics attempts to describe, within the framework of quantum mechanics, the states and the dynamics of these two subsystems whose evolutions are coupled. The charged particles are actually the sources of the field and the field exerts forces on the particles.

We first introduce the variables used to describe the field (§1) and the particles (§2) in classical and quantum theories. Then we introduce the Coulomb-gauge Hamiltonian which governs the dynamics of the global system (§3). We then review a few important quantum states of the field (§4), and last we introduce the electric dipole representation currently used to describe localized systems of charges such as atoms or molecules (§5).

1. Description of the Electromagnetic Field

a) ELECTRIC FIELD **E** AND MAGNETIC FIELD **B**

In classical electrodynamics, the fields $\mathbf{E}(\mathbf{r}, t)$ and $\mathbf{B}(\mathbf{r}, t)$ obey the Maxwell equations

$$\boldsymbol{\nabla} \cdot \mathbf{E}(\mathbf{r}, t) = \frac{1}{\varepsilon_0} \rho(\mathbf{r}, t) \tag{1.a}$$

$$\boldsymbol{\nabla} \cdot \mathbf{B}(\mathbf{r}, t) = 0 \tag{1.b}$$

$$\boldsymbol{\nabla} \times \mathbf{E}(\mathbf{r}, t) = -\frac{\partial}{\partial t} \mathbf{B}(\mathbf{r}, t) \tag{1.c}$$

$$\boldsymbol{\nabla} \times \mathbf{B}(\mathbf{r}, t) = \frac{1}{c^2} \frac{\partial}{\partial t} \mathbf{E}(\mathbf{r}, t) + \frac{1}{\varepsilon_0 c^2} \mathbf{j}(\mathbf{r}, t) \tag{1.d}$$

(*) It is, of course, impossible to discuss in a short appendix all the aspects of the quantization of the electromagnetic field or to prove all the results cited here. The reader can refer to *Photons and Atoms—Introduction to Quantum Electrodynamics* for a more detailed presentation.

where the charge density $\rho(\mathbf{r}, t)$ and current density $\mathbf{j}(\mathbf{r}, t)$ are those associated with the particles. After a spatial Fourier transformation, the Maxwell equations become (*)

$$i\mathbf{k} \cdot \boldsymbol{\mathscr{E}}(\mathbf{k}, t) = \frac{1}{\varepsilon_0}\rho(\mathbf{k}, t) \tag{2.a}$$

$$i\mathbf{k} \cdot \boldsymbol{\mathscr{B}}(\mathbf{k}, t) = 0 \tag{2.b}$$

$$i\mathbf{k} \times \boldsymbol{\mathscr{E}}(\mathbf{k}, t) = -\frac{\partial}{\partial t}\boldsymbol{\mathscr{B}}(\mathbf{k}, t) \tag{2.c}$$

$$i\mathbf{k} \times \boldsymbol{\mathscr{B}}(\mathbf{k}, t) = \frac{1}{c^2}\frac{\partial \boldsymbol{\mathscr{E}}(\mathbf{k}, t)}{\partial t} + \frac{1}{\varepsilon_0 c^2}\boldsymbol{\mathscr{J}}(\mathbf{k}, t) \tag{2.d}$$

where $\mathbf{E}(\mathbf{r}, t)$ and $\boldsymbol{\mathscr{E}}(\mathbf{k}, t)$, for example, are related by:

$$\mathbf{E}(\mathbf{r}, t) = \frac{1}{(2\pi)^{3/2}}\int d^3k \, \boldsymbol{\mathscr{E}}(\mathbf{k}, t) \, e^{i\mathbf{k} \cdot \mathbf{r}}. \tag{3}$$

The two equations (2.a) and (2.b) fix the longitudinal parts $\boldsymbol{\mathscr{E}}_{\parallel}$ and $\boldsymbol{\mathscr{B}}_{\parallel}$ of the electric and magnetic fields, i.e., the projections onto $\mathbf{k}/k$ of $\boldsymbol{\mathscr{E}}$ and $\boldsymbol{\mathscr{B}}$:

$$\boldsymbol{\mathscr{E}}_{\parallel}(\mathbf{k}, t) = -\frac{i\mathbf{k}}{\varepsilon_0 k^2}\rho(\mathbf{k}, t) \tag{4.a}$$

$$\boldsymbol{\mathscr{B}}_{\parallel}(\mathbf{k}, t) = \mathbf{0} \tag{4.b}$$

In real space, these relations become

$$\mathbf{E}_{\parallel}(\mathbf{r}, t) = \frac{-1}{4\pi\varepsilon_0}\int d^3r' \, \rho(\mathbf{r}', t)\nabla_r\frac{1}{|\mathbf{r} - \mathbf{r}'|} \tag{5.a}$$

$$\mathbf{B}_{\parallel}(\mathbf{r}, t) = \mathbf{0}. \tag{5.b}$$

The magnetic field is purely transverse, whereas the longitudinal electric field coincides with the Coulomb field associated with the distribution of charge $\rho(\mathbf{r}, t)$ at the same time. The longitudinal fields are thus not really

---

(*) We systematically use the following notation: a scalar or vectorial field in real space is written with a Roman letter, whereas its spatial Fourier transform is designated by the same cursive or italic letter.

independent variables for the field. They are either zero, or they can be expressed through ρ as a function of the particle variables. By contrast, the transverse fields $\mathscr{E}_\perp$ and $\mathscr{B}_\perp = \mathscr{B}$, which are the projections of $\mathscr{E}$ and $\mathscr{B}$ in the plane perpendicular to $\mathbf{k}$, are independent variables whose equations of motion can be deduced from (2.c) and (2.d)

$$\frac{\partial}{\partial t}\mathscr{B}(\mathbf{k}, t) = -i\mathbf{k} \times \mathscr{E}_\perp(\mathbf{k}, t) \tag{6.a}$$

$$\frac{\partial}{\partial t}\mathscr{E}_\perp(\mathbf{k}, t) = c^2 i\mathbf{k} \times \mathscr{B}(\mathbf{k}, t) - \frac{1}{\varepsilon_0}\mathbf{j}_\perp(\mathbf{k}, t). \tag{6.b}$$

b) VECTOR POTENTIAL A AND SCALAR POTENTIAL U

In quantum theory, it is necessary to consider the potentials $\mathbf{A}$ and U related to the fields $\mathbf{E}$ and $\mathbf{B}$ by the equations

$$\mathbf{E}(\mathbf{r}, t) = -\nabla U(\mathbf{r}, t) - \frac{\partial \mathbf{A}(\mathbf{r}, t)}{\partial t} \tag{7.a}$$

$$\mathbf{B}(\mathbf{r}, t) = \nabla \times \mathbf{A}(\mathbf{r}, t) \tag{7.b}$$

which, in reciprocal space, become

$$\mathscr{E}(\mathbf{k}, t) = -i\mathbf{k}\mathscr{U}(\mathbf{k}, t) - \frac{\partial \mathscr{A}(\mathbf{k}, t)}{\partial t} \tag{8.a}$$

$$\mathscr{B}(\mathbf{k}, t) = i\mathbf{k} \times \mathscr{A}(\mathbf{k}, t). \tag{8.b}$$

The fields $\mathbf{E}$ and $\mathbf{B}$ are invariant in the gauge transformation associated with the function $F(\mathbf{r}, t)$

$$\mathbf{A}(\mathbf{r}, t) \rightarrow \mathbf{A}'(\mathbf{r}, t) = \mathbf{A}(\mathbf{r}, t) + \nabla F(\mathbf{r}, t) \tag{9.a}$$

$$U(\mathbf{r}, t) \rightarrow U'(\mathbf{r}, t) = U(\mathbf{r}, t) - \frac{\partial F(\mathbf{r}, t)}{\partial t} \tag{9.b}$$

which can also be written

$$\mathscr{A}(\mathbf{k}, t) \rightarrow \mathscr{A}'(\mathbf{k}, t) = \mathscr{A}(\mathbf{k}, t) + i\mathbf{k}\mathscr{F}(\mathbf{k}, t) \tag{10.a}$$

$$\mathscr{U}(\mathbf{k}, t) \rightarrow \mathscr{U}'(\mathbf{k}, t) = \mathscr{U}(\mathbf{k}, t) - \frac{\partial \mathscr{F}(\mathbf{k}, t)}{\partial t}. \tag{10.b}$$

It is clear from Equations (10) that only $\mathscr{A}_\parallel$ and $\mathscr{U}$ change in a gauge transformation, whereas $\mathscr{A}_\perp$ is gauge invariant

$$\mathscr{A}'_\perp(\mathbf{k}, t) = \mathscr{A}_\perp(\mathbf{k}, t). \tag{11}$$

Equations (8) prove also that the transverse fields $\mathscr{E}_\perp$ and $\mathscr{B}$ depend only on $\mathscr{A}_\perp$

$$\mathscr{E}_\perp(\mathbf{k}, t) = -\frac{\partial}{\partial t}\mathscr{A}_\perp(\mathbf{k}, t) \tag{12.a}$$

$$\mathscr{B}(\mathbf{k}, t) = i\mathbf{k} \times \mathscr{A}_\perp(\mathbf{k}, t). \tag{12.b}$$

c) COULOMB GAUGE

The Coulomb gauge ($\nabla \cdot \mathbf{A} = 0$) corresponds to the choice

$$\mathscr{A}_\|(\mathbf{k}, t) = \mathbf{0} = \mathbf{A}_\|(\mathbf{r}, t). \tag{13}$$

By comparing the longitudinal part of (8.a) with (4.a), we then obtain

$$\mathscr{U}(\mathbf{k}, t) = \frac{1}{\varepsilon_0 k^2}\rho(\mathbf{k}, t) \tag{14}$$

or, equivalently,

$$U(\mathbf{r}, t) = \frac{1}{4\pi\varepsilon_0}\int d^3r' \frac{\rho(\mathbf{r}', t)}{|\mathbf{r} - \mathbf{r}'|}. \tag{15}$$

In the Coulomb gauge, the longitudinal vector potential is zero and the scalar potential coincides with the Coulomb potential associated with the charge distribution $\rho(\mathbf{r}, t)$ at the same time. Therefore, in this gauge the independent variables of the field are the transverse vector potential $\mathscr{A}_\perp(\mathbf{k}, t)$ and its velocity $\dot{\mathscr{A}}_\perp(\mathbf{k}, t) = -\mathscr{E}_\perp(\mathbf{k}, t)$.

d) NORMAL VARIABLES

The simple form for the equations of motion (6) of the transverse fields suggests the introduction of the following linear combination of $\mathscr{B}$ and $\mathscr{E}_\perp$ [or, using (12), of $\mathscr{A}_\perp$ and $\dot{\mathscr{A}}_\perp$]

$$\boldsymbol{\alpha}(\mathbf{k}, t) = \lambda(k)\left[\mathscr{E}_\perp(\mathbf{k}, t) - c\frac{\mathbf{k}}{k} \times \mathscr{B}(\mathbf{k}, t)\right]$$

$$= \lambda(k)\left[-\dot{\mathscr{A}}_\perp(\mathbf{k}, t) + i\omega\mathscr{A}_\perp(\mathbf{k}, t)\right] \tag{16}$$

where $\lambda(k)$ is a normalization constant that we will later on take to be equal to $-i\sqrt{\varepsilon_0/2\hbar\omega}$. The evolution equation for $\boldsymbol{\alpha}$ is then quite simple:

$$\dot{\boldsymbol{\alpha}}(\mathbf{k}, t) + i\omega\boldsymbol{\alpha}(\mathbf{k}, t) = \frac{i}{\sqrt{2\varepsilon_0\hbar\omega}}\boldsymbol{j}_{\perp}(\mathbf{k}, t). \tag{17}$$

In the absence of sources ($\mathbf{j}_{\perp} = 0$), the variables $\boldsymbol{\alpha}(\mathbf{k}, t)$ corresponding to the different possible values for $\mathbf{k}$ evolve independently of each other, with a time dependence $\exp(-i\omega t)$ where $\omega = ck$. The variables $\boldsymbol{\alpha}$ thus describe the normal modes of vibration of the free field and are for this reason called normal variables.

Equation (16) shows that $\boldsymbol{\alpha}$ is, like $\mathscr{E}_{\perp}$ and $\mathscr{B}$, a transverse field. For each value of $\mathbf{k}$, we can introduce two unitary vectors $\boldsymbol{\varepsilon}$ and $\boldsymbol{\varepsilon}'$, orthogonal to each other and both perpendicular to $\mathbf{k}$. Each ensemble $\mathbf{k}, \boldsymbol{\varepsilon}$ defines a normal vibrational mode of the field and the normal variable associated with this mode

$$\alpha_{\varepsilon}(\mathbf{k}) = \boldsymbol{\varepsilon} \cdot \boldsymbol{\alpha}(\mathbf{k}) \tag{18.a}$$

obeys, according to (17), the evolution equation

$$\dot{\alpha}_{\varepsilon}(\mathbf{k}, t) + i\omega\alpha_{\varepsilon}(\mathbf{k}, t) = \frac{i}{\sqrt{2\varepsilon_0\hbar\omega}}\boldsymbol{\varepsilon} \cdot \boldsymbol{j}(\mathbf{k}, t). \tag{18.b}$$

By using the reality conditions for the fields $\mathbf{E}_{\perp}$, $\mathbf{B}$, and $\mathbf{A}_{\perp}$, which, for example, for $\mathbf{E}_{\perp}$ are written

$$\mathscr{E}_{\perp}(\mathbf{k}, t) = \mathscr{E}_{\perp}^{*}(-\mathbf{k}, t), \tag{19}$$

we can invert Equations (16) and express $\mathscr{E}_{\perp}(\mathbf{k}, t)$, $\mathscr{B}(\mathbf{k}, t)$, and $\mathscr{A}_{\perp}(\mathbf{k}, t)$ as a function of $\alpha_{\varepsilon}(\mathbf{k}, t)$ and $\alpha_{\varepsilon}^{*}(-\mathbf{k}, t)$. A Fourier transformation then gives the expansions of the various transverse fields as functions of the normal variables. Later on, these expansions are written directly as a function of the operators $a_{\varepsilon}(\mathbf{k})$ and $a_{\varepsilon}^{+}(\mathbf{k})$ which are associated, in quantum theory, with the normal variables $\alpha_{\varepsilon}(\mathbf{k})$ and $\alpha_{\varepsilon}^{*}(\mathbf{k})$. The set of degrees of freedom associated with the transverse field is usually designated as "radiation". The state of the radiation is thus defined at time t by the given normal variables $\alpha_{\varepsilon}(\mathbf{k}, t)$ for all $\mathbf{k}$ and all $\boldsymbol{\varepsilon}$.

e) PRINCIPLE OF CANONICAL QUANTIZATION IN THE COULOMB GAUGE

The canonical quantization procedure requires pairs of conjugated dynamical variables to be identified, which, after quantization, become operators whose commutators equal $i\hbar$.

For the electromagnetic field, we can introduce a Lagrangian which contains only the really independent variables of the field ($\mathscr{A}_\perp$ and $\dot{\mathscr{A}}_\perp$) and the variables of the particles (standard Lagrangian in the Coulomb gauge) and which leads to the Maxwell equations for the field and to the Newton–Lorentz equations for the particles. With regard to this Lagrangian, the conjugate moment of the generalized coordinate $\mathscr{A}_\varepsilon(\mathbf{k})$ is found to be equal to $\pi_\varepsilon(\mathbf{k}) = \varepsilon_0 \dot{\mathscr{A}}_\varepsilon(\mathbf{k})$ and the canonical commutation relations are written

$$[\mathscr{A}_\varepsilon(\mathbf{k}), \pi_{\varepsilon'}^+(\mathbf{k}')] = i\hbar\delta_{\varepsilon\varepsilon'}\delta(\mathbf{k} - \mathbf{k}'). \qquad (20.\text{a})$$

The operator $a_\varepsilon(\mathbf{k})$ associated with the normal variable $\alpha_\varepsilon(\mathbf{k})$ is expressed as a function of the operators $\mathscr{A}_\varepsilon(\mathbf{k})$ and $\pi_\varepsilon(\mathbf{k}) = \varepsilon_0 \dot{\mathscr{A}}_\varepsilon(\mathbf{k})$ by an equation analogous to (16). By choosing an appropriate normalization constant $\lambda(k)$ $[\lambda(k) = -i\sqrt{\varepsilon_0/2\hbar\omega}\,]$, we then find that relation (20.a) is equivalent to

$$[a_\varepsilon(\mathbf{k}), a_{\varepsilon'}^+(\mathbf{k}')] = \delta_{\varepsilon\varepsilon'}\delta(\mathbf{k} - \mathbf{k}') \qquad (20.\text{b})$$

with all the other commutators being zero.

To follow the quantum electrodynamic calculations presented in this book, it is sufficient to know the commutation relations (20.b) and the expressions for the physical variables as a function of the operators a and a^+ which are discussed in the next subsection.

f) Quantum Fields in the Coulomb Gauge

As we explained above, it is possible to invert Equations (16) between operators and to use the Hermiticity conditions of these operators to obtain, by Fourier transformation, the expansions of the field operators in a_ε and a_ε^+. We find in this way that

$$\mathbf{A}_\perp(\mathbf{r}) = \int d^3k \sum_\varepsilon \mathscr{A}_\omega[\boldsymbol{\varepsilon} a_\varepsilon(\mathbf{k}) e^{i\mathbf{k}\cdot\mathbf{r}} + \boldsymbol{\varepsilon} a_\varepsilon^+(\mathbf{k}) e^{-i\mathbf{k}\cdot\mathbf{r}}] \qquad (21)$$

$$\mathbf{E}_\perp(\mathbf{r}) = \int d^3k \sum_\varepsilon i\mathscr{E}_\omega[\boldsymbol{\varepsilon} a_\varepsilon(\mathbf{k}) e^{i\mathbf{k}\cdot\mathbf{r}} - \boldsymbol{\varepsilon} a_\varepsilon^+(\mathbf{k}) e^{-i\mathbf{k}\cdot\mathbf{r}}] \qquad (22)$$

$$\mathbf{B}(\mathbf{r}) = \int d^3k \sum_\varepsilon i\mathscr{B}_\omega[(\boldsymbol{\kappa} \times \boldsymbol{\varepsilon}) a_\varepsilon(\mathbf{k}) e^{i\mathbf{k}\cdot\mathbf{r}} - (\boldsymbol{\kappa} \times \boldsymbol{\varepsilon}) a_\varepsilon^+(\mathbf{k}) e^{-i\mathbf{k}\cdot\mathbf{r}}]$$
$$(23)$$

where

$$\omega = ck \tag{24.a}$$

$$\kappa = k/k \tag{24.b}$$

$$\mathscr{A}_\omega = \left[\hbar/2\varepsilon_0\omega(2\pi)^3\right]^{1/2} \qquad \mathscr{E}_\omega = \omega\mathscr{A}_\omega \qquad \mathscr{B}_\omega = \mathscr{E}_\omega/c. \tag{24.c}$$

The total electric field $\mathbf{E}(\mathbf{r})$ is written

$$\mathbf{E}(\mathbf{r}) = \mathbf{E}_\perp(\mathbf{r}) + \mathbf{E}_\|(\mathbf{r}) \tag{24.d}$$

where $\mathbf{E}_\|(\mathbf{r})$ is given in (5.a).

It is often convenient to consider the field as being contained in a cubic box with periodic boundary conditions. The dimension L of this box is taken as being large compared with all the characteristic dimensions of the problem under consideration. The components of the wave vectors are then multiples of $2\pi/L$ and the modes form a discrete ensemble designated by the subscript j. The Fourier integrals are replaced by series following the rule

$$\int d^3k \sum_\varepsilon f(\mathbf{k}, \varepsilon) \leftrightarrow \sum_j \left(\frac{2\pi}{L}\right)^3 f(\mathbf{k}_j, \varepsilon_j). \tag{25}$$

The creation and annihilation operators are redefined by

$$a_j = \left(\frac{L}{2\pi}\right)^{3/2} \int_{C_j} d^3k\, a_\varepsilon(\mathbf{k}) \tag{26}$$

where C_j is the elementary cell of volume $(2\pi/L)^3$ about $\mathbf{k}_j$. The a_i and a_j^+ satisfy the simple commutation relation

$$\left[a_i, a_j^+\right] = \delta_{ij}. \tag{27}$$

The fields are expressed as a function of the a_j in the form

$$\mathbf{A}_\perp(\mathbf{r}) = \sum_j \sqrt{\frac{\hbar}{2\varepsilon_0\omega_j L^3}} \left[a_j \boldsymbol{\varepsilon}_j\, e^{i\mathbf{k}_j\cdot\mathbf{r}} + a_j^+ \boldsymbol{\varepsilon}_j\, e^{-i\mathbf{k}_j\cdot\mathbf{r}}\right] \tag{28}$$

$$\mathbf{E}_\perp(\mathbf{r}) = \sum_j i\sqrt{\frac{\hbar\omega_j}{2\varepsilon_0 L^3}} \left[a_j \boldsymbol{\varepsilon}_j\, e^{i\mathbf{k}_j\cdot\mathbf{r}} - a_j^+ \boldsymbol{\varepsilon}_j\, e^{-i\mathbf{k}_j\cdot\mathbf{r}}\right] \tag{29}$$

$$\mathbf{B}(\mathbf{r}) = \sum_j \frac{i}{c}\sqrt{\frac{\hbar\omega_j}{2\varepsilon_0 L^3}} \left[a_j \boldsymbol{\kappa}_j \times \boldsymbol{\varepsilon}_j\, e^{i\mathbf{k}_j\cdot\mathbf{r}} - a_j^+ \boldsymbol{\kappa}_j \times \boldsymbol{\varepsilon}_j\, e^{-i\mathbf{k}_j\cdot\mathbf{r}}\right]. \tag{30}$$

Last, to finish this subsection, we give several useful formulas. In the bilinear expressions with respect to the fields, the following sums involving the Cartesian components of the transverse polarization vectors $\boldsymbol{\varepsilon}$ and $\boldsymbol{\varepsilon}'$, perpendicular to $\mathbf{k}$ are often encountered:

$$\sum_{\varepsilon} \varepsilon_l \varepsilon_m = \delta_{lm} - \kappa_l \kappa_m \tag{31}$$

$$\sum_{\varepsilon} \varepsilon_l (\boldsymbol{\kappa} \times \boldsymbol{\varepsilon})_m = \sum_n e_{lmn} \kappa_n \tag{32}$$

$$\sum_{\varepsilon} (\boldsymbol{\kappa} \times \boldsymbol{\varepsilon})_l (\boldsymbol{\kappa} \times \boldsymbol{\varepsilon})_m = \delta_{lm} - \kappa_l \kappa_m \tag{33}$$

where l and $m = x, y, z$, e_{lmn} is the antisymmetric tensor, and where $\boldsymbol{\kappa}$ is defined in (24.b). Expression (31) represents, in reciprocal space, the projector onto the subspace of transverse fields. In real space, this projector is represented by the transverse delta function

$$
\begin{aligned}
\delta_{lm}^{\perp}(\mathbf{r} - \mathbf{r}') &= \frac{1}{(2\pi)^3} \int d^3k \, e^{i\mathbf{k}\cdot(\mathbf{r}-\mathbf{r}')}(\delta_{lm} - \kappa_l \kappa_m) \\
&= \frac{2}{3}\delta_{lm}\delta(\mathbf{r} - \mathbf{r}') + \frac{Y(|\mathbf{r}-\mathbf{r}'|)}{4\pi|\mathbf{r}-\mathbf{r}'|^5} \times \\
&\quad \times \left[3(r_l - r_l')(r_m - r_m') - (\mathbf{r} - \mathbf{r}')^2 \delta_{lm} \right]
\end{aligned} \tag{34}
$$

where $Y(|\mathbf{r} - \mathbf{r}'|)$ is a regularization function equal to 1 everywhere except inside a small sphere around $|\mathbf{r} - \mathbf{r}'| = 0$, where it tends to zero. Finally, from (27)–(29), we obtain the following commutation relations for the fields in real space:

$$[A_{\perp l}(\mathbf{r}), A_{\perp m}(\mathbf{r}')] = 0 \qquad [E_{\perp l}(\mathbf{r}), E_{\perp m}(\mathbf{r}')] = 0 \tag{35.a}$$

$$[A_{\perp l}(\mathbf{r}), E_{\perp m}(\mathbf{r}')] = \frac{-i\hbar}{\varepsilon_0}\delta_{lm}^{\perp}(\mathbf{r} - \mathbf{r}'). \tag{35.b}$$

2. Particles

Particles are described within a nonrelativistic framework: their velocity is assumed to be small compared with c, and their number is invariant. In this limit, we can describe each particle α by the conjugate variables $\mathbf{r}_\alpha$ (position) and $\mathbf{p}_\alpha$ (momentum), rather than using a field theory. In quantum theory these variables become observables obeying the canonical commutation relations

$$[r_{\alpha i}, p_{\beta j}] = i\hbar \delta_{\alpha\beta}\delta_{ij}. \tag{36}$$

In the presence of the vector potential, the velocity $\mathbf{v}_\alpha$ of the particle is related to $\mathbf{p}_\alpha$ by

$$m_\alpha \mathbf{v}_\alpha = \mathbf{p}_\alpha - q_\alpha \mathbf{A}_\perp(\mathbf{r}_\alpha) \tag{37}$$

(m_α and q_α are, respectively, the mass and the charge of the particle α). The charge density and the current density are expressed as a function of the preceding variables

$$\rho(\mathbf{r}) = \sum_\alpha q_\alpha \delta(\mathbf{r} - \mathbf{r}_\alpha) \tag{38.a}$$

$$\mathbf{j}(\mathbf{r}) = \sum_\alpha q_\alpha \mathbf{v}_\alpha \delta(\mathbf{r} - \mathbf{r}_\alpha). \tag{38.b}$$

We will also use their spatial Fourier transforms

$$\rho(\mathbf{k}) = \sum_\alpha \frac{q_\alpha}{(2\pi)^{3/2}} e^{-i\mathbf{k}\cdot\mathbf{r}_\alpha} \tag{39.a}$$

$$\mathbf{\not{j}}(\mathbf{k}) = \sum_\alpha \frac{q_\alpha \mathbf{v}_\alpha}{(2\pi)^{3/2}} e^{-i\mathbf{k}\cdot\mathbf{r}_\alpha} \tag{39.b}$$

3. Hamiltonian and Dynamics in the Coulomb Gauge

a) HAMILTONIAN

The Hamiltonian H describing the dynamics of the system formed by the transverse field and the particles can be written as

$$H = \sum_\alpha \frac{1}{2m_\alpha} [\mathbf{p}_\alpha - q_\alpha \mathbf{A}_\perp(\mathbf{r}_\alpha)]^2 +$$

$$+ \sum_\alpha \left(-g_\alpha \frac{q_\alpha}{2m_\alpha}\right) \mathbf{S}_\alpha \cdot \mathbf{B}(\mathbf{r}_\alpha) + V_{\text{Coul}} + H_R. \tag{40}$$

The first term of H is the kinetic energy of the particles [see expression (37) for the velocity].

The second term represents the interaction energy of the spin magnetic moments possibly carried by the particles (g_α is the g factor for the particle α) with the magnetic field $\mathbf{B}(\mathbf{r}_\alpha)$.

The third term, V_{Coul}, is the energy of the longitudinal field (Coulomb energy)

$$V_{\text{Coul}} = \frac{\varepsilon_0}{2} \int d^3r\, E_\parallel^2(\mathbf{r}) = \frac{\varepsilon_0}{2} \int d^3k\, |\boldsymbol{\mathcal{E}}_\parallel(\mathbf{k})|^2 \tag{41}$$

which, using (4.a) and (39.a), equals

$$V_{\text{Coul}} = \frac{1}{2\varepsilon_0} \int d^3k\, \frac{\rho^*(\mathbf{k})\rho(\mathbf{k})}{k^2} = \sum_\alpha \varepsilon_{\text{Coul}}^\alpha + \frac{1}{8\pi\varepsilon_0} \sum_{\alpha \neq \beta} \frac{q_\alpha q_\beta}{|\mathbf{r}_\alpha - \mathbf{r}_\beta|}. \tag{42}$$

$\varepsilon_{\text{Coul}}^\alpha$ is the Coulomb self-energy of particle α, which is expressed in reciprocal space in the form of a divergent integral, unless a cutoff k_c is introduced

$$\varepsilon_{\text{Coul}}^\alpha = \frac{q_\alpha^2}{2\varepsilon_0} \int \frac{d^3k}{(2\pi)^3 k^2} = \frac{q_\alpha^2}{4\varepsilon_0\pi^2} \int_0^{k_c} dk = \frac{q_\alpha^2 k_c}{4\varepsilon_0\pi^2}. \tag{43}$$

The fourth term of H represents the energy of the transverse field

$$H_R = \frac{\varepsilon_0}{2} \int d^3r\big[E_\perp^2(\mathbf{r}) + c^2 B^2(\mathbf{r})\big] \tag{44}$$

which, using expressions (29) and (30) for $E_\perp$ and B, can be put in the form

$$H_R = \sum_i \hbar\omega_i\big(a_i^+ a_i + \tfrac{1}{2}\big). \tag{45}$$

In the presence of external fields described by the potentials $A_e(\mathbf{r}, t)$ and $U_e(\mathbf{r}, t)$, the Hamiltonian (40) must be modified as follows:

$$A_\perp(\mathbf{r}_\alpha) \to A_\perp(\mathbf{r}_\alpha) + A_e(\mathbf{r}_\alpha, t) \qquad V_{\text{Coul}} \to V_{\text{Coul}} + \sum_\alpha q_\alpha U_e(\mathbf{r}_\alpha, t).$$
$$\tag{46}$$

Finally, note that the momentum of the field + particles global system has a simple expression in the Coulomb gauge:

$$P = \sum_\alpha p_\alpha + P_R. \tag{47}$$

The first term represents the momentum of the particles and the momentum of the longitudinal field associated with them; the second term is the momentum of the transverse field

$$\mathbf{P}_R = \varepsilon_0 \int d^3r \, \mathbf{E}_\perp(\mathbf{r}) \times \mathbf{B}(\mathbf{r})$$

$$= \sum_j \hbar \mathbf{k}_j a_j^+ a_j \tag{48}$$

where the fields have been replaced by their expressions (29) and (30).

b) UNPERTURBED HAMILTONIAN AND INTERACTION HAMILTONIAN

It is interesting to split the Hamiltonian H of the global system into three parts:

$$H = H_P + H_R + H_I \tag{49}$$

where H_P depends only on the variables $\mathbf{r}_\alpha$ and $\mathbf{p}_\alpha$ of the particles (particle Hamiltonian), H_R depends only on the variables a_j and a_j^+ of the field (radiation Hamiltonian), H_I depends both on $\mathbf{r}_\alpha, \mathbf{p}_\alpha$ and a_j, a_j^+ (interaction Hamiltonian). Starting with expression (40) for H, we obtain, beside H_R given by (45),

$$H_P = \sum_\alpha \frac{\mathbf{p}_\alpha^2}{2m_\alpha} + V_{\text{Coul}} \tag{50}$$

$$H_I = H_{I1} + H_{I2} + H_{I1}^S \tag{51}$$

where H_{I1} and H_{I1}^S are linear with respect to the fields

$$H_{I1} = -\sum_\alpha \frac{q_\alpha}{m_\alpha} \mathbf{p}_\alpha \cdot \mathbf{A}_\perp(\mathbf{r}_\alpha) \tag{52}$$

$$H_{I1}^S = -\sum_\alpha g_\alpha \frac{q_\alpha}{2m_\alpha} \mathbf{S}_\alpha \cdot \mathbf{B}(\mathbf{r}_\alpha) \tag{53}$$

and where H_{I2} is quadratic

$$H_{I2} = \sum_\alpha \frac{q_\alpha^2}{2m_\alpha} \mathbf{A}_\perp^2(\mathbf{r}_\alpha). \tag{54}$$

For systems of bound particles, the relative orders of magnitude of the different interaction terms are the following:

$$\frac{H_{I2}}{H_{I1}} \simeq \frac{q^2 A^2/m}{qAp/m} = \frac{qAp/m}{p^2/m} \simeq \frac{H_{I1}}{H_P}. \tag{55}$$

For low radiation intensities, the ratio H_{I1}/H_P is small, which results in the ratio H_{I2}/H_{I1} also being small. The ratio H_{I1}^S/H_{I1} is on the order of

$$\frac{H_{I1}^S}{H_{I1}} \simeq \frac{q\hbar B/m}{qAp/m} \simeq \frac{\hbar k A}{pA} = \frac{\hbar k}{p} \tag{56}$$

which is the ratio between the momentum $\hbar k$ of the photon and the momentum p of the particle. For low-energy photons (for example, in the optical or microwave domain) and a bound electron, this ratio is very small compared with 1.

c) Equations of Motion

In the Heisenberg representation, the equations of motion can be deduced from expression (40) for H and the canonical commutation relations (20.b) and (36).

For the position and the velocity of the particles, we find, respectively, the relation (37) between the velocity $\mathbf{v}_\alpha$ of the particle and the momentum $\mathbf{p}_\alpha$, and the Newton–Lorentz equation, appropriately symmetrized, giving the acceleration of the particle in the presence of the fields $\mathbf{E}$ and $\mathbf{B}$.

For the transverse fields, we recover the Maxwell equations between operators. Because the transverse fields are linear functions of a_j and a_j^+, these equations are equivalent to the equations of motion for the a_j:

$$\dot{a}_j = \frac{1}{i\hbar}\left[a_j, H\right]$$

$$= -i\omega_j a_j + \frac{i}{\sqrt{2\varepsilon_0 \hbar \omega L^3}} \int d^3 r\, e^{-i\mathbf{k}_j \cdot \mathbf{r}}\, \boldsymbol{\varepsilon}_j \cdot \mathbf{j}(\mathbf{r}) \tag{57}$$

which are the quantum equivalents of Equations (18.b), and which have the structure of harmonic oscillator equations with source terms. In general, it is not possible to explicitly calculate their solutions, because the source term depends on the motion of the particles, which are themselves affected by the transverse field that we are looking for. However, in the absence of particles, the evolutions of the operators a_j are decoupled and Equation (57) can be integrated to give

$$a_j(t) = a_j(0) \exp(-i\omega_j t). \tag{58}$$

The evolution of the free fields can be immediately deduced from (58). They appear as sums of traveling plane waves with wave vector $\mathbf{k}_j$, frequency $\omega_j = ck_j$ and polarization $\boldsymbol{\varepsilon}_j$. For example:

$$\mathbf{E}_{\text{free}}(\mathbf{r}, t) = \sum_j i \sqrt{\frac{\hbar\omega_j}{2\varepsilon_0 L^3}} \times$$

$$\times \left\{ a_j(0)\boldsymbol{\varepsilon}_j \exp\left[i(\mathbf{k}_j \cdot \mathbf{r} - \omega_j t)\right] - a_j^+(0)\boldsymbol{\varepsilon}_j \exp\left[-i(\mathbf{k}_j \cdot \mathbf{r} - \omega_j t)\right] \right\}. \tag{59}$$

In certain calculations, particularly calculations of photodetection signals, it is necessary to isolate the components of fields with positive and negative frequencies. For the free fields, formulas of the same type as (59) give an explicit expression for these components. For example,

$$\mathbf{E}_{\text{free}}^{(+)}(\mathbf{r}, t) = \sum_j i \sqrt{\frac{\hbar\omega_j}{2\varepsilon_0 L^3}} \, a_j(0)\boldsymbol{\varepsilon}_j \exp\left[i(\mathbf{k}_j \cdot \mathbf{r} - \omega_j t)\right] \tag{60.a}$$

$$\mathbf{E}_{\text{free}}^{(-)}(\mathbf{r}, t) = \left[\mathbf{E}_{\text{free}}^{(+)}(\mathbf{r}, t)\right]^+. \tag{60.b}$$

In the Schrödinger representation, observables are time independent and the state vector evolves in state space according to the Schrödinger equation.

4. State Space

In the Coulomb gauge, the dynamics of the global system is equivalent to the dynamics of an ensemble of nonrelativistic particles and an infinite collection of harmonic oscillators representing the modes of the transverse field. In quantum theory, the state space $\mathscr{E}$ of the system is the tensor product of the state spaces $\mathscr{E}_P$ and $\mathscr{E}_R$ associated with each of these subsystems. The spaces $\mathscr{E}_P$ and $\mathscr{E}_R$ are themselves tensor products of spaces relative to each of the particles, and to each of the modes of the transverse field:

$$\mathscr{E}_P = \cdots \otimes \mathscr{E}_\alpha \otimes \cdots \tag{61}$$

$$\mathscr{E}_R = \cdots \otimes \mathscr{E}_j \otimes \cdots. \tag{62}$$

We now consider $\mathscr{E}_R$ in more detail. An orthonormal basis of each of the spaces $\mathscr{E}_j$ is composed of the basis $\{|n_j\rangle\}$ of the energy eigenstates of the oscillator j. Using expressions (45) and (48) for H_R and $\mathbf{P}_R$, the state $|n_1\rangle \cdots |n_j\rangle \cdots$, written more concisely as $|\{n_j\}\rangle$, is an eigenstate of these two observables:

$$H_R|\{n_j\}\rangle = \left[\sum_j \left(n_j + \tfrac{1}{2}\right)\hbar\omega_j\right]|\{n_j\}\rangle \qquad (63.a)$$

$$\mathbf{P}_R|\{n_j\}\rangle = \left(\sum_j n_j\hbar\mathbf{k}_j\right)|\{n_j\}\rangle. \qquad (63.b)$$

It represents a state of the field containing n_1 *photons* of the mode $1, \ldots, n_j$ *photons* of the mode j, with each photon of a mode j contributing to the total energy and momentum by the elementary "quanta" $\hbar\omega_j$ and $\hbar\mathbf{k}_j$.

The vacuum is the ground state of H_R corresponding to $n_1 = \cdots = n_j = \cdots = 0$. It is written more concisely as $|0\rangle$ and is characterized by the property

$$a_j|0\rangle = 0 \qquad \text{for each } j. \qquad (64)$$

In each of the spaces $\mathscr{E}_j$, the $|n_j\rangle$ are not the only interesting states. The "coherent" states $|\alpha_j\rangle$ play a particular role in the discussion of quasi-classical situations. They can be deduced from the vacuum $|0\rangle$ by a unitary transformation

$$|\alpha_j\rangle = T^+(\alpha_j)|0\rangle \qquad (65)$$

defined by

$$T(\alpha_j) = \exp\left[\alpha_j^* a_j - \alpha_j a_j^+\right] \qquad (66)$$

whose action is a translation of the operator a_j by the quantity α_j

$$T(\alpha_j)a_j T^+(\alpha_j) = a_j + \alpha_j. \qquad (67)$$

The state $|\alpha_j\rangle$ is an eigenstate of the annihilation operator a_j having the eigenvalue α_j

$$a_j|\alpha_j\rangle = \alpha_j|\alpha_j\rangle \qquad (68)$$

and its expansion onto the basis $\{|n_j\rangle\}$ is given by

$$|\alpha_j\rangle = e^{-|\alpha_j|^2/2} \sum_{n_j=0}^{\infty} \frac{(\alpha_j)^{n_j}}{\sqrt{n_j!}} |n_j\rangle. \tag{69}$$

5. The Long-Wavelength Approximation and the Electric Dipole Representation

Atoms and molecules are composed of charged particles (electrons and nuclei) forming bound states whose size a_0 is typically on the order of a few Bohr radii. Assume that such a system interacts with radio frequency, infrared, visible, or ultraviolet radiation. The wavelength λ of this radiation is large compared with a_0 and it is legitimate to neglect the spatial variations of the electromagnetic field over the size of the system of particles: all the particles see the same field. This is the long-wavelength approximation.

For the sake of simplicity, consider an atom (or a molecule) that is globally neutral and located close to the origin **0**. To lowest order in a_0, its electrical properties are characterized by its electric dipole moment

$$\mathbf{d} = \sum_{\alpha} q_\alpha \mathbf{r}_\alpha \tag{70}$$

A unitary transformation on the Hamiltonian (40) can cause the coupling between the atom and the field to appear explicitly in the form of an electric dipole interaction between the atomic dipole **d** and the radiation. To higher orders in a_0/λ, the same procedure would result in the appearance of electric quadrupole and magnetic dipole interactions, etc. Because we neglect them here, we also omit in H the spin magnetic coupling H_{I1}^S, so that we start with the approximate Hamiltonian:

$$H = \sum_{\alpha} \frac{1}{2m_\alpha} \left[\mathbf{p}_\alpha - q_\alpha \mathbf{A}_\perp(\mathbf{0}) \right]^2 + V_{\text{Coul}} + \sum_{j} \hbar\omega_j \left(a_j^+ a_j + \tfrac{1}{2} \right). \tag{71}$$

a) The Unitary Transformation

The transformation

$$T = \exp\left[-\frac{i}{\hbar} \mathbf{d} \cdot \mathbf{A}_\perp(\mathbf{0}) \right] = \exp\left\{ \sum_{j} \left(\lambda_j^* a_j - \lambda_j a_j^+ \right) \right\} \tag{72}$$

where

$$\lambda_j = \frac{i}{\sqrt{2\varepsilon_0 \hbar \omega_j L^3}} \, \boldsymbol{\varepsilon}_j \cdot \mathbf{d} \tag{73}$$

is a translation concerning both the operators $\mathbf{p}_\alpha$ and the operators a_j and a_j^+. The fundamental operators are in fact transformed according to the following rules

$$T\mathbf{r}_\alpha T^+ = \mathbf{r}_\alpha \tag{74.a}$$

$$T\mathbf{p}_\alpha T^+ = \mathbf{p}_\alpha + q_\alpha \mathbf{A}(\mathbf{0}) \tag{74.b}$$

$$Ta_j T^+ = a_j + \lambda_j \tag{74.c}$$

$$Ta_j^+ T^+ = a_j^+ + \lambda_j^*. \tag{74.d}$$

Because the transformation is time independent, the new Hamiltonian is written

$$H' = THT^+$$

$$= \sum_\alpha \frac{\mathbf{p}_\alpha^2}{2m_\alpha} + V_{\text{Coul}} + \varepsilon_{\text{dip}} + \sum_j \hbar\omega_j \left(a_j^+ a_j + \tfrac{1}{2} \right) -$$

$$- \mathbf{d} \cdot \sum_j \mathcal{E}_{\omega_j} \left[i a_j \boldsymbol{\varepsilon}_j - i a_j^+ \boldsymbol{\varepsilon}_j \right] \tag{75}$$

where

$$\varepsilon_{\text{dip}} = \sum_j \frac{1}{2\varepsilon_0 L^3} (\boldsymbol{\varepsilon}_j \cdot \mathbf{d})^2. \tag{76}$$

b) THE PHYSICAL VARIABLES IN THE ELECTRIC DIPOLE REPRESENTATION

The physical variables in the new representation are represented by the transforms G' of the observables G which represent them in the original representation

$$G' = TGT^+. \tag{77}$$

Starting with expressions (37), (21), (23), and (22) for the operators representing, respectively, the velocity of the particle α, the transverse vector potential, the magnetic field and the transverse electric field, we obtain expressions for the new observables which represent these variables:

$$v'_\alpha = T v_\alpha T^+ = \frac{p_\alpha}{m_\alpha} \tag{78}$$

$$A'_\perp(r) = T A_\perp(r) T^+ = A_\perp(r)$$
$$= \sum_j \mathscr{A}_{\omega_j} \left[a_j \varepsilon_j \, e^{i k_j \cdot r} + a_j^+ \varepsilon_j \, e^{-i k_j \cdot r} \right] \tag{79}$$

$$B'(r) = B(r) \tag{80}$$

$$E'_\perp(r) = T E_\perp(r) T^+$$
$$= \sum_j \mathscr{E}_{\omega_j} \left[i(a_j + \lambda_j) \varepsilon_j \, e^{i k_j \cdot r} + \text{h.c.} \right]$$

$$= E_\perp(r) - \frac{1}{\varepsilon_0} P_\perp(r) \tag{81}$$

where $P_\perp(r)$ is the transverse part of the polarization density $P(r)$ associated with the atom when it is considered as a pointlike dipole:

$$P(r) = d \, \delta(r) \tag{82}$$

$$P_\perp(r) = \sum_j \frac{\varepsilon_j (\varepsilon_j \cdot d)}{L^3} e^{i k_j \cdot r}. \tag{83}$$

Note that, according to (74.a), $r'_\alpha = r_\alpha$, so that the position of the particles, the atomic dipole and the polarization density are represented by the same operators in the two representations.

c) THE DISPLACEMENT FIELD

Starting with the total electric field $E(r)$ and the polarization density $P(r)$, we introduce the displacement field

$$D(r) = \varepsilon_0 E(r) + P(r). \tag{84}$$

We study its properties in reciprocal space, where it is written:

$$\mathscr{D}(k) = \varepsilon_0 \mathscr{E}(k) + \frac{d}{(2\pi)^{3/2}}. \tag{85}$$

[Relation (82) has been used to calculate the Fourier transform of $P(r)$].

With expression (39.a) for $\rho(\mathbf{k})$, the Maxwell equation (2.a) is written:

$$i\varepsilon_0\mathbf{k}\cdot\mathscr{E}(\mathbf{k}) = \frac{1}{(2\pi)^{3/2}}\sum_\alpha q_\alpha\, e^{-i\mathbf{k}\cdot\mathbf{r}_\alpha}$$

$$= \frac{1}{(2\pi)^{3/2}}\left[\sum_\alpha q_\alpha - i\mathbf{k}\cdot\sum_\alpha q_\alpha\mathbf{r}_\alpha + \cdots\right]. \qquad (86)$$

The expansion in powers of $\mathbf{k}\cdot\mathbf{r}_\alpha$ made in formula (86) is justified in the long-wavelength approximation because $kr_\alpha \simeq ka_0 \ll 1$. The first term inside the brackets of formula (86) is zero because the system of charges is assumed to be globally neutral. The second term is expressed as a function of the electric dipole. We then have, in the dipole approximation,

$$\varepsilon_0\mathbf{k}\cdot\mathscr{E}(\mathbf{k}) + \frac{\mathbf{k}\cdot\mathbf{d}}{(2\pi)^{3/2}} = 0. \qquad (87)$$

From (85) and (87), it is clear that $\mathbf{k}\cdot\mathscr{D}(\mathbf{k})$ is zero so that the displacement field is transverse for a globally neutral system.

Moreover, the polarization $\mathbf{P}(\mathbf{r})$ is zero outside the system of charges $(r > a_0)$, and (84) is reduced to

$$\mathbf{E}(\mathbf{r}) = \mathbf{D}(\mathbf{r})/\varepsilon_0 \qquad \text{for } |\mathbf{r}| > a. \qquad (88)$$

The displacement field is thus the transverse field which coincides, except for the factor ε_0, with the *total* electric field outside the system of charges. One advantage of the new representation is that a very simple operator describes $\mathbf{D}(\mathbf{r})/\varepsilon_0$, and thus the total electric field. Indeed, according to (84), the transversality of $\mathbf{D}$ results in the fact that $\mathbf{D} = \varepsilon_0\mathbf{E}_\perp + \mathbf{P}_\perp$. As a consequence of (81),

$$\mathbf{D}'(\mathbf{r})/\varepsilon_0 = \mathbf{E}'_\perp(\mathbf{r}) + \frac{1}{\varepsilon_0}\mathbf{P}'_\perp(\mathbf{r})$$

$$= \mathbf{E}_\perp(\mathbf{r})$$

$$= i\sum_j \mathscr{E}_{\omega j}\left(a_j\boldsymbol{\varepsilon}_j\, e^{i\mathbf{k}_j\cdot\mathbf{r}} - a_j^+\boldsymbol{\varepsilon}_j\, e^{-i\mathbf{k}_j\cdot\mathbf{r}}\right). \qquad (89)$$

d) ELECTRIC DIPOLE HAMILTONIAN

We now return to the Hamiltonian H'. The last term of (75) is the new interaction Hamiltonian. Using (89), it is written

$$H_I' = -\mathbf{d}\cdot\mathbf{D}'(0)/\varepsilon_0 \qquad (90)$$

and represents the interaction energy between the electric dipole and the displacement field. It is frequently written in the form

$$H_I' = -\mathbf{d} \cdot \mathbf{E}_\perp(\mathbf{0}) \tag{91}$$

it being understood that $\mathbf{E}_\perp(\mathbf{r})$ is the mathematical operator defined by (29) or (89) as a function of a_j and a_j^+. It is usually designated by the term "electric field", although it coincides with this variable only outside the system of charges. H' is thus written

$$H' = H_P' + H_R + H_I'. \tag{92}$$

H_R is given by (45), H_I' by (90), and H_P' by

$$H_P' = H_P + \varepsilon_{\text{dip}}. \tag{93}$$

The new particle Hamiltonian is obtained simply by adding to H_P given in (50) a dipolar self-energy described by expression (76). However, it should be noted that, in H_P', $\mathbf{p}_\alpha^2/2m_\alpha$ is indeed the kinetic energy of the particle α, according to (78), whereas this is not the case in (50), because the velocity is given by (37) in this representation.

In all the foregoing, the atom is assumed to be at rest at the origin of the coordinate system. In certain problems, it is important to take into account the motion of the center of mass $\mathbf{R}$ of the atom. If the atom is globally neutral, it is sufficient, in the electric dipole representation, to replace $\mathbf{0}$ by $\mathbf{R}$ in H_I':

$$H_I' = -\mathbf{d} \cdot \frac{\mathbf{D}'(\mathbf{R})}{\varepsilon_0} = -\mathbf{d} \cdot \mathbf{E}_\perp(\mathbf{R}). \tag{94}$$

REFERENCES

For more details, see *Photons and Atoms—Introduction to Quantum Electrodynamics*.

Refer also to the work of Kroll, Heitler, Power, Loudon, and Haken.

References

ABRAGAM, A., *The Principles of Nuclear Magnetism*, Clarendon Press, Oxford (1961).

AGARWAL, G. S., *Quantum Statistical Theories of Spontaneous Emission and Their Relation to Other Approaches*, Springer Tracts in Modern Physics, Vol. 70, Springer Verlag, Berlin (1974).

ALLEN, L. and EBERLY, J. H., *Optical Resonance and Two-Level Atoms*, Wiley Interscience, New York (1975).

BETHE, H. A. and SALPETER, E. E., *Quantum Mechanics of One- and Two-Electron Atoms*, Plenum, New York (1977).

BLOEMBERGEN, N., *Non-Linear Optics*, Benjamin, New York (1965).

BROSSEL, J., *Pompage Optique*, in *Quantum Optics and Electronics*, *Les Houches*, *1964*, ed. by C. de Witt, A. Blandin, and C. Cohen-Tannoudji, Gordon and Breach, New York (1965), p. 187.

CARTAN, H., *Théorie Elémentaire des Fonctions Analytiques d'une ou Plusieurs Variables Complexes*, Hermann, Paris (1975).

COHEN-TANNOUDJI, C., *Atoms in Strong Resonant Fields*, in *Les Houches XXVII*, *1975*, *Frontiers in Laser Spectroscopy*, Vol. 1, ed. by R. Balian, S. Haroche, and S. Liberman, North-Holland, Amsterdam (1977), p. 3.

COHEN-TANNOUDJI, C., DIU, B., and LALOË, F., *Mécanique Quantique*, Vols. 1 et 2, Hermann, Paris (1973). English translation: *Quantum Mechanics*, Vols. 1 and 2, Wiley and Hermann, Paris (1977).

FEYNMAN, R. P. and HIBBS, A. R., *Quantum Mechanics and Path Integrals*, MacGraw-Hill, New York (1965).

FEYNMAN, R. P., LEIGHTON, R. B., and SANDS, M., *The Feynman Lectures on Physics*, Vol. III: *Quantum Mechanics*. Addison-Wesley, Reading, Mass. (1966).

GLAUBER, R. J., *Coherence and Quantum Detection*, in *Quantum Optics, Proceedings of the International School of Physics Enrico Fermi*, Course XLVII, ed. by R. J. Glauber, Academic Press, New York (1969), p. 15.

GOLDBERGER, M. L. and WATSON, K. M., *Collision Theory*, Wiley, New York (1964).

GRYNBERG, G., CAGNAC, B. and BIRABEN, F., *Multiphoton Resonant Processes in Atoms*, in *Coherent Nonlinear Optics*, ed. by M. S. Feld and V. S. Letokhov, Springer Verlag, Berlin (1980), p. 111.

HAKEN, H., *Light*, Vol. 1, North-Holland, Amsterdam (1981).

HANNA, D. C., YURATICH, M. A., and COTTER, D., *Nonlinear Optics of Free Atoms and Molecules*, Springer Verlag, Berlin (1979).

HERZBERG, G., *Spectra of Diatomic Molecules*, D. Van Nostrand, Toronto (1963).

HEITLER, W., *The Quantum Theory of Radiation*, 3rd ed., Clarendon Press, Oxford (1954).

JACKSON, J. D., *Classical Electrodynamics*, 2nd ed., Wiley, New York (1975).

KROLL, N. M., *Quantum Theory of Radiation*, in *Quantum Optics and Electronics, Les Houches, 1964*, ed. by C. de Witt, A. Blandin, and C. Cohen-Tannoudji, Gordon and Breach, New York (1965), p. 1.

LANDAU, L. D. and LIFSHITZ, E. M., *Statistical Physics*, Pergamon Press, London (1963).

LANDAU, L. D. and LIFSHITZ, E. M., *The Classical Theory of Fields*, Addison-Wesley, Reading, Mass. (1951); Pergamon Press, London (1951).

LAX, M., *Fluctuations and Coherence Phenomena in Classical and Quantum Physics*, in *Brandeis Summer Institute Lectures, 1966*, Vol. II, ed. by M. Chrétien, E. P. Gross, and S. Dreser, Gordon and Breach, New York (1968), p. 269.

LETOKHOV, V. S. and CHEBOTAYEV, V. P., *Nonlinear Laser Spectroscopy*, Springer Verlag, Berlin (1977).

LEVY-LEBLOND, J. M. and BALIBAR, F., *Quantique, Rudiments*, InterEditions, Paris (1984).

LOUDON, R., *The Quantum Theory of Light*, Clarendon Press, Oxford (1973).

LOUISELL, W. H., *Quantum Statistical Properties of Radiation*, Wiley, New York (1973).

MARTIN, P., *Measurements and Correlation Functions*, in *Many Body Physics, Les Houches, 1967*, ed. by C. De Witt and R. Balian, Gordon and Breach, New York (1968), p. 37.

MERZBACHER, E., *Quantum Mechanics*, Wiley, New York (1970).

MESSIAH, A., *Mécanique Quantique*, Vols. 1 et 2, Dunod, Paris (1964). English translation: *Quantum Mechanics*, Vols. 1 and 2, North Holland, Amsterdam (1962).

MOLLOW, B. R., *Theory of Intensity Dependent Resonance Light Scattering and Resonance Fluorescence*, in *Progress in Optics*, Vol. XIX, ed. by E. Wolf, North-Holland, Amsterdam (1981), p. 1.

NUSSENZVEIG, H. M., *Introduction to Quantum Optics*, Gordon and Breach, New York (1973).

POWER, E. A., *Introductory Quantum Electrodynamics*, Longman, London (1964).

ROMAN, P., *Advanced Quantum Theory*, Addison-Wesley, Reading, Mass. (1965).

SARGENT, M., SCULLY, M. O., and LAMB, W. E. Jr., *Laser Physics*, Addison-Wesley, Reading, Mass. (1974).

SHEN, Y. R., *The Principles of Nonlinear Optics*, Wiley, New York (1984).

SCHIFF, L. I., *Quantum Mechanics*, 2nd ed., McGraw-Hill, New York (1955).

VAN KAMPEN, N. G., *Stochastic Processes in Physics and Chemistry*, North-Holland, Amsterdam (1981).

Index